T0188812

Lecture Notes in Computer Science 12903

More information about this subseries at http://www.springer.com/series/7412

Marleen de Bruijne · Philippe C. Cattin ·
Stéphane Cotin · Nicolas Padoy ·
Stefanie Speidel · Yefeng Zheng ·
Caroline Essert (Eds.)

Medical Image Computing and Computer Assisted Intervention – MICCAI 2021

24th International Conference
Strasbourg, France, September 27 – October 1, 2021
Proceedings, Part III

 Springer

Editors
Marleen de Bruijne ⓘ
Erasmus MC - University Medical Center
Rotterdam
Rotterdam, The Netherlands

University of Copenhagen
Copenhagen, Denmark

Stéphane Cotin ⓘ
Inria Nancy Grand Est
Villers-lès-Nancy, France

Stefanie Speidel ⓘ
National Center for Tumor Diseases
(NCT/UCC)
Dresden, Germany

Caroline Essert ⓘ
ICube, Université de Strasbourg, CNRS
Strasbourg, France

Philippe C. Cattin ⓘ
University of Basel
Allschwil, Switzerland

Nicolas Padoy ⓘ
ICube, Université de Strasbourg, CNRS
Strasbourg, France

Yefeng Zheng ⓘ
Tencent Jarvis Lab
Shenzhen, China

ISSN 0302-9743   ISSN 1611-3349   (electronic)
Lecture Notes in Computer Science
ISBN 978-3-030-87198-7   ISBN 978-3-030-87199-4   (eBook)
https://doi.org/10.1007/978-3-030-87199-4

LNCS Sublibrary: SL6 – Image Processing, Computer Vision, Pattern Recognition, and Graphics

This Springer imprint is published by the registered company Springer Nature Switzerland AG
The registered company address is: Gewerbestrasse 11, 6330 Cham, Switzerland

Preface

The 24th edition of the International Conference on Medical Image Computing and Computer Assisted Intervention (MICCAI 2021) has for the second time been placed under the shadow of COVID-19. Complicated situations due to the pandemic and multiple lockdowns have affected our lives during the past year, sometimes perturbing the researchers work, but also motivating an extraordinary dedication from many of our colleagues, and significant scientific advances in the fight against the virus. After another difficult year, most of us were hoping to be able to travel and finally meet in person at MICCAI 2021, which was supposed to be held in Strasbourg, France. Unfortunately, due to the uncertainty of the global situation, MICCAI 2021 had to be moved again to a virtual event that was held over five days from September 27 to October 1, 2021. Taking advantage of the experience gained last year and of the fast-evolving platforms, the organizers of MICCAI 2021 redesigned the schedule and the format. To offer the attendees both a strong scientific content and an engaging experience, two virtual platforms were used: Pathable for the oral and plenary sessions and SpatialChat for lively poster sessions, industrial booths, and networking events in the form of interactive group video chats.

These proceedings of MICCAI 2021 showcase all 531 papers that were presented at the main conference, organized into eight volumes in the Lecture Notes in Computer Science (LNCS) series as follows:

- Part I, LNCS Volume 12901: Image Segmentation
- Part II, LNCS Volume 12902: Machine Learning 1
- Part III, LNCS Volume 12903: Machine Learning 2
- Part IV, LNCS Volume 12904: Image Registration and Computer Assisted Intervention
- Part V, LNCS Volume 12905: Computer Aided Diagnosis
- Part VI, LNCS Volume 12906: Image Reconstruction and Cardiovascular Imaging
- Part VII, LNCS Volume 12907: Clinical Applications
- Part VIII, LNCS Volume 12908: Microscopic, Ophthalmic, and Ultrasound Imaging

These papers were selected after a thorough double-blind peer review process. We followed the example set by past MICCAI meetings, using Microsoft's Conference Managing Toolkit (CMT) for paper submission and peer reviews, with support from the Toronto Paper Matching System (TPMS), to partially automate paper assignment to area chairs and reviewers, and from iThenticate to detect possible cases of plagiarism.

Following a broad call to the community we received 270 applications to become an area chair for MICCAI 2021. From this group, the program chairs selected a total of 96 area chairs, aiming for diversity — MIC versus CAI, gender, geographical region, and a mix of experienced and new area chairs. Reviewers were recruited also via an open call for volunteers from the community (288 applications, of which 149 were selected by the program chairs) as well as by re-inviting past reviewers, leading to a total of 1340 registered reviewers.

We received 1630 full paper submissions after an original 2667 intentions to submit. Four papers were rejected without review because of concerns of (self-)plagiarism and dual submission and one additional paper was rejected for not adhering to the MICCAI page restrictions; two further cases of dual submission were discovered and rejected during the review process. Five papers were withdrawn by the authors during review and after acceptance.

The review process kicked off with a reviewer tutorial and an area chair meeting to discuss the review process, criteria for MICCAI acceptance, how to write a good (meta-)review, and expectations for reviewers and area chairs. Each area chair was assigned 16–18 manuscripts for which they suggested potential reviewers using TPMS scores, self-declared research area(s), and the area chair's knowledge of the reviewers' expertise in relation to the paper, while conflicts of interest were automatically avoided by CMT. Reviewers were invited to bid for the papers for which they had been suggested by an area chair or which were close to their expertise according to TPMS. Final reviewer allocations via CMT took account of reviewer bidding, prioritization of area chairs, and TPMS scores, leading to on average four reviews performed per person by a total of 1217 reviewers.

Following the initial double-blind review phase, area chairs provided a meta-review summarizing key points of reviews and a recommendation for each paper. The program chairs then evaluated the reviews and their scores, along with the recommendation from the area chairs, to directly accept 208 papers (13%) and reject 793 papers (49%); the remainder of the papers were sent for rebuttal by the authors. During the rebuttal phase, two additional area chairs were assigned to each paper. The three area chairs then independently ranked their papers, wrote meta-reviews, and voted to accept or reject the paper, based on the reviews, rebuttal, and manuscript. The program chairs checked all meta-reviews, and in some cases where the difference between rankings was high or comments were conflicting, they also assessed the original reviews, rebuttal, and submission. In all other cases a majority voting scheme was used to make the final decision. This process resulted in the acceptance of a further 325 papers for an overall acceptance rate of 33%.

Acceptance rates were the same between medical image computing (MIC) and computer assisted interventions (CAI) papers, and slightly lower where authors classified their paper as both MIC and CAI. Distribution of the geographical region of the first author as indicated in the optional demographic survey was similar among submitted and accepted papers.

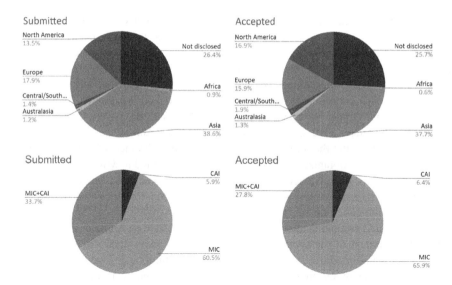

New this year, was the requirement to fill out a reproducibility checklist when submitting an intention to submit to MICCAI, in order to stimulate authors to think about what aspects of their method and experiments they should include to allow others to reproduce their results. Papers that included an anonymous code repository and/or indicated that the code would be made available were more likely to be accepted. From all accepted papers, 273 (51%) included a link to a code repository with the camera-ready submission.

Another novelty this year is that we decided to make the reviews, meta-reviews, and author responses for accepted papers available on the website. We hope the community will find this a useful resource.

The outstanding program of MICCAI 2021 was enriched by four exceptional keynote talks given by Alyson McGregor, Richard Satava, Fei-Fei Li, and Pierre Jannin, on hot topics such as gender bias in medical research, clinical translation to industry, intelligent medicine, and sustainable research. This year, as in previous years, high-quality satellite events completed the program of the main conference: 28 workshops, 23 challenges, and 14 tutorials; without forgetting the increasingly successful plenary events, such as the Women in MICCAI (WiM) meeting, the MICCAI Student Board (MSB) events, the 2nd Startup Village, the MICCAI-RSNA panel, and the first "Reinforcing Inclusiveness & diverSity and Empowering MICCAI" (or RISE-MICCAI) event.

MICCAI 2021 has also seen the first edition of CLINICCAI, the clinical day of MICCAI. Organized by Nicolas Padoy and Lee Swanstrom, this new event will hopefully help bring the scientific and clinical communities closer together, and foster collaborations and interaction. A common keynote connected the two events. We hope this effort will be pursued in the next editions.

We would like to thank everyone who has contributed to making MICCAI 2021 a success. First of all, we sincerely thank the authors, area chairs, reviewers, and session chairs for their dedication and for offering the participants and readers of these proceedings content of exceptional quality. Special thanks go to our fantastic submission platform manager Kitty Wong, who has been a tremendous help in the entire process from reviewer and area chair selection, paper submission, and the review process to the preparation of these proceedings. We also thank our very efficient team of satellite events chairs and coordinators, led by Cristian Linte and Matthieu Chabanas: the workshop chairs, Amber Simpson, Denis Fortun, Marta Kersten-Oertel, and Sandrine Voros; the challenges chairs, Annika Reinke, Spyridon Bakas, Nicolas Passat, and Ingerid Reinersten; and the tutorial chairs, Sonia Pujol and Vincent Noblet, as well as all the satellite event organizers for the valuable content added to MICCAI. Our special thanks also go to John Baxter and his team who worked hard on setting up and populating the virtual platforms, to Alejandro Granados for his valuable help and efficient communication on social media, and to Shelley Wallace and Anna Van Vliet for marketing and communication. We are also very grateful to Anirban Mukhopadhay for his management of the sponsorship, and of course many thanks to the numerous sponsors who supported the conference, often with continuous engagement over many years. This year again, our thanks go to Marius Linguraru and his team who supervised a range of actions to help, and promote, career development, among which were the mentorship program and the Startup Village. And last but not least, our wholehearted thanks go to Mehmet and the wonderful team at Dekon Congress and Tourism for their great professionalism and reactivity in the management of all logistical aspects of the event.

Finally, we thank the MICCAI society and the Board of Directors for their support throughout the years, starting with the first discussions about bringing MICCAI to Strasbourg in 2017.

We look forward to seeing you at MICCAI 2022.

September 2021

Marleen de Bruijne
Philippe Cattin
Stéphane Cotin
Nicolas Padoy
Stefanie Speidel
Yefeng Zheng
Caroline Essert

Organization

General Chair

Caroline Essert   Université de Strasbourg, CNRS, ICube, France

Program Chairs

Marleen de Bruijne   Erasmus MC Rotterdam, The Netherlands,
                                     and University of Copenhagen, Denmark
Philippe C. Cattin   University of Basel, Switzerland
Stéphane Cotin   Inria, France
Nicolas Padoy   Université de Strasbourg, CNRS, ICube, IHU, France
Stefanie Speidel   National Center for Tumor Diseases, Dresden, Germany
Yefeng Zheng   Tencent Jarvis Lab, China

Satellite Events Coordinators

Cristian Linte   Rochester Institute of Technology, USA
Matthieu Chabanas   Université Grenoble Alpes, France

Workshop Team

Amber Simpson   Queen's University, Canada
Denis Fortun   Université de Strasbourg, CNRS, ICube, France
Marta Kersten-Oertel   Concordia University, Canada
Sandrine Voros   TIMC-IMAG, INSERM, France

Challenges Team

Annika Reinke   German Cancer Research Center, Germany
Spyridon Bakas   University of Pennsylvania, USA
Nicolas Passat   Université de Reims Champagne-Ardenne, France
Ingerid Reinersten   SINTEF, NTNU, Norway

Tutorial Team

Vincent Noblet   Université de Strasbourg, CNRS, ICube, France
Sonia Pujol   Harvard Medical School, Brigham and Women's
                                     Hospital, USA

Clinical Day Chairs

Nicolas Padoy   Université de Strasbourg, CNRS, ICube, IHU, France
Lee Swanström   IHU Strasbourg, France

Sponsorship Chairs

Anirban Mukhopadhyay   Technische Universität Darmstadt, Germany
Yanwu Xu   Baidu Inc., China

Young Investigators and Early Career Development Program Chairs

Marius Linguraru   Children's National Institute, USA
Antonio Porras   Children's National Institute, USA
Daniel Racoceanu   Sorbonne Université/Brain Institute, France
Nicola Rieke   NVIDIA, Germany
Renee Yao   NVIDIA, USA

Social Media Chairs

Alejandro Granados   King's College London, UK
   Martinez
Shuwei Xing   Robarts Research Institute, Canada
Maxence Boels   King's College London, UK

Green Team

Pierre Jannin   INSERM, Université de Rennes 1, France
Étienne Baudrier   Université de Strasbourg, CNRS, ICube, France

Student Board Liaison

Éléonore Dufresne   Université de Strasbourg, CNRS, ICube, France
Étienne Le Quentrec   Université de Strasbourg, CNRS, ICube, France
Vinkle Srivastav   Université de Strasbourg, CNRS, ICube, France

Submission Platform Manager

Kitty Wong   The MICCAI Society, Canada

Virtual Platform Manager

John Baxter   INSERM, Université de Rennes 1, France

Program Committee

Ehsan Adeli	Stanford University, USA
Iman Aganj	Massachusetts General Hospital, Harvard Medical School, USA
Pablo Arbelaez	Universidad de los Andes, Colombia
John Ashburner	University College London, UK
Meritxell Bach Cuadra	University of Lausanne, Switzerland
Sophia Bano	University College London, UK
Adrien Bartoli	Université Clermont Auvergne, France
Christian Baumgartner	ETH Zürich, Switzerland
Hrvoje Bogunovic	Medical University of Vienna, Austria
Weidong Cai	University of Sydney, Australia
Gustavo Carneiro	University of Adelaide, Australia
Chao Chen	Stony Brook University, USA
Elvis Chen	Robarts Research Institute, Canada
Hao Chen	Hong Kong University of Science and Technology, Hong Kong SAR
Albert Chung	Hong Kong University of Science and Technology, Hong Kong SAR
Adrian Dalca	Massachusetts Institute of Technology, USA
Adrien Depeursinge	HES-SO Valais-Wallis, Switzerland
Jose Dolz	ÉTS Montréal, Canada
Ruogu Fang	University of Florida, USA
Dagan Feng	University of Sydney, Australia
Huazhu Fu	Inception Institute of Artificial Intelligence, United Arab Emirates
Mingchen Gao	University at Buffalo, The State University of New York, USA
Guido Gerig	New York University, USA
Orcun Goksel	Uppsala University, Sweden
Alberto Gomez	King's College London, UK
Ilker Hacihaliloglu	Rutgers University, USA
Adam Harrison	PAII Inc., USA
Mattias Heinrich	University of Lübeck, Germany
Yi Hong	Shanghai Jiao Tong University, China
Yipeng Hu	University College London, UK
Junzhou Huang	University of Texas at Arlington, USA
Xiaolei Huang	The Pennsylvania State University, USA
Jana Hutter	King's College London, UK
Madhura Ingalhalikar	Symbiosis Center for Medical Image Analysis, India
Shantanu Joshi	University of California, Los Angeles, USA
Samuel Kadoury	Polytechnique Montréal, Canada
Fahmi Khalifa	Mansoura University, Egypt
Hosung Kim	University of Southern California, USA
Minjeong Kim	University of North Carolina at Greensboro, USA

Ender Konukoglu	ETH Zürich, Switzerland
Bennett Landman	Vanderbilt University, USA
Ignacio Larrabide	CONICET, Argentina
Baiying Lei	Shenzhen University, China
Gang Li	University of North Carolina at Chapel Hill, USA
Mingxia Liu	University of North Carolina at Chapel Hill, USA
Herve Lombaert	ÉTS Montréal, Canada, and Inria, France
Marco Lorenzi	Inria, France
Le Lu	PAII Inc., USA
Xiongbiao Luo	Xiamen University, China
Dwarikanath Mahapatra	Inception Institute of Artificial Intelligence, United Arab Emirates
Andreas Maier	FAU Erlangen-Nuremberg, Germany
Erik Meijering	University of New South Wales, Australia
Hien Nguyen	University of Houston, USA
Marc Niethammer	University of North Carolina at Chapel Hill, USA
Tingying Peng	Technische Universität München, Germany
Caroline Petitjean	Université de Rouen, France
Dzung Pham	Henry M. Jackson Foundation, USA
Hedyeh Rafii-Tari	Auris Health Inc, USA
Islem Rekik	Istanbul Technical University, Turkey
Nicola Rieke	NVIDIA, Germany
Su Ruan	Laboratoire LITIS, France
Thomas Schultz	University of Bonn, Germany
Sharmishtaa Seshamani	Allen Institute, USA
Yonggang Shi	University of Southern California, USA
Darko Stern	Technical University of Graz, Austria
Carole Sudre	King's College London, UK
Heung-Il Suk	Korea University, South Korea
Jian Sun	Xi'an Jiaotong University, China
Raphael Sznitman	University of Bern, Switzerland
Amir Tahmasebi	Enlitic, USA
Qian Tao	Delft University of Technology, The Netherlands
Tolga Tasdizen	University of Utah, USA
Martin Urschler	University of Auckland, New Zealand
Archana Venkataraman	Johns Hopkins University, USA
Guotai Wang	University of Electronic Science and Technology of China, China
Hongzhi Wang	IBM Almaden Research Center, USA
Hua Wang	Colorado School of Mines, USA
Qian Wang	Shanghai Jiao Tong University, China
Yalin Wang	Arizona State University, USA
Fuyong Xing	University of Colorado Denver, USA
Daguang Xu	NVIDIA, USA
Yanwu Xu	Baidu, China
Ziyue Xu	NVIDIA, USA

Zhong Xue	Shanghai United Imaging Intelligence, China
Xin Yang	Huazhong University of Science and Technology, China
Jianhua Yao	National Institutes of Health, USA
Zhaozheng Yin	Stony Brook University, USA
Yixuan Yuan	City University of Hong Kong, Hong Kong SAR
Liang Zhan	University of Pittsburgh, USA
Tuo Zhang	Northwestern Polytechnical University, China
Yitian Zhao	Chinese Academy of Sciences, China
Luping Zhou	University of Sydney, Australia
S. Kevin Zhou	Chinese Academy of Sciences, China
Dajiang Zhu	University of Texas at Arlington, USA
Xiahai Zhuang	Fudan University, China
Maria A. Zuluaga	EURECOM, France

Reviewers

Alaa Eldin Abdelaal
Khalid Abdul Jabbar
Purang Abolmaesumi
Mazdak Abulnaga
Maryam Afzali
Priya Aggarwal
Ola Ahmad
Sahar Ahmad
Euijoon Ahn
Alireza Akhondi-Asl
Saad Ullah Akram
Dawood Al Chanti
Daniel Alexander
Sharib Ali
Lejla Alic
Omar Al-Kadi
Maximilian Allan
Pierre Ambrosini
Sameer Antani
Michela Antonelli
Jacob Antunes
Syed Anwar
Ignacio Arganda-Carreras
Mohammad Ali Armin
Md Ashikuzzaman
Mehdi Astaraki
Angélica Atehortúa
Gowtham Atluri

Chloé Audigier
Kamran Avanaki
Angelica Aviles-Rivero
Suyash Awate
Dogu Baran Aydogan
Qinle Ba
Morteza Babaie
Hyeon-Min Bae
Woong Bae
Junjie Bai
Wenjia Bai
Ujjwal Baid
Spyridon Bakas
Yaël Balbastre
Marcin Balicki
Fabian Balsiger
Abhirup Banerjee
Sreya Banerjee
Shunxing Bao
Adrian Barbu
Sumana Basu
Mathilde Bateson
Deepti Bathula
John Baxter
Bahareh Behboodi
Delaram Behnami
Mikhail Belyaev
Aicha BenTaieb

Camilo Bermudez
Gabriel Bernardino
Hadrien Bertrand
Alaa Bessadok
Michael Beyeler
Indrani Bhattacharya
Chetan Bhole
Lei Bi
Gui-Bin Bian
Ryoma Bise
Stefano B. Blumberg
Ester Bonmati
Bhushan Borotikar
Jiri Borovec
Ilaria Boscolo Galazzo
Alexandre Bousse
Nicolas Boutry
Behzad Bozorgtabar
Nathaniel Braman
Nadia Brancati
Katharina Breininger
Christopher Bridge
Esther Bron
Rupert Brooks
Qirong Bu
Duc Toan Bui
Ninon Burgos
Nikolay Burlutskiy
Hendrik Burwinkel
Russell Butler
Michał Byra
Ryan Cabeen
Mariano Cabezas
Hongmin Cai
Jinzheng Cai
Yunliang Cai
Sema Candemir
Bing Cao
Qing Cao
Shilei Cao
Tian Cao
Weiguo Cao
Aaron Carass
M. Jorge Cardoso
Adrià Casamitjana
Matthieu Chabanas

Ahmad Chaddad
Jayasree Chakraborty
Sylvie Chambon
Yi Hao Chan
Ming-Ching Chang
Peng Chang
Violeta Chang
Sudhanya Chatterjee
Christos Chatzichristos
Antong Chen
Chang Chen
Cheng Chen
Dongdong Chen
Geng Chen
Hanbo Chen
Jianan Chen
Jianxu Chen
Jie Chen
Junxiang Chen
Lei Chen
Li Chen
Liangjun Chen
Min Chen
Pingjun Chen
Qiang Chen
Shuai Chen
Tianhua Chen
Tingting Chen
Xi Chen
Xiaoran Chen
Xin Chen
Xuejin Chen
Yuhua Chen
Yukun Chen
Zhaolin Chen
Zhineng Chen
Zhixiang Chen
Erkang Cheng
Jun Cheng
Li Cheng
Yuan Cheng
Farida Cheriet
Minqi Chong
Jaegul Choo
Aritra Chowdhury
Gary Christensen

Daan Christiaens
Stergios Christodoulidis
Ai Wern Chung
Pietro Antonio Cicalese
Özgün Çiçek
Celia Cintas
Matthew Clarkson
Jaume Coll-Font
Toby Collins
Olivier Commowick
Pierre-Henri Conze
Timothy Cootes
Luca Corinzia
Teresa Correia
Hadrien Courtecuisse
Jeffrey Craley
Hui Cui
Jianan Cui
Zhiming Cui
Kathleen Curran
Claire Cury
Tobias Czempiel
Vedrana Dahl
Haixing Dai
Rafat Damseh
Bilel Daoud
Neda Davoudi
Laura Daza
Sandro De Zanet
Charles Delahunt
Yang Deng
Cem Deniz
Felix Denzinger
Hrishikesh Deshpande
Christian Desrosiers
Blake Dewey
Neel Dey
Raunak Dey
Jwala Dhamala
Yashin Dicente Cid
Li Ding
Xinghao Ding
Zhipeng Ding
Konstantin Dmitriev
Ines Domingues
Liang Dong

Mengjin Dong
Nanqing Dong
Reuben Dorent
Sven Dorkenwald
Qi Dou
Simon Drouin
Niharika D'Souza
Lei Du
Hongyi Duanmu
Nicolas Duchateau
James Duncan
Luc Duong
Nicha Dvornek
Dmitry V. Dylov
Oleh Dzyubachyk
Roy Eagleson
Mehran Ebrahimi
Jan Egger
Alma Eguizabal
Gudmundur Einarsson
Ahmed Elazab
Mohammed S. M. Elbaz
Shireen Elhabian
Mohammed Elmogy
Amr Elsawy
Ahmed Eltanboly
Sandy Engelhardt
Ertunc Erdil
Marius Erdt
Floris Ernst
Boris Escalante-Ramírez
Maria Escobar
Mohammad Eslami
Nazila Esmaeili
Marco Esposito
Oscar Esteban
Théo Estienne
Ivan Ezhov
Deng-Ping Fan
Jingfan Fan
Xin Fan
Yonghui Fan
Xi Fang
Zhenghan Fang
Aly Farag
Mohsen Farzi

Lina Felsner
Jun Feng
Ruibin Feng
Xinyang Feng
Yuan Feng
Aaron Fenster
Aasa Feragen
Henrique Fernandes
Enzo Ferrante
Jean Feydy
Lukas Fischer
Peter Fischer
Antonio Foncubierta-Rodríguez
Germain Forestier
Nils Daniel Forkert
Jean-Rassaire Fouefack
Moti Freiman
Wolfgang Freysinger
Xueyang Fu
Yunguan Fu
Wolfgang Fuhl
Isabel Funke
Philipp Fürnstahl
Pedro Furtado
Ryo Furukawa
Jin Kyu Gahm
Laurent Gajny
Adrian Galdran
Yu Gan
Melanie Ganz
Cong Gao
Dongxu Gao
Linlin Gao
Siyuan Gao
Yixin Gao
Yue Gao
Zhifan Gao
Alfonso Gastelum-Strozzi
Srishti Gautam
Bao Ge
Rongjun Ge
Zongyuan Ge
Sairam Geethanath
Shiv Gehlot
Nils Gessert
Olivier Gevaert

Sandesh Ghimire
Ali Gholipour
Sayan Ghosal
Andrea Giovannini
Gabriel Girard
Ben Glocker
Arnold Gomez
Mingming Gong
Cristina González
German Gonzalez
Sharath Gopal
Karthik Gopinath
Pietro Gori
Michael Götz
Shuiping Gou
Maged Goubran
Sobhan Goudarzi
Dushyant Goyal
Mark Graham
Bertrand Granado
Alejandro Granados
Vicente Grau
Lin Gu
Shi Gu
Xianfeng Gu
Yun Gu
Zaiwang Gu
Hao Guan
Ricardo Guerrero
Houssem-Eddine Gueziri
Dazhou Guo
Hengtao Guo
Jixiang Guo
Pengfei Guo
Xiaoqing Guo
Yi Guo
Yulan Guo
Yuyu Guo
Krati Gupta
Vikash Gupta
Praveen Gurunath Bharathi
Boris Gutman
Prashnna Gyawali
Stathis Hadjidemetriou
Mohammad Hamghalam
Hu Han

Liang Han
Xiaoguang Han
Xu Han
Zhi Han
Zhongyi Han
Jonny Hancox
Xiaoke Hao
Nandinee Haq
Ali Hatamizadeh
Charles Hatt
Andreas Hauptmann
Mohammad Havaei
Kelei He
Nanjun He
Tiancheng He
Xuming He
Yuting He
Nicholas Heller
Alessa Hering
Monica Hernandez
Carlos Hernandez-Matas
Kilian Hett
Jacob Hinkle
David Ho
Nico Hoffmann
Matthew Holden
Sungmin Hong
Yoonmi Hong
Antal Horváth
Md Belayat Hossain
Benjamin Hou
William Hsu
Tai-Chiu Hsung
Kai Hu
Shi Hu
Shunbo Hu
Wenxing Hu
Xiaoling Hu
Xiaowei Hu
Yan Hu
Zhenhong Hu
Heng Huang
Qiaoying Huang
Yi-Jie Huang
Yixing Huang
Yongxiang Huang

Yue Huang
Yufang Huang
Arnaud Huaulmé
Henkjan Huisman
Yuankai Huo
Andreas Husch
Mohammad Hussain
Raabid Hussain
Sarfaraz Hussein
Khoi Huynh
Seong Jae Hwang
Emmanuel Iarussi
Kay Igwe
Abdullah-Al-Zubaer Imran
Ismail Irmakci
Mobarakol Islam
Mohammad Shafkat Islam
Vamsi Ithapu
Koichi Ito
Hayato Itoh
Oleksandra Ivashchenko
Yuji Iwahori
Shruti Jadon
Mohammad Jafari
Mostafa Jahanifar
Amir Jamaludin
Mirek Janatka
Won-Dong Jang
Uditha Jarayathne
Ronnachai Jaroensri
Golara Javadi
Rohit Jena
Rachid Jennane
Todd Jensen
Won-Ki Jeong
Yuanfeng Ji
Zhanghexuan Ji
Haozhe Jia
Jue Jiang
Tingting Jiang
Xiang Jiang
Jianbo Jiao
Zhicheng Jiao
Amelia Jiménez-Sánchez
Dakai Jin
Yueming Jin

Bin Jing
Anand Joshi
Yohan Jun
Kyu-Hwan Jung
Alain Jungo
Manjunath K N
Ali Kafaei Zad Tehrani
Bernhard Kainz
John Kalafut
Michael C. Kampffmeyer
Qingbo Kang
Po-Yu Kao
Neerav Karani
Turkay Kart
Satyananda Kashyap
Amin Katouzian
Alexander Katzmann
Prabhjot Kaur
Erwan Kerrien
Hoel Kervadec
Ashkan Khakzar
Nadieh Khalili
Siavash Khallaghi
Farzad Khalvati
Bishesh Khanal
Pulkit Khandelwal
Maksim Kholiavchenko
Naji Khosravan
Seyed Mostafa Kia
Daeseung Kim
Hak Gu Kim
Hyo-Eun Kim
Jae-Hun Kim
Jaeil Kim
Jinman Kim
Mansu Kim
Namkug Kim
Seong Tae Kim
Won Hwa Kim
Andrew King
Atilla Kiraly
Yoshiro Kitamura
Tobias Klinder
Bin Kong
Jun Kong
Tomasz Konopczynski

Bongjin Koo
Ivica Kopriva
Kivanc Kose
Mateusz Kozinski
Anna Kreshuk
Anithapriya Krishnan
Pavitra Krishnaswamy
Egor Krivov
Frithjof Kruggel
Alexander Krull
Elizabeth Krupinski
Serife Kucur
David Kügler
Hugo Kuijf
Abhay Kumar
Ashnil Kumar
Kuldeep Kumar
Nitin Kumar
Holger Kunze
Tahsin Kurc
Anvar Kurmukov
Yoshihiro Kuroda
Jin Tae Kwak
Yongchan Kwon
Francesco La Rosa
Aymen Laadhari
Dmitrii Lachinov
Alain Lalande
Tryphon Lambrou
Carole Lartizien
Bianca Lassen-Schmidt
Ngan Le
Leo Lebrat
Christian Ledig
Eung-Joo Lee
Hyekyoung Lee
Jong-Hwan Lee
Matthew Lee
Sangmin Lee
Soochahn Lee
Étienne Léger
Stefan Leger
Andreas Leibetseder
Rogers Jeffrey Leo John
Juan Leon
Bo Li

Chongyi Li
Fuhai Li
Hongming Li
Hongwei Li
Jian Li
Jianning Li
Jiayun Li
Junhua Li
Kang Li
Mengzhang Li
Ming Li
Qing Li
Shaohua Li
Shuyu Li
Weijian Li
Weikai Li
Wenqi Li
Wenyuan Li
Xiang Li
Xiaomeng Li
Xiaoxiao Li
Xin Li
Xiuli Li
Yang Li
Yi Li
Yuexiang Li
Zeju Li
Zhang Li
Zhiyuan Li
Zhjin Li
Gongbo Liang
Jianming Liang
Libin Liang
Yuan Liang
Haofu Liao
Ruizhi Liao
Wei Liao
Xiangyun Liao
Roxane Licandro
Gilbert Lim
Baihan Lin
Hongxiang Lin
Jianyu Lin
Yi Lin
Claudia Lindner
Geert Litjens

Bin Liu
Chi Liu
Daochang Liu
Dong Liu
Dongnan Liu
Feng Liu
Hangfan Liu
Hong Liu
Huafeng Liu
Jianfei Liu
Jingya Liu
Kai Liu
Kefei Liu
Lihao Liu
Mengting Liu
Peng Liu
Qin Liu
Quande Liu
Shengfeng Liu
Shenghua Liu
Shuangjun Liu
Sidong Liu
Siqi Liu
Tianrui Liu
Xiao Liu
Xinyang Liu
Xinyu Liu
Yan Liu
Yikang Liu
Yong Liu
Yuan Liu
Yue Liu
Yuhang Liu
Andrea Loddo
Nicolas Loménie
Daniel Lopes
Bin Lou
Jian Lou
Nicolas Loy Rodas
Donghuan Lu
Huanxiang Lu
Weijia Lu
Xiankai Lu
Yongyi Lu
Yueh-Hsun Lu
Yuhang Lu

Imanol Luengo
Jie Luo
Jiebo Luo
Luyang Luo
Ma Luo
Bin Lv
Jinglei Lv
Junyan Lyu
Qing Lyu
Yuanyuan Lyu
Andy J. Ma
Chunwei Ma
Da Ma
Hua Ma
Kai Ma
Lei Ma
Anderson Maciel
Amirreza Mahbod
S. Sara Mahdavi
Mohammed Mahmoud
Saïd Mahmoudi
Klaus H. Maier-Hein
Bilal Malik
Ilja Manakov
Matteo Mancini
Tommaso Mansi
Yunxiang Mao
Brett Marinelli
Pablo Márquez Neila
Carsten Marr
Yassine Marrakchi
Fabio Martinez
Andre Mastmeyer
Tejas Sudharshan Mathai
Dimitrios Mavroeidis
Jamie McClelland
Pau Medrano-Gracia
Raghav Mehta
Sachin Mehta
Raphael Meier
Qier Meng
Qingjie Meng
Yanda Meng
Martin Menten
Odyssée Merveille
Islem Mhiri

Liang Mi
Stijn Michielse
Abhishek Midya
Fausto Milletari
Hyun-Seok Min
Zhe Min
Tadashi Miyamoto
Sara Moccia
Hassan Mohy-ud-Din
Tony C. W. Mok
Rafael Molina
Mehdi Moradi
Rodrigo Moreno
Kensaku Mori
Lia Morra
Linda Moy
Mohammad Hamed Mozaffari
Sovanlal Mukherjee
Anirban Mukhopadhyay
Henning Müller
Balamurali Murugesan
Cosmas Mwikirize
Andriy Myronenko
Saad Nadeem
Vishwesh Nath
Rodrigo Nava
Fernando Navarro
Amin Nejatbakhsh
Dong Ni
Hannes Nickisch
Dong Nie
Jingxin Nie
Aditya Nigam
Lipeng Ning
Xia Ning
Tianye Niu
Jack Noble
Vincent Noblet
Alexey Novikov
Jorge Novo
Mohammad Obeid
Masahiro Oda
Benjamin Odry
Steffen Oeltze-Jafra
Hugo Oliveira
Sara Oliveira

Arnau Oliver
Emanuele Olivetti
Jimena Olveres
John Onofrey
Felipe Orihuela-Espina
José Orlando
Marcos Ortega
Yoshito Otake
Sebastian Otálora
Cheng Ouyang
Jiahong Ouyang
Xi Ouyang
Michal Ozery-Flato
Danielle Pace
Krittin Pachtrachai
J. Blas Pagador
Akshay Pai
Viswanath Pamulakanty Sudarshan
Jin Pan
Yongsheng Pan
Pankaj Pandey
Prashant Pandey
Egor Panfilov
Shumao Pang
Joao Papa
Constantin Pape
Bartlomiej Papiez
Hyunjin Park
Jongchan Park
Sanghyun Park
Seung-Jong Park
Seyoun Park
Magdalini Paschali
Diego Patiño Cortés
Angshuman Paul
Christian Payer
Yuru Pei
Chengtao Peng
Yige Peng
Antonio Pepe
Oscar Perdomo
Sérgio Pereira
Jose-Antonio Pérez-Carrasco
Fernando Pérez-García
Jorge Perez-Gonzalez
Skand Peri

Matthias Perkonigg
Mehran Pesteie
Jorg Peters
Jens Petersen
Kersten Petersen
Renzo Phellan Aro
Ashish Phophalia
Tomasz Pieciak
Antonio Pinheiro
Pramod Pisharady
Kilian Pohl
Sebastian Pölsterl
Iulia A. Popescu
Alison Pouch
Prateek Prasanna
Raphael Prevost
Juan Prieto
Sergi Pujades
Elodie Puybareau
Esther Puyol-Antón
Haikun Qi
Huan Qi
Buyue Qian
Yan Qiang
Yuchuan Qiao
Chen Qin
Wenjian Qin
Yulei Qin
Wu Qiu
Hui Qu
Liangqiong Qu
Kha Gia Quach
Prashanth R.
Pradeep Reddy Raamana
Mehdi Rahim
Jagath Rajapakse
Kashif Rajpoot
Jhonata Ramos
Lingyan Ran
Hatem Rashwan
Daniele Ravì
Keerthi Sravan Ravi
Nishant Ravikumar
Harish RaviPrakash
Samuel Remedios
Yinhao Ren

Yudan Ren
Mauricio Reyes
Constantino Reyes-Aldasoro
Jonas Richiardi
David Richmond
Anne-Marie Rickmann
Leticia Rittner
Dominik Rivoir
Emma Robinson
Jessica Rodgers
Rafael Rodrigues
Robert Rohling
Michal Rosen-Zvi
Lukasz Roszkowiak
Karsten Roth
José Rouco
Daniel Rueckert
Jaime S. Cardoso
Mohammad Sabokrou
Ario Sadafi
Monjoy Saha
Pramit Saha
Dushyant Sahoo
Pranjal Sahu
Maria Sainz de Cea
Olivier Salvado
Robin Sandkuehler
Gianmarco Santini
Duygu Sarikaya
Imari Sato
Olivier Saut
Dustin Scheinost
Nico Scherf
Markus Schirmer
Alexander Schlaefer
Jerome Schmid
Julia Schnabel
Klaus Schoeffmann
Andreas Schuh
Ernst Schwartz
Christina Schwarz-Gsaxner
Michaël Sdika
Suman Sedai
Anjany Sekuboyina
Raghavendra Selvan
Sourya Sengupta

Youngho Seo
Lama Seoud
Ana Sequeira
Maxime Sermesant
Carmen Serrano
Muhammad Shaban
Ahmed Shaffie
Sobhan Shafiei
Mohammad Abuzar Shaikh
Reuben Shamir
Shayan Shams
Hongming Shan
Harshita Sharma
Gregory Sharp
Mohamed Shehata
Haocheng Shen
Li Shen
Liyue Shen
Mali Shen
Yiqing Shen
Yiqiu Shen
Zhengyang Shen
Kuangyu Shi
Luyao Shi
Xiaoshuang Shi
Xueying Shi
Yemin Shi
Yiyu Shi
Yonghong Shi
Jitae Shin
Boris Shirokikh
Suprosanna Shit
Suzanne Shontz
Yucheng Shu
Alberto Signoroni
Wilson Silva
Margarida Silveira
Matthew Sinclair
Rohit Singla
Sumedha Singla
Ayushi Sinha
Kevin Smith
Rajath Soans
Ahmed Soliman
Stefan Sommer
Yang Song

Youyi Song
Aristeidis Sotiras
Arcot Sowmya
Rachel Sparks
William Speier
Ziga Spiclin
Dominik Spinczyk
Jon Sporring
Chetan Srinidhi
Anuroop Sriram
Vinkle Srivastav
Lawrence Staib
Marius Staring
Johannes Stegmaier
Joshua Stough
Robin Strand
Martin Styner
Hai Su
Yun-Hsuan Su
Vaishnavi Subramanian
Gérard Subsol
Yao Sui
Avan Suinesiaputra
Jeremias Sulam
Shipra Suman
Li Sun
Wenqing Sun
Chiranjib Sur
Yannick Suter
Tanveer Syeda-Mahmood
Fatemeh Taheri Dezaki
Roger Tam
José Tamez-Peña
Chaowei Tan
Hao Tang
Thomas Tang
Yucheng Tang
Zihao Tang
Mickael Tardy
Giacomo Tarroni
Jonas Teuwen
Paul Thienphrapa
Stephen Thompson
Jiang Tian
Yu Tian
Yun Tian

Aleksei Tiulpin
Hamid Tizhoosh
Matthew Toews
Oguzhan Topsakal
Antonio Torteya
Sylvie Treuillet
Jocelyne Troccaz
Roger Trullo
Chialing Tsai
Sudhakar Tummala
Verena Uslar
Hristina Uzunova
Régis Vaillant
Maria Vakalopoulou
Jeya Maria Jose Valanarasu
Tom van Sonsbeek
Gijs van Tulder
Marta Varela
Thomas Varsavsky
Francisco Vasconcelos
Liset Vazquez Romaguera
S. Swaroop Vedula
Sanketh Vedula
Harini Veeraraghavan
Miguel Vega
Gonzalo Vegas Sanchez-Ferrero
Anant Vemuri
Gopalkrishna Veni
Mitko Veta
Thomas Vetter
Pedro Vieira
Juan Pedro Vigueras Guillén
Barbara Villarini
Satish Viswanath
Athanasios Vlontzos
Wolf-Dieter Vogl
Bo Wang
Cheng Wang
Chengjia Wang
Chunliang Wang
Clinton Wang
Congcong Wang
Dadong Wang
Dongang Wang
Haifeng Wang
Hongyu Wang

Hu Wang
Huan Wang
Kun Wang
Li Wang
Liansheng Wang
Linwei Wang
Manning Wang
Renzhen Wang
Ruixuan Wang
Sheng Wang
Shujun Wang
Shuo Wang
Tianchen Wang
Tongxin Wang
Wenzhe Wang
Xi Wang
Xiaosong Wang
Yan Wang
Yaping Wang
Yi Wang
Yirui Wang
Zeyi Wang
Zhangyang Wang
Zihao Wang
Zuhui Wang
Simon Warfield
Jonathan Weber
Jürgen Weese
Dong Wei
Donglai Wei
Dongming Wei
Martin Weigert
Wolfgang Wein
Michael Wels
Cédric Wemmert
Junhao Wen
Travis Williams
Matthias Wilms
Stefan Winzeck
James Wiskin
Adam Wittek
Marek Wodzinski
Jelmer Wolterink
Ken C. L. Wong
Chongruo Wu
Guoqing Wu

Ji Wu
Jian Wu
Jie Ying Wu
Pengxiang Wu
Xiyin Wu
Ye Wu
Yicheng Wu
Yifan Wu
Tobias Wuerfl
Pengcheng Xi
James Xia
Siyu Xia
Wenfeng Xia
Yingda Xia
Yong Xia
Lei Xiang
Deqiang Xiao
Li Xiao
Yiming Xiao
Hongtao Xie
Lingxi Xie
Long Xie
Weidi Xie
Yiting Xie
Yutong Xie
Xiaohan Xing
Chang Xu
Chenchu Xu
Hongming Xu
Kele Xu
Min Xu
Rui Xu
Xiaowei Xu
Xuanang Xu
Yongchao Xu
Zhenghua Xu
Zhoubing Xu
Kai Xuan
Cheng Xue
Jie Xue
Wufeng Xue
Yuan Xue
Faridah Yahya
Ke Yan
Yuguang Yan
Zhennan Yan

Changchun Yang
Chao-Han Huck Yang
Dong Yang
Erkun Yang
Fan Yang
Ge Yang
Guang Yang
Guanyu Yang
Heran Yang
Hongxu Yang
Huijuan Yang
Jiancheng Yang
Jie Yang
Junlin Yang
Lin Yang
Peng Yang
Xin Yang
Yan Yang
Yujiu Yang
Dongren Yao
Jiawen Yao
Li Yao
Qingsong Yao
Chuyang Ye
Dong Hye Ye
Menglong Ye
Xujiong Ye
Jingru Yi
Jirong Yi
Xin Yi
Youngjin Yoo
Chenyu You
Haichao Yu
Hanchao Yu
Lequan Yu
Qi Yu
Yang Yu
Pengyu Yuan
Fatemeh Zabihollahy
Ghada Zamzmi
Marco Zenati
Guodong Zeng
Rui Zeng
Oliver Zettinig
Zhiwei Zhai
Chaoyi Zhang

Daoqiang Zhang
Fan Zhang
Guangming Zhang
Hang Zhang
Huahong Zhang
Jianpeng Zhang
Jiong Zhang
Jun Zhang
Lei Zhang
Lichi Zhang
Lin Zhang
Ling Zhang
Lu Zhang
Miaomiao Zhang
Ning Zhang
Qiang Zhang
Rongzhao Zhang
Ru-Yuan Zhang
Shihao Zhang
Shu Zhang
Tong Zhang
Wei Zhang
Weiwei Zhang
Wen Zhang
Wenlu Zhang
Xin Zhang
Ya Zhang
Yanbo Zhang
Yanfu Zhang
Yi Zhang
Yishuo Zhang
Yong Zhang
Yongqin Zhang
You Zhang
Youshan Zhang
Yu Zhang
Yue Zhang
Yueyi Zhang
Yulun Zhang
Yunyan Zhang
Yuyao Zhang
Can Zhao
Changchen Zhao
Chongyue Zhao
Fenqiang Zhao
Gangming Zhao

He Zhao

Jun Zhao

Li Zhao

Qingyu Zhao

Rongchang Zhao

Shen Zhao

Shijie Zhao

Tengda Zhao

Tianyi Zhao

Wei Zhao

Xuandong Zhao

Yiyuan Zhao

Yuan-Xing Zhao

Yue Zhao

Zixu Zhao

Ziyuan Zhao

Xingjian Zhen

Guoyan Zheng

Hao Zheng

Jiannan Zheng

Kang Zheng

Shenhai Zheng

Yalin Zheng

Yinqiang Zheng

Yushan Zheng

Jia-Xing Zhong

Zichun Zhong

Bo Zhou

Haoyin Zhou

Hong-Yu Zhou

Kang Zhou

Sanping Zhou

Sihang Zhou

Tao Zhou

Xiao-Yun Zhou

Yanning Zhou

Yuyin Zhou

Zongwei Zhou

Dongxiao Zhu

Hancan Zhu

Lei Zhu

Qikui Zhu

Xinliang Zhu

Yuemin Zhu

Zhe Zhu

Zhuotun Zhu

Aneeq Zia

Veronika Zimmer

David Zimmerer

Lilla Zöllei

Yukai Zou

Lianrui Zuo

Gerald Zwettler

Reyer Zwiggelaar

Outstanding Reviewers

Neel Dey	New York University, USA
Monica Hernandez	University of Zaragoza, Spain
Ivica Kopriva	Rudjer Boskovich Institute, Croatia
Sebastian Otálora	University of Applied Sciences and Arts Western Switzerland, Switzerland
Danielle Pace	Massachusetts General Hospital, USA
Sérgio Pereira	Lunit Inc., South Korea
David Richmond	IBM Watson Health, USA
Rohit Singla	University of British Columbia, Canada
Yan Wang	Sichuan University, China

Honorable Mentions (Reviewers)

Mazdak Abulnaga	Massachusetts Institute of Technology, USA
Pierre Ambrosini	Erasmus University Medical Center, The Netherlands
Hyeon-Min Bae	Korea Advanced Institute of Science and Technology, South Korea
Mikhail Belyaev	Skolkovo Institute of Science and Technology, Russia
Bhushan Borotikar	Symbiosis International University, India
Katharina Breininger	Friedrich-Alexander-Universität Erlangen-Nürnberg, Germany
Ninon Burgos	CNRS, Paris Brain Institute, France
Mariano Cabezas	The University of Sydney, Australia
Aaron Carass	Johns Hopkins University, USA
Pierre-Henri Conze	IMT Atlantique, France
Christian Desrosiers	École de technologie supérieure, Canada
Reuben Dorent	King's College London, UK
Nicha Dvornek	Yale University, USA
Dmitry V. Dylov	Skolkovo Institute of Science and Technology, Russia
Marius Erdt	Fraunhofer Singapore, Singapore
Ruibin Feng	Stanford University, USA
Enzo Ferrante	CONICET/Universidad Nacional del Litoral, Argentina
Antonio Foncubierta-Rodríguez	IBM Research, Switzerland
Isabel Funke	National Center for Tumor Diseases Dresden, Germany
Adrian Galdran	University of Bournemouth, UK
Ben Glocker	Imperial College London, UK
Cristina González	Universidad de los Andes, Colombia
Maged Goubran	Sunnybrook Research Institute, Canada
Sobhan Goudarzi	Concordia University, Canada
Vicente Grau	University of Oxford, UK
Andreas Hauptmann	University of Oulu, Finland
Nico Hoffmann	Technische Universität Dresden, Germany
Sungmin Hong	Massachusetts General Hospital, Harvard Medical School, USA
Won-Dong Jang	Harvard University, USA
Zhanghexuan Ji	University at Buffalo, SUNY, USA
Neerav Karani	ETH Zurich, Switzerland
Alexander Katzmann	Siemens Healthineers, Germany
Erwan Kerrien	Inria, France
Anitha Priya Krishnan	Genentech, USA
Tahsin Kurc	Stony Brook University, USA
Francesco La Rosa	École polytechnique fédérale de Lausanne, Switzerland
Dmitrii Lachinov	Medical University of Vienna, Austria
Mengzhang Li	Peking University, China
Gilbert Lim	National University of Singapore, Singapore
Dongnan Liu	University of Sydney, Australia

Bin Lou	Siemens Healthineers, USA
Kai Ma	Tencent, China
Klaus H. Maier-Hein	German Cancer Research Center (DKFZ), Germany
Raphael Meier	University Hospital Bern, Switzerland
Tony C. W. Mok	Hong Kong University of Science and Technology, Hong Kong SAR
Lia Morra	Politecnico di Torino, Italy
Cosmas Mwikirize	Rutgers University, USA
Felipe Orihuela-Espina	Instituto Nacional de Astrofísica, Óptica y Electrónica, Mexico
Egor Panfilov	University of Oulu, Finland
Christian Payer	Graz University of Technology, Austria
Sebastian Pölsterl	Ludwig-Maximilians Universität, Germany
José Rouco	University of A Coruña, Spain
Daniel Rueckert	Imperial College London, UK
Julia Schnabel	King's College London, UK
Christina Schwarz-Gsaxner	Graz University of Technology, Austria
Boris Shirokikh	Skolkovo Institute of Science and Technology, Russia
Yang Song	University of New South Wales, Australia
Gérard Subsol	Université de Montpellier, France
Tanveer Syeda-Mahmood	IBM Research, USA
Mickael Tardy	Hera-MI, France
Paul Thienphrapa	Atlas5D, USA
Gijs van Tulder	Radboud University, The Netherlands
Tongxin Wang	Indiana University, USA
Yirui Wang	PAII Inc., USA
Jelmer Wolterink	University of Twente, The Netherlands
Lei Xiang	Subtle Medical Inc., USA
Fatemeh Zabihollahy	Johns Hopkins University, USA
Wei Zhang	University of Georgia, USA
Ya Zhang	Shanghai Jiao Tong University, China
Qingyu Zhao	Stanford University, China
Yushan Zheng	Beihang University, China

Mentorship Program (Mentors)

Shadi Albarqouni	Helmholtz AI, Helmholtz Center Munich, Germany
Hao Chen	Hong Kong University of Science and Technology, Hong Kong SAR
Nadim Daher	NVIDIA, France
Marleen de Bruijne	Erasmus MC/University of Copenhagen, The Netherlands
Qi Dou	The Chinese University of Hong Kong, Hong Kong SAR
Gabor Fichtinger	Queen's University, Canada
Jonny Hancox	NVIDIA, UK

Nobuhiko Hata	Harvard Medical School, USA
Sharon Xiaolei Huang	Pennsylvania State University, USA
Jana Hutter	King's College London, UK
Dakai Jin	PAII Inc., China
Samuel Kadoury	Polytechnique Montréal, Canada
Minjeong Kim	University of North Carolina at Greensboro, USA
Hans Lamecker	1000shapes GmbH, Germany
Andrea Lara	Galileo University, Guatemala
Ngan Le	University of Arkansas, USA
Baiying Lei	Shenzhen University, China
Karim Lekadir	Universitat de Barcelona, Spain
Marius George Linguraru	Children's National Health System/George Washington University, USA
Herve Lombaert	ETS Montreal, Canada
Marco Lorenzi	Inria, France
Le Lu	PAII Inc., China
Xiongbiao Luo	Xiamen University, China
Dzung Pham	Henry M. Jackson Foundation/Uniformed Services University/National Institutes of Health/Johns Hopkins University, USA
Josien Pluim	Eindhoven University of Technology/University Medical Center Utrecht, The Netherlands
Antonio Porras	University of Colorado Anschutz Medical Campus/Children's Hospital Colorado, USA
Islem Rekik	Istanbul Technical University, Turkey
Nicola Rieke	NVIDIA, Germany
Julia Schnabel	TU Munich/Helmholtz Center Munich, Germany, and King's College London, UK
Debdoot Sheet	Indian Institute of Technology Kharagpur, India
Pallavi Tiwari	Case Western Reserve University, USA
Jocelyne Troccaz	CNRS, TIMC, Grenoble Alpes University, France
Sandrine Voros	TIMC-IMAG, INSERM, France
Linwei Wang	Rochester Institute of Technology, USA
Yalin Wang	Arizona State University, USA
Zhong Xue	United Imaging Intelligence Co. Ltd, USA
Renee Yao	NVIDIA, USA
Mohammad Yaqub	Mohamed Bin Zayed University of Artificial Intelligence, United Arab Emirates, and University of Oxford, UK
S. Kevin Zhou	University of Science and Technology of China, China
Lilla Zollei	Massachusetts General Hospital, Harvard Medical School, USA
Maria A. Zuluaga	EURECOM, France

Contents – Part III

Machine Learning - Domain Adaptation

Machine Learning - Federated Learning

Machine Learning - Interpretability/Explainability

Machine Learning – Uncertainty

Machine Learning - Advances in Machine Learning Theory

Towards Robust General Medical Image Segmentation

Laura Daza[1]([✉]), Juan C. Pérez[1,2], and Pablo Arbeláez[1]

[1] Universidad de los Andes, Bogotá, Colombia
la.daza10@uniandes.edu.co
[2] King Abdullah University of Science and Technology (KAUST),
Thuwal, Saudi Arabia

Abstract. The reliability of Deep Learning systems depends on their accuracy but also on their robustness against adversarial perturbations to the input data. Several attacks and defenses have been proposed to improve the performance of Deep Neural Networks under the presence of adversarial noise in the natural image domain. However, robustness in computer-aided diagnosis for volumetric data has only been explored for specific tasks and with limited attacks. We propose a new framework to assess the robustness of general medical image segmentation systems. Our contributions are two-fold: *(i)* we propose a new benchmark to evaluate robustness in the context of the Medical Segmentation Decathlon (MSD) by extending the recent *AutoAttack* natural image classification framework to the domain of volumetric data segmentation, and *(ii)* we present a novel lattice architecture for RObust Generic medical image segmentation (ROG). Our results show that ROG is capable of generalizing across different tasks of the MSD and largely surpasses the state-of-the-art under sophisticated adversarial attacks.

Keywords: Robustness assessment · Adversarial training · Adversarial attacks · General medical segmentation

1 Introduction

The observation that imperceptible changes in the input can mislead Deep Neural Networks (DNN) [6,29] has attracted great interest in the deep learning community. This behaviour has been studied for various tasks, such as classification [19,29], object detection [34] and semantic segmentation [2,7,32], highlighting the importance of *reliably* assessing robustness. To improve this dimension, adversarial robustness has been studied both from the side of the attacks [3,9,14,22,24] and the defenses [17,23,27,35], obtaining sizable progress towards DNN models resistant to adversarial perturbations; however, still much is left to advance. In particular, recognition tasks in medical domains are of

Electronic supplementary material The online version of this chapter (https:// doi.org/10.1007/978-3-030-87199-4_1) contains supplementary material, which is available to authorized users.

© Springer Nature Switzerland AG 2021
M. de Bruijne et al. (Eds.): MICCAI 2021, LNCS 12903, pp. 3–13, 2021.
https://doi.org/10.1007/978-3-030-87199-4_1

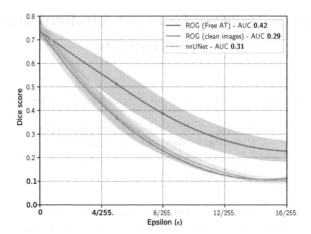

Fig. 1. Medical segmentation robustness benchmark. We introduce a new benchmark for studying adversarial robustness on the MSD [28]. We propose a novel method for robust general segmentation that significantly outperforms the state-of-the-art in our benchmark [13], providing a strong baseline for future reference.

utmost importance for robustness, as these tasks aim at safety-critical applications in which brittleness could have dramatic consequences.

Most semantic segmentation methods for computer aided-diagnosis are designed for specific anatomical structures [12,20,30,38]. This level of specialization has been attributed to the large variability in structures, acquisition protocols, and image modalities, the limited annotated datasets, and the computational cost of processing 3D data. The Medical Segmentation Decathlon (MSD) [28], an experimental framework that combines ten diagnostic tasks, aims at addressing this over-specialization. Top-performing techniques in the MSD solve each task by combining multiple models and automatically adjusting task-specific parameters [13], or conducting Neural Architecture Search (NAS) [33,37]. The MSD, thus, is a promising benchmark to develop general methods for medical image segmentation. However, adversarial robustness, a critical dimension of deep learning systems, remains uncharted territory in the MSD.

In this paper, we propose a new experimental framework to evaluate the adversarial robustness of medical segmentation methods. With this aim, we build on the AutoAttack benchmark [9], an ensemble of adversarial attacks to evaluate recognition models in natural images. We extend AutoAttack from the image classification domain to volumetric multi-channel semantic segmentation. Our results suggest that the adversarial vulnerability of methods in the medical domain, while previously evidenced [16,18], has most likely been *underestimated*. We find that subjecting segmentation methods to our extension of AutoAttack exposes their brittleness against adversarial perturbations.

As a strong baseline for RObust General (ROG) medical segmentation, we propose an efficient lattice architecture that segments organs and lesions on MRI and CT scans. Further, we leverage Adversarial Training (AT) [10,19] to protect our models against adversarial perturbations. Since AT's great computational demands can render training of 3D models prohibitive, we equip ROG with "Free" AT [27], an efficient version of AT. Figure 1 shows the variation in Dice score as the attack's strength, ϵ, increases, comparing ROG trained with clean images, trained with Free AT and nnUnet [13], the winner of the MSD challenge.

Our main contributions can be summarized as follows: *(i)* we extend AutoAttack to the domain of 3D segmentation to introduce an experimental framework for evaluating adversarial robustness in medical segmentation tasks, and *(ii)* we introduce ROG, the first generic medical segmentation model that is robust against adversarial perturbations. We will make our code publicly available to ensure reproducibility and to encourage robustness assessments as a fundamental dimension in the evaluation of 3D DNNs in safety-critical applications[1].

2 Methodology

2.1 Adversarial Robustness

Extending *AutoAttack* to 3D Segmentation. Assessing adversarial robustness has proven to be a complex task [5], hence, many approaches have been proposed [6,22]. In this work, we build on AutoAttack [9], a parameter-free ensemble of attacks that has shown remarkable capacity for breaking adversarial defenses and has emerged as the leading benchmark to assess adversarial robustness. AutoAttack is composed of: *(i)* AutoPGD-CE and *(ii)* AutoPGD-DLR, two Projected Gradient Descent-based (PGD) adversaries that optimize Cross Entropy and Difference-of-Logits Ratio [9] respectively, *(iii)* targeted Fast Adaptive Boundary (FAB-T) [8] which minimizes a decision-changing perturbation, and *(iv)* Square Attack [1], a score-based black-box attack.

Image classification is the most studied setting for adversarial robustness [19,21]. Thus, AutoAttack was designed to fool models that assign a probability distribution to *each image*. However, we are interested in attacking models that assign a probability distribution to *each spatial location*. Therefore, we generalize the definition of the attacks to consider both input and output as four-dimensional tensors: three spatial dimensions plus the channels and the spatial dimensions plus the probability distribution over the classes. In addition, every attack requires a notion of "success", *i.e.* the state in which the attack has fooled the model. While the success of an attack is readily defined for classification, it is not clearly defined for segmentation. Thus, we re-define this notion as a drop in the Dice score below $\mu_i/2$, where μ_i is the average performance in task i over clean images. With this definition, we modify each attack to compute voxel-wise functions across all spatial locations. Finally, we preserve the definition of "Robust Accuracy" as the per-instance worst case across the four attacks.

Adversarial Training. Given the critical applications of medical image segmentation, it is crucial to develop methods that are both accurate and robust to perturbations. To improve ROG's robustness, we apply "Free" Adversarial

---

[1] https://github.com/BCV-Uniandes/ROG.

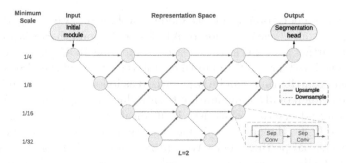

Fig. 2. Overview of ROG. Our lattice architecture for generic medical segmentation preserves high resolution features while also exploiting multiple image scales. The initial module adjusts the input by reducing its scale to a minimum of 1/4 of the original size, while the segmentation head adapts to produce a mask of the same size as the input.

Training [27] in our task domain to fine-tune our models, following the spirit of [4, 11]. For Free AT, we set ϵ, the strength of the attack, to $8/255$, and m, the number of times an image is replayed for gradient computation, to 5.

2.2   Generic Medical Segmentation

We aim to study the unexplored dimension of adversarial robustness in general medical image segmentation. However, existing methods lack publicly available resources [33] or require training multiple models for each task [13]. Combined with the high computational demands of adversarial training, these limitations prevent their adoption for our primary goal. Hence, we propose ROG, an efficient single-architecture model for *RObust Generic* segmentation that exploits increasingly larger receptive fields while preserving high-resolution features. ROG comprises an initial module with four convolutions, the main lattice of processing nodes, and a segmentation head with two convolutions. We organize the nodes in a triangular lattice with four scales, as shown in Fig. 2. Unlike UNet++ [36], we connect each node with both upper and lower level nodes, and we remove the dense connections. As a result, ROG has $10\times$ fewer parameters ($2.6M$ vs. $26.2M$) than [36]. We find that our topology is significantly more efficient than a squared lattice since it reduces the number of layers while preserving the advantages of simultaneous multi-scale processing. We also observe that varying the length L of the lattice, *i.e.* the number of nodes in the last scale, benefits different types of objects. In particular, lower lengths tend to benefit small objects but are detrimental for large objects, while longer lengths favor the opposite. We set $L = 2$ as a trade-off between accuracy and model size.

Within each node, we use two separable convolutions, instance normalization layers, and residual connections. We also use swish activations [26], given the benefit of its smooth gradient for adversarial training [31]. To connect nodes at different levels, we use $1 \times 1 \times 1$ convolutions to adjust the number of feature maps and change the spatial resolution by either using a stride of 2 or adding a trilinear interpolation. Finally, we automatically adjust ROG for each task using the average size of the volumes in the dataset to select the input patch size, strides in the initial module, and upsampling factors in the segmentation head. This selection is driven by two main factors: the target objects' spatial dimensions and the model's memory consumption. Also, we make sure that the minimum possible spatial resolution is $1/32$ of the input resolution.

3 Experiments

3.1 Adversarial Robustness Assessment

We train two variants of ROG for each task: using only clean images and using Free AT. We then use our extended version of AutoAttack to attack the models. For the AutoPGD and FAB attacks, we fix the number of iterations to 5 and vary ϵ between $5/255$ and $16/255$ to assess the adversarial robustness of each model to perturbations of different magnitudes. However, since our experimental framework differs significantly from natural image classification, we also explore the attacks using between 5 and 20 iterations with $\epsilon = 8/255$ and report these results in the supplemental material. For the Square attack, we vary the number of queries between 500 and 5000. We train our models on 80% of the data and validate on the remaining 20%. To reduce bias during the evaluation, all design choices are made based on the results over the tasks of pancreas and prostate segmentation of the MSD. For reference, we evaluate the robustness of nnU-Net, the highest performing method on the MSD with publicly available resources.

3.2 General Medical Segmentation

We evaluate the generalization capacity of ROG on the MSD. Each task has a target that can be an organ or a tumor and source data that can be CT or MRI with one or multiple modalities. The number of patients varies between 30 and 750 for the tasks, with a total of 2,633 volumes [28]. We use Adam [15] with a weight decay of 10^{-5}, and an initial learning rate of 10^{-3} that is reduced by a factor of 0.5 when the validation loss has stagnated for the previous 50 epochs. We optimize a combination of the Dice loss and the Cross-Entropy loss. For data augmentation, we use random rotations, scaling, mirroring, and gamma correction. We sample cubic patches with a 50% probability of being centered at a foreground category to address data imbalance. For pre-processing, we resample the volumes to ensure a homogeneous voxel spacing. Next, we follow [13] and clip the intensities to the $[0.5, 99.5]$ percentiles of the foreground values and perform z-score normalization.

For inference, we uniformly sample patches and combine the predictions to generate the volume. We combine the predictions of the same voxel in multiple patches via weights inversely proportional to the distance to the patch's center. To obtain the final predictions for the MSD test set, we perform simple noise reduction: for organs, we preserve only the largest element; for tumors, we fuse all elements smaller than a threshold with the surrounding category. The threshold is defined based on the average object size in the dataset.

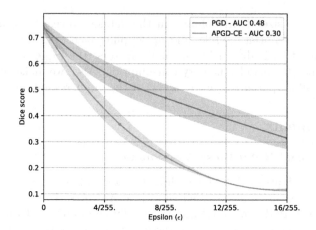

Fig. 3. PGD vs. AutoPGD-CE. We compare the performance of PGD and AutoPGD-CE by attacking ROG trained on clean images averaging the performance over the 10 tasks in the MSD. The results demonstrate that the AutoPGD-CE attack is significantly stronger than regular PGD.

4   Results

4.1   Adversarial Robustness

Adversarial Attacks. Previous assessments of adversarial robustness in segmentation tasks rely on the well-known PGD attack [23]. To elucidate the need for stronger adversaries, we compare the performances of AutoPGD-CE and PGD. We run both attacks on our model trained with clean images in all the datasets, vary the attacks' strength (ϵ), and report the average results in Fig. 3. In line with [9], we find that AutoPGD-CE is dramatically stronger for *every* ϵ. In particular, the AUC for PGD is 0.48, while it drops to 0.30 for AutoPGD-CE.

We test the individual attacks *within* AutoAttack following the same configuration and report the average Dice scores in Table 1. We find that both PGD-based adversaries are the strongest attacks. Further, we note that FAB-T's performance is deficient: FAB-T is unable to find adversaries. While studies on FAB [8,9] often report lower performance for FAB when compared to PGD-based adversaries, our results present a stark contrast. While this observation calls for further experimentation, our initial hypothesis is that FAB's formulation, based on decision-boundary approximation through linearization near the input, may be ill-suited for segmentation. Thus, a re-definition of FAB's inner workings to account for the particularities of segmentation methods may be called for. Regarding the Square attack, despite the simplicity of the random search-based optimization, it consistently finds adversaries. However, since the search space for variable-sized cubes for volumetric inputs is rather large, the computational demands are notably higher than the other attacks.

Table 1. Comparison of individuals attacks of AutoAttack. We report the average performance over all datasets in the MSD with the four attacks in AutoAttack.

	Clean	APGD-CE	APGD-DLR	FAB-T	Square
ROG	0.7375	0.1222	0.3581	0.7375	0.5791

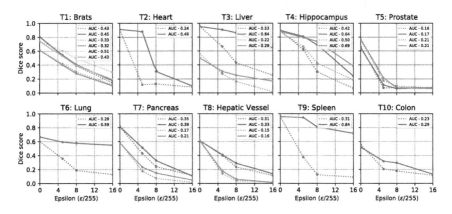

Fig. 4. Adversarial robustness for the 10 tasks in the MSD. We train ROG on clean images (dotted lines) and with Free AT (solid lines), and evaluate their adversarial robustness across various ϵ. Every color represents a different category within each task. In general, training with Free AT improves adversarial robustness against attacks.

Adversarial Robustness. We conduct attacks on every task of the MSD, vary the attack strength (ϵ), and report the Dice score in Fig. 4. Dotted lines represent the performance of ROG with regular training, solid lines show ROG with Free AT, and categories within each task are color-coded. The effect on the Dice score with regular training demonstrates the susceptibility of segmentation models to adversarial attacks. While the ranges for Dice score vary greatly across tasks and categories, our results suggest a coherent finding consistent with the robustness literature: training with Free AT is a tractable approach to improving the adversarial robustness of DNNs. In addition, in most tasks, Free AT enhances robustness *without* major costs to performance on *clean* samples. We identify that the robustness improvement is greater for large structures, such as the liver, and for definite structures like the left atrium of the heart.

Qualitative Evaluation. We show qualitative results in Fig. 5, where the task is to segment liver and liver tumors. We compare ROG with nnUNet over clean and adversarial images computed through APGD-CE with $\epsilon = 8/255$ and $m = 5$. For more results, please refer to the supplemental material. Our results show that, on clean images, ROG provides similar segmentations to those of nnUNet. However, on adversarial images, both methods strongly diverge in their performance. In particular, ROG preserves the segmentations rather untouched, while nnUNet's segmentations are shattered, and the tumor completely disappears.

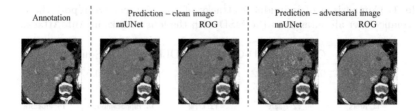

Fig. 5. Qualitative results. We compare the results of ROG and nnUNet on liver images. The adversarial examples are obtained using APGD-CE with $\epsilon = 8/255$ and $m = 5$. Both methods have comparable performance on clean images, while ROG is vastly superior in the presence of adversarial noise.

Table 2. MSD test set results. Comparison of Dice score with top performing methods of the MSD. The results were retrieved from the MSD official leaderboard.

	T1	T2	T3	T4	T5	T6	T7	T8	T9	T10	Mean	Std
MPUNet [25]	0.6	0.89	0.75	0.89	0.77	0.59	0.48	0.48	0.95	0.28	0.66	0.22
nnUNet [13]	0.61	0.93	0.84	0.89	0.83	0.69	0.66	0.66	0.96	0.56	0.75	0.15
C2FNAS [33]	0.62	0.92	0.84	0.88	0.82	0.70	0.67	0.67	0.96	0.59	0.75	0.14
ROG (Ours)	0.56	0.91	0.75	0.89	0.79	0.36	0.66	0.62	0.88	0.46	0.69	0.18

We highlight a visually appealing phenomenon: nnUNet's segmentations appear to follow a pattern of segmenting elements whose boundaries usually correspond with high contrast boundaries. These patterns suggest that nnUNet, and perhaps other segmentation methods, are subject to strong inductive bias that may come from the training data or the convolutional layers with small kernels.

All in all, we observe that nnUNet's vulnerability to adversarial noise is high, failing both on the side of false positives and false negatives. In real-life clinical scenarios, the consequences of such mistakes, while largely unpredictable, could be appalling. Since ROG suffers from these effects to a minor degree, it appears as a better alternative for reliable medical image segmentation.

4.2   Results on the Medical Segmentation Decathlon

To verify the generalization capability of ROG in a clean setup, we evaluate our results in the test server of the MSD. Table 2 summarizes our results compared against all published methods that are evaluated over the entire MSD. ROG surpasses the performance of MPUNet in most tasks with nearly $24\times$ fewer parameters ($2.6M$ vs. $62M$). When we compare to more complex models, we see a trade-off between performance and efficiency. While nnUNet uses an ensemble of models, ROG relies on a *single* architecture. In contrast, C2FNAS uses a single fixed model for most tasks, but requires $6.5\times$ more parameters and almost $5\times$ more GFLOPs (150.78 vs. 30.55). Note that, since ROG adapts to each dataset, we compute the total number of FLOPs by averaging the results of all methods using a maximum input size of $96 \times 96 \times 96$, as in [33]. Under this scheme, our most computationally expensive method has only 79.32 GFLOPs. Considering the large computational demands of studying adversarial robustness, specially in the domain of 3D segmentation, these results demonstrate the advantage of adopting ROG as a strong baseline for the analysis of this dimension in the task of generic medical segmentation.

5   Conclusion

We introduce a benchmark for evaluating the adversarial robustness of volumetric segmentation systems in the medical domain. In particular, we extend the state-of-the-art AutoAttack approach from the image-recognition domain to the volumetric semantic segmentation domain. Further, we propose ROG, a robust and efficient segmentation method for the MSD. We find that ROG provides competitive results over clean images in most tasks of the MSD, while vastly outperforming one of the top-performing methods in this benchmark under adversarial attacks. Our study provides further evidence that the adversarial robustness dimension has been profoundly under-explored in medical image segmentation. Hence, we hope that our work can serve as a tool for developing accurate and robust segmentation models in the clinical domain.

Acknowledgements. We thank Amazon Web Services (AWS) for a computational research grant used for the development of this project.

References

1. Andriushchenko, M., Croce, F., Flammarion, N., Hein, M.: Square attack: a query-efficient black-box adversarial attack via random search. In: Vedaldi, A., Bischof, H., Brox, T., Frahm, J.-M. (eds.) ECCV 2020. LNCS, vol. 12368, pp. 484–501. Springer, Cham (2020). https://doi.org/10.1007/978-3-030-58592-1_29
2. Arnab, A., Miksik, O., Torr, P.H.: On the robustness of semantic segmentation models to adversarial attacks. In: CVPR (2018)
3. Brendel, W., Rauber, J., Bethge, M.: Decision-based adversarial attacks: reliable attacks against black-box machine learning models. In: ICLR (2018)
4. Cai, Q.Z., Liu, C., Song, D.: Curriculum adversarial training. In: International Joint Conference on Artificial Intelligence (IJCAI) (2018)
5. Carlini, N., et al.: On evaluating adversarial robustness. arXiv preprint arXiv:1902.06705 (2019)
6. Carlini, N., Wagner, D.: Towards evaluating the robustness of neural networks. In: 2017 IEEE Symposium on Security and Privacy (SP) (2017)
7. Cisse, M., Adi, Y., Neverova, N., Keshet, J.: Houdini: fooling deep structured prediction models. arXiv preprint arXiv:1707.05373 (2017)
8. Croce, F., Hein, M.: Minimally distorted adversarial examples with a fast adaptive boundary attack. In: International conference on machine learning (ICML) (2020)
9. Croce, F., Hein, M.: Reliable evaluation of adversarial robustness with an ensemble of diverse parameter-free attacks. In: International Conference on Machine Learning (ICML) (2020)

10. Goodfellow, I., Shlens, J., Szegedy, C.: Explaining and harnessing adversarial examples. In: ICLR (2015)
11. Gupta, S., Dube, P., Verma, A.: Improving the affordability of robustness training for DNNs. In: CVPRW (2020)
12. Isensee, F., Jaeger, P.F., Full, P.M., Vollmuth, P., Maier-Hein, K.H.: nnU-net for brain tumor segmentation. arXiv preprint arXiv:2011.00848 (2020)
13. Isensee, F., et al.: nnU-net: self-adapting framework for u-net-based medical image segmentation. CoRR abs/1809.10486 (2018)
14. Joel, M.Z., et al.: Adversarial attack vulnerability of deep learning models for oncologic images. medRxiv (2021). https://doi.org/10.1101/2021.01.17.21249704
15. Kingma, D.P., Ba, J.: Adam: a method for stochastic optimization. In: ICLR (2015)
16. Li, Y., et al.: Volumetric medical image segmentation: a 3D deep coarse-to-fine framework and its adversarial examples. In: Lu, L., Wang, X., Carneiro, G., Yang, L. (eds.) Deep Learning and Convolutional Neural Networks for Medical Imaging and Clinical Informatics. ACVPR, pp. 69–91. Springer, Cham (2019). https://doi.org/10.1007/978-3-030-13969-8_4
17. Liu, Q., et al.: Defending deep learning-based biomedical image segmentation from adversarial attacks: a low-cost frequency refinement approach. In: Martel, A.L., et al. (eds.) MICCAI 2020. LNCS, vol. 12264, pp. 342–351. Springer, Cham (2020). https://doi.org/10.1007/978-3-030-59719-1_34
18. Ma, X., et al.: Understanding adversarial attacks on deep learning based medical image analysis systems. Pattern Recogn. **110**, 107332 (2020)
19. Madry, A., Makelov, A., Schmidt, L., Tsipras, D., Vladu, A.: Towards deep learning models resistant to adversarial attacks. In: ICLR (2018)
20. Milletari, F., Navab, N., Ahmadi, S.A.: V-net: fully convolutional neural networks for volumetric medical image segmentation. In: International Conference on 3D vision (3DV) (2016)
21. Moosavi-Dezfooli, S.M., Fawzi, A., Fawzi, O., Frossard, P.: Universal adversarial perturbations. In: CVPR (2017)
22. Moosavi-Dezfooli, S.M., Fawzi, A., Frossard, P.: DeepFool: a simple and accurate method to fool deep neural networks. In: CVPR (2016)
23. Mummadi, C.K., Brox, T., Metzen, J.H.: Defending against universal perturbations with shared adversarial training. In: ICCV (2019)
24. Ozbulak, U., Van Messem, A., De Neve, W.: Impact of adversarial examples on deep learning models for biomedical image segmentation. In: Shen, D., et al. (eds.) MICCAI 2019. LNCS, vol. 11765, pp. 300–308. Springer, Cham (2019). https://doi.org/10.1007/978-3-030-32245-8_34
25. Perslev, M., Dam, E.B., Pai, A., Igel, C.: One network to segment them all: a general, lightweight system for accurate 3D medical image segmentation. In: Shen, D., et al. (eds.) MICCAI 2019. LNCS, vol. 11765, pp. 30–38. Springer, Cham (2019). https://doi.org/10.1007/978-3-030-32245-8_4
26. Ramachandran, P., Zoph, B., Le, Q.: Searching for activation functions. In: ICLR (2018)
27. Shafahi, A., et al.: Adversarial training for free! In: NeurIPS (2019)
28. Simpson, A.L., et al.: A large annotated medical image dataset for the development and evaluation of segmentation algorithms. CoRR abs/1902.09063 (2019)
29. Szegedy, C., et al.: Intriguing properties of neural networks. In: ICLR (2014)
30. Tang, H., Zhang, C., Xie, X.: NoduleNet: decoupled false positive reduction for pulmonary nodule detection and segmentation. In: Shen, D., et al. (eds.) MICCAI 2019. LNCS, vol. 11769, pp. 266–274. Springer, Cham (2019). https://doi.org/10.1007/978-3-030-32226-7_30

31. Xie, C., Tan, M., Gong, B., Yuille, A.L., Le, Q.V.: Smooth adversarial training. CoRR abs/2006.14536 (2020)
32. Xie, C., Wang, J., Zhang, Z., Zhou, Y., Xie, L., Yuille, A.: Adversarial examples for semantic segmentation and object detection. In: ICCV (2017)
33. Yu, Q., et al.: C2FNAS: coarse-to-fine neural architecture search for 3D medical image segmentation. In: CVPR (2020)
34. Zhang, H., Wang, J.: Towards adversarially robust object detection. In: ICCV (2019)
35. Zhang, H., Yu, Y., Jiao, J., Xing, E.P., Ghaoui, L.E., Jordan, M.I.: Theoretically principled trade-off between robustness and accuracy. In: International Conference on Machine Learning (ICML) (2019)
36. Zhou, Z., Siddiquee, M.M.R., Tajbakhsh, N., Liang, J.: Unet++: redesigning skip connections to exploit multiscale features in image segmentation. Trans. Med. Imaging (2019)
37. Zhu, Z., Liu, C., Yang, D., Yuille, A., Xu, D.: V-NAS: neural architecture search for volumetric medical image segmentation. In: 2019 International Conference on 3D Vision (3DV) (2019)
38. Zhu, Z., Xia, Y., Shen, W., Fishman, E., Yuille, A.: A 3D coarse-to-fine framework for volumetric medical image segmentation. In: International Conference on 3D Vision (3DV) (2018)

Joint Motion Correction and Super Resolution for Cardiac Segmentation via Latent Optimisation

Shuo Wang[1,2,3(✉)], Chen Qin[4,5], Nicolò Savioli[4,6], Chen Chen[4],
Declan P. O'Regan[6], Stuart Cook[7], Yike Guo[3], Daniel Rueckert[4],
and Wenjia Bai[3,8]

[1] Digital Medical Research Center, Fudan University, Shanghai, China
shuowang@fudan.edu.cn
[2] Shanghai Key Laboratory of MICCAI, Shanghai, China
[3] Data Science Institute, Imperial College London, London, UK
[4] Department of Computing, Imperial College London, London, UK
[5] Institute for Digital Communications, University of Edinburgh, Edinburgh, UK
[6] MRC London Institute of Medical Sciences, Imperial College London, London, UK
[7] National Heart Research Institute, Singapore, Singapore
[8] Department of Brain Sciences, Imperial College London, London, UK

Abstract. In cardiac magnetic resonance (CMR) imaging, a 3D high-resolution segmentation of the heart is essential for detailed description of its anatomical structures. However, due to the limit of acquisition duration and respiratory/cardiac motion, stacks of multi-slice 2D images are acquired in clinical routine. The segmentation of these images provides a low-resolution representation of cardiac anatomy, which may contain artefacts caused by motion. Here we propose a novel latent optimisation framework that jointly performs motion correction and super resolution for cardiac image segmentations. Given a low-resolution segmentation as input, the framework accounts for inter-slice motion in cardiac MR imaging and super-resolves the input into a high-resolution segmentation consistent with input. A multi-view loss is incorporated to leverage information from both short-axis view and long-axis view of cardiac imaging. To solve the inverse problem, iterative optimisation is performed in a latent space, which ensures the anatomical plausibility. This alleviates the need of paired low-resolution and high-resolution images for supervised learning. Experiments on two cardiac MR datasets show that the proposed framework achieves high performance, comparable to state-of-the-art super-resolution approaches and with better cross-domain generalisability and anatomical plausibility. The codes are available at https://github.com/shuowang26/SRHeart.

Keywords: Super-resolution · Motion correction · Cardiac MR

Electronic supplementary material The online version of this chapter (https://doi.org/10.1007/978-3-030-87199-4_2) contains supplementary material, which is available to authorized users.

M. de Bruijne et al. (Eds.): MICCAI 2021, LNCS 12903, pp. 14–24, 2021.
https://doi.org/10.1007/978-3-030-87199-4_2

1   Introduction

In cardiac imaging, a high-resolution geometric representation of the heart is desired for accurate assessment of its anatomical structure and function. However, this is not easily available in clinical practice or research. For example, cardiac magnetic resonance (CMR) imaging is the current gold standard imaging modality of the heart. Although high-resolution 3D volumes may be acquired with a research acquisition protocol (Fig. 1A, top row), it is not applicable to the clinical routine, as the research protocol requires a long breath-hold that is not feasible for patients with severe cardiac diseases. Instead, a stack of 2D short-axis slices are usually acquired at multiple breath-holds (Fig. 1A, bottom row). Segmentations built from these 2D slices suffer from three types of degradations: anisotropic low-resolution, motion shifts and potential topological defects (Fig. 1B). To address these issues, we propose a novel latent optimisation framework for joint motion correction and super resolution of 3D cardiac image segmentations.

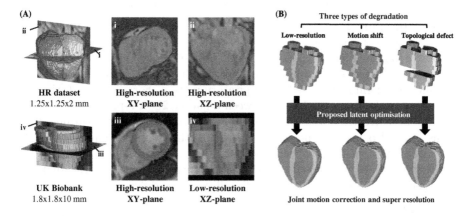

Fig. 1. (A) Illustration of the difference between a high-resolution volume acquired by a research protocol (HR dataset, top) and a low-resolution stack of slices acquired by a standard clinical protocol (UK Biobank, bottom). The UK Biobank data is of low resolution along the Z-axis (subfigure iv: XZ-plane). (B) Illustration of typical degradations in cardiac segmentation.

Related Work: There is a large corpus of works on image super-resolution, including both deep learning-based methods [25] and regularised optimisation methods [26]. Here, we focus on reviewing methods based on convolutional neural networks (CNNs) and related to cardiac imaging enhancement.

Supervised Learning for Super-Resolution: Since the first attempt using CNN for single image super-resolution (SR) [6], deep learning-based SR algorithms have successfully transformed the state-of-the-art performance [5,8,17, 22,27]. In general, these SR networks are trained in a supervised way where pairs of low-resolution (LR) and high-resolution (HR) images are provided. The network is trained to approximate the inverse process of image degradation. However, a large dataset of real paired LR-HR images is often not available but to be simulated, imposing over-fitting risks. Another drawback is their potential failure in cross-domain applications. A well-trained neural network may generate implausible results if applied on a dataset from an unseen domain.

Latent Space Exploration: Instead of approximating the LR→HR mapping through supervised learning, the inverse problem also can be solved via optimisation. Given an LR input, optimisation can be performed in the HR space till the best-fit HR image is found. However, optimisation in the high-dimensional HR image space is challenging and manifold learning has been introduced to shrink the dimensionality [10,23,28]. Menon et al. proposed the PULSE framework to learn the manifold of HR face images [11]. This method does not require paired LR-HR images for supervised training but explores the latent space. In parallel, the concept of latent space exploration has been introduced in medical imaging applications including reconstruction [21], quality control [24] and refinement [15]. These works share the view that images or segmentations are distributed on a low-dimensional manifold which can be parameterised by latent codes.

Cardiac Imaging Enhancement: The main aim of cardiac super-resolution is to improve the resolution along the Z-axis (Fig. 1A). Many studies have investigated the cardiac image enhancement using either traditional methods or deep learning-based methods [4,13]. A particular challenge in CMR imaging is the inter-slice misalignment caused by respiratory motion. For a stacked CMR volume, there exists a potential shift between adjacent 2D XY-plane image, which needs to be corrected for [2,2,12,20]. Besides images, machine learning algorithms have also been proposed for enhancing the segmentations [7,9,14]. In this paper, we focus on the enhancement of cardiac segmentations. Working with segmentations provides us with the convenience to constrain the anatomical geometry and allows better generalisability across different domains.

Contributions: There are three major contributions of this work. Firstly, we propose a novel framework for joint motion correction and super resolution, which alleviates the need of paired samples and achieves competitive results. Secondly, the proposed approach analytically integrates multiple views without retraining the network. Finally, since the framework is based on segmentations, it is agnostic to imaging modalities, devices or contrasts.

2 Methods

2.1 Problem Formulation and Preliminaries

A typical CMR segmentation is a 3D volume that consists of a stack of 2D segmentations for each short-axis slices. Denote the segmentation by $S = \{y_i\}_{i \in N}$, where $y_i \in T = \{1, 2, ...C\}$ denotes the segmentation at i-th voxel, C denotes the number of tissue types, N denotes the total number of voxels, $N = D \times H \times W$, with D, H, W being the depth, height and width. The low-resolution segmentation is denoted by $S_{LR} \in T^{D_{LR} \times H_{LR} \times W_{LR}}$. The corresponding high-resolution segmentation is $S_{HR} \in T^{D_{HR} \times H_{HR} \times W_{HR}}$. The degradation of cardiac segmentation, caused by low-resolution acquisition and motion, is formulated as:

$$S_{LR} = F(S_{HR}) = M_d(\downarrow_s S_{HR}) + \epsilon \qquad (1)$$

where F represents the degradation process of cardiac segmentation, including three components: a down-sampling operator $\downarrow_s$, an inter-slice motion operator M_d and an error term ϵ. $s = [D_{HR}/D_{LR}, H_{HR}/H_{LR}, W_{HR}/W_{LR}]$ is the down-scaling factor. The motion operator M_d is parameterised by $d = [(d_H^1, d_W^1), (d_H^2, d_W^2), ..., (d_H^D, d_W^D)]$, which describes the displacements of 2D slices. $\epsilon \in T^{D_{LR} \times H_{LR} \times W_{LR}}$ represents voxel-wise segmentation errors. Equation (1) describes the HR→LR degradation process with parameters s, d and ϵ. The task of joint motion correction and super resolution is that, given a degraded segmentation S_{LR}, restore a super-resolved segmentation S_{SR} as close to the ground-truth S_{HR} as possible. We briefly describe two commonly used techniques, direct interpolation and supervised learning-based interpolation, before presenting our proposed framework.

Direct Interpolation. The simplest way is to apply a nearest neighbour (NN) up-sampling operation on the low-resolution segmentation, i.e. $S_{NN} = \uparrow_s S_{LR}$. Alternatively, shape-based interpolation (SBI) [18] can be used, which accounts for the shape context.

Learning-Based Interpolation. Similar to image super-resolution, we can utilise neural networks to approximate the inverse process F^{-1} from low-resolution to high-resolution segmentations, i.e., LR→HR mapping. Paired LR and HR samples are needed to train the networks.

2.2 Proposed Latent Optimisation Framework

The proposed framework is composed of two stages. Firstly, we utilise a deep generative model to learn the latent space of plausible high-resolution segmentations (Fig. 2A). Then we perform iterative optimisation for the latent variable z and the motion parameter d, named as 'latent optimisation' (LO). The search is guided by the gradient backpropagated from a consistency loss comparing the degraded high-resolution segmentation by the model with the input low-resolution segmentation (Fig. 2B).

Anatomical Prior. A key component of the proposed framework is the generative model G which learns the manifold Σ_{HR} of plausible high-resolution cardiac segmentations (Fig. 2A), such that:

$$G_\theta(z) : \mathbb{R}^M \ni z \mapsto S = G_\theta(z) \in \Sigma_{HR} \qquad (2)$$

where z denotes the latent representation in a low-dimensional space $\mathbb{R}^M (M \ll N)$ and θ denotes the parameters of the generative model. The generative model can be implemented with different architectures, e.g. VAEs and GANs. With a well-trained generative model, we expect a plausible segmentation via sampling in the latent space, imposing the anatomical prior for latent optimisation.

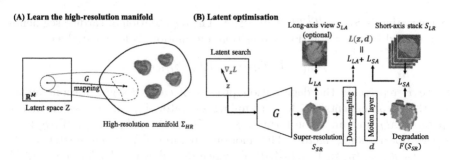

Fig. 2. The latent optimisation framework for joint motion correction and super resolution. (A) A manifold of plausible high-resolution cardiac segmentations is learnt with a generative model G that maps a low-dimensional space Z to the manifold of high-resolution segmentation space Σ_{HR}. (B) Given a low-resolution segmentation S_{LR}, we search in the latent space via gradient descent to find a plausible high-resolution segmentation S_{SR}. The latent variable z and motion displacement d (annotated in red) are optimised to minimise the loss defined on short-axis and long-axis views.

Manifold Traversing. Given a degraded segmentation S_{LR}, the aim of latent optimisation is to traverse the latent space and find the optimal latent variable corresponding to the high-resolution result. The optimisation is formulated as a maximum likelihood problem minimising the consistency loss L.

Multi-view Loss Function. A typical CMR scan provides a stack of short-axis (SA) segmentations as the low-resolution segmentation S_{LR}. The loss function for short-axis view is defined as:

$$L_{SA}(z, d) = CE(M_d(\downarrow_s G(z)), S_{LR}) \qquad (3)$$

where z and d are the latent representation and inter-slice motion shift to be optimised and CE denotes the cross-entropy loss, which describes the consistency between the degraded segmentation and input segmentation. In addition to the short-axis view, auxiliary views such as long-axis (LA) view segmentation S_{LA} may be available. This provides an optional loss term L_{LA}:

$$L_{LA}(z, d) = CE(R \odot G(z), S_{LA}) \qquad (4)$$

where R denotes the slicing operation that maps a volume to the LA view plane. The orientation of LA view is available from the DICOM header. The overall multi-view loss function is:

$$L(z, d) = L_{SA}(z, d) + \gamma \cdot L_{LA}(z, d) \qquad (5)$$

where γ is a hyper-parameter controlling the fidelity weight between SA and LA views. γ is 0 if LA view is not available. To traverse the data manifold, the gradient of the loss function is backpropagated with regard to z and d. We adopt a fully convolutional network architecture. The inter-slice motion shift and down-sampling are implemented as two differentiable layers (Fig. 2B).

2.3   Implementation

For the generative model G, we implemented 3D versions of β-VAE and deep convolutional generative adversarial networks (DCGAN). It turned out that β-VAE with 64 latent dimensions and appropriate regularisation ($\beta = 0.001$) achieved the best performance. The models were implemented in PyTorch and trained using the Adam optimiser with a learning rate of 0.0001 and a batch size of 16. The architecture and loss functions as well as hyper-parameter studies are detailed in *Supple. Material*. The weight γ was set to 1 if long-axis view is available.The trained generative model and deployment code are available at https:// github.com/shuowang26/SRHeart.

For latent optimisation, we used the Adam optimiser with a learning rate of 0.0001. The initial values of z and d were set to 0. The iteration stops if the mean relative change of the loss is smaller than 5%. The degradation process consists of two differentiable layers: the inter-slice motion layer and the down-sampling layer. Specifically, the motion layer was implemented by adding the slice-wise rigid motion onto the mesh grids, followed by bi-linear interpolation. The down-sampling layer was implemented with mesh grids scaled by the scaling factor and with bi-linear interpolation.

For comparison, we implemented interpolation-based and supervised learning-based methods. For direct interpolation, we implemented the NN and SBI up-sampling [18]. For supervised learning, we implemented a 3D version of the enhanced deep residual network (EDSR) which is one of the state-of-the-art super-resolution CNNs [1].

3   Experiments

3.1   Datasets

HR Dataset: This is a research cohort [14] consisting of CMR images from 1,200 healthy volunteers, randomly split into three subsets: training (1,000), validation (100) and test (100). The spatial resolution is $1.25 \times 1.25 \times 2$ mm which is regarded as 'high-resolution' in this work. Cardiac structures of the left ventricle (LV), myocardium (MYO) and right ventricle (RV) were segmented using an atlas-based method, followed by manual correction and quality control.

UK Biobank Dataset: This low-resolution dataset [16] includes short-axis image stacks of 200 subjects, randomly selected from the whole UK Biobank study population ($\sim$100,000). The spatial resolution is $1.8 \times 1.8 \times 10$ mm. The 4-chamber long-axis view is also available, which is a single-slice snapshot with a resolution of 1.8×1.8 mm. The segmentations of LV, MYO and RV on both short-axis and long-axis views were performed automatically with a publicly available fully-convolutional network (FCN) [3].

3.2   Experimental Design

Simulated Degradation. We first simulate the degradation of high-resolution segmentations on the HR dataset using Eq. (1). Technically, we set the down-sampling factor $s = [5, 1, 1]$ to simulate the low-resolution images acquired in the clinical routine. The direction and amplitude of inter-slice displacement of each slice is assumed independent and follow a Gaussian distribution with mean $\mu = 2.3$ mm and standard variance $\sigma = 0.87$ mm, which were fitted from previous studies on UK Biobank [19]. To reflect the fact that respiratory motion could vary among scanning sites and populations, we also simulated scenarios with no inter-slice misalignment and with 4-times misalignment. Overall, we have curated three simulated HR-LR datasets corresponding to the different degradation patterns, named as the No-motion (down-sampling only), Normal-motion (down-sampling+normal respiratory motion), and Severe-motion (down-sampling + 4 × normal respiratory motion) datasets.

Real-World Deployment. We evaluate the real-world performance on the UK Biobank dataset. As ground-truth high-resolution segmentations are not available for evaluation, we slice the super-resolved segmentation along the LA view and compare it to segmentation from LA view. We also investigate the benefits of multi-view loss by incorporating LA view segmentation into the framework.

4   Results and Discussion

4.1   Benchmark on Simulated Degradation

Examples of the super-resolution results from NN, SBI, EDSR and the proposed LO model are demonstrated in Fig. 3. We used Dice coefficient as the metric for evaluating the segmentations. The quantitative results on the simulated datasets of different degradation patterns are reported in *Supple. Material*. When respiratory motion is small (No-motion degradation), all models show good results. With Normal-motion degradation, the performance of all models dropped (*Supple. Material*), especially for the most challenging MYO structure. In this scenario, our LO method showed comparable performance to EDSR and outperformed NN and SBI. Finally, if the respiration motion is high (Severe-motion), our method outperformed other models by a large margin (Table 1). This suggests the robustness of the proposed framework. It is of clinical importance because patients with an acute disease could not follow the breath-holding instruction well, resulting in severe respiratory motions.

We also demonstrate the advantage of our model over supervised learning-based EDSR method in terms of generalisation ability. As shown in Table 2, the performance of EDSR dropped largely if degradation patterns of the training set and test set are different. Under these scenarios, the proposed model outperforms the EDSR models by a large margin (>4%) with a better generalisation ability to unseen degradations.

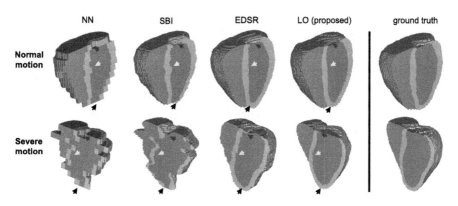

Fig. 3. Illustration of model performance on the simulated Normal- and Severe-motion degradation. Enhancement near basal (red arrow), mid-ventricle (yellow arrow) and apical (black arrow) regions are compared. (Color figure online)

Table 1. Comparison of the model performance under Severe-motion degradation. The dice of different tissues (mean ± std) are evaluated between the ground truth and super-resolved 3D segmentations. Top 2 performance of each tissue are in bold.

	LV	MYO	RV
NN	0.74 ± 0.04	0.32 ± 0.06	0.72 ± 0.04
SBI	0.75 ± 0.03	0.31 ± 0.07	0.74 ± 0.04
EDSR	$\mathbf{0.89 \pm 0.02}$	$\mathbf{0.70 \pm 0.07}$	$\mathbf{0.89 \pm 0.03}$
LO (proposed)	$\mathbf{0.92 \pm 0.02}$	$\mathbf{0.75 \pm 0.08}$	$\mathbf{0.91 \pm 0.02}$

Table 2. Model generalisation under different simulated degradations. Mean Dice of three cardiac structures is evaluated. The first three rows demonstrate the domain-dependence of EDSR performance. Top 2 performance of each column are in bold.

		Test domain		
	Training domain	No-motion	Normal-motion	Severe-motion
EDSR	No-motion	$\mathbf{0.98 \pm 0.00}$	0.88 ± 0.02	0.62 ± 0.04
	Normal-motion	0.91 ± 0.01	$\mathbf{0.93 \pm 0.01}$	0.70 ± 0.05
	Severe-motion	0.68 ± 0.03	0.75 ± 0.03	$\mathbf{0.83 \pm 0.04}$
LO (proposed)		$\mathbf{0.95 \pm 0.01}$	0.92 ± 0.01	$\mathbf{0.86 \pm 0.04}$

4.2   Real-World Performance

We deployed the models on the UK Biobank dataset. The EDSR model was trained on the simulated Normal-motion degradation. The Dice metric between the sliced super-resolved segmentation and the LA view segmentation (considered as ground truth) is reported in Table 3. It shows that the LO method consistently outperformed the other models for all cardiac structures.

Table 3. Real-world performance on UK Biobank dataset evaluated on the 4-chamber long-axis view. LO-multi-view is the LO model incorporating the long-axis segmentation. Top performance of each column is in bold.

	LV	MYO	RV
NN	0.88 ± 0.07	0.57 ± 0.15	0.85 ± 0.12
SBI	0.89 ± 0.07	0.58 ± 0.16	0.85 ± 0.12
EDSR	0.90 ± 0.12	0.64 ± 0.12	0.87 ± 0.13
LO (proposed)	$\mathbf{0.91 \pm 0.08}$	$\mathbf{0.68 \pm 0.12}$	$\mathbf{0.87 \pm 0.12}$
LO-multi-view	$\mathbf{0.92 \pm 0.05}$	$\mathbf{0.72 \pm 0.10}$	$\mathbf{0.89 \pm 0.12}$

In addition, incorporating the long-axis view information further improves the super-resolution performance.

5   Conclusion

We proposed a latent optimisation framework for joint motion correction and super resolution for cardiac segmentation. It achieved high performance on both simulated and real-world datasets. It is a flexible framework, which is able to incorporate multi-view information and robust to domain shift.

Acknowledgements. This research has been conducted using the UK Biobank Resource under Application Number 18545. The authors also acknowledge funding by EPSRC Programme (EP/P001009/1).

References

1. Anwar, S., Khan, S., Barnes, N.: A Deep journey into super-resolution: a survey. ACM Comput. Surv. **53**(3), 1–34 (2020)
2. Avendi, M., Kheradvar, A., Jafarkhani, H.: A combined deep-learning and deformable-model approach to fully automatic segmentation of the left ventricle in cardiac MRI. Med. Image Anal. **30**, 108–119 (2016)
3. Bai, W., Sinclair, M., Tarroni, G., Oktay, O., et al.: Automated cardiovascular magnetic resonance image analysis with fully convolutional networks. J. Cardiovasc. Magn. Reson. **20**(1), 65 (2018)
4. Bhatia, K.K., Price, A.N., Shi, W., Hajnal, J.V., Rueckert, D.: Super-resolution reconstruction of cardiac MRI using coupled dictionary learning. In: IEEE International Symposium on Biomedical Imaging (ISBI), pp. 947–950 (2014)
5. Bulat, A., Yang, J., Tzimiropoulos, G.: To learn image super-resolution, use a GAN to learn how to do image degradation first. In: Proceedings of the European conference on computer vision (ECCV), pp. 185–200 (2018)
6. Dong, C., Loy, C.C., He, K., Tang, X.: Image super-resolution using deep convolutional networks. IEEE Trans. Pattern Anal. Mach. Intell. **38**(2), 295–307 (2015)
7. Duan, J., Bello, G., Schlemper, J., Bai, W., et al.: Automatic 3D bi-ventricular segmentation of cardiac images by a shape-refined multi-task deep learning approach. IEEE Trans. Med. Imaging **38**(9), 2151–2164 (2019)

8. He, K., Zhang, X., Ren, S., Sun, J.: Deep residual learning for image recognition. In: Proceedings of the IEEE Conference on Computer Vision and Pattern Recognition, pp. 770–778 (2016)
9. Larrazabal, A.J., Martínez, C., Glocker, B., Ferrante, E.: Post-DAE: anatomically plausible segmentation via post-processing with denoising autoencoders. IEEE Trans. Med. Imaging **39**, 3813–3820 (2020)
10. Law, M.H., Jain, A.K.: Incremental nonlinear dimensionality reduction by manifold learning. IEEE Trans. Pattern Anal. Mach. Intell. **28**(3), 377–391 (2006)
11. Menon, S., Damian, A., Hu, S., Ravi, N., Rudin, C.: PULSE: self-supervised photo upsampling via latent space exploration of generative models. In: IEEE Conference on Computer Vision and Pattern Recognition (CVPR) (2020)
12. Odille, F., Bustin, A., Chen, B., Vuissoz, P.-A., Felblinger, J.: Motion-corrected, super-resolution reconstruction for high-resolution 3D cardiac cine MRI. In: Navab, N., Hornegger, J., Wells, W.M., Frangi, A.F. (eds.) MICCAI 2015. LNCS, vol. 9351, pp. 435–442. Springer, Cham (2015). https://doi.org/10.1007/978-3-319-24574-4_52
13. Oktay, O., et al.: Multi-input cardiac image super-resolution using convolutional neural networks. In: Ourselin, S., Joskowicz, L., Sabuncu, M.R., Unal, G., Wells, W. (eds.) MICCAI 2016. LNCS, vol. 9902, pp. 246–254. Springer, Cham (2016). https://doi.org/10.1007/978-3-319-46726-9_29
14. Oktay, O., Ferrante, E., Kamnitsas, K., Heinrich, M., et al.: Anatomically constrained neural networks (ACNNs): application to cardiac image enhancement and segmentation. IEEE Trans. Med. Imaging **37**(2), 384–395 (2017)
15. Painchaud, N., Skandarani, Y., Judge, T., Bernard, O., et al.: Cardiac segmentation with strong anatomical guarantees. IEEE Trans. Med. Imaging **39**(11), 3703–3713 (2020)
16. Petersen, S.E., Matthews, P.M., Francis, J.M., Robson, M.D., et al.: UK Biobank's cardiovascular magnetic resonance protocol. J. Cardiovasc. Magn. Reson. **18**(1), 8 (2015)
17. Qiu, Y., Wang, R., Tao, D., Cheng, J.: Embedded block residual network: a recursive restoration model for single-image super-resolution. In: Proceedings of the IEEE International Conference on Computer Vision, pp. 4180–4189 (2019)
18. Raya, S.P., Udupa, J.K.: Shape-based interpolation of multidimensional objects. IEEE Trans. Med. Imaging **9**(1), 32–42 (1990)
19. Tarroni, G., Bai, W., Oktay, O., Schuh, A., et al.: Large-scale quality control of cardiac imaging in population Studies: application to UK Biobank. Sci. Rep. **10**(1), 1–11 (2020)
20. Tarroni, G., Oktay, O., Bai, W., Schuh, A., et al.: Learning-based quality control for cardiac MR images. IEEE Trans. Med. Imaging **38**(5), 1127–1138 (2018)
21. Tezcan, K.C., Baumgartner, C.F., Luechinger, R., Pruessmann, K.P., Konukoglu, E.: MR image reconstruction using deep density priors. IEEE Trans. Med. Imaging **38**(7), 1633–1642 (2018)
22. Tong, T., Li, G., Liu, X., Gao, Q.: Image super-resolution using dense skip connections. In: IEEE International Conference on Computer Vision (ICCV), pp. 4799–4807 (2017)
23. Ulyanov, D., Vedaldi, A., Lempitsky, V.: Deep image prior. In: Proceedings of the IEEE Conference on Computer Vision and Pattern Recognition, pp. 9446–9454 (2018)
24. Wang, S., et al.: Deep generative model-based quality control for cardiac MRI segmentation. In: Martel, A.L., et al. (eds.) MICCAI 2020. LNCS, vol. 12264, pp. 88–97. Springer, Cham (2020). https://doi.org/10.1007/978-3-030-59719-1_9

25. Wang, Z., Chen, J., Hoi, S.C.: Deep learning for image super-resolution: a survey. IEEE Trans. Pattern Anal. Mach. Intell. (2020)
26. Yue, L., Shen, H., Li, J., Yuan, Q., et al.: Image super-resolution: the techniques, applications, and future. Signal Process. **128**, 389–408 (2016)
27. Zhang, Y., Li, K., Li, K., Wang, L., et al.: Image super-resolution using very deep residual channel attention networks. In: European Conference on Computer Vision (ECCV), pp. 286–301 (2018)
28. Zhu, B., Liu, J.Z., Cauley, S.F., Rosen, B.R., Rosen, M.S.: Image reconstruction by domain-transform manifold learning. Nature **555**(7697), 487–492 (2018)

Targeted Gradient Descent: A Novel Method for Convolutional Neural Networks Fine-Tuning and Online-Learning

Junyu Chen[1,2], Evren Asma[3], and Chung Chan[3(✉)]

[1] Department of Radiology and Radiological Science, Johns Hopkins University,
Baltimore, MD, USA
`jchen245@jhmi.edu`
[2] Department of Electrical and Computer Engineering, Johns Hopkins University,
Baltimore, MD, USA
[3] Canon Medical Research USA, Inc., Vernon Hills, IL, USA
{`easma,cchan`}`@mru.medical.canon`

Abstract. A convolutional neural network (ConvNet) is usually trained and then tested using images drawn from the same distribution. To generalize a ConvNet to various tasks often requires a complete training dataset that consists of images drawn from different tasks. In most scenarios, it is nearly impossible to collect every possible representative dataset as a priori. The new data may only become available after the ConvNet is deployed in clinical practice. ConvNet, however, may generate artifacts on out-of-distribution testing samples. In this study, we present Targeted Gradient Descent (TGD), a novel fine-tuning method that can extend a pre-trained network to a new task without revisiting data from the previous task while preserving the knowledge acquired from previous training. To a further extent, the proposed method also enables online learning of patient-specific data. The method is built on the idea of reusing a pre-trained ConvNet's redundant kernels to learn new knowledge. We compare the performance of TGD to several commonly used training approaches on the task of Positron emission tomography (PET) image denoising. Results from clinical images show that TGD generated results on par with training-from-scratch while significantly reducing data preparation and network training time. More importantly, it enables online learning on the testing study to enhance the network's generalization capability in real-world applications.

Keywords: Fine-tuning · Online-learning · Image denoising

Electronic supplementary material The online version of this chapter (https://doi.org/10.1007/978-3-030-87199-4_3) contains supplementary material, which is available to authorized users.

© Springer Nature Switzerland AG 2021
M. de Bruijne et al. (Eds.): MICCAI 2021, LNCS 12903, pp. 25–35, 2021.
https://doi.org/10.1007/978-3-030-87199-4_3

1 Introduction

A Convolutional neural network (ConvNet) is usually trained and tested on datasets where images were sampled from the same distribution. However, if training and testing images are drawn from different distributions, ConvNets would suffer from performance degradation. This is a commonly observed scenario in medical imaging applications due to variations among patients, image acquisition, and reconstruction protocols [8,9,12]. For example, when applying denoising ConvNets on unseen features, it may cause artifacts in the denoised images as demonstrated in both Ultrasound [14] and Positron Emission Tomography (PET) applications [5]. To generalize a trained ConvNet to different image distributions, one has to include images sampled from the new distribution (task) in the training dataset and retrain the ConvNet. However, in medical imaging, generating labeled datasets is often tedious, time-consuming, and expensive. In most scenarios, it is nearly impossible to collect every possible representative dataset as a priori. *In denoising applications, the new data without high quality label may only become available after the ConvNet is deployed (as the high quality label usually requires extra radiation dose or prolonged scan duration).* Moreover, in product development, there is often a need to improve the imaging protocols (i.e., scan or reconstruction protocol) during a product development phase. The change of the image properties, i.e., local pixel correlations, would require regenerating all the training datasets with the updated protocol followed by retraining the denoising network. This recurring process is inevitably time and resource-consuming. Thus, it is more desirable to develop methods that can adapt to various image distributions with minimum need for additional training data and training time. Ultimately, the goal is to develop an online learning algorithm that can quickly retrain and adapt a pre-trained ConvNet to each testing dataset.

Fine-tuning is a promising approach to avoid training a ConvNet from scratch. During fine-tuning, a pre-trained network, usually trained using a large number of datasets from a different application, is used to continue the back-propagation on a smaller dataset from a new task [1,10]. However, fine-tuning the network on a new task does not guarantee retaining the useful knowledge acquired from the previous training. If the number of training dataset from the new task is much less than that used in the old task, the fine-tuned network will overfit to the datasets in the new task with degraded generalization capability [11], which may not be suitable for the applications in which both tasks are of interest during testing. Another approach is using joint training (e.g., [3,21]) or incremental learning (e.g., [4,19,20,22]). They try to adapt a pre-trained network to new tasks while preserving the network's original capabilities. Joint training requires revisiting data from previous tasks while learning the new task [3,17,21] or modifying the network's architecture [19]. Continual learning is used to continuously adapt a ConvNet to a constantly arriving data stream, enabling the ConvNet to learn new tasks incrementally and without forgetting the ones already learned. McClure et al. proposed a continual learning method that consolidates the weights of separate neural networks [18]. The method necessitates that the networks to be trained on the complete datasets. However, obtaining such data may not always be possible. Another example is Elastic Weight Consolidation (EWC) [2,13], which uses Fisher Information Matrix to regularize the penalty term when fine-tuning an existing network using new datasets. Although this method does not require the old training dataset, it might be difficult to fine-tune the hyper-parameter to balance the strength of the weight regularizer and the loss of the new task, especially when only a single testing dataset without label is available.

Instead of blindly fine-tuning all the kernels in the specific layers or retraining the entire network with a mixture of new and old labels, it might be more sensible to precisely retrain the "meaningless" kernels to make them adapt to the new tasks while the "useful" kernels are preserved so they can retain the knowledge acquired from the prior training with a larger training dataset (a wider coverage of data distribution). This work proposes a novel fine-tuning method, the Targeted Gradient Descent (TGD) layer, that can be inserted into any ConvNet architecture. The novel contributions of the proposed method are 2-fold: 1. TGD can extend a pre-trained network to a new task without revisiting data from the previous task while preserving the knowledge acquired from previous training; 2. It enables online learning that adapts a pre-trained network to each testing dataset to avoid generating artifacts on unseen features. We demonstrate the proposed method's effectiveness in denoising tasks for PET images.

2   Methodology

In this study, the pre-trained PET denoising ConvNet was built on the basis of the denoising convolutional neural network (DnCNN) [23]. The architecture of the DnCNN is the same as in Fig. 1 but *without* the TGD layers. It is a 2.5-dimensional network that takes three consecutive 2D image slices as its input. The network consists of eight 3×3 convolutional layers and a single residual layer at the end of the network. Each convolutional layer is followed by a batch normalization and a rectified linear unit (ReLU), except for the first and the last convolutional layers. The first convolution layer is followed by a ReLU, whereas the last convolution layer is not followed by any activation.

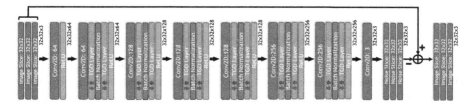

Fig. 1. The architecture of an eight-layer denoising residual network with proposed TGD layers (orange blocks). (Color figure online)

To update the specific kernels in the fine-tuning training, we first need to determine the information richness in the feature maps. The corresponding network kernels can then be identified and updated in the retraining stage to generate new feature maps, such that if the kernels can produce meaningful features, which are identified as "useful" kernels, while the kernels producing "meaningless" features are identified as "meaningless" kernels. However, it is hard to determine a feature map's information richness based solely on some particular input images because different input images may activate different feature maps. Here we used Kernel Sparsity and Entropy (KSE) metric proposed by Li et al. [16]. The KSE quantifies the sparsity and information richness in a kernel to evaluate a feature map's importance to the network. The KSE contains two parts: the kernel sparsity, s_c, and the kernel entropy, e_c, and they are briefly described here. We refer readers to [16] for details. The **kernel sparsity** for the cth input feature map is defined as:

$$s_c = \sum_{n=1}^{N} |W_{n,c}|, \tag{1}$$

where N denotes the total number output feature maps, $W_{n,c}$ denotes the 2D kernels, n and c are, respectively, the indices of the output and input feature maps. The **kernel entropy** is calculated as the entropy of the density metrics (i.e., $dm(\cdot)$):

$$e_c = -\sum_{i=1}^{N} \frac{dm(W_{i,c})}{\sum_{i=1}^{N} dm(W_{i,c})} \log_2 \frac{dm(W_{i,c})}{\sum_{i=1}^{N} dm(W_{i,c})}, \tag{2}$$

where $dm(W_{i,c}) = \sum_j A_{c_{ij}}$, and $A_{c_{ij}}$ is a nearest neighbor distance matrix for the cth convolutional kernel [16]. A small e_c indicates diverse convolution kernels. Thus, the corresponding input feature map provides more information to the ConvNet. KSE is then defined as:

$$KSE = \sqrt{\frac{s_c}{1 + \alpha e_c}}, \tag{3}$$

where KSE, s_c, and e_c are normalized into $[0, 1]$, and α is a parameter for controlling weight between s_c and e_c, which is set to 1 according to [16].

2.1   Targeted Gradient Descent Layer

The KSE score indicates the meaningfulness of feature maps to the ConvNet. Our goal is to retrain the kernels that generate redundant feature maps and keep the "useful" kernels unchanged. In this paper, we denote X and Y, respectively, to be the input and output feature maps of a convolutional layer. As illustrated in Fig. 2, we first calculate KSE for the input feature maps of layer i using the corresponding kernel weights from the ith convolutional layer. The feature maps with KSE scores below a certain user-defined threshold, ϕ, are marked as

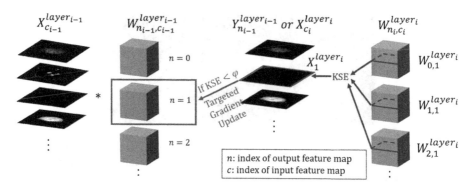

Fig. 2. The framework of identifying kernels that generate redundant feature maps. The kernel weights in layer i (i.e., $W_{n_i,c_i}^{layer_i}$) were used to calculate KSE scores for the input feature maps in layer i (i.e., $X_{c_i}^{layer_i}$), then the kernels in layer $i-1$ (e.g., the green box: $W_{1,c_{i-1}}^{layer_{i-1}}$) that generated the input feature maps in layer i (i.e., $X_{c_i}^{layer_i}$) were identified and would be retrained in the proposed TGD method.

meaningless. We then identify and record the indices of the convolution kernels that generate the "meaningless" feature maps from the $(i-1)^{th}$ layer. The indices were used for creating a binary mask, M:

$$M_n = \begin{cases} \mathbf{1}, & \text{if } KSE(Y_n) < \phi \\ \mathbf{0}, & \text{if } KSE(Y_n) \geq \phi \end{cases},$$

where ϕ is the user-defined KSE threshold. M_n zeros out the gradients for the "useful" kernels (i.e., $KSE(Y_n) \geq \phi$), so that these kernels will not be modified during retraining. The back-propagation formula is then adjusted to incorporate M_n as:

$$W_{n,c}^{(t+1)} = W_n^{(t)} - \eta \frac{\partial \mathcal{L}}{\partial Y_n^{(t)}} M_n X_c^{(t)} - \frac{\partial \mathcal{R}(W_n^{(t)})}{\partial Y_n^{(t)}} M_n X_c^{(t)}, \tag{4}$$

where $\mathcal{L}$ and $\mathcal{R}$ denote, respectively, the loss function and weight regularization. We embedded the gradient zeroing process (i.e., $\frac{\partial \mathcal{L}}{\partial Y_n^{(t)}} M_n$) into a novel layer, named Targeted Gradient Descent layer (the orange blocks in Fig. 1). Notice that the batch normalization layers contain trainable weights as well (i.e., γ, β, and σ^2), where their gradients, $\frac{\partial \mathcal{L}}{\partial \gamma^{(t)}}$, $\frac{\partial \mathcal{L}}{\partial \beta^{(t)}}$, and $\frac{\partial \mathcal{L}}{\partial \sigma^{2(t)}}$ at iteration t can also be expressed as a function of $\frac{\partial \mathcal{L}}{\partial Y_n^{(t)}}$. As a result, the TGD layer was inserted after each convolutional layer as well as each batch normalization layer. Note that *the TGD layers are disabled during forward pass which means all kernels are activated*. During back-propagation, the TGD layers are activated and only the targeted kernels are updated. The final architecture of the TGD-net is shown in Fig. 1.

The TGD-net adapted from a first task to the second task can further be adapted to a third task. The same TGD retraining process can be applied to the

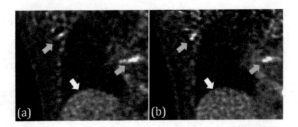

Fig. 3. The PET images of the right lung of a human subject reconstructed with (a) v1 reconstruction and (b) v2 reconstruction methods, respectively. The yellow arrows denote the structures that become elongated in v1 images due to the sub-optimal resolution modeling, which are corrected in the v2 reconstruction. The optimized resolution modeling also produced the liver noise texture (denoted by the white arrows) with a finer grain size (which represents a small correlation between neighboring pixels) in the (b) than (a). (Color figure online)

retrained TGD-net again, i.e., calculate KSE scores for the feature maps in the TGD-net, form new gradient masks, M_n, for the TGD layers, and then retrain the TGD-net with images from a third task. We name this recurring process TGD^n, where n represents the number of TGD retraining processes applied to a single network. In this work, we evaluated the cases in which $n = 1\&2$ (i.e., TGD and TGD^2).

3   TGD in PET Denoising

We demonstrate the proposed TGD method on the task of PET image denoising in two applications. We first use TGD to fine-tune an existing denoising ConvNet to make it adapt to a new reconstruction protocol using substantially fewer training studies. We then further use TGD in an *online-learning* approach to avoid the ConvNet generating artifacts (hallucinations) on unseen features during testing.

3.1   Fine-Tuning

The pre-trained network was trained using FDG PET images acquired on a commercial SiPM PET/CT scanner reconstructed from a prior version of the ordered subset expectation maximization (OSEM) algorithm. For simplicity, we denote these images as the v1 images and the denoising ConvNet trained using these images as the v1 network. We denote the PET images reconstructed by an updated OSEM algorithm as the v2 images and the corresponding denoising ConvNet as the v2 network. The system resolution modeling and scatter estimation in v2 reconstruction were optimized over the v1 reconstruction. Therefore, the noise texture in v2 images is finer, indicating a smaller correlation among neighboring pixels as shown in Fig. 3. The application of the v1 network on v2 images produced over-smoothed results and suppressed activity in small lesions, which could potentially lead to misdiagnosis.

Conventionally, whenever the reconstruction algorithm is updated, the entire training datasets have to be re-reconstructed, and the denoising network has to be retrained using the updated images for optimal performance, followed by qualitative and quantitative assessments on a cohort of testing studies. This process is extremely tedious and time-consuming.

The v1 network was trained using 20 × v1 whole-body FDG-PET human studies with a mixture of low (<23) and high BMI (>28). These studies were acquired for 10-min/bed, which were used as the target images. We uniformly subsampled the list mode data into 6 noise levels as 30, 45, 60, 90, 120, 180 sec/bed as the noisy inputs for noise adaptive training [7]. All these studies consist of 30,720 training slices in total. This v1 network was adapted using the TGD method to denoising v2 PET images. During the TGD's retraining stage, we used only 7 training datasets that consist of PET scans from patients with low BMI (<23). However, the retrained network retained the knowledge on how to denoise PET scans of high BMI patients learned from the previous task (images of high BMI subjects are commonly substantially noisier than those of low BMI subjects). It is important to emphasize that the amount of v1 images used in v1 network training was significantly more than the amount of v2 images used in TGD fine-tuning. Based on this fact, we kept the weights of the noise classifier layer (i.e., the last convolutional layer) in the TGD-net unchanged during the retraining, thus avoiding the last layer from being biased by the v2 image data.

3.2 Online-Learning

In the second experiment, we showed that TGD enables online-learning that further optimize the network's performance on each testing study and prevents artifacts (hallucination) from occurring on out-of-distribution features. This is achieved by using TGD with Noise-2-Noise (N2N) training scheme [6,15]. Specifically, we rebinned a testing study list-mode data acquired with 120-s into 2 noise realizations with equal count levels (60-s). We used TGD to fine-tune the denoising network by using noise samples 1 and 2 as the inputs and noise samples 2 and 1 as the targets. We denote the online-learning network as TGD_{N2N}-net. To a greater extent, this procedure was also applied to the TGD-net from the first experiment (the network was TGD fine-tuned twice), and we denote the resulting network as TGD_{N2N}^2-net for convenience.

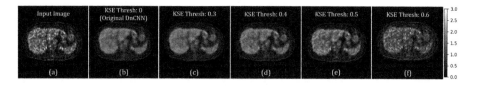

Fig. 4. The network's outputs produced by varying KSE threshold.

4   Experiments

The optimal KSE threshold was first studied. During the prediction, we dropped the kernels (i.e., setting the weights to zeros) identified as "meaningless" by the KSE threshold to examine whether these kernels indeed contributed less to the network. As shown in Fig. 4, where (a) shows an example slice of a PET scan, and (b) shows the denoised PET image from the v1 DnCNN (this can be interpreted as having a KSE threshold = 0, because no kernel was dropped). We then arbitrarily tested four thresholds: 0.3, 0.4, 0.5, and 0.6. The larger the thresholds, the more the kernels were dropped. The percentage of the parameters that were dropped using the four thresholds are, respectively, 51.4%, 51.6%. 53.0%, and 67.3%, and the corresponding denoised results are shown in (c), (d), (e), and (f) of Fig. 4, respectively. The result from $\phi = 0.3$ is almost identical to the original DnCNN's result. Whereas, when $\phi > 0.4$, some severe artifacts begin to occur in the resulting images. Therefore, the KSE threshold, ϕ, was set to be 0.3 in this work.

TGD-net was compared to several baseline methods, including: 1. v1-net: A DnCNN trained using 20× v1 PET images; 2. v2-net: A DnCNN trained using the same 20 studies but reconstructed with v2 algorithm; 3. FT-net: Fine-tuning the last three convolutional blocks [10] of v1-net using only 7× v2 images; 4. TGD-net: v1-net fine-tuned using the TGD layers with 7× v2 images (same studies as used in the FT-net). All networks were trained with 500 epochs.

The proposed $\mathrm{TGD}^2_{\mathrm{N2N}}\text{-net}^{\phi=0.3,0.4}$ and $\mathrm{TGD}_{\mathrm{N2N}}\text{-net}^{\phi=0.4}$ were built based on the previous TGD-net and v2-net, respectively. These networks were retrained using two noise realizations from a single study (i.e., N2N training). They were compared to: 1. v2-net: Same as above. 2. TGD-net$^{\phi=0.3}$: The TGD-net obtained from the previous task. The $\mathrm{TGD}_{\mathrm{N2N}}$ models were trained with 150 epochs.

All these methods were compared in terms of denoising on 3 FDG patient studies (2 are shown in the results, and 1 is shown in the supplementary materials) reconstructed with v2 algorithm (v2 images). One of the studies was acquired with 600-s/bed with a simulated tumor that was inserted in the liver. We rebinned the list-mode study into 10 × 60-s/bed image i.i.d noise realizations to assess the ensemble bias on the tumor and liver coefficient of variation (CoV) by using the 600-s/bed image as the ground truth. We then further evaluated the methods on a second 60-s/bed study.

Table 1. Ensemble bias and CoV comparison results of the different training methods on the PET scan of the low-BMI patient.

	v1-net	v2-net	FT-net	TGD-net$^{0.3}$
Lesion bias (%)	−6.30	−4.07	−4.71	**−3.77**
Liver CoV (%)	6.02	8.56	7.87	6.46

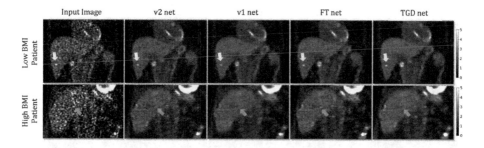

Fig. 5. Qualitative comparisons between the proposed TGD method and other methods on denoising two FDG patient studies. The red arrows indicate the synthetic lesion, and the yellow arrows highlight the difference. (Color figure online)

4.1   Evaluation of TGD on Whole-Body PET Denoising

Figure 5 shows the denoised results of the example cropped slices of the v2 PET images, where the figures in the first column represent the input image. Qualitatively, v1-net (the third column of Fig. 5) over-smoothed the v2 image that led to piece-wise smoothness in the liver and reduced uptake in the synthetic lesion compared to the results from other methods. In contrast, the result from v2-net (the second column) exhibited a higher lesion contrast with more natural noise texture (fine grain size) in liver regions. The fine-tuned networks (FT-net) yielded good performances on denoising the low-BMI-patient PET scans (the top figure of the fourth column) with higher lesion contrast. However, the speckle noise (denoted by the yellow arrow) in the high-BMI-patient PET scans was also preserved. The proposed TGD-net yielded good lesion contrast but also low variations in the liver for both low- and high-BMI patient scans. The quantitative evaluations are shown in Table 1. The best performance is highlighted in bold. For the low-BMI patient study, the proposed method ($\phi = 0.3$) achieved the best lesion quantification with a small ensemble bias of -3.77% while maintaining a low-noise level of 6.45% in terms of CoV. In addition, fine-tuning a TGD net from the v1-net saved 64% of computational time compared to the training-from-scratch v2-net.

Figure 6 shows the denoised results of the example cropped slices of the v2 PET images. Prior to TGD-N2N training, v2-net and TGD-net created artifactual features (hallucination) around the bladder region (denoted by the red arrows). In contrast, the networks fine-tuned using the TGD-N2N online learning scheme did not produce any artifacts, where the bladder's shape is nearly the same as that of the input image. To a greater extend, $\text{TGD}_{\text{N2N}}^2\text{-net}^{\phi=0.3,0.4}$ and $\text{TGD}_{\text{N2N}}\text{-net}^{\phi=0.4}$ retained the denoising performances of their base networks (i.e., v2-net and TGD-net). An additional sample patient study with urinary catheter is shown in the Suppl. Material Fig. 1.

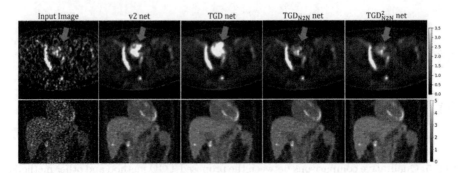

Fig. 6. Qualitative comparisons between the proposed TGD$_{N2N}$ online learning and other methods. The red arrows indicate the artifactual feature generated by the v2-net and TGD-net around the bladder, which was not included in any training datasets. Both TGD$_{N2N}$ and TGD$^2_{N2N}$ yielded images which are in high fidelity to the input image on the bladder, while retaining similar denoising performance as v2-net and TGD-net. (Color figure online)

5   Conclusion

This study introduced Targeted Gradient Descent, a novel incremental learning scheme that effectively reuses the redundant kernels in a pre-trained network. The proposed method can be easily inserted as a layer into an existing network and *does not require revisiting the data from the previous task*. More importantly, it may enable *online learning* on the testing study to enhance the network's generalization capability in real-world applications.

References

1. Amiri, M., Brooks, R., Rivaz, H.: Fine tuning U-net for ultrasound image segmentation: which layers? In: Wang, Q., et al. (eds.) DART/MIL3ID -2019. LNCS, vol. 11795, pp. 235–242. Springer, Cham (2019). https://doi.org/10.1007/978-3-030-33391-1_27
2. Baweja, C., Glocker, B., Kamnitsas, K.: Towards continual learning in medical imaging. arXiv preprint arXiv:1811.02496 (2018)
3. Caruana, R.: Multitask learning. Mach. Learn. **28**(1), 41–75 (1997)
4. Castro, F.M., Marín-Jiménez, M.J., Guil, N., Schmid, C., Alahari, K.: End-to-end incremental learning. In: Proceedings of the European conference on computer vision (ECCV), pp. 233–248 (2018)
5. Chan, C., Yang, L., Asma, E.: Estimating ensemble bias using Bayesian convolutional neural network. In: 2020 IEEE Nuclear Science Symposium and Medical Imaging Conference (NSS/MIC). IEEE (2020)
6. Chan, C., Zhou, J., Yang, L., Qi, W., Asma, E.: Noise to noise ensemble learning for pet image denoising. In: 2019 IEEE Nuclear Science Symposium and Medical Imaging Conference (NSS/MIC), pp. 1–3. IEEE (2019)
7. Chan, C., Zhou, J., Yang, L., Qi, W., Kolthammer, J., Asma, E.: Noise adaptive deep convolutional neural network for whole-body pet denoising. In: 2018 IEEE Nuclear Science Symposium and Medical Imaging Conference Proceedings (NSS/MIC), pp. 1–4. IEEE (2018)

8. Chen, C., Dou, Q., Chen, H., Heng, P.-A.: Semantic-aware generative adversarial nets for unsupervised domain adaptation in chest X-ray segmentation. In: Shi, Y., Suk, H.-I., Liu, M. (eds.) MLMI 2018. LNCS (LNAI and LNB), vol. 11046, pp. 143–151. Springer, Cham (2018). https://doi.org/10.1007/978-3-030-00919-9_17

9. Ghafoorian, M., et al.: Transfer learning for domain adaptation in MRI: application in brain lesion segmentation. In: Descoteaux, M., Maier-Hein, L., Franz, A., Jannin, P., Collins, D.L., Duchesne, S. (eds.) MICCAI 2017. LNCS (LNAI and LNB), vol. 10435, pp. 516–524. Springer, Cham (2017). https://doi.org/10.1007/978-3-319-66179-7_59. https://ncigt.org/files/ncigt/files/ghafoorian-miccai2017.pdf

10. Gong, K., Guan, J., Liu, C.C., Qi, J.: Pet image denoising using a deep neural network through fine tuning. IEEE Trans. Radiat. Plasma Med. Sci. 3(2), 153–161 (2018)

11. Hinton, G., Vinyals, O., Dean, J.: Distilling the knowledge in a neural network. arXiv preprint arXiv:1503.02531 (2015)

12. Kamnitsas, K., et al.: Unsupervised domain adaptation in brain lesion segmentation with adversarial networks. In: Niethammer, M., et al. (eds.) IPMI 2017. LNCS (LNAI and LNB), vol. 10265, pp. 597–609. Springer, Cham (2017). https://doi.org/10.1007/978-3-319-59050-9_47

13. Kirkpatrick, J., et al.: Overcoming catastrophic forgetting in neural networks. Proc. Natl. Acad. Sci. 114(13), 3521–3526 (2017)

14. Laves, M.-H., Tölle, M., Ortmaier, T.: Uncertainty estimation in medical image denoising with Bayesian deep image prior. In: Sudre, C.H., et al. (eds.) UNSURE/GRAIL -2020. LNCS, vol. 12443, pp. 81–96. Springer, Cham (2020). https://doi.org/10.1007/978-3-030-60365-6_9

15. Lehtinen, J., et al.: Noise2noise: learning image restoration without clean data. In: ICML (2018)

16. Li, Y., et al.: Exploiting kernel sparsity and entropy for interpretable CNN compression. In: Proceedings of the IEEE Conference on Computer Vision and Pattern Recognition, pp. 2800–2809 (2019)

17. Li, Z., Hoiem, D.: Learning without forgetting. IEEE Trans. Pattern Anal. Mach. Intell. 40(12), 2935–2947 (2017)

18. McClure, P., et al.: Distributed weight consolidation: a brain segmentation case study. arXiv preprint arXiv:1805.10863 (2018)

19. Rusu, A.A., et al.: Progressive neural networks. arXiv preprint arXiv:1606.04671 (2016)

20. Tasar, O., Tarabalka, Y., Alliez, P.: Incremental learning for semantic segmentation of large-scale remote sensing data. IEEE J. Sel. Top. Appl. Earth Observ. Remote Sens. 12(9), 3524–3537 (2019)

21. Wu, C., Herranz, L., Liu, X., van de Weijer, J., Raducanu, B., et al.: Memory replay GANs: learning to generate new categories without forgetting. In: Advances in Neural Information Processing Systems, pp. 5962–5972 (2018)

22. Xiao, T., Zhang, J., Yang, K., Peng, Y., Zhang, Z.: Error-driven incremental learning in deep convolutional neural network for large-scale image classification. In: Proceedings of the 22nd ACM International Conference on Multimedia, pp. 177–186 (2014)

23. Zhang, K., Zuo, W., Chen, Y., Meng, D., Zhang, L.: Beyond a gaussian denoiser: residual learning of deep CNN for image denoising. IEEE Trans. Image Process. 26(7), 3142–3155 (2017)

A Hierarchical Feature Constraint to Camouflage Medical Adversarial Attacks

Qingsong Yao[1], Zecheng He[2], Yi Lin[3], Kai Ma[3], Yefeng Zheng[3],
and S. Kevin Zhou[1,4(✉)]

[1] Key Lab of Intelligent Information Processing of Chinese Academy of Sciences
(CAS), Institute of Computing Technology, CAS, Beijing 100190, China
yaoqingsong19@mails.ucas.edu.cn
[2] Princeton University, Princeton, USA
zechengh@princeton.edu
[3] Tencent Jarvis Lab, Shenzhen, China
[4] Medical Imaging, Robotics, and Analytic Computing Laboratory and Engineering
(MIRACLE) School of Biomedical Engineering and Suzhou Institute for Advanced
Research, University of Science and Technology of China, Suzhou 215123, China

Abstract. Deep neural networks for medical images are extremely vulnerable to adversarial examples (AEs), which poses security concerns on clinical decision-making. Recent findings have shown that existing medical AEs are easy to detect in feature space. To better understand this phenomenon, we thoroughly investigate the characteristic of traditional medical AEs in feature space. Specifically, we first perform a stress test to reveal the vulnerability of medical images and compare them to natural images. Then, we theoretically prove that the existing adversarial attacks manipulate the prediction by continuously optimizing the vulnerable representations in a fixed direction, leading to outlier representations in feature space. Interestingly, we find this vulnerability is a double-edged sword that can be exploited to help hide AEs in the feature space. We propose a novel hierarchical feature constraint (HFC) as an add-on to existing white-box attacks, which encourages hiding the adversarial representation in the normal feature distribution. We evaluate the proposed method on two public medical image datasets, namely Fundoscopy and Chest X-Ray. Experimental results demonstrate the superiority of our HFC as it bypasses an array of state-of-the-art adversarial medical AEs detector more efficiently than competing adaptive attacks. Our code is available at https://github.com/qsyao/Hierarchical_Feature_Constraint.

Keywords: Adversarial attack · Adversarial defense

Electronic supplementary material The online version of this chapter (https://doi.org/10.1007/978-3-030-87199-4_4) contains supplementary material, which is available to authorized users.

1 Introduction

Deep neural networks (DNNs) are vulnerable to adversarial examples (AEs) [35]. AEs are maliciously generated by adding human-imperceptible perturbations to clean examples, compromising a network to produce the attacker-desired incorrect predictions [6]. The adversarial attack in medical image analysis is disastrous as it can manipulate patients' disease diagnosis and cause serious subsequent problems. More disturbingly, recent studies have shown that DNNs for medical image analysis, including disease diagnosis [9,32], organ segmentation [17,18,20,21,29], and landmark detection [40,41,43,44], are more vulnerable to AEs than natural images.

On the other hand, recent defenses [22,25] have shown that medical AEs can be easily detected in feature space. We plot the 2D t-SNE [26] in Fig. 1 to illustrate the differences between clean and adversarial features from the penultimate layer of a well-trained pneumonia classifier, revealing that adversarial attacks move the deep representations from the original distribution to extreme outlier positions. As a result, a defender can easily take advantage of this characteristic to distinguish AEs.

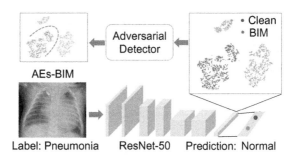

Fig. 1. We craft AEs by the iterative basic method (BIM) [16] to manipulate the prediction and visualize the penultimate layer's features of the adversarial and clean examples by 2D t-SNE.

Given this phenomenon, two key questions are investigated in this paper. The first one is: *What causes medical AEs easier to be detected, compared to AEs of natural images?* To better understand this, we conduct both empirical and theoretical analyses. Firstly, we discover that medical features are more vulnerable than natural ones in a stress test, which aims to change the features by the adversarial attack. Then, we theoretically prove that to generate medical AEs, the adversarial features are pushed in a consistent direction towards outlier regions in the feature space where the clean features rarely reside. In consequence, medical adversarial features become outliers.

The second question is: *Can we hide a medical AE from being spotted in the feature space?* A representative adaptive attack selects a guide example and forces the AE feature to be close to the guide feature [33]. However, this does not directly apply to medical AES. Different medical images have different backgrounds and lesions, making it difficult to manipulate the AE features to be the same as the guide one in all layers within limited perturbation. To hide the adversarial representation at a low price, we propose a novel hierarchical feature constraint (HFC) as an add-on term, which can be plugged into *existing white-box attacks*. HFC first models the normal feature distributions for each layer with a Gaussian mixture model, then encourages the AEs to move to the high-density area by maximizing the log-likelihood of the AE features.

We perform extensive experiments on two public medical diagnosis datasets to validate the effectiveness of HFC. HFC helps medical AE bypass state-of-the-art adversarial detectors, while keeping the perturbations small and unrecognizable to humans. Furthermore, we demonstrate that HFC significantly outperforms other adaptive attacks [1,2,33] on manipulating adversarial features. Our experiments support that **the greater vulnerability of medical representations allows an attacker more room for malicious manipulation**.

Overall, we highlight the following contributions:

- We investigate the feature space of medical images and shed light on why medical AEs can be more easily detected, compared with those of natural images.
- We propose a hierarchical feature constraint (HFC), a novel plug-in that can be applied to existing white-box attacks to avoid being detected.
- Extensive experiments validate the effectiveness of our HFC to help existing adversarial attacks bypass state-of-the-art adversarial detectors simultaneously with small perturbations.

2   Related Work

Preliminaries. Given a clean image x with its label $y \in [1, 2, \ldots, Y]$ and a DNN classifier h, the classifier predicts the class of the input example y' via:

$$y' = \arg\max_{k} \ p(k|x) \equiv \frac{\exp(l_k(x))}{\sum_{j=1}^{K} \exp(l_j(x))}. \tag{1}$$

where the logits $l_k(x)$ (with respect to class k) is given as $l_k(x) = \sum_{n=1}^{N} w_{nk} * z_n(x) + b_k$, in which $z_n(x)$ is the nth activation of the penultimate layer that has N dimensions; w_{nk} and b_k are the weights and the bias from the final dense layer, respectively.

Adversarial Attack. A common way is to manipulate the classifier's prediction by minimizing the classification error between the prediction and target class c, while keeping the AE x_{adv} within a small ϵ-ball of the L_p-norm [27] centered at the original sample x, i.e., $\|x_{adv} - x\|_p \leq \epsilon$. In this paper, we focus on typical L_∞ adversarial attacks, which are most commonly used due to its consistency with respect to human perception [27]. The existing L_∞ approaches can be categorized into two parts. The first one is gradient-based approaches, such as the fast gradient sign method (FGSM) [10], basic iterative method (BIM) [16], the momentum iterative method (MIM) [6] and projected gradient descent (PGD) [27]. BIM computes AE as follows:

$$x_0^* = x; \qquad x_{t+1}^* = \Pi_\epsilon(x_t^* - \alpha \cdot \text{sign}(\nabla_x J(h(x_t^*), c))). \qquad (2)$$

where J is often chosen as the cross-entropy loss; ϵ is the L_∞ norm bound; α is the step size; and $\Pi(\cdot)$ is the project function. Differently from BIM, PGD use a random start $x_0 = x + U^d(-\epsilon, \epsilon)$, where $U^d(-\epsilon, \epsilon)$ is the uniform noise between $-\epsilon$ and ϵ.

The second category is optimization-based methods, among which one of the representative approach is the Carlini and Wanger (CW) attack [3]. According to [27], the L_∞ version of CW attack can be solved by using the following objective function:

$$\hat{J} = \max(l_c(x_t^*) - l_{y_{max} \neq c}(x_t^*), -\kappa). \qquad (3)$$

where l_c is the logits with respect to the target class; $l_{y_{max} \neq c}$ is the maximum logits of the remaining classes; and κ is a parameter managing the confidence.

Adversarial Defenses. Various proactive defenses [12,36] have been proposed to defend adversarial attacks, such as feature squeezing [39], distillation network [31], JPEG compression [7], gradient masking [30]. Per [5], the adversarial training [10,27,38] is the most robust defense, which augments the training set with AEs but consumes too much training time. However, these defenses can be bypassed either completely or partially by adaptive attacks [1,2,37]. Different from the challenging proactive defense, reactive defenses have been developed to detect AEs from clean examples with high accuracy [42]. For example, learning-based methods (e.g., RBF-SVM [23], DNN [28]) train a decision boundary between clean and adversarial features, while anomaly-detection based methods directly reject the outlier feature of the AEs. Specifically, kernel density estimation (KDE) [8] and multivariate Gaussian model (MGM) [22] modeled the normal feature distribution. Local intrinsic dimensionality (LID) [24] characterized the dimensional properties of the adversarial subspaces. The degree of the outlier was measured by Mahalanobis distance (MAHA) [19]. Especially, medical AEs were proven to be easy to detect with nearly *100%* accuracy by these approaches [22,24].

3   Why Are Medical AEs Easy to Detect?

Vulnerability of Representations. A stress test is performed to evaluate the vulnerability of the features of medical and natural images. Specifically, we aim to manipulate the features as much as possible by adversarial attacks. In the test, we decrease ($\downarrow$) and increase ($\uparrow$) the features by replacing the loss function J in BIM (Eq. 2) by $J_\downarrow^* = \text{mean}(f^l(x))$ and $J_\uparrow^* = -\text{mean}(f^l(x))$, respectively, where $f^l(x)$ is the feature from the lth activation layer. We execute the stress test on the medical image dataset (Fundoscopy [13]) and natural image dataset (CIFAR-10 [15]). The results in Table 1 demonstrate that the features of the medical images can be altered more drastically, i.e., **the medical image representations are much more vulnerable than natural images.**

Table 1. Comparison of the robustness between medical images and natural images. We calculate the mean values of the activation layers from ResNet-50 before and after the adversarial attacks. The stress test uses BIM with perturbations under the constraint $L_\infty = 8/256$.

Dataset	Fundoscopy			CIFAR-10		
Layer index	36	45	48	36	45	48
Normal	.0475	.1910	.3750	.0366	.1660	.1900
Adversarial ($\downarrow$)	.0322	.0839	.0980	.0312	.1360	.1480
Adversarial ($\uparrow$)	.0842	.7306	2.0480	.0432	.2030	.2640
Difference ($\downarrow$)	.0153	.1071	.2770	.0054	.0300	.0420
Difference ($\uparrow$)	.0367	.5396	1.6730	.0066	.0370	.0740

Consistency of Gradient Direction. To investigate how the feature vulnerability can be exploited by adversarial attacks, we then focus on the loss function and the corresponding gradient on the final logits output $l_k(x)$. In each iteration of the L_∞ attack introduced above, J and $\hat{J}$ increase the logits of the target class and decrease the others simultaneously. As a result, gradients point to a similar direction across iteration, which will be back-propagated according to the chain rule.

Theorem 1. *Consider a binary disease diagnosis network and its representations from the penultimate layer, the directions of the corresponding gradients are fixed during each iteration under adversarial attack.*[1]

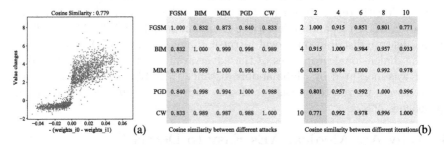

Fig. 2. (a) The similarity between the change values (under adversarial attack) from the penultimate layer and $(w_{i0} - w_{i1})$. (b) The similarity of the changes of features from the penultimate layer between different attacks and the different iterations from basic iterative method (BIM).

Implication. The partial derivative of cross-entropy loss J with respect to the activation value z_i of i-th node in the penultimate layer is computed as:

$$\nabla_{z_i} J(h(x), y_1) = (1 - p_1)(w_{i0} - w_{i1}), \tag{4}$$

---

[1] We provide the proof in the supplementary material.

where p_1 denotes the probability of class 1 and w_{ij} denotes the weight between i-th node in penultimate layer and j-th node in the last layer. Accordingly, the activation z_i with larger $w_{i0} - w_{i1}$ will increase more (guided by the gradient) under attack. We plot the similarity between the value changes and $w_{i0} - w_{i1}$ in Fig. 2(a). Similar conclusions can be derived for other attacks. As shown in Fig. 2(b), **the features are pushed toward a similar direction during different iterations of different attacks**.

4   Adversarial Attack with a Hierarchical Feature Constraint

We investigate the strategies that the attackers can leverage to bypass detection. Our key intuition is to exploit the vulnerability of medical features in the opposite way, i.e., pushing and hiding it in the distribution of normal medical features.

Modeling the Distribution of Normal Features. We model the distribution of normal features using a Gaussian mixture model (GMM).

$$p(f^l(x)) = \sum_{k=1}^{K} \pi_k \mathcal{N}(f^l(x)|\mu_k, \Sigma_k), \tag{5}$$

where p is the probability density of sample x in the target class c; $f^l(\cdot)$ denotes the feature of the lth activation layer; π_k is the mixture coefficient subject to $\sum_{k=1}^{K} \pi_k = 1$; μ_k and Σ_k are the mean and covariance matrix of the k-th Gaussian component in the GMM. These parameters are trained by the expectation-maximization (EM) algorithm [4] on the normal features belonging to the target class c.

Hierarchical Feature Constraint (HFC). We propose a hierarchical feature constraint as a simple-yet-effective add-on that can be applied to existing attacks to avoid being detected. First, we compute the log-likelihood of the AE feature under the distribution of normal features. Specifically, for a given input x, we separately compute the log-likelihood of an AE feature relative to each component and find the most likely Gaussian component:

$$k' = \arg\max_{k} \ \ln(\pi_k \mathcal{N}(f^l(x)|\mu_k, \Sigma_k)). \tag{6}$$

To hide the adversarial representation, we maximize the log-likelihood of this chosen component. Conceptually, it encourages adversarial features in the high-density regions of the distribution of normal features. Then we encourage hiding in all DNN layers. The hierarchical feature constraint induces a loss (J_{HFC}) that is formulated as Eq. 7. It can be used as an add-on to existing attacks by directly adding J_{HFC} to any original attack object function $J_{original}$, i.e., $J = J_{original} + J_{\text{HFC}}$.

$$J_{\text{HFC}} = \sum_{l=1}^{L} \frac{\lambda^l}{2}(f^l(x) - \mu_{k'}^l)^\top (\Sigma_{k'}^l)^{-1}(f^l(x) - \mu_{k'}^l), \tag{7}$$

where λ^l weights the contribution of constraint in layer l. The pseudo-code of the adversarial attack boosted with HFC can be found in supplemental material.

5 Experiments

5.1 Setup

Datasets. Following the literature [9], we use two public datasets on typical medical disease diagnosis tasks. The first one is the Kaggle Fundoscopy dataset [13] on the diabetic retinopathy (DR) classification task, which consists of 3,663 fundus images labeled to one of the five levels from "no DR" to "mid/moderate/severe/proliferate DR". Following [9], we consider all fundoscopies with DR as the same class. The other one is the Kaggle Chest X-Ray [14] dataset on the pneumonia classification task, which consists of 5,863 X-Ray images labeled with "Pneumonia" or "Normal". We split both datasets into three subsets: *Train, AdvTrain* and *AdvTest*. For each dataset, we randomly select 80% of the samples as *Train set* to train the DNN classifier, and treat the left samples as the *Test set*. The incorrectly classified (by the diagnosis network) test samples are discarded. Then we use 70% of the samples (*AdvTrain*) in the *Test set* to train the adversarial detectors and evaluate their effectiveness with the remaining ones (*AdvTest*).

DNN Models. We choose the ResNet-50 [11] and VGG-16 [34] models pretrained on ImageNet. Both models achieve high Area Under Curve (AUC) scores on Fundoscopy and Chest X-Ray datasets: ResNet-50 achieves 99.5% and 97.0%, while VGG-16 achieves 99.3% and 96.5%, respectively.

Adversarial Attacks and Detectors. Following [25], we choose three representative attacks, i.e., BIM, PGD and CW, against our models. For the adversarial detectors, we use kernel density (KD) [8], bayesian uncertainty (BU) [8], local intrinsic dimensionality (LID) [24], Mahalanobis distance (MAHA) [19], RBF-SVM [23], and deep neural network (DNN) [28]. The parameters for KD, LID, BU and MAHA are set per the original papers. For KD, BU, and RBF-SVM, we extract features from the penultimate layers. For DNN, we train a classifier for each layer and ensemble by summing up their logits.

Metrics. We choose three metrics to evaluate the effectiveness of the adversarial detector and the proposed HFC bypassing method: 1) *True positive rate at 90% true negative rate (TPR@90):* The detector will drop 10% of the normal samples to reject more adversarial attacks; 2) *Area Under Curve (AUC) score;* 3) *Adversarial accuracy (Adv. Acc):* The success rate of the targeted adversarial attack against diagnosis network.

Hyperparameters. We set[2] $K = 64$ and 1 for Fundoscopy and Chest X-Ray datasets, respectively. For the lth layer, we compute the mean value of each

---

[2] The hyperparameter analysis can be found in the supplementary material.

channel separately, and set λ_l to $\frac{1}{C_l}$, where C_l is the number of channels. As tiny perturbations in medical images can cause drastic increase in the loss [25], we set $\alpha = 0.02/256$ and $T = 2\epsilon/\alpha$.

Table 2. The point-wise performances of HFC. The metrics scores on the *left* and *right* of the slash are the performances (%) of the adversarial detectors under the attack *without* and *with* HFC, respectively. All of the attacks are evaluated on ResNet-50 with constraint $L_\infty = 1/256$.

Fundoscopy	BIM (Adv. Acc = 99.5)		PGD (Adv. Acc = 99.5)		CW (Adv. Acc = 99.5)	
	AUC	TPR@90	AUC	TPR@90	AUC	TPR@90
KD	99.0/74.2	96.8/20.5	99.4/73.4	98.6/13.2	99.5/74.7	99.1/19.6
MAHA	99.6/6.4	99.5/0.0	100/4.2	100/0.0	99.8/33.0	99.5/0.0
LID	99.8/78.3	100/40.6	99.6/73.2	98.6/35.5	98.8/73.4	97.7/33.3
SVM	99.5/28.6	99.1/0.0	99.8/23.1	99.5/0.0	99.8/27.0	99.5/0.0
DNN	100/60.0	100/12.8	100/58.6	100/8.2	100/62.6	100/15.1
BU	58.9/37.4	9.1/0.0	61.9/35.9	9.1/0.0	93.0/32.8	73.1/5.0
Chest X-Ray	BIM (Adv. Acc = 90.9)		PGD (Adv. Acc = 90.9)		CW (Adv. Acc = 98.9)	
	AUC	TPR@90	AUC	TPR@90	AUC	TPR@90
KD	100/73.1	100/6.8	100/82.3	100/50.5	99.2/71.5	98.4/15.7
MAHA	100/0.0	100/0.0	100/0.0	100/0.0	100/22.4	100/0.0
LID	100/48.6	100/1.8	100/49.1	100/1.5	99.2/64.5	98.4/14.4
SVM	100/16.7	100/6.9	100/5.8	100/0.0	100/21.2	100/0.0
DNN	100/31.8	100/0.7	100/33.7	100/0.0	100/61.6	100/5.2
BU	49.9/26.1	19.2/0.0	49.2/26.2	22.7/0.0	98.3/26.2	94.8/0.0

5.2 Experimental Results

Bypassing Adversarial Detectors. We compare the performances of the existing adversarial attacks *with* and *without* HFC on different DNN classifiers, datasets, and perturbation constraints. The visualization in Fig. 3(a) shows that HFC successfully moves the AE feature (orange) from outlier regions to the regions (cyan) where normal features (purple) reside. Quantitative results in Table 2 show that the HFC boosts all evaluated adversarial attacks to bypass all evaluated detectors simultaneously, with high accuracy and a small perturbation constraint $L_\infty = 1/256$. Furthermore, as shown in Fig. 3(b), a larger perturbation budget provides HFC more room to manipulate the representations, compromising the detectors more drastically (as the dotted lines decrease).

Comparison to Other Adaptive Attacks. We compare the proposed HFC to other adaptive attacks designed to manipulate the deep representations and bypass the detectors: 1) Generate AEs with internal representation similar to a random guide image [33]; 2) Choose a guide image with its representation closest to the input [33]; 3) Minimize the loss terms of KDE and cross-entropy

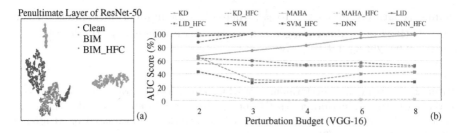

Fig. 3. (a) Visualization of 2D t-SNE of clean and adversarial features generated from BIM and HFC, extracted from ResNet-50 on Chest X-Ray. (b) The AUC scores in the solid lines and dotted lines are the performances of the adversarial detectors against BIM *with* and *without* HFC.

simultaneously [2]; 4) Minimize the loss term of LID and cross-entropy at the same time [1]. As shown in Table 3, our proposed HFC bypasses the five detectors and greatly outperforms other competitive adaptive attacks.

Table 3. AUC scores (%) of the proposed HFC and other adaptive attacks. The AEs are generated on ResNet-50 under constraint $L_\infty = 1/256$.

Fundoscopy	KD	MAHA	LID	SVM	DNN	Chest X-Ray	KD	MAHA	LID	SVM	DNN
Random	75.1	86.1	91.7	48.2	93.7	Random	77.0	64.0	91.0	13.0	79.3
Closest	77.0	64.0	91.0	13.0	79.3	Closest	80.1	38.3	71.3	9.9	87.7
KDE	51.6	86.5	90.9	45.3	95.0	KDE	58.2	66.9	71.7	15.3	95.6
LID	87.6	85.4	93.4	61.2	96.2	LID	84.0	66.6	77.1	28.6	96.6
HFC	74.2	6.4	78.3	28.6	60.0	HFC	70.8	0.0	53.6	16.7	32.6

Table 4. The performance of BIM *w/* and *w/o* HFC under the semi-white-box setting. The AEs are generated under constraint $L_\infty = 4/256$. AUC scores (%) are used as metrics.

DNN model	Fundoscopy	KD	MAHA	LID	SVM	DNN	Adv. Acc
ResNet-50	BIM	98.7	100.0	99.5	92.1	100.0	83.2
	BIM_HFC	78.0	9.1	68.0	16.9	43.8	68.2
VGG-16	BIM	72.3	89.6	81.5	48.1	95.2	88.6
	BIM_HFC	50.9	18.2	64.6	28.8	16.7	73.2

Semi-white-box Attack. We evaluate the proposed method in a more difficult scenario: only the architecture information of DNN models is available. The attacker tries to confuse the victim model and bypass its adversarial detectors simultaneously, without knowing model parameters. The results in Table 4 show that our HFC can help BIM compromise most of the detectors and manipulate the DNN models, which poses more disturbing concerns to the safety of DNN-based diagnosis networks.

6   Conclusion

In this paper, we investigate the characteristics of medical AEs in feature space. A stress test is performed to reveal the greater vulnerability of medical image features, compared to the natural images. Then we theoretically prove that existing adversarial attacks tend to alter the vulnerable features in a fixed direction. As a result, the adversarial features become outliers and easy to detect. However, an attacker can exploit this vulnerability to hide the adversarial features. We propose a novel hierarchical feature constraint (HFC), a simple-yet-effective add-on that can be applied to existing attacks, to avoid AEs from being detected. Extensive experiments validate the effectiveness of HFC, which also significantly outperforms other adaptive attacks. It reveals the limitation of the current methods for detecting medical AEs in the feature space. We hope it can inspire more defenses in future work.

References

1. Athalye, A., Carlini, N., Wagner, D.: Obfuscated gradients give a false sense of security: Circumventing defenses to adversarial examples. In: ICLR (2018)
2. Carlini, N., Wagner, D.: Adversarial examples are not easily detected: bypassing ten detection methods. In: Proceedings of the 10th ACM Workshop on Artificial Intelligence and Security, pp. 3–14 (2017)
3. Carlini, N., Wagner, D.: Towards evaluating the robustness of neural networks. In: IEEE Symposium on Security and Privacy, pp. 39–57 (2017)
4. Dempster, A.P., Laird, N.M., Rubin, D.B.: Maximum likelihood from incomplete data via the EM algorithm. J. R. Stat. Soc.: Ser. B (Methodol.) **39**(1), 1–22 (1977)
5. Dong, Y., et al.: Benchmarking adversarial robustness. In: CVPR (2020)
6. Dong, Y., et al.: Boosting adversarial attacks with momentum. In: CVPR, pp. 9185–9193 (2018)
7. Dziugaite, G.K., Ghahramani, Z., Roy, D.M.: A study of the effect of JPG compression on adversarial images. arXiv preprint arXiv:1608.00853 (2016)
8. Feinman, R., Curtin, R.R., Shintre, S., Gardner, A.B.: Detecting adversarial samples from artifacts. arXiv preprint arXiv:1703.00410 (2017)
9. Finlayson, S.G., Chung, H.W., Kohane, I.S., Beam, A.L.: Adversarial attacks against medical deep learning systems. Science **363**(6433), 1287–1289 (2018)
10. Goodfellow, I.J., Shlens, J., Szegedy, C.: Explaining and harnessing adversarial examples. In: ICLR (2015)
11. He, K., Zhang, X., Ren, S., Sun, J.: Deep residual learning for image recognition. In: CVPR, pp. 770–778 (2016)
12. He, X., Yang, S., Li, G., Li, H., Chang, H., Yu, Y.: Non-local context encoder: Robust biomedical image segmentation against adversarial attacks. In: AAAI, vol. 33, pp. 8417–8424 (2019)
13. Kaggle: APTOS 2019 Blindness Detection (2019). https://www.kaggle.com/c/aptos2019-blindness-detection
14. Kaggle: Chest X-Ray Images (Pneumonia) (2019). https://www.kaggle.com/paultimothymooney/chest-xray-pneumonia
15. Krizhevsky, A.: Learning multiple layers of features from tiny images. University of Toronto, May 2012

16. Kurakin, A., Goodfellow, I., Bengio, S.: Adversarial machine learning at scale. In: ICLR (2017)
17. Ji, W., et al.: Uncertainty quantification for medical image segmentation using dynamic label factor allocation among multiple raters. In: MICCAI on QUBIQ Workshop (2020)
18. Ji, W., et al.: Learning calibrated medical image segmentation via multi-rater agreement modeling. In: CVPR, pp. 12341–12351, June 2021
19. Lee, K., Lee, K., Lee, H., Shin, J.: A simple unified framework for detecting out-of-distribution samples and adversarial attacks. In: ICLR, pp. 7167–7177 (2018)
20. Li, H., et al.: High-resolution chest x-ray bone suppression using unpaired CT structural priors. IEEE Trans. Med. Imaging **39**, 3053–3063 (2020)
21. Li, H., Han, H., Zhou, S.K.: Bounding maps for universal lesion detection. In: Martel, A.L., et al. (eds.) MICCAI 2020. LNCS, vol. 12264, pp. 417–428. Springer, Cham (2020). https://doi.org/10.1007/978-3-030-59719-1_41
22. Li, X., Zhu, D.: Robust detection of adversarial attacks on medical images. In: IEEE International Symposium on Biomedical Imaging, pp. 1154–1158. IEEE (2020)
23. Lu, J., Issaranon, T., Forsyth, D.: SafetyNet: detecting and rejecting adversarial examples robustly. In: ICCV, October 2017
24. Ma, X., et al.: Characterizing adversarial subspaces using local intrinsic dimensionality. In: ICLR (2018)
25. Ma, X., et al.: Understanding adversarial attacks on deep learning based medical image analysis systems. Pattern Recogn. **110**, 107332 (2020)
26. Maaten, L.V.D., Hinton, G.: Visualizing data using t-SNE. J. Mach. Learn. Res. **9**(Nov), 2579–2605 (2008)
27. Madry, A., Makelov, A., Schmidt, L., Tsipras, D., Vladu, A.: Towards deep learning models resistant to adversarial attacks. In: ICLR (2018)
28. Metzen, J.H., Genewein, T., Fischer, V., Bischoff, B.: On detecting adversarial perturbations. In: ICLR (2017)
29. Ozbulak, U., Van Messem, A., De Neve, W.: Impact of adversarial examples on deep learning models for biomedical image segmentation. In: Shen, D., et al. (eds.) MICCAI 2019. LNCS, vol. 11765, pp. 300–308. Springer, Cham (2019). https://doi.org/10.1007/978-3-030-32245-8_34
30. Papernot, N., McDaniel, P., Goodfellow, I., Jha, S., Celik, Z.B., Swami, A.: Practical black-box attacks against machine learning. In: ASIA Computer and Communications Security, pp. 506–519 (2017)
31. Papernot, N., McDaniel, P., Wu, X., Jha, S., Swami, A.: Distillation as a defense to adversarial perturbations against deep neural networks. In: IEEE Symposium on Security and Privacy, pp. 582–597. IEEE (2016)
32. Paschali, M., Conjeti, S., Navarro, F., Navab, N.: Generalizability vs. robustness: investigating medical imaging networks using adversarial examples. In: Frangi, A.F., Schnabel, J.A., Davatzikos, C., Alberola-López, C., Fichtinger, G. (eds.) MICCAI 2018. LNCS, vol. 11070, pp. 493–501. Springer, Cham (2018). https://doi.org/10.1007/978-3-030-00928-1_56
33. Sabour, S., Cao, Y., Faghri, F., Fleet, D.J.: Adversarial manipulation of deep representations. In: IEEE Symposium on Security and Privacy (2016)
34. Simonyan, K., Zisserman, A.: Very deep convolutional networks for large-scale image recognition. In: ICLR (2015)
35. Szegedy, C., et al.: Intriguing properties of neural networks. In: ICLR (2014)
36. Taghanaki, S.A., Abhishek, K., Azizi, S., Hamarneh, G.: A kernelized manifold mapping to diminish the effect of adversarial perturbations. In: CVPR, pp. 11340–11349 (2019)

37. Tramer, F., Carlini, N., Brendel, W., Madry, A.: On adaptive attacks to adversarial example defenses. In: ICLR (2020)
38. Tramèr, F., Kurakin, A., Papernot, N., Goodfellow, I., Boneh, D., McDaniel, P.: Ensemble adversarial training: attacks and defenses. In: ICLR (2018)
39. Xu, W., Evans, D., Qi, Y.: Feature squeezing: detecting adversarial examples in deep neural networks. In: Network and Distributed System Security Symposium (2017)
40. Yao, Q., He, Z., Han, H., Zhou, S.K.: Miss the point: targeted adversarial attack on multiple landmark detection. In: Martel, A.L., et al. (eds.) MICCAI 2020. LNCS, vol. 12264, pp. 692–702. Springer, Cham (2020). https://doi.org/10.1007/978-3-030-59719-1_67
41. Yao, Q., Xiao, L., Liu, P., Zhou, S.K.: Label-free segmentation of COVID-19 lesions in lung CT. IEEE Trans. Med. Imaging (2020)
42. Zheng, Z., Hong, P.: Robust detection of adversarial attacks by modeling the intrinsic properties of deep neural networks. In: Advances in Neural Information Processing Systems, pp. 7913–7922 (2018)
43. Zhou, S.K., et al.: A review of deep learning in medical imaging: imaging traits, technology trends, case studies with progress highlights, and future promises. Proc. IEEE 109(5), 820–838 (2021)
44. Zhou, S.K., Rueckert, D., Fichtinger, G.: Handbook of Medical Image Computing and Computer Assisted Intervention. Academic Press, Cambridge (2019)

Group Shift Pointwise Convolution for Volumetric Medical Image Segmentation

Junjun He[1,2,3,4], Jin Ye[3,4], Cheng Li[5], Diping Song[3,4], Wanli Chen[6], Shanshan Wang[5,7,8], Lixu Gu[1,2], and Yu Qiao[3,4(✉)]

[1] School of Biomedical Engineering, Shanghai Jiao Tong University, Shanghai, China
[2] Institute of Medical Robotics, Shanghai Jiao Tong University, Shanghai, China
[3] Shenzhen Key Lab of Computer Vision and Pattern Recognition,
SIAT-SenseTime Joint Lab, Shenzhen Institute of Advanced Technology,
Chinese Academy of Sciences, Shenzhen, Guangdong, China
yu.qiao@siat.ac.cn
[4] Shanghai AI Lab, Shanghai, China
[5] Paul C. Lauterbur Research Center for Biomedical Imaging,
Shenzhen Institute of Advanced Technology, Chinese Academy of Sciences,
Shenzhen, Guangdong, China
[6] The Chinese University of Hong Kong, Hong Kong, China
[7] Peng Cheng Laboratory, Shenzhen, Guangdong, China
[8] Pazhou Lab, Guangzhou, Guangdong, China

Abstract. Recent studies have witnessed the effectiveness of 3D convolutions on segmenting volumetric medical images. Compared with the 2D counterparts, 3D convolutions can capture the spatial context in three dimensions. Nevertheless, models employing 3D convolutions introduce more trainable parameters and are more computationally complex, which may lead easily to model overfitting especially for medical applications with limited available training data. This paper aims to improve the effectiveness and efficiency of 3D convolutions by introducing a novel Group Shift Pointwise Convolution (GSP-Conv). GSP-Conv simplifies 3D convolutions into pointwise ones with $1 \times 1 \times 1$ kernels, which dramatically reduces the number of model parameters and FLOPs (e.g. $27\times$ fewer than 3D convolutions with $3 \times 3 \times 3$ kernels). Naïve pointwise convolutions with limited receptive fields cannot make full use of the spatial image context. To address this problem, we propose a parameter-free operation, Group Shift (GS), which shifts the feature maps along different spatial directions in an elegant way. With GS, pointwise convolutions can access features from different spatial locations, and the limited receptive fields of pointwise convolutions can be compensated. We evaluate the proposed method on two datasets, PROMISE12 and BraTS18. Results

---

J. He, J. Ye and C. Li—These authors contributed equally to this work.

Electronic supplementary material The online version of this chapter (https://doi.org/10.1007/978-3-030-87199-4_5) contains supplementary material, which is available to authorized users.

M. de Bruijne et al. (Eds.): MICCAI 2021, LNCS 12903, pp. 48–58, 2021.
https://doi.org/10.1007/978-3-030-87199-4_5

show that our method, with substantially decreased model complexity, achieves comparable or even better performance than models employing 3D convolutions.

Keywords: Vomumetric medical image segmentation · Pointwise convolution · Group shift

1 Introduction

Semantic segmentation, which is essential for various applications, is a challenging task in medical imaging. Accurate volumetric medical image segmentation can not only quantitatively assess the volumes of interest (VOIs), but also contribute to the precise disease diagnosis, computer-aided interventions, and surgical planning [9,19]. Manually annotating volumetric medical images (with hundreds of slices and complicate structures) is tedious, time-consuming, and error-prone. Thus, automatic volumetric medical image segmentation methods are highly desired.

Two-dimensional fully convolutional neural network (2D FCN)-based methods have been widely adopted for medical image segmentation [16,21]. However, medical images are commonly in 3D with rich spatial information. Meanwhile, large variations exist in structural appearance, size, and shape among patients. Thus, exploiting 3D structural and anatomical information is critical for accurate volumetric medical image segmentation. Recent works extended 2D FCNs to 3D FCNs by directly adding an operation in the extra dimension [3,4,6,13,15]. Although satisfactory performances were obtained, the parameters and floating-point-operations (FLOPs) increased extremely compared with the 2D counterparts. As a result, increased demands for large training datasets and advanced computational resources arise.

To reduce model parameters and FLOPs and at the same time, maintain the segmentation performance, convolutional kernel factorization-based methods have been extensively investigated for Deep Convolutional Neural Networks (DCNNs) [5,17,18,20,22]. In the earliest DCNNs, filters with large kernels were designed to enlarge the receptive field (RF) and make full use of the spatial context [10]. Later studies found that by decomposing a large filter into several consecutive small filters, the same RF could be obtained and superior performance with fewer parameters and FLOPs could be achieved [17,18]. For example, a 7×7 filter can be decomposed into three 3×3 filters. Decomposing a high dimensional filter into several low dimensional filters along the different dimensions is another method of convolutional kernel factorization. Depthwise Separable Convolutions (DSCs) decompose filters along the spatial and channel dimensions [5]. DSCs treat pointwise convolutions (1×1 for 2D networks and $1 \times 1 \times 1$ for 3D networks) as the endpoint of convolution factorization. Pointwise convolutions are the most efficient convolutions in DCNNs with the fewest parameters and FLOPs. Nonetheless, the severely limited RF of pointwise convolutions makes it difficult to construct a working neural network with pure pointwise convolutions.

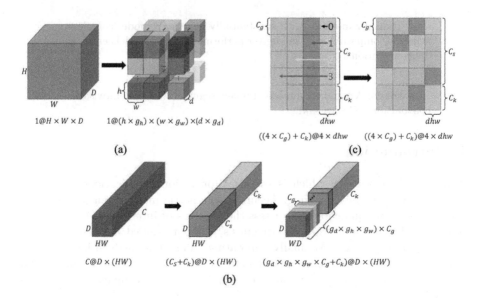

Fig. 1. The proposed Group Shift (GS) operation. GS consists of two steps, grouping and shift. (a) Spatial grouping refers to the operation of grouping features in the three spatial dimensions. (b) Channel grouping represents the operation of grouping features in the channel dimension. (c) Shift is the operation of shifting the grouped features according to the spatial group and channel group indexes.

In this paper, we attempt to build a novel DCNN for volumetric medical image segmentation by answering the following question: Can we replace all the convolutions in DCNNs with pointwise convolutions while keeping the segmentation performance? To achieve the objective, we need to solve the following problems of FCNs with only stacked pointwise convolutions: (1) The receptive field never enlarges. (2) The 3D spatial image context cannot be utilized. (3) Long-term dependencies in images are not exploited. To address these issues, we propose Group Shift (GS), a parameter-free operation. Equipped with GS, our final model with only pointwise convolutions (pointwise FCNs) can achieve comparable or even better performances than the corresponding 3D FCNs with significantly reduced parameters and FLOPs.

2   Method

The major innovation of our proposed method lies in the design of GS. GS is developed to compensate for the limited RF of pointwise convolutions in a parameter-free manner and construct long-term regional dependencies. GS consists of two key steps, grouping and shift. In this section, we will describe the two steps as well as the formulation of GS in detail.

2.1 Grouping

Spatial Grouping. Given the input and output feature maps of GS as $\boldsymbol{F} \in \mathbb{R}^{D \times H \times W \times C}$ and $\boldsymbol{F}_s \in \mathbb{R}^{D \times H \times W \times C}$. D, H, and W are the three spatial dimensions. C is the number of channels. We first divide the images equally into g_d, g_h, and g_w groups along the three spatial dimensions as shown in Fig. 1a, resulting in $g_d \times g_h \times g_w$ image groups in total. The dimension of each spatial group is $d \times h \times w$, and we have $D = d \times g_d$, $H = h \times g_h$, $W = w \times g_w$. So after spatial grouping, the input feature maps are transformed to $\boldsymbol{F}_{sg} \in \mathbb{R}^{(d \times g_d) \times (h \times g_h) \times (w \times g_w) \times C}$.

Channel Grouping. Empirically, we want to shift only a part of the features. The un-shifted features contain the original localization information that is also important for the final segmentation performance. Supposing the number of channels to be shifted is C_s and the number of channels to keep un-shifted is C_k, and $C = C_s + C_k$. Then, we split C_s into $g_d \times g_h \times g_w$ groups (same as the spatial groups). Each channel group contains C_g channels, and $C_s = g_d \times g_h \times g_w \times C_g$. After channel grouping, the output feature map is $\boldsymbol{F}_{cg} \in \mathbb{R}^{D \times H \times W \times (C_g \times g_d \times g_h \times g_w + C_k)}$. Channel grouping is illustrated in Fig. 1b.

Therefore, the input feature maps $\boldsymbol{F} \in \mathbb{R}^{D \times H \times W \times C}$ are transformed to $\boldsymbol{F}_{scg} \in \mathbb{R}^{(d \times g_d) \times (h \times g_h) \times (w \times g_w) \times (C_g \times g_d \times g_h \times g_w + C_k)}$ after spatial and channel grouping. $\boldsymbol{F}_{scg}$ can proceed to the subsequent shift operation.

2.2 Shift

To force the pointwise convolutions into extracting more spatial information, we elaborately design a shift operation. Figure 1c is an example to illustrate how the shift operation works. We assume that the feature maps are divided into four spatial groups ($g_d \times g_h \times g_w = 4$) (corresponding to the four columns with different colors) and rearrange the spatial groups in a column-wise manner (Fig. 1c, left figure). The channels C are divided into shift channels C_s and un-shift channels C_k. The shift channels C_s are further grouped into four groups (corresponding to the upper four rows in Fig. 1c). Then, we shift each channel group in C_s with a step equals to the index of the channel group (Fig. 1c, right figure). Shifting one step means that moving one spatial group in the specific channel group to the neighbor spatial group. All the channel groups shift in the same direction and shifting happens only within the specific channel group without channel shifting.

From Fig. 1c, we can observe that after shifting, every spatial group (i.e. every column) contains one channel group of all the other spatial groups. In other words, one voxel in a specific location in a spatial group contains one group of channels of the corresponding voxel with the same location in all the other spatial groups. Thus, the elaborately designed shift operation can not only increase the RF but also make full advantage of the spatial context, especially long-term dependence. Ideally, it can effectively solve the raised three problems.

2.3   Formulation of Group Shift

Let $\{x', y', z', c'\}$ be the coordinates of a specific voxel in the shifted feature map $\boldsymbol{F}_s \in \mathbb{R}^{D \times H \times W \times C}$ and $\{x, y, z, c\}$ be the corresponding coordinates of the same voxel in the input feature map $\boldsymbol{F} \in \mathbb{R}^{D \times H \times W \times C}$. Specifically, we should find:

$$\boldsymbol{F}_s(x', y', z', c') = \boldsymbol{F}(x, y, z, c) \tag{1}$$

where $x, x' \in [0, D-1]$, $y, y' \in [0, H-1]$, $z, z' \in [0, W-1]$, and $c, c' \in [0, C-1]$. The spatial groups along three dimensions are g_d, g_h, and g_w. The spatial size of each spatial group is $d \times h \times w$, and $D = d \times g_d, H = h \times g_h, W = w \times g_w$. The number of channels to be shifted is C_s. Suppose the current spatial group index of $\{x, y, z, c\}$ in the input feature map is cur_ind, shift step is sft_step, and the shifted spatial group index of $\{x', y', z', c'\}$ in the shifted feature map is $sfted\_ind$. The relationships of the coordinates between the shifted feature map and input feature map are defined as follows:

$$cur\_ind = \lfloor \frac{x}{d} \rfloor + \lfloor \frac{y}{h} \rfloor \times g_d + \lfloor \frac{z}{w} \rfloor \times g_d \times g_h \tag{2}$$

$$sft\_step = \lfloor \frac{c}{c_g} \rfloor \tag{3}$$

$$sfted\_ind = \mod(cur\_ind + sft\_step, g_d \times g_h \times g_w) \tag{4}$$

$$\begin{cases} x' = \mod(sfted\_ind, g_d) \times d + \mod(x, d) & c \in [0, C_s), \\ x' = x & c \in [C_s, C). \end{cases} \tag{5}$$

$$\begin{cases} y' = \lfloor \frac{\mod(sfted\_ind, g_d \times g_h)}{g_d} \rfloor \times h + \mod(y, h) & c \in [0, C_s), \\ y' = y & c \in [C_s, C). \end{cases} \tag{6}$$

$$\begin{cases} z' = \lfloor \frac{sfted\_ind}{g_d \times g_h} \rfloor \times w + \mod(z, w) & c \in [0, C_s), \\ z' = z & c \in [C_s, C). \end{cases} \tag{7}$$

$$c' = c \tag{8}$$

3   Experiments

Extensive experiments are conducted on two benchmark datasets, PROMISE12 [11] and BraTS18 [1,2,12]. PROMISE12 released 50 transversal T2-weighted MR images of the prostate and corresponding segmentation ground truths as the training set and 30 MR images without ground truths as the validation set. The input size of this dataset is set to $128 \times 128 \times 16$ through random cropping. BraTS18 provides multimodal MR scans (T1, T1ce, T2, and FLAIR) for brain tumor segmentation. In the training set, there are 285 images with segmentation labels. All provided volumes have the same matrix size of $240 \times 240 \times 155$. The input size of BraTS18 is set to $128 \times 128 \times 64$ through random cropping.

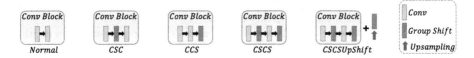

Fig. 2. The basic "Conv Block" in the tiny 3D U-Net and four insert positions of Group Shift. One "Conv Block" contains two convolutional operations. "UpShift" means add Group Shift after "Upsampling".

Table 1. Results of the two baselines, pointwise FCN without GS and 3D FCN with $3 \times 3 \times 3$ convolutions, on the two datasets.

Baselines	BraTS18				PROMISE12
	WT	TC	ET	mDice (%)	mDice (%)
Pointwise FCN without GS	86.4	**79.7**	72.7	79.6	65.5
3D FCN	**89.0**	78.7	**73.7**	**80.5**	**87.3**

The network architecture of the 3D FCN adopted in this study is a tiny 3D U-Net [6] (See supplementary material for the detailed structure). When all convolutions in the network are replaced by pointwise convolutions, the 3D FCN becomes our pointwise FCN. The proposed GS can be inserted to any position of the pointwise FCN. In this paper, we investigate four GS-related configurations, "CSC", "CCS", "CSCS", and "CSCSUpShift" as shown in Fig. 2. The numbers of spatial and channel groups of GS are determined by the size of the input feature maps.

Two baselines are investigated, 3D FCNs with $3 \times 3 \times 3$ convolutions and pointwise FCNs without GS. We randomly split the two datasets into two groups with a ratio of 8:2 for network training and validation. For preprocessing, we normalize each 3D image independently with the mean and standard deviation calculated from the corresponding foreground regions. The poly learning rate policy is adopted with an initial learning rate of 0.01 and a power of 0.9. The optimizer utilized is stochastic gradient descent (SGD), and the loss function is Dice loss [13]. All our models are implemented with PyTorch on a Titan XP GPU (12G) with a batch size of 4. Two evaluation metrics, "dice" and "mDice", are reported. Here, "dice" is the Dice score calculated for each foreground class, and "mDice" is the average "dice" of all foreground classes.

3.1 Results on PROMISE12

Results of the two baselines on PROMISE12 are shown in Table 1. As expected, when all $3 \times 3 \times 3$ convolutions in 3D FCNs are replaced with pointwise convolutions, the network performance drops dramatically. The mDice value is decreased by more than 20%. This reflects that large effective RFs and long-term dependencies in images are important for large foreground object segmentation, such as the prostate.

Table 2. Results of different spatial groups (SG) and different GS positions (IP) on PROMISE12. Stages 1–5 indicate network stages. GS positions refer to Fig. 2. $C_s = C_k = \frac{1}{2}C$. For SG setting of (n1, n2, n3), n1, n2, and n3 are group numbers in depth, height, and width directions, respectively. Results are characterized by mDice (%).

| | Different SG settings | | | | | Results under different SG and IP | | | |
SG	Stage 1	Stage 2	Stage 3	Stage 4	Stage 5	CSC	CCS	CSCS	CSCSUpShift
ProSGv1	(2, 2, 2)	(2, 2, 2)	(2, 4, 4)	(1, 8, 8)	(1, 8, 8)	**84.9**	84.0	83.1	83.0
ProSGv2	(1, 2, 2)	(1, 4, 4)	(2, 4, 4)	(1, 8, 8)	(1, 8, 8)	84.5	**85.4**	85.2	84.5
ProSGv3	(2, 2, 2)	(1, 4, 4)	(1, 4, 4)	(1, 8, 8)	(1, 8, 8)	85.6	84.3	84.8	83.6
ProSGv4	(1, 2, 2)	(2, 2, 2)	(2, 4, 4)	(1, 8, 8)	(1, 8, 8)	85.0	84.3	**85.3**	84.6

Considering the matrix sizes of the input images and the feature maps at different network stages, four settings of spatial groups (ProSGv1 to ProSGv4 in Table 2) are investigated. Specifically, we test different spatial group numbers at different stages. Basically, more spatial groups at deeper stages and more spatial groups in the in-plane dimensions are utilized. Together with the four GS configurations ("CSC", "CCS", "CSCS", and "CSCSUpShift"), there are 16 experimental conditions in total.

Overall, the segmentation results of pointwise FCNs adding GS (Table 2) are better than that without GS (65.5% in Table 1) with a large margin. Among the four spatial group settings, "ProSGv2" achieves the best average results (84.9%) under the four GS configurations. Among the four GS configurations, "CSC" achieves the best average results (85.0%) under the four spatial group settings. Nevertheless, "ProSGv3" with "CSC" achieves the best result with a mDice value of 85.6%, which is only slightly worse than that obtained with normal 3D FCNs (87.3%) utilizing computational intensive 3D convolutions.

With the best configuration of our pointwise FCN ("ProSGv3" with "CSC"), we further investigate the influence of the ratio of the shifted channels on the network performance. When all the input feature channels are allowed to shift ($C_s = C$ and $C_k = 0$), the segmentation results (mDice = 81.4%) are much worse than that obtained when we only shift half of the input features (mDice = 85.6%). Therefore, we conclude that both local (preserved by the un-shifted channel groups) and spatial information (extracted through the shifted channel groups) are important for the final prostate segmentation.

3.2   Results on BraTS18

Surprisingly, for the two baselines, the results of pointwise FCNs (mDice = 79.6%) are only slightly worse than those of 3D FCNs (mDice = 80.5%) on BraTS18 as shown in Table 1, which is quite different from the results achieved on PROMISE12. We suspect that this phenomenon is caused by the different properties of the two datasets. The target objects of BraTS18 data (brain tumors) are much smaller than those of PROMISE12 data (prostate regions). The local information within the limited RF of pointwise FCNs is enough to achieve satisfactory segmentation results on BraTS18.

Table 3. Results on the BraTS18 test set obtained through the official online evaluation server. "CSC" is one of the insert positions of group shift (GS) as shown in Fig. 2. "Dec" means adding GS to the decoder of the network only. $C_s = C_k = \frac{1}{2}C$. ET, WT, and TC refer to enhancing tumor, whole tumor, and tumor core.

Model	Params(M)	FLOPs	Dice (%)			Hausdorff95		
			ET	WT	TC	ET	WT	TC
CSC+Dec (Ours)	**0.25**	**7.91**	78.1	90.2	83.2	4.01	5.22	6.53
S3D-UNet [4]	3.32	75.20	74.93	89.35	83.09	–	–	–
3D-ESPNet [15]	3.63	76.51	73.70	88.30	81.40	–	–	–
Kao et al. [8]	9.45	203.96	78.75	90.47	81.35	3.81	4.32	7.56
No New-Net [7]	10.36	202.25	81.01	**90.83**	85.44	**2.41**	**4.27**	6.52
NVDLMED [14]	40.06	1495.53	**81.73**	90.68	**86.02**	3.82	4.52	6.85

We investigate the influence of insert positions of GS on the final performance with BraTS18 data when utilizing a spatial group setting of (Stage 1–5: spatial groups of (2, 2, 2), (2, 2, 2), (2, 2, 2), (4, 4, 4), and (5, 5, 5)) (See supplementary material). A similar conclusion can be drawn that "CSC" achieves the best average result (mDice = 81.2%) among the four GS configurations (80.7%, 80.1%, and 80.2% for CCS, CSCS, and CSCSUpShift), which is even slightly better than that given by the 3D FCN (80.5%). This indicates the effectiveness of our pointwise FCNs with GS (GSP-Conv) for small object segmentation tasks.

With this dataset, we treat the encoder and the decoder of the network differently and add the GS operations to one of them at a time. Results reflect that adding GS to the decoder (82.6%) is more effective for the brain tumor segmentation task than adding GS to the encoder (81.5%) or to both (81.2%). We speculate that when adding GS only to the decoder, we can keep more local detailed information un-shifted, which is essential for small object segmentation.

Comparisons to state-of-the-art methods [4,7,8,14,15], including factorization-based methods [4,15], are performed on the test set of BraTS18 through the online server (Table 3). Following the best practices, we use the same data preprocessing, training strategies, and training hyper-parameters as [7]. Overall, our method achieves competitive results when compared to these methods with much fewer parameters and FLOPs. With less than 8% parameters and less than 11% FLOPs, our methods can still generate very accurate brain tumor segmentation, which is crucial for acute situations when fast diagnoses are important.

4   Conclusion

Two major limitations exist with our current experimental design. First, we only experimented with the tiny 3D U-Net architecture. Second, our model contains a number of hyper-parameters that might need to be tuned for different applications. Therefore, we believe that we have not made the most of the capability of the proposed GS operation. In our following work, we will investigate the effects of the data (imaging modality, spacing, volume size, and target object size) on the choice of the best model configurations. We will also try to design dedicated network architecture according to these properties of the data. Particularly, the number of stages, the number of channels in each stage, the number of convolution operations in each "Conv Block" (Fig. 2), and the number of "Conv Block" in both the encoder and the decoder will be accordingly optimized. Adding the different settings of the proposed GS operation, all these factors will build a large search space. We are considering introducing the neural architecture search (NAS) method to automate the process.

Nevertheless, equipped with the current version of our proposed GS, the pointwise FCNs can already achieve comparable or even better performance than the corresponding 3D FCNs. To the best of our knowledge, this is the first attempt to segment volumetric images with only pointwise convolutions. We provide a new perspective on model compression. Our proposed GSP-Conv operation can be of high application value when fast and accurate imaging diagnoses are needed. In addition, we believe that the proposed method can be easily extended to other image processing tasks, including image classification, object detection, image synthesis, and image super-resolution.

Acknowledgements. This research is partially supported by the National Key Research and Development Program of China (No. 2020YFC2004804 and 2016YFC0106200), the Scientific and Technical Innovation 2030-"New Generation Artificial Intelligence" Project (No. 2020AAA0104100 and 2020AAA0104105), the Shanghai Committee of Science and Technology, China (No. 20DZ1100800 and 21DZ1100100), Beijing Natural Science Foundation-Haidian Original Innovation Collaborative Fund (No. L192006), the funding from Institute of Medical Robotics of Shanghai Jiao Tong University, the 863 national research fund (No. 2015AA043203), the National Natural Science Foundation of China (No. 61871371 and 81830056), the Key-Area Research and Development Program of GuangDong Province (No. 2018B010109009), the Key Laboratory for Magnetic Resonance and Multimodality Imaging of Guangdong Province (2020B1212060051), the Basic Research Program of Shenzhen (No. JCYJ20180507182400762), and the Youth Innovation Promotion Association Program of Chinese Academy of Sciences (No. 2019351).

References

1. Bakas, S., et al.: Advancing the cancer genome atlas glioma MRI collections with expert segmentation labels and radiomic features. Sci. Data **4**, 170117 (2017)
2. Bakas, S., et al.: Identifying the best machine learning algorithms for brain tumor segmentation, progression assessment, and overall survival prediction in the BRATS challenge. arXiv preprint https://arxiv.org/abs/1811.02629arXiv:1811. 12506 (2018)
3. Chen, C., Liu, X., Ding, M., Zheng, J., Li, J.: 3D dilated multi-fiber network for real-time brain tumor segmentation in MRI. In: Shen, D., et al. (eds.) MICCAI 2019. LNCS, vol. 11766, pp. 184–192. Springer, Cham (2019). https://doi.org/10. 1007/978-3-030-32248-9_21

4. Chen, W., Liu, B., Peng, S., Sun, J., Qiao, X.: S3D-UNet: separable 3D U-net for brain tumor segmentation. In: Crimi, A., Bakas, S., Kuijf, H., Keyvan, F., Reyes, M., van Walsum, T. (eds.) BrainLes 2018. LNCS, vol. 11384, pp. 358–368. Springer, Cham (2019). https://doi.org/10.1007/978-3-030-11726-9_32

5. Chollet, F.: Xception: deep learning with depthwise separable convolutions. In: Proceedings of the IEEE Conference on CVPR, pp. 1251–1258 (2017)

6. Çiçek, Ö., Abdulkadir, A., Lienkamp, S.S., Brox, T., Ronneberger, O.: 3D U-net: learning dense volumetric segmentation from sparse annotation. In: Ourselin, S., Joskowicz, L., Sabuncu, M.R., Unal, G., Wells, W. (eds.) MICCAI 2016. LNCS, vol. 9901, pp. 424–432. Springer, Cham (2016). https://doi.org/10.1007/978-3-319-46723-8_49

7. Isensee, F., Kickingereder, P., Wick, W., Bendszus, M., Maier-Hein, K.H.: No new-net. In: Crimi, A., Bakas, S., Kuijf, H., Keyvan, F., Reyes, M., van Walsum, T. (eds.) BrainLes 2018. LNCS, vol. 11384, pp. 234–244. Springer, Cham (2019). https://doi.org/10.1007/978-3-030-11726-9_21

8. Kao, P.-Y., Ngo, T., Zhang, A., Chen, J.W., Manjunath, B.S.: Brain tumor segmentation and tractographic feature extraction from structural MR images for overall survival prediction. In: Crimi, A., Bakas, S., Kuijf, H., Keyvan, F., Reyes, M., van Walsum, T. (eds.) BrainLes 2018. LNCS, vol. 11384, pp. 128–141. Springer, Cham (2019). https://doi.org/10.1007/978-3-030-11726-9_12

9. Khened, M., Kollerathu, V.A., Krishnamurthi, G.: Fully convolutional multi-scale residual DenseNets for cardiac segmentation and automated cardiac diagnosis using ensemble of classifiers. Med. Image Anal. **51**, 21–45 (2019)

10. Krizhevsky, A., Sutskever, I., Hinton, G.E.: ImageNet classification with deep convolutional neural networks. In: NIPS, pp. 1097–1105 (2012)

11. Litjens, G., et al.: Evaluation of prostate segmentation algorithms for MRI: the PROMISE12 challenge. Med. Image Anal. **18**(2), 359–373 (2014)

12. Menze, B.H., et al.: The multimodal brain tumor image segmentation benchmark (BRATS). IEEE Trans. Med. Imaging **34**(10), 1993–2024 (2014)

13. Milletari, F., Navab, S., Ahmadi, S.-A.: V-net: fully convolutional neural networks for volumetric medical image segmentation. In: Proceedings of the 4th International Conference on 3DV, pp. 565–571 (2016)

14. Myronenko, A.: 3D MRI brain tumor segmentation using autoencoder regularization. In: Crimi, A., Bakas, S., Kuijf, H., Keyvan, F., Reyes, M., van Walsum, T. (eds.) BrainLes 2018. LNCS, vol. 11384, pp. 311–320. Springer, Cham (2019). https://doi.org/10.1007/978-3-030-11726-9_28

15. Nuechterlein, N., Mehta, S.: 3D-ESPNet with pyramidal refinement for volumetric brain tumor image segmentation. In: Crimi, A., Bakas, S., Kuijf, H., Keyvan, F., Reyes, M., van Walsum, T. (eds.) BrainLes 2018. LNCS, vol. 11384, pp. 245–253. Springer, Cham (2019). https://doi.org/10.1007/978-3-030-11726-9_22

16. Ronneberger, O., Fischer, P., Brox, T.: U-net: convolutional networks for biomedical image segmentation. In: Navab, N., Hornegger, J., Wells, W.M., Frangi, A.F. (eds.) MICCAI 2015. LNCS, vol. 9351, pp. 234–241. Springer, Cham (2015). https://doi.org/10.1007/978-3-319-24574-4_28

17. Simonyan, K., Zisserman, A.: Very deep convolutional networks for large-scale image recognition. In: ICLR (2015)

18. Szegedy, C., Vanhoucke, V., Ioffe, S., Shlens, J., Wojna, Z.: Rethinking the inception architecture for computer vision. In: Proceedings of the IEEE Conference on CVPR, pp. 2818–2826 (2016)

19. Tang, H., et al.: Clinically applicable deep learning framework for organs at risk delineation in CT images. Nat. Mach. Intell. **1**(10), 480–491 (2019)

20. Tran, D., Wang, H., Torresani, L., Ray, J., Lecun, Y., Paluri, M.: A closer look at spatiotemporal convolutions for action recognition. In: Proceedings of the IEEE Conference on CVPR, pp. 6450–6459 (2018)
21. Xian, M., et al.: Automatic breast ultrasound image segmentation: a survey. Pattern Recogn. **79**, 340–355 (2018)
22. Xie, S., Sun, C., Huang, J., Tu, Z., Murphy, K.: Rethinking spatiotemporal feature learning: speed-accuracy trade-offs in video classification. In: Proceedings of ECCV, pp. 305–321 (2018)

Machine Learning - Attention Models

UTNet: A Hybrid Transformer Architecture for Medical Image Segmentation

Yunhe Gao[1], Mu Zhou[1,2], and Dimitris N. Metaxas[1(✉)]

[1] Department of Computer Science, Rutgers University, Piscataway, USA
`dnm@cs.rutgers.edu`
[2] SenseBrain and Shanghai AI Laboratory and Centre for Perceptual and Interactive Intelligence, Shanghai, China

Abstract. Transformer architecture has emerged to be successful in a number of natural language processing tasks. However, its applications to medical vision remain largely unexplored. In this study, we present UTNet, a simple yet powerful hybrid Transformer architecture that integrates self-attention into a convolutional neural network for enhancing medical image segmentation. UTNet applies self-attention modules in both encoder and decoder for capturing long-range dependency at different scales with minimal overhead. To this end, we propose an efficient self-attention mechanism along with relative position encoding that reduces the complexity of self-attention operation significantly from $O(n^2)$ to approximate $O(n)$. A new self-attention decoder is also proposed to recover fine-grained details from the skipped connections in the encoder. Our approach addresses the dilemma that Transformer requires huge amounts of data to learn vision inductive bias. Our hybrid layer design allows the initialization of Transformer into convolutional networks without a need of pre-training. We have evaluated UTNet on the multi-label, multi-vendor cardiac magnetic resonance imaging cohort. UTNet demonstrates superior segmentation performance and robustness against the state-of-the-art approaches, holding the promise to generalize well on other medical image segmentations.

1 Introduction

Convolutional networks have revolutionized the computer vision field with outstanding feature representation capability. Currently, the convolutional encoder-decoder architectures have made substantial progress in position-sensitive tasks, like semantic segmentation [6,11,14,17,20]. The used convolutional operation captures texture features by gathering local information from neighborhood pixels. To aggregate the local filter responses globally, these models stack multiple convolutional layers and expand the receptive field through down-samplings.

Electronic supplementary material The online version of this chapter (https://doi.org/10.1007/978-3-030-87199-4_6) contains supplementary material, which is available to authorized users.

© Springer Nature Switzerland AG 2021
M. de Bruijne et al. (Eds.): MICCAI 2021, LNCS 12903, pp. 61–71, 2021.
https://doi.org/10.1007/978-3-030-87199-4_6

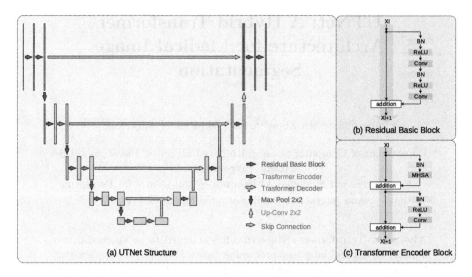

Fig. 1. **(a)** The hybrid architecture of the proposed UTNet. The proposed efficient self-attention mechanism and relative positional encoding allow us to apply Transformer to aggregate global context information from multiple scales in both encoder and decoder. **(b)** Pre-activation residual basic block. **(c)** The structure of Transformer encoder block.

Despite the advances, there are two inherent limitations of this paradigm. First, the convolution only gathers information from neighborhood pixels and lacks the ability to capture long-range (global) dependency explicitly [5,25,26]. Second, the size and shape of convolution kernels are typically fixed thus they can not adapt to the input content [15].

Transformer architecture using the self-attention mechanism has emerged to be successful in natural language processing (NLP) [18] with its capability of capturing long-range dependency. Self-attention is a computational primitive that implements pairwise entity interactions with a context aggregation mechanism, which has the ability to capture long-range associative features. It allows the network to aggregate relevant features dynamically based on the input content. Preliminary studies with simple forms of self-attention have shown its usefulness in segmentation [4,16], detection [24] and reconstruction [9].

Although the application of image-based Transformer is promising, training and deploying of Transformer architecture has several daunting challenges. First, the self-attention mechanism has $O(n^2)$ time and space complexity with respect to sequence length, resulting in substantial overheads of training and inference. Previous works attempt to reduce the complexity of self-attention [10,28], but are still far from perfection. Due to the time complexity, the standard self-attention can be only applied patch-wise, e.g. [3,27] encode images using 16×16 flattened image patches as input sequences, or on top of feature maps from CNN backbone, which are already down-sampled into low-resolution [4,22]. However, for position-sensitive tasks like medical image segmentation, high-resolution feature plays a vital role since most mis-segmented areas are located around the boundary of the region-of-interest. Second, Transformers do not have inductive bias for images and can not perform well on a small-scale dataset [3]. For example, Transformer can be beneficial from pre-training through a large-scale dataset like full JFT-300M [3]. But even with pre-training on ImageNet, Transformer is still worse than the ResNet [7,12], not to mention medical image datasets with much less available amounts of medical data.

In this paper, we propose a **U**-shape hybrid **T**ransformer **Net**work: **UTNet**, integrating the strength of convolution and self-attention strategies for medical image segmentation. The major goal is to apply convolution layers to extract local intensity features to avoid large-scale pretraining of Transformer, while using self-attention to capture long-range associative information. We follow the standard design of UNet, but replace the last convolution of the building block in each resolution (except for the highest one) to the proposed Transformer module. Towards enhanced quality of segmentation, we seek to apply self-attention to extract detailed long-range relationships on high-resolution feature maps. To this end, we propose an efficient self-attention mechanism, which reduces the overall complexity significantly from $O(n^2)$ to approximate $O(n)$ in both time and space. Furthermore, we use a relative position encoding in the self-attention module to learn content-position relationships in medical images. Our UTNet demonstrates superior segmentation performance and robustness in the multi-label, multi-vendor cardiac magnetic resonance imaging cohort. Given the design of UTNet, our framework holds the promise to generalize well on other medical image segmentations.

2   Method

2.1   Revisiting Self-attention Mechanism

The Transformer is built upon the multi-head self-attention (MHSA) module [18], which allows the model to jointly infer attention from different representation subspaces. The results from multiple heads are concatenated and then transformed with a feed-forward network. In this study, we use 4 heads and the dimension of multi-head is not presented for simplicity in the following formulation and in the figure. Consider an input feature map $X \in \mathcal{R}^{C \times H \times W}$, where H,W are the spatial height, width and C is the number of channels. Three 1×1 convolutions are used to project X to query, key, value embeddings: $\mathbf{Q}, \mathbf{K}, \mathbf{V} \in \mathcal{R}^{d \times H \times W}$, where d is the dimension of embedding in each head. The $\mathbf{Q}, \mathbf{K}, \mathbf{V}$ is then flatten and transposed into sequences with size $n \times d$, where $n = HW$. The output of the self-attention is a scaled dot-product:

$$\text{Attention}(\mathbf{Q}, \mathbf{K}, \mathbf{V}) = \text{softmax}(\underbrace{\frac{\mathbf{QK}^{\mathsf{T}}}{\sqrt{d}}}_{P}) \mathbf{V} \tag{1}$$

Note that $P \in \mathcal{R}^{n \times n}$ is named context aggregating matrix, or similarity matrix. To be specific, the i-th query's context aggregating matrix is $P_i =$

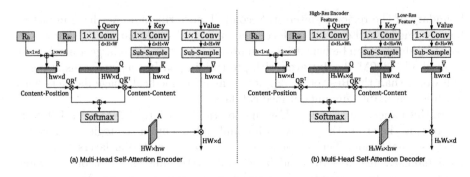

Fig. 2. The proposed efficient multi-head self-attention (MHSA). **(a)** The MHSA used in the Transformer encoder. **(b)** The MHSA used in the Transformer decoder. They share similar concepts, but (b) takes two inputs, including the high-resolution features from skip connections of the encoder, and the low-resolution features from the decoder.

softmax($\frac{\mathbf{q}_i\mathbf{K}^\mathsf{T}}{\sqrt{d}}$), $P_i \in \mathcal{R}^{1\times n}$, which computes the normalized pair-wise dot production between q_i and each element in the keys. The context aggregating matrix is then used as the weights to gather context information from the values. In this way, self-attention intrinsically has the global receptive field and is good at capturing long-range dependence. Also, the context aggregating matrix is adaptive to input content for better feature aggregation. However, the dot-product of $n \times d$ matrices leads to $O(n^2d)$ complexity. Typically, n is much larger than d when the resolution of feature map is large, thus the sequence length dominates the self-attention computation and makes it infeasible to apply self-attention in high-resolution feature maps, e.g. $n = 256$ for 16×16 feature maps, and $n = 16384$ for 128×128 feature maps.

2.2   Efficient Self-attention Mechanism

As images are highly structured data, most pixels in high-resolution feature maps within local footprint share similar features except for the boundary regions. Therefore, the pair-wise attention computation among all pixels is highly inefficient and redundant. From a theoretical perspective, self-attention is essentially low rank for long sequences [21], which indicates that most information is concentrated in the largest singular values. Inspired by this finding, we propose an efficient self-attention mechanism for our task as seen in Fig. 2.

The main idea is to use two projections to project key and value: $\mathbf{K}, \mathbf{V} \in \mathcal{R}^{n\times d}$ into low-dimensional embedding: $\overline{\mathbf{K}}, \overline{\mathbf{V}} \in \mathcal{R}^{k\times d}$, where $k = hw \ll n$, h and w are the reduced size of feature map after sub-sampling. The proposed efficient self-attention is now:

$$\text{Attention}(\mathbf{Q}, \overline{\mathbf{K}}, \overline{\mathbf{V}}) = \underbrace{\text{softmax}(\frac{\mathbf{Q}\overline{\mathbf{K}}^\mathsf{T}}{\sqrt{d}})}_{\overline{P}:n\times k}\underbrace{\overline{\mathbf{V}}}_{k\times d} \tag{2}$$

By doing so, the computational complexity is reduced to $O(nkd)$. Notably, the projection to low-dimensional embedding can be any down-sampling operations, such as average/max pooling, or strided convolutions. In our implementation, we use 1×1 convolution followed by a bilinear interpolation to down-sample the feature map, and the reduced size is 8.

2.3 Relative Positional Encoding

Standard self-attention module totally discards the position information and is perturbation equivariant [1], making it ineffective for modeling image contents that are highly structured. The sinusoidal embedding in previous works [13] does not have the property of translation equivariance in convolutional layers. Therefore, we use the 2-dimensional relative position encoding by adding relative height and width information [1]. The pair-wise attention logit before softmax using relative position encoding between pixel $i = (i_x, i_y)$ and pixel $j = (j_x, j_y)$:

$$l_{i,j} = \frac{q_i^{\mathsf{T}}}{\sqrt{d}}(k_j + r_{j_x-i_x}^W + r_{j_y-i_y}^H) \tag{3}$$

where q_i is the query vector of pixel i, k_i is the key vector for pixel j, $r_{j_x-i_x}^W$ and $r_{j_y-i_y}^H$ are learnable embeddings for relative width $j_x - i_x$ and relative height $j_y - i_y$ respectively. Similar to the efficient self-attention, the relative width and height are computed after low-dimensional projection. The efficient self-attention including relative position embedding is:

$$\text{Attention}(\mathbf{Q}, \overline{\mathbf{K}}, \overline{\mathbf{V}}) = \underbrace{\text{softmax}(\frac{\mathbf{Q}\overline{\mathbf{K}}^{\mathsf{T}} + \mathbf{S}_H^{rel} + \mathbf{S}_W^{rel}}{\sqrt{d}})}_{\overline{P}:n \times k} \underbrace{\overline{\mathbf{V}}}_{k \times d} \tag{4}$$

where $\mathbf{S}_H^{rel}, \mathbf{S}_W^{rel} \in \mathcal{R}^{HW \times hw}$ are matrics of relative position logits along height and width dimensions that satisfy $\mathbf{S}_H^{rel}[i,j] = q_i^{\mathsf{T}} r_{j_y-i_y}^H$, $\mathbf{S}_W^{rel}[i,j] = q_i^{\mathsf{T}} r_{j_x-i_x}^W$.

2.4 Network Architecture

Figure 1 highlights the architecture of UTNet. We seek to combine the strength from both convolution and self-attention mechanism. Therefore, the hybrid architecture can leverage the inductive bias of image from convolution to avoid large-scale pretraining, as well as the capability of Transformer to capture long-range relationships. Because the mis-segmented region usually locates at the boundary of region-of-interest, the high-resolution context information could play a vital role in segmentation. As a result, our focus is placed on the proposed self-attention module, making it feasible to handle large-size feature maps efficiently. Instead of naively integrating the self-attention module on top of the feature maps from the CNN backbone, we apply the Transformer module to each level of the encoder and decoder to collect long-range dependency from multiple scales. Note that we do not apply Transformer on the original resolution, as adding Transformer module in the very shallow layers of the network does not help in experiments but introduces additional computation. A possible reason is that the shallow layers of the network focus more on detailed textures, where gathering global context may not be informative. The building block of UTNet is shown in Fig. 1 (b) and (c), including residual basic block and Transformer block. For both blocks, we use the pre-activation setting for identity mapping in the short cut. This identity mapping has been proven to be effective in vision [8] and NLP tasks [19].

3   Experiment

3.1   Experiment Setup

We systematically evaluate the UTNet on the multi-label, multi-vendor cardiac magnetic resonance imaging (MRI) challenge cohort [2], including the segmentation of left ventricle (LV), right ventricle (RV), and left ventricular myocardium (MYO). In the training set, we have 150 annotated images from two different MRI vendors (75 images of each vendor), including A: Siemens; B: Philips. In the testing set, we have 200 images from 4 different MRI vendors (50 images of each vendor), including A: Siemens; B: Philips; C: GE; D: Canon, where vendor C and D are completely absent in the training set (we discard the unlabeled data). The MRI scans from different vendors have marked differences in appearance, allowing us to measure model robustness and compare with other models under different settings. Specifically, we have performed two experiments to highlight the performance and robustness of UTNet. First, we report primary results with training and testing data are both from the same vendor A. Second, we further measure the cross-vendor robustness of models. Such setting is more challenging since the training and testing data are from independent vendors. We report Dice score and Hausdorff distance of each model to compare the performance.

3.2   Implementation Detail

For data preprocessing, we resample the in-plane spacing to 1.2×1.2 mm, while keeping the spacing along the z-axis unchanged. We train all models from scratch for 150 epochs. We use the exponentially learning rate scheduler with a base learning rate of 0.05. We use the SGD optimizer with a batch size of 16 on one GPU, momentum and weight decay are set to 0.9 and $1e - 4$ respectively. Data augmentation is applied on the fly during model training, including random rotation, scaling, translation, additive noise and gamma transformation. All images are randomly cropped to 256×256 before entering the models. We use the combine of Dice loss and cross-entropy loss to train all networks.

3.3   Segmentation Results

We compare the performance of UTNet with multiple state-of-the-art segmentation models. UNet [14] builds on top of the fully convolutional networks with

Table 1. Segmentation performance in term of Dice score and efficiency comparison. All models are trained and tested using data from vendor A. The Hausdorff distant result is reported in the supplementary.

	UNet	ResUNet	CBAM	Dual-Attn	UTNet
LV	91.8	92.2	92.2	92.4	**93.1**
MYO	81.7	82.5	82.1	82.3	**83.5**
RV	85.6	86.2	87.7	86.4	**88.2**
Average	86.4	86.9	87.3	87.0	**88.3**
Params/M	7.07	9.35	9.41	9.69	9.53
Inference Time/s	0.085	0.115	0.149	0.118	0.145

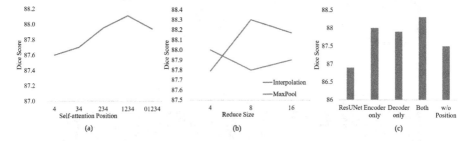

Fig. 3. Ablation study. (a) Effect of different self-attention position. (b) Effect of reduce size and projection of efficient self-attention. (c) Effect of Transformer encoder, Transformer decoder, and the relative positional encoding.

a U-shaped architecture to capture context information. The ResUNet is similar to UNet in architecture, but it uses residual blocks as the building block. CBAM [23] uses two sequential convolutional modules to infer channel and spatial attention to refine intermediate feature maps adaptively. Dual attention network [4] uses two kinds of self-attention to model the semantic inter-dependencies in spatial and channel dimensions, respectively. We have implemented CBAM and dual attention in ResUNet backbone for better comparison. The dual attention is only applied in the feature maps after 4 down-samplings due to its quadratic complexity.

As seen in Table 1, UTNet demonstrates leading performance in all segmentation outcomes (LV, MYO and RV). By introducing residual connections, ResUNet is slightly improved than the original UNet. The spatial and channel attention from CBAM are inferred from convolutional layers, it still suffers from limited receptive field. Thus CBAM only has limited improvement compared with ResUNet. We also recognize that dual-attention approach was almost the same as ResUNet, as it suffers from quadratic complexity that can not process higher resolution feature maps to fix errors in the segmentation boundary. Meanwhile, our UTNet presents less parameters than dual-attention approach and it can capture global context information from high-resolution feature maps.

Table 2. Robustness comparison, measured with Dice score. All models are trained on data from vendor A,B, and are tested on data from vendor A,B,C,D. The number in brackets of C and D indicates the performance drop compared with the average of A and B.

Vendor	ResUNet				CBAM				UTNet			
	A	B	C	D	A	B	C	D	A	B	C	D
LV	92.5	90.1	88.7 (↓2.6)	87.2 (↓4.1)	**93.3**	91.0	89.4 (↓2.8)	88.8 (↓3.4)	93.1	**91.4**	89.8 (↓2.5)	90.5 (↓1.8)
MYO	83.6	85.3	82.8 (↓1.7)	80.2 (↓4.3)	**83.9**	85.8	82.6 (↓2.3)	80.8 (↓4.1)	83.7	**85.9**	83.7 (↓1.1)	82.6 (↓2.2)
RV	87.4	87.5	85.9 (↓1.6)	85.3 (↓2.2)	88.4	88.4	85.3 (↓3.1)	86.4 (↓2.0)	**89.4**	**88.8**	86.3 (↓2.8)	87.3 (↓1.8)
AVG	87.9	87.6	85.7 (↓2.0)	84.2 (↓3.5)	88.5	88.4	85.5 (↓2.7)	85.3 (↓3.2)	**88.7**	**88.7**	86.6 (↓2.1)	86.2 (↓2.5)

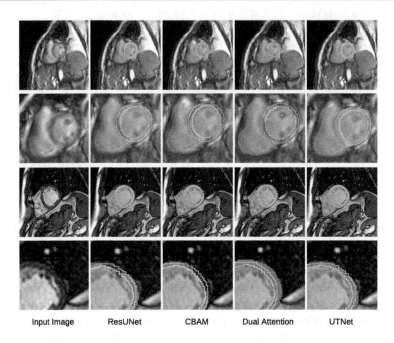

Input Image   ResUNet   CBAM   Dual Attention   UTNet

Fig. 4. Hard cases visualization on unseen testing data from vendor C and D. First two rows and the bottom two rows present the results and a zoom-in view of vendor C and D, respectively. The outline indicates the ground-truth annotation. Best viewed in color with LV(green), MYO(yellow), and RV(red). The test case from vendor C is blur due to motion artifacts, while the test case from vendor D is noisy and has low contrast in the boundary. Only UTNet provides consistent segmentation, which demonstrates its robustness. More visualization of segmentation outcomes are presented in the supplementary. (Color figure online)

Ablation Study. Figure 3 (a) shows the performance of different self-attention positions. The number in the x-axis indicates the level where self-attention is places, e.g., '34' means the level where 3 and 4 times down-samplings are performed. As the level goes up, the self-attention can gather more fine-grained detail information with increased performance. However, the curve saturates when adding to the original resolution. We reason this as the very shallow layer tends to be more focused on local texture, where global context information is not informative anymore. Figure 3 (b) shows the result of efficient self-attention's reduced size of 4, 8, 16. The reduced size 8 results in the best performance. The interpolation down-sampling is slightly better than using max-pooling. Figure 3 (c) shows the effect of the Transformer encoder, decoder, and the relative positional encoding using the optimal hyper-parameter from (a) and (b). The combination of the Transformer encoder and decoder gives the optimal performance. The relative positional encoding also plays a vital role, as removing it causes a large performance drop.

For a head-to-head comparison with standard self-attention on space and time complexity, we further apply dual attention in four resolutions (1, 2, 3, 4, same as UTNet), and use the same input image size and batch size ($256 \times 256 \times 16$) to test the inference time and memory consumption. UTNet gains superior advantage over dual attention with quadratic complexity, where GPU memory: 3.8 GB vs 36.9 GB and time: 0.146 s vs 0.243 s.

Robustness Analysis. Table 2 shows results on training models with data from vendor A and B, and then test the models on vendor A, B, C, and D, respectively. When viewing results on C and D vendors, competing approaches suffer from vendor differences while UTNet retains competitive performance. This observation can probably be attributed to the design of self-attention on multiple levels of feature maps and the content-position attention, allowing UTNet to be better focused on global context information instead of only local textures. Figure 4 further shows that UTNet displays the most consistent results of boundaries, while the other three methods are unable to capture subtle characteristics of boundaries, especially for RV and MYO regions in cardiac MRI.

4 Conclusion

We have proposed a U-shape hybrid Transformer network (UTNet) to merge advances of convolutional layers and self-attention mechanism for medical image segmentation. Our hybrid layer design allows the initialization of Transformer into convolutional networks without a need of pre-training. The novel self-attention allows us to extend operations at different levels of the network in both encoder and decoder for better capturing long-range dependencies. We believe that this design will help richly-parameterized Transformer models become more accessible in medical vision applications. Also, the ability to handle long sequences efficiently opens up new possibilities for the use of the UTNet on more downstream medical image tasks.

Acknowledgement. This research was supported in part by NSF: IIS 1703883, NSF IUCRC CNS-1747778 and funding from SenseBrain, CCF-1733843, IIS-1763523, IIS-1849238, MURI- Z8424104 -440149 and NIH: 1R01HL127661-01 and R01HL127661-05. and in part by Centre for Perceptual and Interactive Intellgience (CPII) Limited, Hong Kong SAR.

References

1. Bello, I., Zoph, B., Vaswani, A., Shlens, J., Le, Q.V.: Attention augmented convolutional networks. In: Proceedings of the IEEE/CVF International Conference on Computer Vision, pp. 3286–3295 (2019)
2. Campello, V.M., Palomares, J.F.R., Guala, A., Marakas, M., Friedrich, M., Lekadir, K.: Multi-Centre, Multi-Vendor & Multi-Disease Cardiac Image Segmentation Challenge (March 2020)
3. Dosovitskiy, A., et al.: An image is worth 16x16 words: transformers for image recognition at scale. arXiv preprint arXiv:2010.11929 (2020)
4. Fu, J., et al.: Dual attention network for scene segmentation. In: Proceedings of the IEEE/CVF Conference on Computer Vision and Pattern Recognition, pp. 3146–3154 (2019)
5. Gao, Y., et al.: Focusnetv 2: imbalanced large and small organ segmentation with adversarial shape constraint for head and neck CT images. Med. Image Anal. **67**, 101831 (2021)
6. Gao, Y., Liu, C., Zhao, L.: Multi-resolution path CNN with deep supervision for intervertebral disc localization and segmentation. In: Shen, D., et al. (eds.) MICCAI 2019. LNCS, vol. 11765, pp. 309–317. Springer, Cham (2019). https://doi.org/10.1007/978-3-030-32245-8_35
7. He, K., Zhang, X., Ren, S., Sun, J.: Deep residual learning for image recognition. In: Proceedings of the IEEE Conference on Computer Vision and Pattern Recognition, pp. 770–778 (2016)
8. He, K., Zhang, X., Ren, S., Sun, J.: Identity mappings in deep residual networks. In: Leibe, B., Matas, J., Sebe, N., Welling, M. (eds.) ECCV 2016. LNCS, vol. 9908, pp. 630–645. Springer, Cham (2016). https://doi.org/10.1007/978-3-319-46493-0_38
9. Huang, Q., Yang, D., Wu, P., Qu, H., Yi, J., Metaxas, D.: MRI reconstruction via cascaded channel-wise attention network. In: 2019 IEEE 16th International Symposium on Biomedical Imaging (ISBI 2019), pp. 1622–1626. IEEE (2019)
10. Huang, Z., Wang, X., Huang, L., Huang, C., Wei, Y., Liu, W.: CCNET: criss-cross attention for semantic segmentation. In: Proceedings of the IEEE/CVF International Conference on Computer Vision, pp. 603–612 (2019)
11. Isensee, F., Jaeger, P.F., Kohl, S.A., Petersen, J., Maier-Hein, K.H.: nnU-Net: a self-configuring method for deep learning-based biomedical image segmentation. Nat. Methods **18**(2), 203–211 (2021)
12. Kolesnikov, A., et al.: Big transfer (bit): General visual representation learning. arXiv preprint arXiv:1912.11370 **6**(2), 8 (2019)
13. Parmar, N., et al.: Image transformer. In: International Conference on Machine Learning, pp. 4055–4064. PMLR (2018)
14. Ronneberger, O., Fischer, P., Brox, T.: U-Net: convolutional networks for biomedical image segmentation. In: Navab, N., Hornegger, J., Wells, W.M., Frangi, A.F. (eds.) MICCAI 2015. LNCS, vol. 9351, pp. 234–241. Springer, Cham (2015). https://doi.org/10.1007/978-3-319-24574-4_28
15. Schlemper, J., et al.: Attention gated networks: learning to leverage salient regions in medical images. Med. Image Anal. **53**, 197–207 (2019)
16. Sinha, A., Dolz, J.: Multi-scale self-guided attention for medical image segmentation. IEEE J. Biomed. Health Inform. **25**(1), 121–130 (2020)
17. Tajbakhsh, N., Jeyaseelan, L., Li, Q., Chiang, J.N., Wu, Z., Ding, X.: Embracing imperfect datasets: a review of deep learning solutions for medical image segmentation. Med. Image Anal. **63**, 101693 (2020)

18. Vaswani, A., et al.: Attention is all you need. In: NIPS (2017)
19. Wang, Q., Li, B., Xiao, T., Zhu, J., Li, C., Wong, D.F., Chao, L.S.: Learning deep transformer models for machine translation. arXiv preprint arXiv:1906.01787 (2019)
20. Wang, S., et al.: Central focused convolutional neural networks: developing a data-driven model for lung nodule segmentation. Med. Image Anal. 40, 172–183 (2017)
21. Wang, S., Li, B., Khabsa, M., Fang, H., Ma, H.: Linformer: Self-attention with linear complexity. arXiv preprint arXiv:2006.04768 (2020)
22. Wang, X., Girshick, R., Gupta, A., He, K.: Non-local neural networks. In: Proceedings of the IEEE Conference on Computer Vision and Pattern Recognition, pp. 7794–7803 (2018)
23. Woo, S., Park, J., Lee, J.Y., Kweon, I.S.: CBAM: convolutional block attention module. In: Proceedings of the European Conference on Computer Vision (ECCV), pp. 3–19 (2018)
24. Yi, J., Wu, P., Jiang, M., Huang, Q., Hoeppner, D.J., Metaxas, D.N.: Attentive neural cell instance segmentation. Med. Image Anal. 55, 228–240 (2019). https://doi.org/10.1016/j.media.2019.05.004
25. Yu, F., Koltun, V.: Multi-scale context aggregation by dilated convolutions. arXiv preprint arXiv:1511.07122 (2015)
26. Zhao, H., Shi, J., Qi, X., Wang, X., Jia, J.: Pyramid scene parsing network. In: Proceedings of the IEEE Conference on Computer Vision and Pattern Recognition, pp. 2881–2890 (2017)
27. Zheng, S., et al.: Rethinking semantic segmentation from a sequence-to-sequence perspective with transformers. arXiv preprint arXiv:2012.15840 (2020)
28. Zhu, Z., Xu, M., Bai, S., Huang, T., Bai, X.: Asymmetric non-local neural networks for semantic segmentation. In: Proceedings of the IEEE/CVF International Conference on Computer Vision, pp. 593–602 (2019)

AlignTransformer: Hierarchical Alignment of Visual Regions and Disease Tags for Medical Report Generation

Di You[1], Fenglin Liu[1], Shen Ge[2], Xiaoxia Xie[3], Jing Zhang[3], and Xian Wu[2(✉)]

[1] School of ECE, Peking University, Beijing, China
{diyou,fenglinliu98}@pku.edu.cn
[2] Tencent Medical AI Lab, Shenzhen, China
{shenge,kevinxwu}@tencent.com
[3] Harbin Chest Hospital, Harbin, China

Abstract. Recently, medical report generation, which aims to automatically generate a long and coherent descriptive paragraph of a given medical image, has received growing research interests. Different from the general image captioning tasks, medical report generation is more challenging for data-driven neural models. This is mainly due to 1) the serious data bias: the normal visual regions dominate the dataset over the abnormal visual regions, and 2) the very long sequence. To alleviate above two problems, we propose an AlignTransformer framework, which includes the Align Hierarchical Attention (AHA) and the Multi-Grained Transformer (MGT) modules: 1) AHA module first predicts the disease tags from the input image and then learns the multi-grained visual features by hierarchically aligning the visual regions and disease tags. The acquired disease-grounded visual features can better represent the abnormal regions of the input image, which could alleviate data bias problem; 2) MGT module effectively uses the multi-grained features and Transformer framework to generate the long medical report. The experiments on the public IU-Xray and MIMIC-CXR datasets show that the AlignTransformer can achieve results competitive with state-of-the-art methods on the two datasets. Moreover, the human evaluation conducted by professional radiologists further proves the effectiveness of our approach.

Keywords: Medical report generation · Data bias · Transformer

1 Introduction

Medical images, e.g., radiology and pathology images, as well as their corresponding reports (see Fig. 1) are widely-used for diagnosis [9,12]. In clinical practice, writing a medical report can be time-consuming and tedious for experienced radiologists, and error-prone for inexperienced radiologists [5,28,29]. As

D. You and F. Liu—Equal Contributions.

Input Image	Aligning Visual Regions and Disease Tags	Ground Truth	R2Gen [8]	AlignTransformer [Ours]
	Cardiomegaly Scoliosis Pleural Effusion Pneumothorax Airspace Disease Hyperlucent Pneumonia Emphysema	The heart is within normal limits in size. Surgical suture material projects over the right lung apex. ¹The lungs are hy-perlucent and hyperinflated compatible with emphysema. ²There is left lower lobe airspace disease identified.³There is moderate left pleural eff-usion and small right pleural effu-sion. No visuali-zed pneumothorax.	The heart is normal in size. Lungs are clear. There is no pneumothorax. No acute bony abnormalities. No airspace disease or significant pleural effusion identified. No evidence of pneumothorax.	The heart and mediastinum are normal. ¹The lungs are hyperlucent. There is no pneumothorax. ²There is left lower lobe airspace disease suspicious for pneumonia. ³There is a moderate left sided pleural effusion.

Fig. 1. An example of ground truth report and reports generated by a state-of-the-art approach R2Gen [7] and our approach. The Red bounding boxes and Red colored text indicate the abnormalities in images and reports, respectively. As we can see, in this example, the R2Gen fails to depict the rare but important abnormalities and generates some error sentences (Blue colored text) and repeated sentences (Purple colored text). (Color figure online)

a result, automatic medical report generation systems can reduce the workload of radiologists by assisting them in clinical decision-making, and are thus in urgent need [4, 7, 16, 17, 21, 22, 24, 28, 29, 38, 42, 44, 45].

Most existing medical report generation models follow the standard image captioning approaches [1, 8, 25–27, 32, 40, 41, 43] and employ the encoder-decoder framework, e.g., a CNN-based image encoder followed by a LSTM-based report decoder. However, directly applying image captioning approaches to medical images has the following problems: 1) **Data bias**: the normal images dominate the dataset over the abnormal ones [36]. Furthermore, for each abnormal image, the normal regions dominate the image over the abnormal ones [24]. As shown in Fig. 1, abnormal regions (Red bounding boxes) only occupy a small part of the entire image. As a result, existing models tend to generate plausible general reports with no prominent abnormal narratives, failing to depict the rare but important abnormalities [16, 22, 44]. 2) **Very long sequence**: Image caption-ing models are designed to generate one single and short sentence, which only describes the most prominent visual contents, while medical report generation requires to generate a long paragraph, including multiple structural sentences with each one focusing on a specific medical observation. For the widely-used LSTM [14] in image captioning, it is hard to model such long sequences or paragraphs due to vanishing or exploding gradients [34]. Recent medical report generation models [16, 17, 44] rely on hierarchical LSTM [20], which have similar long sequence modeling problems. As shown in Fig. 1, even a state-of-the-art model R2Gen [7] still generates some repeated sentences of normalities.

To alleviate above problems, we propose the AlignTransformer, which includes two major modules: 1) Align Hierarchical Attention (AHA) works as the encoder to extract visual features. To focus on the abnormal regions, AHA first predicts the disease tags from the input image and then aligns these tags with the corresponding visual regions. To fit for both coarse- and fine-grained visual features, AHA introduces a hierarchically alignment model [25]. In detail, the disease tags are firstly used to find the most relevant visual regions, extracting disease-grounded visual features for each disease tag. Then the extracted visual features are further used to find the most relevant disease tags and filter out the irrelevant disease tags for each visual feature. By conducting the above process at different granularity level, the visual receptive fields gradually concentrate on salient abnormal regions under the guidance of the disease tags. In this way, we can obtain the features of abnormal regions, which can be used to alleviate data bias; 2) Multi-Grained Transformer (MGT) works as the decoder to generate final reports. MGT adaptively exploit coarse- and fine-grained disease-grounded visual features through a learned gating mechanism. In addition, MGT introduces the Transformer [7,8,28,39] to generate the reports. Benefiting from the visual features from multiple granularities and the Transformer, as shown in Fig. 1, our approach are a better fit for long report.

Overall, the contributions of this paper are as follows:

- In this paper, we propose the AlignTransformer framework, which aims to alleviate the data bias problem and model the very long sequence for medical report generation.
- The experiments on the benchmark IU-Xray [10] and MIMIC-CXR [18] datasets prove the effectiveness of our approach, which achieves results competitive with the existing state-of-the-art methods. We also verify the advantage of AlignTransformer from the manual evaluation of radiologists.

2   Related Works

The related works are introduced from: 1) Image Captioning and Visual Paragraph Generation and 2) Medical Report Generation.

Image Captioning and Visual Paragraph Generation. Image captioning [6] aims to understand the given images and generate corresponding descriptive sentences [1,8,25–27,32,40,41]. However, the sentence generated by AlignTransformer is usually short and describes the most prominent visual contents, which cannot fully convey the rich feature information of the image. Recently, visual paragraph generation [20], which aims to generate long and coherent reports or stories to describe visual contents, has attracted increasing research interests. To conduct the task, the Hierarchical LSTM structure (HLSTM) [20] is widely-used. However, in paragraph generation for medical images, the correctness of generating abnormalities should be emphasized more than other normalities, while in paragraphs of natural images each sentence has equal importance. Besides, due to the data bias in the medical domain, the widely-used HLSTM in the visual paragraph generation does not perform very well in medical report generation and is tend to produce normal reports [16,22,24,28,29,42,44].

Medical Report Generation. Inspired by the success of deep learning models on image captioning, a lot of encoder-decoder based frameworks have been proposed [7,16,17,21,22,24,28–30,42,44,45] to automatically generate a fluent medical report. However, due to the data bias and the very long sequence, some errors occur in the generated reports of the existing methods, like duplicate reports, inexact descriptions, etc. [17,42,44]. To this end, [16,22,30] and [21,28,45] introduced the reinforcement learning and medical knowledge graph, respectively. [29]

introduced the contrastive attention to compare the abnormal image with the normal images to better capture the abnormalities. Meanwhile, [7,28] further introduced the Transformer [39] to model the very long sequence. In our work, different from [7,28], we conduct the alignments between visual regions and disease tags in *an iterative manner* and propose *a multi-grained transformer* to combine disease-grounded visual features at different depths.

3 Approach

We first formulate the problem; Next, we introduce our framework, which includes the Align Hierarchical Attention and Multi-Grained Transformer.

3.1 Problem Formulation

Given a medical image encoded as visual features V and disease tags T, the goal of our framework is to generate a coherent report R that describes the observations in details of both normal and abnormal regions. In implementations, for the visual features V, we follow [16,21,22,24,28,29] to adopt the ResNet-50 [13] pre-trained on ImageNet [11] and fine-tuned on CheXpert dataset [15] to extract the 2,048 7×7 feature maps, which are further projected into 512 7×7 feature maps, denoted as $V = \{v_1, v_2, \ldots, v_{N_V}\} \in \mathbb{R}^{N_V \times d}$ ($N_V = 49, d = 512$). Moreover, we follow [17] to further predict the disease tags T of the input image. Specifically, we further feed the extracted visual features V into a multi-label classification network, which is pre-trained as multi-label classification task on the downstream datasets to generate a distribution over all of the pre-defined tags. Finally, the embeddings of the $N_T = 10$ most likely tags $T = \{t_1, t_2, \ldots, t_{N_T}\} \in \mathbb{R}^{N_T \times d}$ are used as disease tags of current input image. Based on the extracted V and T, our framework first adopt the AHA to align the visual regions V and disease tags T in a coarse-to-fine manner, acquiring multi-grained disease-grounded visual features $\hat{V}$. Next, MGT is introduced to generate final report based on the $\hat{V}$. In brief, our AlignTransformer framework is formulated as:

$$\text{AHA} : \{V, T\} \to \hat{V}; \quad \text{MGT} : \hat{V} \to R. \tag{1}$$

Through the above process, our framework gets the ability to alleviate the data bias problem and model the very long sequence.

3.2 Align Hierarchical Attention

Basic Module. To align the visual regions V and the disease tags T, we adopt the Multi-Head Attention (MHA) from [39], which allows probabilistic many-to-many relations instead of monotonic relations, as in [25,41]. Thus, the MHA can compute the association weights between different features. The MHA consists of n parallel heads and each head is defined as a scaled dot-product attention:

$$\text{Att}_i(X, Y) = \text{softmax}\left(X\text{W}_i^\text{Q}(Y\text{W}_i^\text{K})^T\right) Y\text{W}_i^\text{V} \tag{2}$$

$$\text{MHA}(X, Y) = [\text{Att}_1(X, Y); \ldots; \text{Att}_n(X, Y)]\text{W}^\text{O} \tag{3}$$

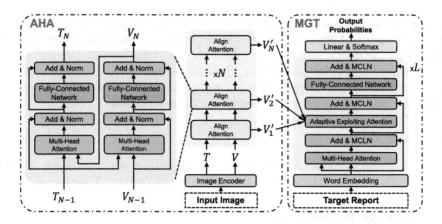

Fig. 2. Illustration of our proposed AlignTransformer framework, which consists of the Align Hierarchical Attention (AHA) and Multi-Grained Transformer (MGT).

where $X \in \mathbb{R}^{l_x \times d}$ and $Y \in \mathbb{R}^{l_y \times d}$ denote the Query matrix and the Key/Value matrix, respectively; the divisor $\sqrt{d_n}$ $(d_n = d/n)$ in Eq. (2) is omitted in equations for conciseness, please see [39] for details; $W_i^Q, W_i^K, W_i^V \in \mathbb{R}^{d \times d_n}$ and $W^O \in \mathbb{R}^{d \times d}$ are learnable parameters. $[\cdot; \cdot]$ stands for concatenation operation.

Conventionally, a Fully-Connected Network (FCN) is followed by the MHA:

$$\text{FCN}(X) = \max(0, X W^f + b^f) W^{ff} + b^{ff} \tag{4}$$

where $\max(0, *)$ represents the ReLU activation function; $W^f \in \mathbb{R}^{d \times 4d}$ and $W^{ff} \in \mathbb{R}^{4d \times d}$ are learnable matrices; b^f and b^{ff} are the bias terms. It is worth noting that both the MHA and the FCN are followed by an operation sequence of dropout [37], skip connection [13] and layer normalization (Norm) [2].

Align Attention. Since the MHA can compute the association weights between different features, if we apply the MHA between visual regions V and the disease tags T, we can learn the correlation and relationship between V and T, resulting in the alignment between visual regions and disease tags. Therefore, based on the MHA, as shown in Fig. 2, the align attention is conducted as:

$$V' = \text{FCN}(\text{MHA}(T, V)); \quad T' = \text{FCN}(\text{MHA}(V', T)) \tag{5}$$

In implementation, the disease tags T are first used as the Query to find the most relevant visual regions in V, generating the disease related visual features V'. Then the V' are further used as the Query to find the most relevant disease tags and filter out the irrelevant disease tags in T. As a result, we can obtain alignment between visual regions and disease tags via above processes. Finally, since V' and T' are aligned, we can add them up to get the disease-grounded visual features $\hat{V}$ that exploits the best of their respective advantages:

$$\hat{V} = \text{LayerNorm}(V' + T') \tag{6}$$

where the LayerNorm denotes the layer normalization [2]. In particular, since T contains the disease tags, the disease-grounded visual features $\hat{V}$ can be referred as the features of abnormal regions, which are useful in alleviating data bias.

Full Model. To improve the performance and acquire the multi-grained disease-grounded visual features, we propose to perform Align Attention hierarchically, which we call Align Hierarchical Attention. Specifically, the process in Eq. (5) that uses the original features is considered as the first round:

$$V_1 = \text{FCN}(\text{MHA}(T_0, V_0)); \quad T_1 = \text{FCN}(\text{MHA}(V_1, T_0)) \tag{7}$$

where $V_0 = V$, $T_0 = T$, V_1 and T_1 denote original visual features, original disease tags, aligned visual features, and aligned disease tags, respectively. By repeating the same process for N times, we obtain the final outputs of the two stacks:

$$V_N = \text{FCN}(\text{MHA}(T_{N-1}, V_{N-1})); \quad T_N = \text{FCN}(\text{MHA}(V_N, T_{N-1})) \tag{8}$$

$$\hat{V}_N = \text{LayerNorm}(V_N + T_N) \tag{9}$$

In this way, we obtain the multi-grained disease-grounded visual features $\{\hat{V}_i\} = \{\hat{V}_1, \hat{V}_2, \ldots, \hat{V}_N\}$.

3.3 Multi-grained Transformer

We propose the Multi-Grained Transformer (MGT), which includes $L = 3$ decoder layers, to exploit the $\{\hat{V}_i\}$ and generate a proper report R. In implementations, for generating each word y_t, given the embedding of current input word $x_t^{(0)} = w_t + e_t$ (w_t: word embedding, and e_t: fixed position embedding). For the lth decoder layer, it takes $x_t^{(l-1)}$ as input to obtain $h_t^{(l)} = \text{MHA}(x_t^{(l-1)}, x_{1:t}^{(l-1)})$. Then, inspired by [8], we introduce the Adaptive Exploiting Attention (AEA) module to adaptively exploit multi-grained disease-grounded visual features:

$$\hat{h}_t^{(l)} = \text{AEA}(h_t^{(l)}, \{\hat{V}_i\}) = \sum_{i=1}^{N} \lambda_i \odot \text{MHA}(h_t^{(l)}, \hat{V}_i) \tag{10}$$

$$\lambda_i = \sigma\left(\left[h_t^{(l)}; \text{MHA}(h_t^{(l)}, \hat{V}_i)\right] W_i + b_i\right) \tag{11}$$

where $W_i \in \mathbb{R}^{2d \times d}$ is a learnable parameter and b_i is a learnable bias vector. $\odot$, σ and $[\cdot; \cdot]$ denote the element-wise multiplication, the sigmoid function and the concatenation, respectively. The computed $\{\lambda_i\}$ weights the expected importance of different grained disease-grounded visual features for each target word. Then, we adopt the FCN to acquire the output of lth decoder layer: $x_t^{(l)} = \text{FCN}(\hat{h}_t^{(l)})$ It is worth noting that both the MHA, the AEA and the FCN are followed by an operation sequence of dropout [37], skip connection [13] and memory-driven conditional layer normalization (MCLN) [7].

Finally, the output of last decoder layer $x_t^{(L)}$ is used to predict the next word: $y_t \sim p_t = \text{softmax}(x_t^{(L)} W^p + b^p)$, where the W^p and b^p are the learnable parameters. Given the ground truth report $R^* = \{y_1^*, y_2^*, \ldots, y_{N_R}^*\}$ provided by the radiologists, we can train our AlignTransformer framework by minimizing the widely-used cross-entropy loss: $L_{CE}(\theta) = -\sum_{i=1}^{N_R} \log \left(p_\theta \left(y_i^* \mid y_{1:i-1}^* \right) \right)$.

4   Experiments

Datasets, Metrics and Settings. In our work, we conduct experiments on two public datasets, i.e., a widely-used benchmark IU-Xray [10] and a recently released large-scale MIMIC-CXR [18]. **IU-Xray** is widely-used to evaluate the performance of medical report generation methods. It contains 7,470 chest X-ray images associated with 3,955 reports. Following previous works [7,16,21,22, 24,28,29], we randomly split the dataset into 70%–10%–20% training-validation-testing splits. **MIMIC-CXR** is the recently released largest dataset to date and consists of 377,110 chest X-ray images and 227,835 reports from 64,588 patients. Following [7,24,28,29], we use the official splits to report our results, resulting in 368,960 in the training set, 2,991 in the validation set and 5,159 in the test set.

To test the performance, we adopt the widely-used BLEU [33], METEOR [3] and ROUGE-L [23], which are reported by the evaluation toolkit [6]. Since our approach is based on the Transformer [39], we follow the original settings in [39] and set $d = 512, n = 8$. Based on the average performance on the validation set, both the N and L in our Align Hierarchical Attention and Multi-Grained Transformer, respectively, are set to 3. We use Adam optimizer [19] with a batch size of 16 and a learning rate of $2e-4$ within maximum 100 epochs for parameter optimization. We also use momentum of 0.8 and weight decay of 0.999.

Automatic Evaluation. Several representative models, including three most recently published state-of-the-art models, i.e., PPKED [28], R2Gen [7] and SentSAT + KG [45], are selected for comparison. The results on the test of two datasets are shown in Table 1. As we can see, our AlignTransformer achieves the competitive results with the previous state-of-the-art models on the two datasets under all metrics, which verify the effectiveness of our approach.

Human Evaluation. Moreover, we invite two professional radiologists to evaluate the perceptual quality of 200 randomly selected reports generated by our AlignTransformer and the R2Gen [7]. The radiologists are unaware of which model generates these reports and are encouraged to select a more accurate report for each pair. The results of last column in Table 1 show that our approach is better than R2Gen in clinical practice with winning pick-up percentages.

Overall, the results verify that our proposed framework can generate accurate and proper reports, which could assist radiologists in making decisions.

Table 1. Results of our AlignTransformer and existing models on the MIMIC-CXR and IU-Xray datasets. Hit represents the human evaluation results and is measured by the pick-up percentages (%). The differences between their sum and 100% represent the percentage of ties. Higher value denotes better performance in all columns.

Datasets	Methods	Year	BLEU-1	BLEU-2	BLEU-3	BLEU-4	METEOR	ROUGE-L	Hit
MIMIC-CXR	CNN-RNN [40]	2015	0.299	0.184	0.121	0.084	0.124	0.263	–
	AdaAtt [31]	2017	0.299	0.185	0.124	0.088	0.118	0.266	–
	Att2in [35]	2017	0.325	0.203	0.136	0.096	0.134	0.276	–
	Up-Down [1]	2018	0.317	0.195	0.130	0.092	0.128	0.267	–
	R2Gen [7]	2020	0.353	0.218	0.145	0.103	0.142	0.277	29
	PPKED [28]	2021	0.360	0.224	0.149	0.106	0.149	**0.284**	–
	AlignTransformer	Ours	**0.378**	**0.235**	**0.156**	**0.112**	**0.158**	0.283	**54**
IU-Xray	HRGR-Agent [22]	2018	0.438	0.298	0.208	0.151	–	0.322	–
	CMAS-RL [16]	2019	0.464	0.301	0.210	0.154	–	0.362	–
	SentSAT [45]	2019	0.445	0.289	0.200	0.143	–	0.359	–
	SentSAT+KG [45]	2020	0.441	0.291	0.203	0.147	–	0.367	–
	R2Gen [7]	2020	0.470	0.304	0.219	0.165	0.187	0.371	22
	PPKED [28]	2021	0.483	**0.315**	0.224	0.168	0.190	0.376	–
	AlignTransformer	Ours	**0.484**	0.313	**0.225**	**0.173**	**0.204**	**0.379**	**63**

Table 2. Ablation study of our method, which includes the AHA and the MGT. The N denotes the number of alignments in AHA. The 'Baseline' and 'w/ AHA+MGT' denote the ResNet-50 image encoder [13] equipped with the Transformer report decoder in [7] and our proposed AlignTransformer, respectively.

Datasets	Methods	N	BLEU-1	BLEU-2	BLEU-3	BLEU-4	METEOR	ROUGE-L
IU-Xray	Baseline	–	0.442	0.285	0.195	0.138	0.176	0.357
	w/ AHA	1	0.467	0.301	0.210	0.159	0.185	0.368
	w/ AHA	2	0.473	0.305	0.217	0.163	0.188	0.372
	w/ AHA	3	0.476	0.308	0.219	0.164	0.192	0.374
	w/ AHA	4	0.471	0.303	0.215	0.160	0.187	0.370
	w/ AHA+MGT	3	**0.484**	**0.313**	**0.225**	**0.173**	**0.204**	**0.379**

5 Analysis

In this section, we conduct analysis on the benchmark IU-Xray dataset [10] from different perspectives to better understand our proposed framework.

Ablation Study. We investigate the contribution of each component in our approach. Table 2 shows that our AHA can significantly promote the performance over all metrics, with up to 18.8% gain in BLEU-4 ($N = 3$). This indicates that the disease-grounded visual features extracted by our AHA contain sufficient accurate abnormal information, which is vital in alleviating the data deviation problem and improving the performance of medical report generation. For the MGT, Table 2 shows that the MGT can further boost the performance under all metrics, which demonstrates the effectiveness of our MGT to adaptively exploit the multi-grained disease-grounded visual features.

Visualization and Example. In Fig. 1, we give the visualization of the AHA and an intuitive example to better understand our approach. As we can see, for the current state-of-the-art model R2Gen [7], it produces some inexact descriptions and repeated sentences of normalities, which is due to the overwhelming normal images in the dataset, i.e., data deviation [16,24,28,36]. Since our AHA can efficiently capture the abnormal regions by aligning visual regions and disease tags, we can generate more accurate report for the abnormal regions. For example, the generated report correctly describes "*The lungs are hyperlucent*", "*There is left lower lobe airspace disease suspicious for pneumonia*" and "*There is a moderate left sided pleural effusion*".

6   Conclusions

In this paper, we propose an AlignTransformer to alleviate the data bias problem and model the very long sequence for medical report generation. The experiments prove the effectiveness of our method, which not only generates meaningful medical reports with accurate abnormal descriptions and regions, but also achieves competitive results with state-of-the-art models on two public datasets with the best human preference. The results justify the usefulness of our framework in assisting radiologists in clinical decision-making, reducing their workload.

Acknowledgments. We sincerely thank all the anonymous reviewers and chairs for their constructive comments and suggestions that substantially improved this paper.

References

1. Anderson, P., et al.: Bottom-up and top-down attention for image captioning and VQA. In: CVPR (2018)
2. Ba, L.J., Kiros, R., Hinton, G.E.: Layer normalization. arXiv preprint arXiv:1607.06450 (2016)
3. Banerjee, S., Lavie, A.: METEOR: an automatic metric for MT evaluation with improved correlation with human judgments. In: IEEvaluation@ACL (2005)
4. Biswal, S., Xiao, C., Glass, L., Westover, M.B., Sun, J.: CLARA: clinical report auto-completion. In: WWW (2020)
5. Brady, A., Laoide, R.Ó., Mccarthy, P., Mcdermott, R.: Discrepancy and error in radiology: concepts, causes and consequences. Ulster Med. J. **81**, 3–9 (2012)
6. Chen, X., et al.: Microsoft COCO captions: data collection and evaluation server. arXiv preprint arXiv:1504.00325 (2015)
7. Chen, Z., Song, Y., Chang, T., Wan, X.: Generating radiology reports via memory-driven transformer. In: EMNLP (2020)
8. Cornia, M., Stefanini, M., Baraldi, L., Cucchiara, R.: Meshed-memory transformer for image captioning. In: CVPR (2020)
9. Delrue, L., Gosselin, R., Ilsen, B., Van Landeghem, A., de Mey, J., Duyck, P.: Difficulties in the interpretation of chest radiography. In: Coche, E., Ghaye, B., de Mey, J., Duyck, P. (eds.) Comparative Interpretation of CT and Standard Radiography of the Chest. Medical Radiology, pp. 27–49. Springer, Heidelberg (2011). https://doi.org/10.1007/978-3-540-79942-9_2

10. Demner-Fushman, D., et al.: Preparing a collection of radiology examinations for distribution and retrieval. J. Am. Med. Inform. Assoc. **23**(2), 304–310 (2016)
11. Deng, J., Dong, W., Socher, R., Li, L., Li, K., Li, F.: ImageNet: a large-scale hierarchical image database. In: CVPR (2009)
12. Goergen, S.K., et al.: Evidence-based guideline for the written radiology report: methods, recommendations and implementation challenges. J. Med. Imaging Radiat. Oncol. **57**(1), 1–7 (2013)
13. He, K., Zhang, X., Ren, S., Sun, J.: Deep residual learning for image recognition. In: CVPR (2016)
14. Hochreiter, S., Schmidhuber, J.: Long short-term memory. Neural Comput. **9**(8), 1735–1780 (1997)
15. Irvin, J., et al.: CheXpert: a large chest radiograph dataset with uncertainty labels and expert comparison. In: AAAI (2019)
16. Jing, B., Wang, Z., Xing, E.P.: Show, describe and conclude: on exploiting the structure information of chest X-ray reports. In: ACL (2019)
17. Jing, B., Xie, P., Xing, E.P.: On the automatic generation of medical imaging reports. In: ACL (2018)
18. Johnson, A.E.W., et al.: MIMIC-CXR: a large publicly available database of labeled chest radiographs. arXiv preprint arXiv:1901.07042 (2019)
19. Kingma, D.P., Ba, J.: Adam: a method for stochastic optimization. In: ICLR (2014)
20. Krause, J., Johnson, J., Krishna, R., Fei-Fei, L.: A hierarchical approach for generating descriptive image paragraphs. In: CVPR (2017)
21. Li, C.Y., Liang, X., Hu, Z., Xing, E.P.: Knowledge-driven encode, retrieve, paraphrase for medical image report generation. In: AAAI (2019)
22. Li, Y., Liang, X., Hu, Z., Xing, E.P.: Hybrid retrieval-generation reinforced agent for medical image report generation. In: NeurIPS (2018)
23. Lin, C.Y.: ROUGE: a package for automatic evaluation of summaries. In: ACL (2004)
24. Liu, F., Ge, S., Wu, X.: Competence-based multimodal curriculum learning for medical report generation. In: ACL (2021)
25. Liu, F., Liu, Y., Ren, X., He, X., Sun, X.: Aligning visual regions and textual concepts for semantic-grounded image representations. In: NeurIPS (2019)
26. Liu, F., Ren, X., Liu, Y., Lei, K., Sun, X.: Exploring and distilling cross-modal information for image captioning. In: IJCAI (2019)
27. Liu, F., Ren, X., Liu, Y., Wang, H., Sun, X.: simNet: stepwise image-topic merging network for generating detailed and comprehensive image captions. In: EMNLP (2018)
28. Liu, F., Wu, X., Ge, S., Fan, W., Zou, Y.: Exploring and distilling posterior and prior knowledge for radiology report generation. In: CVPR (2021)
29. Liu, F., Yin, C., Wu, X., Ge, S., Zhang, P., Sun, X.: Contrastive attention for automatic chest X-ray report generation. In: ACL (Findings) (2021)
30. Liu, G., et al.: Clinically accurate chest X-ray report generation. In: MLHC (2019)
31. Lu, J., Xiong, C., Parikh, D., Socher, R.: Knowing when to look: adaptive attention via a visual sentinel for image captioning. In: CVPR (2017)
32. Pan, Y., Yao, T., Li, Y., Mei, T.: X-linear attention networks for image captioning. In: CVPR (2020)
33. Papineni, K., Roukos, S., Ward, T., Zhu, W.: BLEU: a method for automatic evaluation of machine translation. In: ACL (2002)
34. Pascanu, R., Mikolov, T., Bengio, Y.: On the difficulty of training recurrent neural networks. In: ICML (2013)

35. Rennie, S.J., Marcheret, E., Mroueh, Y., Ross, J., Goel, V.: Self-critical sequence training for image captioning. In: CVPR (2017)

36. Shin, H., Roberts, K., Lu, L., Demner-Fushman, D., Yao, J., Summers, R.M.: Learning to read chest X-rays: recurrent neural cascade model for automated image annotation. In: CVPR (2016)

37. Srivastava, N., Hinton, G.E., Krizhevsky, A., Sutskever, I., Salakhutdinov, R.: Dropout: a simple way to prevent neural networks from overfitting. J. Mach. Learn. Res. **5**, 1929–1958 (2014)

38. Syeda-Mahmood, T., et al.: Chest X-ray report generation through fine-grained label learning. In: Martel, A.L., et al. (eds.) MICCAI 2020. LNCS, vol. 12262, pp. 561–571. Springer, Cham (2020). https://doi.org/10.1007/978-3-030-59713-9_54

39. Vaswani, A., et al.: Attention is all you need. In: NIPS (2017)

40. Vinyals, O., Toshev, A., Bengio, S., Erhan, D.: Show and tell: a neural image caption generator. In: CVPR (2015)

41. Xu, K., et al.: Show, attend and tell: neural image caption generation with visual attention. In: ICML (2015)

42. Xue, Y., et al.: Multimodal recurrent model with attention for automated radiology report generation. In: Frangi, A.F., Schnabel, J.A., Davatzikos, C., Alberola-López, C., Fichtinger, G. (eds.) MICCAI 2018. LNCS, vol. 11070, pp. 457–466. Springer, Cham (2018). https://doi.org/10.1007/978-3-030-00928-1_52

43. You, Q., Jin, H., Wang, Z., Fang, C., Luo, J.: Image captioning with semantic attention. In: CVPR (2016)

44. Yuan, J., Liao, H., Luo, R., Luo, J.: Automatic radiology report generation based on multi-view image fusion and medical concept enrichment. In: Shen, D., et al. (eds.) MICCAI 2019. LNCS, vol. 11769, pp. 721–729. Springer, Cham (2019). https://doi.org/10.1007/978-3-030-32226-7_80

45. Zhang, Y., Wang, X., Xu, Z., Yu, Q., Yuille, A.L., Xu, D.: When radiology report generation meets knowledge graph. In: AAAI (2020)

Continuous-Time Deep Glioma Growth Models

Jens Petersen[1(✉)], Fabian Isensee[2], Gregor Köhler[1], Paul F. Jäger[3],
David Zimmerer[1], Ulf Neuberger[4], Wolfgang Wick[5,6], Jürgen Debus[7,8,9],
Sabine Heiland[4], Martin Bendszus[4], Philipp Vollmuth[4],
and Klaus H. Maier-Hein[1]

[1] Division of Medical Image Computing, German Cancer Research Center,
Heidelberg, Germany
`jens.petersen@dkfz.de`
[2] HIP Applied Computer Vision Lab, Division of Medical Image Computing,
German Cancer Research Center, Heidelberg, Germany
[3] Interactive Machine Learning Group, German Cancer Research Center,
Heidelberg, Germany
[4] Department of Neuroradiology, Heidelberg University Hospital,
Heidelberg, Germany
[5] Neurology Clinic, Heidelberg University Hospital, Heidelberg, Germany
[6] DKTK CCU Neurooncology, German Cancer Research Center,
Heidelberg, Germany
[7] Division of Molecular and Translational Radiation Oncology, Heidelberg Institute
of Radiation Oncology (HIRO), Heidelberg, Germany
[8] Heidelberg Ion-Beam Therapy Center (HIT), Heidelberg University Hospital,
Heidelberg, Germany
[9] Clinical Cooperation Unit Radiation Oncology, German Cancer Research Center,
Heidelberg, Germany

Abstract. The ability to estimate how a tumor might evolve in the
future could have tremendous clinical benefits, from improved treatment
decisions to better dose distribution in radiation therapy. Recent work
has approached the glioma growth modeling problem via deep learning and variational inference, thus learning growth dynamics entirely
from a real patient data distribution. So far, this approach was constrained to predefined image acquisition intervals and sequences of fixed
length, which limits its applicability in more realistic scenarios. We overcome these limitations by extending Neural Processes, a class of conditional generative models for stochastic time series, with a hierarchical
multi-scale representation encoding including a spatio-temporal attention mechanism. The result is a learned growth model that can be conditioned on an arbitrary number of observations, and that can produce a
distribution of temporally consistent growth trajectories on a continuous
time axis. On a dataset of 379 patients, the approach successfully captures both global and finer-grained variations in the images, exhibiting
superior performance compared to other learned growth models.

Keywords: Glioma growth · Stochastic time series · Generative
modeling

© Springer Nature Switzerland AG 2021
M. de Bruijne et al. (Eds.): MICCAI 2021, LNCS 12903, pp. 83–92, 2021.
https://doi.org/10.1007/978-3-030-87199-4_8

1   Introduction

Glioma growth modeling refers to the problem of estimating how the tumor burden in a patient might evolve in the future given the current state of the disease and possibly its history. In the past, this was often approached with so-called reaction-diffusion models, where tumor growth is described by the physical diffusion of an assumed tumor cell density. There is a large body of related literature, so we only refer to [16,17] as overviews. Newer works not included therein are [4,20], among others. All of these approaches have in common that they depend on the quality of the chosen forward model, which always includes simplifying assumptions when compared to the actual biological processes.

A recent alternative was proposed in [18] in the form of fully learned growth models. The authors used a probabilistic U-Net to map a fixed-length input sequence of MRI scans to a distribution of possible spatial tumor configurations in the future. The model thus learns a growth model directly from data and doesn't rely on explicit assumptions and simplifications with respect to the underlying biological process. At the same time, the approach suffers from several limitations, requiring a fixed number of input observations, requiring fixed time intervals between observations, and being able to make predictions only for a single predefined future time point. In order to overcome all of these limitations, we build upon the Neural Process framework [5,6], a family of conditional generative models specifically conceived to model continuous-valued stochastic processes. They work by encoding so-called *context* observations into a joint representation and then predicting a desired *target* output from it. In particular, we extend the work from [3,13], who apply Neural Processes in the image domain, by introducing a convolutional encoder-decoder structure with representations at multiple scales. The single-scale encoder-decoder setups used in [3,13] require extremely large decoders to produce outputs with sufficient spatial resolution. As a result, they require vast amounts of data to be trained successfully, and the authors only demonstrate their approaches on synthetic images. Ours is one of the first Neural Processes to be applied to real data in the image domain[1], with [8] the only other work we're aware of. To allow our approach to model tumor changes with high spatial resolution, we also introduce an attention mechanism [21] that aggregates the context information at different scales. In contrast to existing work with attention in Neural Processes [10,19], we design the mechanism so that it can model spatial and temporal dependencies simultaneously.

In summary, this work makes the following contributions: 1) We propose the first fully learned tumor growth model that can make predictions at arbitrary points on the time axis, conditioned on an arbitrary number of available observations. 2) In doing so, we propose a Neural Process variant for image-based problems that is efficient enough to work successfully on real-world data. 3) We show that the proposed approach significantly outperforms prior work on the task of learning-based modeling of glioma growth. We compare our approach with both the model proposed in [18] and a naive convolutional Neural Process.

---

[1] *Image domain* refers to the fact that the observations are entire images. The original Neural Processes [5,6] work on images by treating individual pixels as observations.

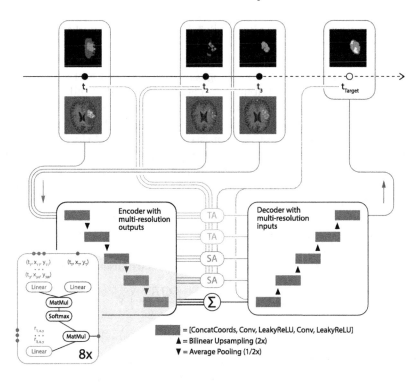

Fig. 1. A schematic of the proposed model. All *context* observations (MRI/ segmentation/time, blue) are encoded separately with the same convolutional encoder that outputs representations at different resolution scales. The lowest scale (black) has no spatial resolution, and the corresponding representations are summed. The result is a global distribution, a sample of which is fed to the decoder together with the *target* input (time, red). For representations with low spatial resolution, we use spatio-temporal attention (SA) to aggregate context information. At higher resolutions we use only temporal attention (TA) to reduce GPU memory requirements. ConcatCo-ords represents an operator that appends pixel locations (-0.5 to 0.5 per axis). The first encoder block also uses InstanceNorm. (Color figure online)

4) We provide a complete implementation at https://github.com/MIC-DKFZ/ deep-glioma-growth.

2   Methods

2.1   Model

The model we propose is shown in Fig. 1. We have a number of observations (MRIs and corresponding tumor segmentations) available at continuous time values t_c. We call those the *context*, with bold indicating a collection henceforth. We are interested in possible tumor segmentations for future *target* times t_t. Each context observation is encoded separately with an encoder that outputs representations hierarchically at multiple scales: $r_c = (r_{c,32}, r_{c,16}, r_{c,8}, r_{c,4}, r_{c,global})$. The second index refers to the scale (the first entry has a resolution of 32^2, etc.) and the global representations $r_{c,global}$ describe mean and standard deviation of diagonal normal distributions. These means and standard deviations are summed over all context encodings (bottom path in Fig. 1) to form a single global distribution. Because it is independent of the target t_t, this allows us to sample growth trajectories that are consistent over time and that can be evaluated at arbitrary new points in time. A sample from the distribution (which is 128-dimensional in our case) along with t_t is fed to the decoder to predict the corresponding segmentation. If only the global representation is used, our model reduces to a Neural Process [5,6] with a convolutional encoder and decoder. We want our approach to be able to model spatial and temporal dependencies simultaneously, so we aggregate the context information at higher scales using a spatio-temporal dot-product attention mechanism [21], conditioned on the target t_{target} and pixel locations (x, y) at the given scale (see inset in Fig. 1):

$$r(t_t, x_t, y_t) = \text{softmax}\left(\frac{QK^T}{\sqrt{d}}\right) V \qquad (1)$$

$$Q = \text{Lin}\left(t_t, x_t, y_t, r_{c,x,y}\right), \quad K = \text{Lin}\left(t_c, x_c, y_c, r_{c,x,y}\right), \quad V = \text{Lin}\left(r_{c,x,y}\right) \quad (2)$$

Q, K, V are linear maps from what is usually referred to as *queries* (target locations (t_t, x_t, y_t)), *keys* (available locations (t_c, x_c, y_c)) and *values* (the representations at the *key* locations). We also use $r_{c,x,y}$ to construct Q and K because of the success of self-attention in transformer architectures, e.g. [1]. This means that the available representations are weighted according to the similarity of the corresponding key to the queries. The similarity is given by the dot product in a higher-dimensional space (with dimension $d = 16$ in our case). Note that in Fig. 1, each attention block represents so-called multihead attention, meaning 8 parallel blocks of the above. The downside of this spatio-temporal attention is that the matrix product QK^T requires GPU memory that scales quadratically with the number of pixels. As a result we only use the full spatio-temporal attention at the two lowest scales (resolutions 4^2 and 8^2) and resort to temporal attention at higher scales, i.e. the same as above but without a dependence on pixel locations.

Our model is trained with the following optimization objective, where I are inputs (MRI scans) while S refers to segmentations. During training the context is absorbed into the target set ($t_t \rightarrow t_{t+c}$, etc.), meaning we let the models reconstruct the inputs as well instead of only predicting a future segmentation:

$$\min_{\theta} \mathop{\mathbb{E}}_{z \sim q_\theta(z|I_t, S_t, t_t)} - \log p_\theta(S_t | I_c, S_c, t_c, t_t, z)$$
$$+ \beta \cdot D_{KL}(q_\theta(z|I_t, S_t, t_t) || q_\theta(z|I_c, S_c, t_c)) \qquad (3)$$

where $q_\theta(z|\cdot)$ is the global representation predicted by the encoder for the given inputs. This is the Neural Process variant of the well-known variational inference objective, with some modifications: 1) There is a direct dependence of the reconstruction likelihood p_θ on $(\boldsymbol{I}_c, \boldsymbol{S}_c, \boldsymbol{t}_c)$, because the attention mechanisms are deterministic (otherwise the influence would only be via z). 2) The predictive likelihood is usually modeled as a factorizing categorical likelihood, leading to a cross-entropy loss. We instead use the sum of cross-entropy and Dice losses [2]. As a result, the likelihood no longer factorizes over pixels, so 3) we introduce a weight β for the KL term, as done for example in [7], which re-balances the two loss terms. We optimized this parameter through line-searching the space of possible values in powers of 10, choosing the one that lead to KL loss values most similar to those obtained during training with only cross-entropy: $\beta = 0.0001$. Note that the objective for the baseline from [18] can be written the same way, but without dependence on $(\boldsymbol{S}_c, \boldsymbol{t}_c, \boldsymbol{t}_t)$. Models are trained for 300 epochs with Adam [11] and a learning rate of 0.0001 at batch size 128. We give between 2 and 5 inputs as context and always have a single target point, but the models predict both context and target during training. For all other hyperparameters we refer to the provided implementation.

2.2   Data and Evaluation

We use a dataset that consists of MRI scans from 379 glioblastoma patients, stemming from a multi-center study that compared treatment using a combination of chemotherapy (Lomustine) and an angiogenesis inhibitor (Bevacizumab) with treatment using chemotherapy alone [23]. The study found no significant difference between test and control arm in terms of overall survival, so we ignore treatment effects in our work and treat it as an additional source of stochasticity. Pre-processing of the data including annotations (segmentations of edema, enhancing tumor and necrosis) was performed by experienced neuroradiologists as described previously [9]. For each patient there are between 3 and 13 (mean 4.85) longitudinal scans, consisting of native T1, contrast-enhanced T1 (with gadolinium agent), T2 and FLAIR. All scans for a patient are skull-stripped, registered to the native T1 image of the first available scan, and normalized by mean and standard deviation on a per-channel basis. We resample the original data to isotropic 128^3 and then extract a region of size 64^3 centered around the tumor. We find that this offers a good trade-off between resolution and compute requirements. Finally, we work on axial slices of the resulting arrays, making extensive use of data augmentation during training. A full list of transforms is given in the implementation. Note that our model only uses skip connections with a resolution of at most 32^2. For the evaluation we perform a random 80/20 train/test split of the data (stratified at the patient level to avoid data leakage). Hyperparameter optimization is done via cross-validation on the training set, for the final results we train all models on the full training set and report scores on the test set, leaving out those instances where the true Dice overlap between the target point and the last context point is 0.

We compare our approach with the one presented in [18] using their official implementation and a hyperparameter configuration that matches our model (U-Net layer size and depth, loss, etc.). We further compare with a regular Neural Process [6], which is a special case of our model where all the attention

Table 1. Results on the test set. ↑ / ↓ indicate that higher/lower is better, bold marks the best-performing method. Errors represent the standard error of the mean. *Test Loss* evaluates Eq. (3) on the test set, *Surprise* measures the distance between prior and posterior distributions. *Query Volume Dice* draws 100 samples from the prior and evaluates the Dice overlap for the sample that is closest to the true future segmentation in terms of tumor volume. Our model outperforms the baselines in all metrics.

	Parameters	Test loss ↓ [1e−2]	Surprise ↓ [nats]	Query vol. dice ↑ [1e−2]
Neural process [6]	5.9M	34.1 ± 0.2	431.9 ± 3.7	57.5 ± 0.2
Learned discrete [18]	11.2M	38.9 ± 0.3	112.1 ± 0.7	60.1 ± 0.2
Learned continuous (ours)	6.4M	**24.1 ± 0.2**	**82.7 ± 1.5**	**71.4 ± 0.2**

skip-connections are removed (see Fig. 1). Because the approach in [18] works with a fixed number of input timesteps, we train 4 separate models (2–5 input timesteps) and evaluate them jointly. We look at three different metrics: 1) The *Test Loss*, which is just the optimization objective (Eq. (3)) evaluated on the test set, but only using the test points and not the context reconstructions. It is by definition the optimal measure of how well the models perform the task they are trained to perform. Note that this measure is only meaningful when all models are trained with the same loss, which we do. 2) The *Surprise* is the KL divergence, i.e. the second summand in Eq. (3). It can be interpreted as a measure of how much the model has to adjust its prior belief when presented with the true future observation (the posterior). 3) The *Query Volume Dice* takes 100 random samples from the predicted prior and measures the Dice overlap for the sample that best matches the observed future in terms of whole tumor volume. It is best interpreted from an application perspective, when clinicians are interested in possible spatial growth estimates *conditioned* on a potential volume increase. Note that we are mainly interested in a comparison with other learned growth models, and the chosen metrics are not suitable for a comparison with diffusion-based models. We leave such an evaluation for future work.

3   Results

An exemplary case is shown in Fig. 2, with predictions from all models. We find that our model produces predictions with high spatial resolution. It can reconstruct the context segmentations very accurately, while the Neural Process is unable to create sufficient spatial detail. The mean prediction from our model overestimates how fast the necrosis (yellow in Fig. 2) in the tumor center expands, but it is important to consider that the observed future isn't necessarily the most likely one, so it is expected to find such differences. More importantly, when we sample from the predicted prior and take the samples that are closest to the observed future in terms of tumor volume, our model matches the observed future closely. The model from [18] can produce spatial detail comparable to our

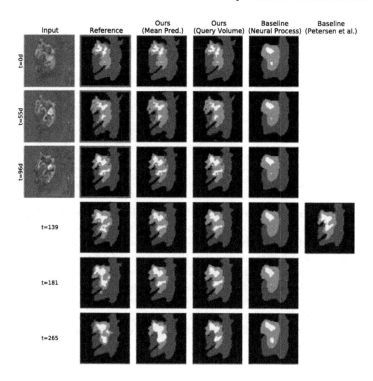

Fig. 2. Example case with predictions from all methods we compare in this work (purple = edema, orange = enhancing tumor, yellow = necrosis). Red borders mark the context, we only show the T1c input channel. Our model can reconstruct the context observations very accurately, while the Neural Process prediction lacks spatial resolution. The mean prediction of our model anticipates an expansion of the necrosis in the tumor center (at $t = 265$ days), which is commonly observed, but does not occur in this particular instance. The query volume prediction is very close to the observed future, indicating that our model does indeed learn meaningful growth trajectories. The model from [18] can only make a prediction at a single fixed point in time. (Color figure online)

model, but it can only make a prediction at a single fixed time relative to the context observations.

The average performance of all methods on the test set is shown in Table 1. We evaluate the *Test Loss* (i.e. Eq. (3), but only on target points, not context reconstructions), the *Surprise*, meaning the KL divergence between prior and posterior, as well as the *Query Volume Dice*, which is the Dice for the sample that best matches the observed future in terms of tumor volume (out of 100 samples from the prior). Our method significantly outperforms the competing approaches in all metrics, with only a little more than half the number of parameters that the model from [18] uses, because ours doesn't require separate prior and posterior networks. Interestingly, the test loss of the Neural Process [6] is actually lower than that of the learned discrete model [18], even though

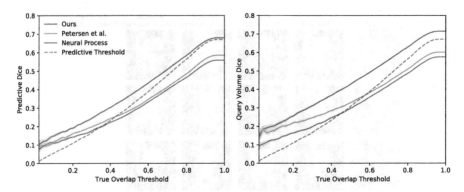

Fig. 3. *Predictive Dice* and *Query Volume Dice* at different thresholds of true overlap. There is often very little change between consecutive observations, so we filter test cases by varying a threshold of their true overlap (Dice) to see how performance changes. As an example, if we select a threshold of 0.6, the corresponding y-values are the average performances on test cases where the overlap of the test timestep with the last context input is less than 0.6. The *predictive threshold* is the average true overlap of those thresholded cases (and the average true Dice for all test cases is 0.67), meaning that if a model falls below that line, it would be better to just assume no change will take place instead of trusting the model's prediction. Only our proposed model stays above that line for all thresholds.

its perceived quality is much lower (see Fig. 2). It should be noted that the numerical results are not comparable to those reported in [18], because we use a slightly different training objective. There are many test cases with only very small changes between the test timestep and the previous context observation. To understand how the performance depends on changes that take place between those times, we iterate a threshold over their true Dice overlap to evaluate the average *Predictive Dice* and Query Volume Dice on the subset of cases below the given threshold. The result is displayed in Fig. 3, where we also include a predictive threshold, which is the average true overlap below the given threshold. Below that line, a model performs worse than one that simply predicts no change between consecutive times. Ours is the only approach that is always above the predictive threshold, while the others actually fall *below* it at higher cutoffs.

4   Discussion

This work proposes a fully learned glioma growth model, first introduced in [18] as an alternative to commonly used biological diffusion models [16,17]. While successful, the approach from [18] has a number of practical limitations: it requires a fixed number of context observations, it requires a fixed time interval between consecutive observations and it can only make a prediction one time interval step into the future. Our proposed model overcomes all of those limitations and can be conditioned on arbitrarily many context observations on a continuous time axis. From these context observations, our model predicts a distribution over growth trajectories, where each sample is temporally consistent and can be evaluated at any desired continuous-valued time. Our model also significantly outperforms the one from [18] in several metrics, which we demonstrate on a dataset ten times larger than the one used in [18]. Our model's main limitation is the high GPU memory requirement of the spatio-temporal attention mechanism we introduce. This is a problem many attention and transformer architectures suffer from, and ways to make them less resource intensive are actively being researched [12,22]. It's also the reason why we performed our experiments on two-dimensional slices rather than full 3D volumes. As a result, one should be careful not to draw any premature conclusions with respect to a possible clinical application. While our results look promising, the value of our model for clinical purposes, for example in radiation therapy, must be validated extensively. We leave such efforts for future work. A comparison with diffusion-based models, particularly in a radiation therapy context [14,15], is another interesting opportunity for future work. Building on the Neural Process framework [5,6], our proposed approach constitutes an efficient Neural Process variant for image time series and is, to the best of our knowledge, only the second time Neural Processes have been demonstrated on real data in the image domain [8]. We believe it can prove useful for both other types of tumor growth as well as any other kind of stochastic time series with image data.

Acknowledgements. Part of this work was funded by the Helmholtz Imaging Platform (HIP), a platform of the Helmholtz Incubator on Information and Data Science.

References

1. Devlin, J., Chang, M.W., Lee, K., Toutanova, K.: BERT: pre-training of deep bidirectional transformers for language understanding. arXiv:1810.04805 [cs] (2018)
2. Drozdzal, M., Vorontsov, E., Chartrand, G., Kadoury, S., Pal, C.: The importance of skip connections in biomedical image segmentation. In: Carneiro, G., et al. (eds.) LABELS/DLMIA -2016. LNCS, vol. 10008, pp. 179–187. Springer, Cham (2016). https://doi.org/10.1007/978-3-319-46976-8_19
3. Eslami, S.M.A., et al.: Neural scene representation and rendering. Science **360**(6394), 1204–1210 (2018)
4. Ezhov, I., et al.: Neural parameters estimation for brain tumor growth modeling. In: Shen, D., et al. (eds.) MICCAI 2019. LNCS, vol. 11765, pp. 787–795. Springer, Cham (2019). https://doi.org/10.1007/978-3-030-32245-8_87
5. Garnelo, M., et al.: Conditional neural processes. In: International Conference on Machine Learning, pp. 1704–1713 (2018)
6. Garnelo, M., et al.: Neural processes. In: International Conference on Machine Learning – Workshop on Theoretical Foundations and Applications of Deep Generative Models (2018)
7. Higgins, I., et al.: β-VAE: learning basic visual concepts with a constrained variational framework. In: International Conference on Learning Representations (2017)
8. Kia, S., Marquand, A.: Neural processes mixed-effect models for deep normative modeling of clinical neuroimaging data. In: International Conference on Medical Imaging with Deep Learning (MIDL)(2019)

9. Kickingereder, P., et al.: Automated quantitative tumour response assessment of MRI in neuro-oncology with artificial neural networks: a multicentre, retrospective study. Lancet Oncol. **20**(5), 728–740 (2019)
10. Kim, H., et al.: Attentive neural processes. In: International Conference on Learning Representations (2019)
11. Kingma, D.P., Ba, J.: Adam: a method for stochastic optimization. In: International Conference on Learning Representations (2015)
12. Kitaev, N., Kaiser, L., Levskaya, A.: Reformer: the efficient transformer. In: International Conference on Learning Representations (2020)
13. Kumar, A., et al.: Consistent jumpy predictions for videos and scenes. In: Advances in Neural Information Processing Systems – Bayesian Deep Learning Workshop (2018)
14. Lipková, J., et al.: Personalized radiotherapy design for glioblastoma: integrating mathematical tumor models, multimodal scans, and Bayesian inference. IEEE Trans. Med. Imaging **38**(8), 1875–1884 (2019)
15. Lê, M., et al.: Personalized radiotherapy planning based on a computational tumor growth model. IEEE Trans. Med. Imaging **36**(3), 815–825 (2017)
16. Mang, A., Bakas, S., Subramanian, S., Davatzikos, C., Biros, G.: Integrated biophysical modeling and image analysis: application to neuro-oncology. Annu. Rev. Biomed. Eng. **22**(1), 309–341 (2020)
17. Menze, B.H., Stretton, E., Konukoglu, E., Ayache, N.: Image-based modeling of tumor growth in patients with glioma. Technical report (2011)
18. Petersen, J., et al.: Deep probabilistic modeling of glioma growth. In: Shen, D., et al. (eds.) MICCAI 2019. LNCS, vol. 11765, pp. 806–814. Springer, Cham (2019). https://doi.org/10.1007/978-3-030-32245-8_89
19. Rosenbaum, D., Besse, F., Viola, F., Rezende, D.J., Eslami, S.M.A.: Learning models for visual 3D localization with implicit mapping. In: Advances in Neural Information Processing Systems - Bayesian Deep Learning Workshop (2018)
20. Subramanian, S., Scheufele, K., Himthani, N., Biros, G.: Multiatlas calibration of biophysical brain tumor growth models with mass effect. In: Martel, A.L., et al. (eds.) MICCAI 2020. LNCS, vol. 12262, pp. 551–560. Springer, Cham (2020). https://doi.org/10.1007/978-3-030-59713-9_53
21. Vaswani, A., et al.: Attention is all you need. In: Advances in Neural Information Processing Systems, pp. 5998–6008 (2017)
22. Wang, S., Li, B.Z., Khabsa, M., Fang, H., Ma, H.: Linformer: self-attention with linear complexity. arXiv:2006.04768 [cs, stat] (2020)
23. Wick, W., et al.: Lomustine and bevacizumab in progressive glioblastoma. N. Engl. J. Med. **377**(20), 1954–1963 (2017)

Spine-Transformers: Vertebra Detection and Localization in Arbitrary Field-of-View Spine CT with Transformers

Rong Tao and Guoyan Zheng$^{(\boxtimes)}$

Institute of Medical Robotics, School of Biomedical Engineering,
Shanghai Jiao Tong University, No. 800, Dongchuan Road,
Shanghai 200240, China
guoyan.zheng@sjtu.edu.cn

Abstract. In this paper, we address the problem of automatic detection and localization of vertebrae in arbitrary Field-Of-View (FOV) Spine CT. We propose a novel transformers-based 3D object detection method that views automatic detection of vertebrae in arbitrary FOV CT scans as an one-to-one set prediction problem. The main components of the new framework, called Spine-Transformers, are an one-to-one set based global loss that forces unique predictions and a light-weighted transformer architecture equipped with skip connections and learnable positional embeddings for encoder and decoder, respectively. It reasons about the relations of different levels of vertebrae and the global volume context to directly output all vertebrae in parallel. We additionally propose an inscribed sphere-based object detector to replace the regular box-based object detector for a better handling of volume orientation variation. Comprehensive experiments are conducted on two public datasets and one in-house dataset. The experimental results demonstrate the efficacy of the present approach. A reference implementation of our method can be found at: https://github.com/gloriatao/Spine-Transformers.

Keywords: Vertebra detection and localization · Transformer · Arbitrary field-of-view spine CT · One-to-one set prediction

1 Introduction

Automatic identification of vertebrae from CT images is a crucial step for various computer aided spinal intervention tasks. The overall morphology of the vertebral column has a basic similarity, with the exception of the first two cervical vertebrae and the sacrum. Thus, compared to other anatomical structures, the spine poses additional challenges to quantitative image analysis. Although the shape

Electronic supplementary material The online version of this chapter (https://doi.org/10.1007/978-3-030-87199-4_9) contains supplementary material, which is available to authorized users.

© Springer Nature Switzerland AG 2021
M. de Bruijne et al. (Eds.): MICCAI 2021, LNCS 12903, pp. 93–103, 2021.
https://doi.org/10.1007/978-3-030-87199-4_9

of the individual vertebrae changes significantly along the spine, most neighboring vertebrae look very similar and are difficult to distinguish. Moreover, the presence of pathological diseases and/or fractures, the arbitrary Field-Of-View (FOV) of scans, and the existence of metal implant artifacts pose additional challenges.

To tackle these challenges, previous studies resort to incorporating prior knowledge. For example, Glocker et al. [1] combined machine learning with a hidden Markov model to find locations of vertebral body. Yang et al. [2] proposed to use a deep image-to-image network to initialize vertebra locations, followed by a refinement with message passing and sparsity regularization. Liao et al. [3] proposed to use 3D Convolutional Neural Networks (CNN) for a rough vertebra localization and then a subsequent bi-directional Recurrent Neural Network (RNN) for refinement. Chen et al. [4] applied 3D fully convolutional networks on down-scaled images to get initial results, followed by a hidden Markov model based post processing. The BtryNet, proposed by Sekuboyina et al. [5], features a 2.5D network design and an energy-based adversarial training regime. In [6], multi-stage convolutional pose machines were combined with regression tree ensembles for key point localization from spine CT. Significant progresses have been achieved with the open dataset released by the Large Scale Vertebrae Segmentation Challenge (VerSe) [7]. For example, Payer et al. [8] designed spatial configuration net (SCN) to identify vertebrae on arbitrary FOV spinal CTs. The SCN implicitly captured contextual correlation of all vertebrae within a patch. Their approach demonstrated superior performance on VerSe 2019 challenge dataset [7]. With recent advances in Graph Convolutional Networks (GCN), GCN-based methods have been adopted by various studies [9,10] featuring a pre-defined adjacency matrix to encode structural information.

Although previous studies achieved promising detection results, they required either complicated steps to handle arbitrary FOV issue [1–3,5,6,8] or were difficult to be applied to scans with arbitrary FOV due to the use of pre-defined adjacency matrix [9,10]. Transformers, which are powerful neural network architectures originally introduced by Vaswani et al. [11] for natural language processing, are recently applied to solve various computer vision problems [12–19]. DEtection TRansformer (DETR) [13] is recently proposed as the first fully end-to-end object detector in 2D images, which offers the advantages of removing the need for hand-designed components like a non-maximum suppression procedure or anchor generation that explicitly encode prior knowledge about the task. Despite significant progresses achieved [13,15,18,19], to the best knowledge of the authors, transformers-based models have never been applied to object detection from 3D images. The main reason is because the core block of the regular transformers [11,13] is the attention module, which identifies complex dependencies between elements of each input data, requires to compute similarity scores of all pairs of positions in the input data. This, however, scales quadratically with the number of positions, which is prohibitively expensive when the number is large. Moreover, based on the Hungarian algorithm [20], Carion et al. [13] designed a loss function to find a bipartite marching between ground-truth and prediction, which enforced permutation-invariance. Such a loss function,

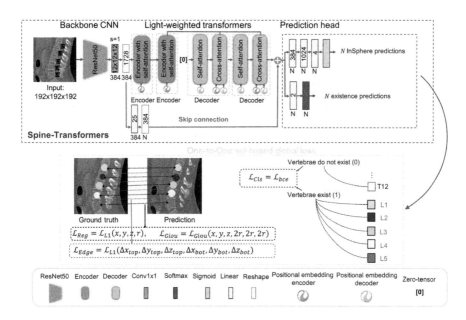

Fig. 1. A schematic illustration of the Spine-Transformers and on how the losses are computed. See text for detailed explanation.

however, cannot be applied to vertebra detection problem as spine consists of a sequence of vertebrae and thus the order of the predictions is important, i.e., vertebra detection is not permutation-invariant.

In this paper, we propose a novel transformers-based 3D object detection method that views automatic detection of vertebrae in arbitrary FOV CT scans as an one-to-one set prediction problem, thus preserving the sequential order of different levels of vertebrae and eliminating the requirement of bipartite matching between ground truth and prediction. The main components of the new framework, called Spine-Transformers, are an one-to-one set based global loss that forces unique prediction and a light-weighted transformer architecture equipped with skip connections and learnable positional embeddings for encoder and decoder, respectively. It reasons about the relations of different levels of vertebrae and the global volume context to directly output all vertebrae in parallel. We additionally propose an inscribed sphere-based object detector to replace the regular box-based object detector for a better handling of volume orientation variation.

Our contributions can be summarized as follows: (1) we are the first to apply a transformers-based model to 3D medical image detection task by designing a light-weighted transformer architecture; (2) we formulate the automatic vertebra detection in arbitrary field-of-view spine CT as a direct one-to-one set prediction problem and introduce a new one-to-one set-based global loss to force unique prediction and to preserve the sequential order of different levels of vertebrae. Our method can reason about the relations of different levels of vertebrae and the global volume context to directly output all vertebrae in parallel; (3)

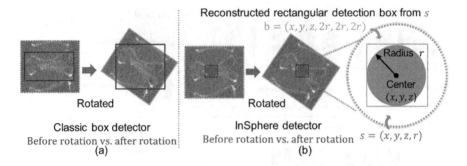

Fig. 2. A comparison of box detector (a) with InSphere detector (b). InSphere detection is not sensitive to vertebral orientation.

we introduce a novel inscribed sphere-based object detector which can handle volume orientation variation better than the regular box-based detector.

2   Method

2.1   Vertebrae Detection as an One-to-one Set Prediction

Patch-based Vertebrae Detection. The input to the Spine-Transformers is fixed-size patches. Although the input patch usually contains part of a spine, the Spine-Transformers infer all N vertebrae in parallel, as shown in Fig. 1, where N is set to be the maximum number of vertebral levels found in the ground truth annotation. The ground truth label $v_i = (c_i, s_i)$ of each vertebra contains two parts: a binary indication tag $c_i \in \{0, 1\}$ to indicate whether a query vertebra exists in the input patch or not, and a geometrical description $s_i = [x_i, y_i, z_i, r_i]$, where r_i is the radius and $[x_i, y_i, z_i]$ is the center as explained below.

One-to-one Set Prediction Loss. Let us denote $v = \{v_i = (c_i, s_i)\}_{i=1}^{n}$ the ground truth set of $n \ll N$ vertebrae in an input patch, and $\hat{v} = \{\hat{v}_i = (\hat{c}_i, \hat{s}_i)\}_{i=1}^{N}$ the set of N predictions. In our problem, we have at most one object for each level of vertebra. Thus, during training, we force one-to-one matching in a single pass through Spine-Transformers. We estimate the binary classification loss for all N objects, but only calculate localization loss for n visible vertebrae by ground truth. During inference, we only keep the coordinates of vertebrae whose binary classification probabilities are above a threshold. We empirically set the threshold to be 0.9.

The loss design takes into account of both the class prediction and the vertebrae coordinate regression. The classic box detectors are not sufficient for vertebra detection, as they are not rotational invariant. Their performance is sensitive to the object orientation [21]. To handle orientation variation, we propose to use a rotational invariant descriptor representing as the center of the sphere inscribed inside the vertebral body and the radius of the inscribed sphere. We name the new rotational invariant detector as InSphere detector. Figure 2 shows the comparison of an InSphere detector with a box detector.

We compute binary cross-entropy loss between ground truth labels $\{c_i\}$ and predicted probabilities $\{\hat{c}_i\}$:

$$L_{Cls} = -\frac{1}{N} \cdot \sum_{i=1}^{N} (c_i log(\hat{c}_i) + (1 - c_i)log(1 - \hat{c}_i)) \tag{1}$$

Given that a patch containing n vertebrae by ground truth, we estimate a regression loss for all n vertebrae:

$$L_{Reg} = \frac{1}{n} \cdot \sum_{i=1}^{n} |(x_i - \hat{x}_i)| + |(y_i - \hat{y}_i)| + |(z_i - \hat{z}_i)| + |(r_i - \hat{r}_i)| \tag{2}$$

We further define an edge loss to enforce the order of the predicted vertebrae. We define $(edges_{i,top} = |s_i - s_{i-1}|, edges_{i,bottom} = |s_i - s_{i+1}|)$ as the $L1$ distances between target vertebral center and its top and bottom neighborhood vertebral centers, respectively. We then estimate the edge loss as:

$$L_{Edges} = \frac{1}{n} \cdot \sum_{i=1}^{n} (|(edge_{i,top} - \hat{edge}_{i,top})| + |(edge_{i,bottom} - \hat{edge}_{i,bottom}|) \tag{3}$$

Area loss is shown to be effective in bounding box detection. We adopt generalized interception-over-union (GIoU) loss as proposed in [22], which is shown in [13] to increase localization accuracy.

To apply GIoU loss, we construct a cube $b_i = (x_i, y_i, z_i, 2r_i, 2r_i, 2r_i)$ from the InSphere descriptor $s_i = (x_i, y_i, z_i, r_i)$. The GIoU loss is estimated as follows:

$$L_{Giou} = \frac{1}{n} \cdot \sum_{i=1}^{n} \{1 - (\frac{b_i \cap \hat{b}_i}{b_i \cup \hat{b}_i} - \frac{B_i \setminus (b_i \cap \hat{b}_i)}{B_i})\} \tag{4}$$

where $\hat{b}_i = (\hat{x}_i, \hat{y}_i, \hat{z}_i, 2\hat{r}_i, 2\hat{r}_i, 2\hat{r}_i)$ is the cube constructed from the predicted values; B_i is the smallest box enclosing b_i and $\hat{b}_i$.

The overall loss function for the Spine-Transformers is:

$$L = \lambda_{Cls}L_{Cls} + \lambda_{Reg}L_{Reg} + \lambda_{Edges}L_{Edges} + \lambda_{Giou}L_{Giou} \tag{5}$$

where λ_{Cls}, λ_{Reg}, λ_{Edges} and λ_{Giou} are parameters controlling the relative weights of different terms.

2.2   The Spine-Transformers

The architecture of the Spine-Transformers, as presented in Fig. 1, consists of a CNN backbone, a light-weighted transformer equipped with skip connections and learnable positional embeddings for encoder and decoder, respectively, and two feed-forwarded branches that aims at predicting the existence of query vertebrae as well as regressing their coordinates. The network design features a skip connection from backbone CNN to the output of transformer for improved performance.

Backbone Network: we use a modified ResNet50 architecture [23] as the backbone network to extract high-level vertebral feature representations. We remove the max pooling layer to obtain feature maps with higher spatial resolution on the bottom level. For an input image patch $x \in \mathbb{R}^{1 \times W \times H \times D}$, the backbone network generates bottom feature map $f \in \mathbb{R}^{C \times w \times h \times d}$, where $h = \frac{H}{16}$, $w = \frac{W}{16}$ and $d = \frac{D}{16}$, and $C = 2048$ is the number of output feature map channels.

Transformer Encoder: the bottom level feature map f is first passed through a $1 \times 1 \times 1$ convolution to reduce channel dimension from C to $3 \times d_{hidden}$, where 3 represents spatial dimension and d_{hidden} is pre-defined hidden dimension for the learnable positional embedding. We then collapse the spatial dimension of $f' \in \mathbb{R}^{3d_{hidden} \times w \times h \times d}$ to obtain feature sequence $z \in \mathbb{R}^{3d_{hidden} \times (w \cdot h \cdot d)}$ as the input to the transformers. Since the attention module in the transformers is permutation-invariant, we supplement it with a learnable positional embedding.

Transformer Decoder: the transformer decoder yields N predictions of vertebrae in a single forward pass. We follow the design of vanilla transformer decoder. In addition to the learnable encoder positional embedding, a learnable decoder positional embedding is also added to each layer of the decoder.

Skip Connection: It has been reported that transformers-based models are data hungry and takes extra-long time (3 days on 16 V100 GPUs as reported in [13]) to converge. To alleviate this problem, we design a skip connection from the backbone CNN to the output of the transformers, as shown in Fig. 1. Such a skip connection facilitates the propagation of context and gradient information in both forward and backward phases during training.

Light-weighted Architecture Design: The Spine-Transformers feature a light-weighted design with only two layers encoder and two layers decoder due to the trade-off between feature resolution and memory constraint. Increasing the resolution of the bottom level feature map significantly increase GPU memory requirement but reducing bottom feature map resolution to save GPU memory is not recommended because higher resolution of bottom feature map facilitates the non-local global context information aggregation in encoder. Empirically we find two layers encoder and two layers decoder, together with the proposed skip connection, is a reasonable choice between detection performance and GPU memory consumption.

Localization Refinement. Spine-Transformers can be trained end-to-end to predict all vertebrae in an input patch. During inference, it can be used to infer all vertebrae in a given spine CT with arbitrary FOV when combining with a sliding-window sampling method. Empirically, we find that the localization accuracy can be further improved by using a 3D CNN-based landmark regression [24]. Specifically, for each detected vertebra, we generate a contextual heatmap from the predicted InSphere using a Gaussian kernel, whose standard deviation equals to 1.5 times of the predicted radius. The contextual heatmap is concatenated to the original image. We then crop a sub-volume of fixed size around the detected center, which is used as the input to train a 3D CNN for refinement.

Table 1. Results when evaluated on the VerSe 2019 challenge dataset. R: Refinement

Methods	Test dataset		Hidden dataset	
	Id-Rate (%)	L-Error (mm)	Id-Rate (%)	L-Error (mm)
Christian payer [7]	95.65	**4.27**	94.25	**4.80**
iFLYTEK [7]	96.94	4.43	86.73	7.13
nlessmann [7]	89.86	14.12	90.42	7.04
Spine-Transformers	97.16	8.39	96.06	7.38
Spine-Transformers+R	**97.22**	4.33	**96.74**	5.31

2.3 Implementation Details

All data are resampled to isotropic resolution of 1 mm. The fixed size of the image patch is [192, 192, 192]. During training, for each epoch, we sample an image patch per volume at a random position. For on-the-fly data augmentation, we employed similar strategy as in [8]. All methods were implemented in Python using Pytorch framework and trained on a workstation with a Tesla V100 graphics card. The learning rates of the backbone CNN and the Spine-Transformers are set to be $1e-5$ and $1e-6$, respectively. We train the network with 200 epochs of warmup steps until the target learning rate, reduced by a factor 10 at every 500 interactions. The AdamW [25] optimizer is adopted, with weight decay $1e-4$. We empirically set $d_{hidden} = 128$, $\lambda_{Cls} = 1$, $\lambda_{Reg} = 2$, $\lambda_{Edges} = 2$, and $\lambda_{Giou} = 1$.

3 Experiments and Results

To demonstrate the performance of the proposed method, we designed and conducted comprehensive experiments on two public datasets and one in-house dataset: the VerSe 2019 challenge dataset [7], the pathological spine CT scans used for the MICCAI CSI 2014 Vertebra Localization and Identification challenge (MICCAI-CSI 2014 challenge) [1] and an in-house lower spine dataset. We adopt the evaluation metrics as proposed in [7]. The mean localization error (L-Error) is the average Euclidean distance between the ground truth and the predicted vertebral centers. The identification rate (Id-Rate) is the ratio between the number of correctly identified vertebrae and the total number of vertebrae present in a scan. A prediction is counted as valid when the distance from the predicted center to the ground-truth center is less than 20 mm.

Results on the VerSe 2019 Challenge Dataset. The VerSe 2019 challenge dataset contains 160 spinal CT with ground truth annotations, among which 80 scans are training data, 40 scans as test data and another 40 scans as hidden data. We trained Spine-Transformers on the training data and evaluated the trained model on the test and the hidden datasets. We compared the performance of our method with the State-Of-The-Art (SOTA) methods described in [7]. The

Table 2. Results on the MICCAI-CSI 2014 challenge dataset. R: Refinement

Methods	Average		Cervical		Thoracic		Lumbar	
	Id-Rate	L-Error	Id-Rate	L-Error	Id-Rate	L-Error	Id-Rate	L-Error
Payer [8]	86.1	6.2	–	–	–	–	–	–
Yang [2]	80.0	9.1 ± 7.2	83.0	6.6 ± 3.9	74.0	9.9 ± 7.5	80.0	10.9 ± 9.1
Liao [3]	88.3	6.5 ± 8.6	95.1	**4.5 ± 4.6**	84.0	7.8 ± 10.2	92.2	5.6 ± 7.7
Glocker [1]	70.0	12.4 ± 11.2	80.0	7.0 ± 4.7	62.0	13.8 ± 11.8	75.0	14.3 ± 12.3
Spine-Transformers	91.8	8.6 ± 6.4	97.6	5.5 ± 4.4	88.9	9.2 ± 6.7	91.8	9.8 ± 6.7
Spine-Transformers+R	**92.2**	**4.8 ± 3.6**	**97.6**	5.5 ± 4.4	**89.3**	**5.3 ± 4.2**	**92.7**	**3.7 ± 2.8**

Table 3. Ablation study results.

Architecture	Components				Results	
	Box Detector	InSphere Detector	Edge Loss	Refinement	Id-Rate	L-Error
Spine-Transformers	✓				96.00	8.21 ± 3.71
Spine-Transformers		✓			96.71	8.10 ± 3.58
Spine-Transformers		✓	✓		96.96	8.04 ± 3.63
Spine-Transformers		✓	✓	✓	**98.70**	**3.41 ± 3.09**

experimental results are shown in Table 1. Without adding the refinement step, the Spine-Transformers achieved an average Id-Rate of 97.16% and an average L-Error of 8.39 mm when evaluated on the test dataset, and an average Id-Rate of 96.06% and an average L-Error of 7.38 mm when evaluated on the hidden dataset. With refinement, these values are improved to 97.22% and 4.33 mm on the test dataset, and 96.74% and 5.31 mm on the hidden dataset, respectively. On both datasets, our method achieved the best Id-Rate but a slightly higher L-Error than the team "christian payer" [7]. This may be caused by the different nature of the two methods. Specifically, their method was based on landmark heatmap regression while our method was based on object detection, which treated the complete object as the target, leading to relatively larger localization uncertainty.

Results on the MICCAI-CSI 2014 Challenge Dataset. The MICCAI-CSI 2014 challenge dataset contains 224 spinal CT scans with varying pathologies [1], including abnormal curvature, fractures and mental implantation. We conducted a two-fold cross-validation study on this dataset using the protocol introduced in [1]. The experimental results are shown in Table 2. With refinement, the Spine-Transformers achieved an average Id-Rate of 92.2% and an average L-Error of 4.8 ± 3.6 mm. As the same protocol was used in [1–3,8] and our work, we can compare our results with theirs. As shown in Table 2, our method achieved better results than state-of-the-art methods.

Results on the In-house Lower Spine Dataset. The in-house lower spine dataset consists of 210 lower spine CT scans, with centroid annotations for 1, 845 vertebrae from T11 to L5. We conducted a two-fold cross-validation study. Without refinement, Spine-Transformers achieved a vertebral Id-Rate of 97.24%

and a L-Error of 6.65 mm. With refinement, the vertebral Id-Rate and the L-Error are improved to 98.82% and 1.12 mm, respectively.

Ablation Study. We conducted ablation study on the 80 training data with ground truth annotation obtained from the VerSe 2019 challenge dataset. The data was randomly split to 80% for training and 20% for testing. The results are shown in Table 3. When box detector was used with the Spine-Transformers, the average Id-Rate is 96.00% and the average L-Error is 8.21 ± 3.71 mm. Replacing the box detector with the InSphere detector, the average Id-Rate and the average L-Error improved to 96.71% and 8.10 ± 3.58 mm, respectively. By adding edge loss and the refinement step, our method achieved an average Id-Rate of 98.70% and an average L-Error of 3.41 ± 3.09 mm, respectively.

4 Conclusion

In this paper, we introduced a transformers-based 3D object detector called Spine-Transformers and applied to the task of automatic detection and localization of vertebrae in arbitrary FOV spine CT. Comprehensive experiments conducted on two public datasets and one in-house dataset demonstrated that the present method achieved equivalent or better results than state-of-the-art methods.

Acknowledgments. This study was partially supported by the Natural Science Foundation of China via project U20A20199 and by Shanghai Municipal S&T Commission via Project 20511105205.

References

1. Glocker, B., Zikic, D., Konukoglu, E., Haynor, D.R., Criminisi, A.: Vertebrae localization in pathological spine CT via dense classification from sparse annotations. In: Mori, K., Sakuma, I., Sato, Y., Barillot, C., Navab, N. (eds.) MICCAI 2013. LNCS, vol. 8150, pp. 262–270. Springer, Heidelberg (2013). https://doi.org/10.1007/978-3-642-40763-5_33
2. Yang, D., et al.: Automatic vertebra labeling in large-scale 3D CT using deep image-to-image network with message passing and sparsity regularization. In: Niethammer, M., et al. (eds.) IPMI 2017. LNCS, vol. 10265, pp. 633–644. Springer, Cham (2017). https://doi.org/10.1007/978-3-319-59050-9_50
3. Liao, H., Mesfin, A., Luo, J.: Joint vertebrae identification and localization in spinal CT images by combining short-and long-range contextual information. IEEE Trans. Med. Imaging **37**(5), 1266–1275 (2018)
4. Chen, Y., Gao, Y., Li, K., Zhao, L., Zhao, J.: vertebrae identification and localization utilizing fully convolutional networks and a hidden Markov model. IEEE Trans. Med. Imaging **39**(2), 387–399 (2020)
5. Sekuboyina, A., et al.: Btrfly Net: vertebrae labelling with energy-based adversarial learning of local spine prior. In: Frangi, A.F., Schnabel, J.A., Davatzikos, C., Alberola-López, C., Fichtinger, G. (eds.) MICCAI 2018. LNCS, vol. 11073, pp. 649–657. Springer, Cham (2018). https://doi.org/10.1007/978-3-030-00937-3_74

6. Mader, A.O., Lorenz, C., von Berg, J., Meyer, C.: Automatically localizing a large set of spatially correlated key points: a case study in spine imaging. In: Shen, D., et al. (eds.) MICCAI 2019. LNCS, vol. 11769, pp. 384–392. Springer, Cham (2019). https://doi.org/10.1007/978-3-030-32226-7_43

7. Sekuboyina, A., et al.: Verse: a vertebrae labelling and segmentation benchmark. arXiv. org e-Print archive (2020)

8. Payer, C., Štern, D., Bischof, H., Urschler, M.: Integrating spatial configuration into heatmap regression based CNNs for landmark localization. Med. Image Anal. **54**, 207–219 (2019)

9. Chang, H., Zhao, S., Zheng, H., Chen, Y., Li, S.: Multi-vertebrae segmentation from arbitrary spine MR images under global view. In: Martel, A.L., et al. (eds.) MICCAI 2020. LNCS, vol. 12266, pp. 702–711. Springer, Cham (2020). https://doi.org/10.1007/978-3-030-59725-2_68

10. Pang, S., et al.: Spineparsenet: spine parsing for volumetric MR image by a two-stage segmentation framework with semantic image representation. IEEE Trans. Med. Imaging **40**(1), 262–273 (2021)

11. Vaswani, A., et al.: Attention is all you need. In: Advances in Neural Information Processing Systems, pp. 5998–6008 (2017)

12. Parmar, N., et al.: Image transformer. In: International Conference on Machine Learning, pp. 4055–4064. PMLR (2018)

13. Carion, N., Massa, F., Synnaeve, G., Usunier, N., Kirillov, A., Zagoruyko, S.: End-to-end object detection with transformers. In: Vedaldi, A., Bischof, H., Brox, T., Frahm, J.-M. (eds.) ECCV 2020. LNCS, vol. 12346, pp. 213–229. Springer, Cham (2020). https://doi.org/10.1007/978-3-030-58452-8_13

14. Li, Z., et al.: Revisiting stereo depth estimation from a sequence-to-sequence perspective with transformers. arXiv preprint arXiv:2011.02910 (2020)

15. Sun, Z., Cao, S., Yang, Y., Kitani, K.: Rethinking transformer-based set prediction for object detection. arXiv preprint arXiv:2011.10881 (2020)

16. Chen, H., et al.: Pre-trained image processing transformer. In: Proceedings of the IEEE/CVF Conference on Computer Vision and Pattern Recognition, pp. 12299–12310 (2021)

17. Dosovitskiy, A., et al.: An image is worth 16x16 words: Transformers for image recognition at scale. arXiv preprint arXiv:2010.11929 (2020)

18. Khan, S., Naseer, M., Hayat, M., Zamir, S.W., Khan, F.S., Shah, M.: Transformers in vision: A survey. arXiv preprint arXiv:2101.01169 (2021)

19. Han, K., et al.: A survey on visual transformer. arXiv preprint arXiv:2012.12556 (2020)

20. Kuhn, H.W.: The Hungarian method for the assignment problem. Nav. Res. Log. Q. **2**(1–2), 83–97 (1955)

21. Wang, Y., Wang, L., Lu, H., He, Y.: Segmentation based rotated bounding boxes prediction and image synthesizing for object detection of high resolution aerial images. Neurocomputing **388**, 202–211 (2020)

22. Rezatofighi, H., Tsoi, N., Gwak, J., Sadeghian, A., Reid, I., Savarese, S.: Generalized intersection over union: a metric and a loss for bounding box regression. In: Proceedings of the IEEE/CVF Conference on Computer Vision and Pattern Recognition, pp. 658–666 (2019)

23. He, K., Zhang, X., Ren, S., Sun, J.: Deep residual learning for image recognition. In: Proceedings of the IEEE/CVF Conference on Computer Vision and Pattern Recognition, pp. 770–778 (2016)

24. Çiçek, Ö., Abdulkadir, A., Lienkamp, S.S., Brox, T., Ronneberger, O.: 3D U-Net: learning dense volumetric segmentation from sparse annotation. In: Ourselin, S., Joskowicz, L., Sabuncu, M.R., Unal, G., Wells, W. (eds.) MICCAI 2016. LNCS, vol. 9901, pp. 424–432. Springer, Cham (2016). https://doi.org/10.1007/978-3-319-46723-8_49

25. Lüscher, C., et al.: Rwth ASR systems for librispeech: Hybrid vs attention-w/o data augmentation. arXiv preprint arXiv:1905.03072 (2019)

Multi-view Analysis of Unregistered Medical Images Using Cross-View Transformers

Gijs van Tulder[✉], Yao Tong, and Elena Marchiori

Data Science Group, Faculty of Science, Radboud University,
Nijmegen, The Netherlands
g.vantulder@cs.ru.nl

Abstract. Multi-view medical image analysis often depends on the combination of information from multiple views. However, differences in perspective or other forms of misalignment can make it difficult to combine views effectively, as registration is not always possible. Without registration, views can only be combined at a global feature level, by joining feature vectors after global pooling. We present a novel cross-view transformer method to transfer information between unregistered views at the level of spatial feature maps. We demonstrate this method on multi-view mammography and chest X-ray datasets. On both datasets, we find that a cross-view transformer that links spatial feature maps can outperform a baseline model that joins feature vectors after global pooling.

Keywords: Multi-view medical images · Transformers · Attention

1 Introduction

Many medical imaging tasks use data from multiple views or modalities, but it can be difficult to combine those effectively. While multi-modal images can usually be registered and treated as multiple input channels in a neural network, images from different views can be difficult to register correctly (e.g., [2]). Therefore, most multi-view models process views separately and only combine them after global pooling, which removes any local correlations between views. If these local correlations are important for the interpretation of the images, models could be improved by linking views at an earlier, spatial feature level.

We discuss two example tasks: mammography classification with craniocaudal (CC) and mediolateral oblique (MLO) views, and chest X-ray classification with frontal and lateral views. In both applications, multi-view models can outperform single-view models (e.g., [3,16]) However, the different perspectives make registration challenging and make a channel-based approach unsuitable.

Electronic supplementary material The online version of this chapter (https://doi.org/10.1007/978-3-030-87199-4_10) contains supplementary material, which is available to authorized users.

© Springer Nature Switzerland AG 2021
M. de Bruijne et al. (Eds.): MICCAI 2021, LNCS 12903, pp. 104–113, 2021.
https://doi.org/10.1007/978-3-030-87199-4_10

We propose a method that can link unregistered views at the level of spatial feature maps. Inspired by the attention-based transformer models [13] that model links between distant parts of a sequence or image, our model uses attention to link relevant areas between views. We apply this transformer to the intermediate feature maps produced by a CNN. Based on a trainable attention mechanism, the model retrieves features from one view and transfers them to the other, where they can be used to add additional context to the original view.

Our approach does not require pixel-wise correspondences – it compares all pixels in the feature maps from view A to all pixels in the feature maps from view B – but combines views using a trainable attention model. By applying this to feature maps instead of directly to the input, we allow the model to link higher-level features and reduce computational complexity. Since linking all pixel pairs can still be expensive, we additionally investigate an alternative implementation that groups pixels with similar features in visual tokens [15].

In this paper, we present these novel pixel-wise and token-based cross-view transformer approaches and apply them to two public datasets. Although combining features after global pooling is a relatively common way to handle multi-view information with unregistered medical images, to our knowledge there are no methods that use a transformer-based approach to do this at the spatial feature level. The proposed model can be easily embedded as a module within baseline multi-view architectures that combine views after global pooling. We evaluate our method on the CBIS-DDSM mammography dataset [5,7] and the CheXpert chest X-ray dataset [6]. Based on our experiments, we think that early combination of features can improve the classification of multi-view images.

2   Related Work

There are many works on multi-view classification of medical images. In this section we focus on methods applied to mammography and chest X-ray data. Most methods combine views at a late stage, usually by concatenating feature vectors obtained from the different views, followed by a fully connected part to make a final prediction. We use this approach in our multi-view baseline.

Combining features at a global level is common for mammography images, which are difficult to register [2]. For example, Bekker et al. [1] combined binary predictions from view-specific classifiers. Carneiro et al. [2] combined feature from view-specific CNN branches after global pooling. Wu et al. [16] discuss multiple ways to combine views in a single network, all with view-specific convolution branches. Similar architectures were proposed elsewhere (e.g., [10,12]).

Other works combine views at a regional level. Wang et al. [14] proposed a region-based three-step method: after extracting mass ROIs (regions of interest) from each view, they used a CNN with an attention-driven approach to extract view-specific features from each ROI. Finally, the features from both views are combined with additional clinical features by an LSTM-based fusion model. Similarly, Ma et al. [9] proposed using Faster RCNNs to detect ROIs in each view, which they then converted to feature vectors and combined in a multi-view network. Liu et al. [8] used a model with bipartite graph convolution to link views based on pseudo landmarks, while satisfying geometric constraints.

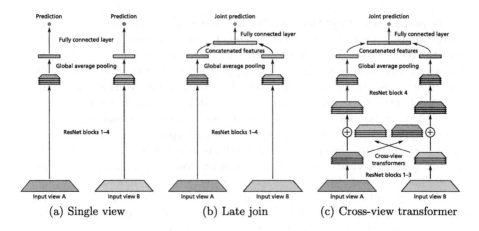

(a) Single view   (b) Late join   (c) Cross-view transformer

Fig. 1. Schematic overview of the three architectures.

Most similar to our approach is the work by Zhao et al. [17], who applied a joint attention mechanism that combined two views or two sides (left and right breasts) to produce channel-wise and spatial attention maps that highlight asymmetric regions. The outputs of the attention-weighted, view-specific branches are pooled and concatenated to produce a final classification. Different from our approach, which transfers feature values between views, Zhao et al. use cross-view information only to compute cross-view attention weights.

For chest X-rays, many datasets only include the frontal view, since the lateral view is harder to read, is mostly useful for specific diagnoses, and is sometimes replaced by a CT scan [3]. Rubin et al. [11] evaluated a model with view-specific convolution branches, global average pooling and a shared fully connected layer, and report that combining frontal and lateral views improved classification performance. Recently, Hashir et al. [3] compared several multi-view models on a large chest X-ray dataset, showing that while multi-view data is useful for some diagnosis tasks, the frontal view can be sufficient for others.

3   Methods

In this section, we describe two baseline models and our cross-view transformer models. All models are designed for a classification task with unregistered, dual-view image pairs. We use a ResNet-based [4] architecture for the view-specific convolution branches, similar to what is used in related work (e.g., [6,16]).

3.1   Baseline Models

Our *single-view* baseline (Fig. 1a) follows the basic ResNet architecture. The network consists of a stack of ResNet blocks with convolution and pooling layers, followed by global average pooling and a fully connected part that computes the final output. Our *late-join* baseline (Fig. 1b) extends this model to multiple unregistered views, by using a separate convolution branch for each view. After global pooling, the feature vectors for all views are concatenated and fed into a shared fully connected part to compute the prediction. This is similar to how multi-view data is commonly combined in other work, such as [16].

Depending on the type of data, the view-specific branches can be linked through weight sharing. This can be a useful regularization if the same low-level features can be used for both views, but might be too restrictive if the views have very different appearances. We do not use this for our models.

3.2   Cross-View Transformer Models

Using view-specific branches and combining feature vectors after global pooling works for unregistered views, but limits the model to learning global correlations. We propose a transformer-based method that links views at feature map level, without requiring pixel-wise correspondences. Instead of the self-attention used in standard transformers [13] to transfer information within a single sequence, we use cross-view attention to transfer information between views. This approach can be used in any dual-view model, such as our late-join baseline.

The *cross-view transformer* (Fig. 1c) works on the intermediate level of the convolutional part of the model. In our case, we apply the module to the feature maps after the third ResNet block, leaving one ResNet block before global pooling. At this level, we use the cross-view transformer to transfer features from the source view to the target view. We make the model bidirectional by applying a second cross-view transformer module in the opposite direction.

We will define two variants of this model: a pixel-wise variant that links pixels from the source and target views, and a token-based variant in which the pixels in the target view are linked to visual tokens in the source view. For conciseness, we will use the term 'pixels' to refer to pixels in the intermediate feature maps.

Cross-View Attention. We use a multi-head attention model [13] with scaled dot-product attention. For each attention head, we use a 1×1 convolution with view-specific weights to compute an embedding for the source and target pixels. We reshape the embedded feature maps for the target view to a query matrix $\mathbf{Q} \in \mathbb{R}^{n \times d}$ and the reshape feature maps for the source view to a key matrix $\mathbf{K} \in \mathbb{R}^{m \times d}$, where d is the size of the embedding and m and n are the number of source and target pixels. We also reshape the original source feature maps to the value matrix $\mathbf{V} \in \mathbb{R}^{m \times f}$, where f is the number of feature maps. Next, we use the scaled dot-product attention function [13] to compute

$$\text{Attention}(Q, K, V) = \text{softmax}\left(\frac{\mathbf{QK}^{\top}}{\sqrt{d}}\right)\mathbf{V} \in \mathbb{R}^{n \times f}. \tag{1}$$

For each target pixel, this computes a weighted sum of the features from the source view, resulting in f new features per attention head. We reshape the output to $m \times f$ feature maps with the shape of the target feature maps and apply 1×1 convolution to reduce these to f attention-based feature maps.

We combine the attention-based feature maps a with the original feature maps x of the target data to obtain the combined feature maps y:

$$y = \text{LayerNorm}(x + \text{Dropout}(\text{Linear}(a))), \tag{2}$$

where LayerNorm is layer normalization, Dropout applies dropout, and Linear a 1×1 convolution that maps the attention features to the feature space of x. The resulting feature maps y are used as the input for the following ResNet block.

Unlike standard transformer networks [13], we do not include a positional encoding. This encoding encodes the relative location of each pixel, allowing the model to distinguish between nearby and faraway pixels within a single image. This is useful in self-attention for natural images, but is less suitable for cross-view attention. However, an alternative positional encoding that works across views might help the network to exploit spatial constraints. We leave this for future work.

Semantic Visual Tokens. Computing the attention between all pairs of source and target pixels can be computationally expensive, even when it is applied to the smaller feature maps at the later stages of the CNN. We therefore evaluate a variant that uses the tokenization method from [15] to replace the source pixels with a smaller number of visual tokens, by grouping semantically related pixels.

We apply a three-layer tokenization procedure to the source feature maps. In the first layer, given flattened feature maps $\mathbf{X} \in \mathbb{R}^{m \times f}$, where m is the number of source pixels and f is the number of feature maps, we compute tokens $\mathbf{T}$:

$$\mathbf{T} = \text{softmax}_m (\mathbf{X}\mathbf{W}_A)^\top \mathbf{X}. \tag{3}$$

The softmax over the spatial dimension uses the tokenizer weights $\mathbf{W}_A \in \mathbb{R}^{f,L}$ to compute a spatial attention map, which is then used to compute a weighted sum of features for each of the L tokens in $\mathbf{T} \in \mathbb{R}^{L \times f}$.

In the second layer, we use the previous tokens $\mathbf{T}_{in}$ to obtain new tokens $\mathbf{T}$:

$$\mathbf{W}_R = \mathbf{T}_{in}\mathbf{W}_{\mathbf{T} \to \mathbf{R}} \tag{4}$$

$$\mathbf{T} = \text{softmax}_m (\mathbf{X}\mathbf{W}_R)^\top \mathbf{X}. \tag{5}$$

where $\mathbf{W}_{\mathbf{T} \to \mathbf{R}} \in \mathbb{R}^{f \times f}$. We repeat this with a different $\mathbf{W}_{\mathbf{T} \to \mathbf{R}}$ in the third tokenization layer to obtain the final set of tokens $\mathbf{T}$. We use these tokens instead of the source pixels in our token-based cross-view transformer.

4   Data

CBIS-DDSM. The CBIS-DDSM [5,7] is a public mammography dataset with craniocaudal (CC) and mediolateral-oblique (MLO) views with manual annotations. We solve a binary classification problem on the scans with mass abnormalities, predicting benign vs. malignant for each CC/MLO pair. We created five subsets for cross-validation, using stratified sampling while ensuring that all scans for a patient remain in the same subset. In total, we used image pairs of 708 breasts (636 unique patients), with approximately 46% labelled malignant.

During preprocessing, we cropped the scans using the method described by Wu et al. [16], using thresholding to position a fixed-size cropping window that includes the breast but excludes most of the empty background. We downsampled the cropped images to 1/16th of the original resolution to obtain images of 305×188 pixels. We normalized the intensities to $\mu = 0$ and $\sigma = 1$, measured on the nonzero foreground pixels of each scan.

CheXpert. The CheXpert dataset [6] is a large public dataset of frontal and lateral chest X-ray scans, annotated for 13 different observations with labels negative, positive, uncertain, or unknown (see supplement). We used the downsampled version of the dataset as provided on the website. We selected the visits with complete frontal and lateral views and divided the patients in random subsets for training (23628 samples for 16810 unique patients), validation (3915 s, 2802p) and testing 3870 s, 2802p). We normalized the images to $\mu = 0$ and $\sigma = 1$ and used zero-padding to obtain a constant size of 390×390 pixels for each view.

To handle the uncertain and unknown labels, we followed [6] and used a single network with a three-class softmax output for each task (negative/uncertain/positive). We excluded the samples with an unknown label from the loss computation for that specific task. At test time, following [6], we remove the uncertain label and compute the softmax only for the negative and positive outputs.

5 Experiments

Models. We compare four models: the single-view model, the late-join model, and the token-based and pixel-based cross-view transformers. All models use the same ResNet-18 architecture [4] for the convolution and pooling blocks up to the global average pooling layer. We use pre-trained weights on ImageNet, as provided by PyTorch. After global average pooling, we concatenate the feature vectors for both views and use this as input for a single fully connected layer that computes the output. (See the supplementary material for a detailed view.)

In the cross-view transformers, we use bidirectional attention and apply the cross-view transformer before the final ResNet block, adding the transformer features to the input for the final convolution and pooling layers. For the CBIS-DDSM dataset, we evaluated models with 12 or 18 attention heads and 16, 32 or 48 tokens, as well as the pixel-based transformer. For the CheXpert dataset, we use token-based transformers with 6 or 12 attention heads and 16 or 32 tokens. In all cases, the embedding size is set to 32 features per head.

Implementation. We implemented the models in PyTorch[1] and trained using the Adam optimizer with cosine learning rate annealing from 0.0001 to 0.000001, with linear warm-up in the first epochs. We used rotation, scaling, translation, flipping and elastic deformations as data augmentation.

---

[1] The code for the experiments is available at https://vantulder.net/code/2021/miccai-transformers/.

Table 1. Area under the ROC curve for the CBIS-DDSM dataset. Mean and standard deviation computed over three runs. p-values for a two-sided Wilcoxon signed-rank test against the late-join baseline model.

Model	Views	ROC-AUC ± std.dev.	p-value
Single view	CC	0.750 ± 0.007	0.005
	MLO	0.763 ± 0.003	0.036
Late join	CC + MLO	0.788 ± 0.008	
Cross-view transformer (tokens)	CC + MLO	0.803 ± 0.007	0.061
Cross-view transformer (pixels)	CC + MLO	0.801 ± 0.003	0.006

Table 2. Area under the ROC curve for tasks in the CheXpert dataset. Mean and standard deviation computed over four runs.

Task	Single view		Late join	Cross-view (token-based)
	Frontal	Lateral		
Overall	0.827 ± 0.007	0.817 ± 0.009	0.829 ± 0.010	0.834 ± 0.002
Atelectasis	0.812 ± 0.009	0.809 ± 0.004	0.812 ± 0.016	0.833 ± 0.009
Cardiomegaly	0.924 ± 0.004	0.902 ± 0.003	0.919 ± 0.003	0.925 ± 0.004
Consolidation	0.863 ± 0.005	0.848 ± 0.006	0.867 ± 0.004	0.867 ± 0.004
Edema	0.882 ± 0.005	0.861 ± 0.005	0.889 ± 0.002	0.889 ± 0.005
Enlarged Cardiomed.	0.812 ± 0.008	0.796 ± 0.003	0.814 ± 0.005	0.810 ± 0.006
Fracture	0.775 ± 0.003	0.764 ± 0.018	0.766 ± 0.019	0.769 ± 0.013
Lung Lesion	0.744 ± 0.013	0.726 ± 0.010	0.747 ± 0.018	0.748 ± 0.007
Lung Opacity	0.808 ± 0.005	0.782 ± 0.006	0.806 ± 0.008	0.805 ± 0.004
Pleural Effusion	0.945 ± 0.001	0.946 ± 0.001	0.955 ± 0.002	0.954 ± 0.001
Pleural Other	0.789 ± 0.025	0.808 ± 0.030	0.786 ± 0.036	0.803 ± 0.030
Pneumonia	0.750 ± 0.004	0.740 ± 0.009	0.766 ± 0.011	0.754 ± 0.008
Pneumothorax	0.869 ± 0.003	0.853 ± 0.004	0.868 ± 0.004	0.872 ± 0.002
Support Devices	0.773 ± 0.006	0.786 ± 0.015	0.786 ± 0.007	0.803 ± 0.013

On the CBIS-DDSM dataset, we used a weighted binary cross-entropy loss to correct for the slight class imbalance. We trained for 300 epochs (30 warm-up), which was sufficient for all models to converge, and used the model of the final epoch. We report the mean of three runs over all five folds.

On the CheXpert dataset, we optimized the unweighted cross-entropy loss averaged over all tasks, training for 60 epochs (6 warm-up) to account for the larger dataset. We computed the AUC-ROC for each task separately and used the performance on the validation set to choose the best epoch for each task. We report the mean performance over four runs.

We ran all experiments on NVIDIA GeForce RTX 2080 Ti GPUs with 11 GB VRAM. For the CBIS-DDSM dataset, median training times were approximately 23 min for a single-view model, 36 min for a late-join model, and 37 min for the cross-view transformers. For the much larger CheXpert dataset, we trained for approximately 5 h per single-view model, versus 10 h for the late-join and cross-view models.

6   Results

On the CBIS-DDSM dataset (Table 1) the late-join baselines outperformed the single-view baselines. Adding the cross-view transformer improved the performance further, both for the token-based and pixel-wise variants. The transformer performance was not very sensitive to the number of heads or tokens: all settings produced similar results (see the table in the supplementary results).

On the CheXpert dataset, the cross-view transformer model also shows an improvement over the baselines (Table 2). The improvement is visible for the average performance over all tasks, with the cross-view model performing better than the late-join and single-view frontal models. The single-view model with the lateral view is less successful.

For individual chest X-ray tasks, the results are quite varied. For some tasks, such as atelectasis, cardiomegaly and support devices, the cross-view models show an improvement over the late-join model. For others, such as consolidation and edema, the models are closer together. This is consistent with observations in other work [3]. In general, with a few exceptions, the cross-view model has a performance that is equal to or better than the late-join models.

7   Discussion and Conclusion

The common multi-view approach to merge views after global pooling restricts models to learning global correlations between views. This may be sufficient for some applications, but a more local approach may be required for others. This is relatively easy if the images are spatially aligned and can be treated as multiple input channels, but is difficult when there are different perspectives or other misalignments that make it impossible to register the images correctly.

In this paper, we proposed a cross-view transformer approach to link unregistered dual-view images based on feature maps. Our experiments on two unregistered multi-view datasets indicate that this approach can outperform a model that links views on a global level. The cross-view transformer module is easy to integrate in any multi-view model with view-specific convolution branches.

Whether a cross-view transformer approach is useful depends on the application, since some tasks will benefit more from multi-view information and local correlations than others. We can see an indication of this in the results for the chest X-ray data (Table 2), where the late-join and cross-view transformer models sometimes do and sometimes do not have an advantage over the single-view models. This is consistent with results from Hashir et al. [3], who made similar observations about multi-view features on a different chest X-ray dataset.

The cross-view transformer mechanism can be computationally expensive. This can be reduced by applying the transformer later in the network, when the number of pixels is smaller. The memory requirements can be reduced with an efficient gradient implementation, by recomputing the pairwise attention scores on the fly. Using the token-based approach further reduces the requirements. In practice, we found that the additional computational and memory requirements were relatively limited, compared with those of the convolution layers.

For this paper we focussed on an evaluation of the cross-view transformer, presenting experiments with downsampled images and with the relatively small CBIS-DDSM dataset. While this allowed us to run more experiments, a better absolute performance could be achieved with higher-resolution data and more training images. For mammography classification, the state-of-the-art methods [16] use similar ResNet-based architectures, trained with different and larger datasets. For chest X-rays, the best-performing methods on the CheXpert dataset [6] use the full-resolution dataset and ensemble methods.

In summary, we presented a novel cross-view transformer approach that can transfer information between views, by linking feature maps before global pooling. On two datasets, we found that combining multi-view information on a spatial level can achieve better results than a model that merges features at an image level. We believe this can be an interesting addition for models that need to learn inter-view correlations in applications with unregistered images.

Acknowledgments. The research leading to these results is part of the project "MARBLE", funded from the EFRO/OP-Oost under grant number PROJ-00887. Some of the experiments were carried out on the Dutch national e-infrastructure with the support of SURF Cooperative.

References

1. Bekker, A.J., Shalhon, M., Greenspan, H., Goldberger, J.: Multi-view probabilistic classification of breast microcalcifications. IEEE Trans. Med. Imaging **35**(2), 645–653 (2015). https://doi.org/10.1109/TMI.2015.2488019
2. Carneiro, G., Nascimento, J., Bradley, A.P.: Deep learning models for classifying mammogram exams containing unregistered multi-view images and segmentation maps of lesions. In: Zhou, S.K., Greenspan, H., Shen, D. (eds.) Deep Learning for Medical Image Analysis, pp. 321–339. Academic Press (January 2017). https://doi.org/10.1016/B978-0-12-810408-8.00019-5
3. Hashir, M., Bertrand, H., Cohen, J.P.: Quantifying the value of lateral views in deep learning for chest X-rays. In: Medical Imaging with Deep Learning, pp. 288–303. PMLR (September 2020)
4. He, K., Zhang, X., Ren, S., Sun, J.: Deep residual learning for image recognition. In: CVPR, pp. 770–780 (2016). https://doi.org/10.3389/fpsyg.2013.00124
5. Heath, M., Bowyer, K., Kopans, D., Moore, R., Kegelmeyer Jr, P.: The Digital Database for Screening Mammography (2001)
6. Irvin, J., et al.: CheXpert: A Large Chest Radiograph Dataset with Uncertainty Labels and Expert Comparison. arXiv:1901.07031 [cs, eess] (January 2019)
7. Lee, R.S., Gimenez, F., Hoogi, A., Miyake, K.K., Gorovoy, M., Rubin, D.L.: A curated mammography data set for use in computer-aided detection and diagnosis research. Sci. Data **4**, 1–9 (2017). https://doi.org/10.1038/sdata.2017.177

8. Liu, Y., Zhang, F., Zhang, Q., Wang, S., Wang, Y., Yu, Y.: Cross-view correspondence reasoning based on bipartite graph convolutional network for mammogram mass detection. In: 2020 IEEE/CVF Conference on Computer Vision and Pattern Recognition (CVPR), pp. 3811–3821. IEEE, Seattle, WA, USA (June 2020). https://doi.org/10.1109/CVPR42600.2020.00387

9. Ma, J., et al.: Cross-view Relation Networks for Mammogram Mass Detection. arXiv:1907.00528 [cs] (June 2019)

10. Nasir Khan, H., Shahid, A.R., Raza, B., Dar, A.H., Alquhayz, H.: Multi-view feature fusion based four views model for mammogram classification using convolutional neural network. IEEE Access **7**, 165724–165733 (2019). https://doi.org/10.1109/ACCESS.2019.2953318

11. Rubin, J., Sanghavi, D., Zhao, C., Lee, K., Qadir, A., Xu-Wilson, M.: Large Scale Automated Reading of Frontal and Lateral Chest X-Rays using Dual Convolutional Neural Networks. arXiv:1804.07839 [cs, stat] (April 2018)

12. Sun, L., Wang, J., Hu, Z., Xu, Y., Cui, Z.: Multi-view convolutional neural networks for mammographic image classification. IEEE Access **7**, 126273–126282 (2019). https://doi.org/10.1109/ACCESS.2019.2939167

13. Vaswani, A., et al.: Attention is all you need. Adv. Neural. Inf. Process. Syst. **30**, 5998–6008 (2017)

14. Wang, H., et al.: Breast mass classification via deeply integrating the contextual information from multi-view data. Pattern Recogn. **80**, 42–52 (2018). https://doi.org/10.1016/j.patcog.2018.02.026

15. Wu, B., et al.: Visual Transformers: Token-based Image Representation and Processing for Computer Vision. arXiv:2006.03677 [cs, eess] (November 2020)

16. Wu, N., et al.: Deep neural networks improve radiologists' performance in breast cancer screening. IEEE Trans. Med. Imaging **39**(4), 1184–1194 (2019). https://doi.org/10.1109/tmi.2019.2945514

17. Zhao, X., Yu, L., Wang, X.: Cross-view attention network for breast cancer screening from multi-view mammograms. In: ICASSP 2020–2020 IEEE International Conference on Acoustics, Speech and Signal Processing (ICASSP), pp. 1050–1054 (May 2020). https://doi.org/10.1109/ICASSP40776.2020.9054612

Machine Learning - Domain Adaptation

Stain Mix-Up: Unsupervised Domain Generalization for Histopathology Images

Jia-Ren Chang[1](✉), Min-Sheng Wu[1], Wei-Hsiang Yu[1], Chi-Chung Chen[1], Cheng-Kung Yang[1], Yen-Yu Lin[2], and Chao-Yuan Yeh[1](✉)

[1] aetherAI, Taipei, Taiwan
{JiaRenChang,vincentwu,seanyu,chenchc,jimmy,joeyeh}@aetherai.com
[2] National Yang Ming Chiao Tung University, Hsinchu, Taiwan
lin@cs.nctu.edu.tw

Abstract. Computational histopathology studies have shown that stain color variations considerably hamper the performance. Stain color variations indicate the slides exhibit greatly different color appearance due to the diversity of chemical stains, staining procedures, and slide scanners. Previous approaches tend to improve model robustness via data augmentation or stain color normalization. However, they still suffer from generalization to new domains with unseen stain colors. In this study, we address the issue of unseen color domain generalization in histopathology images by encouraging the model to adapt varied stain colors. To this end, we propose a novel data augmentation method, **stain mix-up**, which incorporates the stain colors of unseen domains into training data. Unlike previous mix-up methods employed in computer vision, the proposed method constructs the combination of stain colors without using any label information, hence enabling unsupervised domain generalization. Extensive experiments are conducted and demonstrate that our method is general enough to different tasks and stain methods, including H&E stains for tumor classification and hematological stains for bone marrow cell instance segmentation. The results validate that the proposed stain mix-up can significantly improves the performance on the unseen domains.

Keywords: Domain generalization · Mix-up · Stain color

1 Introduction

Computer-aided diagnosis based on histopathology images, such as whole slide images (WSIs) and field of views (FoVs) of tissue sections, gains significant progress owing to the great success of machine learning algorithms in digital pathology. Tissue sections are typically stained with various stains to make tissues visible under the microscope. However, tissue manipulation, staining,

Electronic supplementary material The online version of this chapter (https://doi.org/10.1007/978-3-030-87199-4_11) contains supplementary material, which is available to authorized users.

© Springer Nature Switzerland AG 2021
M. de Bruijne et al. (Eds.): MICCAI 2021, LNCS 12903, pp. 117–126, 2021.
https://doi.org/10.1007/978-3-030-87199-4_11

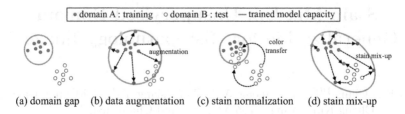

(a) domain gap   (b) data augmentation   (c) stain normalization   (d) stain mix-up

Fig. 1. Comparisons among different methods for model capacity generalization for histological images analysis. (a) There exists the domain gap between domain A (red dots) and domain B (purple cycles) due to the color variation; (b) Data augmentation is adopted to increase the color variations of domain A; (c) Color normalization transfers domain B to the color patterns of domain A to avoid color variation; (d) In this paper, we propose stain mix-up for randomly augmenting the domain A according to the stain color matrices of domain B and thus generalize the model to domain B. Moreover, the proposed domain generalization technique is unsupervised where data labels on domain B are not required. (Color figure online)

and even scanning often result in substantial color appearance variations in histopathology images, and degrade machine learning algorithms due to the domain gap of colors. Thus, it is crucial to take color appearance variations into account when developing machine learning algorithms for histopathology image analysis. Specifically, two strategies are widely used, including 1) augmenting color patterns of training data to enhance model robustness; and 2) normalizing all histopathology images to a single color pattern so that the unfavorable impact of color variations in the subsequent process can be alleviated.

To augment color patterns of training data, most techniques conduct brightness, contrast and hue perturbations [7,19]. Specifically, Bug *et al.* [3] utilize principle component analysis (PCA) to decompose images into a low-dimensional space spanned by a few principle components where augmentation of H&E images is carried out by perturbing the main components. Tellez *et al.* [17] divide H&E images into the hematoxylin, eosin, and residual color channels by color deconvolution [14]. Then, stain-specific transformation by perturbing each channel is used to complete augmentation.

Another way for addressing color variations is to normalize all images to have similar color patterns. Several color normalization methods have been proposed to achieve it. Given a source image and a target image, Reinhard *et al.* [13] convert images to the $l\alpha\beta$ space and normalize the source image by aligning the mean and standard deviation of the source and target images. Other methods decompose stain colors into a stain color matrix and a stain density map, and apply the stain color matrix of the target image onto the source image. Macenko *et al.* [10] use singular value decomposition while Vahadane *et al.* [18] employ sparse non-negative matrix factorization (SNMF) to decompose the stain color matrix. Nadeem *et al.* [12] adopt Wasserstein barycenter to normalize images.

We observe that highly effective methods [17,18] eliminate color variations by decomposing stain color matrices for further transferring or augmenting. It implies that stain color matrices can reliably encode the color patterns. However, current studies [13,17,18] consider stain color matrices decomposed from only a single domain and thus may restrict their generalization abilities.

In this paper, we propose a novel method **stain mix-up** for data augmentation. It randomly interpolates a new stain color matrix between different domains during training and can improve the generalization performance accordingly. The Mix-up technique [20] has become essential to data augmentation for recognition [6] or domain adaption [9,11] in computer vision. In contrast to Mix-up [20] which mixes images and labels, the proposed method is *label-free*. It mixes stain color matrices between different domains and can synthesize various types of stain colors for learning color-invariant representations. In Fig. 1, we describe the concept of different approaches for generalizing histology image analysis, including data augmentation, stain normalization and the proposed stain mix-up. Extensive experiments are conducted on two kinds of stains and tasks, *i.e.*, tumor classification on H&E stained images and bone marrow cell instance segmentation on hematological stained images. Since both tasks have multiple sources (domains) of images, we train the model on one domain where data were collected from one medical center, and test it on others. To this end, the training center is denoted as the source domain which consists of images and labels whereas other centers are denoted as the target domain which only have images but no labels. The stain color matrices of the target domain are mixed with those of the source domain to synthesize new training samples in the source domain for on-the-fly augmentation. The results show that the proposed stain mix-up achieves the-state-of-art generalization performance on both tasks.

The main contributions of this work are summarized as follows. First, we propose a novel data augmentation approach, namely stain mix-up, to achieve unsupervised domain generalization for histology image analysis. Second, we perform extensive experiments to demonstrate the effectiveness of the proposed method. It turns out that our method consistently achieves the state-of-the-art performance on different tasks and stains. To the best of our knowledge, the proposed method is the first work for unsupervised domain generalization in histology image analysis.

2   Method

This section describes the proposed method, which is composed of two stages: stain separation and stain mix-up augmentation. The former is conducted to extract color characteristics from histology images of different domains. It estimates stain color matrices that represent chromatic components and stain density maps of each domain. The latter uses the estimated matrices of different domains to augment training images on-the-fly through the proposed stain mix-up, enabling unsupervised domain generalization. Details of the two stages are elaborated as follows. Figure 2 illustrates the pipeline of our method.

2.1   Stain Separation via SNMF

Stains are optical absorption materials that occlude certain spectra of light, making tissues visible in the complementary colors. They help visualize tissues for

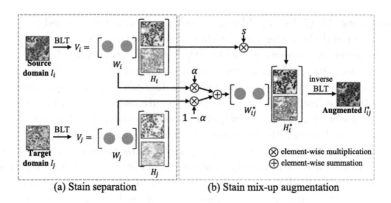

(a) Stain separation   (b) Stain mix-up augmentation

Fig. 2. Pipeline of the proposed stain mix-up augmentation. (a) Given a labeled image I_i from the source domain and an unlabeled image I_j from the target domain, stain separation decomposes the optical density of I_i derived via Beer-Lambert transformation (BLT), *i.e.* V_i, into its stain color matrix W_i and density map H_i. Similarly, we have W_j and H_j from V_j. (b) Stain mix-up augmentation is carried out by applying inverse BLT to a mixed stain color matrix W_{ij}^* and a perturbed density map H_i^*.

medical diagnosis. Stained tissue colors result from light attenuation, depending on the type and amount of stains that tissues have absorbed. This property can be mathematically formulated by Beer-Lambert law [2] defined as follows:

$$V = -\log \frac{I}{I_0} = WH , \qquad (1)$$

where $I \in \mathbb{R}^{3 \times n}$ is a histology image in the RGB color space, I_0 is the illuminating light intensity of sample with $I_0 = 255$ for 8-bit images in our cases, $W \in \mathbb{R}^{3 \times m}$ is the stain color matrix to encode the color appearance of each stain, and $H \in \mathbb{R}^{m \times n}$ is the density map of each stain, for an m-stained n-pixel image. Given a histology image I, its optical density V can be derived via Beer-Lambert transformation (BLT) in Eq. 1. Stain separation aims to estimate the corresponding stain color matrix W and density map H. In this work, we follow [18] and solve W and H of a histology image I through SNMF in the experiments.

2.2   Stain Mix-Up Augmentation

The proposed stain mix-up for data augmentation aims to reduce the domain gaps caused by color appearance variations between different sources. It synthesizes diversified images for augmentation and can increase the potential data coverage. As shown in Fig. 2, we carry out this task by mixing the stain color characteristics of both the source and target domains. Specifically, we randomly sample a pair of histological images I_i and I_j from the source domain and the target domain respectively, followed by decomposing them into the stain color matrices W_i and W_j and the stain density maps H_i and H_j through BLT and

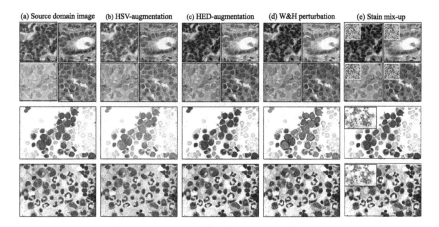

Fig. 3. Comparisons among different augmentation methods. The **top two rows** show images from CAMELYON17 [1] and the **bottom two rows** show images from Hema. (a) shows the training images from the source domain. (b), (c), and (d) are the augmented training images generated via three baseline methods which cannot include target domain information. The results of our proposed stain mix-up are demonstrated in (e). By mixing the stain matrices from (a) source domain image and target domain images (upper-left corners in (e)), the stain mix-up yields more realistic stain color compared with the other augmentations. Therefore, our generated images can effectively help accommodate the model to the target domain.

SNMF. A mixed stain color matrix W_{ij}^* is a linear interpolation between W_i and W_j with a coefficient α randomly sampled from a uniform distribution, namely,

$$W_{ij}^* = \alpha W_i + (1 - \alpha)W_j, \quad \text{where } \alpha \sim U(0,1). \tag{2}$$

Random interpolation between stain color matrices increases the diversity of stain color appearance while keeping the mixed stain color matrices realistic, thus improving the generalization ability to the target domain.

Similar to [17], we perturb the stain density map H_i to simulate the extents of stain concentrations and color fading,

$$H_i^* = sH_i, \quad \text{where } s \sim U(1 - \delta, 1 + \delta), \tag{3}$$

where s is a scaling factor randomly drawn from a uniform distribution controlled by $\delta \in [0,1]$. By referring to the interpolated stain color matrix W_{ij}^* in Eq. 2 and the perturbed map H_i^* in Eq. 3, the resulting augmented image I_{ij}^* is generated by the inverse BLT,

$$I_{ij}^* = I_0 \exp\left(-W_{ij}^* H_i^*\right). \tag{4}$$

Figure 3 shows several examples of the augmented images.

In this study, the histological images are collected from multiple medical centers, and their stained color appearances may vary considerably due to different staining processes. We aim to generalize the model trained on labeled data of one center (source domain) to unlabeled data of other centers (target domains). To this end, the proposed stain mix-up is applied for augmentation. Unlike existing augmentation methods such as [13,17,18] where only source domain data are considered, our method leverages data from both source and target domains to synthesize augmented data that are more consistent with the target domain. Compared with existing domain generalization methods such as [20], our method make use of image data without labels in the target domain, and hence enables unsupervised domain generalization. The properties described above make the stain mix-up a simple yet efficient module that can augment images on-the-fly to achieve a state-of-the-art performances on various tasks.

3   Experiments

In this section, after describing materials, implementation details, and evaluation metrics of our proposed method, we present and elaborate the results of the experiments. Two datasets, namely CAMELYON17 and Hema, are adopted to experiment with different augmentation methods on different types of computer vision tasks in histology image analysis. To better understand how stain matrix augmenting affects the model generalization, we also perform an ablation study on CAMELYON17 for validating the effect from perturbing W and H.

3.1   Datasets

CAMELYON17. We use the CAMELYON17 [1] dataset to evaluate the performance of the proposed method on tumor/normal classification. In this dataset, a total of 500 H&E stained WSIs are collected from *five* medical centers (denoted by C_1, C_2, ... C_5 respectively), 50 of which include lesion-level annotations. All positive and negative WSIs are randomly split into training/validation/test sets with the following distributions: $C_1 : 37/22/15$, $C_2 : 34/20/14$, $C_3 : 43/24/18$, $C_4 : 35/20/15$, $C_5 : 36/20/15$. We extract image tiles in a size of 256×256 pixels from the annotated tumors for positive patches and from tissue regions of WSIs without tumors for negative patches.

Hema. We evaluate the proposed method on a custom hematology dataset for bone marrow cell instance segmentation. In the Hema dataset, a total of 595 WSIs of hematological stained bone marrow smears are collected from *two* medical centers, denoted by M_1 and M_2 respectively. We sample 21,048 FoVs from M_1 as training data and 311 FoVs from M_2 as testing data. All FoVs are in a size of $1,149 \times 1,724$. This dataset has a total of 662,988 blood cell annotations, which are annotated by a cohort of ten annotators, consisting of senior hematologists and medical technicians with an average of over ten years of clinical experience.

3.2   Implementation Details

For the CAMELYON17 dataset, we train five ResNet-50 [5] classifiers on each center individually and test each classifier on the test data of all centers to

Table 1. (Top) Mean AUC of different methods for tumor classification on the CAME-LYON17 [1] dataset. (**Bottom**) Ablation studies on the components of our method.

Method	C_1	C_2	C_3	C_4	C_5	Average (95% CI)
No manipulation	0.947	0.753	0.854	0.777	0.861	0.838 (0.824, 0.852)
Vahadane normalization [18]	0.943	0.826	0.898	0.853	0.707	0.846 (0.829, 0.861)
HSV-aug.	0.984	0.844	0.954	0.937	0.845	0.913 (0.902, 0.924)
HED-aug. [17]	0.984	0.925	0.903	0.861	0.908	0.916 (0.909, 0.924)
Stain mix-up (ours)	0.979	0.948	0.946	0.965	0.942	**0.956** (0.953, 0.959)
Ablation study						
H perturb	0.984	0.925	0.903	0.861	0.908	0.916 (0.909, 0.924)
W, H perturb	0.979	0.948	0.911	0.965	0.939	0.948 (0.944, 0.953)
W mix-up, H perturb	0.979	0.948	0.946	0.965	0.942	**0.956** (0.953, 0.959)

evaluate the effectiveness of generalization. Since the CAMELYON17 dataset contains H&E stained images, we decompose each image into a stain color matrix $W \in \mathbb{R}^{3 \times 2}$ and a density map $H \in \mathbb{R}^{2 \times n}$. The parameter δ in Eq. 3 is set to 0.2. All models are trained with AdamW [8], a learning rate of 0.001, and a batch size of 32 for 100,000 iterations with an Nvidia Quadro RTX8000.

For the Hema dataset, we adopt Mask R-CNN [4] with the ResNet-50 backbone pre-trained on ImageNet [15] for instance segmentation. The stain of WSIs in Hema is composed of three chemicals, namely methylene blue, eosin, and azur. Thereby, we factorize each image into a stain color matrix $W \in \mathbb{R}^{3 \times 3}$ and a density map $H \in \mathbb{R}^{3 \times n}$. The parameter δ is set to 0.5. The model is trained on M_1 with SGD, a learning rate of 0.02, a momentum of 0.9, a batch size of 4, and weight decay 0.0001 for 12 epochs with an Nvidia V100. After data augmentation and model fine-tuning, we evaluate the generalization performance on M_2.

Please note that the stain matrices are calculated using SNMF before training for saving computational time. That is, we only compute stain matrices once and use them repeatedly during training. The computational time of SNMF decomposition for a single image in CAMELYON17 and Hema takes 1.14 and 2.40 s, respectively, measured on an Intel Xeon CPU E5-2697 v3.

3.3   Results on the CAMELYON17 Dataset

In Table 1, we compare the proposed stain mix-up with existing augmentation methods for tumor classification on the CAMELYON17 dataset. It is consistent with previous findings [16,21] that models trained without color augmentations result in weaker performance and a larger performance fluctuation when testing on images from other centers (AUC = 0.838, 95% CI 0.824 − 0.852), which reveals the domain gaps among different centers. The models trained with data augmented by the proposed stain mix-up achieves significant performance gains

Table 2. Performance in mAP of bone marrow cell instance segmentation using different augmentation methods on the Hema dataset.

Method	Box			Mask		
	AP	AP_{50}	AP_{75}	AP	AP_{50}	AP_{75}
No manipulation	48.3	52.5	51.2	49.6	52.5	52.0
HSV-augmentation	48.7	53.4	52.1	50.2	53.4	52.9
HED-augmentation [17]	48.1	52.4	51.5	49.5	52.5	52.0
Stain mix-up (ours)	**50.3**	**55.4**	**54.5**	**52.1**	**55.5**	**55.1**

over HSV-augmentation and HED-augmentation [17]. In addition, the stain mix-up method helps reach stable performance when evaluated on images of different centers, while other competing methods show larger performance variations. We attribute these advantages to the cross-domain interpolation of the stain color matrices in the proposed stain mix-up, while the competing methods such as HSV-augmentation and HED-augmentation refer to only images of the source domain. The augmented images by our method are realistic and more consistent with those in the target domain, leading to a better generalization ability.

Cross-domain interpolation is the key component of the proposed stain mix-up for unsupervised and diversified stain color matrix generation. While the stain color matrix W can be interpolated between the source and target domains, it can be self-perturbed with random degrees sampled from a uniform distribution. The implementation details of the self-perturbed W is described in the supplementary material. In the ablation study, we explore how different perturbation methods contribute to the model generalization. Some example patches generated by using random W, H perturbation are visualized in Fig. 3d. As shown in the bottom of Table 1, stochastic fluctuations in W achieve the AUC of 0.948 (95% CI $0.944 - 0.953$), which is inferior to models trained with the stain mix-up. This result suggests that: 1) Models can benefit from perturbing interaction between color channels, and 2) with the identified stain matrix of centers in advance, interpolating combinations of matrices can be more effective for model adaptation across different centers.

3.4   Results on the Hema Dataset

In addition to tumor classification, we evaluate the proposed stain mix-up for cell instance segmentation on the Hema dataset. As shown in Table 2, our method consistently outperforms baseline methods by substantial margins, more than 2.0% in box mAP and mask mAP in most cases. We observe that the baseline methods, HSV-augmentation and HED-augmentation, make no improvement on this dataset. The major reason is that the large domain gap makes augmentation based on only the source domain, which is irrelevant to the target domain. By taking the stain color matrices of unlabeled data of the target domain into consideration, our method can effectively accomplish domain generalization in an unsupervised manner. The results validate that our model can alleviate the domain gaps between different histology image collections even on the challenging instance segmentation task.

4   Conclusions

We have presented stain mix-up, a simple yet effective data augmentation method for unsupervised domain generalization in histological image analysis. Our stain mix-up constructs various virtual color patterns by random linear interpolation of two stain color matrices, one from the source domain and one from the target domain. Cross-domain interpolation refers to color distributions of both domains, and color patterns that are realistic and more consistent to the target patterns can be synthesized, facilitating model adaptation to the target domain. Since accessing only stain color matrices is label-free, the proposed method carries out unsupervised domain generalization. Through extensive experiments, we have shown that the proposed stain mix-up significantly improves the generalization ability on diverse tasks and stains, such as tumor classification on the H&E stained images and bone marrow cell segmentation on the hematological stained images. We believe the proposed stain mix-up can advance the community of digit pathology for practical usage.

Acknowledgement. We thank Wen-Chien Chou M.D.(National Taiwan University Hospital), Ta-Chuan Yu M.D.(National Taiwan University Hospital Yunlin Branch) and Poshing Lee M.D.(Department of Hematopathology, BioReference) for Hema dataset construction. This paper was supported in part by the Ministry of Science and Technology, Taiwan, under Grants MOST 110-2634-F-007-015 and MOST 109-2221-E-009-113-MY3.

References

1. Bandi, P., et al.: From detection of individual metastases to classification of lymph node status at the patient level: the camelyon17 challenge. IEEE Trans. Med. Imaging **38**(2), 550–560 (2018)
2. Beer, A., Beer, P.: Determination of the absorption of red light in colored liquids. Annalen der Physik und Chemie **86**(5), 78–88 (1852)
3. Bug, D., et al.: Context-based normalization of histological stains using deep convolutional features. In: Cardoso, M.J., et al. (eds.) DLMIA/ML-CDS -2017. LNCS, vol. 10553, pp. 135–142. Springer, Cham (2017). https://doi.org/10.1007/978-3-319-67558-9_16
4. He, K., Gkioxari, G., Dollár, P., Girshick, R.: Mask r-cnn. In: Proceedings of the IEEE/CVF International Conference on Computer Vision, pp. 2961–2969 (2017)
5. He, K., Zhang, X., Ren, S., Sun, J.: Deep residual learning for image recognition. In: Proceedings of the IEEE/CVF Conference on Computer Vision and Pattern Recognition, pp. 770–778 (2016)
6. He, T., Zhang, Z., Zhang, H., Zhang, Z., Xie, J., Li, M.: Bag of tricks for image classification with convolutional neural networks. In: Proceedings of the IEEE/CVF Conference on Computer Vision and Pattern Recognition, pp. 558–567 (2019)

7. Liu, Y., et al.: Detecting cancer metastases on gigapixel pathology images. arXiv preprint arXiv:1703.02442 (2017)
8. Loshchilov, I., Hutter, F.: Decoupled weight decay regularization. In: International Conference on Learning Representations (2019)
9. Luo, C., Song, C., Zhang, Z.: Generalizing person re-identification by camera-aware invariance learning and cross-domain mixup. In: Vedaldi, A., Bischof, H., Brox, T., Frahm, J.-M. (eds.) ECCV 2020. LNCS, vol. 12360, pp. 224–241. Springer, Cham (2020). https://doi.org/10.1007/978-3-030-58555-6_14
10. Macenko, M., et al.: A method for normalizing histology slides for quantitative analysis. In: 2009 IEEE International Symposium on Biomedical Imaging: From Nano to Macro, pp. 1107–1110. IEEE (2009)
11. Mao, X., Ma, Y., Yang, Z., Chen, Y., Li, Q.: Virtual mixup training for unsupervised domain adaptation. arXiv preprint arXiv:1905.04215 (2019)
12. Nadeem, S., Hollmann, T., Tannenbaum, A.: Multimarginal Wasserstein barycenter for stain normalization and augmentation. In: Martel, A.L., et al. (eds.) MICCAI 2020. LNCS, vol. 12265, pp. 362–371. Springer, Cham (2020). https://doi.org/10.1007/978-3-030-59722-1_35
13. Reinhard, E., Adhikhmin, M., Gooch, B., Shirley, P.: Color transfer between images. IEEE Comput. Gr. Appl. 21(5), 34–41 (2001)
14. Ruifrok, A.C., Johnston, D.A., et al.: Quantification of histochemical staining by color deconvolution. Anal. Quant. Cytol. Histol. 23(4), 291–299 (2001)
15. Russakovsky, O., et al.: ImageNet large scale visual recognition challenge. Int. J. Comput. Vis. 115(3), 211–252 (2015). https://doi.org/10.1007/s11263-015-0816-y
16. Shaban, M.T., Baur, C., Navab, N., Albarqouni, S.: Staingan: stain style transfer for digital histological images. In: 2019 IEEE 16th International Symposium on Biomedical Imaging (Isbi 2019), pp. 953–956. IEEE (2019)
17. Tellez, D., et al.: Whole-slide mitosis detection in h&e breast histology using phh3 as a reference to train distilled stain-invariant convolutional networks. IEEE Trans. Med. Imaging 37(9), 2126–2136 (2018)
18. Vahadane, A., et al.: Structure-preserving color normalization and sparse stain separation for histological images. IEEE Trans. Med. Imaging 35(8), 1962–1971 (2016)
19. Veta, M., Van Diest, P.J., Jiwa, M., Al-Janabi, S., Pluim, J.P.: Mitosis counting in breast cancer: object-level interobserver agreement and comparison to an automatic method. PloS one 11(8), e0161286 (2016)
20. Zhang, H., Cisse, M., Dauphin, Y.N., Lopez-Paz, D.: Mixup: beyond empirical risk minimization. In: International Conference on Learning Representations (2018)
21. Zhou, N., Cai, D., Han, X., Yao, J.: Enhanced cycle-consistent generative adversarial network for color normalization of h&e stained images. In: Shen, D., et al. (eds.) MICCAI 2019. LNCS, vol. 11764, pp. 694–702. Springer, Cham (2019). https://doi.org/10.1007/978-3-030-32239-7_77

A Unified Hyper-GAN Model for Unpaired Multi-contrast MR Image Translation

Heran Yang[1,2], Jian Sun[1,2(✉)], Liwei Yang[1], and Zongben Xu[1,2]

[1] School of Mathematics and Statistics, Xi'an Jiaotong University, Xi'an, China
jiansun@xjtu.edu.cn
[2] Pazhou Lab, Guangzhou, China

Abstract. Cross-contrast image translation is an important task for completing missing contrasts in clinical diagnosis. However, most existing methods learn separate translator for each pair of contrasts, which is inefficient due to many possible contrast pairs in real scenarios. In this work, we propose a unified Hyper-GAN model for effectively and efficiently translating between different contrast pairs. Hyper-GAN consists of a pair of hyper-encoder and hyper-decoder to first map from the source contrast to a common feature space, and then further map to the target contrast image. To facilitate the translation between different contrast pairs, contrast-modulators are designed to tune the hyper-encoder and hyper-decoder adaptive to different contrasts. We also design a common space loss to enforce that multi-contrast images of a subject share a common feature space, implicitly modeling the shared underlying anatomical structures. Experiments on two datasets of IXI and BraTS 2019 show that our Hyper-GAN achieves state-of-the-art results in both accuracy and efficiency, e.g., improving more than 1.47 and 1.09 dB in PSNR on two datasets with less than half the amount of parameters.

Keywords: Multi-contrast MR · Unpaired image translation · Unified hyper-GAN

1 Introduction

Magnetic resonance (MR) imaging has been widely utilized in clinical diagnosis, as it has a range of imaging contrasts and largely increases the diversity of diagnostic information. However, due to practical limits, e.g., long scan time [23], image corruption [27], etc., it is often hard to collect all multi-contrast MR images of one subject. To solve this problem, a large variety of synthesis methods [4, 6,12,13,17,18,20,24,25] try to synthesize missing contrast from the available contrast, and most of them are one-to-one cross-contrast synthesis methods, i.e.,

Electronic supplementary material The online version of this chapter (https://doi.org/10.1007/978-3-030-87199-4_12) contains supplementary material, which is available to authorized users.

M. de Bruijne et al. (Eds.): MICCAI 2021, LNCS 12903, pp. 127–137, 2021.
https://doi.org/10.1007/978-3-030-87199-4_12

one model is trained for each specific pair of contrasts. For example, Dar et al. [6] proposed a conditional generative adversarial network to translate between T1w and T2w images. However, it is impractical to train each network for each pair of contrasts due to a range of commonly used MR contrasts in real scenarios. In this work, we focus on tackling the multi-contrast MR image translation problem by a more efficient and effective way in an unpaired training setting, i.e., the training multi-contrast MR images are not required to be captured from same subjects.

There are already several unpaired multi-contrast image translation methods in literature, e.g., [2,5,14,21], in recent years. Firstly, CycleGAN [29], as a one-to-one cross-contrast synthesis method, could be extended to learn multi-contrast mappings and require $N \times (N-1)$ generators for N contrasts, which is impractical in real scenarios. Furthermore, ComboGAN [2] and DomainBank [14] decouple generator into encoder/decoder and reduce the requirement to N contrast-specific encoders/decoders for N contrasts, while the parameter size and training time of them scale linearly with the contrast number. In addition, StarGAN [5] and SUGAN [21] share a generator and discriminator for all contrasts, and rely on a contrast indicator to specify the desired output contrast. However, since the contrast indicator is simply concatenated with input image, it might be insufficient to control the translation process [1,15]. In summary, existing methods either depend on multiple encoders/decoders, or insufficiently control/modulate the generator to be adaptive to contrast.

In this work, we aim to design a unified deep network for unpaired multi-contrast MR image translation in an efficient and effective way. We design shared encoder and decoder to translate between different contrast pairs based on a common feature space constraint, and the encoding and decoding processes are respectively modulated by the source and target contrast codes. This is inspired by MR imaging [3] that multi-contrast MR images are determined by human intrinsic tissue and scanner imaging parameters. The common feature space implicitly models the intrinsic tissue parameters, and the scanner imaging parameters are encoded to modulate the encoder and decoder in our network design.

Specifically, we first define contrast-specific information as one-hot code indicating MR contrast, and construct two contrast modulators as hyper-network [11] to respectively modulate the encoder and decoder to be adaptive to different MR contrasts. To enforce the common feature space shared by different contrasts, we further design a common space loss to enforce extracted deep features from different contrasts within a common feature space, implicitly modeling the shared underlying anatomical structures, besides traditional adversarial [10] and cycle-consistency [29] losses. This unified multi-contrast MR image translation model, dubbed *Hyper-GAN*, can effectively and efficiently translate between different contrast pairs using a single network. Experiments on two multi-contrast brain MR datasets of IXI and BraTS 2019 show that Hyper-GAN achieves start-of-the-art results in both accuracy and efficiency, e.g., improving more than 1.47 and 1.09 dB in PSNR on two datasets with less than half the amount of parameters.

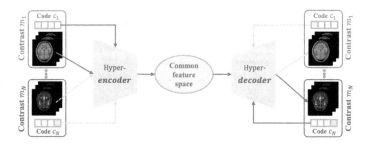

Fig. 1. Hyper-GAN utilizes shared hyper-encoder and hyper-decoder to translate between different contrast pairs, and the encoding and decoding processes are respectively modulated by the source and target contrast codes. Multi-contrast images of a subject are constrained to be within a common feature space after encoding.

2   Method

As shown in Fig. 1, our *Hyper-GAN* utilizes a shared pair of *hyper-encoder* and *hyper-decoder* to translate between different contrast pairs. The source and target contrasts are each represented by a one-hot code, dubbed *contrast code*, with the value of 1 representing the corresponding contrast in a given list of multiple contrasts. These codes are further utilized to adaptively tune the encoding and decoding processes in our hyper-encoder and hyper-decoder. Specifically, our hyper-encoder/hyper-decoder is respectively an encoder/decoder with parameters modulated by a *contrast modulator* with contrast code as input, in order to make the hyper-encoder/hyper-decoder adaptive to different contrasts. The extracted deep features by hyper-encoder are constrained to be within a common feature space shared by different contrast images of a subject. In this way, our Hyper-GAN is an encoder-decoder network adaptive to different source and target contrast pairs. We next introduce the details.

2.1   Network Architecture

As shown in Fig. 2(a) and (b), hyper-encoder E and hyper-decoder G respectively consist of two subnets, i.e., a backbone encoder/decoder and a contrast modulator. The encoder extracts deep features from a source contrast image, while the decoder estimates the target contrast image from the extracted features. The encoder and decoder are respectively paired with a contrast modulator, achieving contrast-adaptive tuning of parameters of encoder and decoder based on the source and target contrast codes. With the contrast code as input, the contrast modulator tunes the parameters using the following two different strategies, i.e., filter scaling or conditional instance normalization.

Filter Scaling (FS) [1]. For each convolutional filter f in encoder/decoder, the modulator produces a corresponding scalar α based on the contrast code and modifies this filter as $f' \triangleq \alpha * f$, where $*$ is scalar multiplication operation.

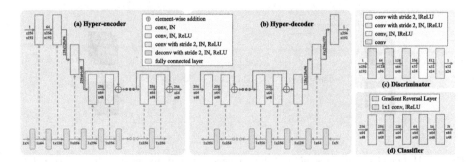

Fig. 2. The illustration of the detailed network architecture in Hyper-GAN.

Conditional Instance Normalization (CIN) [8]. For each instance normalization (IN) layer in encoder/decoder, the modulator estimates its affine parameters γ' and β' based on contrast code, and then IN layer becomes $y = \gamma'\frac{x-\mu(x)}{\sigma(x)} + \beta'$, where $\mu(x)$ and $\sigma(x)$ are mean and standard deviation of the input features x.

Network Architecture. Encoder and decoder are respectively taken as the first five and the remaining four residual blocks of the generator in [29], and contrast modulators are set to multilayer perceptrons. Please refer to Fig. 2(a) and (b) for the detailed definitions of hyper-encoder and hyper-decoder. As shown in Fig. 2(c), discriminators, as in [29], are fully convolutional networks [19] with five 4×4 convolutional layers to classify whether 70×70 overlapping image patches are real or synthetic. The contrast-classifier is composed of a gradient reversal layer [9] and four 1×1 convolutional layers to classify which contrast the features extracted by hyper-encoder belong to and outputs contrast probabilities.

2.2   Training Loss

For a N-contrast translation task with contrasts $\{m_k\}_{k=1}^N$ and one-hot contrast codes $\{c_k \in \{0,1\}^N\}_{k=1}^N$, our Hyper-GAN consists of a pair of hyper-encoder E and hyper-decoder G, a contrast-classifier C and contrast-specific discriminators $\{D_k\}_{k=1}^N$. Note that C and $\{D_k\}_{k=1}^N$ are only utilized in training.

Our Hyper-GAN is trained on unpaired multi-contrast MR images, i.e., different contrasts are acquired from different subjects without requiring accurate registration. When training the network, two random training images I_{m_i}, I_{m_j} in distinct contrasts $m_i, m_j (i \neq j)$ are fed into Hyper-GAN as the source and target contrasts for updating network parameters. For simplicity, we next present the training loss for the translation from contrast m_i to m_j.

Adversarial Loss [10]. The hyper-encoder E and hyper-decoder G are required to generate a synthetic image $G(E(I_{m_i}, c_i), c_j)$ close to a real image, while the discriminator D_j is to distinguish between this synthetic image $G(E(I_{m_i}, c_i), c_j)$ and a real image I_{m_j}. The adversarial loss is defined as

$$\mathcal{L}_{adv}(E, G, D_j) = D_j(G(E(I_{m_i}, c_i), c_j))^2 + (1 - D_j(I_{m_j}))^2 , \qquad (1)$$

where $E(\cdot, c_i)$ and $G(\cdot, c_j)$ respectively denote the outputs of hyper-encoder E and hyper-decoder G tuned by contrast codes c_i and c_j. $G(E(I_{m_i}, c_i), c_j)$ represents the synthetic m_j contrast image translated from an m_i contrast input.

Cycle-consistency Loss [29]. A cycle-consistency loss for E and G is to force the reconstructed image $G(E(G(E(I_{m_i}, c_i), c_j), c_j), c_i)$ (mapping from contrast m_i to m_j and back) to be identical to the input I_{m_i}, which is written as

$$\mathcal{L}_{cyc}(E, G) = \|G(E(G(E(I_{m_i}, c_i), c_j), c_j), c_i) - I_{m_i}\|_1 . \qquad (2)$$

Common Space Loss. The hyper-encoder is adaptive to the contrast of input image, and we expect that the extracted features by hyper-encoder should be in a common space shared by multi-contrast images for each subject, implicitly modeling the intrinsic tissue parameters of the subject. As paired training multi-contrast images are not given, we design the following common space constraints.

(i) Identical loss. As the images I_{m_i} and $G(E(I_{m_i}, c_i), c_j)$ are of the same subject but different contrasts, an identical loss is designed to enforce their extracted features by hyper-encoder E to be identical, which is written as

$$\mathcal{L}_{id}(E, G) = \|E(G(E(I_{m_i}, c_i), c_j), c_j) - E(I_{m_i}, c_i)\|_1 . \qquad (3)$$

(ii) Reconstruction loss. A reconstruction loss constrains that image I_{m_i} could reconstruct itself by hyper-encoder and hyper-decoder, which is defined as

$$\mathcal{L}_{rec}(E, G) = \|G(E(I_{m_i}, c_i), c_i) - I_{m_i}\|_1 . \qquad (4)$$

(iii) Contrast-classification loss. Contrast-classification loss is to force classifier C to predict the contrast of extracted features by hyper-encoder E. We adversarially train the classifier to make deep features of multiple contrasts extracted by E to be same distributed, i.e., within a common feature space, using a gradient reversal layer [9] in C, which flips gradient sign during backpropagation to force extracted deep features unable to be classified by C. This loss is defined as

$$\mathcal{L}_{cla}(E, G, C) = L_{CE}(C(E(G(E(I_{m_i}, c_i), c_j), c_j)), c_j) \\ + L_{CE}(C(E(I_{m_i}, c_i)), c_i) , \qquad (5)$$

where L_{CE} computes the cross entropy between estimated contrast probability by C and real contrast code. Then the common space loss is defined as

$$\mathcal{L}_{com}(E, G, C) = \lambda_{id}\mathcal{L}_{id} + \lambda_{rec}\mathcal{L}_{rec} + \lambda_{cla}\mathcal{L}_{cla} . \qquad (6)$$

Table 1. Accuracies of different methods for arbitrary cross-contrast MR image translation on IXI dataset, which are averaged over test set and all 6 translation tasks. Each cell is formatted as "mean (standard deviation)".

Method	Ablated version	MAE	PSNR	SSIM
CycleGAN		0.0175 (0.0045)	27.18 (2.17)	0.857 (0.055)
StarGAN		0.0174 (0.0052)	27.77 (2.81)	0.856 (0.068)
DGGAN		0.0175 (0.0049)	27.51 (2.59)	0.858 (0.066)
SUGAN		0.0182 (0.0055)	27.30 (2.88)	0.845 (0.068)
ComboGAN		0.0163 (0.0040)	28.17 (2.34)	0.876 (0.045)
Ours (CIN)	$\mathcal{L}_{adv+cyc}$	0.0145 (0.0039)	28.89 (2.45)	0.897 (0.044)
	$\mathcal{L}_{adv+cyc+id}$	0.0141 (0.0037)	29.12 (2.46)	0.906 (0.040)
	$\mathcal{L}_{adv+cyc+rec}$	0.0144 (0.0043)	29.09 (2.76)	0.900 (0.046)
	$\mathcal{L}_{adv+cyc+cla}$	0.0145 (0.0037)	28.86 (2.32)	0.902 (0.038)
	$\mathcal{L}_{adv+cyc+id+cla}$	0.0138 (0.0037)	29.36 (2.54)	0.909 (0.037)
	$\mathcal{L}_{adv+cyc+id+rec+cla}$	0.0138 (0.0043)	29.45 (3.03)	**0.910 (0.042)**
Ours (FS)	$\mathcal{L}_{adv+cyc}$	0.0154 (0.0032)	28.28 (1.74)	0.897 (0.030)
	$\mathcal{L}_{adv+cyc+id}$	0.0141 (0.0040)	29.24 (2.73)	0.904 (0.041)
	$\mathcal{L}_{adv+cyc+rec}$	0.0140 (0.0047)	29.39 (3.22)	0.908 (0.045)
	$\mathcal{L}_{adv+cyc+cla}$	0.0141 (0.0036)	29.04 (2.47)	0.903 (0.041)
	$\mathcal{L}_{adv+cyc+id+cla}$	0.0135 (0.0033)	29.49 (2.27)	**0.910 (0.034)**
	$\mathcal{L}_{adv+cyc+id+rec+cla}$	**0.0133 (0.0038)**	**29.64 (2.70)**	**0.910 (0.040)**

We heuristically set λ_{id}, λ_{rec} and λ_{cla} to 0.5, 10 and 0.2 to make each term in a similar range of loss values as adversarial loss.

Total Loss. The total training loss of Hyper-GAN is summation of above training losses over all training image pairs in distinct contrasts, which is defined as

$$\mathcal{L}(E, G, C, D_j) = \mathcal{L}_{adv} + \lambda_{cyc}\mathcal{L}_{cyc} + \mathcal{L}_{com}, \tag{7}$$

where λ_{cyc} is set to 10 as per [29]. To optimize $\mathcal{L}$, the networks are divided into two groups, i.e., $\{D_j\}_{j=1}^{N}$ and $\{E, G, C\}$, which are alternately updated, and the networks are totally optimized in 100 epochs using an Adam optimizer with betas of $(0.5, 0.999)$. As in [29], the learning rate is set to 0.0002 with a batch size of 1. Our source code will be released on GitHub.

3   Experiments

3.1   Data Sets

IXI Dataset[1]. We utilize all 319 subjects from Guy's Hospital, and randomly split them into 150, 5 and 164 subjects for training, validation and testing. Each subject contains three contrasts (T1w, T2w and PDw), and only one of three contrasts per subject is used for training to generate unpaired data.

BraTS 2019 Dataset[2]. We use all 150 subjects from CBICA institution, and split them into 100, 5 and 45 subjects for training, validation and testing. Each

---

[1] https://brain-development.org/ixi-dataset/.
[2] https://www.med.upenn.edu/cbica/brats2019.html.

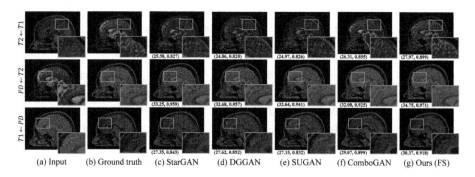

Fig. 3. Visual comparison of cross-contrast MR image translation results on IXI dataset. Top to bottom: T1w-to-T2w, T2w-to-PDw and PDw-to-T1w translation results on a test subject. The values under each sub-image are PSNR and SSIM scores.

subject contains four contrasts (T1w, T1Gd, T2w, FLAIR), and only one of four contrasts per subject is used for training. All volumes of both datasets are N4 corrected and peak normalized, and the intensities are linearly scaled to $[0, 1]$.

3.2 Experimental Results

We compare our Hyper-GAN with the state-of-the-art unpaired multi-contrast image translation methods, including StarGAN [5], DGGAN [22], SUGAN [21] and ComboGAN [2], for arbitrary cross-contrast MR image translation. Our Hyper-GAN includes two versions, i.e., the contrast modulator using filter scaling (FS) or conditional instance normalization (CIN). All experiments are performed on 2D saggital slices. Quantitatively, we compute mean absolute error (MAE), peak signal-to-noise ratio (PSNR), and structural similarity (SSIM) between 3D volumes of ground truth and image translation results. The final accuracy is averaged over test set and all translation tasks between all possible source/target contrast pairs, i.e., 6 tasks for IXI and 12 tasks for BraTS 2019. A paired two-sided Wilcoxon signed-rank test is conducted to compare the performance.

Results on IXI Dataset. Table 1 reports the results on IXI dataset of three contrasts. It shows that, compared with StarGAN, DGGAN, SUGAN and ComboGAN, our Hyper-GAN (i.e., $\mathcal{L}_{adv+cyc+id+rec+cla}$) achieves significantly better performance in all metrics ($p < .001$) and produces 29.45/29.64 (using CIN/FS respectively) in PSNR, comparing favorably with 27.77 of StarGAN, 27.51 of DGGAN, 27.30 of SUGAN and 28.17 of ComboGAN. Specifically, our method achieves the highest accuracies for all 6 translation tasks in all metrics.

Effectiveness of Training Losses. In Table 1, we also compare Hyper-GAN trained with different losses defined in Sect. 2.2. Interestingly, even our baseline (i.e., $\mathcal{L}_{adv+cyc}$) performs better than compared methods ($p < .005$) and obtains 28.89/28.28 in PSNR, justifying effectiveness of our network design with hyper-encoder/hyper-decoder using contrast modulators. Starting from this baseline,

Table 2. Accuracies of different methods for arbitrary cross-contrast MR image translation on BraTS 2019 dataset, which are averaged over test set and all 12 translation tasks. Each cell is formatted as "mean (standard deviation)".

Method	MAE	PSNR	SSIM	# of parameters
StarGAN	0.0083 (0.0029)	30.75 (2.64)	0.905 (0.028)	50.72M
DGGAN	0.0081 (0.0026)	30.81 (2.46)	0.908 (0.023)	58.74M
SUGAN	0.0087 (0.0021)	30.37 (1.97)	0.909 (0.013)	82.63M
ComboGAN	0.0081 (0.0025)	30.55 (2.17)	0.913 (0.015)	71.02M
Ours (CIN)	0.0072 (0.0022)	31.66 (2.46)	0.922 (0.018)	23.71M
Ours (FS)	**0.0070 (0.0026)**	**31.90 (2.74)**	**0.930 (0.018)**	22.57M

(a) Input   (b) Ground truth   (c) StarGAN   (d) DGGAN   (e) SUGAN   (f) ComboGAN   (g) Ours (FS)

Fig. 4. Visual comparison of T1Gd-to-FLAIR image translation results on BraTS 2019 dataset. The values under each sub-image are PSNR and SSIM scores.

each extra loss enforcing common feature space constraint improves results in all metrics except that contrast-classification loss in CIN version produces higher SSIM but comparable MAE and PSNR scores. Specifically, the identical and reconstruction losses respectively improve PSNR from 28.89/28.28 of baseline to 29.12/29.24 and 29.09/29.39. In addition, the PSNR scores are further improved to 29.36/29.49 when both identical and contrast-classification losses are utilized and our whole model performs best. This indicates the effectiveness of our defined common space loss implicitly modeling the shared underlying anatomical structures of each subject. Both FS and CIN versions achieve good performance while FS works generally better. Figure 3 shows visual comparison results.

Results on BraTS 2019 Dataset. Table 2 reports the results on BraTS 2019 dataset of four contrasts. Our Hyper-GAN obtains significantly higher accuracies than compared methods in all metrics ($p < .001$) and achieves 31.66/31.90 (using CIN/FS respectively) in PSNR, comparing favorably with 30.75 of StarGAN, 30.81 of DGGAN, 30.37 of SUGAN and 30.55 of ComboGAN. Specifically, our method works better than all compared methods for all 12 tasks in all metrics, except MAE score of StarGAN for FLAIR-to-T2 translation (0.0103 of StarGAN vs. 0.0105 of ours). Figure 4 shows the T1Gd-to-FLAIR image translation results, and our method performs best even with lesions.

Comparison with CycleGAN. We also compare our Hyper-GAN with Cycle-GAN [29] in Table 1. Note that concatenation of backbones of hyper-encoder and hyper-decoder is identical to generator in CycleGAN as discussed in Sect. 2.1. As CycleGAN is for one-to-one contrast translation, we train $\frac{N \times (N-1)}{2}$ CycleGAN models for a N-contrast translation task, each of which is for a specific pair of contrasts. We average the results over different contrast pairs as final accuracy of CycleGAN. The results show that our Hyper-GAN significantly outperforms CycleGAN ($p < .005$), demonstrating the effectiveness of our network design based on contrast modulator and common feature space constraint.

Computational Efficiency. As shown in Table 2, our parameter size is 2 and 3 times smaller than StarGAN/DGGAN and SUGAN/ComboGAN respectively. Also for BraTS 2019 dataset, CycleGAN requires 169.56M parameters that is 7 times larger than ours. This is because Hyper-GAN utilizes only a single pair of hyper-encoder and hyper-decoder, while the number of encoders/decoders in ComboGAN or CycleGAN scales linearly or quadratically with contrast number.

4 Conclusion

We have proposed a unified GAN for unpaired multi-contrast MR image translation. It can flexibly translate between different contrast pairs using a unified network consisting of hyper-encoder/hyper-decoder and common feature space constraint. This design enables the network to fully investigate the common anatomical structures by common feature space modeling, and contrast-specific imaging by hyper-network design of hyper-encoder and hyper-decoder. It achieves state-of-the-art performance, and outperforms previous GAN-based image translation models that depend on multiple encoders/decoders for different contrast pairs. In the future, we are interested in the extension to multi-contrast, multi-institute setting [7], as well as the combination with segmentation task [16,26,28].

Acknowledgments. This work is supported by the NSFC (12026603, 12026605, 12090021, 61721002, 11690011, U1811461, U20B2075, 11971373).

References

1. Alharbi, Y., Smith, N., Wonka, P.: Latent filter scaling for multimodal unsupervised image-to-image translation. In: IEEE Conference on Computer Vision and Pattern Recognition, pp. 1458–1466 (2019)
2. Anoosheh, A., Agustsson, E., Timofte, R., Van Gool, L.: Combogan: unrestrained scalability for image domain translation. In: IEEE Conference on Computer Vision and Pattern Recognition Workshops, pp. 783–790 (2018)
3. Bernstein, M.A., King, K.F., Zhou, X.J.: Handbook of MRI Pulse Sequences. Elsevier, Amsterdam (2004)
4. Bui, T.D., Nguyen, M., Le, N., Luu, K.: Flow-based deformation guidance for unpaired multi-contrast MRI image-to-image translation. In: Martel, A.L., et al. (eds.) MICCAI 2020. LNCS, vol. 12262, pp. 728–737. Springer, Cham (2020). https://doi.org/10.1007/978-3-030-59713-9_70
5. Choi, Y., Choi, M., Kim, M., Ha, J.W., Kim, S., Choo, J.: Stargan: unified generative adversarial networks for multi-domain image-to-image translation. In: IEEE Conference on Computer Vision and Pattern Recognition, pp. 8789–8797 (2018)
6. Dar, S.U., Yurt, M., Karacan, L., Erdem, A., Erdem, E., Çukur, T.: Image synthesis in multi-contrast MRI with conditional generative adversarial networks. IEEE Trans. Med. Imaging **38**(10), 2375–2388 (2019)

7. Dewey, B.E., et al.: A disentangled latent space for cross-site MRI harmonization. In: Martel, A.L., et al. (eds.) MICCAI 2020. LNCS, vol. 12267, pp. 720–729. Springer, Cham (2020). https://doi.org/10.1007/978-3-030-59728-3_70
8. Dumoulin, V., Shlens, J., Kudlur, M.: A learned representation for artistic style. In: International Conference on Learning Representations (2017)
9. Ganin, Y., Lempitsky, V.: Unsupervised domain adaptation by backpropagation. In: International Conference on Machine Learning, pp. 1180–1189 (2015)
10. Goodfellow, I., et al.: Generative adversarial nets. In: Advances in Neural Information Processing Systems, pp. 2672–2680 (2014)
11. Ha, D., Dai, A.M., Le, Q.V.: Hypernetworks. In: International Conference on Learning Representations (2017)
12. Huang, Y., Shao, L., Frangi, A.F.: DOTE: dual convolutional filter learning for super-resolution and cross-modality synthesis in MRI. In: Descoteaux, M., Maier-Hein, L., Franz, A., Jannin, P., Collins, D.L., Duchesne, S. (eds.) MICCAI 2017. LNCS, vol. 10435, pp. 89–98. Springer, Cham (2017). https://doi.org/10.1007/978-3-319-66179-7_11
13. Huang, Y., Zheng, F., Cong, R., Huang, W., Scott, M.R., Shao, L.: MCMT-GAN: multi-task coherent modality transferable GAN for 3D brain image synthesis. IEEE Trans. Image Process. **29**, 8187–8198 (2020)
14. Hui, L., Li, X., Chen, J., He, H., Yang, J.: Unsupervised multi-domain image translation with domain-specific encoders/decoders. In: International Conference on Pattern Recognition, pp. 2044–2049 (2018)
15. Isola, P., Zhu, J.Y., Zhou, T., Efros, A.A.: Image-to-image translation with conditional adversarial networks. In: IEEE Conference on Computer Vision and Pattern Recognition, pp. 1125–1134 (2017)
16. Jiang, J., Veeraraghavan, H.: Unified cross-modality feature disentangler for unsupervised multi-domain MRI abdomen organs segmentation. In: Martel, A.L., et al. (eds.) MICCAI 2020. LNCS, vol. 12262, pp. 347–358. Springer, Cham (2020). https://doi.org/10.1007/978-3-030-59713-9_34
17. Jog, A., Carass, A., Roy, S., Pham, D.L., Prince, J.L.: Random forest regression for magnetic resonance image synthesis. Med. Image Anal. **35**, 475–488 (2017)
18. Liu, X., et al.: Dual-cycle constrained bijective VAE-GAN for tagged-to-cine magnetic resonance image synthesis. In: International Symposium on Biomedical Imaging (2021)
19. Long, J., Shelhamer, E., Darrell, T.: Fully convolutional networks for semantic segmentation. In: IEEE Conference on Computer Vision and Pattern Recognition, pp. 3431–3440 (2015)
20. Roy, S., Chou, Y.Y., Jog, A., Butman, J.A., Pham, D.L.: Patch based synthesis of whole head MR images: application to EPI distortion correction. In: International Workshop on Simulation and Synthesis in Medical Imaging, pp. 146–156 (2016)
21. Sohail, M., Riaz, M.N., Wu, J., Long, C., Li, S.: Unpaired multi-contrast MR image synthesis using generative adversarial networks. In: International Workshop on Simulation and Synthesis in Medical Imaging, pp. 22–31 (2019)
22. Tang, H., Xu, D., Wang, W., Yan, Y., Sebe, N.: Dual generator generative adversarial networks for multi-domain image-to-image translation. In: Asian Conference on Computer Vision, pp. 3–21 (2018)
23. Vranic, J., Cross, N., Wang, Y., Hippe, D., de Weerdt, E., Mossa-Basha, M.: Compressed sensing-sensitivity encoding (CS-SENSE) accelerated brain imaging: reduced scan time without reduced image quality. Am. J. Neuroradiol. **40**(1), 92–98 (2019)

24. Yu, B., Zhou, L., Wang, L., Shi, Y., Fripp, J., Bourgeat, P.: Ea-GANs: edge-aware generative adversarial networks for cross-modality MR image synthesis. IEEE Trans. Med. Imaging **38**(7), 1750–1762 (2019)
25. Yu, B., Zhou, L., Wang, L., Shi, Y., Fripp, J., Bourgeat, P.: Sample-adaptive GANs: linking global and local mappings for cross-modality MR image synthesis. IEEE Trans. Med. Imaging **39**(7), 2339–2350 (2020)
26. Yuan, W., Wei, J., Wang, J., Ma, Q., Tasdizen, T.: Unified attentional generative adversarial network for brain tumor segmentation from multimodal unpaired images. In: Shen, D., et al. (eds.) MICCAI 2019. LNCS, vol. 11766, pp. 229–237. Springer, Cham (2019). https://doi.org/10.1007/978-3-030-32248-9_26
27. Zaitsev, M., Maclaren, J., Herbst, M.: Motion artifacts in MRI: a complex problem with many partial solutions. J. Magn. Reson. Imaging **42**(4), 887–901 (2015)
28. Zhou, T., Canu, S., Vera, P., Ruan, S.: Brain tumor segmentation with missing modalities via latent multi-source correlation representation. In: Martel, A.L., et al. (eds.) MICCAI 2020. LNCS, vol. 12264, pp. 533–541. Springer, Cham (2020). https://doi.org/10.1007/978-3-030-59719-1_52
29. Zhu, J.Y., Park, T., Isola, P., Efros, A.A.: Unpaired image-to-image translation using cycle-consistent adversarial networks. In: IEEE International Conference on Computer Vision, pp. 2223–2232 (2017)

Generative Self-training
for Cross-Domain Unsupervised
Tagged-to-Cine MRI Synthesis

Xiaofeng Liu[1]($\boxtimes$), Fangxu Xing[1], Maureen Stone[2], Jiachen Zhuo[2],
Timothy Reese[3], Jerry L. Prince[4], Georges El Fakhri[1], and Jonghye Woo[1]

[1] Gordon Center for Medical Imaging, Department of Radiology,
Massachusetts General Hospital and Harvard Medical School,
Boston, MA 02114, USA
{xliu61,JWOO}@mgh.harvard.edu
[2] Department of Neural and Pain Sciences,
University of Maryland School of Dentistry, Baltimore, MD, USA
[3] Athinoula A. Martinos Center for Biomedical Imaging, Department of Radiology,
Massachusetts General Hospital and Harvard Medical School, Boston, MA, USA
[4] Department of Electrical and Computer Engineering, Johns Hopkins University,
Baltimore, MD, USA

Abstract. Self-training based unsupervised domain adaptation (UDA) has shown great potential to address the problem of domain shift, when applying a trained deep learning model in a source domain to unlabeled target domains. However, while the self-training UDA has demonstrated its effectiveness on discriminative tasks, such as classification and segmentation, via the reliable pseudo-label selection based on the softmax discrete histogram, the self-training UDA for generative tasks, such as image synthesis, is not fully investigated. In this work, we propose a novel generative self-training (GST) UDA framework with continuous value prediction and regression objective for cross-domain image synthesis. Specifically, we propose to filter the pseudo-label with an uncertainty mask, and quantify the predictive confidence of generated images with practical variational Bayes learning. The fast test-time adaptation is achieved by a round-based alternative optimization scheme. We validated our framework on the tagged-to-cine magnetic resonance imaging (MRI) synthesis problem, where datasets in the source and target domains were acquired from different scanners or centers. Extensive validations were carried out to verify our framework against popular adversarial training UDA methods. Results show that our GST, with tagged MRI of test subjects in new target domains, improved the synthesis quality by a large margin, compared with the adversarial training UDA methods.

1 Introduction

Deep learning has advanced state-of-the-art machine learning approaches and excelled at learning representations suitable for numerous discriminative and

© Springer Nature Switzerland AG 2021
M. de Bruijne et al. (Eds.): MICCAI 2021, LNCS 12903, pp. 138–148, 2021.
https://doi.org/10.1007/978-3-030-87199-4_13

generative tasks [14,21,22,29]. However, a deep learning model trained on labeled data from a source domain, in general, performs poorly on unlabeled data from unseen target domains, partly because of discrepancies between source and target data distributions, i.e., domain shift [15]. The problem of domain shift in medical imaging arises, because data are often acquired from different scanners, protocols, or centers [17]. This issue has motivated many researchers to investigate unsupervised domain adaptation (UDA), which aims to transfer knowledge learned from a labeled source domain to different but related unlabeled target domains [30,33].

There has been a great deal of work to alleviate the domain shift using UDA [30]. Early methods attempted to learn domain-invariant representations or to take instance importance into consideration to bridge the gap between the source and target domains. In addition, due to the ability of deep learning to disentangle explanatory factors of variations, efforts have been made to learn more transferable features. Recent works in UDA incorporated discrepancy measures into network architectures to align feature distributions between source and target domains [18,19]. This was achieved by either minimizing the distribution discrepancy between feature distribution statistics, e.g., maximum mean discrepancy (MMD), or adversarially learning the feature representations to fool a domain classifier in a two-player minimax game [18].

Recently, self-training based UDA presents a powerful means to counter unknown labels in the target domain [33], surpassing the adversarial learning-based methods in many discriminative UDA benchmarks, e.g., classification and segmentation (i.e., pixel-wise classification) [23,26,31]. The core idea behind the deep self-training based UDA is to iteratively generate a set of one-hot (or smoothed) pseudo-labels in the target domain, followed by retraining the network based on these pseudo-labels with target data [33]. Since outputs of the previous round can be noisy, it is critical to only select the high confidence prediction as reliable pseudo-label. In discriminative self-training with softmax output unit and cross-entropy objective, it is natural to define the confidence for a sample as the max of its output softmax probabilities [33]. Calibrating the uncertainty of the regression task, however, can be more challenging. Because of the insufficient target data and unreliable pseudo-labels, there can be both *epistemic* and *aleatoric* uncertainties [3] in self-training UDA. In addition, while the self-training UDA has demonstrated its effectiveness on classification and segmentation, via the reliable pseudo-label selection based on the softmax discrete histogram, the same approach for generative tasks, such as image synthesis, is underexplored.

In this work, we propose a novel generative self-training (GST) UDA framework with continuous value prediction and regression objective for tagged-to-cine magnetic resonance (MR) image synthesis. More specifically, we propose to filter the pseudo-label with an uncertainty mask, and quantify the predictive confidence of generated images with practical variational Bayes learning. The fast test-time adaptation is achieved by a round-based alternative optimization scheme. Our contributions are summarized as follows:

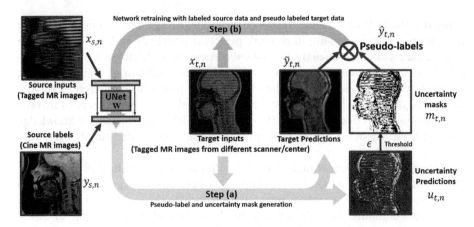

Fig. 1. Illustration of our generative self-training UDA for tagged-to-cine MR image synthesis. In each iteration, two-step alternative training is carried out.

- We propose to achieve cross-scanner and cross-center test-time UDA of tagged-to-cine MR image synthesis, which can potentially reduce the extra cine MRI acquisition time and cost.
- A novel GST UDA scheme is proposed, which controls the confident pseudo-label (continuous value) selection with a practical Bayesian uncertainty mask. Both the aleatoric and epistemic uncertainties in GST UDA are investigated.
- Both quantitative and qualitative evaluation results, using a total of 1,768 paired slices of tagged and cine MRI from the source domain and tagged MR slices of target subjects from the cross-scanner and cross-center target domain, demonstrate the validity of our proposed GST framework and its superiority to conventional adversarial training based UDA methods.

2   Methodology

In our setting of the UDA image synthesis, we have paired resized tagged MR images, $\mathbf{x}_s \in \mathbb{R}^{256 \times 256}$, and cine MR images, $\mathbf{y}_s \in \mathbb{R}^{256 \times 256}$, indexed by $s = 1, 2, \cdots, S$, from the source domain $\{\mathbf{X}_S, \mathbf{Y}_S\}$, and target samples $\mathbf{x}_t \in \mathbb{R}^{256 \times 256}$ from the unlabeled target domain $\mathbf{X}_T$, indexed by $t = 1, 2, \cdots, T$. In both training and testing, the ground-truth target labels, i.e., cine MR images in the target domain, are inaccessible, and the pseudo-label $\hat{\mathbf{y}}_t \in \mathbb{R}^{256 \times 256}$ of $\mathbf{x}_t$ is iteratively generated in a self-training scheme [16,33]. In this work, we adopt the U-Net-based Pix2Pix [9] as our translator backbone, and initialize the network parameters $\mathbf{w}$ with the pre-training using the labeled source domain $\{\mathbf{X}_S, \mathbf{Y}_S\}$. In what follows, alternative optimization based self-training is applied to gradually update the U-Net part for the target domain image synthesis by training on both $\{\mathbf{X}_S, \mathbf{Y}_S\}$ and $\mathbf{X}_T$. Figure 1 illustrates the proposed algorithm flow, which is detailed below.

2.1   Generative Self-training UDA

The conventional self-training regards the pseudo-label $\hat{y}_t$ as a learnable latent variable in the form of a categorical histogram, and assigns all-zero vector label for the uncertain samples or pixels to filter them out for loss calculation [16, 33]. Since not all pseudo-labels are reliable, we define a confidence threshold to progressively select confident pseudo-labels [32]. This is akin to self-paced learning that learns samples in an easy-to-hard order [12,27]. In classification or segmentation tasks, the confidence can be simply measured by the maximum softmax output histogram probability [33]. The output of a generation task, however, is continuous values and thus setting the pseudo-label as 0 cannot drop the uncertain sample in the regression loss calculation.

Therefore, we first propose to formulate the generative self-training as a unified regression loss minimization scheme, where pseudo-labels can be a pixelwise continuous value and indicate the uncertain pixel with an uncertainty mask $\mathbf{m}_t = \{m_{t,n}\}_{n=1}^{256 \times 256}$, where n indexes the pixel in the images, and $m_{t,n} \in \{0,1\}, \forall t, n$:

$$\min_{\mathbf{w},\mathbf{m}_t} \underbrace{\sum_{s \in S}\sum_{n=1}^{N}||y_{s,n} - \tilde{y}_{s,n}||_2^2}_{\mathcal{L}_{reg}^s(\mathbf{w})} + \underbrace{\sum_{t \in T}\sum_{n=1}^{N}||(\hat{y}_{t,n} - \tilde{y}_{t,n})m_{t,n}||_2^2}_{\mathcal{L}_{reg}^t(\mathbf{w},\mathbf{m}_t)} \tag{1}$$

$$s.t. \ \ m_{t,n} = \begin{cases} 1 & u_{t,n} < \epsilon \\ 0 & u_{t,n} > \epsilon \end{cases}; \ \ \epsilon = \min\{\text{top } p\% \text{ sorted } u_{t,n}\}, \tag{2}$$

where $x_{s,n}, y_{s,n}, x_{t,n}, \hat{y}_{t,n} \in [0, 255]$. For example, $y_{s,n}$ indicates the n-th pixel of the s-th source domain ground-truth cine MR image $\mathbf{y}_s$. $\tilde{y}_{s,n}$ and $\tilde{y}_{t,n}$ represent the generated source and target images, respectively. $\mathcal{L}_{reg}^s(\mathbf{w})$ and $\mathcal{L}_{reg}^t(\mathbf{w},\mathbf{m}_t)$ are the regression loss of the source and target domain samples, respectively. Notably, there is only one network parameterized with $\mathbf{w}$, which is updated with the loss in both domains. $u_{t,n}$ is the to-be estimated uncertainty of a pixel and determines the value of the uncertainty mask $m_{t,n}$ with a threshold ϵ. ϵ is a critical parameter to control pseudo-label learning and selection, which is determined by a single meta portion parameter p, indicating the portion of pixels to be selected in the target domain. Empirically, we define ϵ in each iteration, by sorting $u_{t,n}$ in increasing order and set ϵ to minimum $u_{t,n}$ of the top $p \in [0, 1]$ percentile rank.

2.2   Bayesian Uncertainty Mask for Target Samples

Determining the mask value $m_{t,n}$ for the target sample requires the uncertainty estimation of $u_{t,n}$ in our self-training UDA. Notably, the lack of sufficient target domain data can result in the *epistemic* uncertainty w.r.t. the model parameters, while the noisy pseudo-label can lead to the *aleatoric* uncertainty [3,8,11].

To counter this, we model the *epistemic* uncertainty via Bayesian neural networks which learn a posterior distribution $p(\mathbf{w}|\mathbf{X}_T, \hat{\mathbf{Y}}_T)$ over the probabilistic model parameters rather than a set of deterministic parameters [25]. In particular, a tractable solution is to replace the true posterior distribution with a variational approximation $q(\mathbf{w})$, and dropout variational inference can be a practical technique. This can be seen as using the Bernoulli distribution as the approximation distribution $q(\mathbf{w})$ [5]. The K times prediction with independent dropout sampling is referred to as Monte Carlo (MC) dropout. We use the mean squared error (MSE) to measure the epistemic uncertainty as in [25], which assesses a one-dimensional regression model similar to [4]. Therefore, the epistemic uncertainty with MSE of each pixel with K times dropout generation is given by

$$u_{t,n}^{epistemic} = \frac{1}{K}\sum_{k=1}^{K}||\tilde{y}_{t,n} - \mu_{t,n}||_2^2; \quad \mu_{t,n} = \frac{1}{K}\sum_{k=1}^{K}\tilde{y}_{t,n}, \quad (3)$$

where $\mu_{t,n}$ is the predictive mean of $\tilde{y}_{t,n}$.

Because of the different hardness and divergence and because the pseudo-label noise can vary for different $\mathbf{x}_t$, the heteroscedastic *aleatoric* uncertainty modeling is required [13,24]. In this work, we use our network to transform $\mathbf{x}_t$, with its head split to predict both $\tilde{\mathbf{y}}_t$ and the variance map $\sigma_t^2 \in \mathbb{R}^{256\times256}$; and its element $\sigma_{t,n}^2$ is the predicted variance for the n-th pixel. We do not need "uncertainty labels" to learn σ_t^2 prediction. Rather, we can learn σ_t^2 implicitly from a regression loss function [11,13]. The masked regression loss can be formulated as

$$\mathcal{L}_{reg}^t(\mathbf{w},\mathbf{m}_t,\sigma_t^2) = \sum_{t\in T}\sum_{n=1}^{N}(\frac{1}{\sigma_{t,n}^2}||(\hat{y}_{t,n} - \tilde{y}_{t,n})m_{t,n}||_2^2 + \beta\log\sigma_{t,n}^2), \quad (4)$$

which consists of a variance normalized residual regression term and an uncertainty regularization term. The second regularization term keeps the network from predicting an infinite uncertainty, i.e., zero loss, for all the data points. Then, the averaged aleatoric uncertainty of K times MC dropout can be measured by $u_{t,n}^{aleatoric} = \frac{1}{K}\sum_{k=1}^{K}\sigma_{t,n}^2$ [11,13].

Moreover, minimizing Eq. (4) can be regarded as the Lagrangian with a multiplier β of $\min_{\mathbf{w}}\sum_{t\in T}\sum_{n=1}^{N}\frac{1}{\sigma_{t,n}^2}||(\hat{y}_{t,n} - \tilde{y}_{t,n})m_{t,n}||_2^2; \quad s.t. \sum_{t\in T}\sum_{n=1}^{N}\log\sigma_{t,n}^2 < C^1$, where $C \in \mathbb{R}^+$ indicates the strength of the applied constraint. The condition term essentially controls the target domain predictive uncertainty, which is helpful for UDA [7]. Our final pixel-wise self-training UDA uncertainty $u_{t,n} = u_{t,n}^{epistemic} + u_{t,n}^{aleatoric}$ is a combination of the two uncertainties [11].

2.3   Training Protocol

As pointed out in [6], directly optimizing the self-training objectives can be difficult and thus the deterministic annealing expectation maximization (EM)

---

[1] It can be rewritten as $\min_{\mathbf{w}} \mathcal{F} = \{\sum_{t\in T}\sum_{n=1}^{N}\frac{1}{\sigma_{t,n}^2}||(\hat{y}_{t,n}-\tilde{y}_{t,n})m_{t,n}||_2^2 + \beta(\sum_{t\in T}\sum_{n=1}^{N}\log\sigma_{t,n}^2 - C)\}$. Since $\beta, C \geq 0$, an upper bound on $\mathcal{F}$ can be obtained as $\mathcal{F} \leq \mathcal{L}_{reg}^t$.

algorithms are often used instead. Specifically, the generative self-training can be solved by alternating optimization based on the following **a)** and **b)** steps.

a) Pseudo-label and uncertainty mask generation. With the current $\mathbf{w}$, apply the MC dropout for K times image translation of each target domain tagged MR image $\mathbf{x}_t$. We estimate the pixel-wise uncertainty $u_{t,n}$, and calculate the uncertainty mask $\mathbf{m}_t$ with the threshold ϵ. We set the pseudo-label of the selected pixel in this round as $\hat{y}_{t,n} = \mu_{t,n}$, i.e., the average value of K outputs.

b) Network w retraining. Fix $\hat{\mathbf{Y}}_T = \{\hat{\mathbf{y}}_t\}_{t=1}^T$, $\mathbf{M}_T = \{\mathbf{m}_t\}_{t=1}^T$ and solve:

$$\min_{\mathbf{w}} \sum_{s \in S} \sum_{n=1}^N ||y_{s,n} - \tilde{y}_{s,n}||_2^2 + \sum_{t \in T} \sum_{n=1}^N (\frac{1}{\sigma_{t,n}^2}||(\hat{y}_{t,n} - \tilde{y}_{t,n})m_{t,n}||_2^2 + \beta\log\sigma_{t,n}^2)$$

(5)

to update $\mathbf{w}$. Carrying out step **a)** and **b)** for one time is defined as one round in self-training. Intuitively, step **a)** is equivalent to simultaneously conducting pseudo-label learning and selection. In order to solve step **b)**, we can use a typical gradient method, e.g. Stochastic Gradient Descent (SGD). The meta parameter p is linearly increasing from 30% to 80% alongside the training to incorporate more pseudo-labels in the subsequent rounds as in [33].

3   Experiments and Results

We evaluated our framework on both cross-scanner and cross-center tagged-to-cine MR image synthesis tasks. For the labeled source domain, a total of 1,768 paired tagged and cine MR images from 10 healthy subjects at clinical center A were acquired. We followed the test time UDA setting [10], which uses only one unlabeled target subject in UDA training and testing.

For fair comparison, we adopted Pix2Pix [9] for our source domain training as in [20], and used the trained U-Net as the source model for all of the comparison methods. In order to align the absolute value of each loss, we empirically set weight $\beta = 1$ and $K = 20$. Our framework was implemented using the PyTorch deep learning toolbox. The GST training was performed on a V100 GPU, which took about 30 min. We note that K times MC dropout can be processed parallel. In each iteration, we sampled the same number of source and target domain samples.

3.1   Cross-Scanner Tagged-to-Cine MR Image Synthesis

In the cross-scanner image synthesis setting, a total of 1,014 paired tagged and cine MR images from 5 healthy subjects in the target domain were acquired at clinical center A with a different scanner. As a result, there was an appearance discrepancy between the source and target domains.

The synthesis results using source domain Pix2Pix [9] without UDA training, gradually adversarial UDA (GAUDA) [2], and our proposed framework are

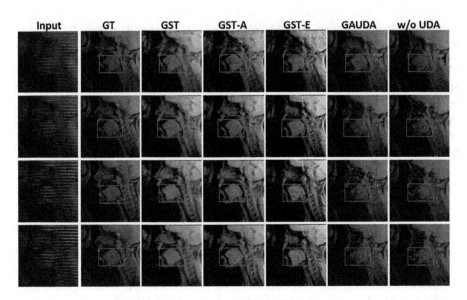

Fig. 2. Comparison of different UDA methods on the cross-scanner tagged-to-cine MR image synthesis task, including our proposed GST, GST-A, and GST-E, adversarial UDA [2]*, and Pix2Pix [9] without adaptation. * indicates the first attempt at tagged-to-cine MR image synthesis. GT indicates the ground-truth.

shown in Fig. 2. Note that GAUDA with source domain initialization took about 2 h for the training, which was four times slower than our GST framework. In addition, it was challenging to stabilize the adversarial training [1], thus yielding checkerboard artifacts. Furthermore, the hallucinated content with the domain-wise distribution alignment loss produced a relatively significant difference in shape and texture within the tongue between the real cine MR images. By contrast, our framework achieved the adaptation with relatively limited target data in the test time UDA setting [10], with faster convergence time. In addition, our framework did not rely on adversarial training, generating visually pleasing results with better structural consistency as shown in Fig. 2, which is crucial for subsequent analyses such as segmentation.

For an ablation study, in Fig. 2, we show the performance of GST without the aleatoric or epistemic uncertainty for the uncertainty mask, i.e., GST-A or GST-E. Without measuring the aleatoric uncertainty caused by the inaccurate label, GST-A exhibited a small distortion of the shape and boundary. Without measuring the epistemic uncertainty, GST-E yielded noisier results than GST.

The synthesized images were expected to have realistic-looking textures, and to be structurally cohesive with their corresponding ground truth images. For quantitative evaluation, we adopted widely used evaluation metrics: mean L1 error, structural similarity index measure (SSIM), peak signal-to-noise ratio (PSNR), and unsupervised inception score (IS) [20]. Table 1 lists numerical comparisons using 5 testing subjects. The proposed GST outperformed GAUDA [2] and ADDA [28] w.r.t. L1 error, SSIM, PSNR, and IS by a large margin.

Table 1. Numerical comparisons of cross-scanner and cross-center evaluations. $\pm$ standard deviation is reported over three evaluations.

Methods	Cross-scanner				Cross-center
	L1 $\downarrow$	SSIM $\uparrow$	PSNR $\uparrow$	IS $\uparrow$	IS $\uparrow$
w/o UDA [9]	176.4 ± 0.1	0.8325 ± 0.0012	26.31 ± 0.05	8.73 ± 0.12	5.32 ± 0.11
ADDA [28]	168.2 ± 0.2	0.8784 ± 0.0013	33.15 ± 0.04	10.38 ± 0.11	8.69 ± 0.10
GAUDA [2]	161.7 ± 0.1	0.8813 ± 0.0012	33.27 ± 0.06	10.62 ± 0.13	8.83 ± 0.14
GST	$\mathbf{158.6} \pm 0.2$	$\mathbf{0.9078} \pm 0.0011$	$\mathbf{34.48} \pm 0.05$	$\mathbf{12.63} \pm 0.12$	$\mathbf{9.76} \pm 0.11$
GST-A	159.5 ± 0.3	0.8997 ± 0.0011	34.03 ± 0.04	12.03 ± 0.12	9.54 ± 0.13
GST-E	159.8 ± 0.1	0.9026 ± 0.0013	34.05 ± 0.05	11.95 ± 0.11	9.58 ± 0.12

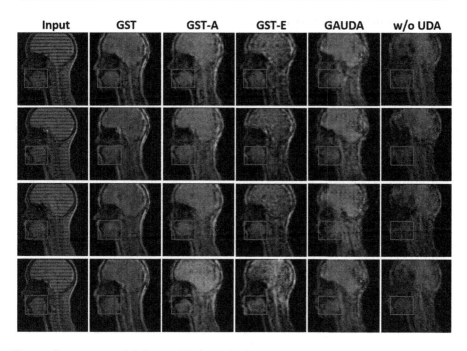

Fig. 3. Comparison of different UDA methods on the cross-center tagged-to-cine MR image synthesis task, including our proposed GST, GST-A, and GST-E, adversarial UDA [2]*, and Pix2Pix [9] without adaptation. * indicates the first attempt at tagged-to-cine MR image synthesis.

3.2   Cross-Center Tagged-to-Cine MR Image Synthesis

To further demonstrate the generality of our framework for the cross-center tagged-to-cine MR image synthesis task, we collected 120 tagged MR slices of a subject at clinical center B with a different scanner. As a result, the data at clinical center B had different soft tissue contrast and tag spacing, compared with clinical center A, and the head position was also different.

The qualitative results in Fig. 3 show that the anatomical structure of the tongue is better maintained using our framework with both the aleatoric and epistemic uncertainties. Due to the large domain gap present in the datasets between the two centers, the overall synthesis quality was not as good as the cross-scanner image synthesis task, as visually assessed. In Table 1, we provide the quantitative comparison using IS, which does not need the paired ground truth cine MR images [20]. Consistently with the cross-scanner setting, our GST outperformed adversarial training methods, including GAUDA and ADDA [2, 28], indicating the self-training can be a powerful technique for the generative UDA task, similar to the conventional discriminative self-training [16,33].

4   Discussion and Conclusion

In this work, we presented a novel generative self-training framework for UDA and applied the framework to cross-scanner and cross-center tagged-to-MR image synthesis tasks. With a practical yet principled Bayesian uncertainty mask, our framework was able to control the confident pseudo-label selection. In addition, we systematically investigated both the aleatoric and epistemic uncertainties in generative self-training UDA. Our experimental results demonstrated that our framework yielded the superior performance, compared with the popular adversarial training UDA methods, as quantitatively and qualitatively assessed. The synthesized cine MRI with test time UDA can potentially be used to segment the tongue and to observe surface motion, without the additional acquisition cost and time.

Acknowledgments. This work is supported by NIH R01DC014717, R01DC018511, and R01CA133015.

References

1. Che, T., et al.: Deep verifier networks: verification of deep discriminative models with deep generative models. In: AAAI (2021)
2. Cui, S., Wang, S., Zhuo, J., Su, C., Huang, Q., Tian, Q.: Gradually vanishing bridge for adversarial domain adaptation. In: Proceedings of the IEEE/CVF Conference on Computer Vision and Pattern Recognition, pp. 12455–12464 (2020)
3. Der Kiureghian, A., Ditlevsen, O.: Aleatory or epistemic? Does it matter? Struct. Saf. **31**(2), 105–112 (2009)
4. Fruehwirt, W., et al.: Bayesian deep neural networks for low-cost neurophysiological markers of Alzheimer's disease severity. arXiv preprint arXiv:1812.04994 (2018)
5. Gal, Y., Ghahramani, Z.: Bayesian convolutional neural networks with Bernoulli approximate variational inference. arXiv preprint arXiv:1506.02158 (2015)
6. Grandvalet, Y., Bengio, Y.: Entropy regularization (2006)
7. Han, L., Zou, Y., Gao, R., Wang, L., Metaxas, D.: Unsupervised domain adaptation via calibrating uncertainties. In: Proceedings of the IEEE Conference on Computer Vision and Pattern Recognition Workshops, pp. 99–102 (2019)

8. Hu, S., Worrall, D., Knegt, S., Veeling, B., Huisman, H., Welling, M.: Supervised uncertainty quantification for segmentation with multiple annotations. In: Shen, D., et al. (eds.) MICCAI 2019. LNCS, vol. 11765, pp. 137–145. Springer, Cham (2019). https://doi.org/10.1007/978-3-030-32245-8_16

9. Isola, P., Zhu, J.Y., Zhou, T., Efros, A.A.: Image-to-image translation with conditional adversarial networks. In: CVPR, pp. 1125–1134 (2017)

10. Karani, N., Erdil, E., Chaitanya, K., Konukoglu, E.: Test-time adaptable neural networks for robust medical image segmentation. Med. Image Anal. **68**, 101907 (2021)

11. Kendall, A., Gal, Y.: What uncertainties do we need in Bayesian deep learning for computer vision? arXiv preprint arXiv:1703.04977 (2017)

12. Kumar, M.P., Packer, B., Koller, D.: Self-paced learning for latent variable models. In: Advances in Neural Information Processing Systems, pp. 1189–1197 (2010)

13. Le, Q.V., Smola, A.J., Canu, S.: Heteroscedastic Gaussian process regression. In: Proceedings of the 22nd International Conference on Machine Learning, pp. 489–496 (2005)

14. Liu, X., et al.: Unimodal regularized neuron stick-breaking for ordinal classification. Neurocomputing **388**, 34–44 (2020)

15. Liu, X., et al.: Domain generalization under conditional and label shifts via variational Bayesian inference. In: IJCAI (2021)

16. Liu, X., Hu, B., Liu, X., Lu, J., You, J., Kong, L.: Energy-constrained self-training for unsupervised domain adaptation. In: ICPR (2020)

17. Liu, X., et al.: Subtype-aware unsupervised domain adaptation for medical diagnosis. In: AAAI (2021)

18. Liu, X., Xing, F., Yang, C., El Fakhri, G., Woo, J.: Adapting off-the-shelf source segmenter for target medical image segmentation. In: de Bruijne, M., et al. (eds.) MICCAI 2021, LNCS 12902, pp. 549–559. Springer, Cham (2021)

19. Liu, X., Xing, F., El Fakhri, G., Woo, J.: A unified conditional disentanglement framework for multimodal brain MR image translation. In: ISBI, pp. 10–14. IEEE (2021)

20. Liu, X., et al.: Dual-cycle constrained bijective VAE-GAN for tagged-to-cine magnetic resonance image synthesis. In: ISBI (2021)

21. Liu, X., Xing, F., Yang, C., Kuo, C.-C.J., El Fakhri, G., Woo, J.: Symmetric-constrained irregular structure inpainting for brain MRI registration with tumor pathology. In: Crimi, A., Bakas, S. (eds.) BrainLes 2020. LNCS, vol. 12658, pp. 80–91. Springer, Cham (2021). https://doi.org/10.1007/978-3-030-72084-1_8

22. Liu, X., Zou, Y., Song, Y., Yang, C., You, J., Kumar, B.V.K.V.: Ordinal regression with neuron stick-breaking for medical diagnosis. In: Leal-Taixé, L., Roth, S. (eds.) ECCV 2018. LNCS, vol. 11134, pp. 335–344. Springer, Cham (2019). https://doi.org/10.1007/978-3-030-11024-6_23

23. Mei, K., Zhu, C., Zou, J., Zhang, S.: Instance adaptive self-training for unsupervised domain adaptation. In: Vedaldi, A., Bischof, H., Brox, T., Frahm, J.-M. (eds.) ECCV 2020. LNCS, vol. 12371, pp. 415–430. Springer, Cham (2020). https://doi.org/10.1007/978-3-030-58574-7_25

24. Nix, D.A., Weigend, A.S.: Estimating the mean and variance of the target probability distribution. In: Proceedings of 1994 IEEE International Conference on Neural Networks (ICNN 1994), vol. 1, pp. 55–60. IEEE (1994)

25. Rasmussen, C.E.: Gaussian processes in machine learning. In: Bousquet, O., von Luxburg, U., Rätsch, G. (eds.) ML -2003. LNCS (LNAI), vol. 3176, pp. 63–71. Springer, Heidelberg (2004). https://doi.org/10.1007/978-3-540-28650-9_4

26. Shin, I., Woo, S., Pan, F., Kweon, I.S.: Two-phase pseudo label densification for self-training based domain adaptation. In: Vedaldi, A., Bischof, H., Brox, T., Frahm, J.-M. (eds.) ECCV 2020. LNCS, vol. 12358, pp. 532–548. Springer, Cham (2020). https://doi.org/10.1007/978-3-030-58601-0_32
27. Tang, K., Ramanathan, V., Fei-Fei, L., Koller, D.: Shifting weights: adapting object detectors from image to video. In: NIPS (2012)
28. Tzeng, E., Hoffman, J., Saenko, K., Darrell, T.: Adversarial discriminative domain adaptation. In: CVPR (2017)
29. Wang, J., et al.: Automated interpretation of congenital heart disease from multi-view echocardiograms. Med. Image Anal. **69**, 101942 (2021)
30. Wang, M., Deng, W.: Deep visual domain adaptation: a survey. Neurocomputing **312**, 135–153 (2018)
31. Wei, C., Shen, K., Chen, Y., Ma, T.: Theoretical analysis of self-training with deep networks on unlabeled data. arXiv preprint arXiv:2010.03622 (2021)
32. Zhu, X.: Semi-supervised learning tutorial. In: ICML Tutorial (2007)
33. Zou, Y., Yu, Z., Liu, X., Kumar, B., Wang, J.: Confidence regularized self-training. In: Proceedings of the IEEE/CVF International Conference on Computer Vision, pp. 5982–5991 (2019)

Cooperative Training and Latent Space Data Augmentation for Robust Medical Image Segmentation

Chen Chen[1(✉)], Kerstin Hammernik[1,2], Cheng Ouyang[1], Chen Qin[3], Wenjia Bai[4,5], and Daniel Rueckert[1,2]

[1] BioMedIA Group, Department of Computing, Imperial College London, London, UK
chen.chen15@imperial.ac.uk
[2] Klinikum rechts der Isar, Technical University of Munich, Munich, Germany
[3] Institute for Digital Communications, University of Edinburgh, Edinburgh, UK
[4] Data Science Institute, Imperial College London, London, UK
[5] Department of Brain Sciences, Imperial College London, London, UK

Abstract. Deep learning-based segmentation methods are vulnerable to unforeseen data distribution shifts during deployment, e.g. change of image appearances or contrasts caused by different scanners, unexpected imaging artifacts etc. In this paper, we present a cooperative framework for training image segmentation models and a latent space augmentation method for generating hard examples. Both contributions improve model generalization and robustness with limited data. The cooperative training framework consists of a fast-thinking network (FTN) and a slow-thinking network (STN). The FTN learns decoupled image features and shape features for image reconstruction and segmentation tasks. The STN learns shape priors for segmentation correction and refinement. The two networks are trained in a cooperative manner. The latent space augmentation generates challenging examples for training by masking the decoupled latent space in both channel-wise and spatial-wise manners. We performed extensive experiments on public cardiac imaging datasets. Using only 10 subjects from a *single* site for training, we demonstrated improved cross-site segmentation performance, and increased robustness against various unforeseen imaging artifacts compared to strong baseline methods. Particularly, cooperative training with latent space data augmentation yields 15% improvement in terms of average Dice score when compared to a standard training method.

1 Introduction

Segmenting anatomical structures from medical images is an important step for diagnosis, treatment planning and clinical research. In recent years, deep

Electronic supplementary material The online version of this chapter (https://doi.org/10.1007/978-3-030-87199-4_14) contains supplementary material, which is available to authorized users.

© Springer Nature Switzerland AG 2021
M. de Bruijne et al. (Eds.): MICCAI 2021, LNCS 12903, pp. 149–159, 2021.
https://doi.org/10.1007/978-3-030-87199-4_14

convolutional neural networks (CNNs) have been widely adopted to automate the segmentation procedure [1,2]. However, a major obstacle for deploying deep learning-based methods to real-world applications is domain shift during clinical deployment, which includes changes of image appearance and contrasts across medical centers and scanners as well as various imaging artefacts. Recent works on domain generalization provide a promising direction to address this issue [3–7]. A majority of them require training data from *multiple* domains to learn domain-invariant features for segmentation. Multi-domain datasets, however, may not always be feasible due to data privacy concerns and collection costs. Learning robust networks from single-domain data and limited data is of great practical value for medical imaging research.

In this work, we propose a novel cooperative training framework for learning a robust segmentation network from *single-domain* data. We make the following contributions. (1) First, to improve model performance on unseen domains, we design a cooperative training framework where two networks collaborate in both training and testing. This is inspired by the two-system model in human behaviour sciences [8], where a fast-thinking system makes intuitive judgment and a slow-thinking system corrects it with logical inference. Such a collaboration is essential for humans to deal with unfamiliar situations. In our framework, a fast-thinking network (FTN) aims to understand the context of images and extracts task-related image and shape features for an initial segmentation. Subsequently, a slow-thinking network (STN) refines the initial segmentation according to a learned shape prior. (2) We introduce a latent space data augmentation (DA) method, which performs channel-wise and spacial-wise masking for the latent code learnt from FTN in random and targeted fashions. Reconstructing images with masked latent codes generates a diverse set of challenging images and corrupted segmentation maps to reinforce the training of both networks. Experimental results on cardiac imaging datasets show the cooperative training mechanism with generated challenging examples can effectively enhance FTN's segmentation capacity and STN's shape correction ability, leading to more robust segmentation. (3) The proposed method alleviates the need for multi-domain data and expertise for data augmentation, making it applicable to a wide range of applications.

Related Work. Our work is conceptually related to DA, multi-task learning (MTL). *a) DA* applies transformations or perturbations to improve the diversity of training data, which is effective for improving model generalization [9]. A large number of the works focuses on image-space DA, including both intensity and geometric transformation functions [10,11] and patch-wise perturbations [12–15]. Adversarial DA has also been explored, which takes the segmentation network into account and generates adversarial examples that can fool the network [16–19]. A major novelty of our work is that we perform DA in the latent space. The latent space contains abstract representation of both image and shape features and challenging examples can be generated by manipulating this space. Different from existing latent DA methods used in metric learning [20], our method is based on feature masking rather than feature interpolation and thus does not

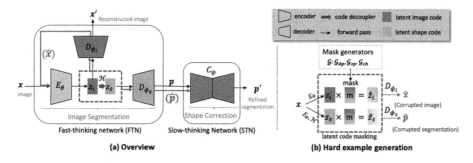

Fig. 1. (a) The proposed cooperative training framework, which consists of a fast-thinking network (FTN) and a slow-thinking network (STN). (b) Hard example generation in latent space. Latent code masking is performed for generating both corrupted images and segmentations for cooperative training.

require paired images from the same/different categories to generate synthetic data. To the best of our knowledge, our work is the first to explore latent space DA for robust segmentation with single domain data. *b) MTL* is extremely beneficial when training data is limited [14,15,21]. MTL enhances network capacity by encouraging the learning of common semantic features across various tasks. Several methods consist of two stages of segmentation: a first network for coarse segmentation from images and a second network for refinement, where two networks are trained independently [22,23]. For example, in [22], manually designed functions are used to generate poor segmentation and a denoising autoencoder is independently trained for segmentation refinement. Another novelty of our work is that we seek the mutual benefits of a segmentation network and a denoising network by training them cooperatively, using hard examples constructed from latent space.

2   Methodology

Given a training dataset from *one, single* domain $D_{tr} = \{(\mathbf{x}_i, \mathbf{y}_i)\}_{i=1}^n$, with pairs of images $\mathbf{x}_i \in \mathbb{R}^{H \times W}$ and one-hot encoded C-class label maps $\mathbf{y}_i \in \{0,1\}^{H \times W \times C}$ as ground truth (GT), our goal is to learn a robust segmentation network across various 'unseen' domains with different image appearance and/or quality. Here, H, W denote image height and width, respectively.

2.1   Overview of the Framework

An overview of the proposed framework is illustrated in Fig. 1(a). At a high level, our framework consists of a fast-thinking network (FTN) and a slow-thinking network (STN). Given an image $\mathbf{x}$, the FTN extracts task-specific shape features $\mathbf{z}_s$ to perform the segmentation task and image contextual features $\mathbf{z}_i$ to perform the image reconstruction task. This network consists of a shared encoder E_θ, a feature decoupler $\mathcal{H}$ and two task-specific decoders $\mathcal{D}_{\phi_s}$ and $\mathcal{D}_{\phi_i}$ for image segmentation and reconstruction tasks. We apply the latent code decoupler $\mathcal{H}$ to $\mathbf{z}_i$, so that task-unrelated information (e.g. image texture information, brightness) is deactivated in $\mathbf{z}_s$. This encourages a sparse latent code $\mathbf{z}_s$, which is beneficial for model robustness [24]. $\mathcal{H}$ employs a stack of two convolutional layers followed by a ReLU activation function. Please find Supple. Fig. 5 for more details. STN is a denoising autoencoder network $\mathcal{C}_\psi$, which corrects the segmentation predicted by FTN by using a learned shape prior encoded in $\mathcal{C}_\psi$. At inference time, we first employ FTN to perform fast segmentation for a given image $\mathbf{x}$: $\mathbf{p} = \mathcal{D}_{\phi_s}(\mathcal{H}(E_\theta(\mathbf{x})))$, and then STN to refine the prediction for improved segmentation quality: $\mathbf{p}' = \mathcal{C}_\psi(\mathbf{p})$.

2.2   Standard Training

To train the two networks, we propose a standard approach which jointly trains the three encoder-decoder pairs with a supervised multi-task loss function for image reconstruction $\mathcal{L}_{rec}$, image segmentation $\mathcal{L}_{seg}$ and shape correction $\mathcal{L}_{shp}$. The loss is defined as:

$$\mathcal{L}_{std} = \mathbb{E}_{(\mathbf{x},\mathbf{y})\in D_{tr}}[\mathcal{L}_{rec}(\mathbf{x}',\mathbf{x}) +   \mathcal{L}_{seg}(\mathbf{p},\mathbf{y}) +   \mathcal{L}_{shp}(\mathbf{p}',\mathbf{y}) + \mathcal{L}_{shp}(\mathbf{y}',\mathbf{y})], \quad (1)$$

where $\mathcal{L}_{rec}$ is the mean squared error (MSE) between the original input image $\mathbf{x}$ and the reconstructed image $\mathbf{x}' = \mathcal{D}_{\phi_i}(E_\theta(\mathbf{x}))$, $\mathcal{L}_{seg}$ and $\mathcal{L}_{shp}$ are cross-entropy loss functions between ground truth $\mathbf{y}$ and predicted segmentation. The predicted segmentation can be the initial prediction $\mathbf{p} = \mathcal{D}_{\phi_s}(\mathcal{H}(E_\theta(\mathbf{x})))$, or the reconstructed prediction $\mathbf{p}' = \mathcal{C}_\psi(\mathbf{p})$ or reconstructed ground-truth map $\mathbf{y}' = \mathcal{C}_\psi(\mathbf{y})$. Different from $\mathcal{L}_{seg}$, optimizing $\mathcal{L}_{shp}(\mathbf{p}',\mathbf{y})$ will trigger gradient flows from STN to FTN. This allows STN to transfer shape knowledge to FTN to improve model generalizability.

2.3   Latent Space Data Augmentation for Hard Example Generation

Standard training is likely to suffer from over-fitting when training data is limited. To solve this problem, a novel latent space DA method is proposed which allows FTN to automatically construct hard examples. As shown in Fig. 1(b), the proposed method requires a mask generator $\mathcal{G}$ to produce a mask $\mathbf{m}$ on the latent code $\mathbf{z}$. The masked latent code $\hat{\mathbf{z}} = \mathbf{z} \cdot \mathbf{m}$ is then fed to the decoders to reconstruct a corrupted image $\hat{\mathbf{x}} = \mathcal{D}_{\phi_i}(\hat{\mathbf{z}}_i)$ and segmentation $\hat{\mathbf{p}} = \mathcal{D}_{\phi_s}(\hat{\mathbf{z}}_s)$. Here, $\cdot$ denotes element-wise multiplication. In our work, we use latent code masking for data augmentation. This differs from existing latent code dropout techniques for explicit regularization [25,26]. By dynamically masking the latent code, the proposed method can generate samples with a wide diversity of image appearances and segmentations, which are not bound to specific image transformation or corruption functions. Below we introduce three latent-code masking schemes: random dropout $\mathcal{G}_{dp}$, and two targeted masking schemes, channel-wise targeted mask generation $\mathcal{G}_{ch}$ and spatial-wise targeted mask generation $\mathcal{G}_{sp}$.

(1) Random Masking with Dropout. A naïve approach for latent code masking is random channel-wise dropout [26], which is an enhanced version of the original dropout method. An entire channel of the latent code can be masked with all zeros at a probability of p at training. Mathematically, this can be viewed as sampling a mask from a Bernoulli distribution:

$$\mathcal{G}_{dp}(\mathbf{m}^{(i)}; p) = \begin{cases} p & \mathbf{m}^{(i)} = \mathbf{0} \in \mathbb{R}^{h \times w} \\ 1 - p & \mathbf{m}^{(i)} = \mathbf{1} \in \mathbb{R}^{h \times w}; \end{cases} \forall i \in 1, ..., c. \tag{2}$$

The masked code at i-th channel is obtained via $\hat{\mathbf{z}}^{(i)} = \mathbf{z}^{(i)} \cdot \mathbf{m}^{(i)}$. In the following, we will use i-j-k to denote the three coordinates of latent code $\mathbf{z} \in \mathbb{R}^{c \times h \times w}$.

(2) Targeted Masking. Inspired by the recent success on latent code masking for domain generalized image classification algorithm [25], we propose targeted latent code masking schemes which takes gradients as a clue to identify 'salient' features to mask. Following the common practice in adversarial DA [27,28], we take task-specific losses (image reconstruction loss and image segmentation loss) to calculate the gradients $\mathbf{g}_{\mathbf{z}_i}, \mathbf{g}_{\mathbf{z}_s}$ for $\mathbf{z}_i$ and $\mathbf{z}_s$ respectively, formulated as: $\mathbf{g}_{\mathbf{z}_i} = \nabla_{\mathbf{z}_i} \mathcal{L}_{rec}(\mathcal{D}_{\phi_i}(\mathbf{z}_i), \mathbf{x})$, $\mathbf{g}_{\mathbf{z}_s} = \nabla_{\mathbf{z}_s} \mathcal{L}_{seg}(\mathcal{D}_{\phi_s}(\mathbf{z}_s), \mathbf{y})$. By ranking the values of task-specific gradients, we can identify most predictive elements in the latent space to attack. We hypothesize that the elements with high response to task-specific loss functions are leading causes to performance drop under unforeseen domain shifts. We therefore focus on attacking these primary elements to simulate strong data distribution shifts. Two types of targeted masking are implemented, which mask features in latent code $\mathbf{z}$ along the channel dimension and spatial dimension.

a) **channel-wise mask generator:**

$$\mathcal{G}_{ch}(\mathbf{m}^{(i)}; \mathbf{g}_{\mathbf{z}}, p) = \begin{cases} \mathbf{m}^{(i)} = a\mathbf{1} \in \mathbb{R}^{h \times w} & \text{if } \mathbb{E}[\mathbf{g}_{\mathbf{z}}^{(i)}] \geq z_p^{ch} \\ \mathbf{m}^{(i)} = \mathbf{1} \in \mathbb{R}^{h \times w} & \text{if } \mathbb{E}[\mathbf{g}_{\mathbf{z}}^{(i)}] < z_p^{ch}; \end{cases} \forall i \in 1, ..., c, \tag{3}$$

b) **spatial-wise mask generator:**

$$\mathcal{G}_{sp}(\mathbf{m}^{(j,k)}; \mathbf{g}_{\mathbf{z}}, p) = \begin{cases} \mathbf{m}^{(j,k)} = a\mathbf{1} \in \mathbb{R}^c & \text{if } \mathbb{E}[\mathbf{g}_{\mathbf{z}}^{(j,k)}] \geq z_p^{sp} \\ \mathbf{m}^{(j,k)} = \mathbf{1} \in \mathbb{R}^c & \text{if } \mathbb{E}[\mathbf{g}_{\mathbf{z}}^{(j,k)}] < z_p^{sp}; \end{cases} \forall j \in [1, h], \forall k \in [1, w]. \tag{4}$$

Thresholds $z_p^{ch}, z_p^{sp} \in \mathbb{R}$ are top p-th value across the channel means and spatial means. a is an annealing factor randomly sampled from $(0, 0.5)$ to create soft masks. Compared to hard-masking ($a = 0$), soft-masking generates more diverse corrupted data (Supple. Fig. 1). Channel-wise masked code at i-th channel is obtained via $\hat{\mathbf{z}}^{(i)} = \mathbf{z}^{(i)} \cdot \mathbf{m}^{(i)}$. Spatial-wise masked code at (j, k) position is obtained via $\hat{\mathbf{z}}^{(j,k)} = \mathbf{z}^{(j,k)} \cdot \mathbf{m}^{(j,k)}$.

2.4   Cooperative Training

During training, we randomly apply one of the three mask generators to both $\mathbf{z}_i, \mathbf{z}_s$. This process generates a rich set of corrupted images $\hat{\mathbf{x}}$ and segmentations $\hat{\mathbf{p}}$ on-the-fly. It allows us to train our dual-network on three hard example pairs, i.e. corrupted images-clean images $(\hat{\mathbf{x}}, \mathbf{x})$, corrupted images-GT $(\hat{\mathbf{x}}, \mathbf{y})$, corrupted prediction-GT $(\hat{\mathbf{p}}, \mathbf{y})$. The final loss for the proposed cooperative training method is a combination of losses defined on easy examples and hard examples: $\mathcal{L}_{cooperative} = \mathcal{L}_{std} + \mathcal{L}_{hard}$, where $\mathcal{L}_{hard}$ is defined as:

$$\mathcal{L}_{hard} = \mathbb{E}_{\hat{\mathbf{x}},\hat{\mathbf{p}},\mathbf{x},\mathbf{y}}[\mathcal{L}_{rec}(\mathcal{D}_{\phi_i}(E_\theta(\hat{\mathbf{x}})), \mathbf{x}) + \mathcal{L}_{seg}(\bar{\mathbf{p}}, \mathbf{y}) + \mathcal{L}_{shp}(\mathcal{C}_\psi(\hat{\mathbf{p}}), \mathbf{y}) + \mathcal{L}_{shp}(\mathcal{C}_\psi(\bar{\mathbf{p}}), \mathbf{y})].$$
(5)

Here, $\bar{\mathbf{p}} = \mathcal{D}_{\phi_i}(\mathcal{H}(E_\theta(\hat{\mathbf{x}})))$ is FTN's predicted segmentation on $\hat{\mathbf{x}}$.

3   Experiments and Results

Datasets. To evaluate the efficacy of the proposed method, we apply it to the cardiac image segmentation task to segment the left ventricle cavity, left ventricular myocardium and right ventricle from MR images. Three datasets are used, ACDC[1] [29], M&Ms[2] [30] and corrupted ACDC, named as ACDC-C. For all experiments, the training set is a **single-site** set of only 10 subjects from ACDC. 10 and 20 subjects from ACDC are used for validation and intra-domain test. The multi-site M&Ms dataset (150 subjects from 5 different sites) is used for cross-domain test. The ACDC-C dataset is used for evaluating the robustness of the method for corrupted images. Challenging scenarios are simulated, where 20 ACDC test subjects are augmented three times with four different types of MR artefacts: bias field, ghosting, motion and spike artifacts [31] using the TorchIO[3] toolkit. This produces 4 subsets with 60 subjects, named as *RandBias, RandGhosting, RandMotion, RandSpike* in experiments.

Implementation and Evaluation. We employed the image pre-processing and default DA pipeline described in [17], including common photo-metric and geometric image transformations. Our encoder and decoder pairs support general structures. Without loss of generality, we used a U-net like structure[32]. Supple. Fig. 5 visualizes detailed structures of encoder-decoder pairs as well as the latent space decoupler. For mask generation, we randomly select one type of proposed masking schemes at training, where p is randomly selected from [0%, 50%]. We use the Adam optimizer with a batch size of 20 to update network parameters, with a learning rate $= 1e^{-4}$. Our code will be available on the Github[4]. For all methods, we trained the same network *three* times using a set of randomly selected 10 ACDC subjects (600 epochs each run, on an Nvidia®, using Pytorch). The average Dice score is reported for segmentation performance evaluation.

Experiment 1: Standard Training vs Cooperative Training. We compared the proposed cooperative training method with the standard training method (using $\mathcal{L}_{standard}$ only) using the same backbone network structure.

---

[1] https://www.creatis.insa-lyon.fr/Challenge/acdc/databases.html.

[2] https://www.ub.edu/mnms/.

[3] https://github.com/fepegar/torchio.

[4] https://github.com/cherise215/Cooperative_Training_and_Latent_Space_Data_Aug mentation.

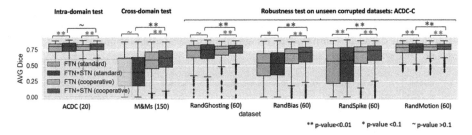

Fig. 2. Compared to standard training, cooperative training with self-generating hard examples greatly improves the segmentation performance on various *unseen, challenging* domains (p-value < 0.01, average improvement: 15%). (Color figure online)

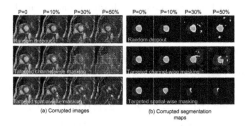

Fig. 3. Visualization of generated hard examples with the three masking schemes ($a =$ 0 for ease of comparison). Best viewed in zoom-in.

Figure 2 shows the box-plots for each method. While both methods achieve comparable performance on the intra-domain test set (p-value > 0.1), it is clear that cooperative training with dual-network (FTN+STN) yields the best performance across out-of domain test sets (see dark green boxes). Consistent improvements made by STN can be clearly observed across all domains. By contrast, STN with standard training fails to provide significant improvements on some datasets (p-value > 0.1). This indicates the superiority of cooperative training with latent space DA.

Experiment 2: Latent Space DA vs Image Space DA. We compared the proposed latent space -based method to other competitive image space DA methods: a) *random multi-window in-and-out masking (Rand MWM)* [14,15], which uses an enhanced variant of Cutout [12] and Patch Gaussian [13] to introduce patch-wise perturbation to images; b) *random convolutional kernels (Rand Conv)* [33], which applies various random convolutional kernels to augment image texture and appearance variations; c) *adversarial noise (Adv Noise)* [16]; d) *adversarial bias field (Adv Bias)* [17], which augments image styles by adding realistic intensity inhomogeneities. We visualize augmented images using above methods in Supple. Fig. 2 for better illustration. For methods under comparison, we used their official code implementation if available and ran experiments using the same backbone network for fairness. Results are shown in Table 1.

Table 1. Comparison to image space data augmentation methods for domain generalization. The proposed latent space augmentation method improves the performance on out-of-domain datasets compared to image space data augmentation methods. AVG: Average Dice scores across six datasets.

Method	ACDC	M&Ms	RandBias	RandGhosting	RandMotion	RandSpike	AVG (FTN)	AVG (FTN+STN)
Standard training	0.7681	0.3909	0.4889	0.6964	0.7494	0.4901	0.5970	0.6018
Rand MWM [14]	0.7515	0.3984	0.4914	0.6685	0.7336	0.5713	0.6024	0.6131
Rand Conv [33]	0.7604	0.4544	0.5538	0.6891	0.7493	0.4902	0.6162	0.6404
Adv Noise [16]	0.7678	0.3873	0.4903	0.6829	0.7543	0.6244	0.6178	0.6276
Adv Bias [17]	0.7573	**0.6013**	**0.6709**	0.6773	0.7348	0.3840	0.6376	0.6604
Proposed w. $\hat{x}$	0.7497	0.5154	0.5921	0.6921	0.7417	**0.6633**	0.6591	0.6709
Proposed w. $\hat{x}, \hat{p}$	**0.7696**	0.5454	0.6174	**0.7073**	**0.7643**	0.6226	**0.6711**	**0.6901**

Table 2. Effectiveness of targeted masking, latent code decoupler $\mathcal{H}$ and cooperative training

Methods	FTN	FTN+STN
w.o. $\mathcal{G}_{ch}, \mathcal{G}_{sp}$	0.6344	0.6584
Share code (a) ($\mathbf{z}_i = \mathbf{z}_i, \mathbf{z}_s = \mathbf{z}_i$)	0.6625	0.6868
Share code (b) ($\mathbf{z}_i = \mathbf{z}_s, \mathbf{z}_s = \mathbf{z}_s$)	0.6343	0.6587
Separate training [22]	0.6020	0.6077
Proposed	**0.6711**	**0.6901**

Surprisingly, with limited training data, both random and adversarial DA methods do not necessarily improve the network generalization on all datasets. While *AdvBias* achieves the best performance on *M&Ms* dataset and *RandBias*, this method has a side effect, making it more sensitive to the spiking artifacts (Dice score 0.4901 vs 0.3840). By contrast, the proposed latent space DA achieves the top average performance across six datasets, without any dramatic failures (Dice score < 0.5). Similar results can be found in a large training setting, see Supple. Fig. 3. Our method can generate not only perturbed images but also realistically corrupted segmentations with increased uncertainty (Supple. Fig. 4). These corrupted segmentations attribute to the increased model generalization (AVG Dice: 0.6709 vs 0.6901). While one may argue that characterizing and combining various image-space DAs and corruptions together could be an interesting direction to improve cross-domain performance, it is time-consuming and computationally inefficient to find the optimal DA policy [34], and has the risk of sacrificing intra-domain performance [35].

Experiment 3: Ablation Study. We further investigate three key contributions: 1) the proposed targeted masking; 2) latent code decoupler $\mathcal{H}$; 3) cooperative training. Results are shown in Table 2. We can see that disabling $\mathcal{G}_{ch}, \mathcal{G}_{sp}$ drops the average Dice score from 0.6901 to 0.6584, highlighting the effectiveness of targeted masking. Figure 3 shows that targeted masking focuses more on attacking cardiac structures, resulting in more challenging images with mixed artifacts and under or over-segmented predictions. We compared the proposed network architecture to its two variants, where $\mathbf{z}_i$ and $\mathbf{z}_s$ are shared in two different ways. Both variants lead to inferior performance. This suggests the benefit of $\mathcal{H}$ for a more sparse $\mathbf{z}_s$ code. Image reconstruction requires low-level information, whereas image segmentation relies on more concentrated high-level information. Introducing $\mathcal{H}$ explicitly defines a hierarchical feature structure to improve model generalization. Lastly, we compared our method to the state-of-the-art denoising auto-encoder-based shape refinement method (Separate Training) [22] where FTN and STN are trained independently. It has been shown that this learning-based method can outperform the commonly used non-learning-based condition random field-based refinement method [36]. Results show that our method can greatly outperform this advanced method by a large margin (Dice score 0.6901 vs. 0.6077), highlighting the benefits of the cooperative training strategy for enhancing learning-based shape refinement and correction.

4   Conclusion

We present a novel cooperative training framework in together with a latent space masking-based DA method. Experiments show that it greatly improves model generalizability and robustness against unforeseen domain shifts. Unlike existing methods which require multi-domain datasets or domain knowledge to specify particular forms of image transformation and corruption functions, our latent space DA method requires *little* human effort, and it has the potential to be applied to other data-driven applications. Although we only demonstrate the performance for cardiac image segmentation, our *generic* framework has the potential to be extended to a wide range of data-driven applications.

Acknowledgment. This work was supported by the SmartHeart EPSRC Programme Grant (EP/P001009/1).

References

1. Shen, D., et al.: Deep learning in medical image analysis. Annu. Rev. Biomed. Eng. **19**, 221–248 (2017)
2. Litjens, G., et al.: A survey on deep learning in medical image analysis. Med. Image Anal. **42**, 60–88 (2017)
3. Dou, Q., et al.: Domain generalization via model-agnostic learning of semantic features. In: Wallach, H.M., et al. (eds.) NeurIPS 2019, pp. 6447–6458 (2019)
4. Albuquerque, I., et al.: Improving out-of-distribution generalization via multi-task self-supervised pretraining. arXiv preprint arXiv:2003.13525 (2020)
5. Chattopadhyay, P., Balaji, Y., Hoffman, J.: Learning to balance specificity and invariance for in and out of domain generalization. In: Vedaldi, A., Bischof, H., Brox, T., Frahm, J.-M. (eds.) ECCV 2020. LNCS, vol. 12354, pp. 301–318. Springer, Cham (2020). https://doi.org/10.1007/978-3-030-58545-7_18

6. Wang, S., Yu, L., Li, C., Fu, C.-W., Heng, P.-A.: Learning from extrinsic and intrinsic supervisions for domain generalization. In: Vedaldi, A., Bischof, H., Brox, T., Frahm, J.-M. (eds.) ECCV 2020. LNCS, vol. 12354, pp. 159–176. Springer, Cham (2020). https://doi.org/10.1007/978-3-030-58545-7_10

7. Shankar, S., et al.: Generalizing across domains via cross-gradient training. In: ICLR (2018). OpenReview.net

8. Daniel, K.: Thinking, fast and slow (2017)

9. Shorten, C., et al.: A survey on image data augmentation for deep learning. J. Big Data **6**(1), 60 (2019)

10. Zhang, L., et al.: Generalizing deep learning for medical image segmentation to unseen domains via deep stacked transformation. IEEE Trans. Med. Imaging **39**(7), 2531–2540 (2020)

11. Chen, C., et al.: Improving the generalizability of convolutional neural network-based segmentation on CMR images. Front. Cardiovasc. Med. **7**, 105 (2020)

12. Devries, T., et al.: Improved regularization of convolutional neural networks with cutout. CoRR, abs/1708.04552 (2017)

13. Lopes, R.G., et al.: Improving robustness without sacrificing accuracy with patch gaussian augmentation. CoRR, abs/1906.02611 (2019)

14. Zhou, Z., et al.: Models genesis: generic autodidactic models for 3D medical image analysis. In: Shen, D., et al. (eds.) MICCAI 2019. LNCS, vol. 11767, pp. 384–393. Springer, Cham (2019). https://doi.org/10.1007/978-3-030-32251-9_42

15. Zhou, Z., et al.: Models genesis. Med. Image Anal. **67**, 101840 (2021)

16. Miyato, T., Maeda, S.-I., Koyama, M., Ishii, S.: Virtual adversarial training: a regularization method for supervised and semi-supervised learning. TPAMI **41**, 1979–1993 (2018)

17. Chen, C., et al.: Realistic adversarial data augmentation for MR image segmentation. In: Martel, A.L., et al. (eds.) MICCAI 2020. LNCS, vol. 12261, pp. 667–677. Springer, Cham (2020). https://doi.org/10.1007/978-3-030-59710-8_65

18. Zhang, X., et al.: Deep adversarial data augmentation for extremely low data regimes. IEEE Trans. Circuits Syst. Video Technol. **31**(1), 15–28 (2021)

19. Zhao, L., et al.: Maximum-entropy adversarial data augmentation for improved generalization and robustness. In: NeurIPS (2020)

20. Zheng, W., Chen, Z., Lu, J., Zhou, J.: Hardness-aware deep metric learning. In: CVPR, pp. 72–81 (2019)

21. Zhang, Y., et al.: A survey on multi-task learning. arXiv preprint arXiv:1707.08114 (2017)

22. Larrazabal, A.J., Martinez, C., Ferrante, E.: Anatomical priors for image segmentation via post-processing with denoising autoencoders. In: Shen, D., et al. (eds.) MICCAI 2019. LNCS, vol. 11769, pp. 585–593. Springer, Cham (2019). https://doi.org/10.1007/978-3-030-32226-7_65

23. Painchaud, N., Skandarani, Y., Judge, T., Bernard, O., Lalande, A., Jodoin, P.-M.: Cardiac MRI segmentation with strong anatomical guarantees. In: Shen, D., et al. (eds.) MICCAI 2019. LNCS, vol. 11765, pp. 632–640. Springer, Cham (2019). https://doi.org/10.1007/978-3-030-32245-8_70

24. Tishby, N., et al.: The information bottleneck method. arXiv preprint physics/0004057 (2000)

25. Huang, Z., Wang, H., Xing, E.P., Huang, D.: Self-challenging improves cross-domain generalization. In: Vedaldi, A., Bischof, H., Brox, T., Frahm, J.-M. (eds.) ECCV 2020. LNCS, vol. 12347, pp. 124–140. Springer, Cham (2020). https://doi.org/10.1007/978-3-030-58536-5_8

26. Tompson, J., et al.: Efficient object localization using convolutional networks. In: CVPR, pp. 648–656. IEEE Computer Society (2015)
27. Madry, A., Makelov, A., Schmidt, L., Tsipras, D., Vladu, A.: Towards deep learning models resistant to adversarial attacks. In: ICLR, June 2017
28. Goodfellow, I.J., et al.: Explaining and harnessing adversarial examples. In: ICLR (2015)
29. Bernard, O., et al.: Deep learning techniques for automatic MRI cardiac multi-structures segmentation and diagnosis: is the problem solved? TMI **0062**(11), 2514–2525 (2018)
30. Campello, V.M., et al.: Multi-centre, multi-vendor and multi-disease cardiac segmentation: the M&Ms challenge. IEEE Trans. Med. Imaging (under review)
31. Pérez-García, F., et al.: TorchIO: a Python library for efficient loading, preprocessing, augmentation and patch-based sampling of medical images in deep learning. arXiv:2003.04696 [cs, eess, stat], March 2020
32. Ronneberger, O., Fischer, P., Brox, T.: U-Net: convolutional networks for biomedical image segmentation. In: Navab, N., Hornegger, J., Wells, W.M., Frangi, A.F. (eds.) MICCAI 2015. LNCS, vol. 9351, pp. 234–241. Springer, Cham (2015). https://doi.org/10.1007/978-3-319-24574-4_28
33. Xu, Z., et al.: Robust and generalizable visual representation learning via random convolutions. In: ICLR (2021)
34. Cubuk, E.D., et al.: Autoaugment: learning augmentation strategies from data. In: CVPR, pp. 113–123 (2019)
35. Shaw, R., et al.: MRI k-space motion artefact augmentation: model robustness and task-specific uncertainty. In: Jorge Cardoso, M., et al. (eds.) Proceedings of Machine Learning Research, MIDL, London, UK, 08–10 July 2019, vol. 102, pp. 427–436. PMLR (2019)
36. Christ, P.F., et al.: Automatic liver and lesion segmentation in CT using cascaded fully convolutional neural networks and 3D conditional random fields. In: Ourselin, S., Joskowicz, L., Sabuncu, M.R., Unal, G., Wells, W. (eds.) MICCAI 2016. LNCS, vol. 9901, pp. 415–423. Springer, Cham (2016). https://doi.org/10.1007/978-3-319-46723-8_48

Controllable Cardiac Synthesis via Disentangled Anatomy Arithmetic

Spyridon Thermos[1]($\boxtimes$), Xiao Liu[1], Alison O'Neil[1,3], and Sotirios A. Tsaftaris[1,2]

[1] School of Engineering, University of Edinburgh, Edinburgh EH9 3FB, UK
{SThermos,Xiao.Liu,S.Tsaftaris}@ed.ac.uk
[2] The Alan Turing Institute, London NW1 2DB, UK
[3] Canon Medical Research Europe, Edinburgh EH6 5NP, UK
Alison.ONeil@mre.medical.canon

Abstract. Acquiring annotated data at scale with rare diseases or conditions remains a challenge. It would be extremely useful to have a method that controllably synthesizes images that can correct such underrepresentation. Assuming a proper latent representation, the idea of a "latent vector arithmetic" could offer the means of achieving such synthesis. A proper representation must encode the fidelity of the input data, preserve invariance and equivariance, and permit arithmetic operations. Motivated by the ability to disentangle images into *spatial* anatomy (tensor) factors and accompanying imaging (vector) representations, we propose a framework termed "disentangled anatomy arithmetic", in which a generative model learns to combine anatomical factors of different input images such that when they are re-entangled with the desired imaging modality (*e.g.* MRI), plausible new cardiac images are created with the target characteristics. To encourage a realistic combination of anatomy factors after the arithmetic step, we propose a localized noise injection network that precedes the generator. Our model is used to generate realistic images, pathology labels, and segmentation masks that are used to augment the existing datasets and subsequently improve post-hoc classification and segmentation tasks. Code is publicly available at https://github.com/vios-s/DAA-GAN.

Keywords: Disentangled anatomy arithmetic · Semantic image synthesis · Cardiac data augmentation

1 Introduction

Whilst large scale public datasets are available for traditional vision tasks, medical data are difficult to acquire. Even in a large-scale medical training dataset, examples of rare diseases and anatomies are scarce. As a result, generalisation to observations that are not seen during training will be reduced. To increase the

Electronic supplementary material The online version of this chapter (https://doi.org/10.1007/978-3-030-87199-4_15) contains supplementary material, which is available to authorized users.

M. de Bruijne et al. (Eds.): MICCAI 2021, LNCS 12903, pp. 160–170, 2021.
https://doi.org/10.1007/978-3-030-87199-4_15

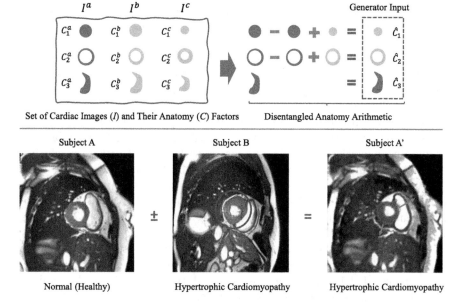

Fig. 1. *Top*: overview of the "disentangled anatomy arithmetic" concept illustrated with 3 factors that represent 3 different anatomical parts of the heart (*e.g.* left/right ventricle and myocardium). *Bottom*: DAA-GAN generated example. Given a healthy Subject A, we aim to generate an image A' which exhibits hypertrophic cardiomyopathy (HCM). We select a Subject B with HCM and remove the anatomical factors from A (i.e. the ones that encode the myocardium and left ventricular cavity) and add the corresponding factors of B (inner part of the red circle). Arrows in A' point to local deformations showing the non-linear abilities of our arithmetic. Arithmetic operations are denoted with $\pm$. (Color figure online)

diversity of training data and for instance, to increase the incidence of rare characteristics, we would like the ability to mix and match factors that encode these variations [4] in a controllable way *i.e.* perform *controllable image synthesis*.

The idea of generating realistic images to augment existing limited data is not new in medical image analysis. Generative Adversarial Networks (GANs) [15] have been used to generate variants for a given input image based on sampling from a random noise vector. In fact, more recent GAN architectures pursue controllability by disentangling existing factors of variation, conditioning the generation process using semantic priors (e.g. segmentation masks) [10, 16,17,21,24,27,33] and class labels [13,19], or learning cross-modality translations [3,8,11,23,28]. An alternative to relying on sampling from a noise vector for new medical images, is the idea of mixing existing information from different populations (*e.g.* patients with different anatomical characteristics) to learn the intermediate latent space and generate more realistic data in a more controllable way. A concept that approximates this idea is "vector arithmetic" [31], where existing vector-based latent representations are combined using simple arithmetic operations to produce new images. However, vector representations do not exploit the spatial equivariance of the image content and the respective task (*e.g.* segmentation) and have shown poor reconstruction quality. An alternative is to use to disentangled representations that use both spatial (tensor) and vector representations to capture factors of variation and permit decomposition of the input in spatially equivariant (and closely to be semantic) and imaging information [7,9,36].

Herein using disentangled representations we propose the concept of "disentangled anatomy arithmetic" (DAA), visualized in Fig. 1, that enables controllable image synthesis of plausible images with a *target pathology*, which we show to be useful for augmenting existing medical training data. We design the DAA-GAN model that learns to combine and transform *spatial* anatomical factors –we provide a visual example of anatomy factors in Fig. 1 of the supplemental– from input images captured by different vendors or from different populations, and then re-entangle them with the chosen imaging factors (*e.g.* MRI) to generate unseen intermediate representations. Inspired by recent findings regarding the advantages of introducing spatial stochasticity in the generation process [1,2,14,22], we propose a convolutional module for the combined anatomical factor representation transformation, in which structured noise is injected at the spatial locations where the arithmetic operations take place.

Our **contributions** are to:

- Introduce the concept of "disentangled anatomy arithmetic" on spatial representations of the anatomy.
- Propose DAA-GAN, a generative model that to the best of our knowledge, is the first to condition image generation using spatial anatomy factors as semantic priors.
- Propose the noise injection module that encourages local deformations to realistically blend the new factors after the arithmetic step.
- Evaluate the impact of using DAA-GAN for cardiac data augmentation in the context of a classification and a semantic segmentation post-hoc task.

2   Generative Model Architecture

Our model assumes disentangled anatomy and imaging representations as inputs. To obtain them, we use SDNet [7] as this model provides binary spatial factors that correspond to the whole anatomy and can be used as semantic priors.

Model Overview. As depicted in Fig. 2, DAA-GAN has 4 distinct steps: 1) We combine anatomy factors, using disentangled anatomy arithmetic, to obtain a new mixed anatomical representation $\hat{\mathbf{C}}$. 2) A noise injection network $\mathcal{J}$ takes $\hat{\mathbf{C}}$ and aims to create a plausible and more refined anatomical representation $\tilde{\mathbf{C}}$. 3) A generator $\mathcal{G}$ reconstructs an image corresponding to $\tilde{\mathbf{C}}$ and a given imaging representation. 4) Two critics, namely a discriminator $\mathcal{D}$ and a pathology classifier $\mathcal{F}$ ensure good image fidelity, but also that the reconstructed image contains the right target characteristics. We proceed detailing these steps.

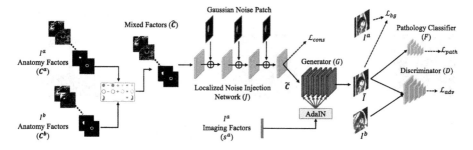

Fig. 2. DAA-GAN overview (from left to right): arithmetic operations are performed between anatomical factors $\mathbf{C}^a$ and $\mathbf{C}^b$ to produce a mixed representation $\hat{\mathbf{C}}$ which is then refined by the noise injection network $\mathcal{J}$. The generator $\mathcal{G}$ receives this refined representation $\tilde{\mathbf{C}}$ and re-entangles it with the input imaging factors to generate the new image $\tilde{I}$. Finally, a discriminator is responsible to judge if $\tilde{I}$ is real or fake, whilst a pathology classifier assesses if $\tilde{I}$ has the desired cardiac pathology.

Disentangled Anatomy Arithmetic. As shown in Fig. 1 (top), we consider an example with 3 cardiac anatomy populations $\mathbf{C}^a, \mathbf{C}^b, \mathbf{C}^c$ (*e.g.* patients from 3 imaging populations a, b, and c) with 3 anatomical factors per population, which –when combined– form a nested structure that corresponds to the heart region. These factors are extracted from I^a, I^b, I^c medical images of dataset $Z_{\sim\{I^a, I^b, I^c\}}$. Based on this setup, we define factor arithmetic between populations any swapping operation between the corresponding factors. Following this example, to create a new $\mathbf{C}^a$ anatomy by mixing factors, we first swap $\mathbf{C}_1^a$ with $\mathbf{C}_1^c$, and then swap $\mathbf{C}_2^a$ with $\mathbf{C}_2^b$. The result is an intermediate $\hat{\mathbf{C}}$ that is used as input to the next module of DAA-GAN. Note that before swapping two factors, assuming upright anatomies, we perform a registration step (warping) to align the swapped-in factor with the center of mass location of the swapped-out one.

Noise Injection. Since cardiac anatomy is a nested structure, the output of the arithmetic step might be non-realistic, *e.g.* have factors that overlap with each other. This can lead to generated images with ambiguous pathology. We tackle this problem with module $\mathcal{J}$, which receives $\hat{\mathbf{C}}$ and produces a refined representation $\tilde{\mathbf{C}}$. Inspired by recent work of Karras *et al.* [22], we introduce stochastic variation (as Gaussian noise) at specific spatial locations of each convolutional (CONV) layer's activations. This variation is exploited by the network to cause local deformations around the added (swapped-in) factor(s) in order to preserve the non-overlapping nested structure of the heart (see Fig. 1(bottom)). $\mathcal{J}$ consists of 4 CONV layers, whilst the noise is injected in the form of noise patches (see Fig. 3) added to each CONV layer's activation features in an element-wise fashion. The last CONV layer is followed by a Gumbel-Softmax operator that bounds $\tilde{\mathbf{C}}$ to $[0, 1]$.

Generator. The generator is responsible for the re-entanglement of anatomy and imaging information, and by extension for the generation of a new image $\tilde{I}$. G consists of 4 CONV-ReLU layers followed by a hyperbolic tangent activation

Anatomy
Factor
Gaussian
Kernel
Blurred
Factor Mask Φ
Gaussian
Noise
Localized Gaussian
Noise Patch

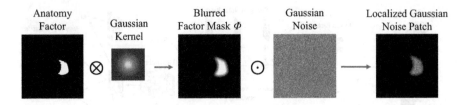

Fig. 3. Visualization of the localized Gaussian noise patch generation process. $\otimes$ and $\odot$ denote convolution and element-wise multiplication, respectively.

function. The first CONV layer receives $\tilde{C}$ as input, while after each CONV-ReLU block, there is an Adaptive Instance Normalization (AdaIN) layer [20] that scales and shifts the activations based on the input imaging factors. Since $\tilde{C}$ has the same dimensions as I and $\tilde{I}$, there is no need for any upsampling.

Critics. Since for $\tilde{I}$ we have no ground truth we use a discriminator to guide reconstruction. We adopt the architecture of the LSGAN discriminator [25] for faster convergence, better stability and ultimately better image quality (compared to a variant with spectral normalization layers adopted from [26]). Additionally, since we want $\tilde{I}$ to have desired properties, we use a VGG16 model [34] $\mathcal{F}$ to classify the pathology represented in $\tilde{I}$. $\mathcal{F}$ has 7 CONV-BN-ReLU blocks and 3 fully-connected layers followed by a softmax function. Note that $\mathcal{F}$ is pretrained on the original data, and is then used as a pathology predictor during training of the generative model.

Having presented each module in detail, we now proceed to describe the 4 losses used to train our model as a whole:

Adversarial Loss ($\mathcal{L}_{adv}$). We use an adversarial loss to encourage our model to generate realistic images. We choose to minimize the LSGAN loss [25] as it is more stable during training and leads to higher quality image generation compared to the traditional GAN loss. Our adversarial loss is defined as:

$$\mathcal{L}_{adv} = \mathcal{L}_{\mathcal{D}} + \mathcal{L}_{\mathcal{G}},$$

$$\mathcal{L}_{\mathcal{D}} = \frac{1}{2}\mathbb{E}_{I' \sim p(Z)}[(\mathcal{D}(M \cdot I') - 1)^2] + \frac{1}{2}\mathbb{E}_{C \sim p(C_Z)}[(\mathcal{D}(M \cdot \mathcal{G}(\hat{C})))^2], \quad (1)$$

$$\mathcal{L}_{\mathcal{G}} = \frac{1}{2}\mathbb{E}_{C \sim p(C_Z)}[(\mathcal{D}(\mathcal{G}(\hat{C})) - 1)^2],$$

where I' is the image that contributes only the anatomy factor(s) C' which are used to form $\hat{C}$ (*e.g.* I^b in Fig. 2). M is a binary mask produced by the union of the anatomical factors that contain information about the heart. Note that $M(j) = 0$ for the remaining (*i.e.* non-heart) pixels.

Pathology Classification ($\mathcal{L}_{path}$). Since we know that we add the anatomical factor C_k to the mixed representation, we expect to be able to recognize the pathology corresponding to C_k in the generated image $\tilde{I}$. To achieve this we minimize the cross entropy loss, defined as $\mathcal{L}_{path} = -\sum_{i=1}^{\Omega} y_i \log(p(x_i))$, where

y_i and $p(x_i)$ are the ground truth and predicted pathology labels, and Ω is the number of pathology classes.

Anatomical and Background Consistency ($\mathcal{L}_{cons}$ and $\mathcal{L}_{bg}$). $\mathcal{L}_{cons}$ encourages the anatomical factors which are not related with the heart to remain unaltered after the arithmetic and noise injection steps. To find which pixels should not be changed, we use the blurred mask produced during the noise patch generation, denoted as $\Phi(j)$, with $\Phi(j) = 0$ for each pixel location j that is not part of the anatomy mixing. We define $\mathcal{L}_{cons} = \frac{1}{N}\sum_j(1-\Phi(j))||\hat{\mathbf{C}}(j)-\tilde{\mathbf{C}}(j)||_1$. $\mathcal{L}_{bg}$ is the same as $\mathcal{L}_{cons}$, but on images. Namely, $\mathcal{L}_{bg} = \frac{1}{N}\sum_j(1-M(j))||(I^a(j)-I\tilde{(j)})||_1$ where M is defined previously, and N is a the total number of pixels.

Total Loss ($\mathcal{L}_{total}$). These losses discussed above, are combined as:

$$\mathcal{L}_{total} = \mathcal{L}_{adv} + \mathcal{L}_{path} + \lambda_1(\mathcal{L}_{cons} + \mathcal{L}_{bg}), \tag{2}$$

where $\lambda_1 = 10$ is a weighting hyperparameter for the consistency losses.

3 Experiments

In this section we present key information on datasets and metrics, and discuss the results of our experiments (see training details in Sec. 1 of the supplemental).

Data. We use the ACDC [5] dataset and data from the M&Ms challenge [6]. **ACDC** consists of cardiac images acquired from MRI scanners across 100 subjects, and provides pathology annotations (5 classes) for all images and pixel-level segmentation masks for end-systole (ES) and end-diastole (ED) per subject. **M&Ms** has cardiac-MR images acquired from 4 different (known) sites (domains), across 345 subjects. It provides ED/ES pixel-level segmentation masks annotations for 320 (of 345) subjects and pathology annotations (4 classes) for all images (for more dataset details see Sec. 2 of the supplemental).

Metrics. We approach this in several levels. To measure image **quality** we compute the Fréchet Inception Distance (FID) [18], which quantifies the distance between feature vectors from real and generated images. To quantify the **utility** of the generated images we study if they help improving the performance of two post-hoc tasks: a) pathology classification, and b) semantic segmentation. For the former, we measure the classification accuracy achieved by a VGG16 model, whilst for the latter, we measure the Dice score [12,35] achieved by a U-Net [32] model. The hardest is to assess **controllability**, *i.e.* the ability to generate images that faithfully create combinations of anatomical factors. We approximate this by examining the existence of the added or removed pathologies. We train two VGG16 classifiers, one 5-class for ACDC data and one 4-class for M&Ms data, and measure the pathology classification accuracy on the generated images.

Setup. We generate data with our DAA-GAN; SPADE [30] a model that conditions synthesis using segmentation masks; and AC-GAN [29] a model that

Table 1. Comparing data augmentation methods in the context of the 4 questions defined in the Results section (see text for details). We report average (standard deviation as subscript) classification accuracy (Acc.) and segmentation performance as average Dice score. "n/a" denotes not applicable experiment. * and ** denote significant improvement over the 2nd best method with $p < 0.05$ and $p < 0.1$ (Wilcoxon non-parametric test), respectively.

Generation	Balanced		Un/ed class		Un/ed vendor		Image qual.	
	ACDC		M&Ms		M&Ms		ACDC	M&Ms
	Acc.	Dice	Acc.	Dice	Acc.	Dice		
Original data	$88.3_{1.6}$	$84.9_{5.0}$	$82.7_{0.6}$	$83.5_{3.7}$	$82.7_{0.6}$	$83.6_{3.6}$	n/a	n/a
Ours	$\mathbf{91.4^{*}_{1.4}}$	$\mathbf{86.5^{**}_{4.7}}$	$\mathbf{86.0^{*}_{0.8}}$	$\mathbf{85.2^{**}_{4.1}}$	$\mathbf{85.8^{*}_{0.7}}$	$\mathbf{84.7^{**}_{4.0}}$	17.2	24.8
/without $\mathcal{F}$	$89.0_{1.6}$	$85.3_{5.5}$	$84.1_{0.8}$	$83.9_{4.0}$	$84.2_{0.5}$	$83.9_{3.7}$	18.8	26.3
/without $\mathcal{J}$	$89.3_{1.5}$	n/a	$84.3_{0.6}$	n/a	$84.5_{0.7}$	n/a	21.5	27.8
/without $\mathcal{F}, \mathcal{J}$	$88.7_{1.6}$	n/a	$83.1_{0.8}$	n/a	$82.9_{0.7}$	n/a	22.5	31.2
AC-GAN [29]	$89.2_{1.2}$	n/a	$83.9_{0.7}$	n/a	$83.0_{0.8}$	n/a	**17.1**	**24.7**
SPADE [30]	$88.3_{1.5}$	n/a	$83.5_{0.5}$	n/a	$83.3_{0.7}$	n/a	20.2	28.7

conditions synthesis on class rather than semantic priors. Then, we train the two posthoc task-specific models (*i.e.* VGG16 and U-Net) with the following training sets (applies to both datasets): i) original data (OD), ii) OD augmented with data from DAA-GAN, iii) OD augmented with data from SPADE, and iv) OD augmented with data from AC-GAN. For (i)-(iv) we further augment the training set using traditional intensity (blur, gamma correction) and geometric (crop, rotate, flip, elastic transformation) augmentations for fairness (validation and test sets remain unaltered). Note that for each generated image, we extract and binarize *only* the heart-related $\tilde{C}$ factors (output of $\mathcal{J}$) and use them as near-ground truth masks for retraining U-Net in the context of post-hoc segmentation.

Results. To demonstrate the effectiveness of our method, we answer several key questions referring to quantitative results presented in Table 1.

Does DAA-GAN Augmentation Improve Learning on a Balanced Dataset? For this experiment we use ACDC, which is balanced in terms of subjects per pathology class. We split the OD into 70%, 15%, 15% subjects for training, validation, and testing, and use DAA-GAN to generate 50 new images for the 5 pathology classes (this corresponds to a 25% samples increase). These images are picked based on $\mathcal{F}$ confidence. As Table 1 shows (column "Balanced") data generated by our model lead to a 3.1% absolute classification accuracy improvement compared to using the OD with traditional augmentations, whilst outperforming both AC-GAN and SPADE. Regarding segmentation, compared to only using the OD, our model improves the Dice score by an absolute 1.6%. Since AC-GAN does not condition on masks and SPADE does not have a mechanism to refine the fused segmentation masks, they cannot be used in this experiment.

Does DAA-GAN augmentation Improve Learning of Underrepresented Classes? The M&Ms dataset is imbalanced in terms of the Abnormal Right Ventricle

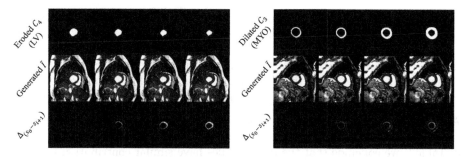

Fig. 4. Two examples of anatomy factor traversals. For each step i increases by 3: a) top row depicts the transformed factor C, b) middle row depicts the generated images $\tilde{I}$, and c) bottom row shows the factor difference between the current generated image $\tilde{I}$ and the input image. LV, MYO denote the left ventricular cavity and the myocardium.

(ARV) pathology class, which represents only 5% of the dataset. We showcase here the impact of our generative model in augmenting underrepresented classes. We split the data in a 70–15–15% fashion (as with ACDC), and use DAA-GAN to generate 168 –high $\mathcal{F}$ confidence– new images (matching the highest represented class) with the ARV pathology by mixing the factor that corresponds to the RV with healthy anatomies from other subjects. We use this new balanced training set that is used to re-train VGG16 and U-Net. As reported in Table 1, column "Un/ed Class", augmentations improve both accuracy by 3.3% and Dice by 1.7% Dice outperforming AC-GAN and SPADE where applicable.

Does DAA-GAN Augmentation Improve Learning of Underrepresented or New Domains? As stated above, M&Ms comprises data captured from 4 different sites (A-D), thus 4 different populations. However, data from site D represent only 14% in the dataset. Here we showcase the augmentation of site D. Thus, in this experiment we aim to balance the training set of M&Ms by augmenting site D data. We adopt the same split percentage with the previous experiments, and augment the original training set by mixing pathological factors from subjects of vendors A-C with anatomies of vendor D. Results in Table 1, column "Un/ed Vendor", show that by augmenting site D with 101 generated images of high $\mathcal{F}$ confidence value, we improve the two tasks' performance by 3.1% in accuracy and 1.2% Dice, respectively, while outperforming AC-GAN and SPADE.

Does DAA-GAN Achieve Good Image Quality? Table 1 (rightmost) reports generated image quality for each model. Our model outperforms SPADE, which is unsurprising since SPADE has no mechanism to mix semantic priors, thus generating images with unusual heart representations due to overlapping anatomical factors (see examples generated using SPADE in Fig. 3 of the supplemental). AC-GAN, has slightly better FID compared to DAA-GAN in both datasets quality but is the worst model in terms of controllability and utility.

Can We Control DAA-GAN Synthesis? We explore this visually by generating images through manipulation of the anatomical factors of the same subject (*i.e.* without mixing between subjects). We erode and dilate only a single factor of each subject and generate images based on the altered anatomy. From the examples in Fig. 4, we observe that the generated cardiac parts do not correspond linearly to the generative factors when approaching extreme cases, *i.e.* for large kernel size. Thus, we argue that our model's controllability is constrained only by the dataset bias, *i.e.* "unseen" extreme (possibly unrealistic) combinations.

Ablations. To evaluate the impact of $\mathcal{J}$ and $\mathcal{F}$ on the augmentation results, we replicate each experiment removing one or both modules. From the results in Table 1, we conclude that the two modules have similar contribution to post-hoc classification improvement, $\mathcal{F}$ slightly improves segmentation, whilst we also observe that $\mathcal{J}$ plays an important role in the quality of the generated images.

4 Conclusions

In this paper we introduced a novel framework for controllable cardiac image synthesis. In particular, we presented a generative model that learns to produce unseen images from existing images using a user-selected combination of spatial anatomy factors. We conducted experiments demonstrating the controllability of the generation process, whilst showcasing the potential of augmenting existing medical data with images generated using the concept of "disentangled anatomy arithmetic". Future work will focus on extending the capability of our model beyond simple mixing and morphology to richer anatomy arithmetic operations.

Acknowledgements. This work was supported by the University of Edinburgh, the Royal Academy of Engineering and Canon Medical Research Europe. This work was partially supported by the Alan Turing Institute under the EPSRC grant EP/N510129/1. We thank Nvidia for donating a Titan-X GPU. S. A. Tsaftaris acknowledges the support of Canon Medical and the Royal Academy of Engineering and the Research Chairs and Senior Research Fellowships scheme (grant RCSRF1819\8\25).

References

1. Alharbi, Y., Smith, N., Wonka, P.: Latent filter scaling for multimodal unsupervised image-to-image translation. In: Proceedings of the CVPR, pp. 1458–1466 (2019)
2. Alharbi, Y., Wonka, P.: Disentangled image generation through structured noise injection. In: Proceedings of the CVPR, pp. 5134–5142 (2020)
3. Ben-Cohen, A., Klang, E., Raskin, S.P., Amitai, M.M., Greenspan, H.: Virtual PET images from CT data using deep convolutional networks: initial results. In: Tsaftaris, S.A., Gooya, A., Frangi, A.F., Prince, J.L. (eds.) SASHIMI 2017. LNCS, vol. 10557, pp. 49–57. Springer, Cham (2017). https://doi.org/10.1007/978-3-319-68127-6_6
4. Bengio, Y., Courville, A., Vincent, P.: Representation learning: a review and new perspectives. IEEE TPAMI **35**(8), 1798–1828 (2013)
5. Bernard, O., Lalande, A., Zotti, C., et al.: Deep learning techniques for automatic MRI cardiac multi-structures segmentation and diagnosis: is the problem solved? IEEE TMI **37**(11), 2514–2525 (2018)

6. Campello, V.M., et al.: Multi-centre, multi-vendor and multi-disease cardiac segmentation: the M&Ms challenge. IEEE TMI (2020, under review)
7. Chartsias, A., et al.: Disentangled representation learning in cardiac image analysis. MIA **58**, 101535 (2019)
8. Chartsias, A., et al.: Multimodal cardiac segmentation using disentangled representation learning. In: Pop, M., et al. (eds.) STACOM 2019. LNCS, vol. 12009, pp. 128–137. Springer, Cham (2020). https://doi.org/10.1007/978-3-030-39074-7_14
9. Chen, C., Dou, Q., Jin, Y., Chen, H., Qin, J., Heng, P.-A.: Robust multimodal brain tumor segmentation via feature disentanglement and gated fusion. In: Shen, D., et al. (eds.) MICCAI 2019. LNCS, vol. 11766, pp. 447–456. Springer, Cham (2019). https://doi.org/10.1007/978-3-030-32248-9_50
10. Costa, P., et al.: End-to-end adversarial retinal image synthesis. IEEE TMI **37**(3), 781–791 (2018)
11. Dar, S.U., Yurt, M., Karacan, L., Erdem, A., Erdem, E., Çukur, T.: Image synthesis in multi-contrast MRI with conditional generative adversarial networks. IEEE TMI **38**(10), 2375–2388 (2019)
12. Dice, L.R.: Measures of the amount of ecologic association between species. Ecology **26**(3), 297–302 (1945)
13. Frid-Adar, M., Diamant, I., Klang, E., Amitai, M., Goldberger, J., Greenspan, H.: GAN-based synthetic medical image augmentation for increased CNN performance in liver lesion classification. Neurocomputing **321**, 321–331 (2018)
14. Gabbay, A., Hoshen, Y.: Demystifying inter-class disentanglement. In: ICLR (2020)
15. Goodfellow, I., et al.: Generative adversarial nets. In: Proceedings of the NeurIPS, pp. 2672–2680 (2014)
16. Guibas, J.T., Virdi, T.S., Li, P.S.: Synthetic medical images from dual generative adversarial networks. In: Advances in Neural Information Processing Systems Workshop (2017)
17. Havaei, M., Mao, X., Wang, Y., Lao, Q.: Conditional generation of medical images via disentangled adversarial inference. Med. Image Anal. **72**, 102106 (2020)
18. Heusel, M., Ramsauer, H., Unterthiner, T., Nessler, B., Hochreiter, S.: GANs trained by a two time-scale update rule converge to a local Nash equilibrium. In: Proceedings of the NeurIPS, pp. 6626–6637 (2017)
19. Hu, X., Chung, A.G., Fieguth, P., Khalvati, F., Haider, M.A., Wong, A.: ProstateGAN: mitigating data bias via prostate diffusion imaging synthesis with generative adversarial networks. In: NeurIPS Workshop (2018)
20. Huang, X., Belongie, S.: Arbitrary style transfer in real-time with adaptive instance normalization. In: Proceedings of the ICCV, pp. 1501–1510 (2017)
21. Jin, D., Xu, Z., Tang, Y., Harrison, A.P., Mollura, D.J.: CT-realistic lung nodule simulation from 3D conditional generative adversarial networks for robust lung segmentation. In: Frangi, A.F., Schnabel, J.A., Davatzikos, C., Alberola-López, C., Fichtinger, G. (eds.) MICCAI 2018. LNCS, vol. 11071, pp. 732–740. Springer, Cham (2018). https://doi.org/10.1007/978-3-030-00934-2_81
22. Karras, T., Laine, S., Aila, T.: A style-based generator architecture for generative adversarial networks. In: Proceedings of the CVPR, pp. 4396–4405 (2019)
23. Li, K., Yu, L., Wang, S., Heng, P.-A.: Unsupervised retina image synthesis via disentangled representation learning. In: Burgos, N., Gooya, A., Svoboda, D. (eds.) SASHIMI 2019. LNCS, vol. 11827, pp. 32–41. Springer, Cham (2019). https://doi.org/10.1007/978-3-030-32778-1_4
24. Li, Q., Yu, Z., Wang, Y., Zheng, H.: TumorGAN: a multi-modal data augmentation framework for brain tumor segmentation. Sensors **20**(15), 4203 (2020)

25. Mao, X., Li, Q., Xie, H., Lau, R.Y.K., Wang, Z., Smolley, S.P.: Least squares generative adversarial networks. In: Proceedings of the ICCV, pp. 2813–2821 (2017)
26. Miyato, T., Kataoka, T., Koyama, M., Yoshida, Y.: Spectral normalization for generative adversarial networks. In: ICLR (2018)
27. Mok, T.C.W., Chung, A.C.S.: Learning data augmentation for brain tumor segmentation with coarse-to-fine generative adversarial networks. In: Crimi, A., Bakas, S., Kuijf, H., Keyvan, F., Reyes, M., van Walsum, T. (eds.) BrainLes 2018. LNCS, vol. 11383, pp. 70–80. Springer, Cham (2019). https://doi.org/10.1007/978-3-030-11723-8_7
28. Nie, D., et al.: Medical image synthesis with context-aware generative adversarial networks. In: Descoteaux, M., Maier-Hein, L., Franz, A., Jannin, P., Collins, D.L., Duchesne, S. (eds.) MICCAI 2017. LNCS, vol. 10435, pp. 417–425. Springer, Cham (2017). https://doi.org/10.1007/978-3-319-66179-7_48
29. Odena, A., Olah, C., Shlens, J.: Conditional image synthesis with auxiliary classifier GANs. In: Proceedings of the ICML, pp. 2642–2651 (2017)
30. Park, T., Liu, M.Y., Wang, T.C., Zhu, J.Y.: Semantic image synthesis with spatially-adaptive normalization. In: Proceedings of the CVPR, pp. 2337–2346 (2019)
31. Radford, A., Metz, L., Chintala, S.: Unsupervised representation learning with deep convolutional generative adversarial networks. In: ICLR (2016)
32. Ronneberger, O., Fischer, P., Brox, T.: U-Net: convolutional networks for biomedical image segmentation. In: Navab, N., Hornegger, J., Wells, W.M., Frangi, A.F. (eds.) MICCAI 2015. LNCS, vol. 9351, pp. 234–241. Springer, Cham (2015). https://doi.org/10.1007/978-3-319-24574-4_28
33. Shin, H.-C., et al.: Medical image synthesis for data augmentation and anonymization using generative adversarial networks. In: Gooya, A., Goksel, O., Oguz, I., Burgos, N. (eds.) SASHIMI 2018. LNCS, vol. 11037, pp. 1–11. Springer, Cham (2018). https://doi.org/10.1007/978-3-030-00536-8_1
34. Simonyan, K., Zisserman, A.: Very deep convolutional networks for large-scale image recognition. In: ICLR (2015)
35. Sørensen, T.: A method of establishing groups of equal amplitude in plant sociology based on similarity of species content and its application to analyses of the vegetation on Danish commons. R. Danish Acad. Sci. Lett. **5**(4), 1–34 (1948)
36. Yang, J., Dvornek, N.C., Zhang, F., Chapiro, J., Lin, M.D., Duncan, J.S.: Unsupervised domain adaptation via disentangled representations: application to cross-modality liver segmentation. In: Shen, D., et al. (eds.) MICCAI 2019. LNCS, vol. 11765, pp. 255–263. Springer, Cham (2019). https://doi.org/10.1007/978-3-030-32245-8_29

CoTr: Efficiently Bridging CNN and Transformer for 3D Medical Image Segmentation

Yutong Xie[1,2], Jianpeng Zhang[1,2], Chunhua Shen[2], and Yong Xia[1(✉)]

[1] National Engineering Laboratory for Integrated Aero-Space-Ground-Ocean Big Data Application Technology, School of Computer Science and Engineering, Northwestern Polytechnical University, Xi'an 710072, China
yxia@nwpu.edu.cn
[2] The University of Adelaide, Adelaide, Australia

Abstract. Convolutional neural networks (CNNs) have been the de facto standard for nowadays 3D medical image segmentation. The convolutional operations used in these networks, however, inevitably have limitations in modeling the long-range dependency due to their inductive bias of locality and weight sharing. Although Transformer was born to address this issue, it suffers from extreme computational and spatial complexities in processing high-resolution 3D feature maps. In this paper, we propose a novel framework that efficiently bridges a **C**onvolutional neural network and a **Tr**ansformer (**CoTr**) for accurate 3D medical image segmentation. Under this framework, the CNN is constructed to extract feature representations and an efficient deformable Transformer (DeTrans) is built to model the long-range dependency on the extracted feature maps. Different from the vanilla Transformer which treats all image positions equally, our DeTrans pays attention only to a small set of key positions by introducing the deformable self-attention mechanism. Thus, the computational and spatial complexities of DeTrans have been greatly reduced, making it possible to process the multi-scale and high-resolution feature maps, which are usually of paramount importance for image segmentation. We conduct an extensive evaluation on the Multi-Atlas Labeling Beyond the Cranial Vault (BCV) dataset that covers 11 major human organs. The results indicate that our CoTr leads to a substantial performance improvement over other CNN-based, transformer-based, and hybrid methods on the 3D multi-organ segmentation task. Code is available at: https://github.com/YtongXie/CoTr.

Keywords: 3D Medical image segmentation · Deformable self-attention · CNN · Transformer

---

Y. Xie—The work was partially done while the first two authors were visiting The University of Adelaide.

M. de Bruijne et al. (Eds.): MICCAI 2021, LNCS 12903, pp. 171–180, 2021.
https://doi.org/10.1007/978-3-030-87199-4_16

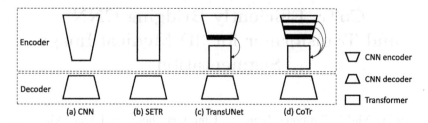

Fig. 1. Comparison of different segmentation architectures. All of them have an encoder-decoder structure, but with different encoders. The encoder in CNN (a) is composed of multiple stacked convolutional layers. The encoder in SETR (b) is purely formed from self-attention layers, *i.e.*, Transformer. The encoder in both TransUNet (c) and our proposed CoTr (d) are the hybrid of CNN and Transformer. Differently, TransUNet only processes the low-resolution feature maps from the last stage due to the high computation and spatial complexities. Thanks to the efficient design of Transformer, CoTr is able to process the multi-scale and high-resolution feature maps.

1   Introduction

Image segmentation is a longstanding challenge in medical image analysis. Since the introduction of U-Net [16], fully convolutional neural networks (CNNs) have become the predominant approach to addressing this task [10,12,23–25,28]. Despite their prevalence, CNNs still suffer from the limited receptive field and fail to capture the long-range dependency, due to the inductive bias of locality and weight sharing [6]. Many efforts have been devoted to enlarge a CNN's receptive field thus improve its ability to context modeling. Yu *et al.* [22] proposed the atrous convolution with an adjustable dilated rate, which shows superior performance in semantic segmentation [5]. More straightforwardly, Peng *et al.* [15] designed large kernels to capture rich global context information. Zhao *et al.* [26] employed the pyramid pooling at multiple feature scales to aggregate multi-scale global information. Wang *et al.* [20] presented the non-local operations which is usually embedded at the end of encoder to capture the long-range dependency. Although improving the context modeling to some extent, these models still have an inevitably limited receptive field, stranded by the CNN architecture.

Transformer, a sequence-to-sequence prediction framework, has a proven track record in machine translation and nature language processing [8,19], due to its strong ability to long-range modeling. The self-attention mechanism in Transformer can dynamically adjust the receptive field according to the input content, and hence is superior to convolutional operations in modeling the long-range dependency.

Recently, Transformer has been considered as an alternative architecture, and has achieved competitive performance on many computer vision tasks, like image recognition [9,17], semantic/instance segmentation [21,27], object detection [2,29], low-level vision [3,14], and image generation [13]. A typical example is the vision Transformer (ViT) [9], which outperforms a ResNet-based CNN on

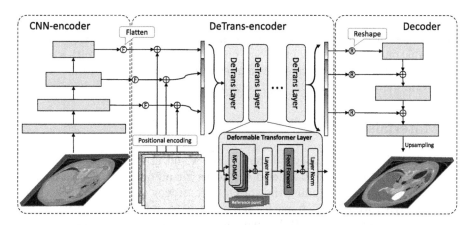

Fig. 2. Diagram of CoTr: A CNN-encoder, a DeTrans-encoder, and a decoder. Gray rectangles: CNN blocks. Yellow rectangles: 3D deformable Transformer layers. The CNN-encoder extracts multi-scale feature maps from an input image. The DeTrans-encoder processes the flattened multi-scale feature maps that embedded with the positional encoding in a sequence-to-sequence manner. The features with long-range dependency are generated by the DeTrans-encoder and fed to the decoder for segmentation. (Color figure online)

recognition tasks but at a cost of using 300M data for training. Since a huge training dataset is not always available, recent studies attempt to combine a CNN and a Transformer into a hybrid model. Carion *et al.* [2] employed a CNN to extract image features and a Transformer to further process the extracted features. Chen *et al.* [4] designed TransUNet, in which a CNN and a Transformer are combined in a cascade manner to make a strong encoder for 2D medical image segmentation. Although the design of TransUNet is interesting and the performance is good, it is challenging to optimize this model due to the existence of self-attention [19]. First, it requires extremely long training time to focus the attention, which was initially cast to each pixel uniformly, on salient locations, especially in a 3D scenario. Second, due to its high computational complexity, a vanilla Transformer [19] can hardly process multi-scale and high-resolution feature maps, which play a critical role in image segmentation.

In this paper, we propose a hybrid framework that efficiently bridges **Co**nvolutional neural network and **Tr**ansformer (**CoTr**) for 3D medical image segmentation. CoTr has an encoder-decoder structure. In the encoder, a concise CNN structure is adopted to extract feature maps and a Transformer is used to capture the long-range dependency (see Fig. 1). Inspired by [7,29], we introduce the deformable self-attention mechanism to the Transformer. This attention mechanism casts attentions only to a small set of key sampling points, and thus dramatically reduces the computational and spatial complexity of Transformer. As a result, it is possible for the Transformer to process the multi-scale feature maps produced by the CNN and keep abundant high resolution information for segmentation. The main contributions of this paper are three-fold: (1) we are the first to explore Transformer for 3D medical image segmentation, particularly in a computationally and spatially efficient way; (2) we introduce the deformable self-attention mechanism to reduce the complexity of vanilla Transformer, and thus enable our CoTr to model the long-range dependency using multi-scale features; (3) our CoTr outperforms the competing CNN-based, Transformer-based, and hybrid methods on the 3D multi-organ segmentation task.

2   Materials

The Multi-Atlas Labeling **B**eyond the **C**ranial **V**ault (BCV) dataset[1] was used for this study. It contains 30 labeled CT scans for automated segmentation of 11 abdominal organs, including the spleen (Sp), kidney (Ki), gallbladder (Gb), esophagus (Es), liver (Li), stomach (St), aorta (Ao), inferior vena cava (IVC), portal vein and splenic vein (PSV), pancreas (Pa), and adrenal gland (AG).

3   Methods

CoTr aims to learn more effective representations for medical image segmentation via bridging CNN and Transformer. As shown in Fig. 2, it consists of a CNN-encoder for feature extraction, a deformable Transformer-encoder (DeTrans-encoder) for long-range dependency modeling, and a decoder for segmentation. We now delve into the details of each module.

3.1   CNN-encoder

The CNN-encoder $\mathcal{F}^{CNN}(\cdot)$ contains a Conv-IN-LeakyReLU block and three stages of 3D residual blocks. The Conv-IN-LeakyReLU block contains a 3D convolutional layer followed by an instance normalization (IN) [18] and leaky rectified linear unit (LeakyReLU) activation. The numbers of 3D residual blocks in three stages are three, three, and two, respectively.

Given an input image x with a height of H, a width of W, and a depth (*i.e.*, number of slices) of D, the feature maps produced by $\mathcal{F}^{CNN}(\cdot)$ can be formally expressed as

$$\{f_l\}_{l=1}^{L} = \mathcal{F}_l^{CNN}(x; \boldsymbol{\Theta}) \in \mathbb{R}^{C \times \frac{D}{2^l} \times \frac{H}{2^{l+1}} \times \frac{W}{2^{l+1}}}, \tag{1}$$

where L indicates the number of feature levels, $\boldsymbol{\Theta}$ denotes the parameters of the CNN-encoder, and C denotes the number of channels.

3.2   DeTrans-Encoder

Due to the intrinsic locality of convolution operations, the CNN-encoder cannot capture the long-range dependency of pixels effectively. To this end, we propose the DeTrans-encoder that introduces the multi-scale deformable self-attention (MS-DMSA) mechanism for efficient long-range contextual modeling.

---

[1] https://www.synapse.org/#!Synapse:syn3193805/wiki/217789.

The DeTrans-encoder is a composition of an input-to-sequence layer and L_D stacked deformable Transformer (DeTrans) layers.

Input-to-Sequence Transformation. Considering that Transformer processes the information in a sequence-to-sequence manner, we first flatten the feature maps produced by the CNN-encoder $\{f_l\}_{l=1}^{L}$ into a 1D sequence. Unfortunately, the operation of flattening the features leads to losing the spatial information that is critical for image segmentation. To address this issue, we supplement the 3D positional encoding sequence $\{p_l\}_{l=1}^{L}$ to the flattened $\{f_l\}_{l=1}^{L}$. For this study, we use sine and cosine functions with different frequencies [19] to compute the positional coordinates of each dimension pos, shown as follows

$$\begin{cases} PE_\#(pos, 2k) = sin(pos \cdot v) \\ PE_\#(pos, 2k+1) = cos(pos \cdot v) \end{cases} \tag{2}$$

where $\# \in \{D, H, W\}$ indicates each of three dimensions, $v = 1/10000^{2k/\frac{C}{3}}$. For each feature level l, we concatenate PE_D, PE_H, and PE_W as the 3D positional encoding p_l and combine it with the flattened f_l via element-wise summation to form the input sequence of DeTrans-encoder.

MS-DMSA Layer. In the architecture of Transformer, the self-attention layer would look over all possible locations in the feature map. It has the drawback of slow convergence and high computational complexity, and hence can hardly process multi-scale features. To remedy this, we design the MS-DMSA layer that focuses only on a small set of key sampling locations around a reference location, instead of all locations.

Let $z_q \in \mathbb{R}^C$ be the feature representation of query q and $\hat{p}_q \in [0,1]^3$ be the normalized 3D coordinate of the reference point. Given the multi-scale feature maps $\{f_l\}_{l=1}^{L}$ that are extracted in the last L stages of CNN-encoder, the feature representation of the i-th attention head can be calculated as

$$\text{head}_i = \sum_l^L \sum_k^K \Lambda(z_q)_{ilqk} \cdot \Psi(f_l)(\sigma_l(\hat{p}_q) + \Delta_{p_{ilqk}}) \tag{3}$$

where K is the number of sampled key points, $\Psi(\cdot)$ is a linear projection layer, $\Lambda(z_q)_{ilqk} \in [0,1]$ is the attention weights, $\Delta_{p_{ilqk}} \in \mathbb{R}^3$ is the sampling offset of the k-th sampling point in the l-th feature level, and $\sigma_l(\cdot)$ re-scales $\hat{p}_q$ to the l-th level feature. Following [29], both $\Lambda(z_q)_{ilqk}$ and $\Delta_{p_{ilqk}}$ are obtained via linear projection over the query feature z_q. Then, the MS-DMSA layer can be formulated as

$$\text{MS} - \text{DMSA}(z_q, \{f_l\}_{l=1}^{L}) = \Phi(\text{Concat}(\text{head}_1, \text{head}_2, ..., \text{head}_H)) \tag{4}$$

where H is the number of attention heads, and $\Phi(\cdot)$ is a linear projection layer that weights and aggregates the feature representation of all attention heads.

DeTrans Layer. The DeTrans layer is composed of a MS-DMSA layer and a feed forward network, each being followed by the layer normalization [1] (see Fig. 2). The skip connection strategy [11] is employed in each sub-layer to avoid gradient vanishing. The DeTrans-encoder is constructed by repeatedly stacking DeTrans layers.

3.3   Decoder

The output sequence of DeTrans-encoder is reshaped into feature maps according to the size at each scale. The decoder, a pure CNN architecture, progressively upsamples the feature maps to the input resolution (*i.e.*, $D \times H \times W$) using the transpose convolution, and then refines the upsampled feature maps using a 3D residual block. Besides, the skip connections between encoder and decoder are also added to keep more low-level details for better segmentation. We also use the deep supervision strategy by adding auxiliary losses to the decoder outputs with different scales. The loss function of our model is the sum of the Dice loss and cross-entropy loss [12,24,28]. More details on the network architecture gare in Appendix.

3.4   Implementation Details

Following [12], we first truncated the HU values of each scan using the range of $[-958, 327]$ to filter irrelevant regions, and then normalized truncated voxel values by subtracting 82.92 and dividing by 136.97. We randomly split the BCV dataset into two parts: 21 scans for training and 9 scans for test, and randomly selected 6 training scans to form a validation set, which just was used to select the hyper-parameters of CoTr. The final results on the test set are obtained by the model trained on all training scans.

In the training stage, we randomly cropped sub-volumes of size $48 \times 192 \times 192$ from CT scans as the input. To alleviate the over-fitting of limited training data, we employed the online data argumentation [12], including the random rotation, scaling, flipping, adding white Gaussian noise, Gaussian blurring, adjusting rightness and contrast, simulation of low resolution, and Gamma transformation, to diversify the training set. Due to the benefits of instance normalization [18], we adopted the micro-batch training strategy with a small batch size of 2. To weigh the balance between training time cost and performance reward, CoTr was trained for 1000 epochs and each epoch contains 250 iterations. We adopted the stochastic gradient descent algorithm with a momentum of 0.99 and an initial learning rate of 0.01 as the optimizer. We set the hidden size in MS-DMSA and feed forward network to 384 and 1536, respectively, and empirically set the hyper-parameters $L_D = 6$, $H = 6$, and $K = 4$. Besides, we formed two variants of CoTr with small CNN-encoders, denoted as CoTr* and CoTr†. In CoTr*, there is only one 3D residual block in each stage of CNN-encoder. In CoTr†, the number of 3D residual blocks in each stage of CNN-encoder is two.

In the test stage, we employed the sliding window strategy, where the window size equals to the training patch size. Besides, Gaussian importance weighting [12] and test time augmentation by flipping along all axes were also utilized to improve the robustness of segmentation. To quantitatively evaluate the segmentation results, we calculated the Dice coefficient scores (Dice) metric that measures the overlapping between a prediction and its ground truth, and the Hausdorff distance (HD) metric that evaluates the quality of segmentation boundaries by computing the maximum distance between the predicted boundaries and ground truth.

Table 1. Dice and HD scores of our CoTr and several competing methods on the BCV test set. **CoTr*** and **CoTr**† are two variants of CoTr with small CNN-encoders.

Methods	Param (M)	Dice scores of different organs											Ave Dice	Ave HD
		Sp	Ki	Gb	Es	Li	St	Ao	IVC	PSV	Pa	AG		
SETR (ViT-B/16-rand) [27]	100.5	95.2	92.3	55.6	71.3	96.2	80.2	89.7	83.9	68.9	68.7	60.5	78.4	7.47
SETR (ViT-B/16-pre) [27]	100.5	94.8	91.7	55.2	70.9	96.2	76.9	89.3	82.4	69.6	70.7	58.7	77.8	8.47
CoTr w/o CNN-encoder	21.9	95.2	92.8	59.2	72.2	96.3	81.2	89.9	85.1	71.9	73.3	61.0	79.8	7.23
CoTr w/o DeTrans	32.6	96.0	92.6	63.8	77.9	97.0	83.6	90.8	87.8	76.7	81.2	72.6	83.6	5.18
APSS [5]	45.5	96.5	93.8	65.6	78.1	97.1	84.0	91.1	87.9	77.0	82.6	73.9	84.3	4.85
PP [26]	33.9	96.1	93.1	64.3	77.4	97.0	85.3	90.8	87.4	77.2	81.9	72.8	83.9	5.10
Non-local [20]	32.8	96.3	93.7	64.6	77.9	97.1	84.1	90.8	87.7	77.2	82.1	73.3	84.1	4.70
TransUnet [4]	43.5	95.9	93.7	63.1	77.8	97.0	86.2	91.0	87.8	77.8	81.6	73.9	84.2	4.77
CoTr*	27.9	96.4	94.0	66.2	76.4	97.0	84.2	90.3	87.6	76.3	80.8	72.9	83.8	5.04
CoTr†	36.9	96.2	93.8	66.5	78.6	97.1	86.9	90.8	87.8	77.7	82.8	73.2	84.7	4.39
CoTr	41.9	96.3	93.9	66.6	78.0	97.1	88.2	91.2	88.0	78.1	83.1	74.1	**85.0**	**4.01**

4   Results

Comparing to Models with only Transformer Encoder. We first evaluated our CoTr against two variants of the state-of-the-art SEgmentation Transformer (SETR) [27], which were formed by using randomly initialized and pretrained ViT-B/16 [9] as the encoder. We also compared to a variant of CoTr that removes the CNN-encoder (CoTr w/o CNN-encoder). To ensure an unprejudiced comparison, all models use the same decoder. The segmentation performance of these models is shown in Table 1, from which three conclusions can be drawn. First, although the Transformer architecture is not limited by the type of input images, the ViT-B/16 pre-trained on 2D natural images does not work well on 3D medical images. The suboptimal performance may be attributed to the domain shift between 2D natural images and 3D medical images. Second, 'CoTr w/o CNN-encoder' has about 22M parameters and outperforms the SETR with about 100M parameters. We believe that a lightweight Transformer may be more friendly for medical image segmentation tasks, where there is usually a small training dataset. Third, our CoTr* with comparable parameters significantly outperforms 'CoTr w/o CNN-encoder', improving the average Dice over 11 organs by 4%. It suggests that the hybrid CNN-Transformer encoder has distinct advantages over the pure Transformer encoder in medical image segmentation.

Comparing to Models with only CNN Encoder. Then, we compared CoTr against a variant of CoTr that removes the DeTrans-encoder (CoTr w/o DeTrans) and three CNN-based context modeling methods, *i.e.*, the Atrous Spatial Pyramid Pooling (ASPP) [5] module, pyramid parsing (PP) [26] module, and Non-local [20] module. For a fair comparison, we used the same CNN-encoder and decoder but replaced our DeTrans-encoder with ASPP, PP, and Non-local modules, respectively. The results in Table 1 shows that our CoTr elevates consistently the segmentation performance over 'CoTr w/o DeTrans' on all organs and improves the average Dice by 1.4%. It corroborates that our CoTr using a hybrid CNN-Transformer encoder has a stronger ability than using a pure

CNN encoder to learn effective representations for medical image segmentation. Moreover, comparing to these context modeling methods, our Transformer architecture contributes to more accurate segmentation.

Comparing to Models with Hybrid CNN-Transformer Encoder. We also compared CoTr to other hybrid CNN-Transformer architectures like TransUNet [4]. To process 3D images directly, we extended the original 2D TransUNet to a 3D version by using 3D CNN-encoder and decoder as done in CoTr. We also set the number of heads and layers of Transformer in 3D TransUNet to be the same as our CoTr. It shows in Table 1 that CoTr steadily beats TransUNet in the segmentation of all organs, particularly for the gallbladder and pancreas segmentation. Even with a smaller CNN-encoder, CoTr† still achieves better performance than TransUnet in the segmentation of seven organs. The superior performance owes to the deformable mechanism in CoTr that makes it possible to process high-resolution and multi-scale feature maps due to the reduced computational and spatial complexities.

Computational Complexity. The proposed CoTr was trained using a workstation with a NVIDIA GTX 2080Ti GPU. It took about 2 days for training, and less than 30ms to segment a volume of size $48 \times 192 \times 192$.

5   Discussion on Hyper-parameter Settings

In the DeTrans-encoder, there are three hyper-parameters, *i.e.*, K, H, and L_D, which represent the number of sampled key points, heads, and stacked DeTrans layers, respectively. To investigate the impact of their settings on the segmentation, we set K to 1, 2, and 4, set H to 2, 4, and 6, and set L_D to 2, 4, and 6. In Fig. 3 (a-c), we plotted the average Dice over all organs obtained on the validation set versus the values of K, H, and L_D. It shows that increasing the number of K, H, or L_D can improve the segmentation performance. To demonstrate the performance gain resulted from the multi-scale strategy, we also attempted to train CoTr with single-scale feature maps from the last stage. The results in Fig. 3 (d) show that using multi-scale feature maps instead of single-scale feature maps can effectively improve the average Dice by 1.2%.

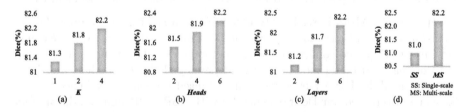

Fig. 3. Average Dice over all organs obtained on the validation set versus (a) the number of sampled key points K, (b) number of heads H, and (c) number of DeTrans layers L_D, and (d) Average Dice obtained by our CoTr using, respectively, single-scale and multi-scale feature maps on the validation set.

6   Conclusion

In this paper, we propose a hybrid model of CNN Transformer, namely CoTr, for 3D medical image segmentation. In this model, we design the deformable Transformer (DeTrans) that employs the deformable self-attention mechanism to reduce the computational and spatial complexities of modelling the long-range dependency on multi-scale and high-resolution feature maps. Comparative experiments were conducted on the BCV dataset. The superior performance of our CoTr over both CNN-based and vanilla Transformer-based models suggests that, via combining the advantages of CNN and Transformer, the proposed CoTr achieves the balance in keeping the details of low-level features and modeling the long-range dependency. As a stronger baseline, our CoTr can be extended to deal with other structures (*e.g.*, brain structure or tumor segmentation) in the future.

Acknowledgment. Y Xie, J Zhang, and Y Xia were supported in part by the National Natural Science Foundation of China under Grants 61771397, in part by the CAAI-Huawei MindSpore Open Fund under Grants CAAIXSJLJJ-2020-005B, and in part by Innovation Foundation for Doctor Dissertation of Northwestern Polytechnical University under Grants CX202010.

References

1. Ba, J.L., Kiros, J.R., Hinton, G.E.: Layer normalization. arXiv preprint arXiv:1607.06450 (2016)
2. Carion, N., Massa, F., Synnaeve, G., Usunier, N., Kirillov, A., Zagoruyko, S.: End-to-end object detection with transformers. In: Vedaldi, A., Bischof, H., Brox, T., Frahm, J.-M. (eds.) ECCV 2020. LNCS, vol. 12346, pp. 213–229. Springer, Cham (2020). https://doi.org/10.1007/978-3-030-58452-8_13
3. Chen, H., et al.: Pre-trained image processing transformer. arXiv preprint arXiv:2012.00364 (2020)
4. Chen, J., et al.: Transunet: Transformers make strong encoders for medical image segmentation. arXiv preprint arXiv:2102.04306 (2021)
5. Chen, L.C., Zhu, Y., Papandreou, G., Schroff, F., Adam, H.: Encoder-decoder with atrous separable convolution for semantic image segmentation. In: Proceedings of the European Conference on Computer Vision (ECCV), pp. 801–818 (2018)
6. Cohen, N., Shashua, A.: Inductive bias of deep convolutional networks through pooling geometry. arXiv preprint arXiv:1605.06743 (2016)
7. Dai, J., et al.: Deformable convolutional networks. In: Proceedings of the IEEE International Conference on Computer Vision, pp. 764–773 (2017)
8. Devlin, J., Chang, M.W., Lee, K., Toutanova, K.: Bert: Pre-training of deep bidirectional transformers for language understanding. arXiv preprint arXiv:1810.04805 (2018)
9. Dosovitskiy, A., et al.: An image is worth 16x16 words: Transformers for image recognition at scale. arXiv preprint arXiv:2010.11929 (2020)
10. Fang, X., Yan, P.: Multi-organ segmentation over partially labeled datasets with multi-scale feature abstraction. IEEE Trans. Med. Imaging **39**(11), 3619–3629 (2020)

11. He, K., Zhang, X., Ren, S., Sun, J.: Deep residual learning for image recognition. In: Proceedings of the IEEE Conference on Computer Vision and Pattern Recognition, pp. 770–778 (2016)
12. Isensee, F., Jäger, P.F., Kohl, S.A., Petersen, J., Maier-Hein, K.H.: Automated design of deep learning methods for biomedical image segmentation. arXiv preprint arXiv:1904.08128 (2019)
13. Jiang, Y., Chang, S., Wang, Z.: Transgan: Two transformers can make one strong gan. arXiv preprint arXiv:2102.07074 (2021)
14. Parmar, N., et al.: Image transformer. In: International Conference on Machine Learning, pp. 4055–4064. PMLR (2018)
15. Peng, C., Zhang, X., Yu, G., Luo, G., Sun, J.: Large kernel matters-improve semantic segmentation by global convolutional network. In: Proceedings of the IEEE Conference on Computer Vision and Pattern Recognition (CVPR), pp. 4353–4361 (2017)
16. Ronneberger, O., Fischer, P., Brox, T.: U-Net: convolutional networks for biomedical image segmentation. In: Navab, N., Hornegger, J., Wells, W.M., Frangi, A.F. (eds.) MICCAI 2015. LNCS, vol. 9351, pp. 234–241. Springer, Cham (2015). https://doi.org/10.1007/978-3-319-24574-4_28
17. Touvron, H., Cord, M., Douze, M., Massa, F., Sablayrolles, A., Jégou, H.: Training data-efficient image transformers & distillation through attention. arXiv preprint arXiv:2012.12877 (2020)
18. Ulyanov, D., Vedaldi, A., Lempitsky, V.: Instance normalization: The missing ingredient for fast stylization. arXiv preprint arXiv:1607.08022 (2016)
19. Vaswani, A., et al.: Attention is all you need. In: Proceedings of the 31st International Conference on Neural Information Processing Systems, pp. 6000–6010 (2017)
20. Wang, X., Girshick, R., Gupta, A., He, K.: Non-local neural networks. In: Proceedings of the IEEE Conference on Computer Vision and Pattern Recognition, pp. 7794–7803 (2018)
21. Wang, Y., et al.: End-to-end video instance segmentation with transformers. arXiv preprint arXiv:2011.14503 (2020)
22. Yu, F., Koltun, V.: Multi-scale context aggregation by dilated convolutions. In: International Conference on Learning Representations (ICLR) (2016)
23. Zhang, J., Xie, Y., Wang, Y., Xia, Y.: Inter-slice context residual learning for 3D medical image segmentation. IEEE Trans. Med. Imaging **40**(2), 661–672 (2020)
24. Zhang, J., Xie, Y., Zhang, P., Chen, H., Xia, Y., Shen, C.: Light-weight hybrid convolutional network for liver tumor segmentation. In: IJCAI, pp. 4271–4277 (2019)
25. Zhang, L., et al.: Block level skip connections across cascaded v-net for multi-organ segmentation. IEEE Trans. Med. Imaging **39**(9), 2782–2793 (2020)
26. Zhao, H., Shi, J., Qi, X., Wang, X., Jia, J.: Pyramid scene parsing network. In: Proceedings of the IEEE Conference on Computer Vision and Pattern Recognition, pp. 2881–2890 (2017)
27. Zheng, S., et al.: Rethinking semantic segmentation from a sequence-to-sequence perspective with transformers. arXiv preprint arXiv:2012.15840 (2020)
28. Zhou, Z., Siddiquee, M.M.R., Tajbakhsh, N., Liang, J.: UNet++: redesigning skip connections to exploit multiscale features in image segmentation. IEEE Trans. Med. Imaging **39**(6), 1856–1867 (2019)
29. Zhu, X., Su, W., Lu, L., Li, B., Wang, X., Dai, J.: Deformable detr: Deformable transformers for end-to-end object detection. arXiv preprint arXiv:2010.04159 (2020)

Harmonization with Flow-Based Causal Inference

Rongguang Wang[1,2(✉)], Pratik Chaudhari[1,3], and Christos Davatzikos[1,2,4]

[1] Department of Electrical and Systems Engineering, University of Pennsylvania, Philadelphia, USA
{rgw,pratikac}@seas.upenn.edu, Christos.Davatzikos@pennmedicine.upenn.edu
[2] Center for Biomedical Image Computing and Analytics (CBICA), Philadelphia, USA
[3] General Robotics, Automation, Sensing and Perception Laboratory (GRASP), Philadelphia, USA
[4] Department of Radiology, Perelman School of Medicine, University of Pennsylvania, Philadelphia, USA

Abstract. Heterogeneity in medical data, e.g., from data collected at different sites and with different protocols in a clinical study, is a fundamental hurdle for accurate prediction using machine learning models, as such models often fail to generalize well. This paper leverages a recently proposed normalizing-flow-based method to perform counterfactual inference upon a structural causal model (SCM), in order to achieve harmonization of such data. A causal model is used to model observed effects (brain magnetic resonance imaging data) that result from known confounders (site, gender and age) and exogenous noise variables. Our formulation exploits the bijection induced by flow for the purpose of harmonization. We infer the posterior of exogenous variables, intervene on observations, and draw samples from the resultant SCM to obtain counterfactuals. This approach is evaluated extensively on multiple, large, real-world medical datasets and displayed better cross-domain generalization compared to state-of-the-art algorithms. Further experiments that evaluate the quality of confounder-independent data generated by our model using regression and classification tasks are provided.

Keywords: Harmonization · Causal inference · Normalizing flows

1 Introduction

Deep learning models have shown great promise in medical imaging diagnostics [11] and predictive modeling with applications ranging from segmentation tasks [19] to more complex decision-support functions for phenotyping brain diseases and personalized prognosis. However deep learning models tend to have

---

Electronic supplementary material The online version of this chapter (https://doi.org/10.1007/978-3-030-87199-4_17) contains supplementary material, which is available to authorized users.

M. de Bruijne et al. (Eds.): MICCAI 2021, LNCS 12903, pp. 181–190, 2021.
https://doi.org/10.1007/978-3-030-87199-4_17

poor reproducibility across hospitals, scanners, and patient cohorts; these high-dimensional models tend to overfit to specific datasets and generalize poorly across training data [6]. One potential solution to the above problem is to train on very large and diverse databases but this can be prohibitive, because data may change frequently (e.g., new imaging devices are introduced) and gathering training labels for medical images is expensive. More importantly, even if it were possible to train a model on data that covers all possible variations across images, such a model would almost certainly sacrifice accuracy in favor of generalization—it would rely on coarse imaging features that are stable across, say imaging devices and patient populations, and might fail to capture more subtle and informative detail. Methods that can tackle heterogeneity in medical data without sacrificing predictive accuracy are needed, including methods for "data harmonization", which would allow training a classifier on, say data from one site, and obtaining similar predictive accuracy on data from another site.

Contributions. We build upon a recently proposed framework [26] for causal inference, by modeling brain imaging data and clinical variables via a causal graph and focus on how causes (site, gender and age) result in the effects, namely imaging measurements (herein we use region of interest (ROI) volumes obtained by preprocessing brain MRI data). This framework uses a normalizing flow parameterized by deep networks to learn the structural assignments in a causal graph. We demonstrate how harmonization of data can be performed efficiently using counterfactual inference on such a flow-based causal model. Given a dataset pertaining to one site (source), we perform a counterfactual query to synthesize the dataset, as if it were from another site (target). Essentially, this amounts to the counterfactual question "what would the scans look like if they had been acquired from the same site". We demonstrate results of such harmonization on regression (age prediction) and classification (predicting Alzheimer's disease) tasks using several large-scale brain imaging datasets. We demonstrate substantial improvement over competitive baselines on these tasks.

2   Related Work

A wide variety of recent advances have been made to remove undesired confounders for imaging data, e.g., pertaining to sites or scanners [3,16,21,22,30,32]. Methods like ComBat [16,30], based on parametric empirical Bayes [20], produce site-removed image features by performing location (mean) and scale (variance) adjustments to the data. A linear model estimates location and scale differences in images features across sites while preserving confounders such as sex and age. In this approach, other unknown variations such as race and disease are removed together with the site variable, which might lead to inaccurate predictions for disease diagnosis. Generative deep learning models such as variational autoencoders (VAEs) [18] and generative adversarial networks (GANs) [12] have been used in many works [3,21,22,32]. These methods typically minimize the mutual information between the site variable and image embedding in the latent space, and learn a site-disentangled representation which can be used to reconstruct images from a different site. Unsupervised image-to-image translation has been used to map scans either between two sites [32] or to a reference domain [3] using models like CycleGAN [37]. Generative models are however challenging to use in practice: VAEs typically suffer from blurry reconstructions while GANs can suffer from mode collapse and convergence issues. These issues are exacerbated for 3D images. In this paper, we focus on regions of interest (ROI) features. We extend the deep structural causal model of [26] which enables tractable counterfactual inference from single-site healthy MR images to multi-center pathology-associated scans for data harmonization. Besides qualitative examination of the counterfactuals performed in [26], we provide extensive quantitative evaluations and compare it with state-of-the-art harmonization baselines.

3 Method

Our method builds upon the causal inference mechanism proposed by Judea Pearl [27] and method of Pawlowski et al. [26] that allows performing counterfactual queries upon causal models parameterized by deep networks. We first introduce preliminaries of our method, namely, structural causal models, counterfactual inference, and normalizing flows, and then describe the proposed harmonization algorithm.

3.1 Building Blocks

Structural Causal Models (SCMs). are analogues of directed probabilistic graphical models for causal inference [29,33]. Parent-child relationships in an SCM denote the effect (child) of direct causes (parents) while they only denote conditional independencies in a graphical model. Consider a collection of random variables $x = (x_1, \ldots, x_m)$, an SCM given by $M = (S, P_\epsilon)$ consists of a collection $S = (f_1, \ldots, f_m)$ of assignments $x_k = f_k(\epsilon_k; \mathrm{pa}_k)$ where pa_k denotes the set of parents (direct causes) of x_k and noise variables ϵ_k are unknown and unmodeled sources of variation for x_k. Each variable x_k is independent of its non-effects given its direct causes (known as the causal Markov condition), we can write the joint distribution of an SCM as $P_M(x) = \prod_{k=1}^{m} P(x_k \mid \mathrm{pa}_k)$; each conditional distribution here is determined by the corresponding structural assignment f_k and noise distribution [27]. Exogenous noise variables are assumed to have a joint distribution $P_\epsilon = \prod_{k=1}^{m} P(\epsilon_i)$, this will be useful in the sequel.

Counterfactual Inference. Given a SCM, a counterfactual query is formulated as a three-step process: abduction, action, and prediction [27–29]. First, we predict exogenous noise ϵ based on observations to get the posterior $P_M(\epsilon \mid x) = \prod_{k=1}^{m} P_M(\epsilon_k \mid x_k, \mathrm{pa}_k)$. Then comes intervention denoted by $\mathrm{do}(\tilde{x}_k)$, where we replace structural assignments of variable x_k. Intervention makes the effect x_k independent of both its causes pa_k and noise ϵ_k and this results in a modified SCM $\tilde{M} = M_{\mathrm{do}(\tilde{x})} \equiv (\tilde{S}, P_M(\epsilon \mid x))$. Note that the noise distribution has

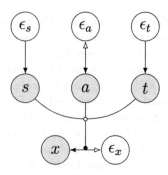

Fig. 1. Causal graph of the structural causal model for brain imaging. Data consists of brain ROIs (x), sex (s), age (a), imaging site (t), and their respective exogenous variables $(\epsilon_x, \epsilon_s, \epsilon_a,$ and $\epsilon_t)$. Bidirectional arrows indicate invertible normalizing flow models and the black dot shows that the flow model associated with x is conditioned on the direct causes (parents) s, a, and t; this follows the notation introduced in [26]. We are interested in answering counterfactual questions of the form "what would the scans look like if they had been acquired from the same site". We first train the flow-based SCM M_θ on the observed data. We then infer the posterior of exogenous variables ϵ_x and ϵ_a with the invertible structural assignments (abduction step). We can now intervene upon the site by replacing site variable t with a specific value τ, this is denoted by $do(t = \tau)$. We sample from the modified flow-based SCM $M_{do(t=\tau)}$ to obtain counterfactual queries.

also been modified, it is now the posterior $P_M(\epsilon \mid x)$ obtained in the abduction step. The third step, namely prediction involves predicting counterfactuals by sampling from the distribution $P_{\tilde{M}}(x)$ entailed by the modified SCM.

Learning a Normalizing Flow-Based SCM. Given the structure of the SCM, learning the model involves learning the structure assignments S from data. Following [26], we next do so using normalizing flows parameterized by deep networks. Normalizing flows model a complex probability density as the result of a transformation applied to some simple probability density [7,10,23,24]; these transformations are learned using samples from the target. Given observed variables x and base density $\epsilon \sim p(\epsilon)$, this involves finding an invertible and differentiable transformation $x = f(\epsilon)$. The density of x is given by $p(x) = p(\epsilon) |\det \nabla f(\epsilon)|^{-1}$ where $\epsilon = f^{-1}(x)$ and $\nabla f(\epsilon)$ is the Jacobian of the flow $f : \epsilon \mapsto x$. The density $p(\epsilon)$ is typically chosen to be a Gaussian. Given a dataset $D = \{x^i \sim p(x)\}_{i=1}^n$ with n samples, a θ-parameterized normalizing flow f_θ can fitted using a maximum-likelihood objective to obtain

$$\theta^* = \operatorname{argmax} \frac{1}{n} \sum_{i=1}^n \log p(\epsilon^i) - \log |\det \nabla f_\theta(\epsilon^i)|.$$

Here $e^i = f_\theta^{-1}(x^i)$. Parameterizing the normalizing flow using a deep network leads to powerful density estimation methods. This approach can be easily extended to conditional densities of the form $p(x_k \mid pa_k)$ in our SCM.

3.2 Harmonization Using Counterfactual Inference in a Flow-Based SCM

Given the structure of a SCM, we fit conditional flows $f_{\theta_k} : \epsilon_k \mapsto x_k$ that map exogenous noise to effect x_k given parents pa_k for all nodes in the SCM. We will denote the combined flow for all nodes in the SCM as f_θ which maps noise $\epsilon^i = (\epsilon_1^i, \dots, \epsilon_m^i)$ to observations $x^i = (x_1^i, \dots, x_m^i)$ in the dataset; the corresponding SCM is denoted by M_θ. Focus on a particular datum x^i in the dataset. The abduction step simply computes $\epsilon^i = f_\theta^{-1}(x^i)$. Formally this corresponds to computing the posterior distribution $P_{M_\theta}(\epsilon \mid x^i)$. Intervention uses the fact that the flow models a conditional distribution and replaces (intervenes) the value of a particular variable, say $x_k^i \leftarrow \tilde{x}_k^i$; this corresponds to the operation $\text{do}(\tilde{x}_k)$. The variable x_k is decoupled from its parents and exogenous noise which corresponds to a modified structural assignment $\tilde{f}_{\theta_k}$ and results in a new SCM $\tilde{M}_\theta$. We can now run the same flow $\tilde{f}_\theta$ forwards using samples ϵ^i from the abduction step to get samples from $P_{\tilde{M}_\theta}(x)$ which are the counterfactuals. Figure 1 shows an example SCM for brain imaging data and shows we perform counterfactual queries to remove site effects.

4 Experimental Results

4.1 Setup

Datasets. We use 6,921 3D T1-weighted brain magnetic resonance imaging (MRI) scans acquired from multiple scanners or sites in Alzheimer's Disease Neuroimaging Initiative (ADNI) [15] and the iSTAGING consortium [13] which consists of Baltimore Longitudinal Study of Aging (BLSA) [2,31], Study of Health in Pomerania (SHIP) [14] and the UK Biobank (UKBB) [34]. Detailed demographic information of the datasets is provided in the supplementary material. We first perform a sequence of preprocessing steps on these images, including bias-filed correction [35], brain tissue extraction via skull-stripping [8], and multi-atlas segmentation [9]. Each scan is then segmented into 145 anatomical regions of interests (ROIs) spanning the entire brain, and finally volumes of the ROIs are taken as the features. We first perform age prediction task using data from the iSTAGING consortium for participants between ages 21–93 years. We then demonstrate our method for classification of Alzheimer's disease (AD) using the ADNI dataset where the diagnosis groups are cognitive normal (CN) and AD; this is a more challenging problem than age prediction.

Implementation. We implement flow-based SCM with three different flows (affine, linear and quadratic autoregressive splines [7,10]) using PyTorch [25] and Pyro [4]. We use a categorical distribution for sex and site, and real-valued normalizing flow for other structural assignments. A linear flow and a conditional flow (conditioned on activations of a fully-connected network that takes age, sex and scanner ID as input) are used as structural assignments for age and ROI features respectively. The density of exogenous noise is standard Gaussian. For

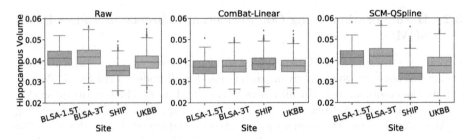

Fig. 2. Comparison of normalized feature (hippocampus volume) distributions for various sites in the iSTAGING consortium data before (raw) and after harmonization using ComBat-Linear and our SCM-QSpline. We observe that ComBat aligns inter-site feature distributions by preserving sex and age effects and removes all other unknown confounders by treating them as site effects. In contrast, the distribution of hippocampus volume is unchanged in our proposed method which takes both known confounders (sex, age, and site) and unknown confounders (exogenous noises) into consideration for harmonization. ComBat removing these useful confounders is detrimental to accuracy (see Table 2).

training, we use Adam [17] with batch-size of 64, initial learning rate 3×10^{-4} and weight decay 10^{-4}. We use a staircase learning rate schedule with decay milestones at 50% and 75% of the training duration. All models are trained for at least 100 epochs. Implementation details for the SCM and the classifier, and the best validation log-likelihood for each model are shown in the supplementary material.

Baselines. We compare with a number of state-of-the-art algorithms: invariant risk minimization (IRM) [1], ComBat [16,30], ComBat++ [36], and CovBat [5] on age regression and Alzheimer's disease classification. IRM learns an invariant representation that the optimal predictor using this representation is simultaneously optimal in all environments. We implement IRM and ComBat algorithms with publicly available code from the original authors. We also show results obtained by training directly on the target data which acts as upper-bound on the accuracy of our harmonization.

4.2   Evaluation of the Learned Flow-Based SCM

We explore three normalizing flow models: affine, linear autoregressive spline [7], and quadratic autoregressive spline [10]. Implementation details and their log-likelihood are in the supplementary material. For both iSTAGING and ADNI datasets, the log-likelihood improves consistently with the model's expressive power. Spline-based autoregressive flow models (17.22 log-likelihood for linear-spline and 17.24 for quadratic-spline) are better for density estimation than an affine flow model (1.88 log-likelihood). A quadratic spline-based model obtains slightly higher log-likelihood than the linear-spline model on the iSTAGING dataset.

Table 1. Mean average error of age prediction for data from the iSTAGING consortium. All experiments were repeated 5 times in cross-validation fashion, and the average performance is reported with the standard errors in the brackets. TarOnly indicates validation MAEs directly trained on each target sites. The hypothesis that our proposed methods achieve a better accuracy than the baselines can be accepted with p-values between 0.06–0.41. This task can be interpreted as a sanity check for our method.

	Study	TarOnly	SrcOnly	IRM	ComBat				Flow-based SCM (ours)		
					Linear	GAM	ComBat++	CovBat	Affine	L-Spline	Q-Spline
Source	BLSA-3T	–	11.74 (0.35)	11.76 (0.35)	11.72 (0.62)	11.74 (0.61)	11.73 (0.62)	11.74 (0.62)	11.74 (0.61)	11.74	11.65 (0.62)
Target	BLSA-1.5T	6.77 (0.82)	7.21 (0.91)	7.16 (0.87)	7.14 (0.99)	7.01 (0.99)	7.00 (1.04)	7.03 (1.08)	7.01 (1.01)	7.00 (1.04)	6.92 (1.09)
Target	UKBB	6.14 (0.16)	7.27 (0.70)	7.18 (0.58)	6.62 (0.46)	6.70 (0.46)	6.71 (0.47)	6.75 (0.49)	6.72 (0.46)	6.75 (0.47)	6.44 (0.28)
Target	SHIP	11.36 (0.31)	17.14 (0.62)	17.05 (0.46)	15.95 (0.61)	16.17 (0.59)	16.21 (0.47)	16.22 (0.65)	16.20	16.25 (0.63)	15.68 (0.80)

We next show the feature (hippocampus volume) distributions of raw data, ComBat [16,30] transformed data and the data generated from the flow-based SCM in Fig. 2. We find that the feature distributions of ComBat are not consistent with those of raw data; ComBat transformed feature distributions show similar means (all shifted to an average value across sites) which arises from removing site-dependent location and scale effects. The third panel shows data generated from counterfactual queries with the flow-based SCM fitted on BLSA-3T. Feature distributions for the SCM-data are similar to those of the raw data. This can be attributed to the fact that our method preserves the unknown confounders (subject-specific information due to biological variability, such as race, gene, and pathology AD/CN) by capturing them as exogenous noise in the SCM.

4.3 Age Prediction

In Table 1, we compare the mean average error (MAE) of age prediction for a regression trained on raw data, site-removed data generated by ComBat [16,30] and its variants [5,36], IRM [1] and counterfactuals generated by our flow-based SCM. All models are trained on BLSA-3T (source site) and then tested on BLSA-1.5T, UKBB, and SHIP separately. We find that model (SrcOnly) trained on the source site with raw data cannot generalize to data from the target site. Models trained with site-removed data generated by ComBat generalize much better compared ones trained on raw data (SrcOnly), whereas IRM shows marginal improvement compared to SrcOnly. All variants (affine, linear-spline, quadratic-spline) of flow-based SCM show substantially smaller MAE; quadratic-spline SCM outperforms the other methods on all target sites.

Table 2. AD classification accuracy (%) comparison on the ADNI dataset and standard deviation (in brackets) across 5-fold cross-validation. TarOnly indicates validation accuracies for training directly on each target site. The hypothesis that our proposed method (Q-Spline) achieves a better accuracy than baselines can be accepted with p-values less than 10^{-5}.

	Study	TarOnly	SrcOnly	IRM	ComBat				Flow-based SCM (ours)		
					Linear	GAM	ComBat++	CovBat	Affine	L-Spline	Q-Spline
Source	ADNI-1	–	76.1 (1.54)	76.2 (2.46)	75.1 (1.37)	75.1 (1.23)	65.1 (6.29)	74.4 (2.29)	76.1 (1.92)	75.3 (1.76)	75.4 (2.45)
Target	ADNI-2	75.8 (3.46)	71.9 (4.88)	73.0 (4.85)	71.4 (4.30)	72.1 (2.83)	56.2 (9.29)	67.4 (5.06)	73.4 (3.52)	72.6 (3.48)	73.7 (4.13)
Source	ADNI-2	–	75.8 (3.46)	76.3 (2.35)	77.5 (2.30)	77.0 (2.74)	67.8 (9.42)	77.9 (2.47)	78.7(1.32)	78.2 (2.80)	77.5(1.76)
Target	ADNI-1	76.1 (1.54)	70.4 (8.80)	72.0 (2.16)	71.1 (4.07)	70.1 (5.67)	58.0 (6.28)	69.1 (5.82)	71.4 (2.41)	71.8 (5.76)	73.3 (3.04)

4.4   Classification of Alzheimer's Disease

In Table 2, we show the accuracy of a classifier trained on raw data, Combat-harmonized data and SCM-generated counterfactuals for the ADNI dataset; this is a binary classification task with classes being CN (cognitive normal) and AD. All classifiers are trained on source sites (ADNI-1 or ADNI-2) and evaluated on target sites (ADNI-2 or ADNI-1 respectively). The classifier works poorly on the target site without any harmonization (SrcOnly). ComBat-based methods show a smaller gap between the accuracy on the source and target site; IRM improves upon this gap considerably. Harmonization using our flow-based SCM, in particular the Q-Spline variant, typically achieves higher accuracies on the target site compared to these methods.

5   Conclusion

This paper tackles harmonization of data from different sources using a method inspired from the literature on causal inference [26]. The main idea is to explicitly model the causal relationship of known confounders such as sex, age, and site, and ROI features in a SCM that uses normalizing flows to model probability distributions. Counterfactual inference can be performed upon such a model to sample harmonized data by intervening upon these variables. We demonstrated experimental results on two tasks, age regression and Alzheimer's Disease classification, on a wide range of real-world datasets. We showed that our method compares favorably to state-of-the-art algorithms such as IRM and ComBat. Future directions for this work include causal graph identification and causal mediation.

Acknowledgements. We thank Ben Glocker, Nick Pawlowski and Daniel C. Castro for suggestions. This work was supported by the National Institute on Aging (grant numbers RF1AG054409 and U01AG068057) and the National Institute of Mental Health (grant number R01MH112070). Pratik Chaudhari would like to acknowledge the support of the Amazon Web Services Machine Learning Research Award.

References

1. Arjovsky, M., Bottou, L., Gulrajani, I., Lopez-Paz, D.: Invariant risk minimization. arXiv preprint arXiv:1907.02893 (2019)
2. Armstrong, N.M., An, Y., Beason-Held, L., Doshi, J., Erus, G., Ferrucci, L., Davatzikos, C., Resnick, S.M.: Predictors of neurodegeneration differ between cognitively normal and subsequently impaired older adults. Neurobiol. Aging **75**, 178–186 (2019)
3. Bashyam, V.M., et al.: Medical image harmonization using deep learning based canonical mapping: Toward robust and generalizable learning in imaging. arXiv preprint arXiv:2010.05355 (2020)
4. Bingham, E., et al.: Pyro: deep universal probabilistic programming. J. Mach. Learn. Res. **20**(1), 973–978 (2019)
5. Chen, A.A., Beer, J.C., Tustison, N.J., Cook, P.A., Shinohara, R.T., Shou, H.: Removal of scanner effects in covariance improves multivariate pattern analysis in neuroimaging data. bioRxiv, p. 858415 (2020)
6. Davatzikos, C.: Machine learning in neuroimaging: progress and challenges. Neuroimage **197**, 652 (2019)
7. Dolatabadi, H.M., Erfani, S., Leckie, C.: Invertible generative modeling using linear rational splines. arXiv preprint arXiv:2001.05168 (2020)
8. Doshi, J., Erus, G., Ou, Y., Gaonkar, B., Davatzikos, C.: Multi-atlas skull-stripping. Acad. Radiol. **20**(12), 1566–1576 (2013)
9. Doshi, J., et al.: Muse: multi-atlas region segmentation utilizing ensembles of registration algorithms and parameters, and locally optimal atlas selection. Neuroimage **127**, 186–195 (2016)
10. Durkan, C., Bekasov, A., Murray, I., Papamakarios, G.: Neural spline flows. In: Advances in Neural Information Processing Systems, pp. 7511–7522 (2019)
11. Esteva, A., et al.: Dermatologist-level classification of skin cancer with deep neural networks. Nature **542**(7639), 115–118 (2017)
12. Goodfellow, I., et al.: Generative adversarial nets. In: Advances in Neural Information Processing Systems, pp. 2672–2680 (2014)
13. Habes, M., et al.: The brain chart of aging: machine-learning analytics reveals links between brain aging, white matter disease, amyloid burden, and cognition in the iSTAGING consortium of 10,216 harmonized MR scans. Alzheimer's Dement. **17**(1), 89–102 (2021)
14. Hegenscheid, K., Kühn, J.P., Völzke, H., Biffar, R., Hosten, N., Puls, R.: Whole-body magnetic resonance imaging of healthy volunteers: pilot study results from the population-based ship study. In: RöFo-Fortschritte auf dem Gebiet der Röntgenstrahlen und der bildgebenden Verfahren, vol. 181, pp. 748–759. Georg Thieme Verlag KG Stuttgart · New York (2009)
15. Jack Jr., C.R., et al.: The Alzheimer's disease neuroimaging initiative (adni): mri methods. J. Magn. Reson. Imaging Off. J. Int. Soc. Magn. Reson. Med. **27**(4), 685–691 (2008)
16. Johnson, W.E., Li, C., Rabinovic, A.: Adjusting batch effects in microarray expression data using empirical bayes methods. Biostatistics **8**(1), 118–127 (2007)
17. Kingma, D.P., Ba, J.: Adam: A method for stochastic optimization. arXiv preprint arXiv:1412.6980 (2014)
18. Kingma, D.P., Welling, M.: Auto-encoding variational bayes. arXiv preprint arXiv:1312.6114 (2013)

19. Menze, B.H., et al.: The multimodal brain tumor image segmentation benchmark (brats). IEEE Trans. Med. Imaging **34**(10), 1993–2024 (2014)
20. Morris, C.N.: Parametric empirical Bayes inference: theory and applications. J. Am. Stat. Assoc. **78**(381), 47–55 (1983)
21. Moyer, D., Gao, S., Brekelmans, R., Galstyan, A., Ver Steeg, G.: Invariant representations without adversarial training. Adv. Neural. Inf. Process. Syst. **31**, 9084–9093 (2018)
22. Moyer, D., Ver Steeg, G., Tax, C.M., Thompson, P.M.: Scanner invariant representations for diffusion MRI harmonization. Magn. Reson. Med. **84**(4), 2174–2189 (2020)
23. Papamakarios, G., Nalisnick, E., Rezende, D.J., Mohamed, S., Lakshminarayanan, B.: Normalizing flows for probabilistic modeling and inference. arXiv preprint arXiv:1912.02762 (2019)
24. Papamakarios, G., Pavlakou, T., Murray, I.: Masked autoregressive flow for density estimation. In: Advances in Neural Information Processing Systems, pp. 2338–2347 (2017)
25. Paszke, A., et al.: Pytorch: an imperative style, high-performance deep learning library. Adv. Neural. Inf. Process. Syst. **32**, 8026–8037 (2019)
26. Pawlowski, N., Castro, D.C., Glocker, B.: Deep structural causal models for tractable counterfactual inference. arXiv preprint arXiv:2006.06485 (2020)
27. Pearl, J.: Causality: Models, Reasoning, and Inference, 2nd edn. Cambridge University Press, Cambridge (2009)
28. Pearl, J., et al.: Causal inference in statistics: an overview. Stat. Surv. **3**, 96–146 (2009)
29. Peters, J., Janzing, D., Schölkopf, B.: Elements of Causal Inference. The MIT Press, Cambridge (2017)
30. Pomponio, R., et al.: Harmonization of large mri datasets for the analysis of brain imaging patterns throughout the lifespan. NeuroImage **208**, 116450 (2020)
31. Resnick, S.M., Pham, D.L., Kraut, M.A., Zonderman, A.B., Davatzikos, C.: Longitudinal magnetic resonance imaging studies of older adults: a shrinking brain. J. Neurosci. **23**(8), 3295–3301 (2003)
32. Robinson, R., et al.: Image-level harmonization of multi-site data using image-and-spatial transformer networks. In: Martel, A.L., et al. (eds.) MICCAI 2020. LNCS, vol. 12267, pp. 710–719. Springer, Cham (2020). https://doi.org/10.1007/978-3-030-59728-3_69
33. Schölkopf, B.: Causality for machine learning. arXiv preprint arXiv:1911.10500 (2019)
34. Sudlow, C., et al.: UK biobank: an open access resource for identifying the causes of a wide range of complex diseases of middle and old age. Plos Med. **12**(3), e1001779 (2015)
35. Tustison, N.J., et al.: N4itk: improved n3 bias correction. IEEE Trans. Med. Imaging **29**(6), 1310–1320 (2010)
36. Wachinger, C., Rieckmann, A., Pölsterl, S., Initiative, A.D.N., et al.: Detect and correct bias in multi-site neuroimaging datasets. Med. Image Anal. **67**, 101879 (2021)
37. Zhu, J.Y., Park, T., Isola, P., Efros, A.A.: Unpaired image-to-image translation using cycle-consistent adversarial networks. In: Proceedings of the IEEE International Conference on Computer Vision, pp. 2223–2232 (2017)

Uncertainty-Aware Label Rectification for Domain Adaptive Mitochondria Segmentation

Siqi Wu[1], Chang Chen[2], Zhiwei Xiong[2,3($\boxtimes$)], Xuejin Chen[2,3], and Xiaoyan Sun[2]

[1] University of Missouri - Kansas City, Kansas City, USA
[2] National Engineering Laboratory for Brain-Inspired Intelligence Technology and Application, University of Science and Technology of China, Hefei, China
zwxiong@ustc.edu.cn
[3] Institute of Artificial Intelligence, Hefei Comprehensive National Science Center, Hefei, China

Abstract. Mitochondria segmentation from electron microscopy images has seen great progress, especially for learning-based methods. However, since the learning of model requires massive annotations, it is time and labour expensive to learn a specific model for each acquired dataset. On the other hand, it is challenging to generalize a learned model to datasets of unknown species or those acquired by unknown devices, mainly due to the difference of data distributions. In this paper, we study unsupervised domain adaptation to enhance the generalization capacity, where no annotation for target datasets is required. We start from an effective solution, which learns the target data distribution with pseudo labels predicted by a source-domain model. However, the obtained pseudo labels are usually noisy due to the domain gap. To address this issue, we propose an uncertainty-aware model to rectify noisy labels. Specifically, we insert Monte-Carlo dropout layers to a UNet backbone, where the uncertainty is measured by the standard deviation of predictions. Experiments on MitoEM and FAFB datasets demonstrate the superior performance of proposed model, in terms of the adaptations between different species and acquisition devices.

Keywords: Mitochondria segmentation · Uncertainty estimation · Domain adaptation · Pseudo-label rectification · Electron microscopy

1 Introduction

Mitochondria Segmentation (MitoSeg) from electron microscopy (EM) images plays an important role in understanding functions of biological systems. With the development of high-throughput EM, a large amount of data can be acquired for investigation. As a representative, Zhang *et al.* propose a Full Adult Fly Brain (FAFB) [26] dataset by using a serial section transmission electron microscopy [6], where 7062 sections of EM images with a resolution of 286720×155648

© Springer Nature Switzerland AG 2021
M. de Bruijne et al. (Eds.): MICCAI 2021, LNCS 12903, pp. 191–200, 2021.
https://doi.org/10.1007/978-3-030-87199-4_18

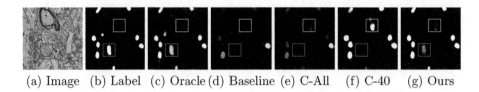

(a) Image   (b) Label   (c) Oracle (d) Baseline   (e) C-All   (f) C-40   (g) Ours

Fig. 1. Examples of the generalization issue and domain adaptation. For a EM image (a) containing human mitochondria (b), a CNN-based model (c) learned from MitoEM-Human [23] performs reasonably well. While the one (d), with the same network structure but learned from MitoEM-Rat [23], fails to predict most human mitochondria, due to the generalization issue. Domain adaptation with All pseudo labels (e) "C-All" cannot well address this issue due to the effect of noisy labels. Label rectification via prediction *confidence* (f), *i.e.*, selecting 40% top-ranked pseudo labels for training ("C-40"), filters out a part of noisy labels. While our proposed method (g) utilizes prediction *uncertainty* for a better label rectification, and thus gives a better performance.

are acquired. This enables the usage of emerging learning-based methods for MitoSeg. However, in addition to the acquisition of EM images, massive annotations are required to facilitate the learning of MitoSeg model. Recently, Wei *et al.* propose an accumulate strategy for the acceleration of annotation [23]. Even so, several rounds of proofreading from neuroscience experts are still required to ensure the accuracy of annotation. Thus, due to the expensive cost of annotation, it is rarely possible to learn a specific model for each acquired dataset.

On the other hand, for a MitoSeg model learned from a specific dataset, it is challenging to generalize this model to other datasets of unknown species or those acquired by unknown devices. As shown in Fig. 1(c), a CNN-based model performs reasonably well when the distribution of the dataset for learning is close to that of the test image. Otherwise, a model with the same network structure fails to segment mitochondria, as shown in Fig. 1(d).

In this paper, we study unsupervised Domain Adaptation (DA) to enhance the generalization capacity of learning-based MitoSeg models. Given a source mitochondria dataset with full annotations, we aim to improve the performance on a target dataset, where no manual annotation is required. Inspired by existing researches for natural image classification tasks [5], we start from an effective DA solution, *i.e.*, pseudo labeling. Specifically, we first train a model on a fully annotated source dataset and generate pseudo labels for a target dataset based on model predictions. Then, we retrain a model on the target dataset by using the generated pseudo label. In this way, we enable the learning of data distribution on the target domain, and thus avoid the generalization issue.

However, the obtained pseudo labels are usually noisy, *i.e.*, containing incorrect ones, due to the gap between source and target domains. These incorrect pseudo labels mislead the model learning, and thus weaken the effectiveness of domain adaptation, as shown in Fig. 1(e). A straightforward solution is to filter out the noisy labels based on the confidence of model predictions. We term this process as label rectification. By selecting the top-ranked pseudo labels for training, we observe the effectiveness of domain adaptation, as shown in Fig. 1(f). Based on this strategy, we make a step forward to utilize prediction *uncertainty* [9] for label rectification. Compared with confidence, the model uncertainty is more suitable to serve as the ranking indicator, since it fundamentally avoids the case when a model predicts incorrect results with high confidence. As shown in Fig. 1(g), our proposed method performs better for domain adaptation.

To enable the estimation of uncertainty, we insert Monte-Carlo dropout [8] layers to a UNet backbone [21], so that the uncertainty of network parameters brings uncertainty in the prediction. To measure the prediction uncertainty, we perform repetitive inference for a given input image and calculate the standard deviation of obtained multiple predictions as the metric of uncertainty level. We conduct extensive experiments for evaluation, in terms of adaptations between different species (*i.e.*, human to rat, rat to human, and human to fly) and different acquisition devices (*i.e.*, SEM [23] to ssTEM [26]). Experimental results demonstrate the significant and consistent advantage of proposed model over competitors, on two representative datasets (*i.e.*, MitoEM [23] and FAFB [26]) and four metrics (*i.e.*, mAP, F1, MCC, and IoU).

Related Works. Domain adaptation (DA) is a hot research topic. Most existing methods are designed for natural images [15,16,25]. While for microscopy images, a representative strategy for DA is to build a dual-branch network structure and perform joint training in either the semi-supervised [3,20] or the adversarial manner [17]. Another strategy focuses on image translation between domains by using adversarial generation techniques [18,24].

Recently, pseudo labeling is introduced for microscopy image DA. To generate pseudo labels, researchers calculate the Normalized Cross Correlation (NCC) [2] to build visual correspondences, so that the corresponding annotations of source-domain images can be aggregated for label generation [1]. Alternatively, pseudo labels can be generated by network prediction [24]. Nevertheless, the importance of label rectification for DA has not been investigated. Following the existing works, we adopt pseudo labeling to achieve DA. Beyond that, we study the importance of label rectification for pseudo label-based DA, for the first time. Moreover, inspired by the analysis of uncertainty in [12], we devise the uncertainty-aware label rectification as an advanced solution.

2   Method

Preliminaries. We aim to achieve a principled Bayesian approximation of uncertainty based on a convolutional neural network (CNN) $f_\omega(\cdot)$. Following the Bayesian equivalent of CNN [7], the parameters of network ω has a prior probability distribution, *i.e.*, $\omega \sim p(\omega)$. Given an input x, the probability of prediction y belonging to a class c can be defined as $p(y = c|x, \omega) = \text{softmax}(f_\omega(x))$, by assuming a standard Gaussian prior $p(\omega)$ [4]. Based on variational Bayesian methods, the inference of Bayesian CNN is equivalent to find a tractable distribution $q(\omega)$ on a training dataset D. This is achieved by minimizing the KL-divergence with the true posterior $p(\omega|D)$. In this way, the uncertainty in

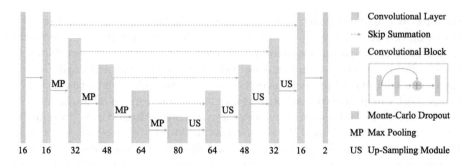

Fig. 2. Illustration of the proposed uncertainty-aware model. The numbers below each block denote the dimension of output channels. We adopt bilinear interpolation to implement the up-sampling module, which is followed by a convolutional layer with 1×1 kernel size to adjust the channel dimension. The ReLU activation layers after each convolutional layer are omitted here for simplification. For the last convolutional layer, we adopt softmax activation to obtain probabilities.

the network parameters ω brings uncertainty in the prediction, which can be represented as

$$\int p(y = c|x, \omega)p(\omega|D)d\omega \approx \int p(y = c|x, \omega)q(\omega)d\omega \approx \frac{1}{T}\sum_{t=1}^{T}p(y = c|x, \hat{\omega}_t), \quad (1)$$

where $q(\omega)$ denotes dropout distribution and the approximate posterior is marginalized by Monte-Carlo integration [8].

Uncertainty-Aware Model. As illustrated in Fig. 2, we build the uncertainty-aware model based on the UNet backbone [21]. Following the design proposed in [14], we upgrade each intermediate convolutional layer to convolutional blocks with a residual connection [11], as shown in the legend of Fig. 2. We insert Monte-Carlo dropout [8] to achieve uncertainty estimation. Different from the common usage of dropout [22], Monte-Carlo dropout enables stochastic sampling in both training and inference phases. Therefore, the confidence of each prediction p is the mean of T times inferences. While the prediction uncertainty is measured by the standard deviation, denoted as $u(p)$. Note that, we adopt p to denote the confidence of positive prediction (or foreground), *i.e.*, mitochondria. A larger p means a model has more prediction confidence for mitochondria. While a smaller p means more prediction confidence for background. For the leaning of model, we adopt binary cross-entropy as the loss function to optimize network parameters in an end-to-end manner.

Rectification of Pseudo Label. Given a model $f_\omega(\cdot)$ learned on a source dataset, we generate the prediction p for a target sample x by T times inferences, *i.e.*, $p = 1/T \sum_{t=1}^{T} f_{\omega_t}(x)$, where ω_t denotes a subset of network parameters ω sampled by stochastic dropout. According to p, we assign the pseudo label y for x as $y = \mathbb{1}(p \geq 0.5)$, where $\mathbb{1}(\cdot)$ denotes the indicator function. To rectify noisy

Table 1. Quantitative comparisons on MitoEM dataset. Compared with the baseline, *i.e.*, direct inference without DA, the pseudo label-based DA is an effective way to narrow the domain gap for a better performance, even for the naive model without label rectification, *i.e.*, C-All. By rectifying noisy pseudo labels, a consistent improvement of performance can be achieved. Compared with the confidence-based rectification, our proposed method utilizes uncertainty for a better DA performance.

Models	Rat to Human (R2H)				Human to Rat (H2R)			
	mAP	F1	MCC	IoU	mAP	F1	MCC	IoU
Oracle[a]	0.9702	0.9160	0.9123	0.8450	0.9820	0.9324	0.9286	0.8734
Baseline	0.7462	0.5675	0.5923	0.3962	0.8850	0.7646	0.7684	0.6189
C-All	0.8050	0.6369	0.6486	0.4672	0.9213	0.7895	0.7905	0.6522
C-20	*0.8204*	0.7429	0.7362	0.5910	0.9104	0.8405	0.8349	0.7248
C-40	0.8155	0.7467	0.7410	0.5958	0.9158	*0.8561*	*0.8482*	*0.7485*
C-60	0.8053	0.7610	0.7547	0.6143	*0.9233*	0.8555	0.8479	0.7475
C-80	0.7896	*0.7653*	*0.7565*	*0.6198*	0.9087	0.8455	0.8353	0.7324
Ours	**0.9072**	**0.8383**	**0.8318**	**0.7216**	**0.9262**	**0.8632**	**0.8551**	**0.7593**

[a] Oracle denotes a model trained on the target dataset with manual annotations, which is the upper-bound of methods trained on pseudo labels.

labels, a straightforward solution is to generate a label-wise mask m based on prediction Confidence, *i.e.*, $m_C = \mathbb{1}(p \geq \tau_p) + \mathbb{1}(p < \tau_n)$, where τ_p and τ_n denote the thresholds for positive and negative labels, respectively. Generally, we have $\tau_p \geq \tau_n$. When $\tau_p = \tau_n = 0.5$, the selection mask m_C degrades to an identity one and all pseudo labels will pass the rectification.

Alternatively, we adopt the prediction Uncertainty to generate a binary mask for selection, *i.e.*, $m_U = \mathbb{1}(u(p) \leq \eta_p) + \mathbb{1}(u(p) \leq \eta_n)$, where η_p and η_n denote the thresholds of uncertainty for positive and negative labels, respectively. This rectification rule is based on an assumption that a pseudo label with lower uncertainty has a higher probability to be the correct one, and vice versa. We exam this assumption qualitatively in Fig. 4 and quantitatively in Table 3, which demonstrates the effectiveness of uncertainty as an indicator for label rectification.

Implementation Details. Theoretically, $\tau_{p,n}$ and $\eta_{p,n}$ can be either hard thresholds, *i.e.*, with fixed values, or soft thresholds determined by a percentile function after the ranking on all alternatives. Yet in the scenario of domain adaptive mitochondria segmentation, the diversity of data distributions make it difficult to set fixed thresholds for different source-target pairs. Thus, both $\tau_{p,n}$ and $\eta_{p,n}$ are implemented as soft thresholds in our experiments. We filter out up to 20% uncertain pseudo labels after the grid search of different $\eta_{p,n}$, and make it as a representative setting of our proposed method. For $\tau_{p,n}$, we traverse different settings (from 20% to 80%) for comparisons.

For the hyper-parameters of network, we set the window size of max pooling as 3×3 with a stride 2. The default kernel size of convolutional layer is set as

Table 2. Quantitative comparison of DA from MitoEM-Human to FAFB. We adopt "C-" to denote the Confidence-based label rectification.

	Baseline	C-All	C-20	C-40	C-60	C-80	Ours
mAP	0.8562	0.8689	0.8459	0.8863	0.8868	*0.8892*	**0.8949**
F1	0.6910	0.7262	0.7259	0.7729	0.7832	*0.7892*	**0.8094**
MCC	0.6905	0.7231	0.7052	0.7533	0.7626	*0.7686*	**0.7890**
IoU	0.5278	0.5701	0.5697	0.6298	0.6437	*0.6519*	**0.6798**

3×3 with a stride 1. The ratio of Monte-Carlo dropout is set as 0.5. We set the times of inferences $T = 10$ for all experiments. For the settings of training, we randomly crop EM images into patches of 256×256 resolution and gather 64 patches into a mini-batch for stochastic gradient descent. We adopt Adam [13] for optimization with the momentum factor set as 0.9. The basic learning rate l_b is set as 0.001. The total iteration for training N is set as 10,000, where a gradually decayed learning rate is adopted for the ith iteration as $l_b \times (1 - i/N)^2$. For random weight initialization, we adopt the method proposed in [10] for all convolutional kernels and set zeros for biases. Experiments are conducted based on PyTorch framework with two 1080Ti GPUs. Code and dataset are available at https://github.com/ngchc/DA-MitoSeg.

3   Experiments and Results

Datasets. We adopt two representative datasets for evaluation, *i.e.*, MitoEM [23] and FAFB [26]. MitoEM is the largest dataset for MitoSeg, where two subsets are imaged by using a multi-beam Scanning EM (SEM). One is collected from an adult human, termed as MitoEM-Human (or MitoEM-H, H for short). Another is collected from an adult rat, termed as MitoEM-Rat (or MitoEM-R, R for short). On each dataset, 500 sections of EM images with a size of 4096×4096 are provided with annotations. We follow the official setting that splits the front 400 sections for training, and the rest 100 sections for validation.

FAFB denotes the first Full imaging for a Adult Fly Brain, by using a high-throughput serial section Transmission EM (ssTEM) [26]. Since there is no corresponding annotation available for MitoSeg, we manually annotate 8 sections with a size of 1736×1736 for quantitative evaluation. We also crop 500 unlabeled sections with a size of 2048×2048 as the resource for domain adaptation.

Metrics. We adopt four metrics for evaluation, *i.e.*, mean Average Precision (mAP), F1 score, Matthews Correlation Coefficient (MCC) [19], and Intersection over Union (IoU, also known as Jaccard index). The values of network prediction, *i.e.*, probabilities, are directly adopted for the calculation of mAP. Since the other metrics work on binary decision maps, we perform binarization for predictions with a fixed threshold of 0.5 before calculation.

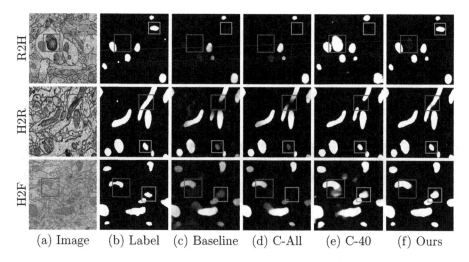

(a) Image   (b) Label   (c) Baseline   (d) C-All   (e) C-40   (f) Ours

Fig. 3. Visual comparisons of domain adaptive MitoSeg, from Rat to Human (R2H), Human to Rat (H2R), and Human to Fly (H2F).

Evaluation on MitoEM. We conduct bi-directional experiments on the two subsets of MitoEM, *i.e.*, from Rat to Human (R2H) and from Human to Rat (H2R). We focus on the effect of label rectification to overall segmentation performance. Take R2H for example, we first train a MitoSeg model on Rat training dataset, where the corresponding annotations are adopted for supervised learning. We exam the direct inference of the obtained model on Human validation dataset, which is termed as "Baseline". Then, based on the learned model, we generate pseudo labels on Human training dataset by using methods described in Sect. 2. Note that, the provided annotations on Human training dataset are *not used* in the setting of R2H. For the label rectification, we abbreviate the confidence-based method as "C-", where the following numbers denote the percentages of top-ranked pseudo labels selected for DA. As listed in Table 1, our proposed method has a significant and consistent advantage over competitors, in terms of different experimental settings and metrics. The corresponding visual comparisons are provided in Fig. 3.

Evaluation on FAFB. Besides the evaluation on MitoEM, we design a more challenging setting, *i.e.*, from MitoEM-Human to Fly mitochondria on FAFB (H2F for short), where not only the species but also the acquisition devices are different between source and target domains. Compared with the experimental settings on MitoEM, H2F is closer to real-world application scenarios since the acquisition devices could be diverse in practice, *e.g.*, ssTEM used on FAFB [26], SEM used on MitoEM [23], and other types of electron microscopes (REM, STEM, and STM). Quantitative results listed in Table 2 and visual comparisons in Fig. 3 demonstrate the superior performance of our proposed model.

Table 3. Comparison for the number of residual noisy labels after rectification, where the pseudo labels for evaluation are generated under the setting of R2H.

Number of noisy labels	Selection by percentages for top-ranked pseudo labels								
	10%	20%	30%	40%	50%	60%	70%	80%	90%
C- ($\times 10^5$)	3.45	8.79	16.51	27.65	44.96	73.48	126.59	238.25	562.33
Ours ($\times 10^5$)	0.30	1.39	3.93	8.68	16.84	28.97	47.65	79.63	157.71

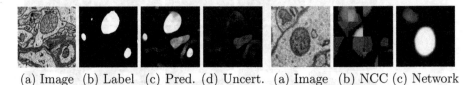

(a) Image   (b) Label   (c) Pred.   (d) Uncert.   (a) Image   (b) NCC   (c) Network

Fig. 4. Model uncertainty indicates false positive errors in predictions, where uncertain degree is colorized from black (low) to red (high). (Color figure online)

Fig. 5. Comparison of pseudo label generation. The network prediction outperforms NCC [2].

4   Analyses and Discussions

Effectiveness of Noisy Label Reduction. As described in Sect. 2, the uncertainty-based rectification is based on an assumption that a pseudo label with lower uncertainty has a higher probability to be the correct label, and vice versa. We exam this assumption by first visualizing the prediction uncertainty in Fig. 4(d). In this example, uncertainty serves as an effect indicator for false positive errors, *i.e.*, the incorrect pseudo labels after the binarization of predictions. Meanwhile, most predictions with low uncertainties match the manual annotations. We also measure the number of residual noisy labels for a quantitative examination. As listed in Table 3, the uncertainty-based method outperforms the confidence-based one, since less noisy labels are left after rectification. This advantage holds whatever percentage of selection is adopted.

Visual Correspondence vs. Network Prediction. Visual correspondence [1] is an alternative method to generate pseudo labels, where the Normalized Cross Correlation (NCC) [2] is adopted to measure the distances between image patches from source and target domains. However, by running the officially provided code, we do not observe the satisfactory robustness of NCC for label generation under the setting of R2H, as shown in Fig. 5. Moreover, to generate pseudo labels for the R2H setting, the running time (dozens of days) required for NCC is not affordable, due to the requirement of greedy correlation search in all possible pairs. In comparison, it only takes a network 5 min.

5   Conclusion

In this paper, we study pseudo labeling for domain adaptive MitoSeg, and reveal the negative effect of noisy labels in this strategy. As a solution, we propose an uncertainty-aware model to rectify noisy labels. Comprehensive experiments validate the effectiveness of our proposed method in various settings.

Acknowledgements. This work was supported by the University Synergy Innovation Program of Anhui Province No. GXXT-2019-025 and the National Natural Science Foundation of China (NSFC) under Grant 62076230.

References

1. Bermúdez-Chacón, R., Altingövde, O., Becker, C., Salzmann, M., Fua, P.: Visual correspondences for unsupervised domain adaptation on electron microscopy images. IEEE Trans. Med. Imaging **39**(4), 1256–1267 (2019)
2. Bermúdez-Chacón, R., Becker, C., Salzmann, M., Fua, P.: Scalable unsupervised domain adaptation for electron microscopy. In: MICCAI (2016)
3. Bermúdez-Chacón, R., Márquez-Neila, P., Salzmann, M., Fua, P.: A domain-adaptive two-stream U-net for electron microscopy image segmentation. In: ISBI (2018)
4. Bishop, C.M.: Pattern Recognition and Machine Learning. Springer, New York (2006)
5. Choi, J., Jeong, M., Kim, T., Kim, C.: Pseudo-labeling curriculum for unsupervised domain adaptation. In: BMVC (2019)
6. Funke, J.: Automatic neuron reconstruction from anisotropic electron microscopy volumes. Ph.D. thesis, ETH Zurich (2014)
7. Gal, Y., Ghahramani, Z.: Bayesian convolutional neural networks with Bernoulli approximate variational inference. In: ICLR (2016)
8. Gal, Y., Ghahramani, Z.: Dropout as a Bayesian approximation: representing model uncertainty in deep learning. In: ICML (2016)
9. Ghahramani, Z.: Probabilistic machine learning and artificial intelligence. Nature **521**(7553), 452–459 (2015)
10. He, K., Zhang, X., Ren, S., Sun, J.: Delving deep into rectifiers: surpassing human-level performance on ImageNet classification. In: ICCV (2015)
11. He, K., Zhang, X., Ren, S., Sun, J.: Deep residual learning for image recognition. In: CVPR (2016)
12. Jungo, A., Reyes, M.: Assessing reliability and challenges of uncertainty estimations for medical image segmentation. In: Shen, D., et al. (eds.) MICCAI 2019. LNCS, vol. 11765, pp. 48–56. Springer, Cham (2019). https://doi.org/10.1007/978-3-030-32245-8_6
13. Kingma, D.P., Ba, J.: Adam: a method for stochastic optimization. In: ICLR (2015)
14. Lee, K., Zung, J., Li, P., Jain, V., Seung, H.S.: Superhuman accuracy on the SNEMI3D connectomics challenge. arXiv preprint arXiv:1706.00120 (2017)
15. Lee, S., Kim, D., Kim, N., Jeong, S.G.: Drop to adapt: learning discriminative features for unsupervised domain adaptation. In: ICCV (2019)
16. Li, G., Kang, G., Liu, W., Wei, Y., Yang, Y.: Content-consistent matching for domain adaptive semantic segmentation. In: Vedaldi, A., Bischof, H., Brox, T., Frahm, J.-M. (eds.) ECCV 2020. LNCS, vol. 12359, pp. 440–456. Springer, Cham (2020). https://doi.org/10.1007/978-3-030-58568-6_26

17. Lin, R., Zeng, X., Kitani, K., Xu, M.: Adversarial domain adaptation for cross data source macromolecule in situ structural classification in cellular electron cryotomograms. Bioinformatics **35**(14), i260–i268 (2019)
18. Liu, D., et al.: Unsupervised instance segmentation in microscopy images via panoptic domain adaptation and task re-weighting. In: CVPR (2020)
19. Matthews, B.W.: Comparison of the predicted and observed secondary structure of T4 phage lysozyme. Biochimica et Biophysica Acta (BBA)-Protein Struct. **405**(2), 442–451 (1975)
20. Roels, J., Hennies, J., Saeys, Y., Philips, W., Kreshuk, A.: Domain adaptive segmentation in volume electron microscopy imaging. In: ISBI (2019)
21. Ronneberger, O., Fischer, P., Brox, T.: U-Net: convolutional networks for biomedical image segmentation. In: Navab, N., Hornegger, J., Wells, W.M., Frangi, A.F. (eds.) MICCAI 2015. LNCS, vol. 9351, pp. 234–241. Springer, Cham (2015). https://doi.org/10.1007/978-3-319-24574-4_28
22. Srivastava, N., Hinton, G., Krizhevsky, A., Sutskever, I., Salakhutdinov, R.: Dropout: a simple way to prevent neural networks from overfitting. J. Mach. Learn. Res. **15**(1), 1929–1958 (2014)
23. Wei, D., et al.: MitoEM dataset: large-scale 3D mitochondria instance segmentation from EM images. In: Martel, A.L., et al. (eds.) MICCAI 2020. LNCS, vol. 12265, pp. 66–76. Springer, Cham (2020). https://doi.org/10.1007/978-3-030-59722-1_7
24. Xing, F., Bennett, T., Ghosh, D.: Adversarial domain adaptation and pseudo-labeling for cross-modality microscopy image quantification. In: Shen, D., et al. (eds.) MICCAI 2019. LNCS, vol. 11764, pp. 740–749. Springer, Cham (2019). https://doi.org/10.1007/978-3-030-32239-7_82
25. Zheng, Z., Yang, Y.: Rectifying pseudo label learning via uncertainty estimation for domain adaptive semantic segmentation. Int. J. Comput. Vision **129**(4), 1106–1120 (2021)
26. Zheng, Z., et al.: A complete electron microscopy volume of the brain of adult drosophila melanogaster. Cell **174**(3), 730–743 (2018)

Semantic Consistent Unsupervised Domain Adaptation for Cross-Modality Medical Image Segmentation

Guodong Zeng[1,5], Till D. Lerch[2,3], Florian Schmaranzer[2,3], Guoyan Zheng[4(✉)],
Jürgen Burger[1], Kate Gerber[1], Moritz Tannast[5], Klaus Siebenrock[2],
and Nicolas Gerber[1]

[1] sitem Center for Translational Medicine and Biomedical Entrepreneurship,
University of Bern, Bern, Switzerland
[2] Department of Orthopaedic Surgery, Inselspital, University of Bern,
Bern, Switzerland
[3] Department of Diagnostic, Interventional and Paediatric Radiology, Inselspital,
University of Bern, Bern, Switzerland
[4] Institute of Medical Robotics, School of Biomedical Engineering,
Shanghai Jiao Tong Unviersity, Shanghai, China
guoyan.zheng@sjtu.edu.cn
[5] Department of Orthopaedic Surgery and Traumatology,
University Hospital of Fribourg, Fribourg, Switzerland

Abstract. Unsupervised domain adaptation (UDA) for cross-modality medical image segmentation has shown great progress by domain-invariant feature learning or image appearance translation. Feature-level adaptation based methods learn good domain-invariant features in classification tasks but usually cannot detect domain shift at the pixel level and are not able to achieve good results in dense semantic segmentation tasks. Image appearance adaptation based methods translate images into different styles with good appearance, but semantic consistency is hard to maintain and results in poor cross-modality segmentation. In this paper, we propose intra- and cross-modality semantic consistency (ICMSC) for UDA and our key insight is that the segmentation of synthesised images in different styles should be consistent. Specifically, our model consists of an image translation module and a domain-specific segmentation module. The image translation module is a standard CycleGAN, while the segmentation module contains two domain-specific segmentation networks. The intra-modality semantic consistency (IMSC) forces the reconstructed image after a cycle to be segmented in the same way as the original input image, while the cross-modality semantic consistency (CMSC) encourages the synthesised images after translation to be segmented exactly the same as before translation. Comprehensive experiments on two different datasets (cardiac and hip) demonstrate that our

---

Electronic supplementary material The online version of this chapter (https://doi.org/10.1007/978-3-030-87199-4_19) contains supplementary material, which is available to authorized users.

© Springer Nature Switzerland AG 2021
M. de Bruijne et al. (Eds.): MICCAI 2021, LNCS 12903, pp. 201–210, 2021.
https://doi.org/10.1007/978-3-030-87199-4_19

proposed method outperforms other UDA state-of-the-art methods by a large margin.

Keywords: Unsupervised domain adaptation · Deep learning · Cardiac segmentation · Hip joint segmentation · Semantic consistency

1 Introduction

Deep Convolutional Neural Network has been successfully applied for medical image analysis [1]. Unfortunately, its performance often drops significantly when tested on medical images from different acquisition protocols or modalities, which is known as domain shift [2]. A naive way to address this challenge is to create new annotations in new domains, but is expensive and sometimes infeasible because of the specialized knowledge in the medical field. Unsupervised Domain Adaptation (UDA) addresses this challenge by transferring knowledge from a domain with labels (source domain) to another domain without labels (target domain) , which requires no additional manual annotations from the target domain and is thus more appealing in the clinical practice [3].

Recent UDA methods for cross-modality medical image segmentation mainly rely on feature alignment or image appearance adaptation, and some works also investigated on uncertainty-aware domain alignment [4] or teacher-student model [5]. Feature alignment based methods force the segmentation network to learn domain invariant features via adversarial training strategy [6–8]. But feature alignment usually cannot detect domain shifts at the pixel level and thus it is not able to achieve good results in dense semantic segmentation tasks. Image appearance adaptation methods (e.g. CycleGAN [9]) implement a pixel-to-pixel image translation between unpaired image domains, and then the translated source-like images from the target domain can be directly tested on the pre-trained source models [10–17]. Though they usually generate source-like or target-like images in good appearance, there is no guarantee to keep semantic consistency after translation and thus leading to unsatisfactory segmentation. CyCADA [18] enforces semantic consistency by constraining the images before and after translation to be segmented the same by a pre-trained source model. However, since the input and translated images are actually from different domains, using the same segmentation model to obtain segmentation results and calculate the semantic consistency loss may be harmful for image translation because the pre-trained source model can hardly perform good segmentation on the target images, especially when the domain shift is dramatic, such as from CT to MRI.

To address the challenge of semantic consistency during image translation for cross-modality medical image segmentation, we propose Intra- and Cross-modality Semantic Consistency (ICMSC) for unsupervised domain adaptation. Our model consists of an image translation module and a domain-specific segmentation module. The image translation module is a standard CycleGAN [9] and the segmentation module consists of two domain-specific segmentation networks, one for the source domain and the other for the target domain. The Intra-modality Semantic Consistency (IMSC) forces the reconstructed image after a

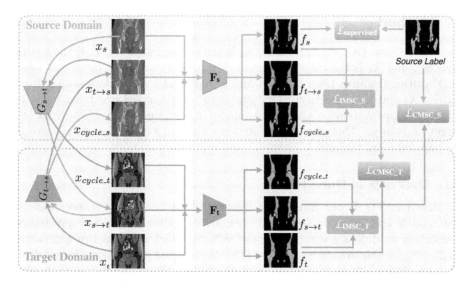

Fig. 1. Method overview. The left side in shadow shows the image appearance translation module (a standard CycleGAN), while the right side shows the domain-specific segmentation module (F_s for the source domain and F_t for the target domain). The blue arrows and green arrows represent the source and target domain, respectively. Our contribution is the proposed Intra and Cross-modality Semantic Consistency loss to encourage semantic consistency during image translation. (Color figure online)

cycle to be segmented in the same way as the input image, while the Cross-modality Semantic Consistency (CMSC) encourages the translated images to be segmented exactly the same as before translation. It should be noted that significant differences exist between our work and CyCADA [18]. Firstly, they simply use one pre-trained source model to segment images from different domains whose appearances may vary significantly. In contrast, we use domain-specific segmentation networks, which is more promising to achieve good segmentation for images from different domains. Secondly, we have IMSC while they don't. Further, our method integrates the image translation and segmentation in one unified model and can be trained in an end-to-end manner, while CyCADA has to be trained in separate stages. Experimental results also demonstrate that our method outperformed theirs.

The main contributions of this paper are summarized as follows: (1) We present an unified UDA framework for cross-modality medical image segmentation, which simultaneously optimizes image translation and segmentation. (2) We propose ICMSC as additional supervisory objectives to enforce semantic consistency during image translation and subsequently improve segmentation performance. (3) Experiment results of challenging cross-modality segmentation between CT and MRI on two different datasets, i.e. cardiac and hip, show that our method outperforms the other state-of-the-art methods by a large margin.

2   Method

Assume we have $\{X_s, Y_s, X_t\}$, where X_s is the source image set, Y_s is the source label set, and X_t is the target image set. We have no access to the target label set Y_t during training. Let x_s, y_s , x_t and y_t represent a single data sample from X_s, Y_s, X_t and Y_t, respectively. Our goal is to learn a model which can perform accurate segmentation on the target images x_t without using labels from the target domain y_t. The network architecture and implementation details are in the supplemental material.

2.1   Overview of Our Method

As shown in Fig. 1, our framework consists of an image appearance translation module and a domain-specific segmentation module. The image appearance translation module is a standard CycleGAN [9], which contains two generators $(G_{s \to t}, G_{t \to s})$ and two discriminators (D_s, D_t, omitted in the figure for clarity). The domain-specific segmentation module includes two segmentation networks, i.e. one for the source domain (F_s) and the other for the target domain (F_t). We proposed the IMSC loss ($\mathcal{L}_{IMSC}$) and CMSC loss ($\mathcal{L}_{CMSC}$) to encourage semantic structure being consistent during the image translation. Our insight is that synthesised image after translation and reconstruction should be segmented in the same way as the original input image. Specifically, $\mathcal{L}_{IMSC}$ forces the input and reconstructed image after a cycle in the same modality to be segmented exactly the same, which can be represented as $F_s(x_s) \approx F_s(x_{cycle\_s})$ and $F_t(x_t) \approx F_t(x_{cycle\_t})$. On the other hand, $\mathcal{L}_{CMSC}$ constrains the same segmentation between the images before and after translation in different modalities, which is formally displayed as $F_s(x_s) \approx F_t(x_{s \to t})$ and $F_t(x_t) \approx F_s(x_{t \to s})$. Since the source label y_s is available, we replace $F_s(x_s)$ directly with y_s. In the testing phase, target images are directly fed into F_t to get segmentation.

2.2   Learning Objective

The overall training objective of our method includes five loss terms, which is defined as:

$$\mathcal{L}_{ICMSC} = \mathcal{L}_{supervised} + \lambda_{adv} \cdot \mathcal{L}_{adv}^{img} + \lambda_{rec} \cdot \mathcal{L}_{rec}^{img}$$
$$+ \lambda_{intra} \cdot \mathcal{L}_{IMSC} + \lambda_{cross} \cdot \mathcal{L}_{CMSC} \tag{1}$$

Supervised Segmentation Loss in the Source Domain. $\mathcal{L}_{supervised}$ is used to train the segmentation network in the source domain (D_s). Let $\mathcal{L}_{seg(x,y,F)}$ be a general supervised loss given input image x, segmentation ground truth y, and a segmentation network F, which is defined as a mixture of cross entropy loss and the Dice coefficient loss in [19]. Then we define $\mathcal{L}_{supervised}$ as:

$$\mathcal{L}_{supervised} = \frac{1}{|X_s|} \sum_{x_s \in X_s} \mathcal{L}_{seg}(x_s, y_s, F_s) \tag{2}$$

Image Domain Adversarial Loss. $\mathcal{L}_{adv}^{img}$ is a combination of two adversarial losses from source and target domains, i.e. $\mathcal{L}_{adv}^{img} = \mathcal{L}_{D_s} + \mathcal{L}_{D_t}$. Two discriminators, D_s and D_t, aim to classify the real input and fake translated images in different domains (label 1 for real and 0 for fake). Let $\mathcal{L}_D$ be the L2 norm loss to train a patchGAN, and then we can then define $\mathcal{L}_{D_s}$ and $\mathcal{L}_{D_t}$ as:

$$
\begin{aligned}
\mathcal{L}_{D_s} &= \frac{1}{|x_s|} \sum_{x_s \in X_s} \mathcal{L}_D(D_s(x_s), 1) + \frac{1}{|x_t|} \sum_{x_t \in X_t} \mathcal{L}_D(D_s(G_{t \to s}(x_t)), 0) \\
\mathcal{L}_{D_t} &= \frac{1}{|x_t|} \sum_{x_t \in X_t} \mathcal{L}_D(D_t(x_t), 1) + \frac{1}{|x_s|} \sum_{x_s \in X_s} \mathcal{L}_D(D_t(G_{s \to t}(x_s)), 0)
\end{aligned}
\tag{3}
$$

Image Reconstruction Loss. $\mathcal{L}_{rec}^{img}$ is used to encourage the preservation of image content between the input and reconstructed image during the cyclic translation process. The motivation lies in that we should achieve the same image after a cycle: $G_{t \to s}(G_{s \to t}(x_s)) \approx x_s$ for any source image x_s and $G_{s \to t}(G_{t \to s}(x_t)) \approx x_t$ for any target image x_t. Formally, the image reconstruction loss is defined as:

$$
\begin{aligned}
\mathcal{L}_{rec}^{img} &= \frac{1}{|x_s|} \sum_{x_s \in X_s} [\|G_{t \to s}(G_{s \to t}(x_s)) - x_s\|_1] \\
&+ \frac{1}{|x_t|} \sum_{x_t \in X_t} [\|G_{s \to t}(G_{t \to s}(x_t)) - x_t\|_1]
\end{aligned}
\tag{4}
$$

Intra-modality Semantic Consistency (IMSC) Loss. $\mathcal{L}_{IMSC}$ enforces the segmentation between input and reconstructed image after a cycle should be exactly the same for both domains. IMSC loss can be separated into two parts: a) $\mathcal{L}_{IMSC\_S}$ in the source domain, b) $\mathcal{L}_{IMSC\_T}$ in the target domain. In other words, we want $F_s(G_{t \to s}(G_{s \to t}(x_s))) \approx y_s$ for any x_s and $F_t(x_t) \approx F_t(G_{s \to t}(G_{t \to s}(x_t)))$ for any x_t, which is defined as:

$$
\begin{aligned}
\mathcal{L}_{IMSC} &= \frac{1}{|x_s|} \sum_{x_s \in X_s} \mathcal{L}_{seg}(G_{t \to s}(G_{s \to t}(x_s)), y_s, F_s) \\
&+ \frac{1}{|x_t|} \sum_{x_t \in X_t} \mathcal{L}_{seg}(G_{s \to t}(G_{t \to s}(x_t)), F_t(x_t), F_t)
\end{aligned}
\tag{5}
$$

Cross-Modality Semantic Consistency (CMSC) Loss. $\mathcal{L}_{CMSC}$ aims to constrain that the translated images will be segmented exactly the same as before translation for both domains. CMSC loss can be separated in two parts: a) $\mathcal{L}_{CMSC\_S}$ in the source domain, and b) $\mathcal{L}_{CMSC\_T}$ in the target domain. In other words, we want $F_t(G_{s \to t}(x_s)) \approx y_s$ for any x_s and $F_t(x_t) \approx F_s(G_{t \to s}(x_t))$ for any x_t, which is defined as:

Table 1. Quantitative comparison on cross-modality hip joint bone segmentation.

Methods	Adaptation	DICE (%) ↑		ASD (mm) ↓	
		Acetabulum	Femur	Acetabulum	Femur
No adaption	Train with source label only	0.08	0.01	29.36	31.27
ADDA [6]	Feature align	50.56	64.80	4.20	4.87
CycleGAN [9]	Image appearance adapt	74.55	72.78	2.05	9.16
CyCADA [18]	Appearance adapt + Feature align	73.80	83.18	3.18	2.24
SIFA [13]	Appearance adapt + Feature align	79.60	83.32	2.93	2.67
Ours	Appearance Adapt + ICMSC	**81.61**	**88.16**	**1.61**	**1.42**
Upper bound	Train with target label directly	95.23	97.29	0.43	0.35

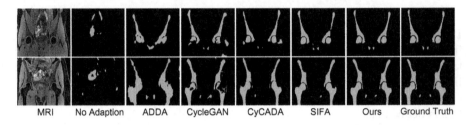

MRI   No Adaption   ADDA   CycleGAN   CyCADA   SIFA   Ours   Ground Truth

Fig. 2. Qualitative comparison results on cross-modality hip joint bone segmentation.

$$\mathcal{L}_{CMSC} = \frac{1}{|x_s|} \sum_{x_s \in X_s} \mathcal{L}_{seg}(G_{s \to t}(x_s), y_s, F_t)$$
$$+ \frac{1}{|x_t|} \sum_{x_t \in X_t} \mathcal{L}_{seg}(x_t, F_s(G_{t \to s}(x_t)), F_t) \tag{6}$$

3 Experiments and Results

3.1 Datasets and Evaluation Metrics

The proposed method was evaluated on cross-modality hip joint bone segmentation (from CT to MRI) and cardiac substructure segmentation (from MRI to CT). The hip dataset consists of unpaired 29 CT and 19 MRI volumes from University Hospital X (anonymized for review). All hip CT and MRI were resampled to the spacing of $1.25 \times 1.25 \times 1\,mm^3$ and manually segmented by experienced clinicians. Eighteen hip CT volumes as source domain, and twelve hip MRI volumes were randomly selected for training and validation. Afterwards, the remaining seven hip MRI volumes were used for testing. The cardiac dataset is a public dataset for cross-modality segmentation [20] and originally comes from the Multi-Modality Whole Heart Segmentation Challenge (MMWHS2017) [21], which consists of unpaired 20 MR and 20 CT volumes. Annotations of ascending aorta (AA), left atrium blood cavity (LAC), left ventricle blood cavity (LVC), and myocardium of the left ventricle (MYO) are provided. Sixteen MR cardiac

Table 2. Quantitative comparison on cross-modality cardiac segmentation.

Methods	DICE (%) ↑					ASD (mm) ↓				
	AA	LAC	LVC	MYO	Average	AA	LAC	LVC	MYO	Average
No adaptation	28.4	27.7	4.0	8.7	17.2	20.6	16.2	N/A	48.4	N/A
ADDA [6]	47.6	60.9	11.2	29.2	37.2	13.8	10.2	N/A	13.4	N/A
CycleGAN [9]	73.8	75.7	52.3	28.7	57.6	11.5	13.6	9.2	8.8	10.8
CyCADA [18]	72.9	77.0	62.4	45.3	64.4	9.6	8.0	9.6	10.5	9.4
SIFA [13]	81.3	79.5	73.8	61.6	74.1	7.9	6.2	5.5	8.5	7.0
Ours	**85.6**	**86.4**	**84.3**	**72.4**	**82.2**	**2.4**	**3.3**	**3.4**	**3.2**	**3.1**
Upper bound	92.7	91.1	91.9	87.7	90.9	1.5	3.5	1.7	2.1	2.2

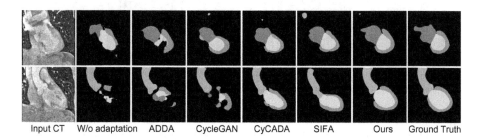

Input CT   W/o adaptation   ADDA   CycleGAN   CyCADA   SIFA   Ours   Ground Truth

Fig. 3. Qualitative comparison on cross-modality cardiac substructure segmentation.

images as source domain, and sixteen CT cardiac images were randomly selected for training and validation, the remaining four CT cardiac volume images were used for testing. For all experiments, we used the Dice Coefficients (DICE) and the Average Surface Distance (ASD) as evaluation metrics.

3.2   Experimental Results

Comparison with State-of-the-art Methods. We compared our method with four state-of-the-art methods: ADDA [6], CycleGAN [9], CyCADA [18] and SIFA [13]. We directly use the code from the paper if that is available, otherwise we reimplement it. We also compared our results with the lower bound, where no adaptation is performed, and with the upper bound, where the model is trained directly with target labels. The quantitative comparison results of cross-modality segmentation on hip and cardiac dataset are displayed in Table 1 and Table 2, respectively. Our method outperformed the four state-of-the-art methods in all metrics by a large margin on both datasets. On the hip dataset, the pretrained source model with no adaptation reported an average DICE of 0.08% for the acetabulum and 0.01% for the proximal femur. This illustrates the dramatic domain shift between hip CT and MRI images. Remarkably, our method yielded an average DICE of 81.61% and 88.16%, an average ASD of 1.61 mm for the acetabulum and 1.41 mm for the proximal femur. On the cardiac dataset, our

Table 3. Ablation study results on cross-modality hip joint bone segmentation.

Methods	$\mathcal{L}_{rec}^{img} + \mathcal{L}_{adv}^{img}$	$\mathcal{L}_{IMSC\_S}$	$\mathcal{L}_{CMSC\_S}$	$\mathcal{L}_{IMSC\_T}$	$\mathcal{L}_{CMSC\_T}$	DICE (%) ↑		
						Acetabulum	Femur	Average
Base, no consistency	✓					74.55	72.78	73.66
+ Intra consistency (source)	✓	✓				74.97	79.32	77.15
+ Cross consistency (source)	✓	✓	✓			78.46	82.57	80.51
+ Intra consistency (target)	✓	✓	✓	✓		76.93	84.94	80.94
+ Cross consistency (target)	✓	✓	✓	✓	✓	**81.61**	**88.16**	**84.88**

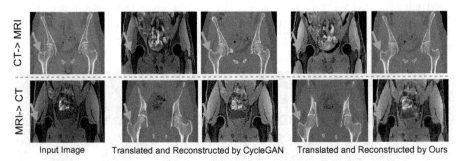

Input Image   Translated and Reconstructed by CycleGAN   Translated and Reconstructed by Ours

Fig. 4. Visual comparison of image translation between CycleGAN and our method. The femur structure to which the arrow point is better preserved in our method.

proposed method achieved superior performance of 82.2% in average Dice and 3.1 mm in average ASD. Notably, our method significantly improves the average Dice of MYO (the most difficult substructure to segment) from 61.6% by the latest state of the art SIFA [13] to 72.4%, achieving a gain of 10.8% in Dice.

Visual comparisons of segmentation results on hip and cardiac dataset are shown in Fig. 2 and Fig. 3, respectively. Due to the dramatic domain shift between CT and MRI, the results from no adaptation are very noisy and cluttered. By using feature alignment or image appearance adaptation, ADDA and CycleGAN output better results but the segmentation is still very poor. CyCADA and SIFA combine feature alignment and image appearance adaptation, and provide more meaningful segmentation in terms of global structure, but hard to maintain semantic consistency in the local structure. For example, in the last row of Fig. 2, the left femur segmentation in CyCADA and the right femur segmentation in SIFA show obvious spurs at greater trochanter while the input MRI and manual segmentation do not. In comparison, our method predicts segmentation that is much closer to the ground truth and keeps better semantic consistency.

Ablation Study. To investigate the role of different semantic consistency losses, we conducted an ablation study and the experimental results are shown in Table 3. The baseline is a standard CycleGAN and then four semantic consistency losses were subsequently integrated. With the integration of each semantic consistency loss term, the performance is consistently improved, which clearly demonstrates the effectiveness of each component of our method. The visual comparison of image translation between the baseline and our method is shown in Fig. 4. The shape of the greater trochanter in the proximal femur of the translated and reconstructed images from the original CycleGAN was altered, whereas our results can better keep the semantic consistency.

4 Discussion and Conclusion

In summary, we presented an UDA framework that unifies image appearance translation and domain-specific segmentation. By incorporating our proposed ICMSC, better semantic consistency can be kept for image translation and thus improve the performance of cross-modality segmentation. Comprehensive experimental results on two different datasets, i.e. cardiac and hip joint, demonstrate the effectiveness of our proposed method. Notably, our method can be elegantly trained end-to-end and easily extended to other UDA tasks. A limitation is that our method is memory intensive, since it involves several sub-networks during training. Therefore, we use a 2D neural network for the 3D MRI segmentation. Our future work will investigate how to reduce the required GPU memory and how to integrate 3D information.

Acknowledgments. This study was partially supported by Shanghai Municipal S&T Commission via Project 20511105205 and by the key program of the medical engineering interdisciplinary research fund of Shanghai Jiao Tong University via project YG2019ZDA22 and YG2019ZDB09 in China.

References

1. Shen, D., Wu, G., Suk, H.I.: Deep learning in medical image analysis. Ann. Rev. Biomed. Eng. **19**, 221–248 (2017)
2. Glocker, B., Robinson, R., Castro, D.C., Dou, Q., Konukoglu, E.: Machine learning with multi-site imaging data: an empirical study on the impact of scanner effects. arXiv preprint arXiv:1910.04597 (2019)
3. Toldo, M., Maracani, A., Michieli, U., Zanuttigh, P.: Unsupervised domain adaptation in semantic segmentation: a review. Technologies **8**(2), 35 (2020)
4. Bian, C., et al.: Uncertainty-aware domain alignment for anatomical structure segmentation. Med. Image Anal. **64**, 101732 (2020)
5. Li, K., Wang, S., Yu, L., Heng, P.A.: Dual-teacher++: exploiting intra-domain and inter-domain knowledge with reliable transfer for cardiac segmentation. IEEE Trans. Med. Imaging (2020)
6. Tzeng, E., Hoffman, J., Saenko, K., Darrell, T.: Adversarial discriminative domain adaptation. In: Proceedings of the IEEE Conference on Computer Vision and Pattern Recognition, pp. 7167–7176 (2017)
7. Kamnitsas, K., et al.: Unsupervised domain adaptation in brain lesion segmentation with adversarial networks. In: Styner, M., et al. (eds.) IPMI 2017. LNCS, vol. 10265, pp. 597–609. Springer, Cham (2017). https://doi.org/10.1007/978-3-319-59050-9_47

8. Zeng, G., et al.: Entropy guided unsupervised domain adaptation for cross-center hip cartilage segmentation from MRI. In: Martel, A.L., et al. (eds.) MICCAI 2020. LNCS, vol. 12261, pp. 447–456. Springer, Cham (2020). https://doi.org/10.1007/978-3-030-59710-8_44

9. Zhu, J.Y., Park, T., Isola, P., Efros, A.A.: Unpaired image-to-image translation using cycle-consistent adversarial networks. In: Proceedings of the IEEE International Conference on Computer Vision, pp. 2223–2232 (2017)

10. Chen, C., Dou, Q., Chen, H., Heng, P.-A.: Semantic-aware generative adversarial nets for unsupervised domain adaptation in chest X-ray segmentation. In: Shi, Y., Suk, H.-I., Liu, M. (eds.) MLMI 2018. LNCS, vol. 11046, pp. 143–151. Springer, Cham (2018). https://doi.org/10.1007/978-3-030-00919-9_17

11. Chen, C., et al.: Unsupervised multi-modal style transfer for cardiac MR segmentation. arXiv preprint arXiv:1908.07344 (2019)

12. Hiasa, Y., et al.: Cross-modality image synthesis from unpaired data using Cycle-GAN. In: Gooya, A., Goksel, O., Oguz, I., Burgos, N. (eds.) SASHIMI 2018. LNCS, vol. 11037, pp. 31–41. Springer, Cham (2018). https://doi.org/10.1007/978-3-030-00536-8_4

13. Chen, C., Dou, Q., Chen, H., Qin, J., Heng, P.: Unsupervised bidirectional cross-modality adaptation via deeply synergistic image and feature alignment for medical image segmentation. IEEE Trans. Med. Imaging (2020)

14. Zhang, Y., Miao, S., Mansi, T., Liao, R.: Task driven generative modeling for unsupervised domain adaptation: application to X-ray image segmentation. In: Frangi, A.F., Schnabel, J.A., Davatzikos, C., Alberola-López, C., Fichtinger, G. (eds.) MICCAI 2018. LNCS, vol. 11071, pp. 599–607. Springer, Cham (2018). https://doi.org/10.1007/978-3-030-00934-2_67

15. Jiang, J., et al.: Tumor-aware, adversarial domain adaptation from CT to MRI for lung cancer segmentation. In: Frangi, A.F., Schnabel, J.A., Davatzikos, C., Alberola-López, C., Fichtinger, G. (eds.) MICCAI 2018. LNCS, vol. 11071, pp. 777–785. Springer, Cham (2018). https://doi.org/10.1007/978-3-030-00934-2_86

16. Yang, J., Dvornek, N.C., Zhang, F., Chapiro, J., Lin, M.D., Duncan, J.S.: Unsupervised domain adaptation via disentangled representations: application to cross-modality liver segmentation. In: Shen, D., et al. (eds.) MICCAI 2019. LNCS, vol. 11765, pp. 255–263. Springer, Cham (2019). https://doi.org/10.1007/978-3-030-32245-8_29

17. Ouyang, C., Kamnitsas, K., Biffi, C., Duan, J., Rueckert, D.: Data efficient unsupervised domain adaptation for cross-modality image segmentation. In: Shen, D., et al. (eds.) MICCAI 2019. LNCS, vol. 11765, pp. 669–677. Springer, Cham (2019). https://doi.org/10.1007/978-3-030-32245-8_74

18. Hoffman, J., et al.: CyCADA: cycle-consistent adversarial domain adaptation. In: International Conference on Machine Learning, pp. 1989–1998. PMLR (2018)

19. Milletari, F., Navab, N., Ahmadi, S.A.: V-net: fully convolutional neural networks for volumetric medical image segmentation. In: 2016 Fourth International Conference on 3D Vision (3DV), pp. 565–571. IEEE (2016)

20. Dou, Q., Ouyang, C., Chen, C., Chen, H., Heng, P.A.: Unsupervised cross-modality domain adaptation of convnets for biomedical image segmentations with adversarial loss. In: Proceedings of the 27th International Joint Conference on Artificial Intelligence, pp. 691–697 (2018)

21. Zhuang, X., Shen, J.: Multi-scale patch and multi-modality atlases for whole heart segmentation of MRI. Med. Image Anal. **31**, 77–87 (2016)

Anatomy of Domain Shift Impact on U-Net Layers in MRI Segmentation

Ivan Zakazov[1,2](✉), Boris Shirokikh[2], Alexey Chernyavskiy[1],
and Mikhail Belyaev[2]

[1] Philips Research, Moscow, Russia
`ivan.zakazov@skoltech.ru`
[2] Skolkovo Institute of Science and Technology, Moscow, Russia
`boris.shirokikh@skoltech.ru`

Abstract. Domain Adaptation (DA) methods are widely used in medical image segmentation tasks to tackle the problem of differently distributed train (source) and test (target) data. We consider the supervised DA task with a limited number of annotated samples from the target domain. It corresponds to one of the most relevant clinical setups: building a sufficiently accurate model on the minimum possible amount of annotated data. Existing methods mostly fine-tune specific layers of the pretrained Convolutional Neural Network (CNN). However, there is no consensus on which layers are better to fine-tune, e.g. the first layers for images with low-level domain shift or the deeper layers for images with high-level domain shift. To this end, we propose SpotTUnet – a CNN architecture that automatically chooses the layers which should be optimally fine-tuned. More specifically, on the target domain, our method additionally learns the policy that indicates whether a specific layer should be fine-tuned or reused from the pretrained network. We show that our method performs at the same level as the best of the non-flexible fine-tuning methods even under the extreme scarcity of annotated data. Secondly, we show that SpotTUnet policy provides a layer-wise visualization of the domain shift impact on the network, which could be further used to develop robust domain generalization methods. In order to extensively evaluate SpotTUnet performance, we use a publicly available dataset of brain MR images (CC359), characterized by explicit domain shift. We release a reproducible experimental pipeline (https://github.com/neuro-ml/domain_shift_anatomy).

Keywords: Domain adaptation · Deep learning · MRI · Segmentation

1 Introduction

Whenever a model, trained on one distribution, is given some data, belonging to another distribution, a detrimental effect called *domain shift* might pop up and

---

I. Zakazov and B. Shirokikh—Equal contribution.

© Springer Nature Switzerland AG 2021
M. de Bruijne et al. (Eds.): MICCAI 2021, LNCS 12903, pp. 211–220, 2021.
https://doi.org/10.1007/978-3-030-87199-4_20

decrease the inference quality [18]. This problem is especially acute in the field of medical imaging since any data collection instance (e.g., MRI apparatus) might appear to sample data, belonging to the domain of its own due to peculiarities of a model or the scanning protocol [5]. Besides, preserving quality on the new data (i.e., *domain adaptation*) is of utter importance because of the industry standards. A realistic set-up is that of *supervised domain adaptation (sDA)*: some data from the new (*target*) domain is labeled and should be utilized for fine-tuning the model, pre-trained on the *source* domain.

The central question of sDA research is *how* should the net be fine-tuned. A great number of works adopt the transfer learning approach of fine-tuning the last layers only, which is underpinned by the notion of feature complexity increasing with depth [19]. It is assumed, that low-level features should be shared across domains, while high-level ones are more prone to domain shift and should be therefore fine-tuned. However, in a number of Domain Adaptation papers the presence of low-level domain shift is demonstrated [3,12,21].

SpotTune [6] is a *Transfer Learning (TL)* approach, allowing for Adaptive Fine-tuning, providing, therefore, a domain shift stamp of each task by learning the corresponding fine-tuning policy. We employ SpotTune approach for getting better insight into the *anatomy of domain shift* in the problem of Domain Adaptation.

Our contribution is threefold:

- To the best of our knowledge, we are the first to propose SpotTUnet: Spot-Tune adapted for supervised DA in medical image segmentation
- We introduce interpretable regularization, with which SpotTune performs on par with the best of the alternative fine-tuning methods across the entire data availability spectrum
- We study the optimal fine-tuning strategies for different data availability scenarios and provide intuition for the obtained results.

2   Related Work

In this paper, we focus on the supervised DA setup in medical image segmentation. Our approach is motivated by SpotTune [6], which learns a policy to choose between fine-tuning and reusing pretrained network layers and at the same time fine-tunes chosen layers on the Target data (see details in Sect. 3). Previous approaches in both supervised and unsupervised medical image DA mostly rely on the explicit choice of layers to fine-tune or to split the adversarial head from. We detail the most relevant approaches below.

The authors of [19] have extensively evaluated features transferability in the image classification task. Their study suggests that features from the first layers of the network could be transferred between tasks, while the last layers should be re-trained. Contrary, convolutional filter reconstruction [1] is designed to tackle the low-level domain shift under unsupervised DA setup showing that the first layers are susceptible to domain shift more than the later ones.

DA methods for medical image segmentation also tackle domain shift problem differently: the earlier approaches follow the motivation of [19], thus fine-tune the later layers of the network. According to the approach of [10] fine-tuning of only the last CNN layer is performed, which yields improvement over transferring without adaptation. However, no comparison is provided with other supervised DA methods or various transferring strategies. Similarly, in [15] the last CNN layer is fine-tuned, but the authors focus more on the training cases selection procedure rather than on the fine-tuning method development. Several works [4,16] provide a comparison between the outcomes of various numbers of the later layers being fine-tuned. Notably, in [4] fine-tuning of the whole network is added to comparison, with better results demonstrated for a smaller number of the later layers fine-tuned. In the unsupervised DA setup, [8] achieves better results adapting features from all the layers except for the first block, but the layers choice strategy remains unlearnable.

In contrast, later approaches follow the motivation of [1], arguing that medical images (e.g. MRI) mostly contain low-level domain shift, which is due to varying intensity profiles but similar high-level structures, thus the first layers should be targeted. In [12] fine-tuning of the first layers is compared to fine-tuning of the last layers and of the whole network. The conclusion is that fine-tuning of the first layers is superior to fine-tuning of the last ones and is even preferable to fine-tuning of the whole network in the case of annotated Target data scarcity. In [9], an adaptive image normalization module is developed for the unsupervised test-time DA, which is conceptually close to fine-tuning of the first layers. Approaches of [3,21] are also motivated by the same hypothesis and adapt the first layers in the unsupervised DA setup.

To this end, we compare SpotTUnet with the best unlearnable layer choice strategies within the supervised DA setup and show it to be a reliable tool for domain shift analysis. While authors of [13] demonstrate SpotTune to perform worse than histogram matching preprocessing in the medical image classification task, we argue that histogram matching is a task-specific method and show its extremely poor segmentation quality in our task. Many approaches competitive to SpotTune have been developed recently, but their focus is more narrow: obtaining the best score on the Target domain rather than analyzing domain shift properties. Therefore, we further study only the SpotTune-based approach.

3 Method

The majority of supervised DA methods in medical image segmentation are based on the explicit choice of layers to fine-tune. However, as indicated in Sect. 2, it is not always clear, whether the first or the last layers are to be targeted. Moreover, state-of-the-art architectures consist of skip-connections and residual paths [2], while residual networks behave like ensembles of *shallow* networks [17]. Therefore, it is also unclear which layers are actually the first across the most meaningful shallow sub-parts of the residual network.

We introduce an extension of SpotTune [6] on the supervised DA for medical image segmentation called **SpotTUnet**. SpotTUnet consists of two copies of the main (segmentation) network and a policy network (see Fig. 1). The main network is pretrained on the Source domain and then duplicated: the first copy has frozen weights (Fig. 1, blue blocks), while the second copy is fine-tuned on the Target domain (Fig. 1, orange blocks). The policy network predicts N pairs of logits for each of N segmentation network blocks (residual blocks or separate convolutions). For each pair of logits, we apply softmax and interpret the result as probabilities in a 2-class classification task: class 0 corresponds to the choice of a frozen block, while class 1 means choosing to fine-tune the unfrozen copy. Then, for the l-th level of the network (frozen block is denoted F_l and fine-tuned block $\tilde{F}_l$) we define its output as $x_l = I_l(x)F_l(x_{l-1}) + (1 - I_l(x))\tilde{F}_l(x_{l-1})$, where $I_l(x)$ is the indicator of choosing the frozen block (i.e. class 0 probability > 0.5). Here, we use Gumbel-Softmax to propagate the gradients through the binary indicator $I_l(x)$ exactly reproducing the methodology of SpotTune [6]. Thus, we simultaneously train the policy network and fine-tune the duplicated layers.

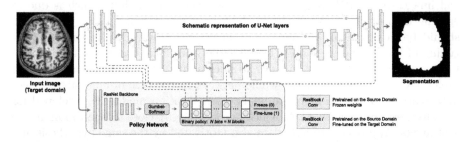

Fig. 1. SpotTUnet architecture for the supervised DA in medical image segmentation. We use U-Net architecture from [12] as the segmentation backbone, which is pretrained on the Source domain. The pretrained segmentation network is frozen (blue blocks) and has a copy (orange blocks) that is fine-tuned on the Target domain. The policy network is simultaneously trained on the Target domain to output binary decisions for each pair of blocks from the segmentation networks: use the frozen block (blue) *vs* use the fine-tuned block (orange). (Color figure online)

Authors of SpotTune also propose a compact global policy (Global-k), which, via additional losses, constrains all the images to fine-tune the same k blocks and aims at reducing the memory and computational costs while possibly decreasing the overall quality. We propose a simplified regularization approach that moreover yields higher quality in the cases of annotated data scarcity. Effectively, we apply $\mathbb{L}_1$ regularization term in the Global-0 SpotTune case, simultaneously minimizing the total number of fine-tuned blocks and achieving a more deterministic policy (exactly 0 or 1, which is due to $\mathbb{L}_1$ properties). The resulting loss is

$$\mathcal{L} = \mathcal{L}_{segm} + \lambda \sum_{l=1}^{N} (1 - I_l(x)), \tag{1}$$

where λ is the balance parameter of the regularization term. Our motivation is different from the original: we assume that fewer blocks should be optimally fine-tuned in case of limited annotated data so that to avoid possible overfitting. Our regularization has only 1 parameter compared to 3 parameters of the original SpotTune regularization, and this parameter (λ) could be optimized during preliminary validation. The intuition behind λ is simple: the less annotated data is available the larger λ value should be.

4   Experiments

4.1   Technical Details

Data. We report our results on a publicly available dataset CC359 [14]. CC359 consists of 359 MR images of head with the task being skull stripping. The dataset is split into 6 equal domains which are shown [12] to contain domain shift resulting in a severe score deterioration. The data preprocessing steps are: interpolation to $1 \times 1 \times 1$ mm voxel spacing and scaling intensities into 0 to 1 interval. All splits and setups are detailed in Sect. 4.2.

Metric. To evaluate different approaches, we use surface Dice Score [11] at the tolerance of 1 mm. While preserving the consistency with the methodology of [12], we also find surface Dice Score to be a more representative metric for the brain segmentation task than the standard Dice Score.

Architecture and Training. The experimental evaluation provided in [12] shows that neither architecture nor training procedure variations, e.g. augmentation, affect the relative performance of conceptually different approaches. Therefore, in all our experiments we consistently use 2D U-Net architecture implementation from [12]. We also use a ResNet architecture [7] as the policy network backbone. In all the experiments we minimize Binary Cross-Entropy loss ($\mathcal{L}_{segm}$) via stochastic gradient descent. Baseline and oracle (see Sect. 4.2) are trained for 100 epochs (100 iterations per epoch) with the learning rate of 10^{-2} reduced to 10^{-3} at the 80-th epoch. All fine-tuning methods are trained for 60 epochs with the learning rate of 10^{-3} reduced to 10^{-4} at the 45-th epoch. We ensure all the models reach the loss plateau. All models are trained with batch size 16. The training takes about 4 h on a 16 GB nVidia Tesla V100 GPU [20].

4.2   Experimental Setup

Baseline and Oracle. The models trained on a single domain form the *baseline* of our study. The transfer of such a model (without fine-tuning) on the other 5 unseen domains results in quality deterioration, which is also shown in [12]. We also obtain scores within each domain (the *oracle*) via 3-fold cross-validation, thereby setting the upper bound for various DA methods.

SpotTUnet Validation. Six domains yield 30 Source-Target pairs, thus, 30 supervised DA experiments. In order to avoid overfitting, we separate one Source domain (Siemens, 1.5T) and, correspondingly, 5 Source-Target pairs for Spot-TUnet validation and use the other 25 pairs for testing various DA approaches. On the 5 validation pairs, we firstly adjust the temperature parameter of Gumbel-Softmax (τ) via grid-search over $\tau \in \{.01, .1, .5, 1, 2, 5\}$. The number of annotated slices from the Target domain, in this case, is 270 (one 3D image). Secondly, we search for the optimal λ for each amount of annotated Target data considered via grid-search over $\lambda \in \{0, 1, 3, 5, 7, 10, 12, 15, 20 \ (\times 10^{-3})\}$. In both validation and testing experiments, we study the same setups of the annotated Target data scarcity: 8, 12, 24, 45, 90, 270, and 800 slices available for fine-tuning. The optimal value of λ is fixed for each data scarcity setup and used for SpotTUnet when testing on the remaining 25 pairs.

Supervised DA Methods. On the rest of the 25 testing pairs, we compare 4 methods: *fine-tuning of the first network layers, fine-tuning of the whole network* from [12], *histogram matching* from [13], and SpotTUnet. We load a *baseline* model pretrained on the corresponding Source domain and then fine-tune it via one of the methods or preprocess the Target data in case of histogram matching. We compare methods by averaging surface Dice Scores over the Target images, separated for test (omitted when fine-tuning).

4.3   Results and Discussion

We firstly find and fix hyperparameters for SpotTUnet through validation: the temperature of Gumbel-Softmax is set to $\tau = 0.1$ and the optimal regularization λ is set to the optimal value for each data scarcity setup (see Fig. 2). We also note that Gumbel-Softmax training stability is extremely sensitive to τ choice, thus suggest to validate different values of τ at one of the first stages of deploying SpotTune-like architectures. Positive regularization term benefits almost all data scarcity setups: surface Dice Score of the optimal λ is significantly higher ($p < 10^{-3}$, one-sided Wilcoxon signed-rank test) than the surface Dice Score of the corresponding model without the regularization. The only exception is the case of 800 available Target slices (or three 3D images), where the optimal λ is close to 0, and the quality drops with the increase of λ (see Fig. 2). We conclude that while SpotTUnet learns the optimal policy without regularization when there is no Target data shortage, regularization improves DA performance significantly in case of Target data scarcity.

We further compare SpotTUnet with *Fine-Tuning All Layers, Fine-Tuning the First Layers,* and *histogram matching.* We present both distributions of the surface Dice Score (violin plots) and their average values (corresponding lines) in Fig. 3. Histogram matching achieves only 0.29 average surface Dice Score, which is even lower than the *baseline* average score of 0.55; we exclude both methods from the comparison in Fig. 3. Here, we show that SpotTune performs at the same level as the best of the other methods regardless of the Target data scarcity severity.

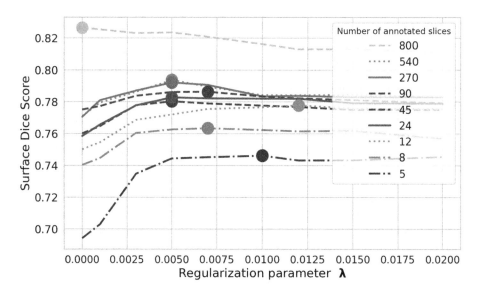

Fig. 2. The validation performance of SpoTUnet dependence on the regularization parameter λ. In each point, the average surface Dice Score over 5 validation experiments is calculated. Each line corresponds to some amount of available annotated data from the Target domain. The bold points indicate the optimal λ values in terms of surface Dice Score.

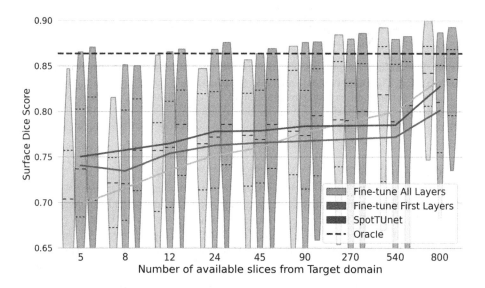

Fig. 3. The methods performance dependence on the amount of available Target data. Baseline and histogram matching yield poor quality, thus not included.

Finally, we track SpotTUnet policy on the test data: for each layer, we calculate the frequency of choosing the fine-tuned one instead of the pretrained and frozen one. The layer-wise visualization is given in Fig. 4. We find that blocks from the encoder part of U-Net are more likely to be fine-tuned, especially in the case of annotated data scarcity. However, these are not exactly the first layers (e.g., preserving the original resolution), which opposes the conclusion of [12]. We further hypothesize, that SpotTUnet policy indicates layers that should be fine-tuned for the optimal solution. Consequently, feature maps preceding these frequently fine-tuned layers might be marked with drastic domain shift. We note that it is worth evaluating if unsupervised DA approaches [8,21] would benefit from passing these SpotTUnet indicated domain shift reach feature maps to the adversarial heads and leave this hypothesis validation for the future research.

We attribute the difference between the policies observed and those presented in the original SpotTune paper [6] (mostly final layers fine-tuned) to the fundamental difference between Transfer Learning (TL) and DA. In TL one deals with data of varying nature, thus the later layers should be addressed; in DA, the datasets contain semantically homogeneous data (e.g., brain MRI scans), thus domain shift is mostly low-level and the first layers should be targeted.

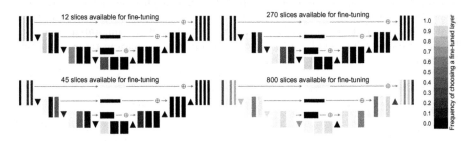

Fig. 4. SpotTUnet learnt policy visualization for the cases of 12 (upper-left), 45 (bottom-left), 270 (upper-right), and 800 (bottom-right) available Target slices. Colored blocks correspond to either residual blocks or convolutions. Triangular blocks are the convolutions that perform ×2 up- or down-sampling.

5   Conclusion

We propose a fine-tuning approach for supervised DA in medical image segmentation called SpotTUnet. Our experiments demonstrate SpotTUnet to preserve the quality of the alternative methods while eliminating the need for switching between various methods depending on the Target data availability. Besides, it learns automatically, which layers are to be optimally fine-tuned on the target domain, therefore providing a policy, indicative of the network layers most susceptible to domain shift. We believe that SpotTUnet generated policy might be used for developing more robust unsupervised DA methods, which is the goal of our future research.

References

1. Aljundi, R., Tuytelaars, T.: Lightweight unsupervised domain adaptation by convolutional filter reconstruction. In: Hua, G., Jégou, H. (eds.) ECCV 2016. LNCS, vol. 9915, pp. 508–515. Springer, Cham (2016). https://doi.org/10.1007/978-3-319-49409-8_43
2. Bakas, S., et al.: Identifying the best machine learning algorithms for brain tumor segmentation, progression assessment, and overall survival prediction in the brats challenge. arXiv preprint arXiv:1811.02629 (2018)
3. Dou, Q., Ouyang, C., Chen, C., Chen, H., Heng, P.A.: Unsupervised cross-modality domain adaptation of convnets for biomedical image segmentations with adversarial loss. In: Proceedings of the 27th International Joint Conference on Artificial Intelligence, IJCAI 2018, pp. 691–697. AAAI Press (2018)
4. Ghafoorian, M., et al.: Transfer learning for domain adaptation in MRI: application in brain lesion segmentation. In: Descoteaux, M., Maier-Hein, L., Franz, A., Jannin, P., Collins, D.L., Duchesne, S. (eds.) MICCAI 2017. LNCS, vol. 10435, pp. 516–524. Springer, Cham (2017). https://doi.org/10.1007/978-3-319-66179-7_59
5. Glocker, B., Robinson, R., Castro, D.C., Dou, Q., Konukoglu, E.: Machine learning with multi-site imaging data: an empirical study on the impact of scanner effects (2019)
6. Guo, Y., Shi, H., Kumar, A., Grauman, K., Rosing, T., Feris, R.: SpotTune: transfer learning through adaptive fine-tuning. In: Proceedings of the IEEE/CVF Conference on Computer Vision and Pattern Recognition, pp. 4805–4814 (2019)
7. He, K., Zhang, X., Ren, S., Sun, J.: Deep residual learning for image recognition. In: Proceedings of the IEEE Conference on Computer Vision and Pattern Recognition, pp. 770–778 (2016)
8. Kamnitsas, K., et al.: Unsupervised domain adaptation in brain lesion segmentation with adversarial networks. In: Niethammer, M., et al. (eds.) IPMI 2017. LNCS, vol. 10265, pp. 597–609. Springer, Cham (2017). https://doi.org/10.1007/978-3-319-59050-9_47
9. Karani, N., Erdil, E., Chaitanya, K., Konukoglu, E.: Test-time adaptable neural networks for robust medical image segmentation. Med. Image Anal. **68**, 101907 (2021)
10. Kushibar, K., et al.: Supervised domain adaptation for automatic sub-cortical brain structure segmentation with minimal user interaction. Sci. Rep. **9**(1), 1–15 (2019)
11. Nikolov, S., et al.: Deep learning to achieve clinically applicable segmentation of head and neck anatomy for radiotherapy. arXiv preprint arXiv:1809.04430 (2018)
12. Shirokikh, B., Zakazov, I., Chernyavskiy, A., Fedulova, I., Belyaev, M.: First U-net layers contain more domain specific information than the last ones. In: Albarqouni, S., et al. (eds.) DART/DCL -2020. LNCS, vol. 12444, pp. 117–126. Springer, Cham (2020). https://doi.org/10.1007/978-3-030-60548-3_12
13. Singh, S., et al.: Adaptation of a deep learning malignancy model from full-field digital mammography to digital breast tomosynthesis. In: Medical Imaging 2020: Computer-Aided Diagnosis, vol. 11314, p. 1131406. International Society for Optics and Photonics (2020)
14. Souza, R., et al.: An open, multi-vendor, multi-field-strength brain MR dataset and analysis of publicly available skull stripping methods agreement. Neuroimage **170**, 482–494 (2018)
15. Valindria, V.V., et al.: Domain adaptation for MRI organ segmentation using reverse classification accuracy. arXiv preprint arXiv:1806.00363 (2018)

16. Valverde, S., et al.: One-shot domain adaptation in multiple sclerosis lesion segmentation using convolutional neural networks. NeuroImage: Clin. **21**, 101638 (2019)

17. Veit, A., Wilber, M.J., Belongie, S.J.: Residual networks behave like ensembles of relatively shallow networks. In: Lee, D.D., Sugiyama, M., von Luxburg, U., Guyon, I., Garnett, R. (eds.) Advances in Neural Information Processing Systems 29: Annual Conference on Neural Information Processing Systems 2016, 5–10 December 2016, Barcelona, Spain, pp. 550–558 (2016)

18. Wang, M., Deng, W.: Deep visual domain adaptation: a survey. Neurocomputing **312**, 135–153 (2018)

19. Yosinski, J., Clune, J., Bengio, Y., Lipson, H.: How transferable are features in deep neural networks? In: Proceedings of the 27th International Conference on Neural Information Processing Systems, NIPS 2014, vol. 2, p. 3320–3328. MIT Press, Cambridge (2014)

20. Zacharov, I., et al.: 'Zhores' - petaflops supercomputer for data-driven modeling, machine learning and artificial intelligence installed in Skolkovo institute of science and technology. Open Eng. **9**, 512–520 (2019)

21. Zhao, X., et al.: Robust white matter hyperintensity segmentation on unseen domain. In: 2021 IEEE 18th International Symposium on Biomedical Imaging (ISBI), pp. 1047–1051 (2021)

FoldIt: Haustral Folds Detection and Segmentation in Colonoscopy Videos

Shawn Mathew[1], Saad Nadeem[2(✉)], and Arie Kaufman[1]

[1] Department of Computer Science, Stony Brook University, Stony Brook, USA
[2] Department of Medical Physics, Memorial Sloan Kettering Cancer Center, New York, USA
nadeems@mskcc.org

Abstract. Haustral folds are colon wall protrusions implicated for high polyp miss rate during optical colonoscopy procedures. If segmented accurately, haustral folds can allow for better estimation of missed surface and can also serve as valuable landmarks for registering pre-treatment virtual (CT) and optical colonoscopies, to guide navigation towards the anomalies found in pre-treatment scans. We present a novel generative adversarial network, FoldIt, for feature-consistent image translation of optical colonoscopy videos to virtual colonoscopy renderings with haustral fold overlays. A new transitive loss is introduced in order to leverage ground truth information between haustral fold annotations and virtual colonoscopy renderings. We demonstrate the effectiveness of our model on real challenging optical colonoscopy videos as well as on textured virtual colonoscopy videos with clinician-verified haustral fold annotations. All code and scripts to reproduce the experiments of this paper will be made available via our Computational Endoscopy Platform at https://github.com/nadeemlab/CEP.

Keywords: Colonoscopy · Haustral folds segmentation

1 Introduction

High polyp miss rates during colonoscopy procedure are mainly attributed to colon wall protrusions, formed by circumferential contraction of the colon inner muscular layer, also known as haustral folds. These folds are extremely difficult to segment in optical colonoscopy (OC) videos due to texture and lighting variations along with specular reflections, fluid motion, and organ movements. If segmented, however, these folds can guide endoscope navigation towards high-occlusion areas, potentially reducing the polyp miss rate. Moreover, even though

---

S. Mathew and S. Nadeem—Equal contribution.

Electronic supplementary material The online version of this chapter (https://doi.org/10.1007/978-3-030-87199-4_21) contains supplementary material, which is available to authorized users.

© Springer Nature Switzerland AG 2021
M. de Bruijne et al. (Eds.): MICCAI 2021, LNCS 12903, pp. 221–230, 2021.
https://doi.org/10.1007/978-3-030-87199-4_21

the colon can stretch and distort shape considerably in OC versus the pre-treatment CT virtual colonoscopy (VC), the haustral folds remain intact across the two modalities and hence can serve as useful landmarks for registration.

Mathew et al. [15] recently introduced a new unsupervised model, XDCy-cleGAN, for inferring scale-consistent depth maps from OC video frames using geometric information from 3D VC data, extracted from abdominal CT scans. The unsupervised model was shown to handle variations in texture, lighting and specular reflections much more effectively than previous supervised approaches that were trained completely on OC or on VC datasets but not both simultaneously. Xu et al. [19] also showed superior performance by using cycle-consistency and optical flow for spatially- and temporally-consistent translation of simulated VC flythroughs to real OC videos.

In this work, we present FoldIt, a new generative adversarial network that can accurately detect and segment haustral folds in OC videos using unpaired image-to-image translation of OC frames to VC haustral fold renderings. *We show that the haustral fold segmentation via our model leads to feature-consistent domain translation for OC video sequences; the feature-consistency refers to consistency of haustral fold annotations between consecutive frames.* FoldIt is available on GitHub via our Computational Endoscopy Platform. The contributions of this work are as follows:

1. A method for haustral fold detection and segmentation in real OC images.
2. A semi-supervised approach to perform image-to-image domain translation via a common domain.
3. A transitive loss to drive the domain translation while retaining/preserving haustral fold features in the common VC domain.

2   Related Works

Deep learning approaches have recently shown promising results in various endoscopic intervention tasks such as depth estimation, 3D reconstruction, and surface coverage [2,6,11,14]. Deep learning models are data driven and the supervised category requires ground truth information. The issue with the supervised approaches, specifically for colonoscopy, is the need for realistic training data that models the specular reflections, fluid motion, and erratic camera and spasm movements. The ground truth data creation is extremely time consuming and even then the network can easily fail since the input data is not representative of the real domain. To overcome this, the supervised methods require additional measures to handle real data. Rau et al. [18] trained a pix2pix variant on textured VC data which needed extra training on OC images to handle real OC data. Mahmood et al. [13] required an additional network to transform their data into a synthetic-like (VC) domain before being able to estimate depth maps. Chen et al. [3] trained on VC input and tested on porcine phantom models, so it is unclear how well their approach performs on real OC data.

In contrast, unsupervised approaches already work on the input domain, so no additional modules are needed to handle real OC images. They, however, require more complicated network design and careful assessment to avoid overfitting. In FoldIt, we present a semi-supervised approach which leverages real OC images. Ground truth haustral fold annotations on 3D VC triangular mesh models were computed using Fiedler vector representation [17] and verified by a clinician. We capture the best aspects of both the unsupervised and supervised approaches in our FoldIt model.

Generative Adversarial Networks (GAN) [7] have shown promising results for image-to-image translation. GANs help the generator output match the training data distribution. Approaches such as pix2pix [10], CycleGAN [20], and StarGAN [4] all utilize adversarial learning in their approaches. Pix2pix does paired image-to-image translation while CycleGAN can handle unpaired data. Both pix2pix and CycleGAN assume a one-to-one mapping between domains. Mathew et al. [15] introduced a one-to-many mapping to handle OC images since there are large variations in texture, lighting, and specular reflections.

Haustral fold annotation requires many-to-many mapping since the depth of the folds can be unclear from a single frame. Travel-GAN [1] introduced a many-to-many approach, however, their translation is not constrained, making it less suitable for our task. To address this, we present a model that performs many-to-many translation via a common domain. This common domain has a one-to-many mapping with the other two domains to help constrain the output. Fang et al. [5] have recently proposed a triple translation loss, similar to our transitive loss, but they focus on face generation based on age in a single domain which does not require translation between multiple domains.

3   Data

The OC and VC data were obtained from 10 patients who underwent a VC procedure followed by an OC procedure. Data from 7 patients were used for training and 3 for testing. The OC frames are cropped and rescaled to 256×256 to remove the borders in the frame. VC meshes are synthesized from abdominal CT scans using a pipeline described in [16]. These meshes do not align with the OC videos, as the shape of the colon changes between the two procedures. Flythroughs are created in the VC mesh along the centerline with random camera rotations, and 2 light sources on left and right side of the camera to replicate the OC procedure. These lights follow the inverse square fall-off property to create realistic lighting [13]. The haustral folds on VC meshes were segmented using Fiedler vector and corresponding levelset representations, as outlined in [17], and verified by a clinician. The segmented folds were overlaid on the VC mesh rendering and used as haustral fold annotation frames. Each video was split into 300 frames for a total of 3000 frames. Example frames from each domain are shown in Fig. 1b.

4   Methods

FoldIt uses 4 generators and discriminators. G_{AB} translates images from domain A to domain B, and G_{BA} acts in the opposite direction. G_{AC} and G_{BC} translate

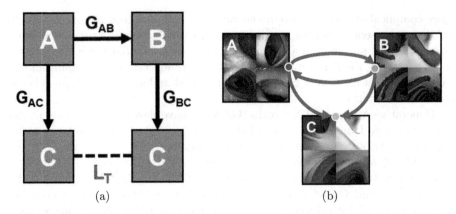

Fig. 1. (a) An image from domain A is converted to domain B and then C through generators G_{AB} and G_{BC}. This resulting image is compared against the image directly translated to domain C via G_{AC} to compute the transitive loss. (b) The relationship between the domains A, B, and C. We can translate between domains A and B, however, the result in domain C must remain the same. For FoldIt, A is the OC domains, B is the Haustral Fold annotation (red overlay) domains, and C is the VC domain. (Color figure online)

from input domains, A and B, to the common domain, C. Our approach can be broken down into four losses: adversarial losses, transitive losses, a ground truth loss, and identity losses. The complete objective function is as follows:

$$\mathcal{L}_{obj} = \lambda_{adv}\mathcal{L}_{adv} + \lambda_T\mathcal{L}_T + \lambda_{GT}\mathcal{L}_{GT} + \lambda_{idt}\mathcal{L}_{idt} \tag{1}$$

Each of the 4 generators have an associated discriminator, D, which applies an adversarial loss to ensure the output image encompasses the features of the output domain. The adversarial loss for each of these GANs is described as:

$$\mathcal{L}_{GAN}(G, D, A, B) = \mathbb{E}_{y \backsim p(B)}\big[\log(D(y))\big] + \mathbb{E}_{x \backsim p(A)}\big[\log(1 - D(G(x)))\big], \tag{2}$$

where $y \backsim p(B)$ represents the output domain B data distribution and $x \backsim p(A)$ represents the input domain A data distribution. An adversarial loss is applied to each generator to create the adversarial component of the objective loss:

$$\begin{aligned}\mathcal{L}_{adv} =&\mathcal{L}_{GAN}(G_{AB}, D_{AB}, A, B) + \mathcal{L}_{GAN}(G_{BA}, D_{BA}, B, A)\\ &+ \mathcal{L}_{GAN}(G_{AC}, D_{AC}, A, C) + \mathcal{L}_{GAN}(G_{BC}, D_{BC}, B, C)\end{aligned} \tag{3}$$

Cycle consistency losses have shown great results for the task of image-to-image translation. Our approach uses a modified cycle consistency loss, which resembles the transitive property, to learn the domain translation via a common domain. The main insight here is that the translation to the common domain, C, should be consistent between our domains A and B. Here, C is the VC domain. When we do the translation, $A \rightarrow B \rightarrow C$, the result should be the same as $A \rightarrow C$ (see Fig. 1a). We can express this transitive loss as ($\|\cdot\|_1$ is $\ell 1$ norm):

$$\mathcal{L}_T(G_{AB}, G_{BC}, G_{AC}, A) = \mathbb{E}_{x \backsim p(A)}\|G_{BC}(G_{AB}(x) - G_{AC}(x))\|_1 \tag{4}$$

This loss is applied on both directions in the total objective loss.

$$\mathcal{L}_T = \mathcal{L}_T(G_{AB}, G_{BC}, G_{AC}, A) + \mathcal{L}_T(G_{BA}, G_{AC}, G_{BC}, B) \tag{5}$$

$\mathcal{L}_{GT}$ is our ground truth loss, meant to utilize the ground truth pairing between the common domain and one of the domains A and B. It simply applies the $\ell1$ norm between the paired domains. Here, we have ground truth correspondence between haustral fold annotations and VC. The ground truth loss is:

$$\mathcal{L}_{GT}(G_{BC}, B, C) = \mathbb{E}_{x,z \sim p(B,C)} \|(G_{BC}(x) - z\|_1 \tag{6}$$

where $x, z \sim p(A, C)$ represents the paired data distribution (A, C).

Lastly, we have the identity loss, which is meant for additional stability during training as described in [15,20]. An image from the output domain is expected to be unchanged when passed as input through the network under an identity loss. This is applied only to G_{AC} and G_{BC} as alterations in the output for G_{AB} and G_{BA} may still be valid. The identity loss is as follows:

$$\mathcal{L}_{idt}(G_{AC}, C) = \mathbb{E}_{z \sim p(C)} \|(G_{AC}(z) - z\|_1 \tag{7}$$

The complete identity loss is:

$$\mathcal{L}_{idt} = \mathcal{L}_{idt}(G_{AC}, C) + \mathcal{L}_{idt}(G_{BC}, C) \tag{8}$$

Each generator used up 18 MB for its ResNet architecture [8] with 9 blocks. We use a PatchGAN discriminator [10] similar to [20], which used 3 MB each. The network was trained for 100 epochs and used the following weights: $\lambda_{adv} = 1, \lambda_T = 10, \lambda_{GT} = 1, \lambda_{idt} = 1$. The inference time for a single image is 0.04 s on an Nvidia RTX 6000 GPU.

5   Results

There is no ground truth correspondence between OC and VC due to the colon shape changes between the two procedures making quantitative analysis difficult. We use textured VC mesh with clinician-verified haustral fold annotations to evaluate our results. We render the VC mesh with two different textures to test the sensitivity of our model to texture and lighting variations. The haustral folds were segmented and overlaid on the VC rendering. Figure 2 shows results from our FoldIt versus XDCycleGAN [15] on these textured VC meshes.

FoldIt results are closer to the ground truth for both textures, while the haustral fold features remain preserved/consistent across frames. The green bounding boxes indicate locations where XDCycleGAN is not feature consistent and removes/adds folds between neighboring frames. As seen in Table 1, FoldIt achieves higher Dice and IoU scores when compared with XDCycleGAN on both textures. Since the textured VC colons have the same underlying geometry, the network should have similar predictions for both videos. The Dice and IoU scores are calculated between Texture 1 and Texture 2 and shown in the third row of Table 1. Again, FoldIt is more consistent in its predictions with varying textures. The complete video sequences are provided in the **supplementary video**[1].

---

[1] Supplementary Video: https://youtu.be/_iWBJnDMXjo.

Table 1. Dice and IoU scores on ground truth VC model with the two textures shown in Fig. 2. Dice and IoU scores between Texture 1 and 2 is shown on the third row.

	Dice		IoU	
	FoldIt	XDCycleGAN	FoldIt	XDCycleGAN
Texture 1	**0.47 ± 0.11**	0.25 ± 0.11	**0.31 ± 0.09**	0.15 ± 0.07
Texture 2	**0.50 ± 0.10**	0.21 ± 0.10	**0.33 ± 0.09**	0.12 ± 0.06
Consistency	**0.77 ± 0.10**	0.64 ± 0.16	**0.64 ± 0.12**	0.49 ± 0.16

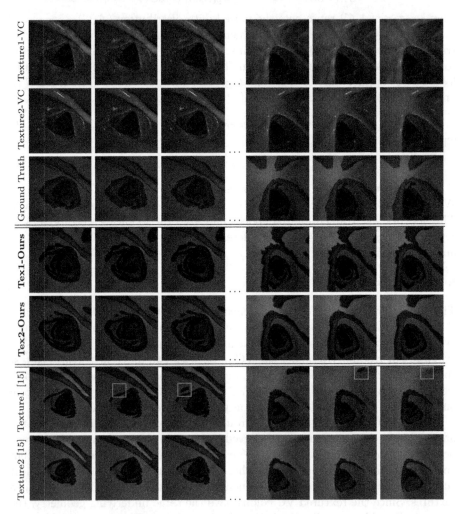

Fig. 2. Evaluation of FoldIt (ours) and XDCycleGAN model [15] on VC with two different textures and ground truth/clinician-verified haustral fold (red) overlays. Green bounding boxes indicate locations where XDCycleGAN is not feature consistent and drops folds between neighboring frames. Complete video sequences are provided in the **supplementary video**. (Color figure online)

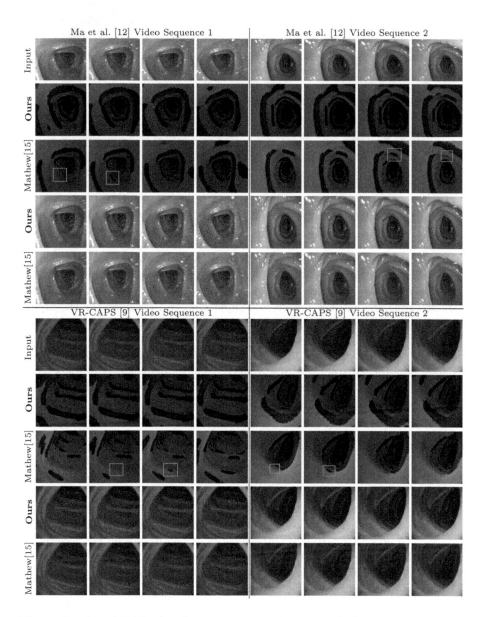

Fig. 3. Results of FoldIt (ours) and XDCycleGAN model [15] are shown on video sequences from Ma et al. [12] and recently released VR-CAPS colon simulator [9]. Complete video sequences are provided in the **supplementary video**. The OC frames are followed by FoldIt and XDCycleGAN output and the fold segmentations overlaid on the OC images. Green bounding boxes indicate examples of XDCycleGAN losing haustral folds information across frames whereas our FoldIt preserves these features.

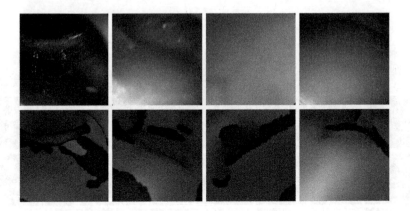

Fig. 4. Examples of our network failure in blurry and fluid-motion frames.

We also show results in Fig. 3 for 11 real OC video sequences from Ma et al. [12] and 2 from recently-released VR-CAPS simulator [9]. For both, FoldIt and XDCycleGAN, we show an overlay view, where the fold segmentation is extracted from the network output and superimposed in blue on the OC input. While XDCycleGAN is capable of doing scale-consistent depth maps for these video sequences, it struggles to retain haustral fold features across frames, as shown by the green bounding boxes and results in flickering in the final output (as shown in the **supplementary video**[2]). FoldIt delineates the folds more accurately and is more consistent in preserving fold features across frames (and does not result in the flickering effect seen with XDCycleGAN).

6   Limitations and Future Work

In this work, we introduced FoldIt, a new model to detect and segment haustral folds in colonoscopy videos. Our model has some limitations. It fails to handle cases where there are large amounts of fluid or blurriness, as shown in Fig. 4. This is due to the fact that our network is trained on clean VC images where there is no fluid or blur. In effect, these types of OC frames do not have valid corresponding frames in our VC training data. In the future, fluid and blur can be introduced in our VC renderings to add a valid correspondence for these failed OC frames. Moreover, even though our current model gives consistent results across video frames, we have not incorporated any explicit temporal constraint. In the future, we will incorporate temporal consistency in our model to track haustral folds across frames; the tracked folds will be used to register OC and VC videos. Finally, we will use additional anatomical features such as taenia coli (three longitudinal muscles that run across colon) to further enrich our model.

---

[2] Supplementary Video: https://youtu.be/_iWBJnDMXjo.

Acknowledgements. This project was supported by MSK Cancer Center Support Grant/Core Grant (P30 CA008748), and NSF grants CNS1650499, OAC1919752, and ICER1940302.

References

1. Amodio, M., Krishnaswamy, S.: Travelgan: image-to-image translation by transformation vector learning. In: Proceedings of the IEEE/CVF Conference on Computer Vision and Pattern Recognition, pp. 8983–8992 (2019)
2. Bae, G., Budvytis, I., Yeung, C.K., Cipolla, R.: Deep multi-view stereo for dense 3D reconstruction from monocular endoscopic video. In: International Conference on Medical Image Computing and Computer-Assisted Intervention, pp. 774–783 (2020)
3. Chen, R.J., Bobrow, T.L., Athey, T., Mahmood, F., Durr, N.J.: Slam endoscopy enhanced by adversarial depth prediction. arXiv preprint arXiv:1907.00283 (2019)
4. Choi, Y., Choi, M., Kim, M., Ha, J.W., Kim, S., Choo, J.: Stargan: unified generative adversarial networks for multi-domain image-to-image translation. In: Proceedings of the IEEE Conference on Computer Vision and Pattern Recognition, pp. 8789–8797 (2018)
5. Fang, H., Deng, W., Zhong, Y., Hu, J.: Triple-GAN: progressive face aging with triple translation loss. In: Proceedings of the IEEE/CVF Conference on Computer Vision and Pattern Recognition Workshops, pp. 804–805 (2020)
6. Freedman, D., et al.: Detecting deficient coverage in colonoscopies. arXiv preprint arXiv:2001.08589 (2020)
7. Goodfellow, I., et al.: Generative adversarial nets. In: Advances in Neural Information Processing Systems, pp. 2672–2680 (2014)
8. He, K., Zhang, X., Ren, S., Sun, J.: Deep residual learning for image recognition. In: Proceedings of the IEEE Conference on Computer Vision and Pattern Recognition, pp. 770–778 (2016)
9. İncetan, K., et al.: VR-Caps: a virtual environment for capsule endoscopy. Med. Image Anal. **70**, 101990 (2021)
10. Isola, P., Zhu, J.Y., Zhou, T., Efros, A.A.: Image-to-image translation with conditional adversarial networks. In: Proceedings of the IEEE Conference on Computer Vision and Pattern Recognition, pp. 1125–1134 (2017)
11. Liu, X., et al.: Reconstructing sinus anatomy from endoscopic video-towards a radiation-free approach for quantitative longitudinal assessment. In: International Conference on Medical Image Computing and Computer-Assisted Intervention, pp. 3–13 (2020)
12. Ma, R., Wang, R., Pizer, S., Rosenman, J., McGill, S.K., Frahm, J.M.: Real-time 3D reconstruction of colonoscopic surfaces for determining missing regions. In: International Conference on Medical Image Computing and Computer-Assisted Intervention, pp. 573–582 (2019)
13. Mahmood, F., Chen, R., Durr, N.J.: Unsupervised reverse domain adaptation for synthetic medical images via adversarial training. IEEE Trans. Med. Imaging **37**(12), 2572–2581 (2018)
14. Mathew, S., Nadeem, S., Kaufman, A.: Visualizing missing surfaces in colonoscopy videos using shared latent space representations. In: IEEE 18th International Symposium on Biomedical Imaging (ISBI), pp. 329–333 (2021)

15. Mathew, S., Nadeem, S., Kumari, S., Kaufman, A.: Augmenting colonoscopy using extended and directional cyclegan for lossy image translation. In: Proceedings of the IEEE/CVF Conference on Computer Vision and Pattern Recognition, pp. 4696–4705 (2020)
16. Nadeem, S., Kaufman, A.: Computer-aided detection of polyps in optical colonoscopy images. SPIE Med. Imaging **9785**, 978525 (2016)
17. Nadeem, S., Marino, J., Gu, X., Kaufman, A.: Corresponding supine and prone colon visualization using eigenfunction analysis and fold modeling. IEEE Trans. Vis. Comput. Gr. **23**(1), 751–760 (2016)
18. Rau, A., et al.: Implicit domain adaptation with conditional generative adversarial networks for depth prediction in endoscopy. Int. J. Comput. Assist. Radiol. Surg. **14**(7), 1167–1176 (2019). https://doi.org/10.1007/s11548-019-01962-w
19. Xu, J., et al.: Ofgan: realistic rendition of synthetic colonoscopy videos. in: international Conference on Medical Image Computing and Computer-Assisted Intervention, pp. 732–741 (2020)
20. Zhu, J.Y., Park, T., Isola, P., Efros, A.: Unpaired image-to-image translation using cycle-consistent adversarial networks. In: IEEE International Conference on Computer Vision, pp. 2223–2232 (2017)

Reference-Relation Guided Autoencoder with Deep CCA Restriction for Awake-to-Sleep Brain Functional Connectome Prediction

Dan Hu, Weiyan Yin, Zhengwang Wu, Liangjun Chen, Li Wang, Weili Lin, Gang Li$^{(\boxtimes)}$, and UNC/UMN Baby Connectome Project Consortium

Department of Radiology and BRIC, University of North Carolina at Chapel Hill, Chapel Hill, NC 27599, USA
gang_li@med.unc.edu

Abstract. The difficulty of acquiring resting-state fMRI of early developing children under the same condition leads to a dedicated protocol, i.e., scanning younger infants during sleep and older children during being awake, respectively. However, the obviously different brain activities of sleep and awake states arouse a new challenge of awake-to-sleep connectome prediction/translation, which remains unexplored despite its importance in the longitudinally-consistent delineation of brain functional development. Due to the data scarcity and huge differences between natural images and geometric data (e.g., brain connectome), existing methods tailored for image translation generally fail in predicting functional connectome from awake to sleep. To fill this critical gap, we unprecedentedly propose a novel reference-relation guided autoencoder with deep CCA restriction (R^2AE-dCCA) for awake-to-sleep connectome prediction. Specifically, 1) A reference-autoencoder (RAE) is proposed to realize a guided generation from the source domain to the target domain. The limited paired data are thus greatly augmented by including the combinations of all the age-restricted neighboring subjects as the references, while the target-specific pattern is fully learned; 2) A relation network is then designed and embedded into RAE, which utilizes the similarity in the source domain to determine the belief-strength of the reference during prediction; 3) To ensure that the learned relation in the source domain can effectively guide the generation in the target domain, a deep CCA restriction is further employed to maintain the neighboring relation during translation; 4) New validation metrics dedicated for connectome prediction are also proposed. Experimental results showed that our proposed R^2AE-dCCA produces better prediction accuracy and well maintains the modular structure of brain functional connectome in comparison with state-of-the-art methods.

Keywords: Functional connectome prediction · rs-fMRI · Autoencoder

1 Introduction

During the first few years of life, the human brain undergoes exceptionally dynamic development that could largely shape later behavioral and cognitive performance [1–3].

© Springer Nature Switzerland AG 2021
M. de Bruijne et al. (Eds.): MICCAI 2021, LNCS 12903, pp. 231–240, 2021.
https://doi.org/10.1007/978-3-030-87199-4_22

Delineating the functional developmental trajectories through this stage with resting-state fMRI (rs-fMRI) is of great importance in understanding the normal brain and diagnosing neurodevelopmental disorders [4, 5]. However, there are unique challenges associated with acquiring rs-fMRI for early developing children under a unified condition (sleep or awake), i.e., 1) it is impossible to request younger infants to be awake while remaining still during scanning; 2) it is difficult to persuade older children (>24 months) to sleep during the daytime and also the brain activity in the deep sleep during the night is significantly different from normal resting state. This dilemma generally leads to a dedicated protocol, i.e., scanning younger infants during sleep, while scanning older children during being awake and keeping them still by watching movies [4, 6]. However, there are big differences lying in the hemodynamic responses of sleep and awake that will be reflected in fMRI and the corresponding brain functional connectome. Thus, to realize meaningful and consistent cross-age studies under different scan conditions, predicting the functional connectome obtained during sleep from that during being awake is critical, which, however, remains unexplored to the best of our knowledge and challenging. Due to the difficulties in image acquisition and recruitment, training the model for awake-to-sleep brain connectome prediction usually confronts the problem of data scarcity. Moreover, although deep generative adversarial networks based domain translation have been successfully developed for image translation [7–9], these methods generally fail in predicting functional connectome because of the huge difference between natural images and geometric data (e.g., brain connectome). In addition, the multi-view brain graph synthesis method [10, 11] may not perform well in awake-to-sleep prediction, as it was designed for connectomes obtained by cortical morphological measures, which usually have smaller distribution differences between the source and target domains.

To address these issues, we unprecedentedly propose a Reference-Relation guided AutoEncoder with deep Canonical Correlation Analysis restriction (R^2AE-dCCA) for awake-to-sleep connectome prediction. First, a reference-autoencoder (RAE) is proposed to realize a guided generation from the source domain to the target domain. During the training stage, reference-couples from the target domain will be used to guide the prediction, which are constructed by the combinations of all the age-restricted neighboring subjects of the ones to be predicted. Merging with the individualized patterns learned from the source domain in the latent space, reference-couples not only provide the sleep-specific patterns but also greatly augment the limited data by random coupling. Then, a relation network is designed and embedded into RAE, which learns the similarity of the reference-couples to the subject to be predicted in the source domain and determine the belief-strength of reference in the process of target connectome generation. To guarantee that that similarity in the source domain will be maintained in the target domain, a deep CCA (Canonical Correlation Analysis) restriction is further employed in the coupled latent space to keep the neighboring relation from the source domain to the target domain. Finally, in the testing stage, all the samples in the target domain will be used as references and the corresponding reference-strengths are co-determined by the relation network and age distance. Our R^2AE-dCCA was implemented on a developing children rs-fMRI dataset and compared with state-of-the-art methods. The superiority of our proposed R^2AE-dCCA was validated not only on the overall accuracy but also on our proposed three new validation metrics dedicated for connectome prediction.

2   Method

2.1   Model Description

The framework of our proposed model, reference-relation guided autoencoder with deep CCA restriction (R^2AE-dCCA), is depicted in Fig. 1 and detailed below.

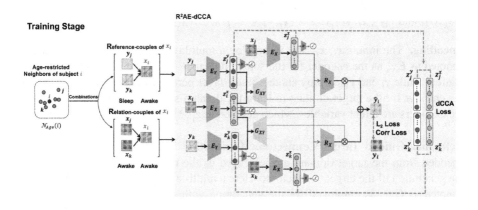

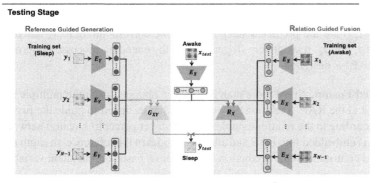

Fig. 1. The framework of our proposed R^2AE-dCCA model.

Our goal is to learn the mapping from the awake domain X to the sleep domain Y, $f : X \rightarrow Y$, given paired samples $\{(x_i, y_i, Age_i) | x_i \in X, y_i \in Y\}$, where $i = 1, \cdots, N$. N is the number of subjects, Age_i is the scan age of subject i, x_i and y_i are brain functional connectome represented by functional connectivity matrix and usually vectorized as the corresponding off-diagonal upper triangular elements for computation.

There are four main steps in **training stage**: (1) Reference-couples and relation-couples construction based on age-restricted neighborhood; (2) Reference guided target connectome generation; (3) Deep-CCA based source domain to target domain correlation; (4) Relation guided fusion.

Reference-Couples and Relation-Couples Construction. For each connectome x_i in the awake domain, except for the paired y_i, other connectomes in the sleep domain Y

could also provide rich information of how the brain connectivity patterns look like during sleep. Since the infant brain undergoes exceptionally dynamic development in both structure and function during early childhood [1–3, 28, 29], we only leverage the subjects within a neighboring age range to guide the learning of prediction. For a subject i, the age-restricted neighborhood of i is defined as $\mathcal{N}_{Age}(i) = \{j \big| |Age_i - Age_j| < \theta\}$, θ is the user-defined threshold and set as 30 days in our experiments. Then, the relation-couples and reference-couples are constructed as $\{(x_j, x_i, x_k) | x_i, x_j, x_k \in X, j, k \in \mathcal{N}_{Age}(i)\}$ and $\{(y_j, x_i, y_k) | x_i \in X, y_j, y_k \in Y, j, k \in \mathcal{N}_{Age}(i)\}$, respectively.

Encoding. The inputs x_i, x_j, and x_k employ a multilayer perceptron neural network, denoted as E_X, as the encoder for the source domain to learn the individualized information, while y_j and y_k employ another multi-layer perceptron neural network, E_Y, as the encoder for the target domain to learn the domain-specific pattern. The outputs of the encoders are latent variables, which denoted as $z_i^x, z_j^x, z_k^x, z_j^y$, and z_k^y, respectively.

Reference Guided Target Connectome Generation. In the latent space, z_j^y and z_k^y encoded from the target domain are leveraged as the reference information to guide the generation of the target connectome. Here a multi-layer perceptron neural network, denoted as G_{XY}, is employed for the generation with the inputs, which are the concatenations of the individualized information z_i^x from the source domain and the sleep-specific information (z_j^y, z_k^y) from the target domain. Thus, $\hat{y}_{ij} = G_{XY}([z_i^x; z_j^y])$ and $\hat{y}_{ik} = G_{XY}([z_i^x; z_k^y])$ are the predicted sleep connectome corresponding to x_i based on the reference of y_j and y_k, respectively. Together with the encoding process, a reference autoencoder (RAE) is designed.

Relation Guided Fusion. Since more than one $\hat{y}_{i*}$ are obtained from the multiple reference couples and the RAE, a relation network is further designed to guide the fusion. Specifically, according to the relation-couples, a multi-layer perceptron neural network, denoted as R_X, is embedded into RAE and employed to learn the reference-strength that y_j and y_k should contribute to the prediction of y_i. That is, based on the latent variables z_i^x, z_j^x, and z_k^x,

$$\hat{y}_i = \frac{e^{R_X(z_i^x, z_j^x)} \cdot \hat{y}_{ij}}{e^{R_X(z_i^x, z_j^x)} + e^{R_X(z_i^x, z_k^x)}} + \frac{e^{R_X(z_i^x, z_k^x)} \cdot \hat{y}_{ik}}{e^{R_X(z_i^x, z_j^x)} + e^{R_X(z_i^x, z_k^x)}} \tag{1}$$

Deep-CCA Based Source Domain to Target Domain Correlation. From Eq. (1), $\hat{y}_i$ is estimated based on the assumption that the similarity relationship in the source domain maintains in the target domain. Therefore, the correlation between the learned embedding of the source domain and the target domain should be maximized during the training, thus ensuring the effectiveness of the fusion by preserving the neighboring relationship cross domains. Suppose $z_j^x, z_k^x, z_j^y, z_k^y \in \mathbb{R}^{l \times s}$, l is the dimension of the latent space, s is the batch size of training, $Z_x = \left[z_j^x, z_k^x\right]$ and $Z_y = [z_j^y, z_k^y]$ are the column-wise concatenation of (z_j^x, z_k^x) and (z_j^y, z_k^y), respectively. Let $\overline{Z}_x = Z_x - \frac{1}{2s}Z_x\mathbf{1}$ and $\overline{Z}_y = Z_y - \frac{1}{2s}Z_y\mathbf{1}$ be the centered matrix, $\mathbf{1}$ is an all-1s matrix, and $\hat{\Sigma}_{xy} = \frac{1}{2s-1}\overline{Z}_x\overline{Z}_y'$,

$\hat{\Sigma}_{xx} = \frac{1}{2s-1}\overline{Z}_x\overline{Z}_x' + \delta_1 I$, $\hat{\Sigma}_{yy} = \frac{1}{2s-1}\overline{Z}_y\overline{Z}_y' + \delta_2 I$, $\delta_1, \delta_2 > 0$ are constants to ensure $\hat{\Sigma}_{xx}$ and $\hat{\Sigma}_{yy}$ to be positive definite. As for the classical CCA, the total correlation of the top m components of Z_x and Z_y is the sum of the top m singular values of the matrix $T = \hat{\Sigma}_{xx}^{-1/2}\hat{\Sigma}_{xy}\hat{\Sigma}_{yy}^{-1/2}$. In our case, m is set as l, then the correlation of Z_x and Z_y is the matrix trace norm of T, i.e.,

$$\mathcal{L}_{dCCA} = corr(Z_x, Z_y) = \|T\|_{tr} = tr(T'T)^{1/2} \tag{2}$$

Taking $\mathcal{L}_{dCCA}$ as one term of the loss function, the embeddings of the source domain and target domain are required to be maximally correlated.

Adversarial Loss. To enforce the stability of the training, a distribution regularization is imposed to the latent space and realized by a shared discriminator D [12]. Let $p(z)$ be the prior distribution imposing on the latent variable z, $q(z|x)$ be the encoding distribution. Training the autoencoder with distribution regularization requires the aggregated posterior distribution $q(z) = \int_{x_i} q(z|x)p_d(x)dx$ matching the predefined prior $p(z)$, where $p_d(x)$ is the distribution of the input data. Here, this regularization is realized by an adversarial procedure, which leads to a $\min_E \max_D \mathcal{L}_{adv}$ problem, where

$$\mathcal{L}_{adv} = \mathcal{L}_{adv\_Ex} + \mathcal{L}_{adv\_Ey} \tag{3}$$

$$\mathcal{L}_{adv\_Ex} = \sum_{x=x_i,x_j,x_k} \mathbb{E}_x \log(1 - D(E_X(x))) + \mathbb{E}_z \log(D(z)) \tag{4}$$

$$\mathcal{L}_{adv\_Ey} = \sum_{y=y_j,y_k} \mathbb{E}_y \log(1 - D(E_Y(y))) + \mathbb{E}_z \log(D(z)) \tag{5}$$

Target Connectome Prediction Loss. L^2 norm and Pearson's correlation are adopted as our generation loss for the target connectome prediction:

$$\mathcal{L}_{element} = \mathbb{E}_{x_i}\|y_i - \hat{y}_i\|_2^2 \tag{6}$$

$$\mathcal{L}_{corr} = \mathbb{E}_{x_i} corr(y_i, \hat{y}_i) \tag{7}$$

Full Objective. The objective functions to optimize E_X, E_Y, G_{XY}, and D are written as:

$$\mathcal{L}_D = \mathcal{L}_{adv} \tag{8}$$

$$\mathcal{L}_{E_X,E_Y,G_{XY}} = -\lambda_1 \mathcal{L}_{corr} + \lambda_2 \mathcal{L}_{element} + \lambda_3 \mathcal{L}_{adv\_E} - \mathcal{L}_{dCCA} \tag{9}$$

where $\mathcal{L}_{adv\_E} = \mathbb{E}_x \log(1 - D(E_X(x))) + \mathbb{E}_y \log(1 - D(E_Y(y)))$, λ_1, λ_2, and λ_3 are trade off parameters. The model alternatively updates E_X, E_Y, G_{XY}, and D with $\mathcal{L}_{E_X,E_Y,G_{XY}}$ and $\mathcal{L}_D$.

Testing Stage. For each x_{test}, with the age of refence connectome being considered into the relation guided fusion, all the connectome y_{Tr} in the training set are used as reference to avoid the lack of variability. With RAE obtaining corresponding sleep connectome $\hat{y}_{test,Tr}$ and relation network providing the reference-strength of each reference, the final prediction of y_{test} is estimated as

$$\hat{y}_{test} = \sum_{y_{Tr}} \sigma\left(W_{y_{Tr}} \cdot R_X\left(E_X\left(x_{test}\right), E_X\left(x_{Tr}\right)\right)\right) \cdot G_{XY}\left(\left[E_X\left(x_{test}\right), E_Y\left(y_{Tr}\right)\right]\right) \quad (10)$$

where $W_{y_{Tr}} = e^{-|Age_{test} - Age_{y_{Tr}}|}$, $\sigma(*) = \frac{\exp(*)}{\sum_{y_{Tr}} \exp(*)}$ is the Softmax function, (x_{Tr}, y_{Tr}) is the paired connectomes obtained during sleep and being awake of the same subject.

2.2 Validation of Functional Connectome Prediction

Although Pearson's correlation coefficient (r) and mean absolute error (MAE) are usually taken as the evaluation metrics, they are general measures without any characteristics of specific applications. Taking the practical requirement of functional connectome prediction into consideration, we propose three new metrics dedicated for the validation of functional connectome prediction, i.e., correlation of top percentile connections (Corr$_{percl}$), normalized variation of information (VIn), and normalized mutual information (MIn) of the induced modular structure.

Correlation of Top Percentile Connections (Corr$_{percl}$). For a functional connectome, the connections with top percentile of strength are usually the focus of functional graph or network construction [13, 14]. Thus, Corr$_{percl}$ is the Pearson's correlation coefficient merely counted within the connections with top percentile of strength, i.e.,

$$\text{Corr}_{percl} = corr\left(y^{percl}, \hat{y}^{percl}\right) \quad (11)$$

where y^{percl} and $\hat{y}^{percl}$ are consisting of connections with top percentile of strength in the expected connectome y. The percentile was set as 95% in our experiments.

Normalized Variation of Information (VIn) and Mutual Information (MIn). The modular structure based on graph theory is one of the most important analyses for functional brain networks [15, 16]. Here we introduce the capability of maintaining the modular structure of the expected connectome y as one metric to validate the predicted connectome. Let $A = \{a_1, a_2, \cdots, a_c\}$ and $B = \{b_1, b_2, \cdots, b_c\}$ be the modular partition induced by y and $\hat{y}$, respectively. The VIn and MIn [17] between A and B are defined as follow:

$$\text{VIn}(A, B) = -\frac{1}{\log \mathbb{N}} \sum_{t_1, t_2 = 1, \dots, c} \frac{|a_{t_1} \cap b_{t_2}|}{\mathbb{N}} \left(\log \frac{|a_{t_1} \cap b_{t_2}|}{|a_{t_1}|} + \log \frac{|a_{t_1} \cap b_{t_2}|}{|b_{t_2}|}\right) \quad (12)$$

$$\text{MIn}(A, B) = 1 + \frac{\text{VIn}(A, B) \cdot \log \mathbb{N}}{\sum_{t_1} \frac{|a_{t_1}|}{\mathbb{N}} \log \frac{|a_{t_1}|}{\mathbb{N}} + \sum_{t_2} \frac{|b_{t_2}|}{\mathbb{N}} \log \frac{|b_{t_2}|}{\mathbb{N}}} \quad (13)$$

where $\mathbb{N}$ is the number of connections in y_i, and $|\cdot|$ in Eqs. (12) and (13) represents the numbers of connections in the module. In our experiments, A and B are obtained by finding the modular structure with the maximal between-class and within-class ratio within 100 repetition of k-means clustering [18]. The number of clusters is set as 10.

3 Experiments

3.1 Data Description

We verified the effectiveness of the proposed R^2AE-dCCA model on a high-resolution resting-state fMRI (rs-fMRI) data including 20 paired sleep and awake scans in the UNC/UMN Baby Connectome Project [6]. All paired rs-fMRI data were acquired during natural sleeping and video watching on a 3T Siemens Prisma MRI scanner using a 32-channel head coil. T1-weighted and T2-weighted MR images were obtained with the resolution $= 0.8 \times 0.8 \times 0.8$ mm^3. The rs-fMRIs scans were acquired with TR/TE $=$ 800/37 ms, FA $= 80°$, FOV $= 220$ mm, resolution $= 2 \times 2 \times 2$ mm^3, and total volumes $= 420$ (5 min 47 s). All structural and functional MR images were preprocessed by a state-of-the-art infant-tailored in-house pipeline [23–27]. Each rs-fMRI was finally parcellated based on the automated anatomical labeling template [19], thus obtaining 116 anatomical regions. The average time series within each ROI was correlated with those from all others. The functional connectivity matrix was derived by calculating the Pearson's correlation coefficient between time series of each pair of ROIs. Fishers r-to-z transformation was conducted to improve the normality of the functional connectivity.

3.2 Validation of R^2AE-dCCA

With the metrics of MAE, r, Corr$_{percl}$, VIn, and MIn, we compared the proposed R^2AE-dCCA model by leave-one-out cross-validation with the following five methods: (1) Connectome prediction with linear model (GLM) [20]; (2) Multi-kernel manifold learning (MKML) [21]; (3) CCA-based MKML (CCA-TSW)[10]; (4) Pixel2Pixel GAN [22]; (5) R^2AE-dCCA without relation network (R^2AE-dCCA no R-Net);

In R^2AE-dCCA, the encoder E_X and E_Y constitute of 3 densely connected layers of dimension (50, 50, 100) with (LeakyReLU, Sigmoid, Linear) as the corresponding activation function. G_{XY} constitutes of 4 densely connected layers of dimension (30, 30, 30, 30) with LeakyReLU as the activation function. The discriminator D constitutes o4 densely connected layers of dimension (50, 50, 25, 1) with LeakyReLU as the activation function of the first 3 layers and Sigmoid as the activation function of the last layer. R^2AE-dCCA was implemented with Pytorch and optimized with Adamax by a fixed learning rate as 0.001. The batch size was set as 400. $\lambda_1 = 0.1$, $\lambda_2 = 0.8$, and $\lambda_3 = 0.1$. Methods (4) and (5) share the similar architecture with R^2AE-dCCA for the fairness of the comparison. The means and standard deviations of the leave-one-out cross-validation are reported in Table 1. Our method achieves lowest MAE, VIn, highest r, Corr$_{percl}$, and MIn among all comparison methods, indicating the superior performance of our method. Figure 2 shows the scatter plots of the expected and predicted connection strength on a representative subject. It can be seen that our method achieved better prediction especially for the connections with strength greater than 1, i.e., the top percentile connections. Figure 3 shows how the predicted connectome maintains the obtained modular structures induced from the expected functional connectome of a representative subject. In Fig. 3, the order of the brain regions are the same in the subfigures, while the values are the corresponding predicted connection strengthes based on different methods. The results obtained by our method shows higher similarity with the groud truth. In summary, our R^2AE-dCCA model outperformed the other five state-of-the-art methods not only on overall prediction accuracy but also on maintaining the modular structure.

Table 1. The comparison of R^2AE-dCCA with other five methods.

Metrics / Methods	Conventional		Connectome Specific		
	MAE	r	Corr$_{percl}$	VIn	MIn
GLM [20]	.272 ± .040	.304 ± .086	.302 ± .090	.553 ± .055	.213 ± .072
MKML [21]	.257 ± .046	.534 ± .081	.542 ± .079	.460 ± .044	.338 ± .066
CCA-TSW [10]	.247 ± .048	.549 ± .083	.546 ± .056	.451 ± .057	.349 ± .085
Pixel2Pixel GAN [22]	.316 ± .088	.426 ± .076	.212 ± .093	.448 ± .051	.330 ± .070
R^2AE-dCCA (proposed)	.239 ± .050	.592 ± .089	.568 ± .095	.442 ± .056	.365 ± .066
R^2AE-dCCA **(proposed)**	**.227 ± .050**	**.614 ± .084**	**.583 ± .010**	**.427 ± 0.51**	**.379 ± .062**

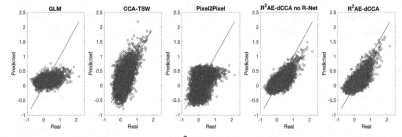

Fig. 2. The scatter plots of our proposed R^2AE-dCCA model in comparison with GLM, CCA-TSW, Pixel2Pixel, and R^2AE-dCCA no R-Net in terms of connection strength.

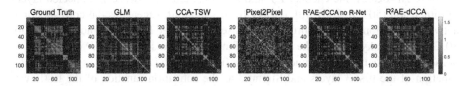

Fig. 3. Comparison of modular structures obtained by different methods.

4 Conclusion

In this paper, to fill the gap of awake-to-sleep connectome prediction for longitudinal study of early brain functional development, we proposed a reference-relation guided autoencoder with deep CCA restriction (R^2AE-dCCA). With the framework of reference guided generation and relation guided fusion, R^2AE-dCCA reaches the superior prediction accuracy by effectively augmenting the severely limited data, utilizing the domain specific pattern from the target domain, and maintaining neighboring relationship from the source domain. As a generalized model for connectome translation, our model and the proposed connectome-dedicated validation metrics have high potential in other connectome prediction related fields.

Acknowledgments. This work was partially supported by NIH grants (MH127544, MH116225, MH117943, MH109773, and MH123202). This work also utilizes approaches developed by an NIH grant(1U01MH110274) and the efforts of the UNC/UMN Baby Connectome Project Consortium.

References

1. Lyall, A.E., et al.: Dynamic development of regional cortical thickness and surface area in early childhood. Cereb. Cortex **25**(8), 2204–2212 (2015)
2. Gilmore, J.H., et al.: Longitudinal development of cortical and subcortical gray matter from birth to 2 years'. Cereb. Cortex **22**(11), 2478–2485 (2012)
3. Li, G., et al.: Cortical thickness and surface area in neonates at high risk for schizophrenia. Brain Struct. Funct. **221**(1), 447–461 (2016). https://doi.org/10.1007/s00429-014-0917-3
4. Zhang, H., Shen, D., Lin, W.: Resting-state functional MRI studies on infant brains: a decade of gap-filling efforts. Neuroimage **185**, 664–684 (2019)
5. Cao, M., Huang, H., He, Y.: Developmental connectomics from infancy through early childhood. Trends Neurosci. **40**(8), 494–506 (2017)
6. Howell, B.R., et al.: The UNC/UMN Baby Connectome Project (BCP): an overview of the study design and protocol development. Neuroimage **185**, 891–905 (2019)
7. Alotaibi, A.: Deep generative adversarial networks for image-to-image translation: a review. Symmetry **12**(10), 1705 (2020)
8. Armanious, K., et al.: MedGAN: medical image translation using GANs. Comput. Med. Imaging Graph. **79**, 101684 (2020)
9. Choi, Y., Choi, M., Kim, M., Ha, J.-W., Kim, S., Choo, J.: StarGAN: unified generative adversarial networks for multi-domain image-to-image translation. In: Proceedings of the IEEE Conference on Computer Vision and Pattern Recognition, pp. 8789–8797 (2018)
10. Zhu, M., Rekik, I.: Multi-view brain network prediction from a source view using sample selection via CCA-based multi-kernel connectomic manifold learning. In: Rekik, I., Unal, G., Adeli, E., Park, S.H. (eds.) PRIME 2018. LNCS, vol. 11121, pp. 94–102. Springer, Cham (2018). https://doi.org/10.1007/978-3-030-00320-3_12
11. Bessadok, A., Mahjoub, M.A., Rekik, I.: Brain graph synthesis by dual adversarial domain alignment and target graph prediction from a source graph. Med. Image Anal. **68**, 101902 (2021)
12. Makhzani, A., Shlens, J., Jaitly, N., Goodfellow, I., Frey, B.: Adversarial autoencoders. In: International Conference on Learning (2015)
13. Van den Heuvel, M.P., de Lange, S.C., Zalesky, A., Seguin, C., Yeo, B.T., Schmidt, R.: Proportional thresholding in resting-state fMRI functional connectivity networks and consequences for patient-control connectome studies: Issues and recommendations. Neuroimage **152**, 437–449 (2017)

14. Garrison, K.A., Scheinost, D., Finn, E.S., Shen, X., Constable, R.T.: The (in)stability of functional brain network measures across thresholds. Neuroimage **118**, 651–661 (2015)
15. Wen, X., Wang, R., Yin, W., Lin, W., Zhang, H., Shen, D.: Development of dynamic functional architecture during early infancy. Cereb. Cortex **30**(11), 5626–5638 (2020)
16. Meunier, D., Achard, S., Morcom, A., Bullmore, E.: Age-related changes in modular organization of human brain functional networks. Neuroimage **44**(3), 715–723 (2009)
17. Meilă, M.: Comparing clusterings-an information based distance. J. Multivar. Anal. **98**(5), 873–895 (2007)
18. Venkataraman, A., Van Dijk, K.R., Buckner, R.L., Golland, P.: Exploring functional connectivity in fMRI via clustering. In: 2009 IEEE International Conference on Acoustics, Speech and Signal Processing, pp. 441–444 (2009)
19. Tzourio-Mazoyer, N., et al.: Automated anatomical labeling of activations in SPM using a macroscopic anatomical parcellation of the MNI MRI single-subject brain. Neuroimage **15**(1), 273–289 (2002)
20. Tavor, I., Jones, O.P., Mars, R.B., Smith, S., Behrens, T., Jbabdi, S.: Task-free MRI predicts individual differences in brain activity during task performance. Science **352**(6282), 216–220 (2016)
21. Wang, B., Zhu, J., Pierson, E., Ramazzotti, D., Batzoglou, S.: Visualization and analysis of single-cell RNA-seq data by kernel-based similarity learning. Nat. Methods **14**(4), 414–416 (2017)
22. Isola, P., Zhu, J.-Y., Zhou, T., Efros, A.A.: Image-to-image translation with conditional adversarial networks. In: Proceedings of the IEEE Conference on Computer Vision and Pattern Recognition, pp. 1125–1134 (2017)
23. Li, G., Wang, L., Yap, P.-T., et al.: Computational neuroanatomy of baby brains: a review. Neuroimage **185**, 906–925 (2019)
24. Wang, L., et al.: Volume-based analysis of 6-month-old infant brain MRI for autism biomarker identification and early diagnosis. In: Frangi, A.F., Schnabel, J.A., Davatzikos, C., Alberola-López, C., Fichtinger, G. (eds.) MICCAI 2018. LNCS, vol. 11072, pp. 411–419. Springer, Cham (2018). https://doi.org/10.1007/978-3-030-00931-1_47
25. Li, G., Wang, L., Shi, F., Gilmore, J.H., Lin, W., Shen, D.: Construction of 4D high-definition cortical surface atlases of infants: methods and applications. Med. Image Anal. **25**(1), 22–36 (2015)
26. Li, G., Wang, L., Shi, F., Lin, W., Shen, D.: Simultaneous and consistent labeling of longitudinal dynamic developing cortical surfaces in infants. Med. Image Anal. **18**(8), 1274–1289 (2014)
27. Yin, W., et al.: The emergence of a functionally flexible brain during early infancy. Proc. Natl. Acad. Sci. **117**(38), 23904–23913 (2020)
28. Hu, D., Zhang, H., Wu, Z., Wang, et al.: Disentangled-multimodal adversarial autoencoder: application to infant age prediction with incomplete multimodal neuroimages. IEEE Trans. Med. Imaging **39**(12), 4137–4149 (2020)
29. Hu, D., et al.: Disentangled intensive triplet autoencoder for infant functional connectome fingerprinting. In: Martel, A.L., et al. (eds.) MICCAI 2020. LNCS, vol. 12267, pp. 72–82. Springer, Cham (2020). https://doi.org/10.1007/978-3-030-59728-3_8

Domain Composition and Attention for Unseen-Domain Generalizable Medical Image Segmentation

Ran Gu[1,3], Jingyang Zhang[2,3], Rui Huang[3], Wenhui Lei[1], Guotai Wang[1(✉)], and Shaoting Zhang[1,3]

[1] School of Mechanical and Electrical Engineering, University of Electronic Science and Technology of China, Chengdu, China
guotai.wang@uestc.edu.cn
[2] School of Biomedical Engineering, Shanghai Jiao Tong University, Shanghai, China
[3] SenseTime Research, Shanghai, China

Abstract. Domain generalizable model is attracting increasing attention in medical image analysis since data is commonly acquired from different institutes with various imaging protocols and scanners. To tackle this challenging domain generalization problem, we propose a Domain Composition and Attention-based network (DCA-Net) to improve the ability of domain representation and generalization. First, we present a domain composition method that represents one certain domain by a linear combination of a set of basis representations (i.e., a representation bank). Second, a novel plug-and-play parallel domain preceptor is proposed to learn these basis representations and we introduce a divergence constraint function to encourage the basis representations are as divergent as possible. Then, a domain attention module is proposed to learn the linear combination coefficients of the basis representations. The result of liner combination is used to calibrate the feature maps of an input image, which enables the model to generalize to different and even unseen domains. We validate our method on public prostate MRI dataset acquired from six different institutions with apparent domain shift. Experimental results show that our proposed model can generalizes well on different and even unseen domains and it outperforms state-of-the-art methods on the multi-domain prostate segmentation task. Code is available at https://github.com/HiLab-git/DCA-Net.

Keywords: Attention · Domain generalization · Segmentation.

1 Introduction

Deep learning with Convolutional Neural Networks (CNNs) have achieved remarkable performance for medical image segmentation [12,17,18]. Their

Ran Gu and Jingyang Zhang contributed equally. The work was done during their internship at SenseTime.

© Springer Nature Switzerland AG 2021
M. de Bruijne et al. (Eds.): MICCAI 2021, LNCS 12903, pp. 241–250, 2021.
https://doi.org/10.1007/978-3-030-87199-4_23

achievements heavily rely on the training process with images from a similar domain distribution for test images. However, for clinical model deployment, the training and test images are commonly acquired from different sites (hospitals) with different imaging protocols and scanner vendors, leading to evident data distribution discrepancy between them and thus substantial performance degradation at test time [6,9,14]. Recently, Domain Adaptation (DA) methods [5,6,16] provide a promising way to solve this problem by aligning the source and target domain distributions. While, most of the DA methods require repeated model re-training with involvement of images from the target domain, which is time-consuming and impractical when the images form a new domain are not available for training in clinical applications [2]. Therefore, it is desirable to enable a model to generalize well to unseen domains, where no images in the new domain are available for retraining the model. Nowadays, domain generalization is still a challenging task but full of practical clinical significance.

Recently, there are some researches have tried to solving the domain generalization problem [1,10,11], and some methods has been applied on medical imaging. From the aspect of data, Zhang et al. [21] proposed a data augmentation-based method assuming that the shift between source and target domains could be simulated by applying extensive data augmentation on a single source domain. However, the configuration of augmentation operations, such as the type and the number of transformation, requires empirical settings and even data-specific modifications. From the aspect of algorithm, Dou et al. [4] proposed a meta-learning paradigm benefited from its model-agnostic manner and providing a flexible way to solve the domain generalization problem. It splits a set of source domains into meta-train and meta-test subsets, and adopts a gradient-based meta-optimization that iteratively updates model parameters to improve performance on both meta-train and meta-test datasets. Subsequently, Liu et al. [13] further introduced shape constraint based on meta-learning for generalizable prostate MRI segmentation. Although meta-learning exhibits remarkable performance for domain generalization [10], it still has two-big limitations. First, the meta-optimization process is highly time-consuming since all potential splitting results of meta-train and meta-test should be considered and involved during training. The process would be even more complicated and hard for convergence when dealing with a larger set of source domains. Second, as meta-learning is used to optimize a unidirectional transfer from meta-train to meta-test domains, it still needs to distinguish distribution discrepancy among source domains by domain-specific statistics, which brings more restricts for data acquirement and storage in clinical practice. And from the aspect of self-ensemble, Wang et al. [19] introduced a domain-oriented feature embedding framework that dynamically enriches the image features with additional domain prior knowledge learned from multi-source domains to make the semantic features more discriminative. However, [19] extracts one domain prior knowledge from each of the K source domains, where they are limited by one-to-one correspondence. Meanwhile, it simply takes an average of features from one domain as its prior knowledge, which may lead to redundancy of prior vectors when two domains are similar and limit the representation power.

Nowadays, attention mechanism is increasingly used because of its powerful self-expression ability, and plays a great role in semantic segmentation [3,7,15]. Inspired by channel attention [8], Wang et al. [20] proposed a domain attention method to construct a universal model for multi-site object detection, leading to a striking performance. However, this method was only validated with domains presented for training, and its generalizablity to unseen domains was not explored. For multi-site images, the segmentation model is expected to achieve high performance not only on diffident domains that have been seen during training, but also on unseen domains that are often involved in the model deployment stage. To deal with this problem, we propose a Domain Composition and Attention-based Network (DCA-Net) for generalizable multi-site medical image segmentation.

The contributions of our work are: 1) we propose a novel Domain Composition and Attention (DCA) method to represent a certain (seen or unseen) domain by a linear combination of a set of basis domain representations, so that the generalizablility of segmentation models are improved. 2) We design a Parallel Domain Preceptor (PDP) to learn the basis domain representations, and propose a divergence constraint function to encourage the basis domain representations to be divergent, which improves the domain composition power for generalization. 3) We propose a network using DCA (i.e., DCA-Net) for generalizable segmentation, and experimental results on multi-site prostate MRI segmentation showed that our DCA-Net outperformed state-of-the-art domain generalization methods in the same experimental settings.

2   Methods

Let $\mathcal{D} = \{\mathcal{D}_1, \mathcal{D}_2, ..., \mathcal{D}_M\}$ be the set of M source domains. The k-th image and its label in domain D_m are denoted as x_m^k and y_m^k, respectively. An unseen target domain for testing is denoted as $\mathcal{D}_t$. To achieve robust segmentation results for both source domains and unseen target domains, we proposed a Domain Composition and Attention (DCA) method to recalibrate the feature maps of an input image to obtain high segmentation performance across different domains. Inspired by shape composition [22], our domain composition represents a certain domain by a linear combination of elements in a domain representation bank consisting of a set of basis representations. As shown in Fig. 1(a), the domain representation bank is learned by a Parallel Domain Preceptor (PDP) and the linear combination coefficients are learned by a domain attention module. The DCA block is combined with a U-Net backbone to achieve generalizability across different domains, which is named as DCA-Net and shown in Fig. 1 (b). In the following, we frist introduce our DCA block and then describe the DCA-Net.

2.1   Domain Composition and Attention Block

As shown in Fig. 1(a), the goal of our DCA block is to calibrate a feature map F of an image from a certain domain adaptively to neutralize its domain bias

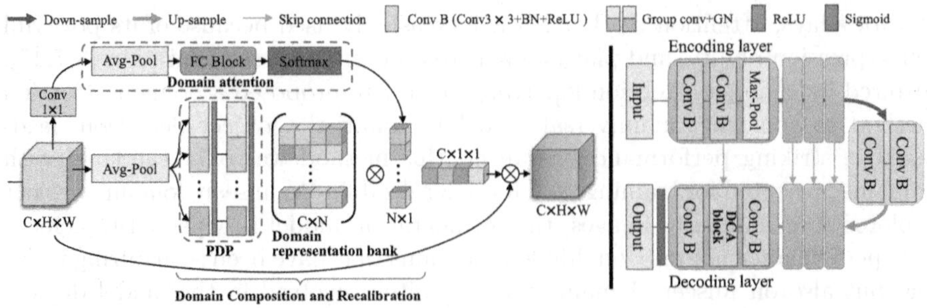

(a) Details of Domain Composition and Attention (DCA) block (b) Structure of proposed DCA-Net

Fig. 1. Overview of the proposed Domain Composition and Attention (DCA) method for multi-site domain generalizable segmentation. (a) Details of the proposed DCA block with a Parallel Domain Preceptor (PDP) and a domain attention module for domain composition. (b) Structure of DCA-Net, which combines the DCA block with the decoder of a U-Net backbone.

for improving the generalizability. DCA includes a PDP module for constructing a domain representation bank and a domain attention module for domain composition, the output of which is used for the feature neutralization.

Parallel Domain Preceptor (PDP). The PDP is used to learn a set of N basis representations in the domain representation bank. We use one basis preceptor for each basis representation, which is designed to capture a specific domain information that are different from the others. For the input feature map F with C channels, its avg-pooled result is denoted as $f \in \mathbb{R}^{C\times 1}$. The n-th basis preceptor p_n converts f into the n-th basis representation $p_n(f) \in \mathbb{R}^{C\times 1}$, thus the domain representation bank can be donated as $\mathcal{B} = \{p_1(f), p_2(f), ..., p_N(f)\}$. As the basis preceptors are independent, they can be implemented by a set of parallel subnetworks. To improve the efficiency, we propose to use a Grouped Convolution (GC) subnetwork as the structure for PDP, where each group corresponds to a basis preceptor. Specifically, the GC subnetwork has two convolutional layers each with N groups, and they are followed by group normalization + ReLU and group normalization + sigmoid, respectively, as illustrated in Fig. 1(a). Thus, the output of group n obtains $p_n(f)$. As there is no interactions between different groups in GC, the N preceptors (groups) are independent to capture different basis representations.

Divergence Constraint for Representation Bank. To encourage the preceptors to capture different domain information, we propose a divergence constraint function $\mathcal{L}_{div}$ that enforces outputs of the N preceptors to be different from each other, which makes DCA be capable of capturing a wide-range domain representations. During training, we randomly choose out three basis preceptors p_i, p_j, p_k and minimize the average consistency between each pair of them.

$$\mathcal{L}_{div} = 1 - \left[\frac{1}{3}\left(||p_i(f) - p_j(f)||_2^2 + ||p_i(f) - p_k(f)||_2^2 + ||p_j(f) - p_k(f)||_2^2\right)\right]^{\frac{1}{2}} \quad (1)$$

where $|| \cdot ||_2$ is the $L2$ norm, and p_i, p_j, p_k ($i \neq j \neq k$) are randomly selected in each iteration. As a result, the basis representations are encouraged to be divergent and $\mathcal{B}$ will have a strong representation ability with its diverse elements.

Domain Attention Module. With the basis domain representations, we represent a corresponding domain calibration vector $\alpha(f) \in \mathbb{R}^{C \times 1}$ for the input feature map F as a linear combination of them:

$$\alpha(f) = \beta_1 p_1(f) + \beta_2 p_2(f) + ... + \beta_N p_N(f) \quad (2)$$

where β_n is the coefficient for $p_n(f)$. The coefficient vector $\beta = (\beta_1, \beta_2, ..., \beta_N)$ is adaptively predicted by a domain attention module, as shown in Fig. 1(a). For the input feature map F, we use a 1×1 convolution followed by avg-pooling in a second branch to convert it to a vector, which is then followed by a Fullly Connected (FC) block and a softmax, where the last layer of FC has a length of N, thus the output of the domain attention module is taken as the coefficient vector β. Given β, we obtain $\alpha(f)$ based on Eq. 2, and the output of DAC is $\hat{F} = F \otimes \alpha(f)$, where $\otimes$ means tensor multiplication with broadcasting.

2.2   DCA-Net

Without loss of generality, we chose the powerfer U-Net [17] as the backbone for its simple structure and good performance for segmentation. We use a DCA in each convolutional block at different levels in the decoder of U-Net, as shown in Fig. 1 (b). Thus, in our DCA-Net, each convolutional block in the decoder consists of two convolutional layers that are connected by the DCA. During training, we combine a standard segmentation loss (i.e., Dice loss) L_{seg} with deep supervision to improve the performance. We also adopt the shape compactness constraint loss $\mathcal{L}_{comp}$ suggested by Liu et al. [13] as regularization to improve the robustness. Considering that we also need a divergence constraint function $\mathcal{L}_{div}$ as explained above, the overall loss function for our DCA-Net is:

$$\mathcal{L}_{all} = \mathcal{L}_{seg} + \lambda_1 \mathcal{L}_{comp} + \lambda_2 \mathcal{L}_{div} \quad (3)$$

where λ_1 and λ_2 are the weights of L_{comp} and L_{div}, respectively.

3   Experiments and Results

Datasets and Implementation. For experiments, we used the well-organized multi-site T2-weighted MRI dataset[1] for prostate segmentation [13] which was acquired from 6 institutes with different patient numbers and acquisition protocols: Data of Site A and B are from NCI-ISBI13 dataset containing 30 samples

---

[1] https://liuquande.github.io/SAML/.

each; data of Site C are from I2CVB dataset containing 19 samples; data of Site D, E and F are from PROMISE12 dataset containing 13, 12 and 12 samples, respectively. The images were preprocessed in the same way as SAML [13]. To assess the generalizability to unseen domains, we follow the leave-one-domain-out strategy in SAML [13], where each time one domain is used as unseen and the others are used for seen domains. All the images in the seen domains are used for training, and we split the unseen domain into 20% and 80% at patient level for validation and testing, respectively.

Our DCA-Net with UNet as backbone [17] was implemented in 2D due to the large variance on through-plane spacing among different sites. The encoder is kept the original settings with the channel numbers of 16, 32, 64, 128 and 256 at five scales, respectively [17]. We concatenate every three slices into an input with three channels, and use DCA-Net to predict the segmentation in the central slice. The loss function weights λ_1 and λ_2 are set as 1.0 and 0.1, respectively. The segmentation model was trained using Adam optimizer and the learning rate was $5e^{-4}$. We trained 20k iterations with batch size of 4 and the basis representation number N was 8 in the domain representation bank. Training was implemented on one NVIDIA Geforce GTX 1080 Ti GPU. For fair comparison, we kept the most parameters be same as those in SAML [13]. We adopt Dice score (Dice) and Average Surface Distance (ASD) as the evaluation metrics.

Generalizability of DCA-Net on Unseen Domains. We used the 'DeepAll' as a baseline which means training all source domains jointly using a standard supervised learning strategy with U-Net [17] and testing on the unseen domain. Furthermore, we compared our DCA-Net with several state-of-the-art generalization methods, including the data-augmentation based method (BigAug) [11], a meta-learning based domain generalization method (MASF) [4], and a shape-aware meta-learning method (SAML) [13].

The 'Unseen' section of Table 1 shows the quantitative evaluation results, where the values of BigAug, MASF and SAML are from [13] due to the same experimental setting as ours. It demonstrates that BigAug and MASF are effective for domain generalization, which shows advantage over the lower bound DeepAll. SAML using shape-aware meta-learning achieved the best performance among existing methods, which gets and average Dice of 87.16% and average ASD of 1.58 mm across the 6 domains. Remarkably, our proposed DCA-Net achieved higher performance in terms of average Dice score and ASD, which are 88.16% and 1.29 mm, respectively. We achieved the highest performance in 5 out of the 6 domains. It is noted that though DCA-Net has a lower Dice score than SAML on the I2CVB site, it has a lower ASD, which means the contour of segmentation obtained by DCA-Net is closer to the ground truth. Figure 2 provides 2D and 3D visual comparisons between SAML and DCA-Net for images form the 6 domains respectively. The 3D visualization shows that the results of our approach are closer to the ground truth. The 2D visualization shows that SAML has more over- and mis-segmented regions than ours.

Table 1. Generalizable segmentation performance of various methods on Dice (%) and ASD (mm). * means training with $\mathcal{L}_{seg}$ and $\mathcal{L}_{comp}$.

Scene	Method	ISBI		ISBI1.5		I2CVB		UCL		BIDMC		HK		Average	
Seen	Intra-site	89.27	1.41	88.17	1.35	88.29	1.56	83.23	3.21	83.67	2.93	85.43	1.91	86.34	2.06
	DeepAll*	91.37	0.77	90.67	0.84	88.39	1.18	89.34	1.19	88.78	1.33	90.56	0.85	89.85	1.03
	DCA-Net	**91.83**	**0.72**	**91.59**	**0.81**	**89.93**	**0.77**	**91.99**	**0.64**	**90.68**	**0.93**	**90.57**	**0.82**	**90.93**	**0.78**
Unseen	DeepAll	87.87	2.05	85.37	1.82	82.49	2.97	86.87	2.25	84.48	2.18	85.58	1.82	85.52	2.18
	BigAug	88.62	1.70	86.22	1.56	83.76	2.72	87.35	1.98	85.53	1.90	85.83	1.75	86.21	1.93
	MASF	88.70	1.69	86.20	1.54	84.16	2.39	87.43	1.91	86.18	1.85	86.57	1.47	86.55	1.81
	SAML	88.89	1.38	87.17	1.46	**85.60**	2.07	86.96	1.56	86.19	1.77	88.12	1.22	87.16	1.58
	DCA-Net*	90.24	1.15	88.12	1.16	82.45	1.76	88.28	1.23	86.14	1.64	88.90	0.97	87.36	1.32
	DCA-Net(N = 4)	90.22	1.19	87.06	1.31	83.62	1.61	88.03	1.29	86.03	1.84	**89.97**	**0.89**	87.49	1.36
	DCA-Net(N = 16)	90.24	**1.11**	87.24	1.31	84.18	1.90	88.64	1.18	86.46	1.64	88.76	1.02	87.59	1.36
	DCA-Net(ours)	**90.61**	1.12	**88.31**	**1.14**	84.89	1.76	**89.22**	**1.09**	**86.78**	1.58	89.17	1.02	**88.16**	**1.29**

Ablation Study. To validate the role of our L_{div}, we removed this term in the loss function and used $\mathcal{L}_{seg}$ with $\mathcal{L}_{comp}$ to train DCA-Net, and the corresponding model is referred to as DCA-Net*. The results in Table 1 show that DCA-Net* already outperformed DeepAll, BigAug, MASF, and SAML, which achieves 0.2% average Dice and 0.26 mm ASD improvements compared with that of SAML, with average Dice and ASD of 87.36% and 1.32 mm, respectively. Introducing L_{div} to DCA-Net further improved the average Dice to 88.16%, and decreased the average ASD to 1.29 mm, demonstrating the importance of learning divergent basis representations for our PDP module. Besides, we investigated the effect of the number of basis preceptors of PDP (i.e., size of domain representation bank) by setting N to 4, 8 and 16 respectively. Results in Table 1 show that 'DCA-Net(ours)' with $N = 8$ achieved the best performance. This demonstrates that a small number of basis preceptors is not enough to represent the different domains on multi-site datesets, while a very large number of basis preceptors or domain representation bank size does not lead to much gain in performance.

Performance on Seen Domains. We also investigated the performance of our method on the seen source domains. We followed [13] to split each source domain dataset at patient level into 70%, 10% and 20% for training, validation and testing. We retrained DeepAll with $\mathcal{L}_{seg}$ and $\mathcal{L}_{comp}$ under the new data setting, which is referred to as DeepAll*, and compared with 'Intra-site' that means training and testing 6 domain-specific models, where each model only uses the data from a single domain. The 'seen' section of Table 1 shows that Intra-site got a good performance by requiring each domain to provide images for training. DeepAll* can improve the performance due to the access to a larger dataset from multiple sites. What's more, DCA-Net outperformed DeepAll*, which demonstrates that the proposed domain composition and attention method also improves the segmentation performance on multiple seen domains.

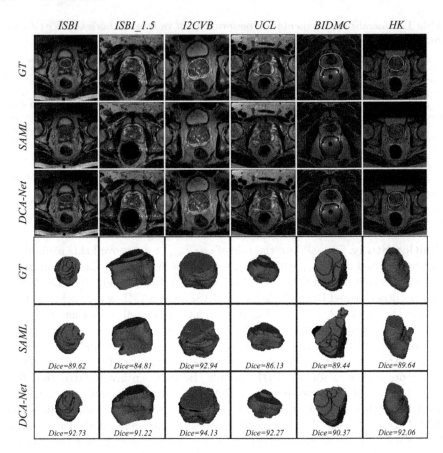

Fig. 2. Visual comparison between DCA-Net and SAML [13]. The first and last three rows are corresponding 2D and 3D visualizations, respectively. Each column shows the results of one domain used for testing and the others for training.

4   Conclusion

We present a Domain Composition and Attention-based model (DCA-Net) to tackle the commonly faced domain generalization problem in medical image segmentation. We propose a Parallel Domain Preceptor (PDP) that synchronously uses domain preceptors to construct a domain representation bank with as set of basis domain representations. Then, the domain attention module learns to predict coefficients for a linear combination of basis representations from the representation bank, which is used to calibrate and neutralize the domain bias of the input feature. Meanwhile, we introduce a novel divergence constraint function to guide the preceptors to capture divergent domain representations, which is important for imroving the generalizability. Experimental results showed the effectiveness of our proposed DCA-Net for achieving robust performance in prostate segmentation from multiple sites and even unseen domains. In the future, it is of interest to apply our DCA to other backbone networks and validate it with other segmentation tasks with more seen and unseen domains.

Acknowledgement. This work was supported by the National Natural Science Foundations of China [61901084 and 81771921] funding, key research and development project of Sichuan province, China [No. 20ZDYF2817].

References

1. Balaji, Y., Sankaranarayanan, S., Chellappa, R.: Metareg: towards domain generalization using meta-regularization. In: Proceedings of the 32nd International Conference on Neural Information Processing Systems, pp. 1006–1016 (2018)
2. Chen, C., Dou, Q., Chen, H., Heng, P.-A.: Semantic-aware generative adversarial nets for unsupervised domain adaptation in chest X-ray segmentation. In: Shi, Y., Suk, H.-I., Liu, M. (eds.) MLMI 2018. LNCS, vol. 11046, pp. 143–151. Springer, Cham (2018). https://doi.org/10.1007/978-3-030-00919-9_17
3. Chen, L.C., Yang, Y., Wang, J., Xu, W., Yuille, A.L.: Attention to scale: scale-aware semantic image segmentation. In: Proceedings of the IEEE Conference on Computer Vision and Pattern Recognition, pp. 3640–3649 (2016)
4. Dou, Q., Castro, D.C., Kamnitsas, K., Glocker, B.: Domain generalization via model-agnostic learning of semantic features. arXiv preprint arXiv:1910.13580 (2019)
5. Dou, Q., Ouyang, C., Chen, C., Chen, H., Heng, P.A.: Unsupervised cross-modality domain adaptation of convnets for biomedical image segmentations with adversarial loss. arXiv preprint arXiv:1804.10916 (2018)
6. Ganin, Y., et al.: Domain-adversarial training of neural networks. J. Mach. Learn. Res. **17**(1), 2096–2030 (2016)
7. Gu, R., et al.: CA-Net: comprehensive attention convolutional neural networks for explainable medical image segmentation. IEEE Trans. Med. Imaging **40**(2), 699–711 (2021)
8. Hu, J., Shen, L., Sun, G.: Squeeze-and-excitation networks. In: Proceedings of the IEEE Conference on Computer Vision and Pattern Recognition (CVPR), June 2018
9. Kamnitsas, K., et al.: Unsupervised domain adaptation in brain lesion segmentation with adversarial networks. In: Niethammer, M., et al. (eds.) IPMI 2017. LNCS, vol. 10265, pp. 597–609. Springer, Cham (2017). https://doi.org/10.1007/978-3-319-59050-9_47
10. Li, D., Yang, Y., Song, Y.Z., Hospedales, T.: Learning to generalize: meta-learning for domain generalization. In: Proceedings of the AAAI Conference on Artificial Intelligence, vol. 32 (2018)
11. Li, H., Pan, S.J., Wang, S., Kot, A.C.: Domain generalization with adversarial feature learning. In: Proceedings of the IEEE Conference on Computer Vision and Pattern Recognition, pp. 5400–5409 (2018)
12. Litjens, G., et al.: Evaluation of prostate segmentation algorithms for MRI: the promise12 challenge. Med. Image Anal. **18**(2), 359–373 (2014)
13. Liu, Q., Dou, Q., Heng, P.-A.: Shape-aware meta-learning for generalizing prostate MRI segmentation to unseen domains. In: Martel, A.L., et al. (eds.) MICCAI 2020. LNCS, vol. 12262, pp. 475–485. Springer, Cham (2020). https://doi.org/10.1007/978-3-030-59713-9_46

14. Liu, Q., Dou, Q., Yu, L., Heng, P.A.: MS-Net: multi-site network for improving prostate segmentation with heterogeneous MRI data. IEEE Trans. Med. Imaging **39**(9), 2713–2724 (2020)
15. Oktay, O., et al.: Attention U-Net: learning where to look for the pancreas. arXiv preprint arXiv:1804.03999 (2018)
16. Perone, C.S., Ballester, P., Barros, R.C., Cohen-Adad, J.: Unsupervised domain adaptation for medical imaging segmentation with self-ensembling. Neuroimage **194**, 1–11 (2019)
17. Ronneberger, O., Fischer, P., Brox, T.: U-Net: convolutional networks for biomedical image segmentation. In: Navab, N., Hornegger, J., Wells, W.M., Frangi, A.F. (eds.) MICCAI 2015. LNCS, vol. 9351, pp. 234–241. Springer, Cham (2015). https://doi.org/10.1007/978-3-319-24574-4_28
18. Shen, D., Wu, G., Suk, H.I.: Deep learning in medical image analysis. Annu. Rev. Biomed. Eng. **19**, 221–248 (2017)
19. Wang, S., Yu, L., Li, K., Yang, X., Fu, C.W., Heng, P.A.: DoFE: domain-oriented feature embedding for generalizable fundus image segmentation on unseen datasets. IEEE Trans. Med. Imaging (2020)
20. Wang, X., Cai, Z., Gao, D., Vasconcelos, N.: Towards universal object detection by domain attention. In: Proceedings of the IEEE/CVF Conference on Computer Vision and Pattern Recognition, pp. 7289–7298 (2019)
21. Zhang, L., et al.: Generalizing deep learning for medical image segmentation to unseen domains via deep stacked transformation. IEEE Trans. Med. Imaging **39**(7), 2531–2540 (2020)
22. Zhang, S., Zhan, Y., Dewan, M., Huang, J., Metaxas, D.N., Zhou, X.S.: Towards robust and effective shape modeling: sparse shape composition. Med. Image Anal. **16**(1), 265–277 (2012)

Fully Test-Time Adaptation for Image Segmentation

Minhao Hu[1,2], Tao Song[2], Yujun Gu[2], Xiangde Luo[4], Jieneng Chen[5],
Yinan Chen[2], Ya Zhang[1,3(✉)], and Shaoting Zhang[2]

[1] Cooperative Medianet Innovation Center, Shanghai Jiao Tong University,
Shanghai, China
ya_zhang@sjtu.edu.cn
[2] SenseTime Research, Shanghai, China
[3] Shanghai AI Laboratory, Shanghai, China
[4] School of Mechanical and Electrical Engineering, University of Electronic Science
and Technology of China, Chengdu, China
[5] College of Electronics and Information Technology, Tongji University,
Shanghai, China

Abstract. When adopting a model from the source domain to the target domain, its performance usually degrades due to the domain shift problem. In clinical practice, the source data usually cannot be accessed during adaptation for privacy policy and the label for the target domain is in shortage because of the high cost of professional labeling. Therefore, it is worth considering how to efficiently adopt a pretrained model with only unlabeled data from the target domain. In this paper, we propose a novel fully test-time unsupervised adaptation method for image segmentation based on Regional Nuclear-norm (RN) and Contour Regularization (CR). The RN loss is specially designed for segmentation tasks to efficiently improve discriminability and diversity of prediction. The CR loss constrains the continuity and connectivity to enhance the relevance between pixels and their neighbors. Instead of retraining all parameters, we modify only the parameters in batch normalization layers with only a few epochs. We demonstrate the effectiveness and efficiency of the proposed method in the pancreas and liver segmentation dataset from the Medical Segmentation Decathlon and CHAOS challenge.

Keywords: Test-time adaptation · Image segmentation · Unsupervised learning

1 Introduction

Deep convolutional neural networks, such as nnUNet [3], have demonstrated state-of-the-art performance in medical image segmentation challenges [5,11], where data are usually finely preprocessed so that the training and testing set share the same distribution. In clinical practice, when the model trained by one medical center (source domain) is applied to the other centers (target domain),

© Springer Nature Switzerland AG 2021
M. de Bruijne et al. (Eds.): MICCAI 2021, LNCS 12903, pp. 251–260, 2021.
https://doi.org/10.1007/978-3-030-87199-4_24

its performance usually degrades if the data are acquired from different equipment with different protocols.

To tackle this problem, it's common to collect the data from target domain and label them for finetuning. It is effective but impractical for all the cases due to the high cost of professional labeling. Without using the labels from target domain, the existing unsupervised domain adaptation (UDA) methods [4] need both source and target data to get the cross-domain loss for training, but the labeled source data may not be available in some real-world scenarios due to data privacy issues [6].

Researches on adaptation with only unlabeled data from target domain are usually called *full test-time adaptation*. [12] proposed a method based on domain adversarial learning and paired consistency, but it needs to train the domain discriminator network for each target domain every time, which is less cost-effective. Test-time entropy minimization (Tent) [13] proposed to optimize the model confidence by minimizing the entropy of its predictions, which is simple yet effective. Batch Nuclear-norm Maximization (BNM) [1] is proposed to improve the discriminability and diversity of prediction measured by Frobenius-norm and rank of batch output matrix, while the nuclear-norm is demonstrated to be the upper-bound of Frobenius-norm and the approximation of the matrix rank [1]. Although Tent and BNM have shown their effectiveness and efficiency for test-time optimization, they are designed specially for classification tasks, without considering the latent regional relation of pixels for segmentation tasks. Furthermore, the existing methods, such as BNM, require a large amount of computational resource if we simply treat the segmentation task as pixel-level classification, especially for three-dimensional medical images.

This paper focuses on the fully test-time adaptation problem for medical image segmentation task. To overcome these difficulties, we propose Regional Nuclear-norm loss and Contour Regularization loss. The Regional Nuclear-norm loss is memory friendly for segmentation task and could improve discriminability and diversity of prediction, while the Contour Regularization loss constrains the continuity and connectivity to enhance the relevance between pixels and their neighbors. Only the parameters of batch normalization layers will be modified during test-time optimization, which could reduce the computational cost and makes it easier to adapt to different target domains for deployment. To the best of our knowledge, this is the first time to adopt the nuclear-norm maximization and contour regularization method to the test-time adaptation for medical image segmentation task. We evaluate our method with different datasets from Medical Segmentation Decathlon [11] and CHAOS [5] challenge and the results show that our method always outperforms state-of-the-art Tent and could obtain the competitive performance as finetuning but with fewer epochs for optimization.

2   Fully Test-Time Adaptation for Image Segmentation

Figure 1 presents the overall flow of our method. Let $(X_s, Y_s) = \{(x_s^i, y_s^i)\}_i^N$ denote the data and corresponding labels from the source domain and $X_t =$

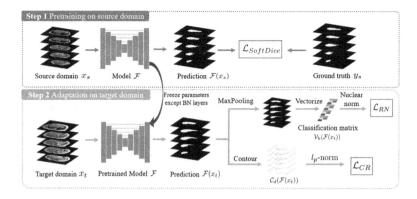

Fig. 1. Illustration of fully test-time adaptation for image segmentation.

$\{x_t^i\}_i^M$ the data from the target domain. We first train the model $\mathcal{F}$ on the source domain with soft dice loss ($\mathcal{L}_{SoftDice}$) and then adapt it on the target domain with an unsupervised loss which is combined by Regional Nuclear-norm term ($\mathcal{L}_{RN}$) and Contour Regularization term ($\mathcal{L}_{CR}$). The Regional Nuclear-norm term improves both discriminability and diversity of prediction, while the Contour Regularization term constrains the continuity and connectivity to enhance the relevance between pixels and their neighbors.

2.1 Objective Function

Regional Nuclear-Norm Loss. Let $\mathcal{F}(x_t)$ with size (C, D, H, W) denote the output of model, where C denotes the number of categories for segmentation including the background and D, H, W denote the depth, height and width separately. We first pass it through a Max Pooling layer with kernel size k and stride k, denoted as $\mathcal{M}_k$, to reduce its spacial size to (C, D', H', W'), where $D' = D/k, H' = H/k, W' = W/k$. For each class $c \in C$, we vectorize the three-dimensional prediction $\mathcal{F}(x_t)_c$ into one-dimensional vector with size of $D' \times H' \times W'$. Then the vector for each class c could be written as $V_c = Vec(\mathcal{M}_k(\mathcal{F}(x_t)_c))$, where Vec denotes the operation of vectorization. Then the classification matrix could be represented as $\mathcal{V}_k(\mathcal{F}(x_t)) = (V_1, V_2, \cdots, V_C)$. We apply the nuclear-norm on this classification matrix to obtain the Regional Nuclear-norm loss:

$$\mathcal{L}_{RN} = -\|\mathcal{V}_k(\mathcal{F}(x_t))\|_* = -\mathbf{tr}(\sqrt{\mathcal{V}_k(\mathcal{F}(x_t))^T \mathcal{V}_k(\mathcal{F}(x_t))}) \tag{1}$$

where $\mathbf{tr}(\cdot)$ denotes the trace of a squared matrix, which is defined to be the sum of elements on the main diagonal. Note that maximizing the Regional Nuclear-norm improves the discriminability and diversity of prediction, so we need to add a minus for optimization. It's also important to remark that if without the operation of Max Pooling, the calculation of nuclear-norm of classification matrix takes up a lot of computing resources especially GPU memory space.

The Max Pooling could greatly reduce the size of the classification matrix by k^3. The value of k is a compromise of the usage of computational resource and the effectiveness of nuclear-norm maximization.

Contour Regularization Loss. Nuclear-norm maximization optimizes the classification prediction for individual pixel but ignores the strong relevance between the pixel and its neighbors in segmentation tasks. In medical scenarios, tumor and organ always have continued and smooth contours, thus we propose to regularize the contour of segmentation output to enhance its continuity and connectivity. For convenient representing, we consider bi-dimensional situation and let $I \in [0,1]^{L \times W}$ denote the predicted probability of each pixel to be the foreground, the contour of prediction on position (i,j) is then *defined* as the largest probability variation surrounding,

$$\mathcal{C}_d(I_{i,j}) = \max_{(i_1,j_1),(i_2,j_2) \in \mathcal{W}_d(i,j)} |I_{i_1,j_1} - I_{i_2,j_2}| \tag{2}$$

where $\mathcal{W}_d(i,j) = \{i \pm k, j \pm l | k,l \in \{0...d\}\}$ is the neighbors pixels surrounding (i,j) with radius d. Since the probability value ranges from 0 to 1, the contour could be *implemented* as following:

$$\mathcal{C}_d(I_{i,j}) = \max_{\substack{(i_c,j_c) \\ \in \mathcal{W}_d(i,j)}} I_{(i_c,j_c)} - \min_{\substack{(i_c,j_c) \\ \in \mathcal{W}_d(i,j)}} I_{(i_c,j_c)} = \mathcal{M}_d(I_{i,j}) + \mathcal{M}_d(-I_{i,j}) \tag{3}$$

where $\mathcal{M}_d$ is the max pooling operation with kernel size $2d+1$. The case in three-dimensional could be easily generalized. Therefore, the Contour Regularization loss can be represented as following:

$$\mathcal{L}_{CR} = \|\mathcal{C}_d(\mathcal{F}(x_t))\|_p \tag{4}$$

where $\|\cdot\|_p$ denotes the l_p-norm. Comparing to the traditional methods such as level-set, our implementation to approximate the contours could back-propagate the gradient of CR loss to the model thanks to the max-pooling operation.

For the test time adaptation, we simultaneously optimize the Regional Nuclear-norm term and the Contour Regularization term. Hence, we combine the two losses with parameter α:

$$\mathcal{L} = \mathcal{L}_{RN} + \alpha \mathcal{L}_{CR} \tag{5}$$

2.2   Optimization Parameters

Since only the target data are available at test, modify all parameters of the model is unstable and inefficient [13]. Instead, we only update the parameters in the Batch normalization layers for the test-time adaptation. According to [7], the normalization statistics for each domain are different, if convolutional layers learn the knowledge of feature extraction, then the batch normalization layers are the key to the domain shift. Furthermore, the batch normalization layers have much less parameters to optimize, which lead to faster adaptation. Therefore, during test-time optimization, the model parameters are frozen except those of batch norm layers. Let $X \in \mathbb{R}^{B \times P}$ denote the input to BN layer, where B denotes the batch size and P the feature dimension. For the feature $j \in \{1...P\}$, the Batch Normalization layer first centers and standardizes the x_j into $\hat{x}_j$ and then transforms it by affine parameters γ and β.

$$\hat{x}_j = \frac{x_j - \mathbb{E}[X_j]}{\sqrt{\mathbb{V}ar[X_j]}}, \quad y_j = \gamma_j \hat{x}_j + \beta_j \tag{6}$$

For the test-time optimization, the mean and standard deviation are estimated from the target data while the affine parameters γ and β are optimized by the gradient back-propagation from the loss function.

3   Experiments

We evaluate our method on pancreas segmentation dataset from Medical Segmentation Decathlon [11] and liver segmentation dataset from CHAOS [5]. Our implementation is in PyTorch [9] and the code will be released for publication.

We first train the model with *training data* from source domain and select models with *validation data*. On the target domain, we use the *adaptation data* for unsupervised adaptation or supervised fine-tuning and use the *testing data* for testing with evaluation metrics.

3.1   Datasets

Pancreas segmentation dataset consists of 281 Portal venous phase CT images with annotated labels. We first resample the spacing of volume to $1.0 \, \text{mm}^3$, then crop it to size of $(128, 128, 128)$ pixel and apply a normalization with fixed windowing $[-125, 275]$ as preprocessing. We plot each data with its mean and variance (Fig. 2), then cluster the data into two groups by Gaussian Mixture Model. Since the two clusters have different statistics, we regard them as source and target domain for the following experiments. The source domain (red) contains 184 subjects and we randomly selected 128 of them as *training set* and the rest are *validation set*. The target domain data (blue, 97 subjects) are randomly divided equally as the *adaptation* and *testing set*.

Liver segmentation dataset consists of 20 subjects with two different modalities: in-phase (IP) and out-of-phase (OOP). IP and OOP correspond to paired MRI gradient echo sequences obtained with the same repetition time but with two different echo time values [10], which are naturally two domains. For source domain, we randomly select 13 subjects as *training set* and 7 as *validation set* and the same for *adaptation set* and *testing set* for target domain. For each instance, we first resample the spacing of volume to $1.5 \, \text{mm}^3$, then crop the MR images to the size of $(256, 256, 64)$ pixel and apply a normalization after replacing the extreme values by 95th percentile value.

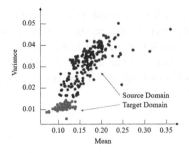
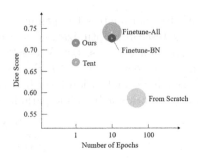

Fig. 2. Illustration of the source domain (blue dots) and target domain (red dots) of pancreas segmentation dataset. (Color figure online)

Fig. 3. Comparison of different methods on pancreas dataset. Each circle represents a method and its diameter shows the number of parameters to optimize.

3.2 Implementation Details

We adopt the state-of-the-art architecture nnUNet [3] and replace its Instance Norm layers with Batch Norm layers. Empirically, the best hyper-parameters for the objective function are $d = 4, p = 2, k = 4$. And we set α to 0.01 so that the values of two loss terms are in the same scale. For all the experiments, we use Adam optimizer for 100 epochs with a learning rate equal to 0.0001. No data augmentation method is adopted during training and testing. The training of nnUNet model on source domain (100 epochs) takes approximately 1 h and the adaptation on target domain (<10 epochs) takes less than 5 min on a NVIDIA GPU 1080Ti.

3.3 Results

We use Dice Score (DSC), Hausdorff Distance (HD) and Average Surface Distance (ASD) as evaluation metrics, which are implemented in SimpleITK [8].

Model Comparison. To demonstrate the effectiveness of the proposed method, we compare it with other five different methods (Fig. 3). First we compare our methods against the state-of-the-art methods for fully test-time adaptation Tent[13]. For both methods, we only modify the parameters in the batch normalization layers as mentioned in Sect. 2.2. The lower bound is provided by the model trained on source domain and directly applied to testing data from target domain without any adaptation, which is denoted as *No Adaptation*. We also compare them with the model that trained from scratch on adaptation data form target domain and refer it as *From Scratch*. The highest performance theoretically is provided by the model trained on source domain and finetuned on target domain, denoted as *Finetune*. We also present the results of finetuning all parameters and that of finetuning only parameters in BN layers. Note that in the problem settings, the adaptation and testing set don't have labels so we won't use them in our method and Tent, but we still need them to evaluate models' results and in addition we could use these labeled data to get the fine-tuned model or train a model from scratch to compare with our method. Table 1 presents the results for each of these methods evaluated on various metrics.

Table 1. Results of the mentioned methods on pancreas dataset. The *Epochs* column refers to the number of epochs that models update to reach the best performance. The *Layers* column refers the layers whose parameters are modified during optimization.

Method	Layers	Epochs	DSC ↑	HD ↓	ASD ↓
From Scratch	All	50	0.5844 ± 0.1922	15.34 ± 9.20	0.9167 ± 0.9282
No Adaptation	None	0	0.2677 ± 0.2312	34.85 ± 14.87	1.2149 ± 1.9217
Finetune	BN	10	0.7268 ± 0.1491	20.78 ± 12.08	0.6293 ± 0.6134
Finetune	All	10	0.7285 ± 0.1423	20.78 ± 15.97	0.7653 ± 0.8325
Tent	BN	1	0.6818 ± 0.1582	18.98 ± 15.26	0.7164 ± 0.9611
Ours	BN	1	$\mathbf{0.7113 \pm 0.1551}$	$\mathbf{18.97 \pm 14.83}$	$\mathbf{0.7066 \pm 0.8231}$

Results from Table 1 show that our method outperforms Tent on all metrics after one-epoch update of BN layer parameters. The poor results of *No Adaptation* indicate that there exits a gap between the source and target domain. Finetuning all layers doesn't have a significantly better performance than fine-tuning only BN layers but needs more parameters to modify. Training from scratch on target domain couldn't reach the same performance as the finetuning because of lack of training samples [2]. It is also worth mentioning that our method needs only one-epoch parameter modulation to get 0.7113 of dice score, while finetuning method needs 10 epochs to get 0.7268. This reveals that the fully test-time adaptation methods could reach a comparable performance of finetuning methods with less computation resource and unlabeled data.

Ablation Study To evaluate the contribution of Contour Regularization term, we make an ablation study by removing the $\mathcal{L}_{CR}$ from the objective function. We also compare the results of models with different choices of parameters p and d defined in Eq. 4.

Table 2. Ablation study of the loss terms and the inference of parameters d and p.

$\mathcal{L}_{RN}$	$\mathcal{L}_{CR}$	p	d	DSC ↑	HD ↓	ASD ↓
✓		–	–	0.6912 ± 0.1545	37.07 ± 21.33	0.8417 ± 0.9532
✓	✓	1	1	0.7099 ± 0.1667	17.99 ± 15.08	0.7029 ± 0.8216
✓	✓	1	2	0.7098 ± 0.1678	$\mathbf{17.96 \pm 15.07}$	0.7023 ± 0.8215
✓	✓	1	4	0.7102 ± 0.1661	17.98 ± 15.08	$\mathbf{0.6980 \pm 0.8074}$
✓	✓	2	4	0.7113 ± 0.1551	18.97 ± 14.83	0.7066 ± 0.8231
✓	✓	inf	4	$\mathbf{0.7158 \pm 0.1571}$	26.97 ± 18.99	0.7610 ± 0.9020

From Table 2, we observe that the Contour Regularization term improves the performance with respect to the model with only $\mathcal{L}_{RN}$. The results of models with different values of d reveals that larger radius of neighbor pixels for contour gets better performance. The results also indicate that higher value of p leads to better Dice score but worse Hausdorff distance and average surface distance, which means the choice of l_p-norm for CR term is a trade-off.

Generalization Results. To demonstrate that our method is generic and effective, we evaluate it on another dataset. The liver segmentation dataset contains three-dimensional MR images with two modalities: in-phase (IP) and out-of-phase (OOP) which are considered as two domains. We make the experiments from in-phase (source) to out-of-phase (target) and the opposite. Table 3 presents the results of seven different methods mentioned above on the liver segmentation dataset.

From Table 3, the results indicate that our method still works and has always better performance than Tent, even the source and target domain are from two different modalities. It worth to mention that although our method needs more epochs than Tent for optimization when OOP is source domain, the increase of Dice score and average surface distance makes the computation worthwhile. It is also important to remark that finetuning methods need about 50 epochs to reach their best performance and if we limit the number of epochs for finetuning to the same as our method, their performance will be closer.

Table 3. Results on the liver segmentation dataset. In the upper part of table, in-phase (IP) is source domain and out-of-phase (OOP) is target domain, the lower part is the reverse.

Method	Source → Target	Layers	Epochs	DSC ↑	HD ↓	ASD ↓
From Scratch	OOP	All	60	0.9075 ± 0.0245	19.86 ± 13.49	0.2236 ± 0.1804
No Adaptation	IP → OOP	None	0	0.7239 ± 0.3004	38.23 ± 24.16	1.0989 ± 1.7215
Finetune	IP → OOP	BN	50	0.8940 ± 0.0276	31.26 ± 19.22	0.2600 ± 0.1703
Finetune	IP → OOP	All	30	0.9175 ± 0.0538	29.65 ± 21.77	0.2437 ± 0.2997
Tent	IP → OOP	BN	5	0.8810 ± 0.0454	**32.11 ± 18.66**	0.2828 ± 0.1815
Ours	IP → OOP	BN	5	**0.8830 ± 0.0038**	32.13 ± 18.64	**0.2789 ± 0.1630**
From Scratch	IP	All	100	0.9269 ± 0.0197	28.05 ± 22.10	0.1742 ± 0.1065
No Adaptation	OOP → IP	None	0	0.6282 ± 0.1128	58.61 ± 19.42	1.1917 ± 0.5830
Finetune	OOP → IP	BN	100	0.8936 ± 0.0189	21.06 ± 10.63	0.2249 ± 0.0963
Finetune	OOP → IP	All	50	0.9281 ± 0.0195	19.44 ± 13.59	0.1798 ± 0.1550
Tent	OOP → IP	BN	5	0.8413 ± 0.0352	24.61 ± 28.62	0.3171 ± 0.1027
Ours	OOP → IP	BN	10	**0.8621 ± 0.0302**	**23.59 ± 10.36**	**0.2857 ± 0.1046**

It's worth mentioning that we also tried to evaluate our method on other dataset with larger domain gaps such as from CT to MR. Results show that our method always outperforms the baseline model (*No adaptation*), but has larger gap to the performance of finetuning. This is reasonable since the characteristics of CT and MR are totally different thus the feature extractor needs to be retrained, which means the modification of parameters in BN layers is no longer enough for adaptation. We will explore this in future work.

4 Conclusion

We present a novel method for fully test-time adaptation on medical image segmentation tasks. We adopt the Regional Nuclear-norm loss to improve discriminability and diversity of prediction. In addition, we employ the Contour Regularization loss to make full use of the strong relevance between the pixels and its neighbors in segmentation tasks. We validate our method on pancreas and liver segmentation datasets, achieving better results than state-of-the-art method Tent. The proposed method is label-free and more efficient than finetuning. The gain in segmentation accuracy and efficiency for fully test-time unsupervised adaptation makes it highly valuable for real-world clinical scenarios.

Acknowledgment. This work is supported partially by SHEITC (No. 2018-RGZN-02046), 111 plan (No. BP0719010), and STCSM (No. 18DZ2270700).

References

1. Cui, S., Wang, S., Zhuo, J., Li, L., Huang, Q., Tian, Q.: Towards discriminability and diversity: batch nuclear-norm maximization under label insufficient situations. In: 2020 IEEE/CVF Conference on Computer Vision and Pattern Recognition, CVPR 2020, Seattle, WA, USA, 13–19 June 2020, pp. 3940–3949. IEEE (2020)
2. Hesamian, M.H., Jia, W., He, X., Kennedy, P.J.: Deep learning techniques for medical image segmentation: achievements and challenges. J. Digit. Imaging **32**(4), 582–596 (2019)
3. Isensee, F., Kickingereder, P., Wick, W., Bendszus, M., Maier-Hein, K.H.: No new-net. In: Crimi, A., Bakas, S., Kuijf, H., Keyvan, F., Reyes, M., van Walsum, T. (eds.) BrainLes 2018. LNCS, vol. 11384, pp. 234–244. Springer, Cham (2019). https://doi.org/10.1007/978-3-030-11726-9_21
4. Kamnitsas, K., et al.: Unsupervised domain adaptation in brain lesion segmentation with adversarial networks. In: Niethammer, M., et al. (eds.) IPMI 2017. LNCS, vol. 10265, pp. 597–609. Springer, Cham (2017). https://doi.org/10.1007/978-3-319-59050-9_47
5. Kavur, A.E., Selver, M.A., Dicle, O., Barış, M., Gezer, N.S.: CHAOS - Combined (CT-MR) Healthy Abdominal Organ Segmentation Challenge Data, April 2019
6. Li, R., Jiao, Q., Cao, W., Wong, H., Wu, S.: Model adaptation: unsupervised domain adaptation without source data. In: 2020 IEEE/CVF Conference on Computer Vision and Pattern Recognition, CVPR 2020, Seattle, WA, USA, 13–19 June 2020, pp. 9638–9647. IEEE (2020)
7. Li, Y., Wang, N., Shi, J., Hou, X., Liu, J.: Adaptive batch normalization for practical domain adaptation. Pattern Recogn. **80**, 109–117 (2018)
8. Lowekamp, B.C., Chen, D.T., Ibáñez, L., Blezek, D.J.: The design of simpleitk. Front. Neuroinform. **7**, 45 (2013)

9. Paszke, A., et al.: Pytorch: an imperative style, high-performance deep learning library. In: Wallach, H.M., Larochelle, H., Beygelzimer, A., d'Alché-Buc, F., Fox, E.B., Garnett, R. (eds.) Advances in Neural Information Processing Systems 32: Annual Conference on Neural Information Processing Systems 2019, NeurIPS 2019, Vancouver, BC, Canada, 8–14 December 2019, pp. 8024–8035 (2019)

10. Ramalho, M., Herédia, V., de Campos, R.O., Dale, B.M., Azevedo, R.M., Semelka, R.C.: In-phase and out-of-phase gradient-echo imaging in abdominal studies: intra-individual comparison of three different techniques. Acta Radiologica **53**(4), 441 (2012)

11. Simpson, A.L., et al.: A large annotated medical image dataset for the development and evaluation of segmentation algorithms. CoRR abs/1902.09063 (2019)

12. Varsavsky, T., Orbes-Arteaga, M., Sudre, C.H., Graham, M.S., Nachev, P., Cardoso, M.J.: Test-time unsupervised domain adaptation. In: Martel, A.L., et al. (eds.) MICCAI 2020. LNCS, vol. 12261, pp. 428–436. Springer, Cham (2020). https://doi.org/10.1007/978-3-030-59710-8_42

13. Wang, D., Shelhamer, E., Liu, S., Olshausen, B., Darrell, T.: Tent: fully test-time adaptation by entropy minimization. In: ICLR 2021 (2021)

OLVA: *O*ptimal *L*atent *V*ector *A*lignment for Unsupervised Domain Adaptation in Medical Image Segmentation

Dawood Al Chanti[(✉)] and Diana Mateus

École Centrale de Nantes, Laboratoire des Sciences du Numérique de Nantes LS2N,
UMR CNRS 6004, Nantes, France
dawood.alchanti@ls2n.fr

Abstract. This paper addresses the domain shift problem for segmentation. As a solution, we propose OLVA, a novel and lightweight unsupervised domain adaptation method based on a Variational Auto-Encoder (VAE) and Optimal Transport (OT) theory. Thanks to the VAE, our model learns a shared cross-domain latent space that follows a normal distribution, which reduces the domain shift. To guarantee valid segmentations, our shared latent space is designed to model the shape rather than the intensity variations. We further rely on an OT loss to match and align the remaining discrepancy between the two domains in the latent space. We demonstrate OLVA's effectiveness for the segmentation of multiple cardiac structures on the public Multi-Modality Whole Heart Segmentation (MM-WHS) dataset, where the source domain consists of annotated 3D MR images and the unlabelled target domain of 3D CTs. Our results show remarkable improvements with an additional margin of 12.5% dice score over concurrent generative training approaches.

Keywords: Unsupervised domain adaptation · Cross modality · Variational Auto-Encoder · Optimal transport · Cardiac segmentation

1 Introduction

Automatic segmentation from multi-modal images is essential for clinical assessment, diagnosis and treatment planning [1,18]. Extensive literature has shown the effectiveness of convolutional neural networks in segmenting accurately cardiac structures [16,19]. Yet, without proper adaptation these models fail when deployed across modalities, new subjects and different clinical sites, due to a domain shift [10], *e.g.* between the modalities' appearance as in Fig. 1. Designing models that can perform well across domains is key in medical applications where labels are scarce and expensive to obtain.

This work has been supported in part by the European Regional Development. Fund, the Pays de la Loire region on the Connect Talent scheme (MILCOM Project) and Nantes Métropole (Convention 2017-10470).

M. de Bruijne et al. (Eds.): MICCAI 2021, LNCS 12903, pp. 261–271, 2021.
https://doi.org/10.1007/978-3-030-87199-4_25

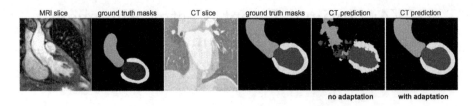

Fig. 1. The appearances of the cardiac structures look significantly different on MR and CT images. Both modalities share the same label space, yellow: left ventricle myocardium (LV-M), blue: left ventricle blood cavity (LV-B), red: left atrium blood cavity (LA-B), and green: ascending aorta (A-A). Bad prediction due to severe domain shift when no adaptation is considered. MM-WHS cardiac public database [26]. (Color figure online)

Semi-supervised and Unsupervised Domain Adaptation (UDA) approaches have been proposed to tackle the domain shift problem. The former assume a few labeled instances in the target domain can be used for joint-training with the source data [20]. The more ambitious UDA strategies [1,3,4,18,24,25] assume no labels are available for the target domain. The core idea of UDA is to go through an adaption phase using a non-linear mapping to find a common domain-invariant representation or a latent space $\mathcal{Z}$. The domain shift in $\mathcal{Z}$ can be reduced by enforcing the two domains distributions to be closer via a certain loss (*e.g.* Maximum Mean Discrepancy [15]). Since $\mathcal{Z}$ is common to all domains who share the same label space, projected labeled source domain samples can be used to train a segmenter for all domains. In this paper, we deal with the problem of UDA for MR-CT cross-modality cardiac structure segmentation.

Related Work. Recent works on UDA for medical image segmentation rely on Generative Adversarial Networks [3,4,8,18,24,25] to translate the appearance from one modality to the other using multiple discriminators and a pixel-wise cycle consistency loss. Despite their success, they: i) suffer from instabilities [2], ii) rely on complex architectures with more than 95 million parameters, iii) are prone to model collapse [17], and iv) may generate images outside the actual target domain [1]. To alleviate some of these limitations, Ouyang *et al.* [18] proposed a Data Efficient Cross-Modality (DECM) model, that combines adversarial networks with VAEs [14]. DECM exploited the VAEs constraint imposed on the latent space to match a prior distribution and experimentally validated that it reduces the domain shift when used as a shared space across domains. To further reduce the domain shift, Ouyang et al. deployed an adversarial loss guided by a cycle-consistency to enforce appearance-invariance. The VAE model in [18] is complex as equipped with three encoders, three decoders, a segmenter, and a domain classier. Its loss function has six trade-off hyper-parameters to tune. Recently, Seg-JDOT framework [1] for UDA based on Optimal Transport (OT) theory and deep learning is proposed. Seg-JDOT adapts a joint OT cost measure combining both the sample distances on the feature space of a deep 3D-Unet and a loss function measuring the discrepancy at the output space between the two domains. Two limitations of Seg-JDOT [1] are i) its success is guaranteed only when the domain shift between the two domains is not severe, and ii) it is deployed on image patches to enable the generation of a higher number of samples to properly estimate the empirical distribution.

Proposal. We present a novel and lightweight domain-invariant variational - segmentation auto-encoder model. We use the latent space of a VAE that is constrained to follow a prior normal distribution as a common space similar to [18] to reduce the domain shift. Then, we exploit the geometry in $\mathcal{Z}$ for matching and aligning probability distributions using OT theory by optimizing for a transport plan γ, similar to [1]. Thereby, further shrinking the remaining domain shift. Different from [18], who maximized the image likelihood, we directly learn a semantic latent representation that maximizes the label likelihood. Our idea is that the prior normal distribution has a limited capacity to handle intensity and shape variations, but it can be efficiently exploited for modelling shapes alone. This claim is supported by [19] who use a VAE as a post-processor on the top of a U-Net output to convert the erroneous U-Net predictions to anatomical plausible outputs. Conversely, we simultaneously perform anatomical plausible segmentation and partial alignment of the label-conditional distributions. Also, different from [1], i) we operate over the full image scale, ii) we bring the source and the target data closer to a normal distribution before solving for γ to guarantee its convergence and iii) we do not require the alignment of the label-conditional distributions at the label space.

Our main contributions are: i) We address the domain shift problem via two principle steps: 1) we learn a common geometrical and continuous space for all domains, that is constrained to follow a prior distribution realized using VAE, to reduce from the severe domain shift; 2) We address the remaining shift by aligning latent vectors from both domains using a transportation plan that minimize an energy function realized using OT theory; ii) We concentrate the limited capacity of the prior normal distribution to model the shape of the segmentation masks rather than pixel intensities; iv) Our model is lightweight with 1.7 million parameters and easy to adapt for other clinical application; v) We validate our model on the MM-WHS public dataset and outperform State of the Art (SOA) methods by a margin of 12.5% dice score.

2 Method

Consider a labeled source domain dataset $\{X^s, Y^s\}_{s=1}^N$ with N images, and a target dataset $\{X^t\}_{t=1}^M$ with M images, but with unknown labels Y^t. The goal of OLVA is to learn a common latent space that conveys aligned structural information for X^s, X^t needed to generate image segmentation masks regardless of the domain. A block diagram of the method is shown in Fig. 2.

2.1 VAEs for Segmentation

The goal of VAEs is to search for the best parameters ϕ^*, θ^* in order to sample a latent variable $z \sim q_{\phi^*}(z|x)$ from a relatively simple distribution such

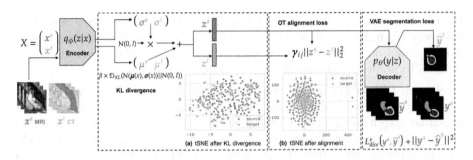

Fig. 2. OLVA: a generative encoder enforces the shared latent vectors of both source and target domain to follow a prior normal distribution (a), further aligned through an OT plan γ (b). A generative decoder is guided with the source domain labels to produce segmentation maps. We use t-distributed stochastic neighbor embedding (t-SNE) to map the latent vectors to a 2D space for visualization purposes only.

as isotropic Gaussian distribution and to generate a new sample $\hat{x} \sim p_{\theta*}(z|x)$ as close as possible to the real observed data x such that $p_\theta^*(x|z) = p^*(x)$. The VAE loss formalized as in Eq. (1) enables an end-to-end training with a first term that maximizes the marginal likelihood so that the generative model becomes better and a regularization term that minimizes the KL-divergence to better approximate $q_\phi(z|x)$ from the posterior $p(z)$. β is a trade-off parameter between the two terms. Commonly, the prior model $p(z)$ is set to the normal distribution $\mathcal{N}(0; I)$ and the re-parametrization trick is applied to facilitate the sampling process as $q_\phi(z|x) = \mathcal{N}(\mu_\phi(x), \sigma_\phi(x)I)$. Thereby, $\mathcal{D}_{KL}$ becomes equivalent to $\frac{1}{2}\sum_{k=1}^{K}\left(1 + \log(\sigma_k^2) - \mu_k^2 - \sigma_k^2\right)$, where K is the latent space dimension and $-\log p_\theta(x|z)$ is conveniently replaced by a reconstruction loss $||x - p_\theta(x|z)||^2$.

$$\mathcal{L}_{vae}(\phi, \theta; x) = \underset{z \sim q_\phi}{E}\left[-\log p_\theta(x|z)\right] + \beta\mathcal{D}_{KL}(q_\phi(z|x)||p(z)), \quad (1)$$

Moreover, we force the VAE to producing valid segmentation masks that respect the anatomical variations of the source domain by constraining z to model a continuous and structured binary shape variation distribution $p_\theta(y|z)$. Unlike the discrete latent space of U-Net-like networks, $p_\theta(y|z)$ facilitates interpolation. The source domain data $\{X^s, Y^s\}$ guides our VAE segmentation Dice loss $\mathcal{L}_{dice}^s$ to provide predictions of the segmentation maps as shown in Eq. (2).

$$\mathcal{L}_{vae}^s = ||y^s - p(y^s|z^s)||^2 + \beta\mathcal{D}_{KL}^s(q(z^s|x^s)||p(z^s)) + \mathcal{L}_{dice}^s(y^s, p_\theta(y^s|z^s)). \quad (2)$$

2.2 Optimal Transport for Latent Vector Alignment (OLVA)

To solve the domain adaptation problem within the segmentation task, we assume the existence of two distinct joint probability distributions $\mathcal{P}^s$ and $\mathcal{P}^t$ defined over a shared latent space $\mathcal{Z}$ and their marginal distributions (ζ^s, ζ^t) are defined over $\Omega \in \mathbb{R}^k$ (a compact input measurable space of dimension k with

$\mathcal{P}(\Omega)$ being the set of all probability measures). We rely on the *Covariate Shift* assumption to solve the underlying domain adaptation problem, which suggests that the conditional distributions of the labels with respect to the data are equal $\mathcal{P}^s(y|z^s) = \mathcal{P}^t(y|z^t)$, despite the data distributions of the two domains being different $\mathcal{P}^s(z^s) \neq \mathcal{P}^t(z^t)$. For the adaptation to be effective using OT theory, this difference needs to be small [7]. Hence, we first minimize the source and the target distribution differences by additionally enforcing the VAE to follow same prior and then further reducing the remaining domain shift in $\mathcal{Z}$ by searching for a nonlinear map of the latent space $\boldsymbol{T} : \Omega^s \to \Omega^t$ has a physical interpretation in our problem as it represent the change in the acquisition modalities.

Following the *Covariate Shift* assumption, $\boldsymbol{T}$ preserves the conditional distribution $\mathcal{P}^t(y|\boldsymbol{T}(z^t)) = \mathcal{P}^s(y|z^s)$. Similar to [1,5], we look at $\boldsymbol{T}$ from a probabilistic point, where $\boldsymbol{T}$ is said to be a transport map or push-forward (denoted as $\sharp$) from ζ^s to ζ^t if satisfying $\boldsymbol{T}\sharp\zeta^s(z) = \zeta^t(\boldsymbol{T}^{-1}(z)), \forall z \subset \Omega^t$. Under this assumption, z^t are drawn from the same probability density function as $\boldsymbol{T}\sharp\zeta^s(z)$.

Now we can solve the adaptation problem based on OT theory using three main steps: i) Estimate ζ^s and ζ^t from Z^s and Z^t; ii) Find a transport map $\boldsymbol{T}$ from ζ^s to ζ^t; iii) Use $\boldsymbol{T}$ to transport labeled samples Z^s and train a segmenter from them. Searching for $\boldsymbol{T}$ in the space of all possible transformations is intractable. We followed similar constraints as those imposed in [1,5,6] and chose $\boldsymbol{T}$ as to minimize a transportation or *energy cost* $C(\boldsymbol{T})$ required to move a probability mass ζ from z to $\boldsymbol{T}(z)$ as:

$$C(\boldsymbol{T}) = \int_{\Omega^s} c(z, \boldsymbol{T}(z)) d\zeta^s(z), \tag{3}$$

where the cost function $c : \Omega^s \times \Omega^t \to \mathbb{R}^+$ is a distance function over the metric space Ω. Monge defined the OT problem Eq. (3) as the minimization of:

$$\boldsymbol{T}_0 = \arg\min_{\boldsymbol{T}} \int_{\Omega^s} c(z, \boldsymbol{T}(z)) d\zeta^s(z), \text{s.t. } \boldsymbol{T}\sharp\zeta^s = \zeta^t \tag{4}$$

The Kantorovitch formulation Eq. (5) [13] of Eq. (4), is a convex relaxation which seeks for a transportation plan γ, that is equivalent to a probabilistic coupling $\gamma_0 \in \prod(\mathcal{P}^s, \mathcal{P}^t)$, to match the distributions $\mathcal{P}^s$ and $\mathcal{P}^t$.

$$\gamma_0 = \arg\min_{\gamma \in \prod(\mathcal{P}^s, \mathcal{P}^t)} \int_{\Omega^s \times \Omega^t} c(z^s, z^t) d\gamma(z^s, z^t), \tag{5}$$

$\prod(\mathcal{P}^s, \mathcal{P}^t)$ is the set of all probabilistic couplings $\in \mathcal{P}(\Omega^s \times \Omega^t)$. Equation (5) allows to define the Wasserstein distance Eq. (6) of order p between ζ^s and ζ^t.

$$W_p \overset{\text{def}}{=} \left(\inf_{\gamma \in \prod} \int d(z^s, z^t)^p d\gamma(z^s, z^t), \right)^{\frac{1}{p}}$$
$$= \left(\inf_{\gamma \in \prod} \left(\underset{z^s \sim \zeta^s, z^t \sim \zeta^t}{\mathbb{E}} d(z^s, z^t)^p \right)^{\frac{1}{p}} \right), \tag{6}$$

where d is a distance and the corresponding cost function $c(\boldsymbol{z}^s, \boldsymbol{z}^t) = d(\boldsymbol{z}^s, \boldsymbol{z}^t)^p$. With $p = 2$, we evaluate distances between measures $\boldsymbol{z}^s$ and $\boldsymbol{z}^t$ according to the squared Wasserstein distance $W_2^2 = d(\boldsymbol{z}^s, \boldsymbol{z}^t)^2$.

In the discrete versions of the problem [21], i.e. when ζ^s and ζ^t are defined as empirical measures based on vectors in $\mathbb{R}^k$, $\prod(\zeta^s, \zeta^t)$ denotes the polytope of matrices $\boldsymbol{\gamma}$ such $\boldsymbol{\gamma}\mathbf{1} = \zeta^s$ and $\boldsymbol{\gamma}^\intercal\mathbf{1} = \zeta^t$ and therefore Eq. (6) read as:

$$\boldsymbol{\gamma}_0(\zeta^s, \zeta^t) = \min_{\gamma \in \prod} <\boldsymbol{\gamma}, \boldsymbol{C}>_F, \tag{7}$$

where $<.,.>_F$ is the Frobenius dot product, $\boldsymbol{C} \geq 0$ is a cost matrix $\in \mathbb{R}^{n \times m}$, representing the pairwise costs of transporting bin i to bin j, and γ is a probabilistic coupling given by a matrix of size $n \times m$, with marginals defined as ζ^s and ζ^t. Solving Eq. (7) requires linear programming problem with equality constraints, but its dimensions scale quadratically with the size of the sample z. Hence, in this paper, we rely on a compact, structured and continuous latent space $\mathcal{Z}$ instead of the original image feature space $\mathcal{X}$ which is high dimensional.

2.3   Learning OLVA with a Stochastic Approximation

The final objective of OLVA Eq. (8) optimizes jointly for: i) an embedding function $q(\boldsymbol{z}|\boldsymbol{x})$ that maps both the source and the target domain to a shared semantic latent space $\mathcal{Z}$ regularized to follow normal distribution; ii) a transportation matrix γ that aligns similar semantic vectors $\boldsymbol{z}$ from both domains in the latent space; and iii) a predictive function $p(\boldsymbol{y}|\boldsymbol{z})$ for masks predictions.

$$\min_{\gamma \in \prod, q_\phi, p_\theta} <\boldsymbol{\gamma}, \boldsymbol{C}>_F + \mathcal{L}_{vae}^s + \mathcal{D}_{KL}^t \tag{8}$$

To solve Eq. (8), we use an Expectation-Maximization alternating method [5]: 1) We fix $q(\boldsymbol{z}|\boldsymbol{x})$ and $p(\boldsymbol{y}|\boldsymbol{z})$ and we optimize for $\boldsymbol{\gamma}$, which reduces to the problem in Eq. (8) to solving a classic OT problem with cost matrix $C_{i,j} = \alpha||\boldsymbol{z}_i^s - \boldsymbol{z}_j^s||_2^2$, α being a hyper parameter to trade-off the focus on the alignment loss; 2) We fix $\boldsymbol{\gamma}$ and optimize for $q(\boldsymbol{z}|\boldsymbol{x})$ and $p(\boldsymbol{y}|\boldsymbol{z})$, this turns the problem in Eq. (8) to a standard deep learning problem, wher the VAE loss is minimized.

Similar to Damodoran $et.al.$ [6], we solve the optimization problem with a stochastic approximation Eq. (9), using mini-batches of size $m + n$ from the source and target domains respectively. Our stochastic approximation yields a computationally feasible solution for both the OT and the VAE. The discrepancy measure and the KL-Divergence regularization are computed at the latent space layer, while the segmentation loss uses the output layer.

$$\min_{q_\phi, p_\theta} \mathbb{E}\left[\frac{1}{m}\sum_{i=1}^{m} \mathcal{L}_{dice}^s(\boldsymbol{y}^s, p(\boldsymbol{y}^s|\boldsymbol{z}^s)) + \frac{1}{m}\sum_{i=1}^{m}\beta\mathcal{D}_{KL}^s + \frac{1}{n}\sum_{i=1}^{n}\beta\mathcal{D}_{KL}^t \right.$$
$$\left. + \frac{1}{m}\sum_{i=1}^{m}||\boldsymbol{y}^s - p(\boldsymbol{y}^s|\boldsymbol{z}^s)||^2 + \min_{\gamma \in \prod(\zeta^s, \zeta^t)}\sum_{i,j}^{m+n}\gamma_{i,j}\alpha||\boldsymbol{z}_i^s - \boldsymbol{z}_j^t||^2\right] \tag{9}$$

Architecture and Implementation Details: OLVA accept batches containing 128 source and 128 target samples. The input dimension is $256 \times 256 \times 3$. The encoder is composed of five convolutional layers, with stride by 2 for downsampling, and with a leaky rectified linear unit (lrelu) activation, with a leakage rate of 0.3. The number of feature maps is successively $32, 32, 64, 64,$ and 64. The last convolutional is flattened and mapped using a linear fully connected layer into two vectors (μ, σ), each composed of $K = 128$ features followed by a dropout of rate 0.3. A latent vector z is generate as $\mu + \sigma \odot \epsilon$, where $\epsilon \sim \mathcal{N}(0, I)$ and given as an input to the decoder. The decoder is composed of five up-convolutional layers, with a lrelu activation, each composed of $64, 64, 32, 32,$ and 4 feature maps. The output layer with a sigmoid activation provides a mask of shape $256 \times 256 \times 4$. A learning rate of 0.0001 is used with Adam optimizer. Using the validation set we experimentally tuned, $\alpha = 10$ to focus more on the alignment loss and $\beta = 0.1$. The total number of iterations is 10,000.

3 Experiments and Results

Dataset. We use the public MM-WHS dataset [26] for cardiac segmentation consisting of 20 MR ($\sim$128 slices) and 20 CT ($\sim$256 slices) unpaired and multi-site images from 40 patients. We followed the SOA data processing, domain adaptation protocol and evaluation metrics [3,4,8,11,18]. For data processing, we use the coronal view slices, cropped to 256×256 and normalized to zero mean and unit variance. To consider contextual information three adjacent slices ($256 \times 256 \times 3$) were stacked at the input and the middle slice label was used as the ground truth. Data augmentation included rotation, scaling, and affine transformations. A total of 11998 MR and 9598 sub-volumes were generated (each $256 \times 256 \times 3$). For domain adaptation, we randomly split each modality into training (16 subject) and testing (4 subjects). We use MR as a source domain, with 9599 sub-volumes for training and 2399 for validation. We set CT as a the target domain, with 8399 sub-volumes for training and 1199 for evaluation. We report the performance in terms of Dice Similarity Coefficient (DSC) and the Average Symmetric Surface Distance (ASSD).

Experimental Settings. We consider four experimental settings: 1) To assess the advantages of using VAE for producing a valid segmentation maps in comparison to a U-Net like model, we consider our VAE model for supervised segmentation without the OT loss, trained and evaluated over CT scans. We refer to this model as oracle VAE. We compare this setting against a U-Net [22] under same data protocol and we refer to it as oracle U-Net. 2) To highlight the severe domain shift between MR and CT modalities, we consider the situation when no adaptation is performed. Thereby, we train our VAE without OT loss over the MR images and evaluated it over CT images. We refer to this setting as VAE-0 and U-Net-0. 3) We consider training OLVA under SOA data protocol setting in which 16 labeled MR and 16 unlabeled CT sequences are used. We refer to this setting as OLVA-16 and we compare against five methods for medical UDA: PnP-AdaNet [8], SIFA [3], Synseg-net [11], DECM [18] and Seg-DJOT [1] and

Table 1. Baseline method with full supervision and with no adaptation.

Methods	DSC score					ASSD score (mm)				
	LV-M	LA-B	LV-B	A-A	**Avg**	LV-M	LA-B	LV-B	A-A	**Avg**
Oracle U-Net	0.83	0.89	0.92	0.93	0.89	0.38	0.39	0.28	0.31	0.34
Oracle VAE	**0.95**	**0.97**	**0.97**	**0.96**	**0.96**	**0.06**	**0.04**	**0.03**	**0.05**	**0.05**
U-Net-0	0.10	0.27	0.02	0.24	0.15	36.0	19.4	48.6	31.9	26.2
VAE-0	**0.41**	**0.51**	**0.60**	**0.48**	**0.49**	**2.51**	**2.21**	**2.44**	**2.95**	**2.53**

two methods for natural image UDA: CycleGAN [12] and AdaOutput [23]. 4) We consider a more ambitious scenario where only one unlabeled CT sequence is available (OLVA-1). Similar to [18], we randomly draw one scan from the target set. To train OLVA-1 and to avoid overfitting, we fix all the model parameters and only the fully connected layer is retrained using the loss evaluated at the latent space of Eq. (9). We also perform an experiment where an auxiliary reconstruction task [9] is integrated (OLVA-R-1).

Discussion: The quantitative and qualitative results of our experiments are presented in Tables 1, 2, 3 and Fig. 3. Table 1 shows that our supervised baseline method outperforms the U-Net, achieving a high DSC and more importantly producing valid cardiac shape predictions as seen in Fig. 3, and as reflected by the ASSD score. When no adaptation is considered, VAE-0 achieves 49% DSC, while U-Net achieved only 15%. As our VAE-0 pushes the latent semantic features to be close to normal distribution, it partially aligns the marginal distributions. In the UDA setting Table 2, OLVA-16 outperforms the SOA's best results by an additional 12.5% in DSC and having minimal erroneous prediction as seen in Fig. 3, with average ASSD of 0.31 mm. Considering the target data scarcity UDA Setting Table 3, OLVA-1 achieved the second best results after DECM-1 with an 8% DSC difference. The results of OLVA-1 when the reconstruction auxiliary task is introduced (OLVA-R-1) reduce the gap to 5% at the price of increasing the model complexity. The second-place is honorable, comparing the 1.7 million parameters of OLVA-1 with the more than 95 million parameters of

Table 2. Performances of UDA: 16 unlabeled CT and 16 labeled MR scans.

Methods	DSC score					ASSD score (mm)				
	LV-M	LA-B	LV-B	A-A	**Avg**	LV-M	LA-B	LV-B	A-A	**Avg**
OLVA-16	**0.79**	**0.87**	**0.88**	**0.88**	**0.85**	**0.56**	**0.53**	**0.37**	**0.45**	**0.31**
Seg-DJOT-16	0.57	0.60	0.57	0.62	0.59	3.64	3.62	3.85	5.20	4.07
SIFA-16	0.58	0.76	0.76	0.81	0.73	3.44	3.83	3.30	2.64	3.32
Pnp-AdaNet-16	0.50	0.77	0.60	0.79	0.66	10.2	4.04	8.60	2.28	6.22
SynSeg-Net-16	0.41	0.69	0.52	0.72	0.58	4.60	3.80	3.40	5.60	4.35
AdaOutput-16	0.43	0.76	0.54	0.65	0.59	4.68	2.89	3.10	6.15	4.20
CycleGAN-16	0.28	0.75	0.52	0.73	0.57	4.85	6.20	3.92	5.54	5.30

Table 3. Performances of UDA: 1 unlabeled CT and 16 labeled MR scans.

Methods	DSC score					ASSD score (mm)				
	LV-M	LA-B	LV-B	A-A	**Avg**	LV-M	LA-B	LV-B	A-A	**Avg**
OLVA-1	0.58	0.69	0.64	<u>0.67</u>	0.64	2.10	1.95	1.85	**2.30**	<u>2.05</u>
OLVA-R-1	**0.68**	<u>0.70</u>	<u>0.78</u>	0.60	<u>0.69</u>	**1.89**	**1.88**	**1.51**	<u>2.43</u>	**1.92**
DECM-1	<u>0.60</u>	**0.78**	0.71	**0.78**	**0.72**	7.37	3.87	6.44	2.77	5.11
Seg-DJOT-1	0.19	0.25	0.21	0.20	0.21	9.64	13.7	8.18	10.3	10.4
SIFA-1	0.39	0.53	**0.80**	0.62	0.62	12.8	4.12	7.70	2.72	6.84
Pnp-AdaNet-1	0.29	0.48	0.33	0.58	0.25	25.1	27.1	27.7	7.14	21.8

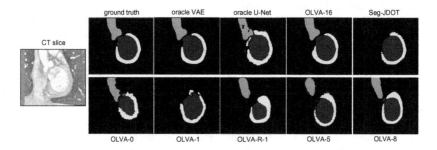

Fig. 3. Qualitative results of adaptation performances on segmentation.

DECM-1, and considering the quality of predictions as reflected by the ASSD. We also examine OLVA's performance when trained with randomly sampled 5 and 8 targets sequences. OLVA-5 achieves similar performance to DECM-1, while OLVA-8 achieves better performance with DSC score of 79%. As for further ablation studies, we change the latent dimension to 64, 256 and 512. With $K = 64$, a degradation in the source domain performance was observed, yielding an average DSC score of 79%. With $K = 256$, similar performances to $K = 128$ is achieved. When $K = 512$, a degradation in the performance over the source and the target domain is observed, leading to 69.6% target DSC score for OLVA-16. This degradation is expected as optimizing for γ requires a reasonable number of samples which grows with K's dimensionality [1].

4   Conclusion

To improve the applicability of deep learning model on new modality where it is expensive to acquire expert annotations, UDA represents a central solution. In this paper, we tackle the problem of unsupervised cross-modality medical image segmentation with a novel framework that jointly integrates VAE and OT theory. OLVA is a simple, efficient and lightweight model. The usability of our method can be integrated within other learning regimes, for instance, a weakly-supervised model where sparse annotation of biomedical volumetric data are available and the aim would be to leverage the rest of the unlabeled data by matching them with the available labeled set. Future work will address the possibility of integrating patient specific information during the inference stage to ease the adaptation process and improve the segmentation performance.

References

1. Ackaouy, A., Courty, N., Vallée, E., Commowick, O., Barillot, C., Galassi, F.: Unsupervised domain adaptation with optimal transport in multi-site segmentation of multiple sclerosis lesions from MRI data. Front. Comput. Neurosci. **14**, 19 (2020)
2. Arjovsky, M., Bottou, L.: Towards principled methods for training generative adversarial networks. In: ICLR (2017)
3. Chen, C., Dou, Q., Chen, H., Qin, J., Heng, P.A.: Synergistic image and feature adaptation: towards cross-modality domain adaptation for medical image segmentation. In: Proceedings of the AAAI Conference on Artificial Intelligence, vol. 33, pp. 865–872 (2019)
4. Chen, C., Dou, Q., Chen, H., Qin, J., Heng, P.A.: Unsupervised bidirectional cross-modality adaptation via deeply synergistic image and feature alignment for medical image segmentation. IEEE TMI **39**, 2494–2505 (2020)
5. Courty, N., Flamary, R., Tuia, D., Rakotomamonjy, A.: Optimal transport for domain adaptation. IEEE TPAMI **39**, 1853–1865 (2017)
6. Damodaran, B.B., Kellenberger, B., Flamary, R., Tuia, D., Courty, N.: DeepJDOT: deep joint distribution optimal transport for unsupervised domain adaptation. In: ECCV, pp. 447–463 (2018)
7. David, S.B., Lu, T., Luu, T., Pál, D.: Impossibility theorems for domain adaptation. In: The Thirteenth AISTATS, pp. 129–136 (2010)
8. Dou, Q., et al.: PnP-AdaNet: plug-and-play adversarial domain adaptation network at unpaired cross-modality cardiac segmentation. IEEE Access **7**, 99065–99076 (2019)
9. Gonzalez Duque, V., Al Chanti, D., Crouzier, M., Nordez, A., Lacourpaille, L., Mateus, D.: Spatio-temporal consistency and negative label transfer for 3D freehand US segmentation. In: Martel, A.L., et al. (eds.) MICCAI 2020. LNCS, vol. 12261, pp. 710–720. Springer, Cham (2020). https://doi.org/10.1007/978-3-030-59710-8_69
10. Heimann, T., Mountney, P., John, M., Ionasec, R.: Learning without labeling: domain adaptation for ultrasound transducer localization. In: Mori, K., Sakuma, I., Sato, Y., Barillot, C., Navab, N. (eds.) MICCAI 2013. LNCS, vol. 8151, pp. 49–56. Springer, Heidelberg (2013). https://doi.org/10.1007/978-3-642-40760-4_7
11. Huo, Y., et al.: SynSeg-Net: synthetic segmentation without target modality ground truth. IEEE TMI **38**(4), 1016–1025 (2018)
12. Isola, P., Zhu, J.Y., Zhou, T., Efros, A.A.: Image-to-image translation with conditional adversarial networks. In: CVPR, pp. 1125–1134 (2017)
13. Kantorovich, L.V.: On the translocation of masses. Dokl. Akad. Nauk. USSR (NS) **37**, 199–201 (1942)
14. Kingma, D.P., Welling, M.: Auto-encoding variational bayes (2014)
15. Kumagai, A., Iwata, T.: Unsupervised domain adaptation by matching distributions based on the maximum mean discrepancy via unilateral transformations. In: AAAI Conference on Artificial Intelligence, vol. 33, pp. 4106–4113 (2019)

16. Li, F., Li, W., Qin, S., Wang, L.: MDFA-Net: multiscale dual-path feature aggregation network for cardiac segmentation on multi-sequence cardiac MR. KBS 106776 (2021)

17. Liu, K., Tang, W., Zhou, F., Qiu, G.: Spectral regularization for combating mode collapse in GANs. In: ICCV, pp. 6382–6390 (2019)

18. Ouyang, C., Kamnitsas, K., Biffi, C., Duan, J., Rueckert, D.: Data efficient unsupervised domain adaptation for cross-modality image segmentation. In: Shen, D., et al. (eds.) MICCAI 2019. LNCS, vol. 11765, pp. 669–677. Springer, Cham (2019). https://doi.org/10.1007/978-3-030-32245-8_74

19. Painchaud, N., Skandarani, Y., Judge, T., Bernard, O., Lalande, A., Jodoin, P.M.: Cardiac segmentation with strong anatomical guarantees. IEEE TMI **39**(11), 3703–3713 (2020)

20. Puybareau, É., et al.: Left atrial segmentation in a few seconds using fully convolutional network and transfer learning. In: Pop, M., et al. (eds.) STACOM 2018. LNCS, vol. 11395, pp. 339–347. Springer, Cham (2019). https://doi.org/10.1007/978-3-030-12029-0_37

21. Redko, I., Courty, N., Flamary, R., Tuia, D.: Optimal transport for multi-source domain adaptation under target shift. In: The 22nd AISTATS, pp. 849–858 (2019)

22. Ronneberger, O., Fischer, P., Brox, T.: U-Net: convolutional networks for biomedical image segmentation. In: Navab, N., Hornegger, J., Wells, W.M., Frangi, A.F. (eds.) MICCAI 2015. LNCS, vol. 9351, pp. 234–241. Springer, Cham (2015). https://doi.org/10.1007/978-3-319-24574-4_28

23. Tsai, Y.H., Hung, W.C., Schulter, S., Sohn, K., Yang, M.H., Chandraker, M.: Learning to adapt structured output space for semantic segmentation. In: CVPR, pp. 7472–7481 (2018)

24. Wu, F., Zhuang, X.: CF distance: a new domain discrepancy metric and application to explicit domain adaptation for cross-modality cardiac image segmentation. IEEE TMI **39**, 4274–4285 (2020)

25. Yang, J., et al.: Domain-agnostic learning with anatomy-consistent embedding for cross-modality liver segmentation. In: ICCV Workshops (2019)

26. Zhuang, X., Shen, J.: Multi-scale patch and multi-modality atlases for whole heart segmentation of MRI. Med. Image Anal. **31**, 77–87 (2016)

Prototypical Interaction Graph for Unsupervised Domain Adaptation in Surgical Instrument Segmentation

Jie Liu[ID], Xiaoqing Guo[ID], and Yixuan Yuan[✉][ID]

Department of Electrical Engineering, City University of Hong Kong,
Kowloon, Hong Kong, China
jliu.ee@my.cityu.edu.hk, yxyuan.ee@cityu.edu.hk

Abstract. Surgical instrument segmentation is fundamental for the advanced computer-assisted system. The variability of the surgical scene, a major obstacle in this task, leads to the domain shift problem. Unsupervised domain adaptation (UDA) technique can be employed to solve this problem and adapt the model to various surgical scenarios. However, existing UDA methods ignore the relationship among different categories, hindering the model learning discriminative features from a global view. Additionally, the adversarial strategy utilized in these methods only narrows down the domain gap at the end of the network, leading to the poor feature alignment. To tackle above mentioned problems, we advance a semantic-prototype interaction graph (SePIG) framework for surgical instrument type segmentation to grasp the category-level relationship and further align the feature distribution. The proposed framework consists of *prototypical inner-interaction graph* (PI-Graph) and *prototypical cross-interaction graph* (PC-Graph). In PI-Graph, EM-Grouping module is designed to generate multi-prototypes representing the semantic information adequately. Then, propagation is performed upon these multi-prototypes to communicate semantic information inner each domain. Aiming at narrowing down the domain gaps, the PC-Graph constructs hierarchical graphs upon multi-prototypes and category centers, and conducts dynamic reasoning to exchange the correlated information among two domains. Extensive experiments on the EndoVis Instrument Segmentation 2017 → 2018 scenarios demonstrate the superiority of our SePIG framework compared with state-of-the-art methods. Code is available at https://github.com/CityU-AIM-Group/SePIG.

Keywords: Surgical instrument type segmentation · Unsupervised domain adaptation · Graph convolution

1 Introduction

Surgical instrument segmentation serves as a crucial cornerstone for various advanced computer-assisted surgery systems, such as instrument tracking and

© Springer Nature Switzerland AG 2021
M. de Bruijne et al. (Eds.): MICCAI 2021, LNCS 12903, pp. 272–281, 2021.
https://doi.org/10.1007/978-3-030-87199-4_26

pose estimation [10,11]. A variety of methods [5,8,13] are proposed to address surgical instrument segmentation and achieve remarkable progress at the cost of abundant annotations. But the pixel-wise annotations conducted by experienced medical experts are expensive and laborious. Moreover, these models trained on one domain may fail in other domains due to poor generalization ability. This situation is even worse in real-world surgical scenarios, such as different procedures and hospitals, where the source data and target data follow the different distributions with evident domain shift. To address these problems, a promising approach is utilizing unsupervised domain adaptation (UDA) to reduce the domain shift and enhance the generalization ability on the target domain only with annotations in the source domain. Existing UDA methods can be primarily categorized into image translation [3,20,21] and feature alignment [7,17,19]. Image translation methods align the image appearance through pixel-to-pixel transformation approaches. Among the other line, feature alignment methods narrow the domain gap among domains in high-level embedding.

In surgical instrument segmentation, existing UDA methods [15,16] just apply image translation to reduce the domain gap in input level. Pfeiffer *et al.* [15] employ the CycleGAN to translate the style of source images to target ones, and train a model using these style translated source images. Sahu *et al.* [16] propose a consistency-based joint-learning framework to combine image translation step and segmentation model training step into an end-to-end way. These methods align domains in image appearance level, highly depending on the image translation qualities. Although these methods achieve impressive progress in *binary segmentation task*, they may generate mismatched visual patterns and hinder the model performance in *instrument type segmentation task*, due to the difficulty in determining suitable visual characterizations for the transferring. Moreover, the existing feature alignment methods [12,19,23] cannot be directly applied to this task owing to the following two reasons. Firstly, most of UDA methods align the feature distribution for each category separately, regardless of the intrinsic relationship among categories. In deed, the cross-category semantic relations among the surgical instruments can enhance the feature representation from a global perspective [4]. Secondly, source domain information is only utilized during the adversarial procedure, i.e., the last step of training procedure, to promote the target performance. Nevertheless, the source domain and target domain are related to each other, and exchanging the complementary information directly can further mitigate the domain gap.

To tackle these two issues, we introduce the graph techniques to model the semantic relationship and propose the semantic-prototype interaction graph (SePIG) framework in feature-level alignment for instrument type segmentation. The proposed SePIG includes prototypical inner-interaction graph (PI-Graph) in each domain and prototypical cross-interaction graph (PC-Graph) cross two domains, representing the first effort in graph guided UDA feature alignment for *instrument type segmentation task*. Specifically, EM-Grouping module is first advanced in PI-Graph, to project the feature map into latent semantic space and generate multi-prototypes for each image. Based on these prototypes, graph convolution is performed to integrate the category level information, efficiently grasping the relationship among categories inner each domain. In PC-Graph, the global multi-prototypes in one domain can be incorporated into the other domain as the complementary information and form a hierarchical structure. Then the dynamic reasoning is proposed to grasp the dynamic relationship in this hierarchical structure and mitigate the domain gap. With these two flexible graph structures, SePIG can fully explore the semantic interaction inner each domain and cross two domains. Extensive experiments and ablation studies validate the effectiveness of our proposed framework.

2   Approach

As illustrated in Fig. 1, we present the SePIG framework for the UDA problem. In the regime of unsupervised domain adaptation, we have labeled source data $\{X_S, Y_S\}$ and unlabeled target data $\{X_T\}$. We first adopt the dilated ResNet pretrained on the ImageNet dataset as the backbone. The input images x_s, x_t are passed through the backbone and produce feature map $f_s, f_t \in \mathcal{R}^{N \times D}$, where $N = W \times H$ and H, W, D represent height, width and channel number, respectively. Then, with the semantic prototypes generated with EM-Grouping Module, we first establish PI-Graph and perform graph convolution upon these prototypes, enabling the prototypes to interact with each other, and enhancing the feature representation in a global view. Afterwards, PC-Graph propagates the information across two domains to incorporate the other domain knowledge and output the refined feature map. This feature map is further fed into a 1×1 convolution layer to get final segmentation result. In the end, the adversarial procedure is adopted to align the source and target feature distribution.

2.1   Prototypical Inner-Interaction Graph (PI-Graph)

In UDA surgical instrument segmentation task, instrument types are easily misclassified due to the similar appearance of instruments. To solve this, the graph structure representing category-level instrument relationship is devised to propagate the semantic information inner each domain, enhancing the feature representation from a global view. It can be formulated as $G_I = (V_I, E_I)$, where vertices set V_I consists of the semantic feature vectors, i.e., multi-prototypes, and edge set E_I encodes the affinity between vertices, as shown in Fig. 1(a).

Vertice Representation via Multi-prototypes. In order to represent the semantic information, prototypes are usually utilized [18, 22]. Considering a single prototype couldn't represent complete category information and may lead to semantic ambiguity for different object parts [6], especially in complex surgical scenario, we propose the EM-Grouping to adaptively generate multi-prototypes and fully explore the semantic information.

Given the extracted feature map f from each image, EM-Grouping produces multi-prototypes. Specifically, EM-Grouping employs probability mixture model to approximate the distribution being $P(f_i|\theta) = \sum_{k=1}^{K} w_k e^{K(f_i, \mu_k)}$, where w_k is

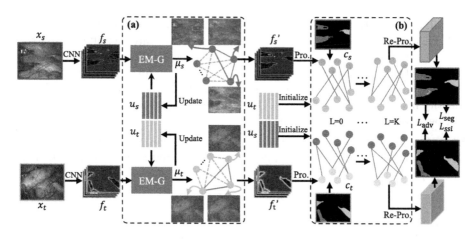

Fig. 1. Overview of SeGIP framework. (a) Prototypical inner-interaction graph. (b) Prototypical cross-interaction graph

the distribution weight and $f_i \in \mathcal{R}^D$ indexes the feature vector for pixel i. The radial basis kernel function is adopted $K(f_i, \mu_k) = -||(s_i - \mu_k)||_2^2$ and $\mu_k \in \mathcal{R}^D$ is the mean vector of the kth distribution. The parameters of probability mixture model $\theta = \{\mu, w\}$ are estimated using EM algorithms [14], which includes iterative E steps and M steps. The iteration is empirically set as 4. μ is first initialized using initial seed u. Then, in the E step, we calculate posterior probability P_{ik} of pixel i and base distribution k:

$$P_{ik} = \frac{w_k e^{K(f_i, \mu_k)}}{\sum_{j=1}^{K} w_j e^{K(f_i, \mu_j)}}. \tag{1}$$

In the M step, parameters are re-estimated with the posterior probability via

$$\mu_k = \frac{\sum_{i=1}^{N} P_{ik} f_i}{\sum_{i=1}^{N} P_{ik}}, \quad w_k = \frac{\sum_{i=1}^{N} P_{ik}}{N}. \tag{2}$$

For each mini-batch, we use weighted moving average to update the initial seed u in the training phase and guide u to a appropriate position for the whole dataset. We update u via $u := \sigma u + (1 - \sigma)\mu$, where σ is the momentum and empirically set as 0.998. This can guarantee the μ converges to the global maximum.

The converged mean vectors μ of probability mixture model are utilized as prototypes. Obviously, each prototype is the mean feature of semantically similar pixels and represents a region in image. It's worth mentioning that the number of multi-prototypes K is larger than the category number and several prototypes represent a single category. This strategy can fully explore the semantic information and mitigate the data imbalanced problem in medical scenario. In addition, with the initial seed u moving towards the centroid feature for the whole dataset in the training process, u can be regarded as the global multi-prototypes and possess global semantic information, which will be utilized in PC-Graph.

Affinity-Based Edge Calculation. Given the calculated multi-prototypes $\mu \in \mathcal{R}^{K \times D}$, we first project μ into affinity space and calculate the affinity matrix, $A = \phi(\mu)\phi'(\mu)^\top$, where $\phi(\cdot)$ and $\phi'(\cdot)$ are linear embedding followed by ReLU function. Then, self-loop and degree normalization are further applied to avoid the imbalance connection problem, $\hat{A} = \widetilde{D}^{-\frac{1}{2}} \widetilde{A} \widetilde{D}^{-\frac{1}{2}}$, where $\widetilde{A} = A + I$. I is the identity matrix and $\widetilde{D}_{ii} = \sum_j \widetilde{A}_{ij}$. This adjacency matrix is data-dependent and can be learned via back propagation.

Relation Reasoning. We perform graph convolution [9] in the constructed PI-Graph to enable the category information interacted inner domain. The reasoning procedure is defined as:

$$\mu' = \mathcal{F}(\hat{A}\mu W), \tag{3}$$

where $\mathcal{F}$ is ReLU activation and $W \in \mathcal{R}^{D \times D}$ is the trainable parameters. After the graph convolution operation, we get the enhanced prototypes μ'. With the affinity matrix modeling the relationship among the prototypes, this enhanced prototypes μ' can model the semantic relations from the global view. Note that the PI-Graph is built in both source and target domains, and the trainable parameters in PI-Graph are shared.

In the end, the prototypes μ' are projected to the 2D feature map f'. This reverse projection is formulated as $f' = P\mu'$, where P is the convergent posterior probability in EM iteration and f' has the same shape as f.

2.2 Prototypical Cross-Interaction Graph (PC-Graph)

Considering the common label space sharing between the source and target domains, the correlated instrument knowledge in one domain can be transferred to the other one. To this end, we propose the PC-Graph building interaction across two domains, so as to reach semantic coherency and narrow down the domain gap. As shown in Fig. 1 (b), we build the PC-Graph as $G_C = (V_C, E_C)$ cross two domains, where vertices set V_C constitutes the category centers c of its own domain and the global multi-prototypes u of the other domain, and the edge set E_C indicates the connection between u and c. Furthermore, as only part of instruments occur in each image and the connection strength varies among the class centers and global multi-prototypes for different images, the fixed adjacency matrix is insufficient to model the dynamic relation between the global multi-prototypes and class centers. Hence, we advance dynamic reasoning to model the dynamic relationship.

Vertice Representation via Category Centers. The category center can be calculated by the weighted average of feature vectors, where the corresponding weight is the probability of the vector belonging to a specific category. Given the enhanced feature map f', class center is generated via $c = \varphi(f')^\top f'$, where $c \in \mathcal{R}^{C \times D}$, $\varphi(f') \in \mathcal{R}^{N \times C}$ is the coarse prediction and C is the class number.

Posterior-Probability-Based Edge Calculation. We utilize the posterior probability P to evaluate the relationship R_{ij} between the ith category center and j^{th} global prototype, which can be formulated as:

$$R_{ij} = \frac{1}{N_D} \sum_{n=1}^{N_D} \frac{\sum_k G_{ki}^n P_{kj}^n}{\sum_k G_{ki}^n},$$ (4)

where N_D is the image number of the whole dataset, $G \in \mathcal{R}^{N \times C}$ represents the ground truth or pseudo label, and P is the convergent posterior probability obtained in E step of EM-Grouping.

Dynamic Reasoning. As the relation among multi-prototypes and category centers change dynamically for different image, dynamic reasoning is proposed in the PC-Graph to promote the model capacity and tolerate the scene variations. Specifically, the adjacency matrix follows a Bernoulli distribution and the relationship value serves as the probability, $A_{ij}^d \sim B(x, R_{ij})$, where B is the Bernoulli distribution. The A_{ij}^d is 1 with the probability R_{ij} and is 0 with probability $1 - R_{ij}$. During training, the adjacency matrix changes dynamically in each iteration, while during the inference stage, it is fixed as R. After obtaining the dynamic adjacency matrix, we aggregate the other domain information as:

$$c_i^{(l+1)} = \mathcal{F}(\frac{1}{\sum_j A_{ij}^d} \sum_{j \in \mathcal{N}(i)} A_{ij}^d \mu_j W) + c_i^{(l)},$$ (5)

where $\mathcal{F}$ is ReLU activation, $W \in \mathcal{R}^{D \times D}$ is trainable parameter defining a linear mapping to encode the message from node j. Besides, we employ the residual connection to facilitate the gradient propagation during training.

2.3   Training Objective

The proposed framework is optimized by three loss functions, i.e., the segmentation loss, the adversarial loss and the self-supervised loss. The segmentation part F is trained with segmentation loss in labeled source data:

$$\mathcal{L}_{seg}(F) = -\mathbb{E}_{(x,y) \sim (X_S, Y_S)}[y \log(F(x))].$$ (6)

Then, domain discriminator D is employed to distinguish between the source and target domain, while F has to fool D. This is achieved by minimaxing the adversarial loss:

$$\mathcal{L}_{adv}(F, D) = -\mathbb{E}_{x \sim X_S}[\log(D(F(x)))] - \mathbb{E}_{x \sim X_T}[\log(1 - D(F(x)))],$$ (7)

where $D(\cdot)$ is the domain prediction of feature map, 0 indicating the target domain and 1 indicating the source domain. The adversarial weight is empirically set as 0.0001. Furthermore, we can't access the corresponding label in the target domain. Thus, we adopt the self-supervised method to dig out the supervised signal inside the unlabeled dataset [24]. We set up different thresholds for different categories considering the data unbalance problem. Specifically, if the predicted probability is higher than the threshold, which is set as [0.95, 0.9, 0.9, 0.9], it could be selected as pseudo-label $\widehat{Y_T}$. By minimizing the self-training loss

Table 1. Performance comparison with different methods and ablation results.

Methods	Background		Forceps		Needle Driver		Scissor		mIoU
	IoU(%)	Dice(%)	IoU(%)	Dice(%)	IoU(%)	Dice(%)	IoU(%)	Dice(%)	
Source only	93.12	96.32	39.86	55.18	12.57	16.96	54.97	62.47	50.13
Oracle	95.84	97.81	62.88	75.64	24.67	35.93	81.30	87.22	66.17
I2I (19′) [15]	94.35	97.15	56.78	69.87	12.47	22.18	55.97	59.41	54.89
FDA (20′) [21]	91.82	94.49	52.13	66.56	18.44	**30.03**	62.31	**72.05**	56.18
CLAN (19′) [12]	94.09	96.92	58.36	72.70	7.77	11.64	60.50	66.14	55.18
ASANet (20′) [23]	**95.32**	**97.43**	60.02	72.24	12.85	22.51	64.25	70.21	58.11
SePIG	94.53	96.83	**61.36**	**74.76**	19.21	29.84	**65.11**	70.86	**60.05**
w/o EM-Grouping	94.37	96.67	55.47	68.32	17.32	25.55	63.66	69.12	57.71
w/o PI-Graph	93.89	96.71	53.68	66.73	14.24	22.23	61.80	66.36	55.90
w/o PC-Graph	94.66	96.90	51.30	64.78	16.83	25.41	63.85	69.93	57.01

over pseudo label, the optimized model can be adapted to the target domain. The self-training loss is defined as:

$$\mathcal{L}_{ssl}(F) = -\mathbb{E}_{(x,\widehat{y}) \sim (X_T, \widehat{Y_T})} [\widehat{y} \log(F(x))]. \tag{8}$$

In unsupervised domain adaptation, we can't access the corresponding label in the target domain. Thus, we adopt the self-supervised method to dig out the supervised signal inside the unlabeled dataset. Specifically, we set up different thresholds for different categories considering the data unbalance problem. If the predicted probability for target data is higher than the threshold, which is set as 0.9, it could be selected as pseudo-label.

3   Experiment

Dataset and Implementation Details. We validate the proposed framework on the surgical instrument type segmentation adaptation task Endovis17 → Endovis18. *Endovis17* [2] consists of 1800 frames with resolution of 1280 × 1024 and instrument-type annotations. *Endovis18* [1] consists of 2384 frames with resolution of 1280 × 1024, including more complex porcine tissue and instrument motion. Following [5], we divide the Endovis18 into 1639 images as training data and the rest as testing data. The common labels include forceps, needle driver and scissor. During training phase, we optimize the segmentation model F using SGD with a momentum of 0.9 and initial learning rate of 2.5e-5, and optimize D using Adam with $\beta_1 = 0.9$, $\beta_2 = 0.99$ and initial learning rate of 1e-4. Both learning rates are decayed by a poly learning rate policy. We set the batch size as 4 and resize the images to 320 × 256. The whole experiments are implemented in RTX 2080Ti with PyTorch.

Benchmark and SePIG Effectiveness Analysis. To observe the domain gap of this benchmark, we first investigate the 'source only' lower bound by directly applying the model trained on the source domain to target domain. Then, the 'oracle' up bound is attained with target annotations. Domain shift

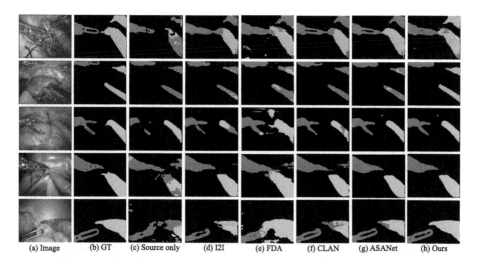

| (a) Image | (b) GT | (c) Source only | (d) I2I | (e) FDA | (f) CLAN | (g) ASANet | (h) Ours |

Fig. 2. Visual comparison of adaptation results produced by different methods. (Color figure online)

degrades the performance from 66.17% (2nd row) to 50.13% (1st row) mIoU, as listed in the Table 1. Remarkably, the proposed SePIG (5th row) presents a great recover with 9.92% mIoU improvement in comparison to 'source only' lower bound, demonstrating our method can leverage unlabeled target domain data to enhance the generalization ability. Moreover, the qualitative results illustrated in Fig. 2 also validate the effectiveness of the proposed method in narrowing the domain gap. Specifically, our segmentation results (Fig. 2 (h)) can successfully localize the instrument and correctly classify the category, while it is difficult to obtain correct predictions without adaptation (Fig. 2 (c)).

Comparison with State-of-the-Art Methods. We compare the SePIG with the state-of-the-art methods [12,15,21,23] in Table 1 ($3^{rd}-7^{th}$ rows). The first two methods [15,21] are image translation based methods, and the last two [12, 23] are feature alignment based methods. The proposed SePIG outperforms these methods [12,15,21,23] with a large margin of 5.16%, 3.87%, 4.87% and 1.94% mIoU. Although the increment in background is marginal, the proposed SePIG ranks first in IoU for all instrument categories, which are more clinically relevant. The qualitative results are shown in Fig. 2. Notably, SePIG can correctly segment the needle driver (yellow part) as shown in $1^{st}-2^{nd}$ rows. This reveals that our framework can model the category relation from a global view and grasp the correct semantic information of each instrument with the help of graph reasoning. Furthermore, as shown in ($3^{rd}-5^{th}$ row), existing methods easily generate piecemeal segmentation patch while our methods segment the complete instrument, which reveals the better feature alignment of our method.

Ablation Study. Furthermore, we conduct the ablation study to demonstrate the effectiveness of each component in our SePIG, as shown in Table 1 ($8^{th}-10^{th}$ rows). Comparing with the 'w/o EM-Grouping' case, i.e., the prototypes are replaced by category centers, SePIG obtains 2.34 % mIoU gain, revealing that the multi-prototypes can explore the complex semantics and fully represent the diverse feature. Then, comparing with 'w/o EM-Grouping' case and 'w/o PI-Graph' case, the performance for three instruments degrades with 1.79%, 3.08% and 1.86% IoU, which quantitatively demonstrates that the category-level information interaction can model the global relation and promote the segmentation accuracy. In contrast with the last row of Table 1, our model promotes the result with 3.04% mIoU, demonstrating the effectiveness in feature alignment for PC-Graph. Overall, the proposed SePIG framework efficiently models the category-level relationship and adequately align the feature distribution.

4 Conclusions

In this work, we propose the SePIG framework to address the shortcomings of UDA in surgical instrument segmentation by introducing graph-based techniques. The PI-Graph with EM-Grouping module efficiently discover the inter-category relationship inner each domain and enhance the feature map from a global view. Then, the PC-Graph dynamically communicate the relative information via hierarchical structure and narrow the domain gap. Experiments demonstrate that SePIG outperforms the state-of-the-art methods with a large margin.

Acknowledgement. This work was supported by National Natural Science Foundation of China (62001410), Hong Kong Research Grants Council (RGC) Early Career Scheme grant 21207420 (CityU 9048179) and Hong Kong RGC Collaborative Research Fund grant C4063-18G (CityU 8739029).

References

1. Allan, M., et al.: 2018 robotic scene segmentation challenge. arXiv preprint arXiv:2001.11190 (2020)
2. Allan, M., et al.: 2017 robotic instrument segmentation challenge. arXiv preprint arXiv:1902.06426 (2019)
3. Chen, Y.C., Lin, Y.Y., Yang, M.H., Huang, J.B.: CrDoCo: pixel-level domain transfer with cross-domain consistency. In: CVPR, pp. 1791–1800 (2019)
4. Dou, Q., Castro, D.C., Kamnitsas, K., Glocker, B.: Domain generalization via model-agnostic learning of semantic features. arXiv preprint arXiv:1910.13580 (2019)
5. González, C., Bravo-Sánchez, L., Arbelaez, P.: ISINet: an instance-based approach for surgical instrument segmentation. In: Martel, A.L., et al. (eds.) MICCAI 2020. LNCS, vol. 12263, pp. 595–605. Springer, Cham (2020). https://doi.org/10.1007/978-3-030-59716-0_57
6. Hao, F., He, F., Cheng, J., Wang, L., Cao, J., Tao, D.: Collect and select: semantic alignment metric learning for few-shot learning. In: ICCV, pp. 8460–8469 (2019)
7. Hoffman, J., Wang, D., Yu, F., Darrell, T.: FCNs in the wild: pixel-level adversarial and constraint-based adaptation. arXiv preprint arXiv:1612.02649 (2016)
8. Islam, M., Li, Y., Ren, H.: Learning where to look while tracking instruments in robot-assisted surgery. In: Shen, D., et al. (eds.) MICCAI 2019. LNCS, vol. 11768, pp. 412–420. Springer, Cham (2019). https://doi.org/10.1007/978-3-030-32254-0_46

9. Kipf, T.N., Welling, M.: Semi-supervised classification with graph convolutional networks. arXiv preprint arXiv:1609.02907 (2016)
10. Kurmann, T., et al.: Simultaneous recognition and pose estimation of instruments in minimally invasive surgery. In: Descoteaux, M., Maier-Hein, L., Franz, A., Jannin, P., Collins, D.L., Duchesne, S. (eds.) MICCAI 2017. LNCS, vol. 10434, pp. 505–513. Springer, Cham (2017). https://doi.org/10.1007/978-3-319-66185-8_57
11. Laina, I., et al.: Concurrent segmentation and localization for tracking of surgical instruments. In: Descoteaux, M., Maier-Hein, L., Franz, A., Jannin, P., Collins, D.L., Duchesne, S. (eds.) MICCAI 2017. LNCS, vol. 10434, pp. 664–672. Springer, Cham (2017). https://doi.org/10.1007/978-3-319-66185-8_75
12. Luo, Y., Zheng, L., Guan, T., Yu, J., Yang, Y.: Taking a closer look at domain shift: category-level adversaries for semantics consistent domain adaptation. In: CVPR, pp. 2507–2516 (2019)
13. Milletari, F., Rieke, N., Baust, M., Esposito, M., Navab, N.: CFCM: segmentation via coarse to fine context memory. In: Frangi, A.F., Schnabel, J.A., Davatzikos, C., Alberola-López, C., Fichtinger, G. (eds.) MICCAI 2018. LNCS, vol. 11073, pp. 667–674. Springer, Cham (2018). https://doi.org/10.1007/978-3-030-00937-3_76
14. Moon, T.K.: The expectation-maximization algorithm. IEEE Signal Process. Mag. **13**(6), 47–60 (1996)
15. Pfeiffer, M., et al.: Generating large labeled data sets for laparoscopic image processing tasks using unpaired image-to-image translation. In: Shen, D., et al. (eds.) MICCAI 2019. LNCS, vol. 11768, pp. 119–127. Springer, Cham (2019). https://doi.org/10.1007/978-3-030-32254-0_14
16. Sahu, M., Strömsdörfer, R., Mukhopadhyay, A., Zachow, S.: Endo-Sim2Real: consistency learning-based domain adaptation for instrument segmentation. In: Martel, A.L., et al. (eds.) MICCAI 2020. LNCS, vol. 12263, pp. 784–794. Springer, Cham (2020). https://doi.org/10.1007/978-3-030-59716-0_75
17. Tsai, Y.H., Hung, W.C., Schulter, S., Sohn, K., Yang, M.H., Chandraker, M.: Learning to adapt structured output space for semantic segmentation. In: CVPR, pp. 7472–7481 (2018)
18. Wang, K., Liew, J.H., Zou, Y., Zhou, D., Feng, J.: PANet: few-shot image semantic segmentation with prototype alignment. In: ICCV, pp. 9197–9206 (2019)
19. Wang, Z., et al.: Differential treatment for stuff and things: a simple unsupervised domain adaptation method for semantic segmentation. In: CVPR, pp. 12635–12644 (2020)
20. Yan, W., et al.: The domain shift problem of medical image segmentation and vendor-adaptation by Unet-GAN. In: Shen, D., et al. (eds.) MICCAI 2019. LNCS, vol. 11765, pp. 623–631. Springer, Cham (2019). https://doi.org/10.1007/978-3-030-32245-8_69
21. Yang, Y., Soatto, S.: FDA: fourier domain adaptation for semantic segmentation. In: CVPR, pp. 4085–4095 (2020)
22. Zhang, F., et al.: ACFNet: attentional class feature network for semantic segmentation. In: ICCV, pp. 6798–6807 (2019)
23. Zhou, W., Wang, Y., Chu, J., Yang, J., Bai, X., Xu, Y.: Affinity space adaptation for semantic segmentation across domains. IEEE Trans. Image Process. **30**, 2549–2561 (2021)
24. Zou, Y., Yu, Z., Kumar, B., Wang, J.: Unsupervised domain adaptation for semantic segmentation via class-balanced self-training. In: ECCV, pp. 289–305 (2018)

Unsupervised Domain Adaptation for Small Bowel Segmentation Using Disentangled Representation

Seung Yeon Shin[(✉)], Sungwon Lee, and Ronald M. Summers

Imaging Biomarkers and Computer-Aided Diagnosis Laboratory,
Radiology and Imaging Sciences, Clinical Center, National Institutes of Health,
Bethesda, MD, USA
{seungyeon.shin,rms}@nih.gov

Abstract. We present a novel unsupervised domain adaptation method for small bowel segmentation based on feature disentanglement. To make the domain adaptation more controllable, we disentangle intensity and non-intensity features within a unique two-stream auto-encoding architecture, and selectively adapt the non-intensity features that are believed to be more transferable across domains. The segmentation prediction is performed by aggregating the disentangled features. We evaluated our method using intravenous contrast-enhanced abdominal CT scans with and without oral contrast, which are used as source and target domains, respectively. The proposed method showed clear improvements in terms of three different metrics compared to other domain adaptation methods that are without the feature disentanglement. The method brings small bowel segmentation closer to clinical application.

Keywords: Small bowel segmentation · Unsupervised domain adaptation · Feature disentanglement · Abdominal computed tomography

1 Introduction

The small bowel is a part of the gastrointestinal tract between the stomach and the large bowel. It ranges from 20 to 30 ft long but is highly convoluted so that it can fit into the abdominal cavity [1]. Apart from its simple tubular structure, it has variable configuration while surrounded by visceral fat and other organs including the large bowel. Also, the appearance may differ locally according to the internal material, e.g., gas and fluid.

Computed tomography (CT) has been considered the first-line imaging modality for the evaluation of small bowel diseases since it is fast and non-invasive compared to other imaging tests such as endoscopy while providing

Electronic supplementary material The online version of this chapter (https://doi.org/10.1007/978-3-030-87199-4_27) contains supplementary material, which is available to authorized users.

M. de Bruijne et al. (Eds.): MICCAI 2021, LNCS 12903, pp. 282–292, 2021.
https://doi.org/10.1007/978-3-030-87199-4_27

essential diagnostic information [11]. Despite the clinical benefit, the interpretation itself is laborious and time-consuming. Automatic segmentation of the small bowel could expedite the interpretation. Specifically, it may help precise localization of diseases, such as inflammatory bowel disease and carcinoid tumors, and preoperative planning by better visualization.

Over the years, there have been attempts to develop automatic methods for small bowel segmentation, especially using deep learning. The small bowel was included in segmenting multiple organs-at-risk for radiotherapy treatment planning of affected tissues, such as pancreatic and cervical cancers, in CT scans [8,9,13]. Although the results obtained for the small bowel are reasonable, some of their data included only the part of the small bowel that is closest to the target area, which needed to be dose-evaluated [8,9]. In [13], the rough bowel location was detected instead of performing pixel-accurate small bowel segmentation. There have been only a few previous works dedicated solely to automatic small bowel segmentation [12,14,21]. While the specific anatomic relationship between the mesenteric vasculature and the small bowel is used to guide the small bowel segmentation in [21], a cylindrical shape constraint is applied during training of the small bowel segmenter in [14]. Although each of the works showed reasonable performance for particular datasets, their generalizability across different datasets was not evaluated. A CT scan is acquired using a specific imaging protocol depending on the purpose of the investigation, which includes the use of different contrast media and scan timing. Thus, the appearance of the small bowel may be different across datasets as exemplified in Fig. 1. It is observed in our experiment that, when trained on one dataset, the model does not generalize well to another dataset due to the domain shift (Sect. 3).

Domain adaptation is a task to address the domain shift problem and has been gaining attention in various fields. In this work, an unsupervised domain adaptation scenario, where ground-truth (GT) labels are not available for the target domain, is considered. Recent unsupervised domain adaptation methods can be categorized into two groups according to which specific space is to be aligned between different domains: 1) input data space [20,22] and 2) output space [3,17,18]. In the first group works, image-to-image translation [22] is used to translate images from target domains to a source domain. Then, the translated images can be tested using a source domain model. Cross-modality liver segmentation was performed by translating between CT and MRI images in [7,20]. On the other hand, adversarial learning is used to encourage the output prediction of the target domain to be similar to the source ones in the second group works. This adaptation can be applied in the feature level [3], multiple output levels [17], or multiple kinds of outputs [18]. Our method falls into the second group. Domain adaptation is even more important for small bowel segmentation since it is very hard to achieve GT labels for multiple datasets due to the high difficulty of the labeling. In practice, relatively small numbers (ten or less) of annotated CT scans were used in recent works [12,14]. It would be beneficial if it is possible to adapt a network to the target dataset without the use of labels.

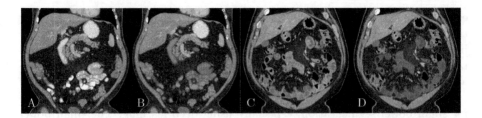

Fig. 1. Example CT scans (A) with and (C) without oral contrast. The respective ground-truth segmentation of the small bowel is shown as red in (B) and (D).

In this paper, we present a novel unsupervised domain adaptation method for small bowel segmentation, which is based on feature disentanglement. Although the absolute intensity values in CT scans (Hounsfield units) carry important information on specific substances of the human body, thus could provide a clue of being specific tissues and organs, they may be variable according to the imaging protocol. Figure 1 shows example CT scans that were acquired with and without oral contrast administration. The absolute intensity values are no longer a strong clue for the small bowel when we train and test across the datasets. Non-intensity features like texture and shape may be more useful. For example, local textures of the valvulae conniventes, which are circular folds on the inner surface of the small bowel, are more recurrent across the datasets.

Disentangling feature representations into desired factors provides not only an understanding of a deep network, but also more controllability on it. However, to achieve it in an unsupervised manner, either some prior assumptions or proper modification on the network architecture is required [2,15]. For example, when decomposing a face image into appearance and deformation components, a smoothness constraint is applied on the inferred deformation in [15]. In our method, feature disentanglement is performed using a unique auto-encoding architecture paired with augmented input images, without any prior assumption.

From the observation that the non-intensity features would be more transferable across the datasets than the intensity features, we first disentangle them within the proposed auto-encoding architecture. Then, only the non-intensity features are guided to be domain-invariant using adversarial learning [6]. Finally, segmentation prediction is performed by aggregating the disentangled features. To the best of our knowledge, this is the first work to develop an unsupervised domain adaptation method for small bowel segmentation. The proposed adaptation method based on the feature disentanglement further increased the adaptability of the segmenter, resulting in clear improvements for all evaluation metrics compared to the alternative methods.

2   Method

2.1   Dataset

Our dataset consists of intravenous contrast-enhanced abdominal CT scans which were done during the portal venous phase. It is composed of two subsets depending on whether or not oral contrast was used. The first subset includes 20 scans that were acquired with oral administration of Gastrografin, which is used as the source domain dataset. Meanwhile, the second subset, which includes 70 scans acquired without any oral contrast, is used as the target domain dataset. We resampled all the volumes to have isotropic voxels of $2\,mm^3$. The images were cropped manually to include from the diaphragm through the pelvis.

GT labels were achieved using "Segment Editor" module in 3DSlicer [5] by a radiologist with 12 years of experience. The GT segmentation includes the duodenum, jejunum, and ileum while not including any mesenteric fat, vessels, colon, and abdominal wall. We note that this annotation took several hours for each scan. We acquired GT segmentations for all 20 scans of the source domain dataset, and for 10 scans of the target domain dataset. All the annotated 10 scans are used as the test set, and the remaining 60 scans are used as unsupervised training samples. We note that this number of GTs is bigger than that of the previous works [12,14], which was ten or less.

2.2 Unsupervised Disentangling of Intensity and Non-intensity Representations

Figure 2 shows the proposed network composed of a sub-network for feature disentanglement and an additional decoder for segmentation prediction based on the disentangled features. The sub-network for feature disentanglement has a two-stream auto-encoding architecture, where the intensity and non-intensity features are first extracted through the separate encoders E_I and E_{NI}, respectively, and then combined in the decoder G_R to reconstruct the input image.

All convolution and pooling layers in the intensity encoder have $1 \times 1 \times 1$ kernels to constrain it to see each voxel independently and learn only the intensity information. While the intensity encoder takes the original image X as input, the non-intensity encoder uses as input the gradient images of X, $\nabla X = \{\frac{\partial X}{\partial x}, \frac{\partial X}{\partial y}, \frac{\partial X}{\partial z}\}$ as shown in Fig. 2. As a simple operation, the gradient still maintains the non-intensity information like texture and shape by keeping the relative values of neighboring voxels while losing the absolute values in a CT scan. Thus, the non-intensity encoder is guided to learn the non-intensity features.

The extracted features from both encoders are finally combined to reconstruct the original image in the reconstruction decoder. To prevent this reconstruction from being done solely from the intensity feature, the following are applied: 1) While the lower level features of the non-intensity encoder are used by skip connections, only the innermost features from the intensity encoder are used for reconstruction. 2) Dropout is applied to the intensity features before concatenation with the non-intensity features. We applied the per-element dropout based on empirical evaluation. The loss function for reconstruction is as follows:

$$L_{recon}(X) = \frac{1}{|X|} \sum_i |x_i - \hat{x}_i|, \tag{1}$$

where $\hat{X} = \{\hat{x}_i\}_{i=1}^{h \times w \times l}$ is the reconstruction from the input X of size $h \times w \times l$.

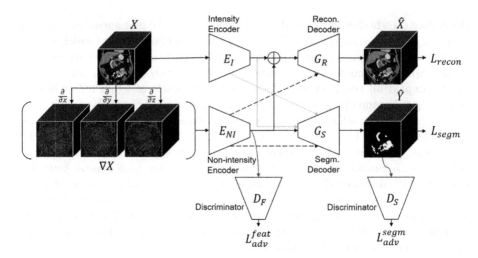

Fig. 2. Network architecture for the proposed method. The network is composed of two encoders E_I, E_{NI}, and two decoders G_R, G_S. The two encoders separately extract the intensity and non-intensity features through the specific network design and augmented input. The two decoders reconstruct the input volume and predict the segmentation, respectively. ∇X represents the gradient images of X, and the concatenation of the gradient along each axis is fed into the non-intensity encoder. The dotted lines represent skip connections, and $\oplus$ means feature concatenation. The gray lines are optional connections, which can be disconnected in some mode. The two discriminator networks D_F, D_S, and four loss functions L_{recon}, L_{segm}, L_{adv}^{feat}, L_{adv}^{segm} involved in training the proposed network are also shown. Refer to the text for the details.

2.3 Unsupervised Domain Adaptation Using Disentangled Representations

The key idea of this work is to perform the domain adaptation on the disentangled representation, but it is also done in the output level.

Feature Level Adaptation. Adversarial learning is applied to the disentangled non-intensity feature to make it domain-invariant while leaving the intensity feature as it is. The involved fully convolutional discriminator D_F takes the innermost features $E_{NI}(\nabla X)$ as input, and outputs $P = D_F(E_{NI}(\nabla X)) = \{p_j\}_{j=1}^{h' \times w' \times l'}$, where the value p_j represents the probability that the image X is drawn from the source domain for the corresponding position j in the feature map. The cross-entropy loss L_{disc}^{feat} for training D_F is defined as:

$$L_{disc}^{feat}(X) = -\frac{1}{|P|} \sum_j \left((1-z) \log p_j + z \log(1 - p_j) \right),$$ (2)

where $z = 0$ or 1 for the source and target domain images, respectively.

While training the discriminator D_F using images from both domains, an adversarial loss is computed for target domain images X^{tar}. Given the discriminator output $P^{tar} = \{p_j^{tar}\}_{j=1}^{h' \times w' \times l'}$, the adversarial loss is defined as:

$$L_{adv}^{feat}(X^{tar}) = -\frac{1}{|P^{tar}|} \sum_j \log p_j^{tar}. \tag{3}$$

The non-intensity encoder E_{NI} is encouraged to learn domain-invariant features in order to fool the discriminator D_F during training.

Output Level Adaptation. The segmentation prediction is performed in an additional decoder G_S by aggregating the disentangled features. Being less transferable across domains, the intensity features are optionally used for segmentation prediction, which is implemented by optional connections as shown in Fig. 2. The effect of this will be evaluated in Sect. 3. We used the generalized Dice loss [16] for L_{segm} to train the segmentation decoder. The calculation of this supervised loss is possible only for source domain images, meaning that the segmentation decoder G_S is fitted by only source domain images.

To adapt the segmentation decoder, another adversarial learning is applied to the segmentation output $\hat{Y}$. The loss L_{disc}^{segm} for training the output level discriminator D_S, and the adversarial loss L_{adv}^{segm} are defined similarly to ones for feature level adaptation but for the segmentation prediction $\hat{Y}$. This encourages the output prediction of the target domain to be similar to the source ones.

Objective Function for Domain Adaptation. Finally, the overall loss function for training the proposed network is:

$$L(X^{src}, X^{tar}) = L_{segm}(X^{src}) + \lambda_{recon} L_{recon}(X^{src}, X^{tar})$$
$$+ \lambda_{adv}^{feat} L_{adv}^{feat}(X^{tar}) + \lambda_{adv}^{segm} L_{adv}^{segm}(X^{tar}), \tag{4}$$

where λ_{recon}, λ_{adv}^{feat}, and λ_{adv}^{segm} are the weight for each loss.

2.4 Evaluation Details

In the proposed network, the structure of each encoder and decoder is based on that of the 3D U-Net [4], but has a smaller number of channels, which are {32, 64, 128, 256}. All convolution layers have $3 \times 3 \times 3$ kernels excepting ones in the intensity encoder and the final inference layer in the decoders, which have $1 \times 1 \times 1$ kernels. Group normalization [19] is used between each convolution layer and non-linearity function. The feature level discriminator network D_F consists of three convolution layers with $3 \times 3 \times 3$ kernels and one final convolution layer with $1 \times 1 \times 1$ kernels, where the numbers of channels are 32, 64, 128, 1, respectively. A pooling layer is added between each convolution layer. The output level discriminator D_S is with a similar structure, but has one more convolution layer, where the numbers of channels are 32, 64, 128, 256, 1.

We implemented all the networks, including ones for the comparable methods, using PyTorch 1.2. For training, we used AdamW optimizers [10] and a weight decay of 5×10^{-4}. Based on the grid search, the learning rates of 10^{-4} and 10^{-6} were used for the encoders and decoders, and for the discriminators, respectively. We used 0.2, 10^{-6}, 10^{-4} for λ_{recon}, λ_{adv}^{feat}, λ_{adv}^{segm}, respectively. Dropout with a probability of 0.3 is applied to the intensity features before concatenation with the non-intensity features as described in Sect. 2.2.

We used a NVIDIA Tesla V100 32GB GPU to conduct experiments. To fit in the memory, sub-volumes of size $112 \times 112 \times 112$ sampled from the original volume are used during training. The mini-batch size was set as 1. We applied image rotation, elastic transformation, random global intensity adjustment for data augmentation. Dice coefficient, 95% Hausdorff distance (HD95), and average symmetric surface distance (ASD) are used as evaluation metrics. Also, paired t-tests are conducted to show the statistical significance of the proposed method.

3   Results

3.1   Quantitative Evaluation

Table 1 shows quantitative results of the proposed method and other unsupervised domain adaptation methods on the target domain dataset. We first provide the segmentation performance without any domain adaptation, which are: 1) performing a 5-fold cross validation within the target domain test set (tar 5-fold CV), 2) applying a model trained from the source domain to the target domain images (w/o DA), and 3) the same with 2) but using the gradient images instead of the original image as input. The use of the gradient images slightly improved the generalizability by preventing the network from seeing absolute intensity values in CT scans. The cross validation within the target domain test set was expected to perform better than other unsupervised domain adaptation methods since it is fully supervised. However, it is worse than most of the presented domain adaptation methods. When only a small amount of labels are accessible in the target domain, rather than attempting to train a 'strictly' supervised model, it may be better to adapt a model from a different domain to the target domain using relatively abundant unlabeled target domain images.

For the other domain adaptation methods, while the same network architecture based on the 3D U-Net [4] is used, adaptation is performed in a different way, i.e., either in the feature level or in the output level, or both. The effect of using the gradient images is evaluated again for the method that is with both feature and output level adaptation (feat. & out. level DA w/ grad.), showing a worse result compared to using the original image. It corresponds to the proposed network without the intensity encoder and reconstruction decoder, which play an important role for feature disentanglement. The proposed method, where domain adaptation is performed on the disentangled representation, shows the best performance in terms of all metrics. We note that, when the segmentation prediction is performed without using the intensity features, our method has the

Table 1. Comparison with other methods on the target domain test set. The first three are without any domain adaptation (DA). The remaining are different domain adaptation methods including the proposed method and its variant. Refer to the text for the explanation on each method. For every metric, the mean and standard deviation are presented. P-values are computed by conducting paired t-tests between the proposed method and the others with the Dice coefficients.

Method	Dice	HD95 (mm)	ASD (mm)	p-value
tar 5-fold CV	0.809 ± 0.081	13.024 ± 5.829	2.998 ± 1.315	0.004
w/o DA	0.725 ± 0.141	15.944 ± 6.707	3.622 ± 1.657	1.380×10^{-5}
gradient input	0.749 ± 0.125	17.405 ± 10.093	4.178 ± 2.077	2.725×10^{-6}
feature level DA [3]	0.807 ± 0.084	13.467 ± 6.053	3.040 ± 1.094	1.006×10^{-5}
output level DA [17]	0.813 ± 0.106	15.355 ± 8.063	3.243 ± 1.864	0.016
multi output level DA [17]	0.816 ± 0.098	14.599 ± 11.562	3.121 ± 2.048	9.270×10^{-5}
feat. & out. level DA	0.814 ± 0.100	12.487 ± 6.547	2.892 ± 1.480	0.006
feat. & out. level DA w/ grad.	0.811 ± 0.063	17.948 ± 9.994	3.527 ± 1.494	0.002
Ours	**0.837 ± 0.084**	**10.290 ± 8.057**	**2.388 ± 1.408**	–
Ours + int. feat.	0.837 ± 0.085	10.559 ± 6.702	2.536 ± 1.207	0.860
Ours w/o output level DA	0.829 ± 0.073	14.024 ± 7.830	2.940 ± 1.059	0.137
Ours w/o feature level DA	0.820 ± 0.083	14.745 ± 8.760	3.085 ± 1.375	0.006

same model complexity with the others since the intensity encoder and reconstruction decoder are no more involved in inference time. The p-values computed by conducting paired t-tests between the proposed method and the others with the Dice coefficients show the statistical significance of the proposed method.

As mentioned, segmentation prediction is performed by aggregating the disentangled features in our network. The variant using the intensity features as well as the non-intensity features (Ours + int. feat.), through the optional connection in Fig. 2, does not improve the performance. It implies that the disentangled non-intensity features are enough to perform the cross-domain small bowel segmentation. To validate the need of feature and output level adaptation in the proposed method, the variants without either of them are also evaluated. While both contribute to achieving the best performance, the feature level adaptation that is applied directly on the disentangled non-intensity features is more important to guide the network to operate as we expect.

3.2 Qualitative Evaluation

Figure 3 shows example segmentation results in 3D. The result corresponding to 'feat. & out. level DA' in Table 1 is compared to ours. We note that the only difference between them is whether the feature disentanglement is involved for the domain adaptation, thus could show its effectiveness. Fewer errors are observed for the proposed method. We believe this is because the proposed method explicitly concentrates on the features more transferable across the datasets, the non-intensity features in this work, by disentangling those features and applying adversarial learning directly to them during the adaptation process. Example

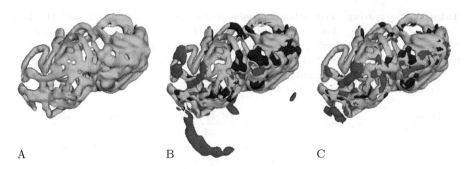

Fig. 3. Example segmentation results in 3D. (A) Ground-truth segmentation. (B) Result corresponding to 'feat. & out. level DA' in Table 1. (C) Result of the proposed method. In (B) and (C), each result is compared with the ground-truth, and false positives and false negatives are marked as red and blue, respectively. (Color figure online)

reconstruction results from the auto-encoding architecture as well as segmentation results in coronal view can be found in supplementary material.

4   Conclusion

We have presented a novel unsupervised domain adaptation method for small bowel segmentation. To increase the adaptability of a trained segmenter across different domains, we disentangle the feature representation into desired factors in an unsupervised way, and concentrate on adapting more transferable features among them. Finally, segmentation prediction is performed by aggregating the disentangled features. We evaluated our method using abdominal CT scans with and without oral contrast as the source and target domains, respectively. The experimental results showed clear improvement compared to other domain adaptation methods that are without the feature disentanglement. Considering the difficulty of labeling the small bowel, the obtained result is encouraging since the proposed method can adapt a model without using any target domain labels.

Acknowledgments. We thank Dr. James Gulley for patient referral and for providing access to CT scans. This research was supported by the National Institutes of Health, Clinical Center.

References

1. Small bowel obstruction (2019). https://my.clevelandclinic.org/health/diseases/ 15850-small-bowel-obstruction
2. Berga, D., Masana, M., de Weijer, J.V.: Disentanglement of color and shape representations for continual learning (2020)
3. Chen, Y., Chen, W., Chen, Y., Tsai, B., Wang, Y.F., Sun, M.: No more discrimination: cross city adaptation of road scene segmenters. In: IEEE International Conference on Computer Vision, pp. 2011–2020 (2017)

4. Çiçek, Ö., Abdulkadir, A., Lienkamp, S.S., Brox, T., Ronneberger, O.: 3D U-Net: learning dense volumetric segmentation from sparse annotation. In: Ourselin, S., Joskowicz, L., Sabuncu, M.R., Unal, G., Wells, W. (eds.) MICCAI 2016. LNCS, vol. 9901, pp. 424–432. Springer, Cham (2016). https://doi.org/10.1007/978-3-319-46723-8_49

5. Fedorov, A., et al.: 3D slicer as an image computing platform for the quantitative imaging network. Magn. Reson. Imaging **30**(9), 1323–1341 (2012). https://doi.org/10.1016/j.mri.2012.05.001

6. Goodfellow, I., et al.: Generative adversarial nets. In: Advances in Neural Information Processing Systems, vol. 27, pp. 2672–2680 (2014)

7. Jiang, J., Veeraraghavan, H.: Unified cross-modality feature disentangler for unsupervised multi-domain MRI abdomen organs segmentation. In: Martel, A.L., et al. (eds.) MICCAI 2020. LNCS, vol. 12262, pp. 347–358. Springer, Cham (2020). https://doi.org/10.1007/978-3-030-59713-9_34

8. Ju, Z., et al.: Automatic segmentation of pelvic organs-at-risk using a fusion network model based on limited training samples. Acta Oncol. **59**(8), 933–939 (2020). https://doi.org/10.1080/0284186X.2020.1775290

9. Liu, Y., et al.: CT-based multi-organ segmentation using a 3D self-attention u-net network for pancreatic radiotherapy. Med. Phys. **47**(9), 4316–4324 (2020). https://doi.org/10.1002/mp.14386

10. Loshchilov, I., Hutter, F.: Decoupled weight decay regularization. In: International Conference on Learning Representations (2019)

11. Murphy, K.P., McLaughlin, P.D., O'Connor, O.J., Maher, M.M.: Imaging the small bowel. Curr. Opin. Gastroenterol. **30**(2), 134–140 (2014)

12. Oda, H., et al.: Visualizing intestines for diagnostic assistance of ileus based on intestinal region segmentation from 3D CT images. arXiv preprint arXiv:2003.01290 (2020)

13. Sartor, H., et al.: Auto-segmentations by convolutional neural network in cervical and anorectal cancer with clinical structure sets as the ground truth. Clin. Transl. Radiat. Oncol. **25**, 37–45 (2020). https://doi.org/10.1016/j.ctro.2020.09.004

14. Shin, S.Y., Lee, S., Elton, D., Gulley, J.L., Summers, R.M.: Deep small bowel segmentation with cylindrical topological constraints. In: Martel, A.L., et al. (eds.) MICCAI 2020. LNCS, vol. 12264, pp. 207–215. Springer, Cham (2020). https://doi.org/10.1007/978-3-030-59719-1_21

15. Shu, Z., Sahasrabudhe, M., Alp Güler, R., Samaras, D., Paragios, N., Kokkinos, I.: Deforming autoencoders: unsupervised disentangling of shape and appearance. In: Ferrari, V., Hebert, M., Sminchisescu, C., Weiss, Y. (eds.) ECCV 2018. LNCS, vol. 11214, pp. 664–680. Springer, Cham (2018). https://doi.org/10.1007/978-3-030-01249-6_40

16. Sudre, C.H., Li, W., Vercauteren, T., Ourselin, S., Jorge Cardoso, M.: Generalised dice overlap as a deep learning loss function for highly unbalanced segmentations. In: Cardoso, M.J., et al. (eds.) DLMIA/ML-CDS -2017. LNCS, vol. 10553, pp. 240–248. Springer, Cham (2017). https://doi.org/10.1007/978-3-319-67558-9_28

17. Tsai, Y., Hung, W., Schulter, S., Sohn, K., Yang, M., Chandraker, M.: Learning to adapt structured output space for semantic segmentation. In: IEEE/CVF Conference on Computer Vision and Pattern Recognition, pp. 7472–7481 (2018)

18. Wang, S., Yu, L., Li, K., Yang, X., Fu, C.-W., Heng, P.-A.: Boundary and entropy-driven adversarial learning for fundus image segmentation. In: Shen, D., et al. (eds.) MICCAI 2019. LNCS, vol. 11764, pp. 102–110. Springer, Cham (2019). https://doi.org/10.1007/978-3-030-32239-7_12

19. Wu, Y., He, K.: Group normalization. arXiv preprint arXiv:1803.08494 (2018)
20. Yang, J., Dvornek, N.C., Zhang, F., Chapiro, J., Lin, M.D., Duncan, J.S.: Unsupervised domain adaptation via disentangled representations: application to cross-modality liver segmentation. In: Shen, D., et al. (eds.) MICCAI 2019. LNCS, vol. 11765, pp. 255–263. Springer, Cham (2019). https://doi.org/10.1007/978-3-030-32245-8_29
21. Zhang, W., et al.: Mesenteric vasculature-guided small bowel segmentation on 3-D CT. IEEE Trans. Med. Imaging **32**(11), 2006–2021 (2013). https://doi.org/10.1109/TMI.2013.2271487
22. Zhu, J., Park, T., Isola, P., Efros, A.A.: Unpaired image-to-image translation using cycle-consistent adversarial networks. In: IEEE International Conference on Computer Vision, pp. 2242–2251 (2017)

Data-Driven Mapping Between Functional Connectomes Using Optimal Transport

Javid Dadashkarimi[1(✉)], Amin Karbasi[1,2], and Dustin Scheinost[3]

[1] Department of Computer Science, Yale University, New Haven, USA
{javid.dadashkarimi,amin.karbasi}@yale.edu
[2] Department of Electrical Engineering, Computer Science, Statistics
and Data Science, Yale University, New Haven, USA
[3] Department of Radiology and Biomedical Imaging, Yale School of Medicine,
New Haven, USA
dustin.scheinost@yale.edu

Abstract. Functional connectomes derived from functional magnetic resonance imaging have long been used to understand the functional organization of the brain. Nevertheless, a connectome is intrinsically linked to the atlas used to create it. In other words, a connectome generated from one atlas is different in scale and resolution compared to a connectome generated from another atlas. Being able to map connectomes and derived results between different atlases without additional pre-processing is a crucial step in improving interpretation and generalization between studies that use different atlases. Here, we use optimal transport, a powerful mathematical technique, to find an optimum mapping between two atlases. This mapping is then used to transform time series from one atlas to another in order to reconstruct a connectome. We validate our approach by comparing transformed connectomes against their "gold-standard" counterparts (*i.e.*, connectomes generated directly from an atlas) and demonstrate the utility of transformed connectomes by applying these connectomes to predictive models based on a different atlas. We show that these transformed connectomes are significantly similar to their "gold-standard" counterparts and maintain individual differences in brain-behavior associations, demonstrating both the validity of our approach and its utility in downstream analyses. Overall, our approach is a promising avenue to increase the generalization of connectome-based results across different atlases.

Keywords: Optimal transport · Functional connectome · fMRI

1 Introduction

Functional connectomics, using functional magnetic resonance imaging (fMRI), are a powerful approach for investigating the functional organization of the brain.

© Springer Nature Switzerland AG 2021
M. de Bruijne et al. (Eds.): MICCAI 2021, LNCS 12903, pp. 293–302, 2021.
https://doi.org/10.1007/978-3-030-87199-4_28

A prerequisite for creating a functional connectome—*i.e.*, a matrix describing the connectivity between any pair of brain regions—is defining an atlas to parcellate the brain into these regions. Given the popularity of this approach, many atlases, for which there is no gold standard, exist [2]. As these atlases divide the brain into a different number of regions, where each vary by size and topology, connectomes created from different atlases are not directly comparable. Thus, results and potential biomarkers generated from one atlas are not readily applicable to connectomes generated from a different atlas. To extend previous results to a connectome generated from a different atlas, additional preprocessing is needed, a barrier to replication and generalization efforts and limiting wider use of potential connectome-based biomarkers.

To overcome these limitations, we propose how to find an optimum mapping between two different atlases, allowing data processed from one atlas to be directly transformed into a connectome based on another atlas. First, in a training sample with time-series data from two different atlases, we find this mapping by solving the Monge–Kantorovich transportation problem [23]. Then, by employing this optimal mapping, time-series data based on the first atlas from novel subjects can be transformed into connectomes based on the second atlas without ever needing to use the second atlas. An overview of our approach is shown in Fig. 1. We validate our approach by comparing transformed connectomes against their "gold-standard" counterparts (*i.e.*, connectomes generated directly from an atlas) and demonstrate the utility of transformed connectomes by applying these connectomes to predictive models based on a different atlas. Overall, our results suggest that data from one atlas can be transformed into a connectome comparable to one generated directly from a different atlas.

2 Methods

2.1 Optimal Transport

The optimal transport problem solves how to transport resources from one location α to another β while minimizing the cost C to do so [12,15,19,26]. It has been used for contrast equalization [9], image matching [20], image watermarking [21], text classification [16], and music transportation [11]. OT is one of the few methods that provides a well-defined distance metric when the support of the distributions is different. Other mappings approaches such as KL divergence do not make this guarantee.

Monge Problem: The original formulation of the optimal transport problem is known as the Monge problem. Lets define some resources $x_1, .., x_n$ in α and some resources $y_1, .., y_m$ in β. Then, we specify weight vectors a and b over these resources and define matrix C as a measure of pairwise distances between points $x_i \in \alpha$ and comparable points $\mathcal{T}(x_i)$; Monge problem aims to solve the following optimizing problem [22]:

$$\min_{\mathcal{T}} \left\{ \sum_i C(x_i, \mathcal{T}(x_i)) : \mathcal{T}_\sharp \alpha = \beta \right\}, \tag{1}$$

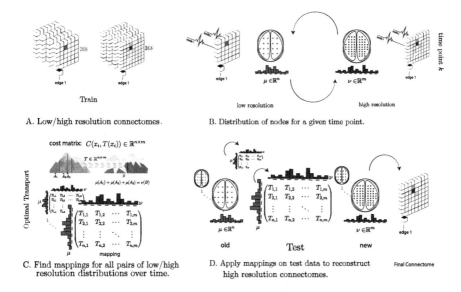

Fig. 1. Optimal transport pipeline to estimate a mapping between different atlases in order to transform connectomes between the atlases. A) Training data includes time series data from two atlases, which can be of different number of brain regions. B) Extracting the empirical distribution of brain activity for each node for both atlases at a given time point. C) Learning the optimal transport mapping between source and target distributions for a pre-defined cost matrix. D) Applying the mappings on testing data and then building transformed connectomes according to the new time series data.

where the push forward operator $\sharp$ indicates that mass from α moves towards β assuming that weights absorbed in $b_j = \sum_{T(x_i)=y_j} a_i$. Assignment problem when the number of elements in the measures are not equal is a special case of this problem, where each point in α can be assigned to several points in β.

Kantorovich Relaxation: As a generalization of the Monge problem, the Kantorvich relaxation solves the mass transportation problem using a probabilistic approach in which the amount of mass located at x_i potentially dispatches to several points in target [18]. Admissible solution for Kantorvich relaxation is defined by a coupling matrix $T \in \mathbb{R}_+^{n \times m}$ indicating the amount of mass being transferred from location x_i to y_j by $T_{i,j}$:

$$U(a,b) = \{T \in \mathbb{R}_+^{n \times m} : T\mathbb{1}_m = a, T^T\mathbb{1}_n = b\}, \tag{2}$$

for vectors of all 1 shown with $\mathbb{1}$. An optimum solution is obtained by solving the following problem for a given "ground metric" matrix $C \in \mathbb{R}^{n \times m}$ [24]:

$$L_c(a,b) = \min_{T \in U(a,b)} <C, T> = \sum_{i,j} C_{i,j} T_{i,j}. \tag{3}$$

which is a linear problem and is not guarantee to have a unique solution [23], but always there exists an optimal solution (see proof in [3,4]). Kantorovich and Monge problems could also be equivalent in some conditions (see proof in [6]).

2.2 Proposed Algorithm for Mapping Atlases Using Optimal Transport

Formulation: For paired time-series data from the same individual but from two different atlases (atlas $\mathscr{P}_n$ with n regions and atlas $\mathscr{P}_m$ with m regions), lets define $\mu_t \in \mathbb{R}^n$ and $\nu_t \in \mathbb{R}^m$ to be the distribution of brain activity at single time point t based on atlases $\mathscr{P}_n$ and $\mathscr{P}_m$, respectively. For a fixed cost matrix $C \in \mathbb{R}^{n \times m}$, we aim to find a mapping $\mathcal{T} \in \mathbb{R}^{n \times m}$ that minimizes transportation cost between μ_t and ν_t:

$$L_c(\mu_t, \nu_t) = \min_{\mathcal{T}} C^T \mathcal{T} \text{ s.t, } A\underline{\mathcal{T}} = \begin{bmatrix} \mu_t \\ \nu_t \end{bmatrix}, \tag{4}$$

in which $\underline{\mathcal{T}} \in \mathbb{R}^{nm}$ is vectorized version of $\mathcal{T}$ such that the $i + n(j-1)$'s element of $\mathcal{T}$ is equal to $\mathcal{T}_{ij}$ and A is defined as:

$$
A = \begin{matrix} & \overset{1}{} & \overset{2}{} & & \overset{n}{} \\ m & \begin{pmatrix} \begin{smallmatrix} 1 & 0 & \cdots & 0 \\ 0 & 1 & \cdots & 0 \\ \vdots & \vdots & \ddots & \vdots \\ 0 & 0 & \cdots & 1 \\ 1 & 1 & \cdots & 1 \end{smallmatrix} & \begin{smallmatrix} 1 & 0 & \cdots & 0 \\ 0 & 1 & \cdots & 0 \\ \vdots & \vdots & \ddots & \vdots \\ 0 & 0 & \cdots & 1 \\ 1 & 1 & \cdots & 1 \end{smallmatrix} & \cdots & \begin{smallmatrix} 1 & 0 & \cdots & 0 \\ 0 & 1 & \cdots & 0 \\ \vdots & \vdots & \ddots & \vdots \\ 0 & 0 & \cdots & 1 \\ \cdots(1 & 1 & \cdots & 1) \end{smallmatrix} \\ \vdots & & \\ (1 & 1 & \cdots & 1) & \cdots & & \cdots(1 & 1 & \cdots & 1) \end{pmatrix} \end{matrix} \tag{5}
$$

The mapping $\mathcal{T}$ represents the optimal way of transforming the brain activity data from n regions into m regions.

Yet, solving a large linear program is computationally hard [8]. Thus, we use the entropy regularization, which gives an approximation solution with complexity of $\mathcal{O}(n^2 \log(n) \eta^{-3})$ for $\epsilon = \frac{4 \log(n)}{\eta}$ [23], and instead solve the following:

$$L_c(\mu_t, \nu_t) = \min_{\mathcal{T}} C^T \mathcal{T} - \epsilon H(\mathcal{T}) \text{ s.t, } A\underline{\mathcal{T}} = \begin{bmatrix} \mu_t \\ \nu_t \end{bmatrix}. \tag{6}$$

Specifically, we use the Sinkhorn algorithm—an iterative solution for Eq. 6 [1]—to find the optimum mapping $\mathcal{T}$ as implemented in the Python Optimal Transport (POT) toolbox [10]. Initial investigations suggest that similar mappings are obtained using the exact linear programming solution or the Sinkhorn approximation. Similarly, results appear stable over a range of ϵ's.

Defining the Cost Matrix: We investigated two different cost matrices. First, we computed the pairwise Euclidean distance between every combination of brain regions between the two atlases by: *(i)* computing cluster centroids for all n regions in $\mathscr{P}_n$ and m regions in $\mathscr{P}_m$ and *(ii)* then, calculating the Euclidean distance between these nodes: $C_{\text{euc}}(p, q) = \sqrt{\sum_{i=1}^{3} (q_i - p_i)^2}$, where p_i and q_i

are coordinates for the arbitrary regions: p in $\mathscr{P}_n$ and q in $\mathscr{P}_m$. Second, we compute functional distance between regions by: (i) calculating the correlation (ρ) between the time series for all pairwise combinations of regions between $\mathscr{P}_n$ and $\mathscr{P}_m$, (ii) normalizing by $\rho_{\mathrm{norm}} = (\rho - \min(\rho))/(\max(\rho) - \min(\rho))$, (iii) converting to distance: $C_{\mathrm{func}} = \mathbb{1}_{n \times m} - \rho_{\mathrm{norm}}$, (iv) and average the cost matrix over participants s to get a more robust estimation for C_{func} (*i.e.*, $C = 1/s \sum_s C_s$).

Estimating Mapping: For training data with S participants and k time points per participant, first, we estimate the optimal mapping T_i, independently, for each time point and participant using Eq. 6. The distributions were normalized using min-max scaling and then dividing by sum. Next, we average all T_i over all participants to produce a single optimal mapping T for one time point in the training data (*e.g.*, $T = \frac{1}{|S|} \sum_{i=1}^{|S|} T_i$). For bigger frames, we use the mapping we learned at the beginning of a frame for the rest time points in the window (i.e., $T[t : t + w] \leftarrow T$, for a frame size of w and time point t).

Estimating Connectomes from Transformed Time Series: Once the n time series from $\mathscr{P}_n$ are transformed to m time series based on the target atlas $\mathscr{P}_m$, we correlate the time series for every pair of regions i and j to build the final, transformed connectomes.

3 Results

3.1 Datasets

To evaluate our approach, we used data from the Human Connectome Project (HCP) [27], starting with the minimally preprocessed data [14]. First, data with a maximum frame-to-frame displacement of 0.15 mm or greater were excluded, resulting in a sample of 876 resting-state scans. Analyses were restricted only to the LR phase encoding, which consisted of 1200 individual time points. Further preprocessing steps were performed using BioImage Suite [17]. These included regressing 24 motion parameters, regressing the mean white matter, CSF, and grey matter time series, removing the linear trend, and low-pass filtering. Regions were delineated according to the Shen 268 and 368 atlases [25]. These atlases, defined in an independent dataset, provide a parcellation of the whole gray matter (including subcortex and cerebellum) into 268 or 368 contiguous, functionally coherent regions. For each scan, the average time series within each region was obtained. To calculate connectomes, the Pearson's correlation between the mean time series of each pair of regions was calculated and converted to be approximately normally distributed using a Fisher transformation.

3.2 Intrinsic Evaluation

Correlation with "Gold-Standard" Connectomes: To validate our approach, we, first, partitioned our sample into 80% training data to estimate the

Table 1. Intrinsic evaluation of the transformed connectomes based on the optimal mapping $\mathcal{T}$. The transformed connectomes exhibited high correlation with the "gold-standard" connectomes for both (*top*) the Euclidean distance and (*bottom*) the functional distance cost matrices. Similarity between connectomes was not affected by sample size and number of time points used to estimate $\mathcal{T}_i$.

Train size (Euclidean Distance)												
		268 → 368						368 → 268					
		100	200	300	400	500	600	100	200	300	400	500	600
Frame size	100	0.489	0.495	0.491	0.498	0.496	0.494	0.461	0.454	0.460	0.445	0.458	0.461
	200	0.496	0.490	0.497	0.502	0.494	0.501	0.456	0.455	0.456	0.454	0.457	0.458
	300	0.500	0.500	0.503	0.499	0.500	0.495	0.447	0.454	0.451	0.454	0.453	0.458
	400	0.490	0.499	0.492	0.496	0.499	0.494	0.454	0.454	0.464	0.461	0.450	0.461
	500	0.491	0.500	0.492	0.496	0.491	0.499	0.461	0.459	0.466	0.457	0.458	0.448
	600	0.503	0.492	0.494	0.495	0.499	0.496	0.454	0.455	0.456	0.457	0.454	0.452
	700	0.491	0.508	0.492	0.500	0.493	0.498	0.455	0.454	0.465	0.462	0.459	0.457
	800	0.493	0.497	0.501	0.495	0.503	0.499	0.460	0.457	0.455	0.456	0.459	0.465
	900	0.505	0.505	0.486	0.498	0.492	0.491	0.451	0.456	0.460	0.461	0.462	0.458
	1000	0.502	0.492	0.489	0.502	0.496	0.503	0.452	0.457	0.469	0.450	0.461	0.452
	1100	0.499	0.496	0.498	0.497	0.503	0.485	0.453	0.460	0.455	0.460	0.453	0.464
Train size (Functional Distance)													
		268 → 368						368 → 268					
		100	200	300	400	500	600	100	200	300	400	500	600
Frame size	100	0.626	0.622	0.630	0.624	0.622	0.624	0.589	0.596	0.587	0.586	0.593	0.591
	200	0.621	0.630	0.623	0.625	0.632	0.621	0.592	0.597	0.590	0.601	0.584	0.591
	300	0.624	0.629	0.629	0.627	0.629	0.630	0.590	0.593	0.591	0.590	0.596	0.600
	400	0.626	0.628	0.631	0.626	0.631	0.625	0.595	0.587	0.590	0.595	0.596	0.590
	500	0.632	0.627	0.629	0.631	0.626	0.624	0.594	0.598	0.596	0.601	0.591	0.593
	600	0.633	0.630	0.631	0.631	0.629	0.631	0.596	0.597	0.595	0.600	0.594	0.593
	700	0.635	0.639	0.636	0.629	0.632	0.627	0.594	0.592	0.589	0.596	0.598	0.597
	800	0.628	0.634	0.631	0.634	0.634	0.634	0.598	0.596	0.590	0.592	0.593	0.601
	900	0.632	0.633	0.634	0.635	0.635	0.638	0.600	0.596	0.597	0.599	0.603	0.595
	1000	0.632	0.635	0.635	0.638	0.643	0.636	0.595	0.593	0.595	0.596	0.600	0.594
	1100	0.638	0.636	0.638	0.634	0.639	0.639	0.598	0.591	0.601	0.594	0.593	0.601

optimal mapping $\mathcal{T}$ between atlases and 20% testing data for evaluating the quality of the transformed connectomes. In the training data, we estimated $\mathcal{T}$ using all 1200 time points and 700 participants for each of the cost matrices (C_{euc} and C_{func}). Next, in the testing data, we applied $\mathcal{T}$ to construct 368×368 connectomes from the 268 atlas data (labeled: 268 → 368) as well as 268×268 connectomes from the 368 atlas data (labeled: 368 → 268). Finally, the transformed connectomes were compared to the "gold-standard" connectomes (*i.e.*, connectomes generated directly from an atlas) using correlation. Using a 12 core processor Intel Xeon Gold 6128 CPU with a 3.40 GHz clock speed, estimating $\mathcal{T}$ took 2,975 s.

For both cost matrices, significant correlations between the transformed connectomes and the "gold-standard" connectomes were observed (for C_{euc},

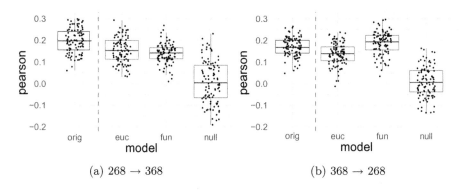

Fig. 2. Box plots for IQ prediction from transformed connectomes. Participants were randomly split into three groups (g_1, g_2, and g_3) with a respective ratio of $\{0.25, 0.5, 0.25\}$. A mapping is trained on g_1, the model is trained on connectomes from g_2, and tested on transformed connectomes from g_3. *orig* shows prediction performance in "gold-standard" connectomes, *euc* and *fun* show prediction performance for transformed connectomes found using either C_{euc} or C_{func}, respectively. A null model (labeled *null*) is obtained by permuting labels.

$268 \rightarrow 368$: $r = 0.508, p < 0.01$; $368 \rightarrow 268$: $r = 0.469, p < 0.01$; for C_{func}, $268 \rightarrow 368$: $r = 0.643, p < 0.01$; $368 \rightarrow 268$: $r = 0.603, p < 0.01$). Notably, transformed connectomes using the C_{func} cost matrix were significantly ($p < 0.01$) more similar to the "gold-standard" connectomes compared to transformed connectomes using the C_{euc} cost matrix. Finally, the $268 \rightarrow 368$ connectomes were more similar to the "gold-standard" connectomes compared to the $368 \rightarrow 268$ connectomes.

Evaluation of Free Parameters: Next, we investigated the sensitivity of our approach to the number of time points and number of participants used to find the mapping between atlases. Using the 80/20 split for training and testing, we varied the number of time points used from 100 to 1100 in 100 increments and varied the number of participants from 100 to 600 in 100 increments. No clear pattern of performance change was observed across either parameter, suggesting that our approach is stable to both the number of frames and participants (Table 1). However, using only 100 participants and 100 time points in a frame significant ($p < 0.05$) reduced the processing time from 2,975 s to 467 s.

3.3   Extrinsic Evaluation

In addition to validating our approach, we demonstrated that the transformed connectomes can be used to elucidate brain-behavior associations. To this aim, **1)** We partitioned our data into three folds g_1, g_2, and g_3 with a respective ratio of $\{0.25, 0.5, 0.25\}$. **2)** Using only participants in g_1, we estimated the optimal mapping $\mathcal{T}$ for both cost matrices. **3)** We applied $\mathcal{T}$ to the participants in g_3 to produce the transformed connectomes ($268 \rightarrow 368$ and $368 \rightarrow 268$). **4)** We

Table 2. Extrinsic evaluation of the transformed connectomes based on the optimal mapping $\mathcal{T}$. Indicator * shows the significance of the results with respect to the null model for $p < 0.05$ using corrected resampled t-tests [5].

	Sex			IQ		
	Null	Euc	Func	Null	Euc	Func
$268 \rightarrow 368$	0.5033	0.6961*	0.7253*	0.0083	0.1553*	0.1376*
$368 \rightarrow 268$	0.5077	0.7312*	0.7243*	0.0036	0.1313*	0.1835*

predicted IQ using ridge regression [13] and classified sex using support vector machine (SVM) with a linear kernel [7] using the connectomes in g_2 for both the 268 and 368, independently. All models were trained with 10-fold cross-validation. 5) We used the predictive models from Step 4) to predict phenotypic information using the transformed matrices from Step 3) (*e.g.*, using the 268 $\rightarrow$ 368 connectomes as inputs to the models trained with the 368 connectomes). We tested the significance of predictions based on the transformed connectomes against a null distribution of prediction based on permuted values using corrected resampled t-tests [5].

Results showed that using transformed connectomes from both cost matrices and both directions (*e.g.*, 368 $\rightarrow$ 268) lead to significantly ($p < 0.05$) better prediction of IQ compared to the null model (see Table 2, Fig. 2a). Similarly, results showed that sex classification achieves up to 72% accuracy and is significantly higher compared to the null distribution for all transformed connectomes (see Table 2). There were no clear differences in prediction performance between the transformed connectomes from either cost matrix or direction.

Finally, we compared the prediction performance of the "gold-standard" connectomes and the transformed connectomes. In this comparison, the prediction performance of the "gold-standard" connectomes was treated as an upper limit of how well the transformed connectomes could perform, as it is unreasonable to expect the transformed connectomes to outperform the 'gold-standard" connectomes. For both IQ and sex prediction, prediction performance of the transformed connectomes overlapped that of the "gold-standard" connectomes, indicating that little information about brain-behavior association is lost when transforming data using the estimated optimal mapping $\mathcal{T}$.

4   Discussion and Conclusions

Atlas selection is a prerequisite for creation of a functional connectome. Yet, any choice of atlas ultimately constrains interpretation and future replication and generalization efforts to that particular atlas. Since there is no single gold-standard atlas, results generated from two distinct atlases must undergo additional processing before comparison. In this work, we propose optimal transport to find optimum mappings between different atlases, which enable data, previously processed with one atlas, to be mapped to a connectome generated from a different atlas, without the need for further prepossessing. We show that these transformed connectomes are significantly similar to their "gold-standard" counterparts and maintain individual differences in brain-behavior associations, demonstrating both the validity of our approach and its utility in downstream analyses. Importantly, our optimal mappings are robust to training parameters, suggesting that a single mapping between any atlas pair could be generated once and used as an off-the-shelf solution by the community.

Our approach is in the spatial domain (*i.e.*, mapping node-to-node), rather than mapping timecourse-to-timecourse or even connectome-to-connectome. Once we have a node-to-node mapping, timecourses (and resulting connectomes) naturally come for free. While our end goal is to generate the transformed connectomes, we chose the node-to-node approach as it is more general. Our method could be used to map other neuroimaging results, such as cortical thickness in the Desikan-Killiany Atlas to the Destrieux Atlas.

Future work will include further validation of our approach in a wider range of atlases and the generation of a publicly available repository of mappings for community use. Overall, our approach is a promising avenue to increase the generalization of connectome-based results across different atlases.

Acknowledgements. Data were provided in part by the Human Connectome Project, WU-Minn Consortium (Principal Investigators: David Van Essen and Kamil Ugurbil; U54 MH091657) and funded by the 16 NIH Institutes and Centers that support the NIH Blueprint for Neuroscience Research; and by the McDonnell Center for Systems Neuroscience at Washington University. We thank Brendan Adkinson for the helpful comments on this work.

References

1. Altschuler, J., Weed, J., Rigollet, P.: Near-linear time approximation algorithms for optimal transport via sinkhorn iteration. arXiv preprint arXiv:1705.09634 (2017)
2. Arslan, S., Ktena, S.I., Makropoulos, A., Robinson, E.C., Rueckert, D., Parisot, S.: Human brain mapping: a systematic comparison of parcellation methods for the human cerebral cortex. NeuroImage **170**, 5–30 (2018). https://doi.org/10.1016/j.neuroimage.2017.04.014. https://www.sciencedirect.com/science/article/pii/S1053811917303026. Segmenting the Brain
3. Bertsimas, D., Tsitsiklis, J.: Introduction to Linear Optimization. Athena Scientific (1997). http://athenasc.com/linoptbook.html
4. Birkhoff, G.: Tres observaciones sobre el algebra lineal. Univ. Nac. Tucuman, Ser. A **5**, 147–154 (1946)
5. Bouckaert, R.R., Frank, E.: Evaluating the replicability of significance tests for comparing learning algorithms. In: Dai, H., Srikant, R., Zhang, C. (eds.) PAKDD 2004. LNCS (LNAI), vol. 3056, pp. 3–12. Springer, Heidelberg (2004). https://doi.org/10.1007/978-3-540-24775-3_3
6. Brenier, Y.: Polar factorization and monotone rearrangement of vector-valued functions. Commun. Pure Appl. Math. **44**(4), 375–417 (1991)
7. Cortes, C., Vapnik, V.: Support-vector networks. Mach. Learn. **20**(3), 273–297 (1995)

8. Dantzig, G.B.: Reminiscences about the origins of linear programming. In: Bachem, A., Korte, B., Grötschel, M. (eds.) Mathematical Programming the State of the Art, pp. 78–86. Springer, Heidelberg (1983). https://doi.org/10.1007/978-3-642-68874-4_4

9. Delon, J.: Midway image equalization. J. Math. Imaging Vis. **21**(2), 119–134 (2004)

10. Flamary, R., Courty, N.: Pot python optimal transport library (2017). https://pythonot.github.io/

11. Flamary, R., Févotte, C., Courty, N., Emiya, V.: Optimal spectral transportation with application to music transcription. arXiv preprint arXiv:1609.09799 (2016)

12. Gangbo, W., McCann, R.J.: The geometry of optimal transportation. Acta Math. **177**(2), 113–161 (1996)

13. Gao, S., Greene, A., Constable, T., Scheinost, D.: Combining multiple connectomes improves predictive modeling of phenotypic measures. Neuroimage (2019, in Press)

14. Glasser, M.F., et al.: The minimal preprocessing pipelines for the human connectome project. Neuroimage **80**, 105–124 (2013)

15. Hitchcock, F.L.: The distribution of a product from several sources to numerous localities. J. Math. Phys. **20**(1–4), 224–230 (1941)

16. Huang, G., Quo, C., Kusner, M.J., Sun, Y., Weinberger, K.Q., Sha, F.: Supervised word mover's distance. In: Proceedings of the 30th International Conference on Neural Information Processing Systems, pp. 4869–4877 (2016)

17. Joshi, A., et al.: Unified framework for development, deployment and robust testing of neuroimaging algorithms. Neuroinformatics **9**(1), 69–84 (2011)

18. Kantorovich, L.: On the transfer of masses. Dokl. Akad. Nauk SSSR **37**, 227–229 (1942). (in Russian)

19. Koopmans, T.C.: Optimum utilization of the transportation system. Econometrica: J. Econometric Soc. 136–146 (1949)

20. Li, P., Wang, Q., Zhang, L.: A novel earth mover's distance methodology for image matching with gaussian mixture models. In: Proceedings of the IEEE International Conference on Computer Vision, pp. 1689–1696 (2013)

21. Mathon, B., Cayre, F., Bas, P., Macq, B.: Optimal transport for secure spread-spectrum watermarking of still images. IEEE Trans. Image Process. **23**(4), 1694–1705 (2014)

22. Monge, G.: Mémoire sur la théorie des déblais et des remblais. Histoire de l'Académie Royale des Sciences de Paris (1781)

23. Peyré, G., Cuturi, M., et al.: Computational optimal transport: with applications to data science. Found. Trends® Mach. Learn. **11**(5–6), 355–607 (2019)

24. Rubner, Y., Tomasi, C., Guibas, L.J.: The earth mover's distance as a metric for image retrieval. Int. J. Comput. Vis. **40**(2), 99–121 (2000)

25. Shen, X., Tokoglu, F., Papademetris, X., Constable, R.T.: Groupwise whole-brain parcellation from resting-state fMRI data for network node identification. Neuroimage **82**, 403–415 (2013)

26. Tolstoi, A.: Methods of finding the minimal total kilometrage in cargo transportation planning in space. TransPress Natl. Commissariat Transp. **1**, 23–55 (1930)

27. Van Essen, D.C., et al.: The WU-Minn human connectome project: an overview. Neuroimage **80**, 62–79 (2013)

EndoUDA: A Modality Independent Segmentation Approach for Endoscopy Imaging

Numan Celik[1,2] , Sharib Ali[1,2,3(✉)] , Soumya Gupta[1,2] ,
Barbara Braden[3,4] , and Jens Rittscher[1,2,3]

[1] Department of Engineering Science, Institute of Biomedical Engineering,
University of Oxford, Oxford, UK
{numan.celik,sharib.ali,jens.rittscher}@eng.ox.ac.uk
[2] Big Data Institute, University of Oxford, Li Ka Shing Centre for Health
Information and Discovery, Oxford, UK
[3] NIHR Oxford Biomedical Research Centre, Oxford, UK
[4] Translational Gastroenterology Unit, Experimental Medicine Division,
John Radcliffe Hospital, University of Oxford, Oxford, UK

Abstract. Gastrointestinal (GI) cancer precursors require frequent monitoring for risk stratification of patients. Automated segmentation methods can help to assess risk areas more accurately, and assist in therapeutic procedures or even removal. In clinical practice, addition to the conventional white-light imaging (WLI), complimentary modalities such as narrow-band imaging (NBI) and fluorescence imaging are used. While, today most segmentation approaches are supervised and only concentrated on a single modality dataset, this work exploits to use a target-independent unsupervised domain adaptation (UDA) technique that is capable to generalize to an unseen target modality. In this context, we propose a novel UDA-based segmentation method that couples the variational autoencoder and U-Net with a common EfficientNet-B4 backbone, and uses a joint loss for latent-space optimization for target samples. We show that our model can generalize to unseen target NBI (target) modality when trained using only WLI (source) modality. Our experiments on both upper and lower GI endoscopy data show the effectiveness of our approach compared to naive supervised approach and state-of-the-art UDA segmentation methods.

Keywords: Barrett's esophagus · Polyp · Endoscopy · Unsupervised domain adaptation · Variational autoencoder · Segmentation

1 Introduction

Most gastrointestinal (GI) cancers are preventable. In 2018, the five major types of GI cancers, which include those of the stomach, liver, oesophagus, and

Electronic supplementary material The online version of this chapter (https:// doi.org/10.1007/978-3-030-87199-4_29) contains supplementary material, which is available to authorized users.

M. de Bruijne et al. (Eds.): MICCAI 2021, LNCS 12903, pp. 303–312, 2021.
https://doi.org/10.1007/978-3-030-87199-4_29

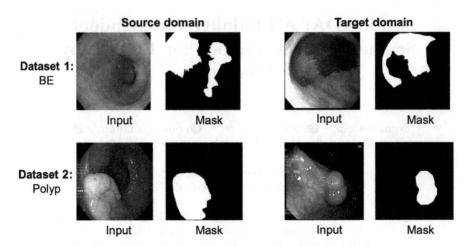

Fig. 1. Example images and masks of the two GI datasets for which source and target domains are separated based on their modality.

colorectum account for 26% of the global cancer incidence and 35% of all cancer deaths [2]. Endoscopy, a vital tool for screening and disease surveillance, is however heavily operator dependent and 12% of cancers are missed [11]. Artificial intelligence-based systems can play a vital role in improving diagnostic quality. However, it is critical that these tools integrate directly into the existing clinical workflow and generalize well to the previously unseen data.

Today, high-definition endoscopes provide a sufficient resolution to allow for a detailed visualisation of the mucosal surface. In addition to the conventional white light imaging (WLI) modality, autofluorescence imaging, and electronic chemoendoscopy techniques such as narrow-band imaging have been developed to improve the early detection of cancer and its precursors [15]. It is important that any computer assisted methods should work seamlessly in these different modalities and avoid the need for any modality-specific training. Developing such a modality agnostic approach for detection and segmentation of pre-cancerous changes in the esophagus, and polyps in the lower GI is the focus of this paper.

Convolutional Neural Network (CNN) based approaches [5,10,14], all well established in biomedical imaging, have been applied to segmentation tasks in GI endoscopy [1,8,19]. Wu et al. [19] proposed a dual neural network based on U-Net with ResNet backbone for automatic esophageal lesion segmentation. Guo et al. [8] utilized a FCN-based architecture with atrous kernels to segment polyps. However, all these supervised approaches are trained on WLI endoscopy and do not support working in different imaging modalities. To date, the problem of quantifying disease related changes in NBI, which can produce a very different appearance of the mucosal surface when compared to WLI (see Fig. 1) has not been addressed. Training methods for different imaging modalities independently is prohibitive. Firstly, extensive training datasets that cover a similar range of conditions would need to be curated to ensure a comparable

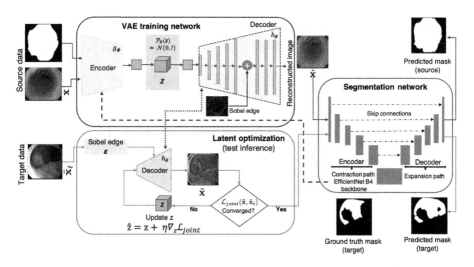

Fig. 2. Complete architecture of the proposed EndoUDA: a target-independent unsupervised domain adaptation.

performance. Secondly, manual annotations by human experts are required which is not only labour intensive but also expensive. To tackle these issues domain adaptation methods have been used to adapt trained models to unseen target data for segmentation [7,9,13,17]. Although these methods promise to generalise, so far there are only a few examples of applying such methods to medical data [7,9]. Additionally, the available methods still under-perform and have not been adapted for extremely diverse modality cases such as that in the GI endoscopy.

To develop a modality-agnostic approach, we propose to use an unsupervised domain adaptation (UDA) technique that can be trained on available single WLI modality dataset and can be applied on an unseen NBI imagery. For such settings, where classical supervised models fail to generalize and perform poorly, we show that our proposed target-independent UDA technique, EndoUDA, has the potential to generalize and to provide improved segmentation on NBI modality for both BE and polyp datasets.

2   Method

Our proposed EndoUDA consists of components for training and test time inference shown in Fig. 2. The training network, utilizes a shared EfficientNet-B4 encoder for our variational autoencoder (VAE) and a U-Net-based segmentation model. The VAE module allows to learn the source domain representations while U-Net module leverages the learned encoded features for semantic segmentation. For inference at test time, learnt latent space embeddings are further optimized to translate to the unseen modality dataset by continuously updating via a joint loss minimization scheme. Below, details of each of these modules are presented.

2.1   VAE Training on Source Domain

VAE is a latent space derived generative model that is based on variational inference distributions. Let $\mathcal{P}_\theta(\mathbf{z}|\mathbf{x})$ be the latent space distribution for sample $\mathbf{x}$ and $\mathcal{Q}_\phi(\mathbf{z}|\mathbf{x})$ be an identical tractable distribution so that it approximates the true posterior $\mathcal{P}_\theta(\mathbf{z}|\mathbf{x})$. Then, the variational lower bound term of the loss $\mathcal{L}(\theta, \phi)$ between probabilistic h_ϕ (encoder, outputs the parameters $\mu_\mathbf{z}$ and $\sigma_\mathbf{z}$ of a distribution) and h_θ (decoder) neural networks. In Eq. (1), approximated $\mathcal{Q}_\phi(\mathbf{z}|\mathbf{x})$ eliminates the non-negative KL-divergence term imposing a lower bound on the data log-likelihood only. This is maximized by the VAE via parameterizing $\mathcal{Q}_\phi(\mathbf{z}|\mathbf{x})$ and $\mathcal{P}_\theta(\mathbf{x}|\mathbf{z})$ which is similar to the minimization of the reconstruction loss $\mathcal{L}_r(\mathbf{x})$ in Eq. (2).

$$\mathcal{L}(\theta, \phi) = \mathbb{E}_{\mathcal{Q}_\phi}[\ln(\mathcal{P}_\theta(\mathbf{x}|\mathbf{z}))] - \mathbb{D}_{KL}[\mathcal{Q}_\phi(\mathbf{z}|\mathbf{x})\|\mathcal{P}_\theta(\mathbf{z}|\mathbf{x})] \tag{1}$$

Compared to the VAE framework in [13], we propose to couple the encoder layer of VAE as the segmentation network backbone. Based on our experiments, we have chosen an EfficientNet which computes the approximate posterior $\mathcal{Q}_\phi(\mathbf{z}|\mathbf{x})$. Unlike Pandey $et\ al.$ [13], we remove the originally used perceptual loss module in the decoder layer used to compute the image likelihood of the observed data $\mathcal{P}_\theta(\mathbf{x}|\mathbf{z})$. The decoder network of the VAE contains 5 convolution layers followed by a LeakyRelu function (Leak=0.2) and a $tanh$ activation function for the reconstructed output in the final convolution layer. A Sobel filter generated edge map is concatenated with the sixth layer of the decoder network of the VAE through a $tanh$ nonlinear function (see Fig. 2). The reconstruction loss $\mathcal{L}_r(\mathbf{x})$ is minimized for the VAE training network given by:

$$\mathcal{L}_r(\mathbf{x}) = \mathcal{L}(\theta, \phi) = \frac{1}{N} \sum_{i=1}^{N} \|\mathbf{x}^i - h_\theta(h_\phi(\mathbf{x}^i))\|^2, \tag{2}$$

where $\mathbf{x}$ is the source domain image, N is total number of samples, h_ϕ and h_θ are encoder and decoder networks, respectively. Here, the reconstructed image is given by $\hat{\mathbf{x}} = h_\theta(h_\phi(\mathbf{x}))$. In addition, as shown in Fig. 2, the edges of the input images are extracted and concatenated into the decoder network in VAE through a skip connection to better generalize the source and target domains. Figure 2 shows a schematic diagram of the complete VAE architecture used for training on the source domain.

2.2   Segmentation Module

The segmentation network consists of a U-Net with an EfficientNet-B4 [16] backbone. The input from the reconstructed image obtained from the decoder of the VAE is fed to the already tuned weights of EfficientNet-B4. The segmentation network is adapted by a supervised loss on the source domain given by a classically used binary cross-entropy loss ($\mathcal{L}_{BCE}$) and dice coefficient loss ($\mathcal{L}_{DSC}$) [12]. $\mathcal{L}_{DSC}$ is represented as $\mathcal{L}_{DSC} = (2\sum \mathbf{pg})/(\sum \mathbf{p}^2 + \sum \mathbf{g}^2)$, where $\mathbf{p}$ is the predicted probability map and $\mathbf{g}$ is the ground truth mask.

2.3   VAE Latent Search Optimization via New Joint Loss

The trained VAE encoding aims to find the nearest point from the source domain, given a sample from the target domain, through an optimization process and updates of the latent space representation $\mathbf{z}$, as can be seen in Fig. 2. Once the decoder h_θ of the VAE is trained on the source distribution $\mathcal{P}_s(\mathbf{x})$, given an image $\tilde{\mathbf{x}}_t$ from the target distribution, the latent search algorithm finds the nearest point by optimizing latent vector over the iterative process. As shown in the Fig. 1, the WLI images are used in the source domain while the NBI are used in the target domain. For faster optimization convergence of the latent space encoding with improved generalization ability of the network, we propose to use a joint loss function that comprises of the normalized cross correlation (NCC) $\mathcal{L}_{NCC}$ and the structural similarity index measure (SSIM) [21] loss $\mathcal{L}_{ssim}$ between the image $\tilde{\mathbf{x}}_t$ from target domain and the reconstructed target image $\tilde{\mathbf{x}}$ from h_θ. The loss function for computing NCC is defined as:

$$\mathcal{L}_{NCC}(\tilde{\mathbf{x}}, \tilde{\mathbf{x}}_t) = \frac{1}{2N} \sum (\frac{\tilde{\mathbf{x}} - \mu_s}{\sqrt{\sigma_s^2 + \epsilon^2}} - \frac{\tilde{\mathbf{x}}_t - \mu_t}{\sqrt{\sigma_t^2 + \epsilon^2}})^2, \tag{3}$$

where $\tilde{\mathbf{x}}$, μ_s and σ_s are reconstructed images from decoder network, mean and standard deviation of the latent encoded image, respectively. Similarly, $\tilde{\mathbf{x}}_t$, μ_t and σ_t correspond to target image input, mean and standard deviation of the target image. The SSIM for a pair of images $(\tilde{\mathbf{x}}, \tilde{\mathbf{x}}_t)$ and loss function $\mathcal{L}_{ssim}$ is defined as:

$$SSIM(\tilde{\mathbf{x}}, \tilde{\mathbf{x}}_t) = l(\tilde{\mathbf{x}}, \tilde{\mathbf{x}}_t)^\alpha * c(\tilde{\mathbf{x}}, \tilde{\mathbf{x}}_t)^\beta * s(\tilde{\mathbf{x}}, \tilde{\mathbf{x}}_t)^\gamma,$$
$$\mathcal{L}_{ssim}(\tilde{\mathbf{x}}, \tilde{\mathbf{x}}_t) = 1 - SSIM(\tilde{\mathbf{x}}, \tilde{\mathbf{x}}_t), \tag{4}$$

here l, c and s denote luminance, contrast and structure similarities of the given pair of images, respectively with the parameters α, β and $\gamma > 0$. Also, the similarity loss function $\mathcal{L}_{ssim}$ is presented. The final joint loss function is given by:

$$\mathcal{L}_{joint} = \lambda \mathcal{L}_{NCC} + (1 - \lambda)\mathcal{L}_{ssim}, \tag{5}$$

where $\lambda = 0.75$ is set empirically. After the convergence of $\mathcal{L}_{joint}$, the optimal latent space representation $\hat{\mathbf{z}} = \mathbf{z} + \eta \mathcal{L}_{joint}$ is used to generate the closest clone of source domain with a learning rate η which is then used as an input to the segmentation network to predict the target mask.

3   Experiments and Results

Implementation Details. For the training of our VAE network we used RMSProp with learning rate (lr) of 0.0001 while for the segmentation module Adam optimizer was used with β_1=0.5, β_2=0.999, and $lr = 0.001$. The initial learning rate was reduced by a factor of 10 for every validation loss that did not improve in the last 3 epochs and a stopping criteria was set for no improvement in validation loss upto 10 epochs. We used 100 epochs for training our EndoUDA training module with batch size of 64. For the test module of EndoUDA, we set

Table 1. Supervised Segmentation analysis: Benchmarking BE and polyp datasets on standard segmentation architectures on 80-20 train-test split.

Models	BE		Polyp	
	IoU	Dice	IoU	Dice
U-Net [14]	0.738	0.846	0.718	0.755
UNet++ [20]	0.742	0.859	0.721	0.778
ResUNet [6]	0.789	0.908	0.734	0.794
EfficientUNet [3]	**0.841**	**0.931**	**0.803**	**0.896**
EfficientUNet++	0.828	0.927	0.767	0.865
DeepLabv3+ [5]	0.802	0.893	0.789	0.869

learning rate of $\eta = 0.001$ for optimization of the latent space representation and a fixed 40 iterations with batch size of 32 was set. The proposed method was implemented using the Tensorflow framework on a machine deployed with NVIDIA RTX 2080Ti graphics card with 12 GB GPU memory. All input images were resized to 128×128 pixels.

Datasets. We have used two gastrointestinal (GI) endoscopy datasets for analyzing two distinct endoscopically found precancerous anomalies, namely, Barrett's esophagus (BE) and polyps. To evaluate the efficacy of the proposed EndoUDA method, we have used clinically acquired white light imaging (WLI) and narrow-band imaging (NBI) modality data for BE and polyps. BE dataset consists of 1182 endoscopy images acquired from 68 unique patients of which 611 WLI images are used as source domain data (train set: 90%, validation set: 10%) and 571 NBI images as target domain data are used for testing. Similarly, for the polyp dataset, we used the publicly available Hyper-Kvasir dataset [4] for training which consists of 1000 images (train set: 80%, validation set: 20%). In addition, we used clinically acquired 42 NBI images of polyps acquired from 20 patients at our local hospital and from collaborators.

3.1   Benchmark Comparison and Ablation Study

Result of supervised segmentation of BE and polyps in our GI dataset are presented in Table 1. Frequently used supervised learning methods have been evaluated with respect to IoU and Dice to establish a baseline. It can be observed that the UNet with an EfficientNet B4 backbone [3] outperformed other baseline methods including DeepLabv3+ [5] by nearly 4% and 2% on both IoU and Dice for BE and polyps, respectively. **Supplementary material Table 1–4** show that our choice of the joint loss function, the Sobel edge operator, the decoder architecture and the computational complexity yield in improved performance compared to alternatives that are often used.

Table 2. Empirical results of EndoUDA along with SOTA UDA methods with mean μ and standard deviation σ are provided. All comparisons are provided for source only trained model and tested on target data. *Paired t-test* between EndoUDA and each SOTA methods are shown in Supplementary Table 5.

Models	BE ($\mu \pm \sigma$)				Polyp ($\mu \pm \sigma$)			
	IoU	Dice	Precision	Recall	IoU	Dice	Precision	Recall
Naive U-Net [3]	0.626	0.718	0.654	0.667	0.492	0.564	0.487	0.496
	±0.02	±0.015	±0.018	±0.021	±0.018	±0.012	±0.011	±0.011
AdaptSegnet [17]	0.658	0.749	0.721	0.698	0.519	0.577	0.601	0.583
	±0.019	±0.016	±0.012	±0.022	±0.017	±0.015	±0.016	±0.012
ADVENT [18]	0.667	0.768	0.739	0.706	0.524	0.591	0.662	0.638
	±0.012	±0.014	±0.015	±0.012	±0.011	±0.019	±0.021	±0.014
CellSegUDA [9]	0.673	0.771	0.733	0.691	0.533	0.629	0.688	0.641
	±0.02	±0.016	±0.018	±0.013	±0.012	±0.021	±0.018	±0.017
GLSS [13]	0.704	0.815	**0.846**	0.735	0.558	0.621	0.675	0.649
	±0.019	±0.022	±0.019	±0.02	±0.017	±0.016	±0.019	±0.012
EndoUDA	**0.733**	**0.854**	0.832	**0.784**	**0.605**	**0.693**	**0.722**	**0.704**
	±0.014	±0.021	±0.017	±0.019	±0.014	±0.02	±0.017	±0.018

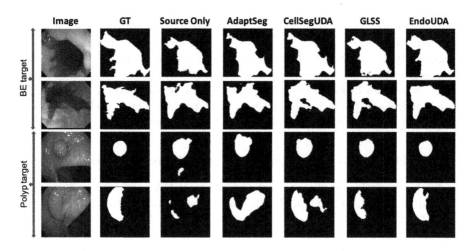

Fig. 3. Qualitative study: Baseline UDA experiments on BE and polyp images.

3.2 Comparison with SOTA Method

We compared the proposed EndoUDA approach with four SOTA UDA methods: AdaptSegnet [17], ADVENT [18], CellSegUDA [9], GLSS [13], along with the naive supervised baseline model. The quantitative results are reported in Table 2 for four different metrics including intersection-over-union (IoU), Dice, precision and recall metrics. It can be observed that our proposed framework EndoUDA outperformed other SOTA UDA methods and naive U-Net method. An

Table 3. IoU and dice comparisons for out-of-distribution target samples (NBI) in the source (WLI) GI endoscopy datasets

Method	BE_{IoU}	$Polyp_{IoU}$	BE_{dice}	$Polyp_{dice}$
U-Net (source-only)	0.626	0.492	0.718	0.564
EndoUDA	0.733	0.605	0.854	0.693
U-Net (source 100% + target 10%)	0.646	0.536	0.738	0.588
EndoUDA (source 100% + target 10%)	0.744	0.613	0.866	0.706
U-Net (source 100% + target 25%)	0.691	0.579	0.761	0.641
EndoUDA (source 100% + target 25%)	0.756	0.624	0.875	0.721
U-Net (source 100% + target 50%)	0.738	0.631	0.827	0.734
EndoUDA (source 100% + target 50%)	0.771	0.658	0.889	0.739
U-Net (target-trained)	0.841	0.802	0.931	0.896

improvement of nearly 4% on BE and 7% on polyp dataset for with respect to the Dice metric compared to the recent SOTA GLSS method. Similar improvements can be seen for other metrics for both GI datasets and also much higher improvements compared to other SOTA UDA methods. Table 5 in the supplementary material shows the statistical significance of the improvement provided by our method when compared to SOTA methods presented in Table 2 with p-values <0.01 for almost all metrics and SOTA approaches for both BE and polyp segmentation.

It can be seen in Fig. 3 that the predicted masks using proposed EndoUDA model for BE and polyp images are very close to the ground truth annotations compared to those predicted by other SOTA UDA methods.

3.3   Effect of Out-of-Distribution Data in Supervised Training

We further investigated the performance of EndoUDA compared to the supervised baseline U-Net model (with EfficientNet backbone) by training the model with 10%, 25% and 50% of target data samples. It can be observed (see Table 3 that the EndoUDA performs consistently better than the baseline approach for a different proportion of target data inclusion in the training. For example, on 10% and 25% of target data mixed in the training sample, EndoUDA still provides nearly a 7–10% improvement on IoU and 8–13% with respect to Dice scores over the baseline method.

4   Conclusion

We have presented a novel target-independent unsupervised domain adaptation VAE-based method that comprises of a shared encoder for both reconstruction and segmentation, and a joint loss function for improved unseen target domain generalization. We have validated our approach on two GI endoscopy datasets that often require additional modalities for screening and intervention. The proposed modality agnostic method can be used in clinical GI endoscopy without imposing any constraints on what imaging modality is being used. Our qualitative and quantitative results demonstrate the effectiveness and promise of the proposed method compared to both the naive baseline supervised model and state-of-the-art UDA segmentation methods.

Acknowledgement. NC is supported by the Emerson Collective Cancer Research Fund, SG is funded by Boston Scientific, SA and JR are supported by National Institute for Health Research (NIHR) Oxford Biomedical Research Centre (BRC). The views expressed are those of the authors and not necessarily those of the NHS, the NIHR or the Department of Health. The authors declare no competing interests.

References

1. Ali, S., Dmitrieva, M., Ghatwary, N., Bano, S., et al.: Deep learning for detection and segmentation of artefact and disease instances in gastrointestinal endoscopy. Med. Image Anal. **70**, 102002 (2021)
2. Arnold, M., et al.: Global burden of 5 major types of gastrointestinal cancer. Gastroenterology **159**(1), 335–349 (2020)
3. Baheti, B., Innani, S., Gajre, S., Talbar, S.: Eff-UNet: a novel architecture for semantic segmentation in unstructured environment. In: IEEE Conference on Computer Vision and Pattern Recognition Workshops (CVPRW), pp. 1473–1481 (2020)
4. Borgli, H., Thambawita, V., Smedsrud, P.H., Hicks, S., et al.: HyperKvasir, a comprehensive multi-class image and video dataset for gastrointestinal endoscopy. Sci. Data **7**(1), 1–14 (2020)
5. Chen, L.-C., Zhu, Y., Papandreou, G., Schroff, F., Adam, H.: Encoder-decoder with atrous separable convolution for semantic image segmentation. In: Ferrari, V., Hebert, M., Sminchisescu, C., Weiss, Y. (eds.) ECCV 2018. LNCS, vol. 11211, pp. 833–851. Springer, Cham (2018). https://doi.org/10.1007/978-3-030-01234-2_49
6. Diakogiannis, F.I., Waldner, F., Caccetta, P., Wu, C.: ResUNet-a: a deep learning framework for semantic segmentation of remotely sensed data. ISPRS J. Photogramm. Remote. Sens. **162**, 94–114 (2020)
7. Dong, N., Kampffmeyer, M., Liang, X., Wang, Z., Dai, W., Xing, E.: Unsupervised domain adaptation for automatic estimation of cardiothoracic ratio. In: Frangi, A.F., Schnabel, J.A., Davatzikos, C., Alberola-López, C., Fichtinger, G. (eds.) MICCAI 2018. LNCS, vol. 11071, pp. 544–552. Springer, Cham (2018). https://doi.org/10.1007/978-3-030-00934-2_61
8. Guo, Y., Bernal, J., Matuszewski, B.J.: Polyp segmentation with fully convolutional deep neural networks-extended evaluation study. J. Imaging **6**(7), 69 (2020)
9. Haq, M.M., Huang, J.: Adversarial domain adaptation for cell segmentation. In: Proceedings of the Third Conference on Medical Imaging with Deep Learning (MIDL), pp. 277–287 (2020)
10. Long, J., Shelhamer, E., Darrell, T.: Fully convolutional networks for semantic segmentation. In: 2015 IEEE Conference on Computer Vision and Pattern Recognition (CVPR), pp. 3431–3440 (2015)
11. Menon, S., Trudgill, N.: How commonly is upper gastrointestinal cancer missed at endoscopy? A meta-analysis. Endosc. Int. Open **2**(2), E46 (2014)

12. Milletari, F., Navab, N., Ahmadi, S.: V-Net: fully convolutional neural networks for volumetric medical image segmentation. In: International Conference on 3D Vision (3DV), pp. 565–571 (2016)

13. Pandey, P., Tyagi, A.K., Ambekar, S., Prathosh, A.P.: Unsupervised domain adaptation for semantic segmentation of NIR images through generative latent search. In: Vedaldi, A., Bischof, H., Brox, T., Frahm, J.-M. (eds.) ECCV 2020. LNCS, vol. 12351, pp. 413–429. Springer, Cham (2020). https://doi.org/10.1007/978-3-030-58539-6_25

14. Ronneberger, O., Fischer, P., Brox, T.: U-Net: convolutional networks for biomedical image segmentation. In: Navab, N., Hornegger, J., Wells, W.M., Frangi, A.F. (eds.) MICCAI 2015. LNCS, vol. 9351, pp. 234–241. Springer, Cham (2015). https://doi.org/10.1007/978-3-319-24574-4_28

15. Subramanian, V., Ragunath, K.: Advanced endoscopic imaging: a review of commercially available technologies. Clin. Gastroenterol. Hepatol. **12**(3), 368–376 (2014)

16. Tan, M., Le, Q.: EfficientNet: rethinking model scaling for convolutional neural networks. In: Proceedings of International Conference on Machine Learning (ICML), pp. 6105–6114 (2019)

17. Tsai, Y., Hung, W., Schulter, S., Sohn, K., Yang, M., Chandraker, M.: Learning to adapt structured output space for semantic segmentation. In: IEEE Conference on Computer Vision and Pattern Recognition (CVPR), pp. 7472–7481 (2018)

18. Vu, T.H., Jain, H., Bucher, M., Cord, M., Pérez, P.: Advent: adversarial entropy minimization for domain adaptation in semantic segmentation. In: 2019 IEEE/CVF Conference on Computer Vision and Pattern Recognition (CVPR), pp. 2512–2521 (2019)

19. Wu, Z., Ge, R., Wen, M., Liu, G., et al.: ELNet: automatic classification and segmentation for esophageal lesions using convolutional neural network. Med. Image Anal. **67**, 101838 (2021)

20. Zhou, Z., Rahman Siddiquee, M.M., Tajbakhsh, N., Liang, J.: UNet++: a nested U-Net architecture for medical image segmentation. In: Stoyanov, D., et al. (eds.) DLMIA/ML-CDS -2018. LNCS, vol. 11045, pp. 3–11. Springer, Cham (2018). https://doi.org/10.1007/978-3-030-00889-5_1

21. Wang, Z., Bovik, A.C., Sheikh, H.R., Simoncelli, E.P.: Image quality assessment: from error visibility to structural similarity. IEEE Trans. Image Process. **13**(4), 600–612 (2004)

Style Transfer Using Generative Adversarial Networks for Multi-site MRI Harmonization

Mengting Liu, Piyush Maiti, Sophia Thomopoulos, Alyssa Zhu, Yaqiong Chai, Hosung Kim, and Neda Jahanshad[✉]

USC Mark and Mary Stevens Neuroimaging and Informatics Institute, Keck School of Medicine of USC, University of Southern California, Los Angeles, CA, USA
Neda.Jahanshad@ini.usc.edu

Abstract. Large data initiatives and high-powered brain imaging analyses require the pooling of MR images acquired across multiple scanners, often using different protocols. Prospective cross-site harmonization often involves the use of a phantom or traveling subjects. However, as more datasets are becoming publicly available, there is a growing need for retrospective harmonization, pooling data from sites not originally coordinated together. Several retrospective harmonization techniques have shown promise in removing cross-site image variation. However, most unsupervised methods cannot distinguish between image-acquisition based variability and cross-site population variability, so they require that datasets contain subjects or patient groups with similar clinical or demographic information. To overcome this limitation, we consider cross-site MRI image harmonization as a style transfer problem rather than a domain transfer problem. Using a fully unsupervised deep-learning framework based on a generative adversarial network (GAN), we show that MR images can be harmonized by inserting the style information encoded from a reference image directly, without knowing their site/scanner labels *a priori*. We trained our model using data from five large-scale multi-site datasets with varied demographics. Results demonstrated that our style-encoding model can harmonize MR images, and match intensity profiles, successfully, without relying on traveling subjects. This model also avoids the need to control for clinical, diagnostic, or demographic information. Moreover, we further demonstrated that if we included diverse enough images into the training set, our method successfully harmonized MR images collected from unseen scanners and protocols, suggesting a promising novel tool for ongoing collaborative studies.

Keywords: MRI harmonization · Style encoding · GAN

1 Introduction

Large-scale multi-site neuroimaging studies allow for high statistical power and are essential for ensuring reliable and robust results. Many multi-site initiatives prospectively aim to harmonize image acquisition protocols to minimize scanner-induced variability due to factors such as magnetic field strength, coil channels, gradient directions, manufacturer, and image resolution. However, in many cases, the need for retrospective

© Springer Nature Switzerland AG 2021
M. de Bruijne et al. (Eds.): MICCAI 2021, LNCS 12903, pp. 313–322, 2021.
https://doi.org/10.1007/978-3-030-87199-4_30

harmonization is often inevitable. Decade-long running studies, such as ADNI undergo scanner upgrades, and many studies of small effects or effects across wide demographic ranges, require retrospective pooling of data.

Several retrospective image harmonization techniques have shown promise in being able to remove cross-site variance from different studies for pooled analyses. Most harmonization methods fall into two broad categories: 1) harmonization of image-derived features using statistical properties of the distribution, for example ComBat [1]; 2) harmonization of the task output of the MR images, i.e. segmentation, classification, and prediction. This second category is largely composed of deep learning-based approaches, namely domain adaptation techniques. In particular, domain transfer learning and domain adversarial learning have been successfully applied for MRI harmonization [2].

However, if a wide range of tasks were to be performed on images, harmonization would need to be performed separately for each task. Furthermore, there are some applications, like cortical surface construction (as opposed to segmentation), that cannot be directly embedded into a deep learning framework [3]. In these scenarios, image translation or a direct image-to-image harmonization for MRIs is needed. Several image translation-based harmonization methods have been proposed. Supervised methods typically require traveling subjects and must be planned prospectively [4]. Unsupervised methods, such as variational auto-encoders [5] or CycleGAN [6], MR images are often separated into well-defined domains in terms of scanners or sites. These methods are prone to overcorrection if data collected at each site are not acquired with different scanning parameters, but are also confounded by differences in population of individuals scanned, such as clinical diagnoses, age ranges, or even race and ethnic groups; the correction for site could then also remove critical biological differences. In these situations, the demographic and clinical conditions need to be strictly controlled and matched. Furthermore, even with the data collected from the same scanners/sites, the images collected may also show slight variance [7]. These intra-site differences make the harmonization less accurate due to the lack of diversity of these domain-based methods.

Recently, deep learning methods have successfully completed diverse image translations by disentangling the image into 'content' and 'style' spaces [8], where contents represent low level information in images like contours and orientations, and styles can be considered high level information such as colors and textures. Images within the same domain share the same content space but may show different styles. In MR images, we can consider the biologically defined anatomical information as the content, and the non-biological information such as intensity variation, SNR, and contrasts as styles. Dewey, et al., [9] have used this breakdown to show promising results for MRI harmonization. However, in [9] paired image modalities from the same subjects were needed to supervise the extraction of the content information, which are not always available. Jiang et al., [10] also used a similar framework to facilitate the cross-domain image translations, where each modality of images was treated as a domain. In [10], styles that span multiple domains must be modeled together using a variational auto-encoder with a universal prior.

Here, we consider image harmonization as a pure style transfer problem rather than a domain transfer problem. We think anatomical patterns (or contents) from MR images collected from different sites share the same latent space, and it is not necessary to separate them into different "domains". These style-irrelevant patterns can be learned using an unsupervised cycle consistency GAN model, and thus any paired modalities are not needed from the same subjects. Due to scanner shifts, and software upgrades, the styles for all the images, even collected from the same site, may be different from each other. We hence consider every single image as a unique "domain" with its own style, and the styles can be robustly learned using an adversarial approach. Furthermore, inspired by [11], we proposed that the style information needed for harmonization can be encoded from a single reference MR image directly, instead of using predefined labels.

2 Method

2.1 The Architecture of Style-Encoding GAN

Let X be the set of MR images. Given an image $x \in X$, our goal is to train a single generator G that can generate diverse images that correspond to the image x with a style code s, where s is associated with the style (non-biological) patterns from another image. The style code s is generated by a mapping network M from sampling a given latent vector z $(s = M(z))$. The reason for using s instead of z has been provided by past studies [12]. Then, the generator G translates an input image x into an output image $G(x, s)$ that reflects the style of s. To validate that the style code s is successfully injected into the output image $G(x, s)$, another style encoding network E was designed to encode the style of s from images. That is, given an image x, the encoder E extracts the style code $s = E(x)$ of x. E can produce diverse style codes using different images. This allows G to synthesize an output image reflecting the style s from different reference images of X. The goal of the network is to train E so that $E(G(x, s)) = s$, meaning that if an image was generated based on style code s, then s can also be encoded when this image was input into the style encoder E. Adaptive instance normalization (AdIN) [13] was used to inject s into G. Finally, the discriminator D learns a binary classification determining whether an image x is a real image or a fake image $G(x, s)$ produced by G. Borrowed by [14], our model includes only one generator, one discriminator, and one style encoder (Fig. 1).

2.2 Network Training

Given an image $x \in X$, we train our framework using the following objectives.

Adversarial Loss. During training, we sample a latent code $z \in Z$ randomly, and the mapping network M learns to generate a target style code $s = M(z)$. The generator G takes an image x and s as inputs and learns to generate an output image $G(x, s)$ that is indistinguishable by the discriminator D from real images via an adversarial loss:

$$L_{GAN} = \mathbb{E}_x\big[logD(x)\big] + \mathbb{E}_{x,z}[\log(1 - D(G(x, s)))]$$

Cycle-Consistency Loss. To guarantee that generated images are meaningful to the original images and properly preserving the style-irrelevant characteristics (e.g. anatomical patterns) of input x, an additional cycle consistency loss [6] is defined as the difference between original and reconstructed images:

$$L_{cyc} = \mathbb{E}_{x,z}[\|x - G(G(x, s), s_x)\|_1]$$

where $s_x = E(x)$ is the estimated style code of the input image x. By encouraging the generator G to reconstruct the input image x with the estimated style code s_x, G learns to preserve the original characteristics of x while changing its style faithfully.

Style Reconstruction Loss. In order to enforce the generator G to use the style code while generating the image $G(x, s)$, we incorporate a style reconstruction loss:

$$L_{sty} = \mathbb{E}_{x,z}[\|s - E(G(x, s))\|_1]$$

Our learned encoder E allows G to transform an input image x, to reflect the style of a reference image.

Style Diversification Loss. To further enable the generator G to produce diverse images, we explicitly regularize G with the diversity sensitive loss [15]:

$$L_{div} = \mathbb{E}_{x,z_1,z_2}[\|G(x, s_1) - G(x, s_2)\|_1]$$

where the target style codes s_1 and s_2 are produced by M conditioned on two random latent codes z_1 and z_2(i.e. $s_i = M(z_i)$ for $i \in \{1, 2\}$). Maximizing the regularization term forces G to explore the image space and discover meaningful style features to generate diverse images.

Put together, our full objective function can be summarized as follows:

$$L(G, M, E, D) = L_{GAN} + \lambda_{cyc}L_{cyc} + \lambda_{sty}L_{sty} - \lambda_{div}L_{div}$$

Where λ_{cyc}, λ_{sty} and λ_{div} are hyperparameters for each term.

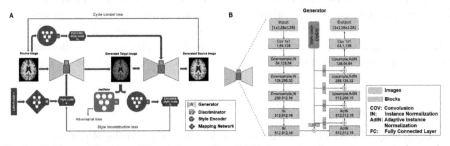

Fig. 1. A) The architecture of the style-encoding GAN. Generators learns to generate image by inputting a source image and a style code. B) The detailed architecture of the generator in network. In each of the block, the three number means number of input channels, number of output channels and the image size.

2.3 Experimental Setup

We obtained T1-weighted brain MR images from five datasets: UK Biobank (UKBB), Parkinson's Progression Markers Initiative (PPMI), and Alzheimer Disease Neuroimaging Initiative (ADNI), Adolescent Brain Cognitive Development (ABCD) and International Consortium for Brain Mapping (ICBM). Scans used in this study were collected from disease-free participants (UKBB: n = 200, age range 45–55 years old; PPMI: n = 76 age range 67–70 years old; ADNI: n = 42, age range 67–70 years old; ABCD: n = 200, age range 9–13 years old; and ICBM: n = 200, age range 19–54 years old), among which 90% were used as training/validation sets and 10% testing sets. We deliberately kept some cohorts overlapping completely in age, while others were non-overlapping. To show that our model generalizes to unseen images, we also applied the trained harmonization model to travelling subjects collected in [7], where three subjects were scanned 12 times at 10 different sites within 13 months respectively. This traveling subject was also used to validate the harmonization quantitatively. All image acquisition information for these public resources can be found elsewhere, but briefly they vary in terms of scanner manufacturer, field strength, voxel size, and more, often within the same study. A list of manufacturers and field strength of the five datasets used in our study can be found in Table 1.

Table 1. Scanner and field strength specifications in 5 datasets utilized

Datasets	Scanner manufacturers				Field strength		
	GE	PHILLIPS	SIEMENS	ELSCENT	1.5T	2T	3T
ADNI			✓				✓
ICBM	✓	✓		✓	✓	✓	✓
UKBB			✓				✓
PPMI	✓	✓	✓		✓		✓
ABCD	✓	✓	✓				✓

All the images were skull-stripped, nonuniformity corrected and registered to the MNI template using a 9 dof linear registration; these simple, standard, and reversible image processing steps allow for some level of concordance between images in advance. MR images from the UKBB and PPMI datasets were further segmented into white matter (WM), gray matter (GM) and cerebrospinal fluid (CSF) using a statistical partial volume model [16]. All the images were resized to $128 \times 128 \times 128$ voxels. The current network may be extended to 3D, yet, to pilot this work, we used 2D slices. We selected 50 axial slices in the middle of each MRI volume (in MNI space) as input as all these slices contain multiple brain tissue types/contrasts to help ensure the model learns style features.

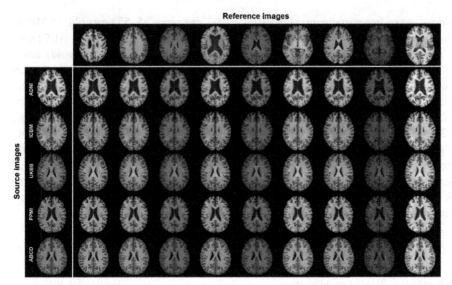

Fig. 2. The style-encoding GAN can harmonize images based on only one single reference image.

3   Results

3.1   Validation of the Removal of Cross-Sites Variances

Histogram Comparison. Figure 2 illustrates the harmonized images among the 5 datasets according to nine randomly selected reference images from across the datasets. Harmonized images are noticeably more similar in contrast and intensity to those of the reference images, and their anatomical structures were well-maintained. To quantitatively compare the intensity histogram of the harmonized images and reference images, we randomly select one participant from ADNI, and one participant from PPMI. Participants were matched for age and sex to ensure approximately similar volumes of each tissue type. We made a slice-matched comparison between the histograms of all 50 slices in MR images from the two participants, and the histogram of the image after the ADNI scan after it was translated to the PPMI domain using Jensen-Shannon divergence (JS). The JS between ADNI and translated ADNI → PPMI image 0.036 ± 0.014) is significantly higher than JS between PPMI and the ADNI → PPMI translated image $(0.017 \pm 0.006;\ p < 0.0001)$, suggesting the histogram of the translated image has an intensity profile more similar to PPMI than ADNI, from which it came.

Segmentation Comparison. Automated brain tissue segmentation into gray matter (GM), white matter (WM), and cerebrospinal fluid (CSF) is generally based on image contrast variations. There may be a systematic over- or under-segmentation of tissue types for different acquisition protocols. Here, we further validate the removal of cross-site variances by testing a segmentation model for images before and after the harmonization. In this example, MR slices from the UKBB dataset (10% testing set as described in the harmonization section) were used to train a deep-learning based U-net [17] to segment the brain into WM, GM, and CSF. The auto-segmentation outputs were used as the ground truth. The trained model was then applied to slices in the PPMI test dataset (testing set in the harmonization section) for segmentation before and after the harmonization using a reference image from the UKBB dataset. The Dice score for WM (before vs. after harmonization: 0.93 ± 0.02 vs. 0.95 ± 0.01), GM (0.80 ± 0.06 vs. 0.86 ± 0.02) and CSF (0.51 ± 0.16 vs. 0.64 ± 0.14) were all significantly higher for images after harmonization (all $p < 0.0001$).

3.2 Validation of the Preservation of Brain Anatomical Information

Ideally, MR image harmonization methods do not only remove the cross-site variances, but also rigorously maintain the anatomical information within subjects. To test whether the anatomical information in MR images were preserved after harmonization, we compared the intra-subject similarity and inter-subject differences before and after the harmonization. In this illustration, to guarantee the similarity between slices, we selected only the middle slices from 10 subjects randomly selected from ADNI as source images, and 100 images randomly selected from other datasets as reference images.

We compared the similarity between images before and after harmonization to different scans using intensity correlation (r) and the structural similarity index measure (SSIM) as metrics. The average intensity correlation between images before and after harmonization is 0.984 and the SSIM is 0.612. Both values were significantly higher than the average value between pairs of images before harmonization ($r = 0.901$; SSIM $= 0.412$), indicating the intra-subject anatomical information was preserved after the harmonization compared to inter-subject variances.

To quantify inter-subject differences, we used the voxelwise intensity differences to compute Euclidean distances [6] between pairs of scans, forming a distance matrix, denoted as $D_{ij}^{n \times n} = \|I_i - I_j\|_2$, where n = 10 is the number of scans, and I the whole-image voxel intensity vectors for scans i and j. The goal was to estimate how the distances were preserved relative to each other before and after harmonization. We computed the correlation, r, between the two distance-matrices before and after harmonization. Our model achieves an average r of 0.963 (range: [0.943, 0.991]) between the distance-matrices before harmonization and the 100 distance-matrices after harmonization, indicating the inter-subject differences were reliably preserved after harmonization.

3.3 Hyperparameter Selection

If the images to be harmonized are confounded by demographic or clinical/pathological differences, such biological differences could also be inadvertently learned during the harmonization. To avoid this, we tuned the λ_{cyc} in our model to preserve the style-irrelevant characteristics. Figure 3 shows an example of the image harmonized using the model with different λ_{cyc} values. In this example, the source image is from ADNI and the reference image is from ABCD. There is approximately a 58-year age difference between these subjects. If $\lambda_{cyc}=0$, meaning none of the style-irrelevant characteristics are needed, the model learns everything from the reference image, generating an image completely identical to the reference. If $\lambda_{cyc} = 1$, then the model learns style from the reference image but also some biological patterns, such as smaller lateral ventricles and thicker

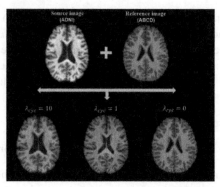

Fig. 3. Small cycle consistency loss coefficients bias the generated images towards the reference images, while larger cycle consistency coefficients rigorously maintain the structure of the source images.

gray matter cortices. If $\lambda_{cyc} = 10$, then the model learns only the style information from the reference and rigorously maintains the style-irrelevant characteristics (ventricles etc.) from the source images.

3.4   Harmonization for MR Images from Unseen Dataset

By design, the model should learn style-invariant features, encouraging better generalization to unseen samples. Figure 4 shows that our model successfully captures styles of the unseen/traveling subject and renders these styles correctly to the source images. Furthermore, we harmonize all the 12 scans from the three traveling subjects using one randomly selected scan (among 12 scans) as the reference. We then compared the slice-matched similarity between all paired scans using SSIM and peak signal to noise ratio (PSNR), and find significant improvements in similarity using SSIM for all three subjects (M = 0.752 before harmonization vs. 0.822 for harmonized images, two-way t-test: p's < 0.031). PSNR also showed an improved similarity but was not statistically significant (23.1 before harmonization and 24.1 after harmonization).

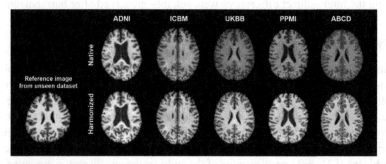

Fig. 4. The trained style-encoding GAN successfully captures styles of reference images from novel acquisition protocols and renders these styles correctly to the source images.

4 Discussion and Conclusion

We have shown in this study that the style-encoding GAN paves a way for multi-site MR images to be synthesized from a single input image from arbitrary sites, taking the style code encoded from an arbitrary reference image directly as extra information, without knowing the acquisition protocols *a priori*. Our model does not rely on traveling subjects or any paired modalities of images from the same subjects. Furthermore, because we consider the cross-site harmonization as a style transfer problem rather than a domain transfer problem, MR images from multiple sites need not be categorized into different domains. Thus, the demographic and pathological conditions do not need to be matched for the harmonization. We showed our model generalizes well even to unseen samples, likely due to the diverse data included in the training set. This is particularly important for individual studies with very small sample sizes using acquisition protocols that are seldomly used by other datasets. One of the remaining questions in applying this model is how to select the best reference image if large-scale harmonization was applied. Furthermore, given the limited availability of ground truth data, direct comparisons with other methods cannot quantifiably confirm improved harmonization. Future work will expand our approach to 3D to allow indirect comparisons with other harmonization methods using features derived from the full image, such as cortical thickness measurements, and tested against established associations with demographic factors including age.

Acknowledgements. This work was supported in part by: R01AG059874, RF1AG057892, U01AG068057, and P41EB015922. BrightFocus Research Grant award (A2019052S). This research has been conducted using the UK Biobank Resource under Application Number '11559'. Data used in the preparation of this article were also obtained from the Parkinson's Progression Markers Initiative (PPMI) database (www.ppmi-info.org/data), the Alzheimer's Disease Neuroimaging Initiative (ADNI) database (adni.loni.usc.edu), and the Adolescent Brain Cognitive Development (ABCD) Study (https://abcdstudy.org), held in the NIMH Data Archive (NDA).

References

1. Fortin, J.P., et al.: Harmonization of cortical thickness measurements across scanners and sites. Neuroimage **167**, 104–120 (2017)
2. Dinsdale, N., Jenkinson, M., Namburete, A.: Deep learning-based unlearning of dataset bias for MRI harmonisation and confound removal. NeuroImage **228**, 117689 (2021)
3. Dong, J., Cong, Y., Sun, G., Zhong, B., Xu, X..: What can be transferred: unsupervised domain adaptation for endoscopic lesions segmentation. In: Proceedings of the IEEE/CVF Conference on Computer Vision and Pattern Recognition, pp. 4023–4032 (2020)
4. Dewey, B.E., et al.: DeepHarmony: a deep learning approach to contrast harmonization across scanner changes. Mag. Reson. Imaging **64**, 160–170 (2019)
5. Moyer, D., et al.: Scanner invariant representations for diffusion MRI harmonization. Mag. Reson. Med. **84**(4), 2174–2189 (2020)
6. Zhao, F., et al.: Harmonization of infant cortical thickness using surface-to-surface cycle-consistent adversarial networks. In: Shen, Dinggang, et al. (eds.) MICCAI 2019. LNCS, vol. 11767, pp. 475–483. Springer, Cham (2019). https://doi.org/10.1007/978-3-030-32251-9_52
7. Tong, Q., et al.: Multicenter dataset of multi-shell diffusion MRI in healthy traveling adults with identical settings. Sci. Data **7**(1), 1–7 (2020)

8. Huang, X., Liu, M.-Y., Belongie, S., Kautz, J.: Multimodal unsupervised image-to-image translation. In: Ferrari, V., Hebert, M., Sminchisescu, C., Weiss, Y. (eds.) ECCV 2018. LNCS, vol. 11207, pp. 179–196. Springer, Cham (2018). https://doi.org/10.1007/978-3-030-01219-9_11

9. Dewey, B.E., et al.: A disentangled latent space for cross-site MRI harmonization. In: Martel, A.L., et al. (eds.) MICCAI 2020. LNCS, vol. 12267, pp. 720–729. Springer, Cham (2020). https://doi.org/10.1007/978-3-030-59728-3_70

10. Jiang, J., Veeraraghavan, H.: Unified cross-modality feature disentangler for unsupervised multi-domain MRI abdomen organs segmentation. In: Martel, A.L., et al. (eds.) MICCAI 2020. LNCS, vol. 12262, pp. 347–358. Springer, Cham (2020). https://doi.org/10.1007/978-3-030-59713-9_34

11. Choi, Y., et al.: Stargan v2: diverse image synthesis for multiple domains. In: Proceedings of the IEEE/CVF Conference on Computer Vision and Pattern Recognition, pp. 8188–8197 (2020)

12. Karras, T., Laine, S., Aila, T.: A style-based generator architecture for generative adversarial networks. In: Proceedings of the IEEE/CVF Conference on Computer Vision and Pattern Recognition, pp. 4401–4410 (2019)

13. Huang, X., Belongie, S.: Arbitrary style transfer in real-time with adaptive instance normalization. In: Proceedings of the IEEE International Conference on Computer Vision, pp. 1501–1510 (2017)

14. Choi, Y., Choi, M., Kim, M., Ha, J.W., Kim, S., Choo, J.: StarGAN: unified generative adversarial networks for multi-domain image-to-image translation. In: Proceedings of the IEEE Conference on Computer Vision and Pattern Recognition, pp. 8789–8797 (2018)

15. Wang, X., et al.: ESRGAN: enhanced super-resolution generative adversarial networks. In: Leal-Taixé, L., Roth, S. (eds.) ECCV 2018. LNCS, vol. 11133, pp. 63–79. Springer, Cham (2019). https://doi.org/10.1007/978-3-030-11021-5_5

16. Tohka, J., Zijdenbos, A., Evans, A.: Fast and robust parameter estimation for statistical partial volume models in brain MRI. Neuroimage 23(1), 84–97 (2004)

17. Ronneberger, O., Fischer, P., Brox, T.: U-net: Convolutional networks for biomedical image segmentation. In: Navab, N., Hornegger, J., Wells, W.M., Frangi, A.F. (eds.) MICCAI 2015. LNCS, vol. 9351, pp. 234–241. Springer, Cham (2015). https://doi.org/10.1007/978-3-319-24574-4_28

Machine Learning - Federated Learning

Federated Semi-supervised Medical Image Classification via Inter-client Relation Matching

Quande Liu[1(✉)], Hongzheng Yang[2], Qi Dou[1], and Pheng-Ann Heng[1,3]

[1] Department of Computer Science and Engineering, The Chinese University of Hong Kong, Kowloon, Hong Kong SAR, China
{qdliu,qdou}@cse.cuhk.edu.hk
[2] Department of Computer Science and Engineering, Beihang University, Beijing, China
[3] Shenzhen Key Laboratory of Virtual Reality and Human Interaction Technology, Shenzhen Institutes of Advanced Technology, Chinese Academy of Sciences, Shenzhen, China

Abstract. Federated learning (FL) has emerged with increasing popularity to collaborate distributed medical institutions for training deep networks. However, despite existing FL algorithms only allow the supervised training setting, most hospitals in realistic usually cannot afford the intricate data labeling due to absence of budget or expertise. This paper studies a practical yet challenging FL problem, named *Federated Semi-supervised Learning* (FSSL), which aims to learn a federated model by jointly utilizing the data from both labeled and unlabeled clients (i.e., hospitals). We present a novel approach for this problem, which improves over traditional consistency regularization mechanism with a new inter-client relation matching scheme. The proposed learning scheme explicitly connects the learning across labeled and unlabeled clients by aligning their extracted disease relationships, thereby mitigating the deficiency of task knowledge at unlabeled clients and promoting discriminative information from unlabeled samples. We validate our method on two large-scale medical image classification datasets. The effectiveness of our method has been demonstrated with the clear improvements over state-of-the-arts as well as the thorough ablation analysis on both tasks (Code will be made available at https://github.com/liuquande/FedIRM).

Keywords: Federated learning · Semi-supervised learning · Medical image classification

1 Introduction

Data collaboration across medical institutions is increasingly desired to mitigate the scarcity and distribution bias of medical images, thereby improving the model performance on important tasks such as disease diagnosis [6,17,28]. Recently, federated learning (FL) has emerged as a privacy-preserving solution for this, which

© Springer Nature Switzerland AG 2021
M. de Bruijne et al. (Eds.): MICCAI 2021, LNCS 12903, pp. 325–335, 2021.
https://doi.org/10.1007/978-3-030-87199-4_31

allows to learn from distributed data sources by aggregating the locally learned model parameters without exchanging the sensitive health data [8,12,14,16,24]. However, despite progress achieved, existing FL algorithms typically only allow the supervised training setting [3,13,15,25,27], which has limited the local clients (i.e., hospitals) without data annotations to join the FL process. Yet, in realistic scenarios, most hospitals usually cannot afford the intricate data labeling due to lack of budget or expertise [23]. How to utilize these widely-existing unlabeled datasets to further improve the FL models is still an open question to be solved.

To this end, we study a practical FL problem which involves only several labeled clients while most of the participating clients are unlabeled, namely *federated semi-supervised learning (FSSL)*. This is also noted by Yang et al. [31] very recently in COVID-19 lesion segmentation, which halts in an extremely simple case containing only one labeled and one unlabeled client. In contrast to their work, we for the first time broaden this problem to a more practical yet complex scenario in which multiple distributed labeled and unlabeled clients are involved. To address this problem, a naive solution is to simply integrate the off-the-rack semi-supervised learning (SSL) methods onto the federated learning paradigm. However, previous SSL methods are typically designed for centralized training setting [2,10,18,30], which *rely heavily on the assumption that the labeled data is accessible to provide necessary assistance for the learning from unlabeled data* [1,4]. In consistency-based methods [5,32], for instance, the regularization of perturbation-invariant model predictions needs the synchronous labeled data supervision, in order to obtain the necessary task knowledge to produce reliable model predictions for unlabeled data where the consistency regularization is imposed on. Unfortunately, such close assistance from labeled data is lost in FSSL scenario, where the local dataset could be completely unlabeled. This will make the local model aloof from original task as the consistency-based training goes on, hence fail to fully exploit the knowledge at unlabeled clients.

Based on the above issues, compared with traditional SSL problem, the main challenge in FSSL lies in how to build the interaction between the learning at labeled and unlabeled clients, given the challenging constraint of data decentralization. In this work, our insight is to communicate their knowledge inherent in disease relationships to achieve this goal. The idea is motivated by an observation that the relationships exist naturally among different categories of disease and reflect the structural task knowledge in the context of medical image classification, as evidenced by disease similarity measure [19,21]. More importantly, such disease relationships are independent of the observed hospitals, i.e., similar disease at one hospital should also be high-related at others. We may consider extracting such client-invariant disease relation information from labeled clients to supervise the learning at unlabeled clients, thereby mitigating the loss of task knowledge at unlabeled clients and effectively exploiting the unlabeled samples.

In this paper, we present *to our knowledge the first FSSL framework* for medical image classification, by exploring the client-independent disease relation information to facilitate the learning at unlabeled clients. Our method roots in the state-of-the-art consistency regularization mechanism, which enforces the prediction consistency under different input perturbations to exploit the unlabeled data. To address the loss of task knowledge at unlabeled clients which would lead to degenerated learning, we introduce a novel *Inter-client Relation Matching* scheme, by explicitly regularizing the unlabeled clients to capture similar disease relationships as labeled clients for preserving the discriminative task knowledge. To this end, we propose to derive the disease relation matrix at labeled clients from pre-softmax features, and devise an uncertainty-based scheme to estimate reliable relation matrix at unlabeled clients by filtering out inaccurate pseudo labels. We validate our method on two large-scare medical image classification datasets, including intracranial hemorrhage diagnosis with 25000 CT slices and skin lesion diagnosis with 10015 dermoscopy images. Our method achieves large improvements by utilizing the unlabeled clients, and clearly outperforms the combination of federated learning with state-of-the-art SSL methods.

2 Method

In FSSL scenario, we denote $\mathcal{D}_L = \{\mathcal{D}^1, \mathcal{D}^2, \ldots, \mathcal{D}^m\}$ be the collection of m labeled clients, where each labeled client l contains N^l data and one-hot label pairs $\mathcal{D}^l = \{(x_i^l, y_i^l)\}_{i=1}^{N^l}$; and let $\mathcal{D}_U = \{\mathcal{D}^{m+1}, \mathcal{D}^{m+2}, \ldots, \mathcal{D}^{m+n}\}$ be the n unlabeled clients, with each unlabeled client u containing N^u data samples $\mathcal{D}^u = \{(x_i^u)\}_{i=1}^{N^u}$. The goal of FSSL is to learn a global federated model f_θ jointly utilizing the data from both labeled and unlabeled clients. Figure 1 gives an overview of our proposed FSSL solution, i.e. FedIRM, in comparison with the naive FSSL solution.

2.1 Backbone Federated Semi-supervised Learning Framework

Our method follows standard FL paradigm which involves the communication between a central server and local clients. Specifically, in each federated round, every client k will receive the same global model parameters θ from the central server, and update the model with local learning objective $\mathcal{L}^k$ for e epochs on its private data $\mathcal{D}^k$. The central server then collects the local model parameters θ^k from all clients and aggregate them to update the global model. Such process repeats until the global model converges stably. In this work, we adopt the well-established federated averaging algorithm [20] (FedAvg) to update the global model, by aggregating the local model parameters with weights in proportional to the size of each local dataset, i.e., $\theta = \sum_{k=1}^{K} \frac{N^k}{N} \theta^k$, where $N = \sum_{k=1}^{K} N^k$.

In our FSSL solution, the local learning objective at labeled clients adopts the cross entropy loss for capturing the discriminative task knowledge. At unlabeled clients, we preserve the state-of-the-art consistency regularization mechanism, which exploits the unlabeled data in an unsupervised manner by enforcing the consistency of model predictions under input perturbations. Formally, this learning objective at each unlabeled client u could be expressed as:

$$\mathcal{L}_c(\mathcal{D}^u, \theta^u) = \sum_{i=1}^{N^u} \mathbb{E}_{\xi, \xi'} ||f_{\theta^u}(x_i^u, \xi), f_{\theta^u}(x_i^u, \xi')||_2^2 \tag{1}$$

where ξ and ξ' denote different input perturbations (e.g., adding Gaussian noise).

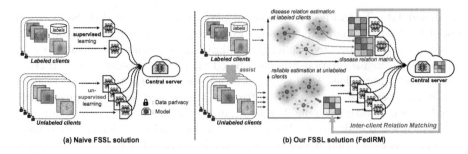

Fig. 1. (a) Naive FSSL solution simply performs unsupervised learning (e.g. consistency regularization) at unlabeled clients, hence the local model is prone to forget the original task knowledge as the training goes on. (b) Our FedIRM explicitly utilizes the knowledge from labeled clients to assist the learning at unlabeled clients by aligning their extracted disease relationships, thereby mitigating the loss of task knowledge at unlabeled clients and promoting discriminative information from unlabeled data.

2.2   Disease Relation Estimation at Labeled and Unlabeled Clients

Without the assistance from labeled data supervision, the local learning at unlabeled clients solely with consistency regularization is prone to forget the original task knowledge, therefore failing to fully exploit the information from unlabeled samples. To tackle this problem, we introduce a novel *inter-client relation matching* (IRM) scheme, which explicitly extracts the knowledge from labeled clients to assist the learning at unlabeled clients, by exploiting the rich information inherent in disease relationships. Specifically, the relationships exist naturally across different categories of disease and reflect structural task knowledge in medical image classification, *independent of the changes of observed hospitals*. In light of this, we aim to enforce the alignment of such disease relations across labeled and unlabeled clients, thereby promoting the learning of discriminative information at unlabeled clients for preserving such structural task knowledge.

Disease Relation Estimation at Labeled Clients. Inspired by knowledge distillation from deep networks, we estimate the disease relationships from the class ambiguity captured by deep models, i.e., per-class soft labels, and enforce them to be consistent between labeled and unlabeled clients. Formally, we first consider the relation estimation at labeled clients. For each labeled client $\mathcal{D}^l$, we summarize the model's knowledge on each class c by computing per-category mean feature vectors $\mathbf{v}_c^l \in \mathbb{R}^C$ (with C denoting total class number):

$$\mathbf{v}_c^l = \frac{1}{N_c^l} \sum_{i=1}^{N^l} \mathbb{1}_{[y_i^l = c]} \hat{f}_{\theta^l}(x_i^l) \tag{2}$$

where N_c^l is the number of samples with class c at labeled client $\mathcal{D}^l$; $\mathbb{1}_{[\cdot]}$ denotes the indicator function, $\hat{f}$ denotes the model without last softmax layer. The obtained $\mathbf{v}_c^l$ is then scaled to a soft label distribution, with a softened softmax function under temperature $\tau > 1$ [7]:

$$\mathbf{s}_c^l = \text{softmax}(\mathbf{v}_c^l/\tau) \tag{3}$$

This distilled knowledge of soft label $\mathbf{s}_c^l$ conveys how the network predictions of samples on certain class generally distribute across all classes, reflecting the relationships across different classes captured by the deep model. Consequently, the collection of soft labels from all classes could form a soft confusion matrix $\mathcal{M}^l = [\mathbf{s}_1^l, \ldots, \mathbf{s}_C^l]$, which encodes the inter-class relationships among different categories of disease hence serve as the disease relation matrix.

Reliable Disease Relation Estimation at Unlabeled Clients. Since the data annotations are unavailable at unlabeled clients, we utilize the pseudo labels generated from model predictions to estimate the disease relation matrix. However, without sufficient task knowledge provided at unlabeled clients, the model predictions on unlabeled data could be noisy and inaccurate. We therefore employ an uncertainty-based scheme to filter out the unreliable model predictions, and only preserve the trustworthy ones to measure the reliable relation matrix.

Specifically, we take the local training at unlabeled client $\mathcal{D}^u$ for instance. Given an input mini-batch $\mathbf{x}^u$ of B image, we denote $\mathbf{p^u}$ as the corresponding predicted probability and $\mathbf{y^u}$ as the pseudo labels, i.e., $\mathbf{y^u} = \text{argmax}(\mathbf{p^u})$. Following the literature on uncertainty estimation, we approximate the uncertainty of model predictions with dropout of Bayesian networks [9]. Concretely, we perform T-time forward propagation for the input mini-batch $\mathbf{x}^u$ under random dropout, obtaining a set of predicted probability vectors $\{\mathbf{q}_t^u\}_{t=1}^T$. The uncertainty $\mathbf{w}^u$ is then estimated as the predictive entropy, which is computed from the averaged probability from the T-time forward passes as:

$$\mathbf{w}^u = -\sum_{c=1}^C \overline{\mathbf{q}}_{(c)}^u \log(\overline{\mathbf{q}}_{(c)}^u), \text{ with } \overline{\mathbf{q}}_{(c)}^u = \frac{1}{T}\sum_{t=1}^T \mathbf{q}_{t(c)}^u \tag{4}$$

where $\mathbf{q}_{t(c)}^u$ is the value of the c-th class of $\mathbf{q}_t^u$. Since the predictive entropy has a fixed range, we can filter out the relatively unreliable predictions and only select the certain ones to compute the disease relation matrix. Hence, the per-category mean feature vectors $\mathbf{v}_c^u$ (c.f. Eq. 2) at unlabeled clients are computed as:

$$\mathbf{v}_c^u = \frac{\sum_{i=1}^B \mathbb{1}_{[(\mathbf{y}_i=c)\cdot(\mathbf{w}_i^u<h)]} \cdot \mathbf{p}_i^u}{\sum_{i=1}^B \mathbb{1}_{[(\mathbf{y}_i=c)\cdot(\mathbf{w}_i^u<h)]}} \tag{5}$$

where h is the threshold to select the certain predictions from $\mathbf{w}^u$. Then, following the same operation as Eq. 3, the disease relation matrix at unlabeled client u is estimated as $\mathcal{M}^u = [\mathbf{s}_1^u, \mathbf{s}_2^u, \ldots, \mathbf{s}_C^u]$.

2.3   Objective of Inter-client Relation Matching

With the above basis, we aim to enforce the unlabeled clients to produce similar disease relationships as labeled clients to preserve such discriminative task knowledge. Specifically, at the end of each federated round, the central server collects the relation matrix $\mathcal{M}^l$ from each labeled client, and average them to compute a matrix representing the general disease relation information captured from all labeled data, i.e., $\mathcal{M} = \frac{1}{m}\sum_{l=1}^{m}\mathcal{M}^l$. This obtained $\mathcal{M}$ is then delivered to unlabeled clients to supervise their next round of local training. To establish the supervision online, the relation matrix at unlabeled clients $\mathcal{M}^u$ is estimated from each mini-batch during training. Finally, the inter-client relation matching loss is designed by minimizing the KL divergence between $\mathcal{M}$ and $\mathcal{M}^u$ as:

$$\mathcal{L}_{\mathrm{IRM}} = \frac{1}{C}\sum_{c=1}^{C}(\mathcal{L}_{\mathrm{KL}}(\mathcal{M}_c||\mathcal{M}_c^u) + \mathcal{L}_{\mathrm{KL}}(\mathcal{M}_c^u||\mathcal{M}_c)),$$

$$\mathrm{with}\ \mathcal{L}_{\mathrm{KL}}(\mathcal{M}_c||\mathcal{M}_c^u)) = \sum_j \mathcal{M}_{c(j)}\log\frac{\mathcal{M}_{c(j)}}{\mathcal{M}_{c(j)}^u} \tag{6}$$

where $\mathcal{M}_c \in \mathbb{R}^C$ denote the relation vector of class c, i.e., $\mathcal{M}_c = \mathbf{s}_c$; and $\mathcal{M}_{c(j)}$ denote its j-th entry. Overall, the local learning objectives at labeled ($\mathcal{L}^l$) and unlabeled ($\mathcal{L}^u$) clients are respectively expressed as:

$$\mathcal{L}^l = \mathcal{L}_{ce}(\mathcal{D}^l, \theta^l)\ \mathrm{and}\ \mathcal{L}^u = \lambda(\omega)(\mathcal{L}_c + \mathcal{L}_{\mathrm{IRM}}) \tag{7}$$

where $\mathcal{L}_{ce}$ is the cross entropy loss; $\mathcal{L}_c$ is the traditional consistency regularization loss (c.f. Eq. 1); $\lambda(\omega)$ is a warming up function regarding federated round ω, which helps to reduce the effect of the learning at unlabeled clients when the model is underfitting at earlier federated rounds.

3   Experiments

3.1   Dataset and Experimental Setup

We validate our method on two important medical image classification tasks, including: intracranial hemorrhage (ICH) diagnosis from brain CT and skin lesion classification from dermoscopy images.

Task 1 - Intracranial Hemorrhage Diagnosis. We perform ICH diagnosis with the RSNA ICH Detection dataset[26], which aims to classify CT slices into 5 subtypes of ICH disease. Since most images in this dataset are healthy without any of the subtypes, we randomly sample 25000 slices from the dataset which contain one of the 5 subtypes of ICH for evaluation. These samples are then randomly divided into 70%, 10% and 20% for training, validation and testing. Since multiple slices may come from the same patient in this dataset, we have ensured no overlapped patients exist across the three split for a valid evaluation.

Table 1. Quantitative comparisons with state-of-the-arts on two different tasks.

Method	Client num		Metrics				
	Label	Unlabel	AUC	Sensitivity	Specificity	Accuracy	F1
Task 1: Intracranial hemorrhage diagnosis							
FedAvg [20]	10	0	90.48 ± 0.31	64.33 ± 1.13	92.68 ± 0.43	89.94 ± 0.92	63.94 ± 1.20
FedAvg [20]	2	0	83.40 ± 0.87	57.88 ± 1.68	90.48 ± 0.79	87.45 ± 1.08	57.10 ± 1.29
Fed-SelfTraining [33]	2	8	84.32 ± 0.82	57.94 ± 1.66	90.22 ± 0.74	87.90 ± 1.81	57.48 ± 1.14
Fed-Consistency [31]	2	8	84.83 ± 0.79	57.26 ± 1.93	90.87 ± 0.62	88.35 ± 1.32	57.61 ± 1.08
FedIRM (ours)	2	8	$\mathbf{87.56 \pm 0.56}$	$\mathbf{59.57 \pm 1.57}$	$\mathbf{91.53 \pm 0.81}$	$\mathbf{88.89 \pm 1.29}$	$\mathbf{59.86 \pm 1.65}$
Task 2: Skin Lesion Diagnosis							
FedAvg [20]	10	0	94.82 ± 0.32	75.11 ± 1.82	94.87 ± 0.35	95.24 ± 0.21	70.16 ± 1.21
FedAvg [20]	2	0	90.65 ± 1.23	65.53 ± 1.76	91.76 ± 0.48	92.53 ± 0.67	52.59 ± 1.42
Fed-SelfTraining [33]	2	8	90.82 ± 0.56	67.03 ± 1.93	$\mathbf{93.61 \pm 0.21}$	92.47 ± 0.34	53.44 ± 1.85
Fed-Consistency [31]	2	8	91.13 ± 0.62	68.55 ± 1.29	93.45 ± 0.94	92.67 ± 0.39	54.25 ± 1.31
FedIRM (ours)	2	8	$\mathbf{92.46 \pm 0.45}$	$\mathbf{69.05 \pm 1.71}$	93.29 ± 0.59	$\mathbf{92.89 \pm 0.25}$	$\mathbf{55.81 \pm 1.49}$

Task 2 - Skin Lesion Diagnosis. We employ ISIC 2018: Skin Lesion Analysis Towards Melanoma Detection[22] dataset for skin lesion diagnosis, which contains 10015 dermoscopy images in the official training set labeled by 7 types of skin lesions. As the ground truth of official validation and testing set was not released, we randomly divide the entire training set to 70% for training, 10% for validation and 20% for testing. We perform the same data pre-processing for the two tasks. Specifically, we first resized the original images from 512×512 to 224×224. To employ the pre-trained model, we then normalized the images with statistic collected from ImageNet dataset before feeding them into the network.

Experiment Setup. To simulate the FL setting, we randomly partition the training set into 10 different subsets serving as 10 local clients. Following the practice in SSL [29], we evaluate the model performance under 20% labeled data setting, i.e., two clients are labeled and the remaining eight are unlabeled in our case. Five metrics are used to extensively evaluate the classification performance, including AUC, Sensitivity, Specificity, Accuracy and F1 score. We report the results in form of average and standard deviation over three independent runs.

Implementation Details. We employ DenseNet121 [11] as the backbone for medical image classification. Two types of perturbations are utilized to drive the consistency regularization, including random transformation on input data (rotation, translation and flip) and dropout layer in the network. The temperature parameter τ is empirically set as 2.0. The forward pass time T used to compute uncertainty is set as 8, and the threshold h to select reliable predictions is set as ln2. We follow [29] to apply a Gaussian warming up function $\lambda(\omega) = 1 * e^{(-5(1-\omega/\Omega))}$, where Ω is set as 30. The local training adopts Adam optimizer with momentum of 0.9 and 0.99, and the batch size is 48 for both labeled and unlabeled clients. We totally train 100 federated rounds when the global model has converged stably, with the local training epoch e set as 1.

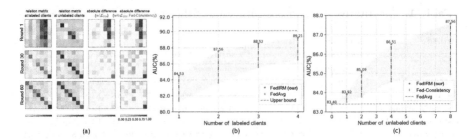

Fig. 2. Ablation analysis. (a) Disease relation matrix at labeled and unlabeled clients under our method, as well as their absolute difference with or without $\mathcal{L}_{IRM}$ (task 2); (b) Model performance under different labeled client number setting, using our approach and FedAvg (task 1); (c) Model performance as the number of unlabeled client increases (with labeled client number fixed), using our approach and Fed-Consistency (task 1).

3.2 Comparison with State-of-the-Arts

We compare with recent FSSL methods, including **Fed-SelfTraining** [33], which performs self training at unlabeled clients by iteratively updating the model parameters and the pseduo labels of unlabeled data with expectation maximization; and **Fed-Consistency** [31], which employs the state-of-the-art SSL strategy, i.e., consistency regularization, to exploit the data at unlabeled clients *(without inter-client relation matching compared with our method)*. We also compare with the **FedAvg** [20] model trained only with labeled clients or with all clients as labeled, which serve as the baseline and upperbound performance in FSSL.

The results on the two tasks are listed in Table 1. As observed, both Fed-SelfTraining and Fed-Consistency performs better than the baseline FedAvg model, which reflects the benefit to integrate the knowledge from additional unlabeled clients to improve FL models. Notably, compared with these methods, our FedIRM achieves higher performance on nearly all metrics, with 2.73% and 1.33% AUC improvements on the two tasks over Fed-Consistency which does not employ our inter-client relation matching scheme. These clear improvements benefit from our FedIRM scheme which explicitly harnesses the discriminative relation information learned from labeled clients to facilitate the learning at unlabeled clients. Without the supervision from $\mathcal{L}_{IRM}$, the local training simply from consistency regularization is prone to forget the original task information, hence fail to fully exploit the discriminative information from unlabeled data.

3.3 Analytical Studies of Our Method

Learning Behavior Under Inter-client Relation Matching. Figure 2(a) displays the disease relation matrix of labeled (first col.) and unlabeled clients (second col.) under our method, as well as their absolute difference under our method with (third col.) and without $\mathcal{L}_{IRM}$ (i.e., Fed-Consistency, forth col.), at different federated rounds. As observed, the relationships between disease at labeled clients become increasingly clear as the federated training goes on, indicating that the model gradually captures such structural knowledge. Notably, the unlabeled clients in our method can well preserve such disease relationships, with highly consistent matrix patterns as labeled clients and low responses in the difference matrix. In contrast, the method without $\mathcal{L}_{IRM}$ (i.e., Fed-Consistency) fails to do so and the responses in difference matrix are relatively high. This observation affirms the benefit to transfer such discriminative knowledge to facilitate the learning at unlabeled clients and also explains our performance gains.

Effectiveness Under Different Labeled Client Number. We investigate the impact of different labeled client number in our method. As shown in the curve of Fig. 2(b), our method shows consistent improvements over the supervised-only FedAvg model under labeled client number from 1 to 4 (corresponds 10% to 40% labeled data setting in SSL). Importantly, using only 40% labeled client, our method achieves 89.21% AUC, which is very close to the upper-bound FedAvg model trained with 10 labeled clients (90.48%). This endorses the capability of our method to leverage the data from unlabeled clients for improving FL models.

Effect of Adding More Unlabeled Clients. We finally analyze the effect of unlabeled client number on the performance of our FSSL method and Fed-Consistency, by fixing the labeled client number as 2 and gradually increasing the unlabeled client number in $[1, 2, 4, 8]$. As shown in Fig. 2(c), an interesting finding is the FSSL performance progresses as the unlabeled client number increases, indicating the potential in realistic scenarios to aggregate more widely-existing unlabeled clients to improve the FL models. Notably, our method consistently outperforms the Fed-Consistency method under different unlabeled client number, highlighting the stable capacity of our proposed FSSL learning scheme.

4 Conclusion

We present a new FSSL framework, which to our knowledge is the first method incorporating unlabeled clients to improve FL models for medical image classification. To address the deficiency of consistency regularization in FSSL, our method includes a novel inter-client relation matching scheme to explicitly utilize the knowledge of labeled clients to assist the learning at unlabeled clients. Experiments on two large-scare datasets demonstrate the effectiveness. Our method is extendable to non-IID scenario in FSSL setting, as the employed disease relations are independent of the observed clients and unaffected by image distributions.

Acknowledgement. The work described in this paper was supported in parts by the following grants: Key-Area Research and Development Program of Guangdong Province, China (2020B010165004), Hong Kong Innovation and Technology Fund (Project No. GHP/110/19SZ), Foundation of China with Project No. U1813204 and Shenzhen-HK Collaborative Development Zone.

References

1. Aviles-Rivero, A.I., et al.: GraphX^{NET} – chest x-ray classification under extreme minimal supervision. In: Shen, D., et al. (eds.) MICCAI 2019. LNCS, vol. 11769, pp. 504–512. Springer, Cham (2019). https://doi.org/10.1007/978-3-030-32226-7_56

2. Bai, W., et al.: Semi-supervised learning for network-based cardiac MR image segmentation. In: Descoteaux, M., Maier-Hein, L., Franz, A., Jannin, P., Collins, D.L., Duchesne, S. (eds.) MICCAI 2017. LNCS, vol. 10434, pp. 253–260. Springer, Cham (2017). https://doi.org/10.1007/978-3-319-66185-8_29

3. Chang, Q., Qu, H., Zhang, Y., Sabuncu, M., Chen, C., Zhang, T., Metaxas, D.N.: Synthetic learning: learn from distributed asynchronized discriminator GAN without sharing medical image data. In: CVPR, pp. 13856–13866 (2020)

4. Cheplygina, V., de Bruijne, M., Pluim, J.P.: Not-so-supervised: a survey of semi-supervised, multi-instance, and transfer learning in Mia. MIA **54**, 280–296 (2019)

5. Cui, W., et al.: Semi-supervised brain lesion segmentation with an adapted mean teacher model. In: Chung, A.C.S., Gee, J.C., Yushkevich, P.A., Bao, S. (eds.) IPMI 2019. LNCS, vol. 11492, pp. 554–565. Springer, Cham (2019). https://doi.org/10.1007/978-3-030-20351-1_43

6. Dhruva, S.S., Ross, J.S., Akar, J.G., Caldwell, B., Childers, K., et al.: Aggregating multiple real-world data sources using a patient-centered health-data-sharing platform. NPJ Digit. Med. **3**(1), 1–9 (2020)

7. Dou, Q., Liu, Q., Heng, P.A., Glocker, B.: Unpaired multi-modal segmentation via knowledge distillation. IEEE TMI **39**(7), 2415–2425 (2020)

8. Dou, Q., So, T.Y., Jiang, M., Liu, Q., Vardhanabhuti, V., Kaissis, G., et al.: Federated deep learning for detecting COVID-19 lung abnormalities in CT: a privacy-preserving multinational validation study. NPJ Digit. Med. **4**(1), 1–11 (2021)

9. Gal, Y., Ghahramani, Z.: Dropout as a Bayesian approximation: representing model uncertainty in deep learning. In: ICML, pp. 1050–1059. PMLR (2016)

10. Gyawali, P.K., Ghimire, S., Bajracharya, P., Li, Z., Wang, L.: Semi-supervised medical image classification with global latent mixing. In: Martel, A.L., et al. (eds.) MICCAI 2020. LNCS, vol. 12261, pp. 604–613. Springer, Cham (2020). https://doi.org/10.1007/978-3-030-59710-8_59

11. Huang, G., Liu, Z., Van Der Maaten, L., Weinberger, K.Q.: Densely connected convolutional networks. In: CVPR, pp. 4700–4708 (2017)

12. Kaissis, G.A., Makowski, M.R., Rückert, D., Braren, R.F.: Secure, privacy-preserving and federated machine learning in medical imaging. Nat. Mach. Intell. 1–7 (2020)

13. Li, D., Kar, A., Ravikumar, N., Frangi, A.F., Fidler, S.: Federated simulation for medical imaging. In: Martel, A.L., et al. (eds.) MICCAI 2020. LNCS, vol. 12261, pp. 159–168. Springer, Cham (2020). https://doi.org/10.1007/978-3-030-59710-8_16

14. Li, W., et al.: Privacy-preserving federated brain tumour segmentation. In: Suk, H.-I., Liu, M., Yan, P., Lian, C. (eds.) MLMI 2019. LNCS, vol. 11861, pp. 133–141. Springer, Cham (2019). https://doi.org/10.1007/978-3-030-32692-0_16

15. Li, X., Gu, Y., Dvornek, N., Staib, L.H., Ventola, P., Duncan, J.S.: Multi-site FMRI analysis using privacy-preserving federated learning and domain adaptation: abide results. MIA **65**, 101765 (2020)

16. Liu, Q., Chen, C., Qin, J., Dou, Q., Heng, P.A.: FedDG: federated domain generalization on medical image segmentation via episodic learning in continuous frequency space. In: CVPR (2021)

17. Liu, Q., Dou, Q., Yu, L., Heng, P.A.: MS-Net: multi-site network for improving prostate segmentation with heterogeneous MRI data. IEEE Trans. Med. Imaging **39**, 2713–2724 (2020)
18. Liu, Q., Yu, L., Luo, L., et al.: Semi-supervised medical image classification with relation-driven self-ensembling model. IEEE TMI **39**(11), 3429–3440 (2020)
19. Mathur, S., Dinakarpandian, D.: Finding disease similarity based on implicit semantic similarity. J. Biomed. Inform. **45**(2), 363–371 (2012)
20. McMahan, B., Moore, E., Ramage, D., Hampson, S., Arcas, B.A.: Communication-efficient learning of deep networks from decentralized data. In: Artificial Intelligence and Statistics, pp. 1273–1282 (2017)
21. Oerton, E., Roberts, I., Lewis, P.S., et al.: Understanding and predicting disease relationships through similarity fusion. Bioinformatics **35**(7), 1213–1220 (2019)
22. Tschandl, P., Rosendahl, C., Kittler, H.: The HAM10000 dataset, a large collection of multi-source dermatoscopic images of common pigmented skin lesions. Sci. Data **5**, 1–9 (2018)
23. Razzak, M.I., Naz, S., Zaib, A.: Deep learning for medical image processing: overview, challenges and the future. In: Dey, N., Ashour, A.S., Borra, S. (eds.) Classification in BioApps. LNCVB, vol. 26, pp. 323–350. Springer, Cham (2018). https://doi.org/10.1007/978-3-319-65981-7_12
24. Rieke, N., Hancox, J., Li, W., Milletari, F., Roth, H.R., et al.: The future of digital health with federated learning. NPJ Digit. Med. **3**(1), 1–7 (2020)
25. Roth, H.R., et al.: Federated learning for breast density classification: a real-world implementation. In: Albarqouni, S., et al. (eds.) DART/DCL -2020. LNCS, vol. 12444, pp. 181–191. Springer, Cham (2020). https://doi.org/10.1007/978-3-030-60548-3_18
26. RSNA: Intracranial hemorrhage detection challenge (2019). https://www.kaggle.com/c/rsna-intracranial-hemorrhage-detection/
27. Sheller, M.J., Reina, G.A., Edwards, B., Martin, J., Bakas, S.: Multi-institutional deep learning modeling without sharing patient data: a feasibility study on brain tumor segmentation. In: Crimi, A., Bakas, S., Kuijf, H., Keyvan, F., Reyes, M., van Walsum, T. (eds.) BrainLes 2018. LNCS, vol. 11383, pp. 92–104. Springer, Cham (2019). https://doi.org/10.1007/978-3-030-11723-8_9
28. Silva, S., Gutman, B.A., Romero, E., Thompson, P.M., Altmann, A., Lorenzi, M.: Federated learning in distributed medical databases: meta-analysis of large-scale subcortical brain data. In: ISBI, pp. 270–274. IEEE (2019)
29. Tarvainen A., V.H.: Mean teachers are better role models: weight-averaged consistency targets improve semi-supervised deep learning results. In: Advances Neural Information Processing Systems (2017)
30. Wang, D., Zhang, Y., Zhang, K., Wang, L.: Focalmix: semi-supervised learning for 3D medical image detection. In: CVPR, pp. 3951–3960 (2020)
31. Yang, D., Xu, Z., Li, W., Myronenko, A., Roth, H.R., Harmon, S., et al.: Federated semi-supervised learning for COVID region segmentation in chest CT using multi-national data from China, Italy, Japan. In: MIA, p. 101992 (2021)
32. Yu, L., Wang, S., Li, X., Fu, C.-W., Heng, P.-A.: Uncertainty-aware self-ensembling model for semi-supervised 3D left atrium segmentation. In: Shen, D., et al. (eds.) MICCAI 2019. LNCS, vol. 11765, pp. 605–613. Springer, Cham (2019). https://doi.org/10.1007/978-3-030-32245-8_67
33. Zhang, Z., Yao, Z., Yang, Y., Yan, Y., Gonzalez, J.E., Mahoney, M.W.: Benchmarking semi-supervised federated learning. arXiv preprint arXiv:2008.11364 (2020)

FedPerl: Semi-supervised Peer Learning for Skin Lesion Classification

Tariq Bdair[1,2](✉)[iD], Nassir Navab[1,3][iD], and Shadi Albarqouni[1,2][iD]

[1] Computer Aided Medical Procedures, Technical University of Munich,
Munich, Germany
{t.bdair,nassir.navab}@tum.de
[2] Helmholtz AI, Helmholtz Zentrum München, Neuherberg, Germany
shadi.albarqouni@helmholtz-muenchen.de
[3] The Whiting School of Engineering, Johns Hopkins University, Baltimore, USA

Abstract. Skin cancer is one of the most deadly cancers worldwide. Yet, it can be reduced by early detection. Recent deep-learning methods have shown a dermatologist-level performance in skin cancer classification. Yet, this success demands a large amount of centralized data, which is oftentimes not available. Federated learning has been recently introduced to train machine learning models in a privacy-preserved distributed fashion demanding annotated data at the clients, which is usually expensive and not available, especially in the medical field. To this end, we propose FedPerl, a semi-supervised federated learning method that utilizes peer learning from social sciences and ensemble averaging from committee machines to build communities and encourage its members to learn from each other such that they produce more accurate pseudo labels. We also propose the peer anonymization (PA) technique as a core component of FedPerl. PA preserves privacy and reduces the communication cost while maintaining the performance without additional complexity. We validated our method on 38,000 skin lesion images collected from 4 publicly available datasets. FedPerl achieves superior performance over the baselines and state-of-the-art SSFL by 15.8%, and 1.8% respectively. Further, FedPerl shows less sensitivity to noisy clients (https://github.com/tbdair/FedPerlV1.0).

Keywords: Semi-supervised federated learning · Skin cancer

1 Introduction

Skin cancer is one of the most costly and deadly cancers in the world. More than 5 million cases are diagnosed with skin cancer in the US every year [17]. In 2021, an estimated 200,000 cases of deadly invasive and in-situ melanoma will be diagnosed in the US [20]. Early detection of skin cancer can reduce the

Electronic supplementary material The online version of this chapter (https://doi.org/10.1007/978-3-030-87199-4_32) contains supplementary material, which is available to authorized users.

M. de Bruijne et al. (Eds.): MICCAI 2021, LNCS 12903, pp. 336–346, 2021.
https://doi.org/10.1007/978-3-030-87199-4_32

treatment cost and death rate substantially [5]. Automated classification of skin lesions using dermoscopic images was investigated in early works [3] demanding exhaustive pre-processing and handcrafted engineered features. Recently, deep learning-based methods have shown dermatologist-level [5,24] or superior performance [8,12,28] in skin cancer classification. Yet, most of these methods rely on a large curated amount of centralized labeled data, which is usually not available due to privacy issues [16].

Federated learning (FL) [13] has been recently introduced to train machine learning models leveraging the abundant amounts of data in a privacy-preserved distributed fashion. FL a.k.a `FedAvg` initiates the training process by sharing initial weight model parameters with the participating clients. Then, each client trains their local model on their local data before sending back the parameters updates to the server (global model). Once all updates are collected, they are aggregated to update the global model parameters. Next, the updated parameters are sent to the participating clients before a new round of training is started. During the training, data is kept locally and only model parameters are shared among the clients. Thus, a couple of works investigated `FedAvg` in the medical domain [2,11,27,29] showing a great potential in training privacy-persevering ML models in real-world scenarios [6,18,19]. Yet, the aforementioned works require highly accurate ground-truth annotated data, e.g., confirmed through histopathology, which oftentimes expensive and not available.

In this paper, we tackle a more realistic scenario, where the clients have a few annotated data besides a large number of unannotated ones (e.g., not confirmed through histopathology). Such a scenario is typically tackled by a semi-supervised learning paradigm. In this context, Yang et al.[26] have proposed semi-supervised federated learning (`SSFL`) for COVID-19 segmentation. The framework has applied a straightforward combination of semi-supervised learning at the clients, e.g. FixMatch [21], and `FedAvg` to coordinate the training among the clients (see Sect. 2.1). While `SSFL` is simple and straightforward, we argue that the knowledge gain for the participating clients is quite limited, particularly in generating pseudo labels. In `SSFL`, clients are only exposed to i) local knowledge, i.e. distilled via the annotated data, and ii) global knowledge, i.e. aggregated through the shared global model parameters. We hypothesize, however, that Peer Learning (PL), i.e. learning from similar clients, is quite essential in empowering self-confidence in the clients by exchanging their knowledge while preserving privacy. Our method is highly inspired by the literature in social science where Peer Learning (PL) is defined as acquiring knowledge through active helping among alike groups [22]. A similar concept has been introduced to the computer science literature known as Committee Machines (CM) [23] which can be defined as an ensemble of estimators (committee members) that collaborate to produce better performance than individuals by ensemble averaging.

To this end, we propose `FedPerl`, an `SSFL` framework, where Peer Learning (PL) and ensemble averaging for the peers anonymization (PA) are the key properties. In contrast to [26], our method employs peer learning such that the clients gain extra knowledge by helping each other in a privacy-preserved way to leverage the unlabeled data, while [26] is limited to the local knowledge by the client itself. Such extra knowledge has been shown to be useful in [9], however, at the cost of communication and privacy. In contrast to [9], our approach is communication efficient, and PA avoids any privacy breaching, model inversion [1,7] and deanonymization attacks [14] by simply employing ensemble before sharing their knowledge to other peers. Our **contributions** are: (i) we propose `FedPerl`, an SSFL framework, inspired by Peer Learning (PL) and Committee Machines (CM), (ii) we propose *peers anonymization* (PA), for the first time in the SSFL, to avoid any privacy breach and reduce the communication cost, while maintains the performance, and most importantly (iii) we evaluate `FedPerl` for skin lesion classification, where more than 38,000 skin lesion images (of which 12% were annotated), distributed in 10 clients, were leveraged. `FedPerl` shows superior performance over the state-of-the-art SSFL and other baselines.

2   Methodology

Given M clients C_j who have access to their own local database $\mathcal{D}_j \in \mathbb{R}^{H \times W \times N_j}$ consisting of labeled $\mathcal{S}_L = \{\mathcal{X}_L, \mathcal{Y}_L\}$ and unlabeled data $\mathcal{S}_U = \{\mathcal{X}_U\}$, where $\{\mathcal{X}_L, \mathcal{X}_U\} = \{x_1, \ldots, x_L, x_{L+1}, \ldots, x_{L+U}\}$ are input images; $x \in \mathbb{R}^{H \times W}$, and $\mathcal{Y}_L = \{y_1, \ldots, y_L\}$; $y \in \mathbb{R}^C$ are the corresponding labels, our objective is to train a global model $f(\cdot)$, in a privacy-preserved fashion, leveraging the unlabeled data to predict the corresponding label $\tilde{y}_q$ for a given query image x_q.

2.1   Semi-Supervised Federated Learning (SSFL)

To meet the aforementioned requirements, one could pick off-the-shelf SoTA SSL models, *e.g.*, FixMatch [21], for the local clients to leverage the unlabeled data, and employ `FedAvg` [13] to train the clients in a federated fashion as [26],

$$\min_{\phi} \mathcal{L}(\mathcal{D}; \phi) \quad \text{with} \quad \mathcal{L}(\mathcal{D}; \phi) = \sum_{j=1}^{M} w_j \, \mathcal{L}_{SSL_j}(\mathcal{D}_j; \phi), \tag{1}$$

where ϕ is the model parameters and $w_j = N_j / \sum_{j=1}^{M} N_j$ is the respective weight coefficient. The SSL objective function, appeared in `FixMatch` [21], can be employed for the clients to train from both labeled and unlabeled data as

$$\mathcal{L}_{SSL_j}(\mathcal{D}_j; \phi) = \arg\min_{\phi} \mathcal{L}_{CE}(\mathcal{Y}_L, f(\alpha(\mathcal{X}_L); \phi)) + \beta \mathcal{L}_{CE}(\tilde{\mathcal{Y}}_U, f(\mathcal{A}(\mathcal{X}_U); \phi)), \tag{2}$$

where $\mathcal{L}_{CE}(\cdot, \cdot)$ is the cross-entropy loss, β is a hyper-parameter, $\alpha(\cdot)$ and $\mathcal{A}(\cdot)$ are weak and strong augmentations respectively, and $\tilde{\mathcal{Y}}_U$ is the pseudo labels for the unlabeled data $\mathcal{X}_U$. The pseudo label $\tilde{y}_i \in \tilde{\mathcal{Y}}_U$, for an unlabeled sample x_i, is generated by applying a confidence threshold τ on the client prediction of an augmented version of x_i, $\tilde{y}_i = \arg\max(\mathbb{I}(f(\alpha(x_i); \phi^*) \geq \tau))$, where $\mathbb{I}(\cdot)$ is the indicator function, and ϕ^* is frozen model parameters.

2.2   Federated Peer Supervised Learning (FedPerl)

Our proposed `FedPerl`, depicted in Fig. 1, consists of three components; namely 1) building communities, 2) peer learning, and 3) peer anonymization.

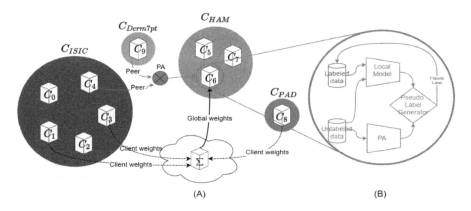

Fig. 1. Illustrative diagram. (A) FedPerl consists of (i) Building communities: similar clients clustered into one community, (ii) Peer Learning: peers are helping in pseudo labeling, and (iii) Peer Anonymization to avoid model inversion & deanonymization. (B) Pseudo labeling *e.g.* an anonymized peer in this diagram.

Building Communities. We borrow the 'peers' definition from social science and call a group of clients as 'peers' if they are similar. To measure the similarity, we represent each client C_j by a feature vector $\mathcal{F}_j = \{(\mu_0, \sigma_0), \dots, (\mu_l, \sigma_l)\} \in \mathbb{R}^{2 \cdot l}$, where (μ_l, σ_l) is the first two statistical moments of the model's layer l parameters, and compute the similarity matrix $\mathcal{W} \in \mathbb{R}^{M \times M}$ where $\omega_{jk} = \frac{\mathcal{F}_j^T \mathcal{F}_k}{\|\mathcal{F}_j\| \cdot \|\mathcal{F}_k\|}$ is the cosine similarity between C_j and C_k. FedPerl starts by standard warm-up rounds (*e.g.* ten rounds in our case). In the next training rounds, the feature vectors for the clients are extracted, and the similarity matrix is computed and updated accordingly. Communities are shaped implicitly based on the similarity matrix where similar members are clustered into one community (see Sect. 3.2).

Peers Learning: In our context, we define 'peer learning' as the process of top T similar clients (peers) help each other by sharing their knowledge (model parameters) to generate pseudo labels. During the training, each peer experiences different data and class distribution. Thus, accumulating and sharing these experiences can help the target peer to generate pseudo labels for the unlabeled data that might never have such experience from its own labeled data. To realize this, the pseudo label is modified to include the predictions of the similar T peers, *i.e.* $f_t(\cdot; \phi)$, according to the similarity matrix $\mathcal{W}$ as

$$\tilde{y}_i = \arg\max \left(\mathbb{I} \left(f_{C_j}(\alpha(x_i); \phi^*) + \sum_{t=0}^{T} f_t(\alpha(x_i); \phi_t^*) \geq \tau \right) \right). \tag{3}$$

Peers Anonymization: Sharing knowledge among peers has to be regulated and anonymized adhering to the privacy regulations. Thus, we propose a simple, yet effective technique, called *peers anonymization* (PA). Specifically, we share an anonymized peer $f_a(\cdot; \phi_a)$ that aggregate the learned knowledge from the top T similar peers $f_a(\cdot; \phi_a) = \frac{1}{|T|} \sum_{i=0}^{T} f_t(\cdot; \phi_t)$. Accordingly, Eq. 3 is modified to

$$\tilde{y}_i = \arg\max \left(\mathbb{I} \left(f_{\mathcal{C}_j}(\alpha(x_i); \phi^*) + f_a(\alpha(x_i); \phi_a^*) \geq \tau \right) \right). \tag{4}$$

Note that sharing the anonymized peer is not equivalent to sharing the peers (Sect. 3.1), i.e. $f_a(\alpha(x_i); \phi_a) \neq \frac{1}{|T|} \sum_{i=0}^{T} f_t(\alpha(x_i); \phi_t))$. Finally, to prevent the local model from deviated from its data, we use an MSE consistency-regularization term that widely used in SSL, $\mathcal{L}_{CON_j} = \| f_{\mathcal{C}_j}(x_i; \phi) - f_a(x_i; \phi_a^*) \|^2$. The overall objective functions is given by $\mathcal{L}_j = \mathcal{L}_{SSL_j} + \gamma \mathcal{L}_{CON_j}$, where the pseudo labels are generated using Eq. 4, and γ is a hyper-parameter.

3 Experiments and Results

Dataset. We validated FedPerl on 38,000 skin lesion images from four publicly available datasets, namely ISIC19 [4] which consists of 25K images with 8 classes of melanoma (MEL), melanocytic nevus (NV), basal cell carcinoma (BCC), actinic keratosis (AK), benign keratosis (BKL), dermatofibroma (DF), the vascular lesion (VASC), and squamous cell carcinoma (SCC); HAM [25] with 10K images (7 classes); Derm7pt [10] with 1K images (6 classes), and PAD-UFES [15] with 2K images (6 classes). The databases are distributed to ten clients besides the global model, cf. Fig. 2. Each client data is divided into labeled (blue), unlabeled (orange), validation (green), and testing (gray) data. The data split is designed to simulate a realistic scenario with severe class imbalance, varying data size, and diverse communities, e.g., clients 0–4 from ISIC19, clients 5–7 from HAM, and client 8 and 9 are PAD-UFES and Derm7pt, respectively.

Baselines. We perform the following baselines, (i) The local clients: lower, upper, and SSL (FixMatch [21]) models; (ii) The federated learning: lower, upper, and SSFL similar to Yang et al.[26] models, and (iii) ablation study for our FedPerl, namely with(out) the PA. In the above settings, the lower bounds are trained only on the labeled data, while FixMatch [21], SSFL, and FedPerl, are trained on both labeled and unlabeled data. The upper bounds are trained similar to SSLs, yet, all labels were revealed.

Implementation. We opt for EfficientNet as our architecture. Adam for optimization for 500 rounds. The batch size and participation rate were set to 16 & 0.3, respectively. The learning rate, β, γ, T were investigated and found best at 0.00005, 0.5, 0.01, and 2, respectively, whereas τ found best at 0.6 & 0.9 for the federated and local models respectively. Further details in the suppl. material.

3.1 Peer Learning Results

Proof-of-Concept. We first validated FedPerl on CIFAR-10 and FMNIST datasets and compare it with all baselines. To have a fair comparison with FedMatch [9], we follow their experimental setup and report the results in Table 1. FedPerl significantly outperforms FedMatch [9] in all experiments

Table 1. Proof-of-concept experiment on CIFAR10 & FMNIST datasets.

PA	Method (SSFL)	CIFAR10 IID Task	CIFAR10 NonIID Task	FMNIST Streaming-NonIID Task
	FedMatch [9]	53.51 ± 0.38	54.26 ± 0.17	75.85 ± 0.14
w/o PA	FedMatch (Our run) [9]	53.12 ± 0.65	53.10 ± 0.99	76.48 ± 0.18
w/ PA	FedMatch	$\mathbf{53.32 \pm 0.59}$	$\mathbf{53.80 \pm 0.39}$	$\mathbf{76.72 \pm 0.44}$
w/o PA	FedPerl	$\mathbf{53.37 \pm 0.11}$	$\mathbf{53.75 \pm 0.40}$	$\mathbf{76.52 \pm 0.08}$
w/ PA	FedPerl	$\mathbf{53.98 \pm 0.06}$	$\mathbf{53.50 \pm 0.71}$	$\mathbf{82.75 \pm 0.44}$

demonstrating the effectiveness of our similarity measure (Sect. 2.2) without additional complexity, e.g., weight decomposition [9]. Besides, *peers anonymization* (PA) boosts the performance while preserving the privacy at low communication cost. Re-implementing FedMatch in PyTorch was not straightforward, so we decided to proceed without it in our next experiments.

Baselines: The classification results (*cf.* Table 2) reconfirms the recent findings that FedAvg significantly outperforms the local models with relative improvement (RI) up to 19.74% (see Local/FixMatch vs. FedAvg). Interestingly, both lower FedAvg and FedAvg ‡ (SSFL) outperform the local SSL, and the local upper bound, respectively. This suggests aggregating knowledge across different clients is more beneficial than exploring local unlabeled or labeled data, respectively.

FedPerl: Next, we present FedPerl results without PA (denoted as w/o PA). Table 2 shows that utilizing peer learning improves local lower model with RI between 14.53% and 15.46%. For example, all FedPerl models outperform the local upper bound by 2.9%, and (SSFL) FedAvg ‡ ($\sim$ Yang *et al.*[26]), which is a special case of FedPerl when $T = 0$, by 1.8%. This enhancement is attributed to the peer learning proposed in our method. Interestingly, the results of FedPerl for different committee size (number of peers T) are comparable, whereas communication cost is increasing proportionally with T over the standard FedAvg (see AC in Table 2). The additional cost is calculated relative to the baseline (SSFL). For simplicity, we assume the initial cost for the SSFL is 0%. The results suggest sharing one similar peer ($T = 1$) is sufficient to achieve remarkable improvement with minimal communication cost, yet, at the cost of the privacy. Next, we present the anonymized peer results in Table 2. In general, all models show a comparable or slightly better performance ($T = 5$) when compared to the FedPerl without PA. Interestingly, we obtain comparable performance while still better in preserving the privacy and reducing the communication cost $O(1)$ regardless the committee size T. This is because while we anonymize the peers, we ensemble the learned knowledge from many peers into a single model. The results confirm the superiority of FedPerl.

3.2 Building Communities Results

In this experiment, we aim to investigate the influence of similarity matrix, used in ranking the similar peers, on clustering the clients into communities. To gain

Table 2. Mean (Median) ± Std. of different evaluation metrics. ‡:~ Yang *et al.* [26]. RI: Relative Improvement. AC: Additional Cost.

Setting	Model	F1-score	Precision	Recall	RI (%)	AC (%)
Lower	Local	0.647(0.632) ± 0.053	0.644(0.622) ± 0.053	0.666(0.650) ± 0.053	–	
	FedAvg [13]	0.698(0.690) ± 0.084	0.711(0.702) ± 0.072	0.709(0.700) ± 0.077	7.88	
SSL	FixMatch [21]	0.664(0.636) ± 0.060	0.666(0.645) ± 0.063	0.692(0.671) ± 0.052	2.63	
SSFL	FedAvg‡	0.734(0.725) ± 0.065	0.744(0.730) ± 0.064	0.739(0.728) ± 0.061	13.44	0
w/o PA	FedPerl(T = 1)	**0.746(0.741) ± 0.071**	**0.753(0.744) ± 0.069**	**0.748(0.744) ± 0.069**	**15.30**	100
w/o PA	FedPerl(T = 2)	**0.747(0.736) ± 0.071**	**0.756(0.741) ± 0.067**	**0.750(0.739) ± 0.069**	**15.46**	200
w/o PA	FedPerl(T = 3)	**0.746(0.741) ± 0.072**	**0.757(0.743) ± 0.066**	**0.747(0.743) ± 0.070**	**15.30**	300
w/o PA	FedPerl(T = 4)	**0.741(0.731) ± 0.077**	**0.751(0.735) ± 0.069**	**0.745(0.736) ± 0.072**	**14.53**	400
w/o PA	FedPerl(T = 5)	**0.744(0.734) ± 0.073**	**0.753(0.744) ± 0.071**	**0.747(0.739) ± 0.069**	**15.00**	500
	FedPerl(T = 2)	**0.746(0.737) ± 0.075**	**0.754(0.741) ± 0.071**	**0.749(0.742) ± 0.073**	**15.30**	100
	FedPerl(T = 3)	**0.746(0.738) ± 0.066**	**0.756(0.743) ± 0.060**	**0.748(0.740) ± 0.065**	**15.30**	100
	FedPerl(T = 4)	**0.746(0.736) ± 0.077**	**0.755(0.745) ± 0.072**	**0.750(0.740) ± 0.074**	**15.30**	100
	FedPerl(T = 5)	**0.749(0.739) ± 0.068**	**0.758(0.744) ± 0.065**	**0.750(0.742) ± 0.066**	**15.77**	100
Upper	Local	0.726(0.701) ± 0.044	0.729(0.705) ± 0.045	0.732(0.710) ± 0.042	12.21	
	FedAvg[13]	0.773(0.757) ± 0.068	0.779(0.765) ± 0.065	0.773(0.759) ± 0.069	19.47	

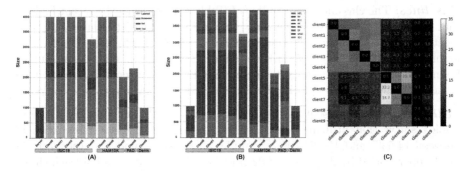

Fig. 2. (A) & (B) Show the clients and classes distribution from our datasets. The data split is carefully designed such that it simulates the medical setting. The clients 5–9 missing one or more classes. (C) FedPerl clustered clients into two main communities, while clients 8 & 9 do not belong to any community. (Color figure online)

more insights, we show the percentage of selecting peers during the training (*cf.* Fig. 2.(C)). For example, the percentage; 33.7% between clients 7 and 5 reflects how often client 7 has chosen client 5 as a peer. Interestingly, the clients were clustered into two main communities matching the clients' distribution we designed in our experiment. Further, the horizontal summations (columns 0–4) for each client in community 1 are 81.6%, 85.6%, 89.5%, 86.4%, and 88.1% respectively. These numbers represent the frequency of selecting peers from the same community. On average 86.24% of the time, first community members learn from each other, while it is 57.77% for community 2.

3.3   The Influence of Peer Learning on Clients

To gain more insights and observe the influence of peer learning on individual clients, we report the results of individual clients in Table 3. `FedPerl` outperforms

Table 3. Client level results (F1-score)

Setting	Model/Client	0	1	2	3	4	C_{ISIC}	5	6	7	C_{HAM}	8	9	$Avg_{/Cs}$	Avg
Lower	Local	0.581	0.618	0.603	0.622	0.596	0.604	0.742	0.738	0.670	0.717	0.656	0.641	0.646	0.647
	FedAvg	0.678	0.687	0.667	0.703	0.692	0.685	0.794	0.796	0.787	0.792	0.492	0.684	0.731	0.698
SSL	FixMatch	0.634	0.635	0.608	0.637	0.626	0.628	0.752	0.783	0.716	0.751	**0.650**	0.602	0.666	0.664
SSFL	FedAvg	0.727	0.718	0.686	0.723	0.735	0.718	0.812	0.831	0.806	0.816	0.602	0.703	0.764	0.734
w/o PA	FedPerl(T = 2)	0.735	**0.731**	0.725	**0.737**	**0.739**	**0.733**	0.805	**0.850**	**0.839**	0.831	0.582	**0.729**	**0.769**	**0.747**
	FedPerl(T = 2)	**0.737**	**0.737**	**0.724**	0.730	**0.751**	0.736	**0.818**	0.846	0.834	**0.833**	0.567	0.717	0.765	0.746
Upper	Local	0.698	0.698	0.677	0.700	0.696	0.694	0.806	0.804	0.752	0.787	0.702	0.722	0.728	0.726
	FedAvg	0.736	0.747	0.735	0.753	0.761	0.746	0.859	0.855	0.861	0.858	0.630	0.789	0.797	0.773

Table 4. Class level results (Mean F1-score)

Setting	Model	MEL	NV	BCC	AK	BKL	DF	VASC	SCC
Lower	Local	0.430	0.811	0.502	0.293	0.357	0.099	0.318	0.124
	FedAvg	0.501	0.834	0.646	0.377	0.507	0.173	0.642	0.111
SSL	FixMatch	0.451	0.831	0.540	0.304	0.374	0.052	0.292	0.135
SSFL	FedAvg	0.565	0.852	0.680	**0.396**	0.570	0.416	0.707	0.253
w/o PA	FedPerl (T = 2)	**0.576**	**0.854**	**0.706**	0.393	**0.589**	**0.552**	0.702	**0.305**
	FedPerl (T = 2)	**0.602**	**0.854**	0.687	0.390	**0.592**	0.493	**0.712**	**0.315**
Upper	Local	0.551	0.853	0.651	0.428	0.520	0.308	0.654	0.308
	FedAvg	0.617	0.867	0.750	0.510	0.637	0.672	0.804	0.282

the baselines, including the local upper bounds, with remarkable margins, e.g., 16.4% for client 7 (Lower Local vs. FedPerl). Surprisingly, FedPerl exceeds the upper FedAvg for client 0. The enhancement also observed at the community level with 13.2% and 11.6% for C_{ISIC} and C_{HAM}, respectively. The boosting in the performance is noticed for all clients except client 8. This can be attributed to the fact that FedPerl did not find suitable peers for client 8 due to the class distribution mismatch (*cf.* Fig. 2.(B)). To study the influence of client 8 further, we experiment excluding it from the training, then report the results as $Avg_{/C8}$ and compare it with the previous results, reported as Avg in Table 3. Results reveal that all FL models (FedAvg and FedPerl) obtain better performance without client 8. Yet, FedPerl exceeds the local upper and the FedAvg (SSFL) models. Note that the degradation in the performance indicates the negative impact of client 8. For instance, it negatively impacts the most of FedAvg (3%), the least on FedPerl (1.9%), and moderate on FedPerl w/o PA (2.2%). Such noisy behavior of client 8 could represent a threat in the federated learning, where an out-of-distribution client might mislead the global model and hurt other clients. The most exciting remark from this experiment that FedPerl shows less sensitivity to such negative impact than SSFL. This is attributed to the learning schema we proposed, where involving peers in training reduces the negative impact and limits client 8 to send more reliable updates.

3.4   Class Level Results

Class level performance is reported in Table 4. `FedPerl` significantly outperforms the performance of local models (`FedPerl` vs. Local/FixMatch) up to 10 times for DF class. In SSL setting, `FedPerl` boosts the accuracy for BCC, BKL, DF, VASC, and SCC classes by 16.6%, 21.8%, 50.4%, 42.0%, and 18.0%, respectively. Insignificant drop in the performance for AK class was observed. The performance boost of `FedPerl` is attributed not only to the local and global knowledge, but also to the knowledge exchanged through peer learning.

4   Discussion and Conclusion

We propose `FedPerl`, an SSFL framework that utilizes peer learning from social science and ensemble averaging from committee machines to overcome the limitations of previous methods. We validated our method on 38,000 skin lesion images collected from 4 public datasets demonstrating superior performance over the baselines and SoTA SSFL by 15.77%, and 1.8%, respectively. Our novel peer anonymization (PA) technique, is simple yet effective to anonymize the peer such that it is less prone to model inversion or deanonymization. PA is designed carefully to reduce the communication cost while maintains performance. Nevertheless, a privacy guarantee for aggregated models (not individuals) is an open issue and has not been thoroughly investigated in the community and mathematical analysis is yet to be proven. Generalization to unseen client is yet to be investigated in future work. This includes investigating different approaches to profile the clients in building the community. Further, a dynamic policy of when and which community to approach should be further investigated.

Acknowledgement. T.B. is financially supported by the German Academic Exchange Service (DAAD).

References

1. Abadi, M., et al.: Deep learning with differential privacy. In: Proceedings of the 2016 ACM SIGSAC Conference on Computer and Communications Security, pp. 308–318 (2016)
2. Albarqouni, S., Bakas, S., Kamnitsas, K., et al.: Domain adaptation and representation transfer, and distributed and collaborative learning (2020)
3. Binder, M., Kittler, H., Seeber, A., Steiner, A., Pehamberger, H., Wolff, K.: Epiluminescence microscopy-based classification of pigmented skin lesions using computerized image analysis and an artificial neural network. Melanoma Res. **8**(3), 261–266 (1998)
4. Codella, N., et al.: Skin lesion analysis toward melanoma detection 2018: a challenge hosted by the international skin imaging collaboration (ISIC). arXiv preprint arXiv:1902.03368 (2019)
5. Esteva, A., et al.: Dermatologist-level classification of skin cancer with deep neural networks. Nature **542**(7639), 115–118 (2017)

6. Flores, M., et al.: Federated learning used for predicting outcomes in SARS-COV-2 patients (2021)
7. Fredrikson, M., Jha, S., Ristenpart, T.: Model inversion attacks that exploit confidence information and basic countermeasures. In: Proceedings of the 22nd ACM SIGSAC Conference on Computer and Communications Security, pp. 1322–1333 (2015)
8. Gessert, N., Nielsen, M., Shaikh, M., Werner, R., Schlaefer, A.: Skin lesion classification using ensembles of multi-resolution EfficientNets with meta data. MethodsX **7**, 100864 (2020)
9. Jeong, W., Yoon, J., Yang, E., Hwang, S.J.: Federated semi-supervised learning with inter-client consistency & disjoint learning (2021). https://openreview.net/forum?id=ce6CFXBh30h
10. Kawahara, J., Daneshvar, S., Argenziano, G., Hamarneh, G.: Seven-point checklist and skin lesion classification using multitask multimodal neural nets. IEEE J. Biomed. Health Inform. **23**(2), 538–546 (2019)
11. Li, D., Kar, A., Ravikumar, N., Frangi, A.F., Fidler, S.: Federated simulation for medical imaging. In: Martel, A.L., et al. (eds.) MICCAI 2020. LNCS, vol. 12261, pp. 159–168. Springer, Cham (2020). https://doi.org/10.1007/978-3-030-59710-8_16
12. Lopez, A.R., Giro-i Nieto, X., Burdick, J., Marques, O.: Skin lesion classification from dermoscopic images using deep learning techniques. In: 2017 13th IASTED international conference on biomedical engineering (BioMed), pp. 49–54. IEEE (2017)
13. McMahan, B., Moore, E., Ramage, D., Hampson, S., Arcas, B.A.: Communication-efficient learning of deep networks from decentralized data. In: Artificial Intelligence and Statistics, pp. 1273–1282. PMLR (2017)
14. Orekondy, T., Oh, S.J., Zhang, Y., Schiele, B., Fritz, M.: Gradient-leaks: understanding and controlling deanonymization in federated learning. arXiv preprint arXiv:1805.05838 (2018)
15. Pacheco, A.G., et al.: PAD-UFES-20: a skin lesion benchmark composed of patient data and clinical images collected from smartphones. arXiv preprint arXiv:2007.00478 (2020)
16. Rieke, N., et al.: The future of digital health with federated learning. NPJ Digit. Med. **3**(1), 1–7 (2020)
17. Rogers, H.W., et al.: Incidence estimate of nonmelanoma skin cancer in the United States, 2006. Arch. Dermatol. **146**(3), 283–287 (2010)
18. Roth, H.R., et al.: Federated learning for breast density classification: a real-world implementation. In: Albarqouni, S., et al. (eds.) DART/DCL -2020. LNCS, vol. 12444, pp. 181–191. Springer, Cham (2020). https://doi.org/10.1007/978-3-030-60548-3_18
19. Sarma, K.V., et al.: Federated learning improves site performance in multicenter deep learning without data sharing. J. Am. Med. Inform. Assoc. **28**, 1259–1264 (2021)
20. Siegel, R.L.: Cancer statistics, 2021. Published early online January 12, 2021 in CA cancer journal for clinicians. MPH, American Cancer Society, Atlanta (2021)
21. Sohn, K., et al.: Fixmatch: simplifying semi-supervised learning with consistency and confidence. arXiv preprint arXiv:2001.07685 (2020)
22. Topping, K.J.: Trends in peer learning. Educ. Psychol. **25**(6), 631–645 (2005)
23. Tresp, V.: Committee machines. In: Handbook for Neural Network Signal Processing, pp. 1–18 (2001)

24. Tschandl, P.: Comparison of the accuracy of human readers versus machine-learning algorithms for pigmented skin lesion classification: an open, web-based, international, diagnostic study. Lancet Oncol. **20**(7), 938–947 (2019)

25. Tschandl, P., Rosendahl, C., Kittler, H.: The HAM10000 dataset, a large collection of multi-source dermatoscopic images of common pigmented skin lesions. Sci. Data **5**(1), 1–9 (2018)

26. Yang, D., et al.: Federated semi-supervised learning for COVID region segmentation in chest CT using multi-national data from China, Italy, Japan. Med. Image Anal. **70**, 101992 (2021)

27. Yeganeh, Y., Farshad, A., Navab, N., Albarqouni, S.: Inverse distance aggregation for federated learning with Non-IID data. In: Albarqouni, S., et al. (eds.) DART/DCL -2020. LNCS, vol. 12444, pp. 150–159. Springer, Cham (2020). https://doi.org/10.1007/978-3-030-60548-3_15

28. Zhang, J., Xie, Y., Xia, Y., Shen, C.: Attention residual learning for skin lesion classification. IEEE Trans. Med. Imaging **38**(9), 2092–2103 (2019)

29. Li, W., et al.: Privacy-preserving federated brain tumour segmentation. In: Suk, H.-I., Liu, M., Yan, P., Lian, C. (eds.) MLMI 2019. LNCS, vol. 11861, pp. 133–141. Springer, Cham (2019). https://doi.org/10.1007/978-3-030-32692-0_16

Personalized Retrogress-Resilient Framework for Real-World Medical Federated Learning

Zhen Chen, Meilu Zhu, Chen Yang, and Yixuan Yuan$^{(\boxtimes)}$

Department of Electrical Engineering, City University of Hong Kong,
Kowloon, Hong Kong, China
yxyuan.ee@cityu.edu.hk

Abstract. Nowadays, deep learning methods with large-scale datasets can produce clinically useful models for computer-aided diagnosis. However, the privacy and ethical concerns are increasingly critical, which make it difficult to collect large quantities of data from multiple institutions. Federated Learning (FL) provides a promising decentralized solution to train model collaboratively by exchanging client models instead of private data. However, the server aggregation of existing FL methods is observed to degrade the model performance in real-world medical FL setting, which is termed as *retrogress*. To address this problem, we propose a personalized retrogress-resilient framework to produce a superior personalized model for each client. Specifically, we devise a Progressive Fourier Aggregation (PFA) at the server to achieve more stable and effective global knowledge gathering by integrating client models from low-frequency to high-frequency gradually. Moreover, with an introduced deputy model to receive the aggregated server model, we design a Deputy-Enhanced Transfer (DET) strategy at the client and conduct three steps of *Recover-Exchange-Sublimate* to ameliorate the personalized local model by transferring the global knowledge smoothly. Extensive experiments on real-world dermoscopic FL dataset prove that our personalized retrogress-resilient framework outperforms state-of-the-art FL methods, as well as the generalization on an out-of-distribution cohort. The code and dataset are available at https://github.com/CityU-AIM-Group/PRR-FL.

Keywords: Federated learning · Skin lesions · Parameters aggregation

1 Introduction

Recent years have witnessed the superior performance of deep learning techniques in the field of computer-aided diagnosis [2,6,27]. By collecting large quantities of data from multiple institutions, tailored deep learning methods achieved

---

Z. Chen, M. Zhu and C. Yang—Equal contribution.

M. de Bruijne et al. (Eds.): MICCAI 2021, LNCS 12903, pp. 347–356, 2021.
https://doi.org/10.1007/978-3-030-87199-4_33

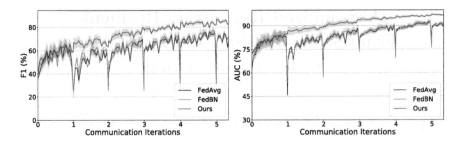

Fig. 1. The training curves of client C in real-world dermoscopic FL dataset. The *retrogress* happens in FedAvg [14] and FedBN [10] after client-server communications.

great success and have been applied in the clinical practice to alleviate the work-load of physicians [5,12,13]. However, this is not a sustainable way to develop future intelligent healthcare systems, where the patient privacy and ethical concerns impede constructing centralized datasets with increasing size. To overcome this challenge, Federated Learning (FL) [8,17] provides a collaborative paradigm to optimize local models at multiple clients, and regularly aggregate these local models together at a global server. After that, the aggregated model is delivered to each client for the next local training. In this way, the client's data is kept private without being exposed to other participants, while the aggregated model with global knowledge can achieve superior performance, compared with the models trained at any single client. Among existing FL works, FedAvg [4,14] proved that averaging local models trained at different clients is a generalization of updating the gradients of each batch when clients are assumed as independent and identically distributed (IID), which can reduce the communication overhead. To solve the practical non-IID distribution among clients, FedProx [9] introduced a proximal term to the objective of clients, regularizing the heterogeneous local updates to be closer to the initial global model. Furthermore, FedBN [10] and SiloBN [1] adopted personalized batch normalization (BN) settings for models at the client side to overcome the inter-client data heterogeneity. Since personalized FL allows each client to choose the aggregated server model or any intermediate client one during the local training [18,19], it is more suitable for medical FL scenarios.

In the scenarios of real-world medical FL, the inter-client data heterogeneity (e.g., caused by different imaging devices, protocols and even regional disease differences) is much more serious than aforementioned studies, thereby leading to enormous discrepancy between different local models. As illustrated in Fig. 1, we observe existing FL methods encounter an abrupt performance drop (termed as *retrogress*) after each communication between the server and clients. The aggregated models with *retrogress* lose the previous knowledge and require re-adaptation to the client task during the local training, which hinders the local training of clients and the knowledge sharing of the server. We suppose the *retrogress* in real-world FL may come from the following two reasons. On the one hand, when the severe data heterogeneity exists among clients, it is irrational to average these parameters in element-wise, because parameters from different clients may represent diverse semantic patterns at the same location [23]. In contrast, representing parameters in the frequency domain can ensure that the components are aligned along the frequency dimension, and also provide the flexibility to select the frequency band for aggregation [11]. On the other hand, replacing the previous local models with the aggregated server model at the communication, obliterates the knowledge learned by the local models and degrades the optimization in the next iteration.

To address the *retrogress* problem in real-world medical FL, we propose a personalized retrogress-resilient framework in both server and client perspectives, aiming to provide a customized model for each client. Specifically, we devise a Progressive Fourier Aggregation (PFA) to integrate client models at the server. Through Fast Fourier Transform (FFT) and inverse FFT (IFFT) to achieve the mutual conversion of client parameters to frequency domain, we average the low-frequency components of client parameters while preserving the remaining high-frequency components. By gradually increasing the frequency threshold of the shared components during the FL, PFA can effectively integrate the client knowledge in a manner consistent with network learning preferences. To the best of our knowledge, this work represents the first attempt to implement parameters aggregation of FL in frequency domain. At the client side, instead of replacing local models, we propose a Deputy-Enhanced Transfer (DET) to introduce a deputy model to receive the updated server model and maintain the personalized local model not contaminated. Considering that the deputy model suffers from the *retrogress* after each communication, our DET conducts three steps, *Recover-Exchange-Sublimate*, to recover the local prior and transfer the global knowledge to improve the personalized local model. Extensive experiments on real-world dermoscopic FL dataset demonstrate the effectiveness and out-of-distribution generalization of our personalized retrogress-resilient framework, which outperforms state-of-the-art FL methods.

2 Personalized Retrogress-Resilient Framework

2.1 Overview

Given a set of K clients with their private data, our personalized retrogress-resilient framework aims to collaboratively generate a personalized model $\boldsymbol{p}$ with dominant performance for each client. These models $\{\boldsymbol{p}_k\}_{k=1}^{K}$ share the same network architecture to benefit from the server aggregation, which is the same as previous FL works. For the k-th client, the personalized model $\boldsymbol{p}_k$ is trained with private data for E epochs locally, which is then uploaded to the server. The server collects client models, and aggregates them into server models with individual high-frequency components of client models using the proposed Progressive Fourier Aggregation (PFA). After that, these server models are delivered to corresponding client as a deputy model. The deputy model $\boldsymbol{d}_k$ can transfer the global knowledge through our Deputy-Enhanced Transfer (DET). Repeat these steps until local training reaches T epochs. The overview of our personalized retrogress-resilient framework is illustrated in Fig. 2.

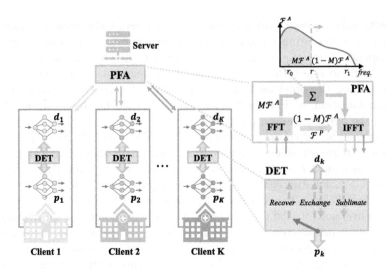

Fig. 2. The personalized retrogress-resilient framework. At the server, Progressive Fourier Aggregation (PFA) integrates global knowledge in the frequency domain. At each client, Deputy-Enhanced Transfer (DET) is employed to improve the local personalized model without being disturbed by the communication.

2.2 Progressive Fourier Aggregation in Server

Previous FL works [1,9,10,14] gathered the global knowledge from different clients by directly element-wise averaging the parameters of local models to generate the aggregated server model. However, this coarse operation in parameter space violently degrades the model performance on clients, as the *retrogress* of FedAvg [14] and FedBN [10] observed in Fig. 1. To alleviate the *retrogress* induced by aggregation, we propose the Progressive Fourier Aggregation (PFA) to stably integrate client models in the frequency domain.

Inspired by the fact that low-frequency components of parameters are the basis for the network capability [11], our PFA aggregates the relatively low-frequency components of parameters to share knowledge from different clients, while retraining their high-frequency components, which may contain specific knowledge for each individual client. Specifically, for a convolutional layer of k-th client model, we first reshape its parameter tensor $w_k \in \mathbb{R}^{N \times C \times d_1 \times d_2}$ into a 2-D matrix $w_k^{'} \in \mathbb{R}^{d_1 N \times d_2 C}$, where N and C represent the output and input channels, and d_1 and d_2 are the spatial shape of the kernel. Then, we obtain the amplitude map $\mathcal{F}^A$ and phase map $\mathcal{F}^P$ through the Fourier transform $\mathcal{F} = \mathcal{F}^A e^{j\mathcal{F}^P}$, as follows:

$$\mathcal{F}(w_k^{'})(m,n) = \sum_{x,y} w_k^{'}(x,y)e^{-j2\pi\left(\frac{x}{d_1 N}m + \frac{y}{d_2 C}n\right)}, j^2 = -1, \tag{1}$$

which can be implemented efficiently using the FFT [7]. To extract the low-frequency components for aggregation, we employ a low-frequency mask M with zero values except for the central region:

$$M(m,n) = \mathbb{1}_{(m,n)\in[-rd_1N:rd_1N,-rd_2C:rd_2C]}, \tag{2}$$

where $r \in (0, 0.5)$ represents the low-frequency threshold, and the center of w'_k is set as the coordinate $(0,0)$. By averaging the low-frequency components over clients, the aggregated frequency components for k-th client are calculated as:

$$\hat{\mathcal{F}}^A(w'_k) = (1 - M) \circ \mathcal{F}^A(w'_k) + \frac{1}{K}\sum_{k=1}^{K} M \circ \mathcal{F}^A(w'_k), \tag{3}$$

where $\circ$ is the element-wise multiplication. Considering the fact that networks are trained to learn the low-frequency knowledge prior to the high-frequency counterpart [16,22,24], we design a progressive strategy to implement our PFA by increasing $r = r_0 + \frac{r_1-r_0}{T}t$ during the FL training, where r_0 and r_1 are the initial and terminated low-frequency threshold.

Finally, after applying the inverse Fourier transform $\mathcal{F}^{-1}$ to convert the amplitude and phase maps back to parameters as $\hat{w}'_k = \mathcal{F}^{-1}([\hat{\mathcal{F}}^A(w'_k), \mathcal{F}^P(w'_k)])$, we obtain the aggregated parameters of k-th client. The same PFA can be applied to the fully connected layers of client models, without the reshape of parameters.

2.3   Deputy-Enhanced Transfer in Client

Our PFA can alleviate the *retrogress* induced by aggregation at the server, however, directly updating client models with the aggregated server parameters still obliterates the learned local knowledge and further degrades the optimization in the next iteration. To solve this problem, we propose the Deputy-Enhanced Transfer (DET) to subtly fuse the global knowledge with the local prior, rather than the direct replacement. In addition to a personalized local model p, each client also contains a deputy model d to receive the aggregated parameters from server. The proposed DET conducts three steps: *Recover*, *Exchange* and *Sublimate*, to smoothly transfer the global knowledge from the deputy d to the personalized local model p.

Recover. When the deputy d is updated with the aggregated model s from the server, its performance encounters a severe deterioration due to the aforementioned *retrogress* problem. Therefore, we firstly regard the personalized local model p as a teacher to recover the deputy model d with the local knowledge. In this step, the personalized local model p is supervised by the cross entropy loss, while the deputy model d is optimized by a joint loss function L_d:

$$L_d = L_{CE1} + \sum_{i=1}^{N} p(x_i) \log \frac{p(x_i)}{q(x_i)}, \tag{4}$$

where x_i is the i-th training sample and N is the number of training samples on the current client. L_{CE1} is the cross entropy loss. The $p(x_i)$ and $q(x_i)$ are the posterior probabilities of the deputy d and personalized local model p. The second term of Eq. (4) is the Kullback-Leibler (KL) divergence that helps the deputy d to quickly recover the adaptability to the client with performance improved. This step is crucial to guarantee the deputy model does not hurt the personalized local model in the next step.

Exchange. Once the recovered performance of the deputy d is close to the teacher p, as $\phi_{val}(d) \geq \lambda_1 \phi_{val}(p)$, where ϕ_{val} represents a specific performance metric (*e.g.*, F1) on the validation set, we conduct the mutual learning [26] between the deputy d and the personalized local model p to exchange the global knowledge and the local knowledge. Here, the deputy model d is supervised by Eq. (4) and the personalized local model p is learned by the following loss function L_p:

$$L_p = L_{CE2} + \sum_{i=1}^{N} q(x_i) \log \frac{q(x_i)}{p(x_i)}. \tag{5}$$

The second term of Eq. (5) transfers the global knowledge of the server to the personalized local model p through the deputy d. In this way, the knowledge exchange can promote the generalization ability of all clients [26].

Sublimate. Finally, when the performance of the deputy d is highly close to the personalized local model p, as $\phi_{val}(d) \geq \lambda_2 \phi_{val}(p)$, where $0 < \lambda_1 < \lambda_2 < 1$, the deputy model d serves as the teacher to further guide p with L_p in Eq. (5), which enables the global knowledge can be transferred to the personalized local model to the greatest extent.

3   Experiment

3.1   Dataset and Implementation Details

Real-World Dermoscopic FL Dataset. Considering that the real-world FL in medical scenarios is involved with complex data heterogeneity, we construct a novel FL benchmark dataset of dermoscopic images to evaluate various FL methods under the actual client settings. Specifically, we collect $8,940$ and $2,000$ images that are diagnosed as nevus, benign keratosis or melanoma from HAM10K [21] and MSK [3] dataset, respectively. To preserve the critical shape information of irregular-sized images for lesion diagnosis, we first crop the central square regions using the shorter side length, and then unify them into 128×128 resolution. With the metadata of each image in HAM10K [21], we segregate

Table 1. Detailed information of real-world dermoscopic FL dataset.

Client	Nevus	Benign Keratosis	Melanoma	Total
A	1832	475	680	2987
B	3720	124	24	3868
C	803	490	342	1635
D	1372	254	374	2000

Table 2. Comparison with state-of-the-art on real-world dermoscopic FL dataset.

Method	F1 (%)					AUC (%)				
	A	B	C	D	Avg	A	B	C	D	Avg
FedAvg [14]	57.44	48.14	56.80	44.20	51.64	81.52	82.75	76.14	71.14	77.89
FedProx [9]	56.70	39.09	54.70	45.95	49.11	81.70	70.09	76.76	74.83	75.84
SiloBN [1]	50.83	63.81	53.98	61.90	57.63	83.17	81.41	77.90	80.56	80.76
IDA [25]	55.62	41.87	55.42	45.64	49.64	81.27	75.95	78.38	73.10	77.18
FedBN [10]	54.96	72.10	54.73	62.07	60.96	83.06	96.35	79.97	81.36	85.18
FML [19]	69.14	75.83	66.02	59.40	67.60	88.38	95.49	85.05	82.81	87.93
Ours *w/o* DET	67.28	75.20	61.44	59.24	65.79	87.95	89.81	83.76	81.39	85.73
Ours *w/o* PFA	69.41	77.90	66.77	61.90	69.00	88.70	97.03	84.25	82.59	88.14
Ours	**72.00**	**79.70**	**68.87**	**62.46**	**70.75**	**88.98**	**97.58**	**85.22**	**83.90**	**88.92**

the images with the same institution and imaging device into one client. Consequently, our FL dataset contains 4 clients, including 3 clients from HAM10K [21] and MSK [3] data as one. Detailed information of the dataset is illustrated in Table 1. We further divide samples of each client into the training, validation and test sets as 7:1:2, ensuring the consistent proportion of categories among these three sets. In this way, our dataset can reflect the difficulty of practical FL deployment, including the data heterogeneity, different imaging devices and various sample numbers among clients.

Implementation Details. We implement our personalized retrogress-resilient framework and state-of-the-art FL methods using VGG-16BN [20] in PyTorch [15]. During the local training at the client side, the networks are optimized using SGD with the batch size of 16. The learning rate is initialized as 1×10^{-2} and halved after every 25 epochs. The communication is conducted after every $E = 5$ epochs in local training until reaching $T = 250$ epochs in total. The frequency thresholds r_0 and r_1 in PFA are set to 0.35 and 0.48, and the performance thresholds λ_1 and λ_2 in DET are set to 0.7 and 0.9, respectively. We choose F1 score as the metric ϕ_{val} to measure the performance of deputy and personalized models. The best model is selected using the hold-out validation set on each client. The training employs a combination of random flip and rotation as data augmentation to avoid overfitting. All experiments are performed on NVIDIA V100 GPU.

Evaluation Metrics. The F1 and AUC are utilized to measure the performance of various FL methods comprehensively. For the 3-category classification, we calculate these two metrics of each category separately, and then perform the macro average to eliminate the impacts of imbalance. The higher scores of both F1 and AUC indicate better performance on the dermoscopic diagnosis task. In addition to showing the performance of FL methods on each client, we also calculate the marco average of 4 clients for a more concise comparison.

3.2 Comparison with State-of-the-art Methods

To evaluate the performance of our personalized retrogress-resilient framework, we perform a comprehensive comparison with the state-of-the-art FL methods,

Table 3. Generalization comparison of different FL works on out-of-distribution data.

Method	FedAvg [14]	FedProx [9]	SiloBN [1]	IDA [25]	FedBN [10]	FML [19]	Ours
F1 (%)	37.76	38.14	42.87	40.11	43.74	37.01	**50.20**
AUC (%)	66.49	67.36	65.65	62.75	70.85	66.81	**73.70**

including FedAvg [14], FedProx [9], SiloBN [1], IDA [25], FedBN [10] and FML [19]. As illustrated in Table 2, our method achieves the best performance, with the overwhelming average F1 of 70.75% and average AUC of 88.92%. Noticeably, our framework outperforms the personalized FL method, FedBN [10], by a large margin, *e.g.*, 9.79% in average F1 and 3.74% in average AUC. This advantage confirms that the proposed PFA and DET can alleviate the *retrogress* in FedBN [10]. Moreover, compared with FML [19] that also employs the deputy model at each client, our method obtains superior performance with a remarkable increase of 3.15% in average F1 and 0.99% in average AUC. These experimental results demonstrate the performance advantage of our method over state-of-the-art FL methods in the scenarios of real-world medical FL.

3.3   Ablation Study

To further validate the effectiveness of PFA and DET, we perform the detailed ablation study in Table 2. Specifically, we implement two ablative baselines of the proposed personalized retrogress-resilient framework, by removing the DET (denoted as *w/o* DET) and the PFA (denoted as *w/o* PFA) individually. As illustrated in Table 2, the proposed PFA and DET can bring a F1 improvement of 1.75% and 4.96%, respectively. Compared with FedBN [10], our method *w/o* DET only replaces the parameter aggregation with PFA, which proves that the PFA results in a 4.83% F1 increase. Moreover, when applying the same parameter aggregation with a deputy at each client, our method *w/o* PFA outperforms FML [19] with a 1.40% increase in F1, which can be attributed to the effect of the tailor-made DET with *Recover-Exchange-Sublimate* three steps. The ablation experiment confirms that the proposed PFA and DET play an important role in solving data heterogeneity in real-world medical FL scenarios, resulting in the performance advantage of our personalized retrogress-resilient framework.

3.4   Out-of-distribution Generalization

To further verify the generalization ability on the out-of-distribution cohort, we test the trained models on an unseen cohort with 427 samples from another institution [21]. As shown in Table 3, our method achieves the best generalization with F1 of 50.20% and AUC of 73.70% in such challenging scenario. Moreover, our method demonstrates superior classification performance in comparison to existing algorithms [1,9,10,14,19,25] with statistically significant increments of 12.44%, 12.06%, 7.33%, 10.09%, 6.46% and 13.19% in average F1 score. These competitive experimental results can indicate the superiority of our method on the generalization ability.

4   Conclusion

In this work, the *retrogress* problem is observed in existing FL methods after every aggregation. To promote the FL in real-world medical scenarios, we propose a personalized retrogress-resilient framework with modification in both the server and clients. Specifically, we conduct PFA to integrate the global knowledge from low-frequency to high-frequency gradually at the server. At the client side, instead of replacing local models, DET conducts three steps of *Recover-Exchange-Sublimate* to transfer global knowledge to improve the personalized model without being disturbed by *retrogress*. Extensive experiments confirm the advantages of our framework over state-of-the-art works on a real-world dermoscopic FL dataset, as well as the generalization on an out-of-distribution cohort.

Acknowledgments. This work is supported by Shenzhen-Hong Kong Innovation Circle Category D Project SGDX2019081623300177 (CityU 9240008) and CityU SRG 7005229.

References

1. Andreux, M., du Terrail, J.O., Beguier, C., Tramel, E.W.: Siloed federated learning for multi-centric histopathology datasets. In: Albarqouni, S., et al. (eds.) DART/DCL -2020. LNCS, vol. 12444, pp. 129–139. Springer, Cham (2020). https://doi.org/10.1007/978-3-030-60548-3_13

2. Bhattacharya, I., et al.: CorrSigNet: learning CORRelated prostate cancer SIGnatures from radiology and pathology images for improved computer aided diagnosis. In: Martel, A.L., et al. (eds.) MICCAI 2020. LNCS, vol. 12262, pp. 315–325. Springer, Cham (2020). https://doi.org/10.1007/978-3-030-59713-9_31

3. Codella, N.C., et al.: Skin lesion analysis toward melanoma detection: a challenge at the 2017 international symposium on biomedical imaging (ISBI), hosted by the international skin imaging collaboration (ISIC). In: ISBI 2018, pp. 168–172. IEEE (2018)

4. Dou, Q., et al.: Federated deep learning for detecting COVID-19 lung abnormalities in CT: a privacy-preserving multinational validation study. NPJ Digit. Med. **4**(1), 1–11 (2021)

5. Esteva, A., et al.: Dermatologist-level classification of skin cancer with deep neural networks. Nature **542**(7639), 115–118 (2017)

6. Esteva, A., et al.: A guide to deep learning in healthcare. Nat. Med. **25**(1), 24–29 (2019)

7. Frigo, M., Johnson, S.G.: FFTW: an adaptive software architecture for the FFT. In: ICASSP, vol. 3, pp. 1381–1384. IEEE (1998)

8. Li, T., Sahu, A.K., Talwalkar, A., Smith, V.: Federated learning: challenges, methods, and future directions. IEEE Signal Process. Mag. **37**(3), 50–60 (2020)

9. Li, T., Sahu, A.K., Zaheer, M., Sanjabi, M., Talwalkar, A., Smith, V.: Federated optimization in heterogeneous networks. In: Proceedings of Machine Learning and Systems, vol. 2, pp. 429–450 (2020)

10. Li, X., Jiang, M., Zhang, X., Kamp, M., Dou, Q.: Fedbn: federated learning on Non-IID features via local batch normalization. In: ICLR (2021)

11. Liu, Z., Xu, J., Peng, X., Xiong, R.: Frequency-domain dynamic pruning for convolutional neural networks. In: NeurIPS, pp. 1051–1061 (2018)
12. Liu, Z., Xiong, R., Jiang, T.: Clinical-inspired network for skin lesion recognition. In: Martel, A.L., et al. (eds.) MICCAI 2020. LNCS, vol. 12266, pp. 340–350. Springer, Cham (2020). https://doi.org/10.1007/978-3-030-59725-2_33
13. Lotter, W., et al.: Robust breast cancer detection in mammography and digital breast tomosynthesis using an annotation-efficient deep learning approach. Nat. Med. **27**, 1–6 (2021)
14. McMahan, B., Moore, E., Ramage, D., Hampson, S., Arcas, B.A.: Communication-efficient learning of deep networks from decentralized data. In: Artificial Intelligence and Statistics, pp. 1273–1282. PMLR (2017)
15. Paszke, A., et al.: Pytorch: an imperative style, high-performance deep learning library. arXiv preprint arXiv:1912.01703 (2019)
16. Rahaman, N., et al.: On the spectral bias of deep neural networks. In: ICML (2019)
17. Rieke, N., et al.: The future of digital health with federated learning. NPJ Digit. Med. **3**(1), 1–7 (2020)
18. Roth, H.R., et al.: Federated learning for breast density classification: a real-world implementation. In: Albarqouni, S., et al. (eds.) DART/DCL -2020. LNCS, vol. 12444, pp. 181–191. Springer, Cham (2020). https://doi.org/10.1007/978-3-030-60548-3_18
19. Shen, T., et al.: Federated mutual learning. arXiv preprint arXiv:2006.16765 (2020)
20. Simonyan, K., Zisserman, A.: Very deep convolutional networks for large-scale image recognition. In: ICLR (2015)
21. Tschandl, P., Rosendahl, C., Kittler, H.: The HAM10000 dataset, a large collection of multi-source dermatoscopic images of common pigmented skin lesions. Sci. Data **5**(1), 1–9 (2018)
22. Ulyanov, D., Vedaldi, A., Lempitsky, V.: Deep image prior. In: CVPR, pp. 9446–9454 (2018)
23. Wang, H., Yurochkin, M., Sun, Y., Papailiopoulos, D., Khazaeni, Y.: Federated learning with matched averaging. In: ICLR (2020)
24. Xu, Z.-Q.J., Zhang, Y., Xiao, Y.: Training behavior of deep neural network in frequency domain. In: Gedeon, T., Wong, K.W., Lee, M. (eds.) ICONIP 2019. LNCS, vol. 11953, pp. 264–274. Springer, Cham (2019). https://doi.org/10.1007/978-3-030-36708-4_22
25. Yeganeh, Y., Farshad, A., Navab, N., Albarqouni, S.: Inverse distance aggregation for federated learning with Non-IID data. In: Albarqouni, S., et al. (eds.) DART/DCL -2020. LNCS, vol. 12444, pp. 150–159. Springer, Cham (2020). https://doi.org/10.1007/978-3-030-60548-3_15
26. Zhang, Y., Xiang, T., Hospedales, T.M., Lu, H.: Deep mutual learning. In: CVPR, pp. 4320–4328 (2018)
27. Zhu, W., Liao, H., Li, W., Li, W., Luo, J.: Alleviating the incompatibility between cross entropy loss and episode training for few-shot skin disease classification. In: Martel, A.L., et al. (eds.) MICCAI 2020. LNCS, vol. 12266, pp. 330–339. Springer, Cham (2020). https://doi.org/10.1007/978-3-030-59725-2_32

Federated Whole Prostate Segmentation in MRI with Personalized Neural Architectures

Holger R. Roth$^{(\boxtimes)}$ ⓘ, Dong Yang ⓘ, Wenqi Li ⓘ, Andriy Myronenko ⓘ,
Wentao Zhu ⓘ, Ziyue Xu ⓘ, Xiaosong Wang ⓘ, and Daguang Xu$^{(\boxtimes)}$ ⓘ

NVIDIA, Bethesda, MD, USA
{hroth,daguangx}@nvidia.com

Abstract. Building robust deep learning-based models requires diverse training data, ideally from several sources. However, these datasets cannot be combined easily because of patient privacy concerns or regulatory hurdles, especially if medical data is involved. Federated learning (FL) is a way to train machine learning models without the need for centralized datasets. Each FL client trains on their local data while only sharing model parameters with a global server that aggregates the parameters from all clients. At the same time, each client's data can exhibit differences and inconsistencies due to the local variation in the patient population, imaging equipment, and acquisition protocols. Hence, the federated learned models should be able to adapt to the local particularities of a client's data. In this work, we combine FL with an AutoML technique based on local neural architecture search by training a "supernet". Furthermore, we propose an adaptation scheme to allow for personalized model architectures at each FL client's site. The proposed method is evaluated on four different datasets from 3D prostate MRI and shown to improve the local models' performance after adaptation through selecting an optimal path through the AutoML supernet.

Keywords: Federated learning · AutoML · Domain adaptation

1 Introduction

The advancements of the last few years in medical image segmentation were dominated by deep learning (DL) approaches. DL mostly eliminated the need for handcrafting image features. However, it has been arguably replaced by the need of domain experts to design application-specific DL models. In particular, the medical image computing field has been dominated by popular hand-engineered network architectures such as 2D and 3D U-Net [2,21], V-Net [20], High-Res-Net [14], DeepMedic [12], and many others. To get a good network design for a particular problem, one promising direction is to automate the time-consuming model designing process via AutoML techniques. As another major challenge in model development, large amounts of data covering sufficient large range of

© Springer Nature Switzerland AG 2021
M. de Bruijne et al. (Eds.): MICCAI 2021, LNCS 12903, pp. 357–366, 2021.
https://doi.org/10.1007/978-3-030-87199-4_34

examples are usually required to train accurate and robust models. To achieve this goal, hospitals and medical institutes often need to collaborate and host centralized databases for the development of clinical-grade DL models. This can become challenging due to data-privacy and various ethical concerns associated with data sharing in healthcare domain. One approach to combat such issues is through federated learning (FL), where only model and/or DL workflow parameters are shared among participating institutes instead of raw medical data. Furthermore, it is well known that global robustness and local accuracy is in many cases conflicting: models trained on large centralized datasets might not always generalize well to the data at a particular imaging site due to various inconsistencies (scanner models, imaging protocols, patient populations, etc.) among the different sites. In this case, domain adaptation (DA) is often needed. In this work, we propose to systematically tackle the three challenges in a unified framework: combining an FL algorithm with AutoML and the capability of global-local model adaptation. In particular, we implement a "supernet" training strategy that can be trained in a federated setting. We believe AutoML and FL technologies are a natural fit for each other because of their complementary nature. By combining the two, we are also able to address the DA problem. For one, FL can circumvent the problem of hosting and accessing large centralized datasets by distributing the learning effort to several clients with their own local data. FL will only communicate the model gradients after a local round of training to a centralized server which aggregates the results and starts the next round of FL. At the same time, AutoML with supernet design allows us to avoid hand-engineering of dataset-specific network architectures and a particular sub-network of the trained supernet can be used as a way of local domain adaptation to handle inconsistencies between the different contributing data sites. Next, we summarize related works.

AutoML: Recently, deep learning is applied for various applications, such as image recognition, semantic segmentation, object detection, natural image generation, etc. However, for each specific task, particular network architectures often need to be hand-designed. Neural architecture search (NAS) [3] is one of the most common approaches to circumvent such hand-design of architectures in AutoML for DL applications. The goal of NAS is to automatically design neural network architectures without any human heuristics or assumptions. In addition to the model weights, after searching, the model architecture itself is optimized for the task at hand, while often still being generalizable to other datasets [28]. A common concept in (one-shot) NAS and AutoML literature is the "supernet" [1,18,27]. The main idea behind supernet is that we can create a large neural network including several candidate modules at each level of the networks. This supernet can be trained jointly, and from the supernet, specific sub-networks can be chosen by selecting a path through the module candidates. At deployment, the final architecture is selected from the supernet by assigning path weights to select particular module candidates. Additional budgeting constraints, such as latency or number of model parameters, can be added to find optimal architectures for a given application. Recent works can achieve state-of-the-art results on computer vision tasks while being computationally efficient [25].

Federated Learning: FL enables collaborative and decentralized DL training without sharing raw patient data [19]. Each client in FL trains locally on their own data and then submits their model parameters to a server that accumulates and aggregates the model updates from each client. Once a certain number of clients have submitted their updates, the aggregated model parameters are redistributed to the clients for local model update, and a new round of local training starts. While out of the scope of this work, FL can also be combined with additional privacy-preserving measures to avoid potential reconstruction of training data through model inversion if the model parameters would be leaked to an adversary [13]. Several works have shown the applicability of FL to medical imaging tasks [13,22,26]. Recent work that combines NAS approaches with FL has been proposed for the mobile phone applications [31]. As such, its focus is on reducing the computational requirements on the local edge devices, making its setting quite different from the "cross-silo" FL [10] medical image segmentation investigated here, where the focus is on model performance and personalization. The closest work in motivation to ours is [7] which focuses on the non-I.I.D. setting but is restricted to using toy datasets for classification tasks, like CIFAR-10, and differs in its implementation details.

Domain Adaptation: Domain adaptation aims to tackle data inconsistencies among different domains, or between training data and unknown data. In its simplest form, fine-tuning, also known as transfer learning [23], can help to adapt a pre-trained model to a particular target domain. More recent approaches for DA typically involve some form of adversarial learning to introduce a specific loss that can minimize the feature-level differences among different domains [11] or through gradient back-propagation using adversarial training [4,5]. An alternative approach is coming from the "image translation" field where generative adversarial networks (GAN) are utilized to translate the image of one domain to mimic another domain. An important part of these approaches is the application of some form of cycle-consistency which is essential to train on un-paired data [9,30,32] The common concept of adversarial training suggests that the gradients from external constraints will help balance various domains and change the model's feature representations.

Contributions: Our proposed approach here is similar in that we will ultimately adapt the model's internal feature representations through the selection of an adapted sub-network of the trained supernet, but without the need to use computationally expensive adversarial learning schemes. Our contributions can be summarized as follows:

1. We show that we can successfully train models through federated learning with comparable or better performance to models trained on centrally hosted data.
2. We extent federated learning by introducing an AutoML approach for supernet model training.

3. We show that finding an optimal path through the supernet can act as a form of local domain adaptation and bring performance gains for each individual client.

2   Method

Here we describe the technical details of the FL and AutoML approach utilized in this work. The proposed method can be separated into two steps: 1) FL with AutoML supernet training and 2) local model adaptation by finding the best path through the supernet with respect to the local data. Both FL and AutoML procedures presented are designed for 3D medical image segmentation tasks.

Client-Server-Based Federated Learning: In its typical form, FL utilizes a client-server setup. Each client trains the same model architecture locally on their own data. Once a certain number of clients finished local training, the updated model weights (or their gradients) are sent to the server for aggregation. After aggregation, the new weights on the server are re-distributed to the clients to execute the next round of local model training. After several FL rounds, the models at each client are converged. Each client can be allowed to select their local best model by monitoring a certain performance metric on a local hold out validation set. In our experiments, we implement the `FederatedAveraging` algorithm proposed in [19]. While there exist variants of this algorithm to address particular learning tasks, in its most general form, FL tries to minimize a global loss function $\mathcal{L}$ which can be a weighted combination of K local losses $\{\mathcal{L}_k\}_{k=1}^{K}$ that each is computed on a client k's local data. Hence, FL can be formulated as the task of finding the model parameters ϕ that minimize L given some local data X.

$$\min_{\phi} \mathcal{L}(X; \phi) \quad \text{with} \quad \mathcal{L}(X; \phi) = \sum_{k=1}^{K} w_k \, \mathcal{L}_k(X_k; \phi), \tag{1}$$

where $w_k > 0$ denote the weight coefficients for each client k, respectively. Note, that the local data X_i is never shared among the different clients. Only the model weights are accumulated and aggregated on the server as shown in Algorithm 1.

AutoML with Supernet: In order to allow for personalized neural architectures, we designed a supernet $\mathcal{S}$ consisting of various DL module candidates $\mathcal{M}$ suitable for 3D medical imaging tasks shown in Fig. 1a. Each candidate $\mathcal{M}$ is a subgraph $s \in \mathcal{S}$, denoted as $\mathcal{N}(s, w)$ with model weights w. These modules are optimized at multiple resolution levels to capture different levels of image features useful for the segmentation task. In general, we follow the popular encoder-decoder structure which has been successfully applied to many medical imaging tasks [2,20,21] as shown in Fig. 1b with skip connections that concatenate features of the encoder with their corresponding layer in the decoder path. During training, we choose arbitrary paths m from the module candidates $\mathcal{M}$ following

Algorithm 1. Client-server federated learning with `FederatedAveraging` [13,19]. T is the number of federated learning rounds and n_k is the number of `LocalTraining` iterations minimizing the local loss $\mathcal{L}_k(X_k; \phi^{(t-1)})$ for a client k.

1: **procedure** FEDERATED LEARNING
2:    Initialize weights: $\phi^{(0)}$
3:    **for** $t \leftarrow 1 \cdots T$ **do**
4:       **for** $client\ k \leftarrow 1 \cdots K$ **do**   ▷ *Executed in parallel*
5:          Send $\phi^{(t-1)}$ to client k
6:          Receive $(\Delta\phi_k^{(t)}, n_k)$ from client's `LocalTraining`$(\phi^{(t-1)})$
7:       **end for**
8:       $\phi_k^{(t)} \leftarrow \phi^{(t-1)} + \Delta\phi_k^{(t)}$
9:       $\phi^{(t)} \leftarrow \frac{1}{\sum_k n_k} \sum_k (n_k \cdot \phi_k^{(t)})$
10:   **end for**
11:   **return** $\phi^{(t)}$

a uniform sampling scheme (see Fig. 1c) to define a sub-network s sampled from the supernet $\mathcal{S}$ as in Eq. 2.

$$W_{\mathcal{S}} = \underset{W}{\mathrm{argmax}}\ \mathcal{L}(\mathcal{N}(\mathcal{S}, W)) \tag{2}$$

In this work, we choose the combination of Dice loss [20] and cross entropy loss as our loss function which is commonly used for segmentation tasks in medical imaging [8]. Dice loss' major advantage is its ability to work well in segmentation tasks with an unbalance in the amount of foreground/background regions. Once the supernet is trained, we can find a sub-network s_0 by identifying a locally optimal path through the supernet, effectively adapting the model to the target domain. During adaptation, the model parameters ϕ stay fixed and only the path weights are optimized for one epoch on the local validation set. This results in an optimal path $m_0 \in \mathcal{M}$ that defines our locally adapted sub-network $s_0 \in \mathcal{S}$ as Eq. 3.

$$s_0 = \underset{s \in \mathcal{S}}{\mathrm{argmax}}\ \mathcal{L}_{\mathrm{val}}(\mathcal{N}(s, w_s)) \tag{3}$$

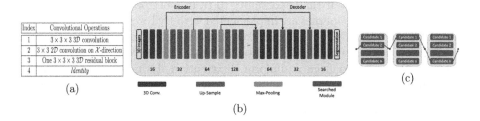

Fig. 1. (a) Searched network module candidates, (b) our supernet architecture $\mathcal{S}$, and (c) a potential path m selecting several module candidates $\mathcal{M}$. Note, *Identity* is only used if input/output sizes are the same.

3   Experiments and Results

Our proposed method is evaluated on the task of 3D whole prostate segmentation in T2-weighted MRI. In particular, MRI has challenges of data inconsistencies due to variations in different imaging protocols and scanners used at each data contributing site, potentially causing drastic variations in contrast and intensity values.

Datasets: We utilize prostate MRI datasets from four different publicly available data sources. MSD-Prostate[1] [24], PROMISE12[2] [17], NCI-ISBI13[3], and ProstateX[4] [16]. For each dataset, we perform three random splits into training, validation, and testing sets at roughly 70%, 10%, and 20% of the total number of cases of each dataset. The resulting number of cases for each dataset are shown in Table 1. We average results across the testing splits of each random split. For reference, we show the results on a centralized dataset where all four datasets have been combined. We also compare the performance for models trained locally and through federated learning an each dataset's testing split. The performance of a standard 3D U-Net [2] which is a subgraph of our supernet (when all candidates are type 1) is shown for a baseline comparison. We resample each image to a constant resolution of $0.5\,\text{mm} \times 0.5\,\text{mm} \times 1.0\,\text{mm}$ and normalize all non-zero image intensities by subtracting their mean and dividing by their standard derivation on a per-image basis.

Implementation: Both U-Net and the supernet are trained using randomly cropped patches of size $160 \times 160 \times 32$ from the input images and labels. We used a mini-batch size of 18 by selecting 3 random crops from any 6 random input image and label pairs. As the optimizer for training the supernet, we chose *NovoGrad* which has typically faster convergence speed than the more commonly used Adam optimizer [6]. The learning rate for supernet training was set to $1e^{-3}$. For finding the optimal path for the final sub-network we use the Adam optimizer with a learning rate of $1e^{-3}$. Augmentation techniques like random intensity shifts, contrast adjustments, and adding Gaussian noise are applied during training to avoid overfitting to the training set. Our supernet has $3^3 \times 4^6 = 110,592$ possible path combinations. Therefore, it is trained $10\times$ longer than 3D U-Net to give it the opportunity to train most paths well. Both 3D U-Net baseline and the supernet are implemented with PyTorch[5] using components from MONAI[6] and NVFlare[7] for FL communication. All models are trained on NVIDIA V100 GPUs with 16 GB memory. We monitor convergence on randomly chosen paths sampled from a uniform distribution during each validation to determine when the supernet is sufficiently trained across clients. The number

---

[1] http://medicaldecathlon.com.

[2] https://promise12.grand-challenge.org.

[3] http://doi.org/10.7937/K9/TCIA.2015.zF0vlOPv.

[4] https://prostatex.grand-challenge.org.

[5] https://pytorch.org.

[6] https://monai.io.

[7] https://pypi.org/project/nvflare.

of training iterations is chosen such that the likelihood of a path being selected during the entire training is at least >1.

Results: Table 1 show the performance for assuming local, centrally hosted, and federated datasets. Table 2 shows the better generalizibilty of supernet models trained in the FL. We show the performance of the proposed supernet training approach and its adaption to the local dataset distribution via path optimization, together with a baseline implementation of 3D U-Net using the same augmentation, optimization and hyperparameters to be comparable. Visualization of the results before and after model adaptation are shown in Fig. 2. In descending order, most commonly chosen operations were 3D conv., 3D residual block, 2D conv., followed by identity.

Table 1. Results for centralized dataset and each dataset trained locally and in federated learning. We show the performance of a baseline 3D U-Net and our proposed supernet (SN) approach. The average Dice of the local model's scores is shown (excluding the scores on centralized data). The highest scores for each dataset are marked in **bold**.

# Cases	Central	NCI	PROMISE12	ProstateX	MSD	
# Training	172	45	35	69	23	
# Validation	23	6	5	9	3	
# Testing	48	12	10	20	6	
# Total	243	63	50	98	32	
Avg. Dice [%]	Central	NCI	PROMISE12	ProstateX	MSD	Avg. (loc.)
U-Net (loc.)	89.72	90.15	83.59	90.73	86.17	87.66
U-Net (fed.)		90.59	85.73	90.35	88.40	88.77
SN (loc.)	90.11	90.42	83.51	90.76	86.76	87.86
SN (fed.)		90.57	85.61	**91.07**	88.39	88.91
SN (loc.) + adapt.	**90.15**	90.50	83.46	90.78	87.83	88.14
SN (fed.) + adapt.		**90.68**	**86.15**	90.65	**88.74**	**89.06**

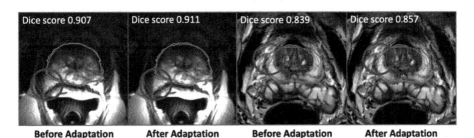

Dice score 0.907   Dice score 0.911   Dice score 0.839   Dice score 0.857

Before Adaptation   **After Adaptation**   **Before Adaptation**   **After Adaptation**

Fig. 2. The visual comparison of results from supernet before and after adaptation. Left two figures show results from PROMISE12, and right two figures are from MSD. Green contours denote ground truth (boundaries), and red contours are predictions from neural networks. We can see that adaptation is not only able to remove false positives, but also capable to correct false negative regions (highlighted by circles). (Color figure online)

4   Discussion and Conclusions

It can be observed from Table 1 that the supernet training with local adaptation in FL *(SN (fed.) + adapt.)* achieves the highest average Dice score on the local datasets. At the same time, the adapted models also show the best generalizability (see Table 2). This illustrates the viability of supernet training with local model adaption to the client's data. We furthermore observe a general improvement of the local supernet models' performance when trained in an FL setting versus local training. This means that in particular supernet model training can benefit from the larger effective training set size made available through FL without having to share any of the raw image data between clients. Overall, we achieve average Dice scores comparable to recent literature on whole prostate segmentation in MRI [16,17,20] and can likely be improved with more aggressive data augmentation schemes [8,29]. Further fine-tuning of the network weights (not the supernet path weights) is likely going to give performance boost on a local client but is also expected to reduce generalizability of the model. Methods of fine-tuning that do not reduce the robustness to other data sources (i.e. generalizability) gained through FL (e.g. learning without forgetting [15]) is still an open research question and was deemed to be out of scope of this work.

Table 2. Generalizability of the supernet models trained locally SN (loc.), in federated learning **SN (fed.)**, and after local adaptation **SN (fed.) + adapt**. Here, client sites A, B, C, and D stand for NCI, PROMISE12, ProstateX, and MSD, respectively. We show the local average performance of the models (**Loc.**, same as in Tab. 1) and the generalizability (**Gen.**), i.e. the Dice score on other clients' test data. The total average generalizability score is denoted as $\overline{\text{Gen.}}$ and the highest scores for each dataset are marked in **bold**.

| | SN (loc.) | | | | | SN (fed.) | | | | | SN (fed.) + adapt | | | | |
| | Test site | | | | | Test site | | | | | Test site | | | | |
Train site	A	B	C	D	Gen.	A	B	C	D	Gen.	A	B	C	D	Gen.
A	**90.4**	81.5	87.7	81.9	83.7	**90.6**	86.1	90.9	85.5	87.5	**90.7**	86.0	90.6	86.2	87.6
B	84.6	**83.5**	85.3	87.3	85.7	87.4	85.6	89.7	88.3	88.5	88.2	**86.2**	90.3	**89.5**	89.3
C	84.9	72.8	**90.8**	84.3	80.7	89.7	**86.4**	**91.1**	87.7	87.9	90.0	86.0	**90.7**	87.7	87.9
D	77.3	66.0	82.9	**86.8**	75.4	83.1	82.3	86.9	**88.4**	84.1	84.5	83.9	87.3	88.7	85.2
	Loc. 87.9			$\overline{\text{Gen.}}$ 81.4		Loc. 88.9			$\overline{\text{Gen.}}$ 87.0		Loc. 89.1			$\overline{\text{Gen.}}$ 87.5	

In conclusion, we proposed to combine the advantages of both federated learning and AutoML. The two techniques are complementary and in combination, they allow for an implicit domain adaptation through the finding of locally optimal model architectures (sub-networks of the supernet) for a client's dataset. We showed that the performances of federated learning are comparable to the model's performance when the dataset is centrally hosted. After local adaptation via choosing the optimal path through the supernet, we can see an additional performance gain on the client's data. In the future, it could be explored if there is a set of optimal sub-networks that could act as an ensemble during inference to further improve performance and provide additional estimates such as model uncertainty. Furthermore, one could adaptively change the path frequencies used during supernet training based on sub-network architectures that work well on each client in order to reduce communication cost and speed-up training.

References

1. Cai, H., Zhu, L., Han, S.: Proxylessnas: direct neural architecture search on target task and hardware. arXiv preprint arXiv:1812.00332 (2018)
2. Çiçek, Ö., Abdulkadir, A., Lienkamp, S.S., Brox, T., Ronneberger, O.: 3D U-net: learning dense volumetric segmentation from sparse annotation. In: Ourselin, S., Joskowicz, L., Sabuncu, M.R., Unal, G., Wells, W. (eds.) MICCAI 2016. LNCS, vol. 9901, pp. 424–432. Springer, Cham (2016). https://doi.org/10.1007/978-3-319-46723-8_49
3. Elsken, T., Metzen, J.H., Hutter, F.: Neural architecture search: a survey. arXiv preprint arXiv:1808.05377 (2018)
4. Ganin, Y., Lempitsky, V.: Unsupervised domain adaptation by backpropagation. In: Proceedings of the 32nd International Conference on International Conference on Machine Learning, vol. 37. pp. 1180–1189. JMLR. org (2015)
5. Ganin, Y., et al.: Domain-adversarial training of neural networks. J. Mach. Learn. Res. **17**(1), 1–35 (2016). 2096-2030
6. Ginsburg, B., et al.: Stochastic gradient methods with layer-wise adaptive moments for training of deep networks. arXiv preprint arXiv:1905.11286 (2019)
7. He, C., Annavaram, M., Avestimehr, S.: Towards Non-IID and invisible data with fednas: federated deep learning via neural architecture search. arXiv preprint arXiv:2004.08546 (2020)
8. Isensee, F., Jaeger, P.F., Kohl, S.A., Petersen, J., Maier-Hein, K.H.: NNU-Net: a self-configuring method for deep learning-based biomedical image segmentation. Nat. Methods **18**(2), 203–211 (2021)
9. Isola, P., Zhu, J.Y., Zhou, T., Efros, A.A.: Image-to-image translation with conditional adversarial networks. In: CVPR, pp. 1125–1134 (2017)
10. Kairouz, P., et al.: Advances and open problems in federated learning. arXiv preprint arXiv:1912.04977 (2019)
11. Kamnitsas, K., et al.: Unsupervised domain adaptation in brain lesion segmentation with adversarial networks. In: Niethammer, M., et al. (eds.) IPMI 2017. LNCS, vol. 10265, pp. 597–609. Springer, Cham (2017). https://doi.org/10.1007/978-3-319-59050-9_47
12. Kamnitsas, K., et al.: Deepmedic for brain tumor segmentation. In: Crimi, A., Menze, B., Maier, O., Reyes, M., Winzeck, S., Handels, H. (eds.) Brainlesion: Glioma, multiple sclerosis, stroke and traumatic brain injuries, pp. 138–149. Springer, Heidelberg (2016). https://doi.org/10.1007/978-3-319-55524-9_14
13. Li, W., et al.: Privacy-preserving federated brain tumour segmentation. In: Suk, H.-I., Liu, M., Yan, P., Lian, C. (eds.) MLMI 2019. LNCS, vol. 11861, pp. 133–141. Springer, Cham (2019). https://doi.org/10.1007/978-3-030-32692-0_16
14. Li, W., Wang, G., Fidon, L., Ourselin, S., Cardoso, M.J., Vercauteren, T.: On the compactness, efficiency, and representation of 3d convolutional networks: brain parcellation as a pretext task. In: Niethammer, M., et al. (eds.) IPMI 2017. LNCS, vol. 10265, pp. 348–360. Springer, Cham (2017). https://doi.org/10.1007/978-3-319-59050-9_28

15. Li, Z., Hoiem, D.: Learning without forgetting. IEEE Trans. Pattern Anal. Mach. Intell. **40**(12), 2935–2947 (2017)
16. Litjens, G., Debats, O., Barentsz, J., Karssemeijer, N., Huisman, H.: Computer-aided detection of prostate cancer in MRI. TMI **33**(5), 1083–1092 (2014)
17. Litjens, G., et al.: Evaluation of prostate segmentation algorithms for MRI: the PROMISE12 challenge. Med. Image Anal. **18**(2), 359–373 (2014)
18. Liu, H., Simonyan, K., Yang, Y.: Darts: differentiable architecture search. arXiv preprint arXiv:1806.09055 (2018)
19. McMahan, H.B., Moore, E., Ramage, D., Hampson, S., et al.: Communication-efficient learning of deep networks from decentralized data. arXiv preprint arXiv:1602.05629 (2016)
20. Milletari, F., Navab, N., Ahmadi, S.A.: V-Net: fully convolutional neural networks for volumetric medical image segmentation. In: 2016 Fourth International Conference on 3D Vision (3DV), pp. 565–571. IEEE (2016)
21. Ronneberger, O., Fischer, P., Brox, T.: U-Net: convolutional networks for biomedical image segmentation. In: Navab, N., Hornegger, J., Wells, W.M., Frangi, A.F. (eds.) MICCAI 2015. LNCS, vol. 9351, pp. 234–241. Springer, Cham (2015). https://doi.org/10.1007/978-3-319-24574-4_28
22. Sheller, M.J., Reina, G.A., Edwards, B., Martin, J., Bakas, S.: Multi-institutional deep learning modeling without sharing patient data: a feasibility study on brain tumor segmentation. In: Crimi, A., Bakas, S., Kuijf, H., Keyvan, F., Reyes, M., van Walsum, T. (eds.) BrainLes 2018. LNCS, vol. 11383, pp. 92–104. Springer, Cham (2019). https://doi.org/10.1007/978-3-030-11723-8_9
23. Shin, H.C., Roth, H.R., Gao, M., Lu, L., Xu, Z., Nogues, I., Yao, J., Mollura, D., Summers, R.M.: Deep convolutional neural networks for computer-aided detection: CNN architectures, dataset characteristics and transfer learning. TMI **35**(5), 1285–1298 (2016)
24. Simpson, A.L., et al.: A large annotated medical image dataset for the development and evaluation of segmentation algorithms. arXiv preprint arXiv:1902.09063 (2019)
25. Tan, M., Le, Q.: EfficientNet: rethinking model scaling for convolutional neural networks. In: International Conference on Machine Learning, pp. 6105–6114 (2019)
26. Yang, D., et al.: Federated semi-supervised learning for COVID region segmentation in chest CT using multi-national data from China, Italy, Japan. Med. Image Anal. **70**, 101992 (2021)
27. You, S., Huang, T., Yang, M., Wang, F., Qian, C., Zhang, C.: GreedyNAS: towards fast one-shot NAS with greedy supernet. In: CVPR, pp. 1999–2008 (2020)
28. Yu, Q., et al.: C2fnas: coarse-to-fine neural architecture search for 3D medical image segmentation. In: CVPR, pp. 4126–4135 (2020)
29. Zhang, L., et al.: Generalizing deep learning for medical image segmentation to unseen domains via deep stacked transformation. TMI (2020)
30. Zhang, Y., Miao, S., Mansi, T., Liao, R.: Task driven generative modeling for unsupervised domain adaptation: application to x-ray image segmentation. In: Frangi, A.F., Schnabel, J.A., Davatzikos, C., Alberola-López, C., Fichtinger, G. (eds.) MICCAI 2018. LNCS, vol. 11071, pp. 599–607. Springer, Cham (2018). https://doi.org/10.1007/978-3-030-00934-2_67
31. Zhu, H., Jin, Y.: Real-time federated evolutionary neural architecture search. arXiv preprint arXiv:2003.02793 (2020)
32. Zhu, J.Y., Park, T., Isola, P., Efros, A.A.: Unpaired image-to-image translation using cycle-consistent adversarial networks. In: Proceedings of the IEEE International Conference on Computer Vision, pp. 2223–2232 (2017)

Federated Contrastive Learning for Volumetric Medical Image Segmentation

Yawen Wu[1]([✉]), Dewen Zeng[2], Zhepeng Wang[1], Yiyu Shi[2], and Jingtong Hu[1]

[1] University of Pittsburgh, Pittsburgh, PA 15260, USA
{yawen.wu,zhepeng.wang,jthu}@pitt.edu
[2] University of Notre Dame, Notre Dame, IN 46556, USA
{dzeng2,yshi4}@nd.edu

Abstract. Supervised deep learning needs a large amount of labeled data to achieve high performance. However, in medical imaging analysis, each site may only have a limited amount of data and labels, which makes learning ineffective. Federated learning (FL) can help in this regard by learning a shared model while keeping training data local for privacy. Traditional FL requires fully-labeled data for training, which is inconvenient or sometimes infeasible to obtain due to high labeling cost and the requirement of expertise. Contrastive learning (CL), as a self-supervised learning approach, can effectively learn from unlabeled data to pre-train a neural network encoder, followed by fine-tuning for downstream tasks with limited annotations. However, when adopting CL in FL, the limited data diversity on each client makes federated contrastive learning (FCL) ineffective. In this work, we propose an FCL framework for volumetric medical image segmentation with limited annotations. More specifically, we exchange the features in the FCL pre-training process such that diverse contrastive data are provided to each site for effective local CL while keeping raw data private. Based on the exchanged features, global structural matching further leverages the structural similarity to align local features to the remote ones such that a unified feature space can be learned among different sites. Experiments on a cardiac MRI dataset show the proposed framework substantially improves the segmentation performance compared with state-of-the-art techniques.

Keywords: Federated learning · Contrastive learning · Self-supervised learning · Image segmentation

This work was supported in part by NSF IIS-2027546.

Electronic supplementary material The online version of this chapter (https://doi.org/10.1007/978-3-030-87199-4_35) contains supplementary material, which is available to authorized users.

M. de Bruijne et al. (Eds.): MICCAI 2021, LNCS 12903, pp. 367–377, 2021.
https://doi.org/10.1007/978-3-030-87199-4_35

1   Introduction

Deep learning (DL) provides state-of-the-art medical image segmentation performance by learning from large-scale labeled datasets [7,18,21,29], without which the performance of DL will significantly degrade [12]. However, medical data exist in isolated medical centers and hospitals [30], and combining a large dataset consisting of very sensitive and private medical data in a single location is impractical and even illegal. It requires multiple medical institutions to share medical patient data such as medical images, which is constrained by the Health Insurance Portability and Accountability Act (HIPAA) [12] and EU General Data Protection Regulation (GDPR) [25]. Federated learning (FL) is an effective machine learning approach in which distributed clients (i.e. individual medical institutions) collaboratively learn a shared model while keeping private raw data local [8,20,22,23]. By applying FL to medical image segmentation, an accurate model can be collaboratively learned and data is kept local for privacy.

Existing FL approaches use supervised learning on each client and require that all data are labeled. However, annotating all the medical images is usually unrealistic due to the high labeling cost and requirement of expertise. The deficiency of labels makes supervised FL impractical. Self-supervised learning can address this challenge by pre-training a neural network encoder with unlabeled data, followed by fine-tuning for a downstream task with limited labels. Contrastive learning (CL), a variant of the self-supervised learning approach, can effectively learn high-quality image representations. By integrating CL to FL as federated contrastive learning (FCL), clients can learn models by first collaboratively learning a shared image-level representation. Then the learned model will be fine-tuned by using limited annotations. Compared with local CL, FCL can learn a better encoder as the initialization for fine-tuning, and provide higher segmentation performance. In this way, a high-quality model can be learned by using limited annotations while data privacy is preserved.

However, integrating FL with CL to achieve good performance is nontrivial. Simply applying CL to each client and then aggregating the models is not the optimal solution for the following two reasons: First, each client only has a small amount of unlabeled data with limited diversity. Since existing contrastive learning frameworks [5,11] rely on datasets with diverse data to learn distinctive representations, directly applying CL on each client will result in an inaccurate learned model due to the lack of data diversity. Second, if each client only focuses on CL on its local data while not considering others' data, each client will have its own feature space based on its raw data and these feature spaces are inconsistent among different clients. When aggregating local models, the inconsistent feature space among local models will degrade the performance of the aggregated model.

To address these challenges, we propose a framework consisting of two stages to enable effective FCL for volumetric medical image segmentation with limited annotations. The first stage is feature exchange (FE), in which each client exchanges the features (i.e. low-dimensional vectors) of its local data with other clients. It provides more diverse data to compare with for better local contrastive learning while avoiding raw data sharing. In the learning process, the improved data diversity in feature space provides more accurate contrastive information in the local learning process on each client and improves the learned representations.

The second stage is global structural matching (GSM), in which we leverage structural similarity of 3D medical images to align similar features among clients for better FCL. The intuition is that the same anatomical region for different subjects has similar content in volumetric medical images such as MRI. By leveraging the structural similarity across volumetric medical images, GSM aligns the features of local images to the shared features of the same anatomical region from other clients. In this way, the learned representations of local models are more unified among clients and they further improve the global model after model aggregation. Experimental results show that the proposed approaches substantially improve the segmentation performance over state-of-the-art techniques.

2   Background and Related Work

Federated Learning. Federated learning (FL) learns a shared model by aggregating locally updated models on clients while keeping raw data accessible on local clients for privacy [14,15,17,32]. In FL, the training data are distributed among clients. FL is performed round-by-round by repeating the local model learning and model aggregation process until convergence.

The main drawback of these works is that fully labeled data are needed to perform FL, which results in high labeling costs. To solve this problem, an FL approach using limited annotations while achieving good performance is needed.

Contrastive Learning. Contrastive learning (CL) is a self-supervised approach to learn useful visual representations by using unlabeled data [10,19,24]. The learned model provides good initialization for fine-tuning on the downstream task with few labels [4–6,11,27]. CL performs a proxy task of instance discrimination [28], which maximizes the similarity of representations from similar pairs and minimizes the similarity of representations from dissimilar pairs [26].

The main drawback of existing CL approaches is that they are designed for centralized learning on large-scale datasets with sufficient data diversity. However, when applying CL to FL on each client, the limited data diversity will greatly degrade the performance of the learned model. Therefore, an approach to increase the local data diversity while avoiding raw data sharing for privacy is needed. Besides, while [4] leverages structural information in medical images for improving centralized CL, it requires accessing raw images of similar pairs for learning. Since sharing raw medical images is prohibitive due to privacy, [4] cannot be applied to FL. Therefore, an approach to effectively leverage similar images across clients without sharing raw images is needed.

Federated Unsupervised Pre-training. Some concurrent works employ federated pre-training on unlabeled data. [1] employs autoencoder in FL for pre-training on time-series data, but the more effective contrastive learning for visual tasks is not explored in FL. FedCA [31] combines contrastive learning with FL.

However, it relies on a shared dataset available on each client, which is impractical for medical images due to privacy concerns.

The proposed work differs from these federated unsupervised pre-training approaches in the following ways. First, we do not share raw data among clients to preserve privacy. Second, we leverage the structural similarity of images across clients to improve the quality of representation learning.

3   Method

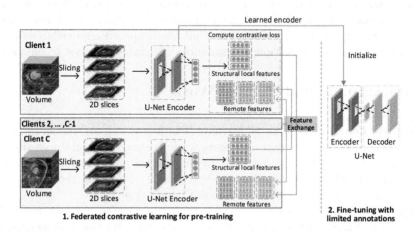

Fig. 1. Federated contrastive learning with structural feature exchange for learning the encoder with unlabeled data. Then the learned encoder initializes the encoder in U-Net for fine-tuning with limited annotations.

Overview of Federated Contrastive Learning. The overview of the proposed FCL process is shown in Fig. 1. Distributed clients first collaboratively learn a shared encoder by FCL with unlabeled data. Then the learned encoder initializes the encoder in U-Net [21] for fine-tuning with limited annotations, either independently on each client by supervised learning or collaboratively by supervised federated learning. Since the supervised fine-tuning can be trivially achieved by using available annotations, in the rest of the paper, we focus on FCL to learn a good encoder as the initialization for fine-tuning.

As shown in Fig. 1, in the FCL stage, given a volumetric 3D image on one client, multiple 2D slices are sampled from the volume while keeping structural order along the slicing axis. Then the ordered 2D images are fed into the 2D encoder to generate feature vectors, one vector for each 2D image.

To improve the data diversity in local contrastive learning, one natural way is to share raw images [32]. However, sharing raw medical images is prohibitive due to privacy concerns. To solve this problem, the proposed FCL framework exchanges the feature vectors instead of raw images among clients, which can improve the data diversity while preserving privacy. As shown in

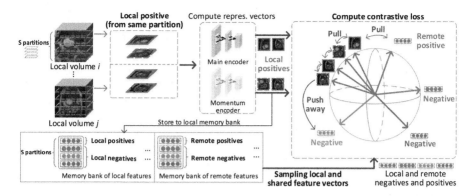

Fig. 2. Contrastive learning on one client with exchanged features. The exchanged features consist of remote negatives and remote positives, in which remote negatives improve the local data diversity and remote positives are used for global structural matching to learn a unified feature space among clients.

Fig. 1, client 1 generates structural local features denoted as blue vectors and shares them with other clients. Meanwhile, client 1 collects structural features from other clients, such as remote features shown in green and gray vectors. After that, the contrastive loss is computed based on both local and remote features.

3.1   Contrastive Learning with Feature Exchange

With feature exchange, each client has both remote and local features and is ready to perform local CL in each round of FCL. The remote features provide more diverse features to compare with and improve the learned representations. As shown in Fig. 2, we use MoCo [11] architecture for local CL since it has a memory bank for negatives, which can leverage local and remote features. There are two encoders, including the main encoder and the momentum encoder. The main encoder will be learned and used as the initialization for fine-tuning, while the momentum encoder is the slowly-evolving version of the main encoder and generates features to contrast with and for sharing. Now the most important steps are to construct negatives and positives from both local and remote features.

Negatives from Local and Remote Features. Local features are generated by the momentum encoder from local images and used as local negatives. Each client has a memory bank of local features and a memory bank of remote features. Let $Q_{l,c}$ be the size-K memory bank of local features on client c, which are used as local negatives. $Q_{l,c}$ is progressively updated by replacing the oldest features with the latest ones. In each round of FCL, the remote negatives from other clients will be shared with client c to form its aggregated memory bank including local and remote negatives as:

$$Q = Q_{l,c} \cup \{Q_{l,i} \mid 1 \leq i \leq |C|, i \neq c\}. \tag{1}$$

where C is the set of all clients and $Q_{l,i}$ is the local memory bank on client i.

Compared with using only local memory bank $Q_{l,c}$, the aggregated memory bank Q provides more data diversity to improve CL. However, Q is $|C|$ times the size of the local memory bank $Q_{l,c}$. More negatives make CL more challenging since for one local feature q, more negatives need to be simultaneously pushed away from it than when using $Q_{l,c}$, which can result in ineffective learning. To solve this problem, instead of using all negatives in Q, for each q we sample a size-K (i.e. the same size as $Q_{l,c}$) subset of Q as negatives, which is defined as:

$$Q' = \{Q_i| \ i \sim \mathcal{U}(|Q|, K)\}. \tag{2}$$

where $i \sim \mathcal{U}(|Q|, K)$ means i is a set of indices sampled uniformly from $[|Q|]$.

Local Positives. We leverage the structural similarity in the volumetric medical images to define the local positives, in which the same anatomical region from different subjects has similar content [4]. Each volume is grouped into S partitions, and one image sampled from partition s of volume i is denoted as x_s^i.

Local positives are features of images from the same partition in different volumes. Given an image x_s^i, its feature q_s^i and corresponding positives $P(q_s^i) = \{k_s^{i+}, k_s^{j+}\}$ are formed as follows. Two transformations (e.g. cropping) are applied to x_s^i to get $\tilde{x}_s^i$ and $\hat{x}_s^i$, which are then fed into the main encoder and momentum encoder to generate two representation vectors q_s^i and k_s^{i+}, respectively. Then another image x_s^j is sampled from partition s of volume j, and its features q_s^j and k_s^{j+} are generated accordingly. In this way, the local positives for both q_s^i and q_s^j are formed as $P(q_s^i) = P(q_s^j) = \{k_s^{i+}, k_s^{j+}\}$.

Loss Function for Local Positives. By using the sampled memory bank Q' consisting of both *local* negatives and *remote* negatives, one local feature q is compared with its local positives $P(q)$ and each negative in Q'. The contrastive loss is defined as:

$$\mathcal{L}_{local} = \ell_{q, P(q), Q'} = -\frac{1}{|P(q)|} \sum_{k^+ \in P(q)} \log \frac{\exp(q \cdot k^+/\tau)}{\exp(q \cdot k^+/\tau) + \sum_{n \in Q'} \exp(q \cdot n/\tau)}. \tag{3}$$

where τ is the temperature and the operator $\cdot$ is the dot product between two vectors. By minimizing the loss, the distance between q and each local positive is minimized, and the distance between q and each negative in Q' is maximized.

3.2   Global Structural Matching

Remote Positives. We use the remote positives from the shared features to further improve the learned representations. On each client, we align the features of one image to the features of images in the same partition from other clients. In this way, the features of images in the same partition across clients will be aligned in the feature space and more unified representations can be learned among clients. To achieve this, for one local feature q, in addition to its local positives

$P(q)$, we define remote positives $\Lambda(q)$ as features in the sampled memory bank Q' which are in the same partition as q.

$$\Lambda(q) = \{p \mid p \in Q', partition(p) = partition(q)\}. \qquad (4)$$

$partition(\cdot)$ is the partition number of one feature and Q' is defined in Eq.(2).

Final Loss Function. By replacing local positives $P(q)$ in Eq. (3) with remote positives $\Lambda(q)$ as $\mathcal{L}_{remote}$, the final loss function for one feature q is defined as:

$$\mathcal{L}_q = \mathcal{L}_{remote} + \mathcal{L}_{local} = \ell_{q,\Lambda(q),Q'} + \ell_{q,P(q),Q'}. \qquad (5)$$

With $\mathcal{L}_q$, the loss for one batch of images is defined as $\mathcal{L}_B = \frac{1}{|B|} \sum_{q \in B} \mathcal{L}_q$, where B is the set of features generated by the encoder from the batch of images.

4   Experiments

Dataset and Preprocessing. We evaluate the proposed approaches on the ACDC MICCAI 2017 challenge dataset [2], which has 100 patients with 3D cardiac MRI images. Each patient has about 15 volumes covering a full cardiac cycle, and only volumes for the end-diastolic and end-systolic phases are annotated by experts for three structures, including left ventricle, myocardium, and right ventricle. Details of preprocessing can be found in the supplementary material.

Federated and Training Setting. Following [32], we use 10 clients. We randomly split 100 patients in ACDC dataset into 10 partitions, each with 10 patients. Then each client is assigned one partition with 10 patients. We use the proposed FCL approaches to pre-train the U-Net encoder on the assigned dataset partition on each client without labels. Then the pre-trained encoder (i.e. the final global encoder after pre-training) is used as the initialization for fine-tuning the U-Net segmentation model by using a small number of labeled samples. The U-Net model follows the standard 2D U-Net architecture [21] with the initial number of channels set to 48. We evaluate with two settings for fine-tuning: *local fine-tuning* and *federated fine-tuning*. In local fine-tuning, each client fine-tunes the model on its local annotated data. In federated fine-tuning, all clients collaboratively fine-tune the model by supervised FL with a small number of annotations. Training details can be found in the supplementary material.

Evaluation. During fine-tuning, we use 5-fold cross validation to evaluate the segmentation performance. In each fold, 10 patients on one client are split into a training set of 8 patients and a validation set of 2 patients. For each fold, we fine-tune with annotations from $N \in \{1, 2, 4, 8\}$ patients in the training set, and validate on the validation set of the same fold on all clients (i.e. 20 patients). Dice similarity coefficient (DSC) is used as the metric for evaluation.

Baselines. We compare the proposed approaches with multiple baselines. *Random init* fine-tunes the model from random initialization. *Local CL* performs

Table 1. Comparison of the proposed approaches and baselines on **local fine-tuning** with limited annotations on the ACDC dataset. N is the number of annotated patients for fine-tuning on each client. The average dice score and standard deviation across 10 clients are reported, in which on each client the dice score is averaged on 5-fold cross validation. The proposed approaches substantially outperform all the baselines with different numbers of annotations.

Methods	$N = 1$	$N = 2$	$N = 4$	$N = 8$
Random init	0.280 ± 0.037	0.414 ± 0.070	0.618 ± 0.026	0.766 ± 0.027
Local CL [4]	0.320 ± 0.106	0.456 ± 0.095	0.637 ± 0.043	0.770 ± 0.029
FedRotation [9]	$\underline{0.357} \pm 0.058$	$\underline{0.508} \pm 0.054$	$\underline{0.660} \pm 0.021$	$\underline{0.783} \pm 0.029$
FedSimCLR [5]	0.288 ± 0.049	0.435 ± 0.046	0.619 ± 0.032	0.765 ± 0.033
FedSwAV [3]	0.323 ± 0.066	0.480 ± 0.067	0.659 ± 0.019	0.782 ± 0.030
FedCA [31]	0.280 ± 0.047	0.417 ± 0.042	0.610 ± 0.030	0.766 ± 0.029
Proposed	$\mathbf{0.506} \pm 0.056$	$\mathbf{0.631} \pm 0.051$	$\mathbf{0.745} \pm 0.017$	$\mathbf{0.824} \pm 0.025$

contrastive learning on each client by the SOTA approach [4] with unlabeled data for pre-training the encoder before fine-tuning. *Rotation* [9] is a self-supervised pre-training approach by predicting the image rotations. *SimCLR* [5] and *SwAV* [3] are the SOTA contrastive learning approaches for pre-training. We combine these three self-supervised approaches with *FedAvg* [17] as their federated variants *FedRotation*, *FedSimCLR*, and *FedSwAV* for pre-training the encoder. *FedCA* is the SOTA federated unsupervised learning approach for pre-training [31].

4.1   Results of Local Fine-Tuning

We evaluate the performance of the proposed approaches by fine-tuning locally on each client with limited annotations. As shown in Table 1, the proposed approaches substantially outperform the baselines. First, with 1, 2, 4, or 8 annotated patients, the proposed approaches outperform the best-performing baseline by 0.149, 0.123, 0.085, and 0.041 dice score, respectively. Second, the proposed approaches significantly improve the annotation efficiency. For example, with 1 annotated patient, the proposed approaches achieve a similar dice score to the best-performing baseline with 2 annotations (0.506 vs. 0.508), which improves labeling-efficiency by 2×.

4.2   Results of Federated Fine-Tuning

We evaluate the performance of the proposed approaches by collaborative federated fine-tuning with limited annotations. Similar to local fine-tuning, the proposed approaches significantly outperform the SOTA techniques as shown in Table 2. First, with 1, 2, 4, or 8 annotated patients per client, the proposed approaches outperform the best-performing baselines by 0.130, 0.107, 0.050,

Table 2. Comparison of the proposed approaches and baselines on **federated fine-tuning** with limited annotations on the ACDC dataset. N is the number of annotated patients for fine-tuning on each client. The proposed approaches significantly outperform all the baselines with different numbers of annotations.

Methods	$N = 1$	$N = 2$	$N = 4$	$N = 8$
Random init	0.445 ± 0.012	0.572 ± 0.061	0.764 ± 0.017	0.834 ± 0.011
Local CL [4]	0.473 ± 0.013	$\underline{0.717} \pm 0.024$	0.784 ± 0.015	0.847 ± 0.009
FedRotation [9]	$\underline{0.516} \pm 0.015$	0.627 ± 0.074	$\underline{0.821} \pm 0.015$	$\underline{0.867} \pm 0.010$
FedSimCLR [5]	0.395 ± 0.023	0.576 ± 0.046	0.788 ± 0.014	0.859 ± 0.011
FedSwAV [3]	0.500 ± 0.015	0.594 ± 0.058	0.815 ± 0.015	0.862 ± 0.010
FedCA [31]	0.397 ± 0.020	0.561 ± 0.047	0.784 ± 0.015	0.858 ± 0.011
Proposed	$\mathbf{0.646 \pm 0.052}$	$\mathbf{0.824 \pm 0.004}$	$\mathbf{0.871 \pm 0.007}$	$\mathbf{0.894 \pm 0.006}$

and 0.027 dice score, respectively. Second, the proposed approaches effectively reduce the annotations needed for fine-tuning. For example, with 2 or 4 annotated patients per client, the proposed approaches achieve better performance than the best-performing baseline with 2× annotated patients per client, which achieve more than 2× labeling-efficiency. Third, compared with local fine-tuning in Table 1, all the approaches achieve a higher dice score.

5   Conclusion and Future Work

This work aims to enable federated contrastive learning for volumetric medical image segmentation with limited annotations. Clients first learn a shared encoder on distributed unlabeled data and then a model is fine-tuned on annotated data. Feature exchange is proposed to improve data diversity for contrastive learning while avoiding sharing raw data. Global structural matching is developed to learn an encoder with unified representations among clients. The experimental results show significantly improved segmentation performance and labeling-efficiency compared with state-of-the-art techniques.

Discussion. Sharing features needs additional communication, and we will explore techniques to reduce the communication cost. Besides, we will explore defenses such as [13] against inversion attacks [16] for improved security.

References

1. van Berlo, B., Saeed, A., Ozcelebi, T.: Towards federated unsupervised representation learning. In: Proceedings of the Third ACM International Workshop on Edge Systems, Analytics and Networking, pp. 31–36 (2020)
2. Bernard, O., et al.: Deep learning techniques for automatic MRI cardiac multistructures segmentation and diagnosis: is the problem solved? IEEE Trans. Med. Imaging **37**(11), 2514–2525 (2018)

3. Caron, M., Misra, I., Mairal, J., Goyal, P., Bojanowski, P., Joulin, A.: Unsupervised learning of visual features by contrasting cluster assignments. arXiv preprint arXiv:2006.09882 (2020)

4. Chaitanya, K., Erdil, E., Karani, N., Konukoglu, E.: Contrastive learning of global and local features for medical image segmentation with limited annotations. arXiv preprint arXiv:2006.10511 (2020)

5. Chen, T., Kornblith, S., Norouzi, M., Hinton, G.: A simple framework for contrastive learning of visual representations. arXiv preprint arXiv:2002.05709 (2020)

6. Chen, T., Kornblith, S., Swersky, K., Norouzi, M., Hinton, G.: Big self-supervised models are strong semi-supervised learners. arXiv preprint arXiv:2006.10029 (2020)

7. Dong, H., Yang, G., Liu, F., Mo, Y., Guo, Y.: Automatic brain tumor detection and segmentation using U-Net based fully convolutional networks. In: Valdés Hernández, M., González-Castro, V. (eds.) MIUA 2017. CCIS, vol. 723, pp. 506–517. Springer, Cham (2017). https://doi.org/10.1007/978-3-319-60964-5_44

8. Dou, Q., et al.: Federated deep learning for detecting COVID-19 lung abnormalities in CT: a privacy-preserving multinational validation study. NPJ Digit. Med. 4(1), 1–11 (2021)

9. Gidaris, S., Singh, P., Komodakis, N.: Unsupervised representation learning by predicting image rotations. arXiv preprint arXiv:1803.07728 (2018)

10. Hadsell, R., Chopra, S., LeCun, Y.: Dimensionality reduction by learning an invariant mapping. In: 2006 IEEE Computer Society Conference on Computer Vision and Pattern Recognition (CVPR'06), vol. 2, pp. 1735–1742. IEEE (2006)

11. He, K., Fan, H., Wu, Y., Xie, S., Girshick, R.: Momentum contrast for unsupervised visual representation learning. In: Proceedings of the IEEE/CVF Conference on Computer Vision and Pattern Recognition, pp. 9729–9738 (2020)

12. Kairouz, P., et al.: Advances and open problems in federated learning. arXiv preprint arXiv:1912.04977 (2019)

13. Li, A., Guo, J., Yang, H., Chen, Y.: DeepObfuscator: adversarial training framework for privacy-preserving image classification. arXiv preprint arXiv:1909.04126 (2019)

14. Li, T., Sahu, A.K., Zaheer, M., Sanjabi, M., Talwalkar, A., Smith, V.: Federated optimization in heterogeneous networks. arXiv preprint arXiv:1812.06127 (2018)

15. Li, T., Sahu, A.K., Zaheer, M., Sanjabi, M., Talwalkar, A., Smith, V.: Federated optimization in heterogeneous networks. In: Proceedings of Machine Learning and Systems, vol. 2, pp. 429–450 (2020)

16. Lyu, L., Yu, H., Yang, Q.: Threats to federated learning: a survey. arXiv preprint arXiv:2003.02133 (2020)

17. McMahan, B., Moore, E., Ramage, D., Hampson, S., Arcas, B.A.: Communication-efficient learning of deep networks from decentralized data. In: Artificial Intelligence and Statistics, pp. 1273–1282. PMLR (2017)

18. Milletari, F., Navab, N., Ahmadi, S.A.: V-Net: fully convolutional neural networks for volumetric medical image segmentation. In: 2016 Fourth International Conference on 3D Vision (3DV), pp. 565–571. IEEE (2016)

19. Misra, I., Maaten, L.V.D.: Self-supervised learning of pretext-invariant representations. In: Proceedings of the IEEE/CVF Conference on Computer Vision and Pattern Recognition, pp. 6707–6717 (2020)

20. Rieke, N.: The future of digital health with federated learning. NPJ Digit. Med. 3(1), 1–7 (2020)

21. Ronneberger, O., Fischer, P., Brox, T.: U-Net: convolutional networks for biomedical image segmentation. In: Navab, N., Hornegger, J., Wells, W.M., Frangi, A.F.

(eds.) MICCAI 2015. LNCS, vol. 9351, pp. 234–241. Springer, Cham (2015). https://doi.org/10.1007/978-3-319-24574-4_28

22. Sheller, M.J., et al.: Federated learning in medicine: facilitating multi-institutional collaborations without sharing patient data. Sci. Rep. **10**(1), 1–12 (2020)

23. Sheller, M.J., Reina, G.A., Edwards, B., Martin, J., Bakas, S.: Multi-institutional deep learning modeling without sharing patient data: a feasibility study on brain tumor segmentation. In: Crimi, A., Bakas, S., Kuijf, H., Keyvan, F., Reyes, M., van Walsum, T. (eds.) BrainLes 2018. LNCS, vol. 11383, pp. 92–104. Springer, Cham (2019). https://doi.org/10.1007/978-3-030-11723-8_9

24. Tian, Y., Krishnan, D., Isola, P.: Contrastive multiview coding. arXiv preprint arXiv:1906.05849 (2019)

25. Truong, N., Sun, K., Wang, S., Guitton, F., Guo, Y.: Privacy preservation in federated learning: An insightful survey from the GDPR perspective. arXiv preprint arXiv:2011.05411 (2020)

26. Wang, T., Isola, P.: Understanding contrastive representation learning through alignment and uniformity on the hypersphere. In: International Conference on Machine Learning, pp. 9929–9939. PMLR (2020)

27. Wu, Y., Wang, Z., Zeng, D., Shi, Y., Hu, J.: Enabling on-device self-supervised contrastive learning with selective data contrast. arXiv preprint arXiv:2106.03796 (2021)

28. Wu, Z., Xiong, Y., Yu, S.X., Lin, D.: Unsupervised feature learning via non-parametric instance discrimination. In: Proceedings of the IEEE Conference on Computer Vision and Pattern Recognition, pp. 3733–3742 (2018)

29. Xu, X., et al.: Whole heart and great vessel segmentation in congenital heart disease using deep neural networks and graph matching. In: Shen, D., et al. (eds.) MICCAI 2019. LNCS, vol. 11765, pp. 477–485. Springer, Cham (2019). https://doi.org/10.1007/978-3-030-32245-8_53

30. Yang, Q., Liu, Y., Chen, T., Tong, Y.: Federated machine learning: concept and applications. ACM Trans. Intell. Syst. Technol. (TIST) **10**(2), 1–19 (2019)

31. Zhang, F., et al.: Federated unsupervised representation learning. arXiv preprint arXiv:2010.08982 (2020)

32. Zhao, Y., Li, M., Lai, L., Suda, N., Civin, D., Chandra, V.: Federated learning with Non-IID data. arXiv preprint arXiv:1806.00582 (2018)

Federated Contrastive Learning for Decentralized Unlabeled Medical Images

Nanqing Dong[✉] and Irina Voiculescu

Department of Computer Science, University of Oxford, Oxford, UK
nanqing.dong@cs.ox.ac.uk

Abstract. A label-efficient paradigm in computer vision is based on self-supervised contrastive pre-training on unlabeled data followed by fine-tuning with a small number of labels. Making practical use of a federated computing environment in the clinical domain and learning on medical images poses specific challenges. In this work, we propose FedMoCo, a robust federated contrastive learning (FCL) framework, which makes efficient use of decentralized unlabeled medical data. FedMoCo has two novel modules: *metadata transfer*, an inter-node statistical data augmentation module, and *self-adaptive aggregation*, an aggregation module based on representational similarity analysis. To the best of our knowledge, this is the first FCL work on medical images. Our experiments show that FedMoCo can consistently outperform FedAvg, a seminal federated learning framework, in extracting meaningful representations for downstream tasks. We further show that FedMoCo can substantially reduce the amount of labeled data required in a downstream task, such as COVID-19 detection, to achieve a reasonable performance.

Keywords: Federated learning · Contrastive representation learning

1 Introduction

Recent studies in self-supervised learning (SSL) [21] have led to a renaissance of research on contrastive learning (CL) [1]. Self-supervised or unsupervised CL aims to learn transferable representations from unlabeled data. In a CL framework, a model is first pre-trained on unlabeled data in a self-supervised fashion via a contrastive loss, and then fine-tuned on labeled data. Utilizing the state-of-the-art (SOTA) CL frameworks [3,8,16,24], a model trained with only unlabeled data plus a small amount of labeled data can achieve comparable performance with the same model trained with a large amount of labeled data on various downstream tasks.

As a data-driven approach, deep learning has fueled many breakthroughs in medical image analysis (MIA). Meanwhile, large-scale fully labeled medical datasets require considerable human annotation cost, which makes data scarcity a major bottleneck in practical research and applications. To leverage unlabeled

© Springer Nature Switzerland AG 2021
M. de Bruijne et al. (Eds.): MICCAI 2021, LNCS 12903, pp. 378–387, 2021.
https://doi.org/10.1007/978-3-030-87199-4_36

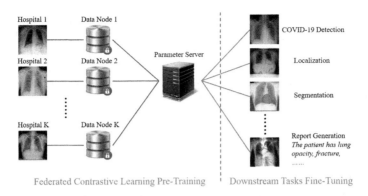

Fig. 1. Illustration of FCL workflow. Each node works on an independent local dataset. The parameter server works with data nodes for periodic synchronization, metadata transfer (Sect. 2.2), and self-adaptive aggregation (Sect. 2.3). For each downstream task, the pre-trained model weights are fine-tuned with small amount of labeled data in either supervised or semi-supervised fashion. Due to data privacy, the labeled data of a downstream task can only be accessed *locally*.

data, CL [4,10,23] has obtained promising results on unlabeled data, yet none of these studies consider federated learning (FL), which can handle sensitive data stored in multiple devices [7,14,20,22]. Data privacy regulations restrict collecting clinical data from different hospitals for conventional data-centralized CL. Under FL protocols, medical images (either raw or encoded) must not be exchanged between data nodes. In this work, we design an FCL framework on decentralized medical data (i.e. medical images stored on multiple devices or at multiple locations) to extract useful representations for MIA.

There are two direct *negative* impacts of the FL environment on CL. First, different imaging protocols in hospitals (nodes) create domain shift [6]. In contrast to CL on centralized data, each node in FCL only has access to its local data, which has a smaller variation in sample distribution. This impairs the CL performance on single nodes. Second, without the supervision of ground truth labels, there is no guarantee for the performance of CL in each node. Applying supervised FL frameworks on CL directly might lead to worse generalization ability and hence poorer outcome as the performance of FCL could be dominated by the performance of a few nodes. How to aggregate CL models across nodes is still an open question.

We introduce FedMoCo, a robust FCL framework for MIA (see Fig. 1). FedMoCo uses MoCo [8] as the intra-node CL model. To mitigate the above hurdles, we propose *metadata transfer*, an inter-node augmentation module utilizing Box-Cox power transformation [2] and Gaussian modeling; we also propose *self-adaptive aggregation*, a module based on representational similarity analysis (RSA) [12]. We empirically evaluate FedMoCo under various simulated scenarios. Our experiments show that FedMoCo can consistently outperform FedAvg [15], a SOTA FL framework; FedMoCo can efficiently reduce annotation cost in downstream tasks, such as COVID-19 detection. By pre-training on unlabeled non-COVID datasets, `FedMoCo` requires only 3% examples of a labeled COVID-19 dataset to achieve 90% accuracy. Our contributions are threefold: (1) to the best of our knowledge, this is the first work of FCL on medical images; (2) we propose `FedMoCo` with two novel modules; (3) our results provide insights into future research on FCL.

2   Method

Problem Formulation. $K > 1$ denotes the number of nodes and $\mathcal{D}_k$ is an *unlabeled* dataset stored in node $k \leq K$. x_i is a medical image in $\mathcal{D}_k = \{x_i\}_{i=1}^{n_k}$, with $n_k = |\mathcal{D}_k|$. As is standard practice in FL, $\mathcal{D}_k \cap \mathcal{D}_l = \emptyset$ for $k \neq l$ and $\{\mathcal{D}_k\}_{k=1}^{K}$ are non-IID data. There is an additional master node, which does not store any clinical data. The master node is implemented as a *parameter server* (PS) [13].

The model of interest is f_θ and its parameter set θ_0 is randomly initialized in PS. At the beginning of the federated training, K copies of θ_0 are distributed into each node as $\{\theta_k\}_{k=1}^{K}$, i.e. we fully synchronize the data nodes with PS. An important FL protocol is data privacy preservation: exchanging training data between nodes is strictly prohibited. Instead, node k updates θ_k by training on $\mathcal{D}_k$ independently. After the same number of local epochs, $\{\theta_k\}_{k=1}^{K}$ are aggregated into θ_0 in PS. Again, we synchronize $\{\theta_k\}_{k=1}^{K}$ with aggregated θ_0. This process is repeated until certain criteria are met. To enforce data privacy, we only ever exchange model parameters $\{\theta_k\}_{k=0}^{K}$ and metadata between the data nodes and PS. The learning outcome is θ_0 for relevant downstream tasks.

2.1   Intra-Node Contrastive Learning

We create a positive pair by generating two random views of the same image through data augmentation. A negative pair is formed by taking two random views from two different images. Given an image x and a family of stochastic image transformations $\mathcal{T}$, we randomly sample two transformations τ and τ' to have two random views $\tau(x)$ and $\tau'(x)$ which form a positive pair. Let z denote the representation of an image extracted by a CNN encoder f_θ. In contrast to previous works [3,8,27], we add a ReLU function between the last fully-connected layer and a L2-normalization layer, which projects all extracted features into a non-negative feature space. The non-negative feature space is a prerequisite for Sect. 2.2. For query image $x \in \mathcal{D}_k$, we have positive pair $z_q = f_{\theta_k^q}(\tau(x))$ and $z_0 = f_{\theta_k^d}(\tau'(x))$. If there are N negative examples $\{x_i\}_{i=1}^{N}$, the contrastive loss InfoNCE [18] is defined as

$$\mathcal{L}_q = -\log \frac{\exp(z_q \cdot z_0/te)}{\sum_{i=0}^{N} \exp(z_q \cdot z_i/te)}, \tag{1}$$

where the temperature parameter is te.

We use the dynamic dictionary with momentum update in MoCo [8] to maintain a large number of negative examples in Eq. 1. In node k, there are two CNNs, a CNN encoder (θ_k^q) for the query image and another (θ_k^d) for the corresponding positive example which are defined as

$$\theta_k^q \leftarrow \theta_k, \qquad \theta_k^d \leftarrow m\theta_k^d + (1-m)\theta_k^q, \tag{2}$$

where $m \in [0,1)$ is the momentum coefficient. For query image x, we have $z_q = f_{\theta_k^q}(\tau(x))$ and $z_0 = f_{\theta_k^d}(\tau'(x))$.

2.2 Metadata Transfer

The dynamic dictionary of node k must not exchange information with other nodes. This limits the sample variation of the dynamic dictionary, which limits CL performance. In addition, a node overfitted to local data may not generalize well to other nodes. We overcome these two hurdles by utilizing metadata from the encoded feature vectors in each node. For node k, after the full synchronization with PS (i.e. before the start of next round of local updates), we extract feature vectors $\{f_{\theta_k}(x_i)\}_{i=1}^{n_k}$ of $\mathcal{D}_k$. In order to enforce a Gaussian-like distribution of the features we use the Box-Cox power transformation (BC). BC is a reversible transformation defined as

$$BC(x) = \begin{cases} \frac{x^\lambda - 1}{\lambda}, & \lambda \neq 0 \\ \log(x), & \lambda = 0 \end{cases} \tag{3}$$

where λ controls the skewness of the transformed distribution. Then, we calculate the mean and covariance[1] of the transformed features

$$\boldsymbol{\mu}_k = \frac{\sum_{i=1}^{n_k} y_i}{n_k}, \qquad \boldsymbol{\Sigma}_k = \frac{\sum_{i=1}^{n_k}(y_i - \boldsymbol{\mu}_k)(y_i - \boldsymbol{\mu}_k)^T}{n_k - 1} \tag{4}$$

where $y_i = BC(f_{\theta_k}(x_i))$. The metadata of the learned representations in each node are collected by PS as $\{(\boldsymbol{\mu}_k, \boldsymbol{\Sigma}_k)\}_{k=1}^{K}$. $\{(\boldsymbol{\mu}_j, \boldsymbol{\Sigma}_j)\}_{j \neq k}$ is sent to node k. We name this operation as *metadata transfer*. Metadata transfer improves the CL performance of node k by augmenting the dynamic dictionary of node k statistically with metadata collected from all the other nodes $j \neq k$.

For each pair of $(\boldsymbol{\mu}_k, \boldsymbol{\Sigma}_k)$, there is a corresponding Gaussian distribution $\mathcal{N}(\boldsymbol{\mu}_k, \boldsymbol{\Sigma}_k)$. In the next round of local updates, we increase the sample variation in each node by sampling new points from the Gaussian distributions when minimizing Eq. 1. Specifically, for node k, we increase the number of negative examples in Eq. 1 from N to $\lfloor N(1 + \eta) \rfloor$, where $\eta \geq 0$ is a hyper-parameter to control the level of interaction between the node k and the other nodes (e.g. $\eta = 0$ means no interaction). We sample $\left\lfloor \frac{\eta N}{K-1} \right\rfloor$ examples from each $\mathcal{N}(\boldsymbol{\mu}_l, \boldsymbol{\Sigma}_l) \ \forall \ l \neq k$. Let $\tilde{y} \sim \mathcal{N}(\boldsymbol{\mu}_l, \boldsymbol{\Sigma}_l) \ \forall \ l \neq k$, we have $\tilde{z} = BC^{-1}(\tilde{y})$ and the new contrastive loss is

$$\mathcal{L}_q = -\log \frac{\exp(z_q \cdot z_0 / \tau)}{\sum_{i=0}^{N} \exp(z_q \cdot z_i / \tau) + \sum_{j=1}^{(K-1)\left\lfloor \frac{\eta N}{K-1} \right\rfloor} \exp(z_q \cdot \tilde{z}_j / \tau)} \tag{5}$$

---

[1] For large-scale datasets, the mean and covariance could be estimated by using random samples from the population. Here, we use the notation n_k for simplicity.

Algorithm 1. FedMoCo. The training in each data node is warmed up for t_w rounds before metadata transfer. A local round could be a few local epochs.

Initialize θ_0^0
for $t = 1, 2, \cdots, T$ do
    for $k = 1, 2, \cdots, K$ do
        $\theta_k^t \leftarrow \theta_0^{t-1}$ ▷ Synchronize with PS
        if $t \leq t_w$ then   ▷ Warm up
            $\theta_k^t \leftarrow local\_update(\theta_k^t)$ ▷ Eq. 1
        else   ▷ Metadata transfer
            Download $\{(\boldsymbol{\mu}_j, \boldsymbol{\Sigma}_j)\}_{j \neq k}$ from PS
            $\theta_k^t \leftarrow local\_update(\theta_k^t, \{(\boldsymbol{\mu}_j, \boldsymbol{\Sigma}_j)\}_{j \neq k})$ ▷ Eq. 5
            Upload $(\boldsymbol{\mu}_k, \boldsymbol{\Sigma}_k)$ to PS
    Compute A ▷ Eq. 6 and Eq. 7
    $\theta_0^t \leftarrow \sum_{k=1}^{K} a_k^t \theta_k^t$
Output θ_0^T

2.3 Self-adaptive Aggregation

Given locally updated $\{\theta_k^t\}_{k=1}^K$, which are collected at the end of round t, the aggregation step of round t can be formulated as $\theta^t = \theta_0^t = \sum_{k=1}^K a_k \theta_k^t$, where a_k denotes k^{th} element of diagonal matrix A. For example, in FedAvg [15], we use $a_k = \frac{n_k}{\sum_{j=1}^K n_j}$ because the number of labels indicates the strength of supervision in each node, which does not work for unsupervised learning. For node k with large n_k and small variation in $\mathcal{D}_k$ (i.e. examples are similar), f_{θ_k} converges faster but θ_k learns less meaningful representations (e.g. overfitted to no findings or to certain diseases). But $a_k = \frac{n_k}{\sum_{j=1}^K n_j}$ gives θ_k^t a larger weight and θ_k^0 is dominated by θ_k^t. Instead, we propose *self-adaptive aggregation* to compute matrix A.

Let $-1 \leq r_k \leq 1$ denote representational similarity analysis (RSA) [12] of node k. In round t we first take n_k' random samples from $\mathcal{D}_k$ as a subset $\mathcal{D}_k'$ for computational and statistical efficiency. For each $x_i \in \mathcal{D}_k'$, we get two representations $f_{\theta_0^{t-1}}(x_i)$ (aggregated weights at the end of round $t-1$, which are also globally synchronized weights at the beginning of round t) and $f_{\theta_k^t}(x_i)$ (locally updated weights at the end of round t). We define ρ_{ij} as Pearson's correlation coefficient between $f_\theta(x_i)$ and $f_\theta(x_j)$ $\forall\, 0 \leq i, j \leq n_k'$ and define $\text{RDM}_{ij} = 1 - \rho_{ij}$, where RDM is representation dissimilarity matrix [12] for f_θ. Then, r_k is defined based on Spearman's rank correlation:

$$r_k = 1 - \frac{6 \sum_{i=1}^n d_i^2}{n(n^2 - 1)} \tag{6}$$

where d_i is the difference between the ranks of ith elements of the lower triangular of RDM for $f_{\theta^{t-1}}$ and RDM for $f_{\theta_k^t}$, and $n = n_k'(n_k' - 1)/2$. We define A for the aggregation step at the end of round t as

$$a_k^t = \frac{1 - r_k}{\sum_{j=1}^K 1 - r_j}. \tag{7}$$

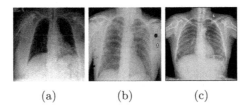

Fig. 2. Pre-training on: (a) CheXpert; (b) ChestX-ray8; (c) VinDr-CXR.

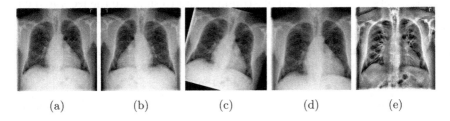

Fig. 3. Stochastic data augmentation: (a) raw; (b) flip; (c) rotate; (d) crop and resize; (e) histogram equalization.

The CL performance at node k in round t is measured by r_k. Intuitively, given the same θ^{t-1} at the start of round t, a small r_k indicates that the change between the representations $f_{\theta^{t-1}}$ and f_{θ^t} is large, i.e. node k is still learning meaningful representations. Equation 7 assigns larger weights to local models with higher potentials in representational power. The complete pseudo-code is given in Algorithm 1.

3 Experiments

3.1 Experimental Setup

For a fair comparison, we use the same set of hyperparameters and the same training strategy for all experiments. We use ResNet18 [9] as the network backbone, initialized with the same random seed. We use the Momentum optimizer with momentum 0.9. Without using any pre-trained weights, we demonstrate that FedMoCo can learn from scratch. The initial learning rate is 0.03 and is multiplied by 0.1 (and 0.01) at 120 (and 160) epochs. The batch size is 64 for each node, the weight decay is 10^{-4}, m in Eq. 2 is 0.999, λ in Eq. 3 is 0.5, τ is 0.2, N is 1024, and η in Eq. 5 is 0.05. We estimate RSA with $n_k' = 100$.

In the absence of FCL for MIA, we compare FedMoCo with a strong baseline which is an integration of MoCo [8] and FedAvg [15], a seminal supervised FL model. All models are implemented with PyTorch on NVIDIA Tesla V100.

Datasets. FCL should work for any type of medical images. We illustrate FCL on anterior and posterior chest X-rays (CXRs). We use three public large-scale

Table 1. Dataset description.

Dataset	Size	# of classes	Multi-label	Multi-view	Balanced	Resolution
CheXpert [11]	371920	14	✓	✓	✗	390×320
ChestX-ray8 [26]	112120	15	✓	✗	✗	1024×1024
VinDr-CXR [17]	15000	15	✓	✗	✗	512×512
COVID-19 [5]	3886	3	✗	✗	✓	256×256

CXR datasets as the unlabeled pre-training data to simulate the federated environment, namely CheXpert [11], ChestX-ray8 [26], and VinDr-CXR [17] (see Table 1). Three datasets are collected and annotated from different sources independently and express a large variety in data modalities (see Fig. 2). The original images are cropped and downsampled to 256×256. Three datasets contain noisy labels and the label distributions are different.

Data Augmentation. Driven by clinical domain knowledge and experimental findings, We provide a augmentation policy as illustrated in Fig. 3. Compared with previous studies of MoCo on medical images [4,10,23], we propose stochastic histogram equalization for CL on medical images. The stochasticity comes from uniformly sampling the parameters of CLAHE [19]. For a fair comparison, we use the same augmentation policy for all models in this work.

Linear Classification Protocol. We evaluate the performance of unsupervised pre-training by following *linear classification protocol* (LCP) [3,8,16,24]. ResNet18 is first pre-trained on the unlabeled dataset. Then a supervised linear classifier (a fully-connected layer) is trained on top of the frozen features extracted from a labeled dataset for 50 epochs with a constant learning rate 0.1. We report the classification accuracy on the validation set of the labeled data as the performance of FCL. We choose a public COVID-19 CXR dataset [5] with 3886 CXRs. Note that COVID-19 has not been seen in three pre-training datasets. We use 50% CXRs for training and 50% for testing.

3.2   Evaluation

The experiments are simulated in a controllable federated environment. To eliminate the effect of n_k, we first evaluate FedMoCo in a situation that each node has the same number of CXRs. We create 3 data nodes by randomly sampling 10000 CXRs from each of three datasets. We create 6 nodes by partitioning each data node equally. We evaluate the pre-training performance on $T = 200$ and $T = 400$, where one round is one local epoch and FedMoCo is warmed up for $t_w = 50$ epochs. We provide the performance of MoCo trained in a single node with the same data as *Oracle* for centralized CL. LCP results with mean and standard deviation of 3 rounds are present in Table 2. We have two empirical findings for FCL: more nodes will decrease the performance, and more training time may not improve the performance of downstream tasks.

To show the isolated contribution of metadata transfer and self-adaptive aggregation, we simulate two common situations in FL with $K = 3$ and $T = 200$.

Table 2. LCP comparison between FCL models on non-IID distributions with equal numbers of examples.

Model	K	$T = 200$ (%)	$T = 400$ (%)
FedAvg	3	94.92 ± 0.16	95.03 ± 0.23
FedMoCo	3	95.73 ± 0.05	95.58 ± 0.26
FedAvg	6	94.60 ± 0.32	94.65 ± 0.94
FedMoCo	6	95.11 ± 0.71	94.92 ± 1.27
Oracle	1	92.80 ± 0.05	92.37 ± 0.06

Table 3. LCP comparison between FCL models on non-IID distributions with imbalanced numbers of examples.

Model	$\gamma = 5$ (%)	$\gamma = 10$ (%)
FedAvg	94.27 ± 1.37	94.77 ± 0.86
FedMoCo-M	94.66 ± 1.57	95.23 ± 0.96
FedMoCo-S	94.37 ± 0.77	95.01 ± 0.45
FedMoCo	94.77 ± 0.80	95.58 ± 0.31
Oracle	95.22 ± 0.05	95.23 ± 0.39

Table 4. LCP comparison between FCL models on non-IID distributions with imbalanced label distribution.

Model	$n_k = 5000$ (%)	$n_k = 10000$ (%)
FedAvg	93.30 ± 1.21	95.25 ± 0.33
FedMoCo-M	93.67 ± 1.19	95.43 ± 0.36
FedMoCo-S	94.11 ± 0.88	95.74 ± 0.29
FedMoCo	94.38 ± 0.68	96.02 ± 0.25
Oracle	95.78 ± 0.22	94.82 ± 0.11

Table 5. Fine-tuning models pre-trained by FCL for COVID-19 detection.

Model	Accuracy (%)
w/o Pre-Training	79.63
FedAvg	87.24
FedMoCo	91.56
Oracle(MoCo)	88.42
Oracle(Supervised)	95.78

First, with the same data as above, we create unbalanced data distributions in terms of the number of examples by only using γ% CXRs in nodes of CheXpert and ChestX-ray8. We use FedMoCo-M to denote FedMoCo with only metadata transfer module and use FedMoCo-S to denote FedMoCo with only self-adaptive aggregation. The results are presented in Table 3. Second, with the same number of examples in each node, we create unbalanced data distributions in terms of the label distribution by only sampling healthy CXRs (no findings) for the nodes of CheXpert and ChestX-ray8 and only sampling CXRs with disease labels for the node of VinDr-CXR. The results are presented in Table 4. In summary the above results, FedMoCo can outperform FedAvg consistently under non-IID challenges while metadata transfer and self-adaptive aggregation both can improve the performance as an individual module. An interesting finding against common sense is that sometimes FCL can outperform centralized CL depending on the data distributions, which may worth further investigation.

Downstream Task Evaluation. To demonstrate the practical value of the representations extracted by FCL under data scarcity, we use COVID-19 detection [5] as the downstream task. We use the same training set and test set in LCP. We fine-tune the ResNets pre-trained by FCL models in Table 2 ($K = 3$ and $T = 200$) with only 3% of the training set. We train a randomly initialized ResNet with the full training set as the *Oracle* for supervised learning. For a fair comparison, we use a fixed learning rate 0.01 to train all models for 100 epochs. We report the highest accuracy in Table 5. FedMoCo outperforms centralized CL and FedAvg. Compared with standard supervised learning, FedMoCo utilizes only 3% of labels to achieve 90% accuracy, which greatly reduces the annotation cost.

4   Conclusion

In this work, we formulate and discuss FCL on medical images and propose FedMoCo. We evaluate the robustness of FedMoCo under a few characteristic non-IID challenges and use COVID-19 detection as the downstream task. More investigations will be conducted, but these initial results already provide insights into future FCL research. In future, we plan to focus on the task affinity [25] between FCL and its corresponding downstream tasks. We will quantitatively analyze how the representations extracted by FCL can influence the performance of different downstream tasks.

Acknowledgements. We would like to thank Huawei Technologies Co., Ltd. for providing GPU computing service for this study.

References

1. Baldi, P., Pineda, F.: Contrastive learning and neural oscillations. Neural Comput. **3**(4), 526–545 (1991)
2. Box, G.E., Cox, D.R.: An analysis of transformations. J. R. Stat. Soc.: Ser. B (Methodol.) **26**(2), 211–243 (1964)
3. Chen, T., Kornblith, S., Norouzi, M., Hinton, G.: A simple framework for contrastive learning of visual representations. In: ICML (2020)
4. Chen, X., Yao, L., Zhou, T., Dong, J., Zhang, Y.: Momentum contrastive learning for few-shot COVID-19 diagnosis from chest CT images. Pattern Recognit. **113**, 107826 (2020)
5. Chowdhury, M.E., et al.: Can AI help in screening viral and COVID-19 pneumonia? IEEE Access **8**, 132665–132676 (2020)
6. Dong, N., Kampffmeyer, M., Liang, X., Wang, Z., Dai, W., Xing, E.: Unsupervised domain adaptation for automatic estimation of cardiothoracic ratio. In: Frangi, A.F., Schnabel, J.A., Davatzikos, C., Alberola-López, C., Fichtinger, G. (eds.) MICCAI 2018. LNCS, vol. 11071, pp. 544–552. Springer, Cham (2018). https://doi.org/10.1007/978-3-030-00934-2_61
7. Dou, Q.: Federated deep learning for detecting COVID-19 lung abnormalities in CT: a privacy-preserving multinational validation study. NPJ Digit. Med. **4**(1), 1–11 (2021)
8. He, K., Fan, H., Wu, Y., Xie, S., Girshick, R.: Momentum contrast for unsupervised visual representation learning. In: CVPR, pp. 9729–9738 (2020)
9. He, K., Zhang, X., Ren, S., Sun, J.: Deep residual learning for image recognition. In: CVPR, pp. 770–778 (2016)
10. He, X., et al.: Sample-efficient deep learning for COVID-19 diagnosis based on CT scans. MedRxiv (2020)

11. Irvin, J., et al.: CheXpert: a large chest radiograph dataset with uncertainty labels and expert comparison. In: AAAI, vol. 33, no. 01, pp. 590–597 (2019)
12. Kriegeskorte, N., Mur, M., Bandettini, P.A.: Representational similarity analysis-connecting the branches of systems neuroscience. Front. Syst. Neurosci. **2**, 4 (2008)
13. Li, M., Andersen, D.G., Smola, A.J., Yu, K.: Communication efficient distributed machine learning with the parameter server. In: NIPS, pp. 19–27 (2014)
14. Li, W., et al.: Privacy-preserving federated brain tumour segmentation. In: Suk, H.-I., Liu, M., Yan, P., Lian, C. (eds.) MLMI 2019. LNCS, vol. 11861, pp. 133–141. Springer, Cham (2019). https://doi.org/10.1007/978-3-030-32692-0_16
15. McMahan, B., Moore, E., Ramage, D., Hampson, S., Aguera Arcas, B.: Communication-efficient learning of deep networks from decentralized data. In: AISTATS, pp. 1273–1282. PMLR (2017)
16. Misra, I., Maaten, L.V.D.: Self-supervised learning of pretext-invariant representations. In: CVPR, pp. 6707–6717 (2020)
17. Nguyen, H.Q., et al.: VinDr-CXR: an open dataset of chest x-rays with radiologist's annotations. arXiv preprint arXiv:2012.15029 (2020)
18. Oord, A.V.D., Li, Y., Vinyals, O.: Representation learning with contrastive predictive coding. arXiv preprint arXiv:1807.03748 (2018)
19. Pizer, S.M., et al.: Adaptive histogram equalization and its variations. Comput. Vis. Graph. Image Process. **39**(3), 355–368 (1987)
20. Rieke, N., et al.: The future of digital health with federated learning. NPJ Digit. Med. **3**(1), 1–7 (2020)
21. de Sa, V.R.: Learning classification with unlabeled data. In: NIPS, pp. 112–119. Citeseer (1994)
22. Sheller, M.J., et al.: Federated learning in medicine: facilitating multi-institutional collaborations without sharing patient data. Sci. Rep. **10**(1), 1–12 (2020)
23. Sowrirajan, H., Yang, J., Ng, A.Y., Rajpurkar, P.: MoCo pretraining improves representation and transferability of chest x-ray models. arXiv preprint arXiv:2010.05352 (2020)
24. Tian, Y., Sun, C., Poole, B., Krishnan, D., Schmid, C., Isola, P.: What makes for good views for contrastive learning. In: NIPS (2020)
25. Vandenhende, S., Georgoulis, S., De Brabandere, B., Van Gool, L.: Branched multi-task networks: deciding what layers to share. In: BMVC (2020)
26. Wang, X., Peng, Y., Lu, L., Lu, Z., Bagheri, M., Summers, R.M.: ChestX-ray8: hospital-scale chest x-ray database and benchmarks on weakly-supervised classification and localization of common thorax diseases. In: CVPR, pp. 2097–2106 (2017)
27. Wu, Z., Xiong, Y., Stella, X.Y., Lin, D.: Unsupervised feature learning via non-parametric instance discrimination. In: CVPR (2018)

Machine Learning - Interpretability/Explainability

Explaining COVID-19 and Thoracic Pathology Model Predictions by Identifying Informative Input Features

Ashkan Khakzar[1], Yang Zhang[1], Wejdene Mansour[1], Yuezhi Cai[1], Yawei Li[1], Yucheng Zhang[1], Seong Tae Kim[2(✉)], and Nassir Navab[1,3]

[1] Technical University of Munich, Munich, Germany
[2] Kyung Hee University, Yongin-si, South Korea
[3] Johns Hopkins University, Baltimore, USA
st.kim@khu.ac.kr

Abstract. Neural networks have demonstrated remarkable performance in classification and regression tasks on chest X-rays. In order to establish trust in the clinical routine, the networks' prediction mechanism needs to be interpretable. One principal approach to interpretation is feature attribution. Feature attribution methods identify the importance of input features for the output prediction. Building on Information Bottleneck Attribution (IBA) method, for each prediction we identify the chest X-ray regions that have high mutual information with the network's output. Original IBA identifies input regions that have *sufficient* predictive information. We propose Inverse IBA to identify *all* informative regions. Thus all predictive cues for pathologies are highlighted on the X-rays, a desirable property for chest X-ray diagnosis. Moreover, we propose Regression IBA for explaining regression models. Using Regression IBA we observe that a model trained on cumulative severity score labels implicitly learns the severity of different X-ray regions. Finally, we propose Multi-layer IBA to generate higher resolution and more detailed attribution/saliency maps. We evaluate our methods using both human-centric (ground-truth-based) interpretability metrics, and human-agnostic feature importance metrics on NIH Chest X-ray8 and BrixIA datasets. The code (https://github.com/CAMP-eXplain-AI/CheXplain-IBA) is publicly available.

Keywords: Explainable AI · Feature attribution · Chest X-rays · Covid

1 Introduction

Deep Neural Network models are the de facto standard in solving classification and regression problems in medical imaging research. Their prominence is specifically more pronounced in chest X-ray diagnosis problems, due to the availability

---

A. Khakzar and Y. Zhang—Shared first authorship.
S. T. Kim and N. Navab—Shared senior authorship.

© Springer Nature Switzerland AG 2021
M. de Bruijne et al. (Eds.): MICCAI 2021, LNCS 12903, pp. 391–401, 2021.
https://doi.org/10.1007/978-3-030-87199-4_37

of large public chest X-ray datasets [6,7,19,30]. Chest X-ray is an economical, fast, portable, and accessible diagnostic modality. A modality with the aforementioned properties is specifically advantageous in worldwide pandemic situations such as COVID-19 where access to other modalities such as Computed Tomography (CT) is limited [16,18,23]. Therefore, diagnostic chest X-ray neural network models can be of great value in large-scale screening of patients worldwide.

However, the black-box nature of these models is of concern. It is crucial for their adoption to know whether the model is relying on features relevant to the medical condition. In pursuit of interpretability of chest X-ray models, a class of works focuses on instilling interpretability into the models during optimization [9,26,29], another class pursues optimization semi-supervised with localization [13], and another class of works provides post-hoc explanations [8,19,30]. Post-hoc explanations have the advantage that they can be applied to any model without changing the objective function.

One principal method for post-hoc explanation is feature attribution (aka saliency methods), i.e. identifying the importance/relevance of input features for the output prediction [4,11,21,22,25,28]. Feature attribution problem remains largely open to this date, however, many branches of solutions are proposed. The question is which attribution solution to use. Attributions are evaluated from several perspectives, and one crucial and *necessary* aspect is to evaluate whether the attributed features are indeed important for model prediction, which is done by feature importance metrics [3,17,20]. One *desirable* property is human interpretability of the results, i.e. if the attribution is interpretable for the user. For example, Class Activation Maps (CAM, GradCAM) [22,32] being a solid method that is adopted by many chest X-ray model interpretation works, satisfies feature importance metrics. However, it generates attributions that are of low resolution, and while accurately highlighting the important features, they do not highlight these regions with *precision*. Such precision is of more importance in chest X-rays where features are subtle. On the other hand, some other methods (e.g. Guided BackPropagation [27], $\alpha 1\beta 0$ [4], Excitation Backprop [31]) have pixel-level resolution and are human-interpretable, but do not satisfy feature important metrics and some do not explain model behavior [1,5,10,12,15].

Information Bottleneck Attribution (IBA) [21] is a recent method proposed in neural networks literature that satisfies feature importance metrics, is more human-interpretable than established methods such as CAMs [32], and is of solid theoretical grounding. The method also visualizes the amount of information each image region provides for the output in terms of bits/pixels, thus its attribution maps (saliency maps) of different inputs are comparable in terms of quantity of the information (bits/pixels). Such properties make IBA a promising candidate for chest X-ray model interpretation.

In this work, we build upon IBA and propose extended methodologies that benefit chest X-ray model interpretations.

1.1   Contribution Statement

Inverse IBA: The original IBA method finds input regions that have sufficient predictive information. The presence of these features is sufficient for the target prediction. However, if sufficient features are removed, some other features can have predictive information. We propose Inverse IBA to find any region that can have predictive information.

Regression IBA: IBA (and many other methods such as CAMs) is only proposed for classification. We propose Regression IBA and by using it we observe that a model trained on cumulative severity score labels implicitly learns the severity of different X-ray regions.

Multi-layer IBA: We investigate approaches to use the information in layers of all resolutions, to generate high-resolution saliency maps that *precisely* highlight informative regions. Using Multi-layer IBA, for instance, we can precisely highlight subtle regions such as Mass, or we observe that the model is using corner regions to classify Cardiomegaly.

Effect of Balanced Training: We also observe that considering data imbalance during training results in learned features being aligned with the pathologies.

2   Methodology

Information Bottleneck for Attribution (IBA) [21] inserts a bottleneck into an existing network to restrict the flow of information during inference given an input. The bottleneck is constructed by adding noise into the feature maps (activations) of a layer. Let F denote the feature maps at layer l, the bottleneck is represented by $Z = \lambda F + (1 - \lambda)\epsilon$, where ϵ is the noise, the mask λ has the same dimension as F and controls the amount of noise added to the signal. Each element in the mask $\lambda_i \in [0, 1]$. Since the goal is post-hoc explanation for an input X, the model weights are fixed and *the mask λ is optimized* such that mutual information between the noise-injected activation maps Z and the input X is minimized, while the mutual information between Z and the target Y is maximized:

$$\max_{\lambda} I(Y, Z) - \beta I(X, Z) \tag{1}$$

The term $I(X, Z)$ is intractable, thus it is (variationally) approximated by

$$I(X, Z) \approx \mathcal{L}_I = E_F[D_{KL}(P(Z|F)||Q(Z))] \tag{2}$$

where $Q(Z) \sim \mathcal{N}(\mu_F, \sigma_F)$ (μ_F and σ_F are the estimated mean and variance of hidden feature F from a batch of data samples). In [21], the mutual information $I(Y, Z)$ is replaced by cross entropy loss L_{CE}. It is proven $-L_{CE}$ is a lower bound for I(Y,Z) [2]. Minimizing L_{CE} corresponds to maximizing $-L_{CE}$ and thus maximizing the lower bound of I(Y,Z). The objective becomes:

$$\mathcal{L} = \beta \mathcal{L}_I + \mathcal{L}_{CE} \tag{3}$$

2.1   Inverse IBA

In IBA formulation (Eq. 3), the first term tries to remove as many features as possible (by setting $\lambda = 0$) and while the second term tries to keep features (by setting $\lambda = 1$) such that mutual information with target Y is kept. Therefore, if only a small region can keep the second term ($\mathcal{L}_{CE}$) minimized (keep the mutual information with target Y), the rest of the features are removed (their $\lambda = 0$). The identified regions ($\lambda = 1$) have *sufficient* predictive information, as their existence is sufficient for the prediction. However, there might exist other regions that have predictive information in the absence of these sufficient regions. From another perspective, IBA is playing a preservation game, which results in preserving features that keep the output close to the target.

To find all regions that have predictive information we change the formulation of IBA such that the optimization changes to a deletion game. I.e. deleting the smallest fraction of features such that there is no predictive information for the output anymore after deletion. In order to change IBA optimization to a deletion game we make two changes: 1) for the second term ($\mathcal{L}_{CE}$) in Eq. 3 we use an inverse mask: $Z_{inv} = \lambda \epsilon + (1 - \lambda)F$, and denote the new term with $\mathcal{L}_{CE}^{inv}$. 2) we maximize the $\mathcal{L}_{CE}^{inv}$ in order for the optimization to remove predictive features. Thus, the objective is:

$$\mathcal{L}_{inv} = \beta \mathcal{L}_I - \mathcal{L}_{CE}^{inv} \qquad (4)$$

Minimizing $\mathcal{L}_I$ corresponds to the feature map becoming noise (similar to IBA) and pushes λ to 0. Minimizing $\mathcal{L}_{CE}^{inv}$ (maximizing $\mathcal{L}_{CE}^{inv}$) in Eq. 4 corresponds to removing *all* predictive information. In the $\mathcal{L}_{CE}^{inv}$ term, we use Z_{inv}, thus removing features corresponds to pushing the λ to 1 (if we instead use Z instead of Z_{inv}, λ moves to 0, and as $\mathcal{L}_I$ also pushes λ to 0, we get 0 everywhere). Therefore, λ is pushed to 1 for *any* predictive feature, and to 0 for the rest. As such, Inverse IBA identifies *any* predictive feature in the image and not just the *sufficiently* predictive features (examples in Fig. 1).

2.2   Regression IBA

Original IBA is proposed for classification setting. In this section, we discuss several variations of IBA for the regression case. We discuss three different regression objectives: 1) MSE Loss defined as $\mathcal{L}_{MSE} = (\mathbf{\Phi}(Z) - \mathbf{y})^2$. MSE loss has the property that if the target score is small, it identifies regions with small brixIA score as informative. Because in this case, the objective is trying to find regions that have information for output to be zero. 2) Regression Maximization (RM) Loss is simply defined as $\mathcal{L}_{RM} = \mathbf{\Phi}(Z)^2$. This loss has the property that it favors regions with high scores as informative. 3) Deviation loss defined as $\mathcal{L}_{DV} = (\mathbf{\Phi}(Z) - X)^2$. We subtract the score of the noisy feature map from the score of the original image. Similar to IBA for classification, this formulation identifies regions with sufficient information for the prediction. We also apply Inverse IBA to regression (see Fig. 1) to identify all predictive features.

2.3   Multi-layer IBA

For original IBA, the bottleneck is inserted in one of the later convolutional layers. As we move towards earlier layers, the variational approximation becomes less accurate. Thus the optimization in Eq. 3 highlights extra regions that do not have predictive information in addition to highlighting the sufficient regions. However, as the resolution of feature maps in earlier layers are higher, the highlighted regions are crisper and more interpretable. In order to derive regions that are crips and have high predictive information we compute IBA for several layers and combine their results, thus introducing Multi-layer IBA:

$$\mathcal{T}(IBA_{L_1}) \cap \mathcal{T}(IBA_{L_2})... \cap \mathcal{T}(IBA_{L_L}) \tag{5}$$

where $\mathcal{T}$ denotes a thresholding operation to binarize the IBA maps.

2.4   Chest X-Ray Models

Classification Model: We denote a neural network function by $\boldsymbol{\Phi}_\Theta(\mathbf{x})$: $\mathbb{R}^{H \times W} \rightarrow \mathbb{R}^C$ where C is the number of output classes. For a dataset $\mathbf{X} = \{\mathbf{x}^{(1)}, ..., \mathbf{x}^{(N)}\}$, and their labels $\mathbf{Y} = \{\mathbf{y}^{(1)}, ..., \mathbf{y}^{(N)}\}$, where $\mathbf{y} = [\mathbf{y_j}]^C$, and $\mathbf{y_j} \in \{0, 1\}$. Chest X-rays can have multiple pathologies. We use Binary Cross Entropy (BCE) loss on each output for multilabel prediction.

$$\mathcal{L}_{BCE} = (\hat{\mathbf{y}}, \mathbf{y}) = -\sum_j \beta \mathbf{y_j} \log(\hat{\mathbf{y_j}}) + (1 - \mathbf{y_j}) \log(1 - \hat{\mathbf{y_j}}) \tag{6}$$

where β is a weighting factor to balance the positive labels.

Regression Model: Consider a neural network $\mathbf{f}_\Theta(\mathbf{x}) : \mathbb{R}^{H \times W} \rightarrow \mathbb{R}$ and a dataset $\mathbf{X} = \{\mathbf{x}^{(1)}, ..., \mathbf{x}^{(N)}\}$ of N X-ray images, and their corresponding labels $\mathbf{Y} = \{\mathbf{y}^{(1)}, ..., \mathbf{y}^{(N)}\}$, where $\mathbf{y}_j \in 0, ..., 18$ is the cumulative severity score on each image. We model the regression problem with a MSE loss:

$$\mathcal{L}_{MSE} = \frac{1}{N} \sum (\boldsymbol{\Phi}_\Theta(\mathbf{x})^{(\mathbf{n})} - \mathbf{y}^{(\mathbf{n})})^2 \tag{7}$$

3   Experiments and Results

Implementation Details. We use three models: 1) NIH ChestX-ray8 classification: Network with 8 outputs for the 8 pathologies. 2) BrixIA regression: Network with one output and predicts the total severity score (sum of severity scores of 6 regions) 3) BrixIA classifier: 3 outputs detecting whether a severity score of 3, 2, and 0/1 exists in the X-rays. We use Densenet 121, and insert the IBA bottleneck on the output of DenseBlock 3. For Multi-layer IBA we insert it on the outputs of DenseBlock 1, 2 and 3.

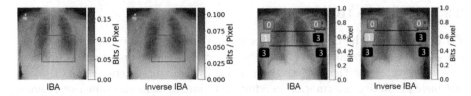

Fig. 1. Inverse IBA: Inverse IBA compared with IBA on a sample from the NIH Chest X-ray8 (left) and a sample from BrixIA (right). **NIH Chest X-ray8 (left):** Inverse IBA is identifying both sides of Cardiomegaly as informative. The bounding box denotes the expert's annotation. **BrixIA (Right):** IBA is identifying two regions with a severity score of 3 as sufficient for predicting the score of 3, however, Inverse IBA is identifying all regions with a severity score of 3, and the region with a score of 1. The horizontal lines denote the 6 regions within the lungs, and the numbers represent the severity score of each region.

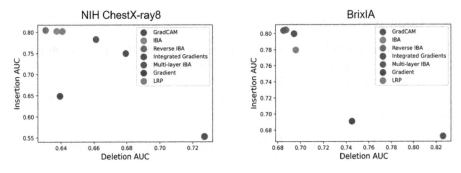

Fig. 2. Insertion/Deletion metric: Comparison of different attribution methods in terms of feature importance. Method with high Insertion AUC and low Deletion AUC is the best (top left corner is the best).

3.1   Feature Importance (Human-Agnostic) Evaluations

Experiments in this section evaluate whether an attribution method is identifying important features for the model prediction.

Insertion/Deletion [17,20]. Insertion: given a baseline image (we use the blurred image) features of the original image are added to the baseline image starting from the most important feature and the output is observed. If the attribution method is correct, after inserting a few features the prediction changes significantly, thus the AUC of output is high. Deletion: deleting important features first. The lower the AUC the better. Results are presented in Fig. 2.

Sensitivity-N [3]. We adapt this metric to regression case (not trivial with Insertion/Deletion) and use it to evaluate Regression-IBA. In Sensitivity-n we mask the input randomly and observe the correlation between the output change and the values of attribution map overlapping with the mask. The higher the correlation the more accurate the attribution map. Results in Fig. 3b.

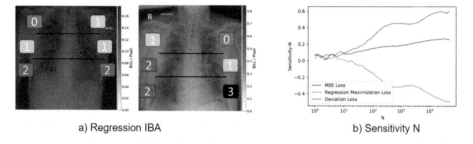

a) Regression IBA   b) Sensitivity N

Fig. 3. Regression IBA: a) Regression IBA (with $\mathcal{L}_{DV}$) applied on a regression model that predicts the total severity score. Using Regression IBA we see that the model has identified the severity scores of different regions implicitly. b) Sensitivity N metric for evaluating feature importance of different Regression IBA losses

Table 1. Mean IOU on NIH ChestX-ray8 (all pathologies) for various methods

GradCAM[22]	InteGrad[28]	LRP[14]	Gradients[25]	IBA	Inverse IBA	Multi-layer IBA
0.077	0.076	0.025	0.114	0.114	0.088	**0.189**

3.2 Ground Truth Based (Human-centric) Evaluations

Experiments in this section evaluate the attribution maps in terms of the human notion of interpretability, i.e. the alignment between what we understand and the map. Moreover, they measure how fine-grained the attribution maps are.

Localization. For NIH ChestX-ray8 dataset the bounding boxes are available. To generate boxes for BrixIA score regions, similar to [24] we use a lung segmentation network with the same architecture in [24]. We divide each lung into 3 regions. We threshold the attribution maps and compute their IoU with these divided regions (over dataset). The results are presented in Tables 1 and 2, respectively.

Correlation Analysis (Regression Models). For the BrixIA dataset, we evaluate the performance of regression models by measuring the correlation between the attribution scores and the severity scores. For each image, we first assign each pixel with its severity score, obtaining a severity score map. We then flatten both the attribution and severity score maps and compute their Pearson correlation coefficient (PCC). The PCC results are as follows: for $\mathcal{L}_{MSE}$, 0.4766, for $\mathcal{L}_{RM}$, 0.4766, for $\mathcal{L}_{MSE}$, 0.4404, and for random heatmaps, 0.0004.

4   Discussion

Inverse IBA: We observe that (Fig. 1) Inverse IBA highlights all regions with predictive information. On BrixIA sample, IBA only identifies two regions with a score of 3 as being predictive, while Inverse IBA identifies all regions with a score of 3. On NIH sample, if we remove the highlighted areas of both methods (equally remove) from the image, the output change caused by the removal of Inverse IBA regions is higher. This is also quantitatively validated across dataset in the Deletion experiment (Fig. 2).

Regression IBA: Using Regression IBA we observe that (Fig. 3) a regression model which only predicts one cumulative severity score (0–18) for each X-ray implicitly identifies the severity scores of different regions.

Multi-layer IBA: We use Multi-layer IBA for obtaining fine-grained attributions. In Fig. 4 we see that such fine-grained attributions allow for identifying subtle features such as Mass. Moreover, Multi-layer IBA also uncovers some hidden insights regarding what features the model is using for the Cardiomegaly example. While IBA highlights the entire region, Multi-layer IBA shows precisely the regions to which IBA is pointing.

Imbalanced Loss: We observe in Table 3 that using weighted BCE results in an increased IoU with the pathologies. This signifies that the contributing features of the weighted BCE model are more aligned with the pathology annotations. The observation is more significant when we consider the AUC of ROC and the Average Precision (AP) of these models. The AUCs (BCE = 0.790, Weighted BCE = 0.788) and APs (BCE = 0.243, Weighted BCE = 0.236) are approxi-

Table 2. Mean IOU on BrixIA for each detector and for various attribution methods

	GradCAM	InteGrad	LRP	Gradients	IBA	Inverse IBA	Multi-layer IBA
Detector 0/1	0.11	0.176	0.0	0.04	0.145	**0.194**	0.171
Detector 2	0.0	0.13	0.0	0.019	0.13	0.245	**0.257**
Detector 3	0.011	0.14	0.0	0.052	0.222	0.243	**0.257**

Table 3. Mean IOU on NIH ChestX-ray8 dataset for BCE and Weighted BCE models, reported for all the pathologies in NIH Chest X-ray8 using Inverse IBA

	Atelec.	Cardio.	Effusion	Infiltrate.	Mass	Nodule	Pneumo.	Pn. thorax	Mean
BCE	0.016	0.071	0.004	0.001	0.102	0.011	0.0	0.003	0.024
W. BCE	0.073	0.131	0.032	0.058	0.097	0.02	0.066	0.016	**0.065**

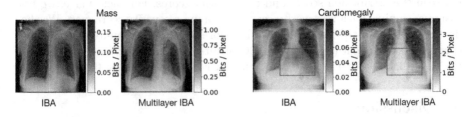

Fig. 4. Multi-layer IBA: Multi-layer IBA generates more fine-grained maps compared to IBA. (Left): Multi-layer IBA precisely highlights subtle features such as Mass (Right) Using Multi-layer we observe that for Cardiomegaly in this X-ray the corner regions of Cardiomegaly are used. IBA highlights the entire region.

mately equivalent. The BCE even archives marginally higher scores in terms of AUC of ROC and AP but its learned features are less relevant to the pathologies.

5   Conclusion

In this work, we build on IBA feature attribution method and come up with different approaches for identifying input regions that have predictive information. Contrary to IBA, our Inverse IBA method identifies *all* regions that can have predictive information. Thus all predictive cues from the pathologies in the X-rays are highlighted. Moreover, we propose Regression IBA for attribution on regression models. In addition, we propose Multi-layer IBA, an approach for obtaining fine-grained attributions which can identify subtle features.

Acknowledgement. This work was partially funded by the **Munich Center for Machine Learning (MCML)** and the **Bavarian Research Foundation** grant AZ-1429-20C. The computational resources for the study are provided by the **Amazon Web Services Diagnostic Development Initiative**. S.T. Kim is supported by the **Korean MSIT**, under the National Program for Excellence in SW (2017-0-00093), supervised by the IITP.

References

1. Adebayo, J., Gilmer, J., Muelly, M., Goodfellow, I., Hardt, M., Kim, B.: Sanity checks for saliency maps. arXiv preprint arXiv:1810.03292 (2018)
2. Alemi, A.A., Fischer, I., Dillon, J.V., Murphy, K.: Deep variational information bottleneck. arXiv preprint arXiv:1612.00410 (2016)
3. Ancona, M., Ceolini, E., Öztireli, C., Gross, M.: Towards better understanding of gradient-based attribution methods for deep neural networks. arXiv preprint arXiv:1711.06104 (2017)
4. Bach, S., Binder, A., Montavon, G., Klauschen, F., Müller, K.R., Samek, W.: On pixel-wise explanations for non-linear classifier decisions by layer-wise relevance propagation. PloS One **10**(7), e0130140 (2015)
5. Hooker, S., Erhan, D., Kindermans, P.J., Kim, B.: A benchmark for interpretability methods in deep neural networks. In: Advances in Neural Information Processing Systems, vol. 32. Curran Associates, Inc. (2019). https://proceedings.neurips.cc/paper/2019/file/fe4b8556000d0f0cae99daa5c5c5a410-Paper.pdf
6. Irvin, J., et al.: Chexpert: a large chest radiograph dataset with uncertainty labels and expert comparison. In: Proceedings of the AAAI Conference on Artificial Intelligence, vol, 33, pp. 590–597 (2019)
7. Johnson, A.E., et al.: MIMIC-CXR-JPG, a large publicly available database of labeled chest radiographs. arXiv preprint arXiv:1901.07042 (2019)
8. Karim, M., Döhmen, T., Rebholz-Schuhmann, D., Decker, S., Cochez, M., Beyan, O., et al.: DeepCOVIDExplainer: explainable covid-19 predictions based on chest x-ray images. arXiv preprint arXiv:2004.04582 (2020)
9. Khakzar, A., Albarqouni, S., Navab, N.: Learning interpretable features via adversarially robust optimization. In: Shen, D., et al. (eds.) MICCAI 2019. LNCS, vol. 11769, pp. 793–800. Springer, Cham (2019). https://doi.org/10.1007/978-3-030-32226-7_88

10. Khakzar, A., Baselizadeh, S., Khanduja, S., Kim, S.T., Navab, N.: Explaining neural networks via perturbing important learned features. arXiv preprint arXiv:1911.11081 (2019)

11. Khakzar, A., Baselizadeh, S., Khanduja, S., Rupprecht, C., Kim, S.T., Navab, N.: Neural response interpretation through the lens of critical pathways. In: Proceedings of the IEEE/CVF Conference on Computer Vision and Pattern Recognition (2021)

12. Khakzar, A., Baselizadeh, S., Navab, N.: Rethinking positive aggregation and propagation of gradients in gradient-based saliency methods. arXiv preprint arXiv:2012.00362 (2020)

13. Li, Z., et al.: Thoracic disease identification and localization with limited supervision. In: Proceedings of the IEEE Conference on Computer Vision and Pattern Recognition, pp. 8290–8299 (2018)

14. Montavon, G., Lapuschkin, S., Binder, A., Samek, W., Müller, K.R.: Explaining nonlinear classification decisions with deep Taylor decomposition. Pattern Recogn. **65**, 211–222 (2017)

15. Nie, W., Zhang, Y., Patel, A.: A theoretical explanation for perplexing behaviors of backpropagation-based visualizations. In: International Conference on Machine Learning, pp. 3809–3818. PMLR (2018)

16. Oh, Y., Park, S., Ye, J.C.: Deep learning COVID-19 features on CXR using limited training data sets. IEEE Trans. Med. Imaging **39**(8), 2688–2700 (2020)

17. Petsiuk, V., Das, A., Saenko, K.: Rise: Randomized input sampling for explanation of black-box models. arXiv preprint arXiv:1806.07421 (2018)

18. Punn, N.S., Agarwal, S.: Automated diagnosis of COVID-19 with limited posteroanterior chest X-ray images using fine-tuned deep neural networks. Appl. Intell. **51**(5), 2689–2702 (2020). https://doi.org/10.1007/s10489-020-01900-3

19. Rajpurkar, P., et al.: CheXNet: radiologist-level pneumonia detection on chest x-rays with deep learning. arXiv preprint arXiv:1711.05225 (2017)

20. Samek, W., Binder, A., Montavon, G., Lapuschkin, S., Müller, K.R.: Evaluating the visualization of what a deep neural network has learned. IEEE Trans. Neural Netw. Learn. Syst. **28**(11), 2660–2673 (2016)

21. Schulz, K., Sixt, L., Tombari, F., Landgraf, T.: Restricting the flow: information bottlenecks for attribution. arXiv preprint arXiv:2001.00396 (2020)

22. Selvaraju, R.R., Cogswell, M., Das, A., Vedantam, R., Parikh, D., Batra, D.: Grad-CAM: visual explanations from deep networks via gradient-based localization. In: Proceedings of the IEEE International Conference on Computer Vision, pp. 618–626 (2017)

23. Signoroni, A., et al.: End-to-end learning for semiquantitative rating of COVID-19 severity on chest X-rays. arXiv preprint arXiv:2006.04603 (2020)

24. Signoroni, A., et al.: BS-Net: learning COVID-19 pneumonia severity on a large chest X-ray dataset. Med. Image Anal. **71**, 102046 (2021)

25. Simonyan, K., Vedaldi, A., Zisserman, A.: Deep inside convolutional networks: visualising image classification models and saliency maps. arXiv preprint arXiv:1312.6034 (2013)

26. Singh, R.K., Pandey, R., Babu, R.N.: COVIDScreen: explainable deep learning framework for differential diagnosis of COVID-19 using chest X-rays. Neural Comput. Appl. **33**(14), 8871–8892 (2021). https://doi.org/10.1007/s00521-020-05636-6

27. Springenberg, J.T., Dosovitskiy, A., Brox, T., Riedmiller, M.: Striving for simplicity: the all convolutional net. arXiv preprint arXiv:1412.6806 (2014)

28. Sundararajan, M., Taly, A., Yan, Q.: Axiomatic attribution for deep networks. In: International Conference on Machine Learning, pp. 3319–3328. PMLR (2017)

29. Taghanaki, S.A., et al.: InfoMask: masked variational latent representation to localize chest disease. In: Shen, D., et al. (eds.) MICCAI 2019. LNCS, vol. 11769, pp. 739–747. Springer, Cham (2019). https://doi.org/10.1007/978-3-030-32226-7_82

30. Wang, X., Peng, Y., Lu, L., Lu, Z., Bagheri, M., Summers, R.M.: ChestX-ray8: hospital-scale chest X-ray database and benchmarks on weakly-supervised classification and localization of common thorax diseases. In: Proceedings of the IEEE Conference on Computer Vision and Pattern Recognition, pp. 2097–2106 (2017)

31. Zhang, J., Bargal, S.A., Lin, Z., Brandt, J., Shen, X., Sclaroff, S.: Top-down neural attention by excitation backprop. Int. J. Comput. Vis. **126**(10), 1084–1102 (2018). https://doi.org/10.1007/s11263-017-1059-x

32. Zhou, B., Khosla, A., Lapedriza, A., Oliva, A., Torralba, A.: Learning deep features for discriminative localization. In: Proceedings of the IEEE Conference on Computer Vision and Pattern Recognition, pp. 2921–2929 (2016)

Demystifying T1-MRI to FDG^{18}-PET Image Translation via Representational Similarity

Chia-Hsiang Kao$^{(\boxtimes)}$, Yong-Sheng Chen, Li-Fen Chen, and Wei-Chen Chiu

National Yang Ming Chiao Tung University, Taipei, Taiwan
{chkao.md04,yschen,lfchen,walon}@nycu.edu.tw

Abstract. Recent development of image-to-image translation techniques has enabled the generation of rare medical images (e.g., PET) from common ones (e.g., MRI). Beyond the potential benefits of the reduction in scanning time, acquisition cost, and radiation exposure risks, the translation models in themselves are inscrutable black boxes. In this work, we propose two approaches to demystify the image translation process, where we particularly focus on the T1-MRI to PET translation. First, we adopt the representational similarity analysis and discover that the process of T1-MR to PET image translation includes the stages of brain tissue segmentation and brain region recognition, which unravels the relationship between the structural and functional neuroimaging data. Second, based on our findings, an Explainable and Simplified Image Translation (ESIT) model is proposed to demonstrate the capability of deep learning models for extracting gray matter volume information and identifying brain regions related to normal aging and Alzheimer's disease, which untangles the biological plausibility hidden in deep learning models.

Keywords: Explainability · Medical image translation

1 Introduction

Recent advances in deep learning have brought the magic leap to various research areas such as computer vision and natural language processing. In these days, we have witnessed the great performance of deep learning models [18,26,27] for making translation from T1-weighted magnetic resonance (T1-MR) images towards 18-fluorodeoxyglucose proton emission tomography images (FDG^{18}-PET) (FDG^{18}-PET is abbreviated to PET unless stated otherwise). In these days, we have witnessed the great performance of deep learning models [18,26,27]

Electronic supplementary material The online version of this chapter (https://doi.org/10.1007/978-3-030-87199-4_38) contains supplementary material, which is available to authorized users.

M. de Bruijne et al. (Eds.): MICCAI 2021, LNCS 12903, pp. 402–412, 2021.
https://doi.org/10.1007/978-3-030-87199-4_38

for making translation from T1-weighted MR (T1-MR) images towards FDG[18]-PET images (FDG[18]-PET is abbreviated to PET unless stated otherwise). Typically, the T1-MR imaging, a type of structural magnetic resonance imaging (MRI) scan, is an anatomical image used to observe the brain structures, thus being known as one of the *structural imaging*; while PET studies the *in vivo* glucose metabolism, which is correlated with neuronal activity and stimulus intensity [24], thus being defined as *functional imaging*. Yet these two imaging techniques are different in terms of their physical and biological characteristics, it remains unexplored to understand the mechanism behind deep-learning-based translation models of bridging the difference between structural and functional images, in which it motivates us to demystify such image translation process, as the main goal of this paper. The basic idea behind our research framework is illustrated in Fig. 1.

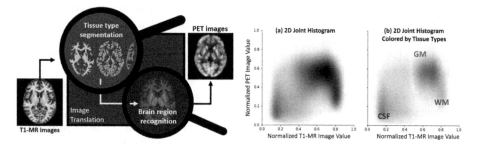

Fig. 1. We propose to lay the explanatory groundwork for the cross-modal medical image translation model by exploring the biological plausibility behind the deep neural net. We experimentally validate our proposed hypotheses that the T1-MR to PET translation comprises successive recognition of the brain tissue types and the brain regions.

Fig. 2. Figure (a) shows the 2D joint histogram of values from corresponding T1-MR and PET image pixels, while it is further colorized by the probability of pixels belonging to different tissue types (i.e., white matter (WM) be purple, gray matter (GM) be cyan, and CSF be yellow) and shown in (b). (Color figure online)

Our study begins with building a 2D joint histogram between the values obtained from the corresponding T1-MR and PET image pixels. In Fig. 2, we observe three clusters which are associated to cerebro-spinal fluid (CSF) and brain tissues. We therefore have our first hypothesis: during image translation, the model would first recognize (or segment) the areas related to various brain tissue types. Nevertheless, as these clusters are not compact and their boundaries are blurry, merely having the tissue regions as the intermediary can not conclusively reason the translation between images. We hence step further and hypothesize: as different brain regions (e.g., thalamus, insula, or putamen) have different baseline uptake of glucose, there would be an in-between stage of brain regions identification followed by performing region-dependent transformation.

We experimentally verify our hypotheses using the representational similarity analysis, thus contributing to decipher the black box translation model. Moreover, we advance to leverage these two hypotheses by building up an **Explainable and Simplified Image Translation (ESIT)** model. With ESIT model, we look into its learnt feature representations and discover their close relationship with the gray matter volume as well as the informativeness on clinical status.

2   Related Works

Image Translation in Medical Domains. Image-to-image translation is the task of transforming the images from one domain to the ones of another domain with distinct characteristics/styles, i.e. mapping between image domains. Several approaches [10,17,18,25,26] have been proposed to translate images from T1-MR to PET, where most of them are based on U-Net architecture [21] with using the paired data (i.e. cross-domain correspondences). Instead of pushing forward the model performance on T1-MR to PET image translation, our primary goal is to explore the underlying mechanism of deep-learning-based translation methods. Therefore, our research focuses on the typical U-Net structure that is widely used as a building block for various translation models.

Analysis of Representational Similarity for Deep Networks. To better characterize deep learning models, various methods have been proposed to analyze the optimization process or the internal organization of neural networks, such as filter visualization [28,29] or saliency maps [1,22]. However, as these approaches are developed mainly for classification models, their direct application on image-to-image translation models is challenging since the high-dimensional structural and semantic information are substantially entangled during translation. Recently, *Canonical Correlation Analysis* (CCA) has emerged as a robust tool to study the similarity of representations within a deep network or across networks, as CCA measures the multivariate joint relationship among variables/representations via maximizing the correlation between their projections. In our study, we take advantage of CCA to calculate the per-layer representational similarity scores with respect to brain tissue maps and brain region templates.

3   Image Translation Model, Dataset, and Analysis Tool

As motivated previously, we propose to analyze the trained U-Net model of T1-MR to PET image translation by exploring its layer-wise representational similarity with respect to medically interpretable concepts, i.e., the brain tissue maps and the brain region templates. In this section, we provide the details of the U-Net model, the dataset in use, and CCA as our analysis tool.

U-Net Model for T1-MR to PET Image Translation. As the basis for our study, we start off with training the U-Net model for performing image translation from T1-MR to PET images. We adopt the typical U-Net architecture proposed in [21] but substitute its batch normalization, ReLU activation, and those convolution layers followed by the maxpooling operation with the instance normalization (IN), Leaky ReLU (LR) activation, and strided convolution layers, respectively. Such U-Net model is trained to minimize the L1 loss with AdamW optimizer [11] for 40 epochs over five-fold cross-validation. The quantitative metrics used for evaluating the translation performance are mean square error (MAE), peak signal-to-noise ratio (PSNR), and structure similarity index (SSIM).

Alzheimer's Disease Neuroimaging Initiative Dataset. The paired T1-MR and PET images are obtained from the Alzheimer's Disease Neuroimaging Initiative dataset (ADNI [9]) including ADNI-1, ADNI-2, ADNI-GO, and ADNI-3 phases. Overall, we include the data of cognitively normal (CN, n = 300) subjects and subjects diagnosed as significant memory concern (SMC, n = 54), mild cognitive impairment (MCI, n = 868) and Alzheimer's disease (AD, n = 219).

Preprocessing. We follow standard procedure to perform image preprocessing using the SPM 12 software [19]. PET images are coregistered to the corresponding T1-MR images. Then, T1-MR images are transformed to Montreal Neurological Institute (MNI) reference space with the voxel size resliced to $1.5 \times 1.5 \times 1.5$ mm^3, and the estimated deformation field is applied to the aligned PET images. Skull-stripping and cerebellum removal are performed to all PET images as indicated in [18]. Min-Max normalization of pixel values to the range $[0, 1]$ is performed to all T1-MR images and PET images.

Brain Tissue Maps and Brain Region Templates. The subject-specific probabilistic maps of gray matter (GM), white matter (WM) and CSF are obtained from the T1-MR image segmentation by SPM. For brain region templates, we utilize Automated Anatomical Labeling (AAL3) [20] and Hammersmith atlas [8], which are resliced to match the spatial resolution of T1-MR images.

Canonical Component Analysis. Canonical Component Analysis (CCA) measures the relationship between two variables by finding a canonical space in which the correlation between their projections onto that space is maximized. Given two centered random vectors $\mathbf{X}$ and $\mathbf{Y}$ of dimensions m and n respectively, where $\mathbf{X} = (x_1, ..., x_m)^T$ and $\mathbf{Y} = (y_1, ..., y_n)^T$, CCA seeks pairs of projection weight vectors $\mathbf{a}_i \in \mathbb{R}^m$ and $\mathbf{b}_i \in \mathbb{R}^n$ iteratively to maximize the correlation coefficient between $\mathbf{a}_i^T\mathbf{X}$ and $\mathbf{b}_i^T\mathbf{Y}$, under the orthogonality constraints that $\forall j < i, \mathbf{a}_i^T\mathbf{X} \perp \mathbf{a}_j^T\mathbf{X}$ and $\mathbf{b}_i^T\mathbf{Y} \perp \mathbf{b}_j^T\mathbf{Y}$. For $1 \leq i \leq \min(m, n)$, the ith canonical correlation coefficient ρ_i is therefore given by:

$$\rho_i = \max \frac{< \mathbf{u}_i, \mathbf{v}_i >}{\|\mathbf{u}_i\|\|\mathbf{v}_i\|} = \max \frac{(\mathbf{a}_i^T\mathbf{X})(\mathbf{b}_i^T\mathbf{Y})^T}{\|\mathbf{a}_i^T\mathbf{X}\|\|\mathbf{b}_i^T\mathbf{Y}\|} \tag{1}$$

where $\mathbf{u}_i = \mathbf{a}_i^T\mathbf{X}$ and $\mathbf{v}_i = \mathbf{b}_i^T\mathbf{Y}$ are called *canonical components*.

In details, the activation maps from a network are concatenated along spatial dimension to form $d \times h \times w$ vectors with dimension c, denoted by $\mathbf{L}$ (where

d, h, w, c are respectively the number of datapoint, height, width, and the number of channels). The brain tissue maps and the brain region templates are flattened into $d \times h \times w$ vectors with dimension 1, denoted by $\mathbf{K}$. We use CCA to find the projection weight vectors $\mathbf{w}_1 \in \mathbb{R}^c$ and $\mathbf{w}_2 \in \mathbb{R}^1$, such that the correlation coefficient $\rho = \max \frac{(\mathbf{w}_1^T \mathbf{L})(\mathbf{w}_2^T \mathbf{K})^T}{\|\mathbf{w}_1^T \mathbf{L}\| \| \mathbf{w}_2^T \mathbf{K}\|}$ is maximized.

We name the correlation coefficient ρ as **CCA similarity**, since such values reflect the degree of correlation between two random variables in a canonical coordinate space. To avoid interference by zero-valued regions (i.e., pixels of air and bone), we remove the data point which is neither brain tissues nor CSF. Due to the expensive computational cost, for each model, we randomly sample 20 subjects from corresponding training data and retrieve their activation maps of every layer (after Leaky ReLU operator) from the translation model. The standard deviation of the CCA similarities is plotted as the shaded region.

4   Demystification from a Medical Perspective

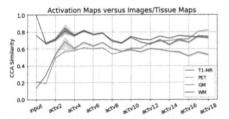 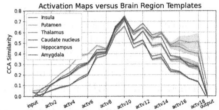

(a) CCA similarity of activation maps against T1-MR images, PET images, and brain tissue maps. Please refer to Section 4.1 for more detailed discussion.

(b) CCA similarities between the per-layer activation maps and the brain region templates. Please refer to Section 4.3 for more detailed discussion.

Fig. 3. Per-layer CCA similarity scores of per-layer activation maps

4.1   Brain Tissues Are Segmented in the Early Encoding Stage

To verify our first hypothesis that, during the translation process, the translation model would first recognize the areas related to brain tissue types, our work begins with showing the CCA similarity scores of per-layer activation maps with respect to the brain tissue maps and the T1-MR/PET images, as provided in Fig. 3a. We have several observations: (1) The increase of CCA scores between the activation maps and the gray matter maps (i.e., the curve in green color) in the early encoding stage (i.e., from layer *actv1* to *actv3*), indicating that the translation model learns to recognize the gray matter distribution; (2) In the encoding stage, the trend of CCA similarity between the activation maps and the gray matter maps is similar to the one between the activation maps and the PET images (i.e., the curve in orange color). Such observation echoes the physical principle [13] that gray matter has higher metabolic activity, which PET imaging aims to capture. To further understand the information captured by the translation model, we show the canonical components of the *actv3* activation maps in Figure S1 of supplement, which highly resemble the tissue maps.

4.2   Brain Tissue Information Is the Key to PET Image Synthesis

Based on our observation that the translation model learns to represent the tissue type information in the early encoding stage, we wonder if the tissue information alone is sufficient for PET image synthesis. As shown in the first two columns in Table S1 of supplement, the performance of the U-Net-based translation model which is directly trained on tissue maps as input is competitive to the original translation model (i.e., trained by taking T1-MR images as input), verifying our assumption that the brain tissue information is the key to PET image synthesis.

4.3   Brain Regions Are Recognized Later in the Translation

In Fig. 3a, we observe a steady increase in the per-layer CCA similarity with PET images (denoted by the orange curve), while the per-layer CCA similarity with the gray matter maps (denoted by the green curve) declines, suggesting that the tissue type information as the intermediary might not be enough to conclusively reason the transformation from T1-MR to PET. Inspired by the findings from previous medical researches [3,23] that certain brain regions (e.g., the basal ganglia, posterior cingulate cortex, and visual cortex) have variations in baseline glucose uptake, we propose our second hypothesis: there could be an in-between stage of identifying brain regions, followed by performing region-dependent transformation to infer the final PET image.

To verify this hypothesis, the per-layer CCA similarity scores with respect to brain region templates are computed, and are provided in Fig. 3b. We observe that the translation model represents most of the brain region information in the bottleneck (i.e., from layer *actv9* to *actv12*). Remarkably, the per-layer CCA similarity scores with respect to the thalamus (denoted by the green curve) and putamen (denoted by the orange curve) are relatively high and sustained even in the late decoding stage, consistent with the finding that the thalamus and putamen have slightly higher metabolism activity [3]. To better understand the encoded brain region information in the translation model, the canonical components learnt from CCA are shown in Figure S2 of supplement. Intriguingly, from the visualization of layer *actv15* regarding the brain regions of caudate nucleus and putamen, we can observe co-occurrence of caudate nucleus and putamen in the feature space of the translation model. This echoes the well-documented fact that the caudate nucleus and putamen (which together are referred to as the striatum) act jointly in function as the major input zone for basal ganglia [5,6]. With the aforementioned analysis, we validate our second hypothesis that in the cross-modal medical image translation, there is likely an in-between stage of identifying and representing brain regions.

5   Explainable and Simplified Image Translation Model

Via CCA analysis, we validate our two main hypotheses that the translation from T1-MR images to PET ones includes the recognition of the brain tissues and regions. Yet, U-Net-based model comprises an entangled process of segmentation and transformation, it is challenging to further analyze the information related to brain tissue types and brain regions from the feature maps of the model.

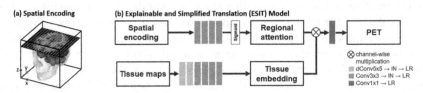

Fig. 4. (a) Illustration of the spatial encoding. (b) Overview of our Explainable and Simplified Image Translation (ESIT) model.

To address such issue, we propose an Explainable and Simplified Image Translation (ESIT) model, where the information of the brain tissue types and brain regions are explicitly unraveled and tackled in our model design. The architecture of our proposed ESIT model is illustrated in Fig. 4. The ESIT model takes both the brain tissue maps and the spatial encoding as inputs. The spatial encoding is an $H \times W \times 3$ map that transforms the pixel coordinates of the T1-MR image (of size $H \times W$) into the universal brain coordinate system (as illustrated in Fig. 4 (a)). In the ESIT model, the input map of spatial encoding is passed through several convolution layers and then becomes the *regional attention* map. On the other hand, the input tissue maps are transformed into the *tissue embeddings* by a shallow network. We integrate the regional attention maps and the tissue embeddings via channel-wise multiplication followed by a 1×1 convolution layer, then we obtain the final PET output. Simple as it may seem, our ESIT model attains competitive performance compared to the U-Net-based translation model (cf. Table S1 of supplement). Our ESIT provides a more straightforward and more explainable way to understand the underlying mechanism of T1-MR to PET image translation, which we will detail later.

5.1   Extraction of Regional Gray Matter Volume Information from Brain Tissue Maps

As our dataset comprises of images of subjects with a wide spectrum of clinical status, the good performance of deep models implies that the information of structural abnormality (e.g., gray matter volume, gray matter thickness) could be well extracted from the T1-MR images or brain tissue maps. Hence, we roll out experiments with a focus on the hippocampus and amygdala, the two brain regions susceptible to structural changes due to AD [4], to explore the associations between the tissue embeddings and the gray matter volume (obtained with Computational Anatomy Toolbox (CAT12) [7]) as well as the clinical status.

We visualize the tissue embeddings of the hippocampus and amygdala regions in Figure S3 of supplement, where we observe that the tissue embeddings are fairly likely to correlate with the gray matter volume as well as the clinical status of subjects in both areas. Such observation has its corresponding medical explanation: as PET is used to measure the brain metabolism (mostly synaptic activities [15]) and our tissue embeddings are contributing to the synthesis of PET images, the decrease in gray matter volume which reflects the loss of synapses and neurons can hence be captured by our tissue embeddings.

5.2   Regional Attention on Metabolic Variation in Aging and AD

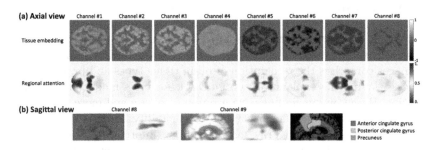

Fig. 5. The visualization of **(a) tissue embedding maps (top) and regional attention maps (bottom) learnt by our ESIT model, from axial view. (b) The tissue embeddings and regional attention maps from sagittal view.**

In this section, we revisit the brain region identification process using our proposed ESIT model, in which the regional information is processed independently such that we can focus more on the learnt regional patterns. In Fig. 5, we show the learnt tissue embeddings as well as regional attention maps from axial and sagittal views. The tissue embeddings mostly follow the distribution of brain tissues and CSF. For regional attention maps, we summarize the four typical brain region patterns:

- *Regions with normal variation.* Channel #2 reveals the pallidum, caudate nucleus and putamen, all of which are the main components of the basal ganglia that have high metabolic activity in healthy population [3].
- *Regions with preserved metabolism during aging.* In channel #1, attention is given to the bilateral occipital lobes, thalamus, pallidum and heschl's gyrus, which are relatively unaffected during aging [3,12]. Likewise, channel #4 focuses on the superior frontal gyrus, superior temporal gyrus, and lingual gyrus, all of which has relatively preserved metabolism in aging population [12].
- *Aging-related regions.* Channel #9 mainly focuses on the anterior cingulate gyrus, the region with profound metabolic decrease during aging [16].

– *AD-related regions.* Channel #8 focuses on the posterior cingulate gyrus and precuneus regions, and both of them have the most reduction in glucose metabolism in subjects with AD [2,14,16].

These results show that our ESIT model identifies four typical patterns of regional hyper-metabolism or hypo-metabolism and thus reinforces our proposed hypothesis that there is region-dependent transformation for inferring the final PET images. We humbly conclude that the deep learning translation models are reasonably plausible from a medical perspective, where they learn to capture the characteristics of the regional metabolic differences such as normal regional variation, age-related change, and dementia-related regional alterations.

6 Conclusion

In this paper, we conduct extensive experiments using representational similarity to verify our proposed hypotheses that the translation from T1-MR to PET images comprises the recognition of brain tissue types and brain regions in its process, laying the explanatory groundwork for cross-modal medical image translation. Based on our findings, we propose a concise and more interpretable model, ESIT, and further demonstrate the capability of deep learning technique in extracting regional gray matter volume information and identifying the regional metabolic variation in normal, aging and dementia population.

References

1. Bach, S., Binder, A., Montavon, G., Klauschen, F., Müller, K.R., Samek, W.: On pixel-wise explanations for non-linear classifier decisions by layer-wise relevance propagation. PloS One **10**, e0130140 (2015)
2. Bailly, M., et al.: Precuneus and cingulate cortex atrophy and hypometabolism in patients with Alzheimer's disease and mild cognitive impairment: MRI and 18F-FDG PET quantitative analysis using FreeSurfer. BioMed Res. Int. (2015)
3. Berti, V., Mosconi, L., Pupi, A.: Brain: normal variations and benign findings in fluorodeoxyglucose-PET/computed tomography imaging. PET Clin. **9**, 129–140 (2014)
4. Chételat, G., et al.: Direct voxel-based comparison between grey matter hypometabolism and atrophy in Alzheimer's disease. Brain **131**, 60–71 (2008)
5. Dale, P., George, A., David, F., Lawrence, K., Anthony-Samuel, L., James, M., S, W.: Neuroscience, 2nd edn. Sinauer Associates, Sunderland (2001)
6. Driscoll, M.E., Bollu, P.C., Tadi, P.: Neuroanatomy, Nucleus Caudate. StatPearls Publishing, Treasure Island (FL) (2020)
7. Gaser, C., Dahnke, R.: Cat-a computational anatomy toolbox for the analysis of structural MRI data. Hum. Brain Mapp. (2016)
8. Hammers, A., et al.: Three-dimensional maximum probability atlas of the human brain, with particular reference to the temporal lobe. Hum. Brain Mapp. **19**, 224–247 (2003)
9. Jack Jr., C.R., et al.: The Alzheimer's disease neuroimaging initiative (ADNI): MRI methods. J. Magn. Reson. Imaging (2008)

10. Lan, H., Toga, A., Sepehrband, F.: SC-GAN: 3D self-attention conditional GAN with spectral normalization for multi-modal neuroimaging synthesis. bioRxiv:2020.06.09.143297 (2020)

11. Loshchilov, I., Hutter, F.: Decoupled weight decay regularization. In: International Conference on Learning Representations (ICLR) (2019)

12. Lowe, V.J., et al.: Association of hypometabolism and amyloid levels in aging, normal subjects. Neurology **82**, 1959–1967 (2014)

13. Manninen, S., et al.: Cerebral grey matter density is associated with neuroreceptor and neurotransporter availability: a combined PET and MRI study. bioRxiv:2020.01.29.924530 (2020)

14. Marcus, C., Mena, E., Subramaniam, R.M.: Brain PET in the diagnosis of Alzheimer's disease. Clin. Nucl. Med. **39**, e413 (2014)

15. Márquez, F., Yassa, M.A.: Neuroimaging biomarkers for Alzheimer's disease. Mol. Neurodegeneration (2019)

16. Mosconi, L.: Glucose metabolism in normal aging and Alzheimer's disease: methodological and physiological considerations for PET studies. Clin. Transl. Imaging **1**, 217–233 (2013)

17. Nie, D., et al.: Medical image synthesis with context-aware generative adversarial networks. In: Descoteaux, M., Maier-Hein, L., Franz, A., Jannin, P., Collins, D.L., Duchesne, S. (eds.) MICCAI 2017. LNCS, vol. 10435, pp. 417–425. Springer, Cham (2017). https://doi.org/10.1007/978-3-319-66179-7_48

18. Pan, Y., Liu, M., Lian, C., Zhou, T., Xia, Y., Shen, D.: Synthesizing missing PET from MRI with cycle-consistent generative adversarial networks for Alzheimer's disease diagnosis. In: Frangi, A.F., Schnabel, J.A., Davatzikos, C., Alberola-López, C., Fichtinger, G. (eds.) MICCAI 2018. LNCS, vol. 11072, pp. 455–463. Springer, Cham (2018). https://doi.org/10.1007/978-3-030-00931-1_52

19. Penny, W.D., Friston, K.J., Ashburner, J.T., Kiebel, S.J., Nichols, T.E.: Statistical Parametric Mapping: The Analysis of Functional Brain Images. Academic Press, Cambridge (2011)

20. Rolls, E.T., Huang, C.C., Lin, C.P., Feng, J., Joliot, M.: Automated anatomical labelling atlas 3. NeuroImage **206**, 116189 (2020)

21. Ronneberger, O., Fischer, P., Brox, T.: U-Net: convolutional networks for biomedical image segmentation. In: Navab, N., Hornegger, J., Wells, W.M., Frangi, A.F. (eds.) MICCAI 2015. LNCS, vol. 9351, pp. 234–241. Springer, Cham (2015). https://doi.org/10.1007/978-3-319-24574-4_28

22. Selvaraju, R.R., Cogswell, M., Das, A., Vedantam, R., Parikh, D., Batra, D.: Grad-CAM: visual explanations from deep networks via gradient-based localization. In: IEEE Conference on Computer Vision and Pattern Recognition (CVPR) (2017)

23. Shamchi, S.P., et al.: Normal patterns of regional brain 18F-FDG uptake in normal aging. Hell. J. Nucl. Med. (2018)

24. Shulman, R.G., Rothman, D.L., Behar, K.L., Hyder, F.: Energetic basis of brain activity: implications for neuroimaging. Trends Neurosci. **27**, 489–495 (2004)

25. Sikka, A., Peri, S.V., Bathula, D.R.: MRI to FDG-PET: cross-modal synthesis using 3D U-Net for multi-modal Alzheimer's classification. In: Gooya, A., Goksel, O., Oguz, I., Burgos, N. (eds.) SASHIMI 2018. LNCS, vol. 11037, pp. 80–89. Springer, Cham (2018). https://doi.org/10.1007/978-3-030-00536-8_9

26. Sun, H., et al.: Dual-glow: conditional flow-based generative model for modality transfer. In: IEEE International Conference on Computer Vision (ICCV) (2019)

27. Wei, W., et al.: Learning myelin content in multiple sclerosis from multimodal MRI through adversarial training. In: Frangi, A.F., Schnabel, J.A., Davatzikos, C., Alberola-López, C., Fichtinger, G. (eds.) MICCAI 2018. LNCS, vol. 11072, pp. 514–522. Springer, Cham (2018). https://doi.org/10.1007/978-3-030-00931-1_59
28. Yosinski, J., Clune, J., Nguyen, A., Fuchs, T., Lipson, H.: Understanding neural networks through deep visualization. ArXiv:1506.06579 (2015)
29. Zeiler, M.D., Fergus, R.: Visualizing and understanding convolutional networks. In: Fleet, D., Pajdla, T., Schiele, B., Tuytelaars, T. (eds.) ECCV 2014. LNCS, vol. 8689, pp. 818–833. Springer, Cham (2014). https://doi.org/10.1007/978-3-319-10590-1_53

Fairness in Cardiac MR Image Analysis: An Investigation of Bias Due to Data Imbalance in Deep Learning Based Segmentation

Esther Puyol-Antón[1(✉)], Bram Ruijsink[1,2], Stefan K. Piechnik[7], Stefan Neubauer[7], Steffen E. Petersen[3,4,5,6], Reza Razavi[1,2], and Andrew P. King[1]

[1] School of Biomedical Engineering & Imaging Sciences, King's College London, London, UK
esther.puyol_anton@kcl.ac.uk
[2] Guy's and St Thomas' Hospital, London, UK
[3] William Harvey Research Institute, NIHR Barts Biomedical Research Centre, Queen Mary University London, Charterhouse Square, London EC1M 6BQ, UK
[4] Barts Heart Centre, St Bartholomew's Hospital, Barts Health NHS Trust, West Smithfield, London EC1A 7BE, UK
[5] Health Data Research UK, London, UK
[6] Alan Turing Institute, London, UK
[7] Division of Cardiovascular Medicine, Radcliffe Department of Medicine, University of Oxford, Oxford, UK

Abstract. The subject of 'fairness' in artificial intelligence (AI) refers to assessing AI algorithms for potential bias based on demographic characteristics such as race and gender, and the development of algorithms to address this bias. Most applications to date have been in computer vision, although some work in healthcare has started to emerge. The use of deep learning (DL) in cardiac MR segmentation has led to impressive results in recent years, and such techniques are starting to be translated into clinical practice. However, no work has yet investigated the fairness of such models. In this work, we perform such an analysis for racial/gender groups, focusing on the problem of training data imbalance, using a nnU-Net model trained and evaluated on cine short axis cardiac MR data from the UK Biobank dataset, consisting of 5,903 subjects from 6 different racial groups. We find statistically significant differences in Dice performance between different racial groups. To reduce the racial bias, we investigated three strategies: (1) stratified batch sampling, in which batch sampling is stratified to ensure balance between racial groups; (2) fair meta-learning for segmentation, in which a DL classifier is trained to classify race and jointly optimized with the segmentation model; and (3) protected group models, in which a different segmentation model is trained for each racial group. We also compared the results to the scenario where we have a perfectly balanced database. To assess fairness we used the standard deviation (SD) and skewed error ratio (SER) of the average Dice values. Our results demonstrate that the racial bias

© Springer Nature Switzerland AG 2021
M. de Bruijne et al. (Eds.): MICCAI 2021, LNCS 12903, pp. 413–423, 2021.
https://doi.org/10.1007/978-3-030-87199-4_39

results from the use of imbalanced training data, and that all proposed bias mitigation strategies improved fairness, with the best SD and SER resulting from the use of protected group models.

Keywords: Fair AI · Segmentation · Cardiac MRI · Inequality

1   Introduction

Fairness in artificial intelligence (AI) is a relatively new but fast-growing research field which deals with assessing and addressing potential bias in AI models. For example, an early landmark paper [2] found differences in performance of a video-based gender classification model for different racial groups. With AI models starting to be deployed in the real world it is essential that the benefits of AI are shared equitably according to race, gender and other demographic characteristics, and so efforts to ensure the fairness of deployed models have generated much interest. Most work so far has focused on computer vision problems but some applications in healthcare are starting to emerge [15,22].

Recently, deep learning (DL) models have shown remarkable success in automating many medical image segmentation tasks. In cardiology, human-level performance in segmenting the main structures of the heart has been reported [1], and researchers have proposed to use these models for tasks such as automating cardiac functional quantification [19]. These methods are now starting to move towards wider clinical translation.

It has long been well understood that cardiac structure and function, as well as the mechanisms leading to cardiovascular disease, vary according to demographic characteristics such as race and gender [11]. For example, the Multi-Ethnic Study of Atherosclerosis (MESA) [24] showed that there are racial differences in regional left ventricular (LV) systolic function in a large cohort study of adults. In addition, there are profound race-associated disparities among those who are affected by and die from cardiovascular disease [13]. Inequalities in detection of disease likely play a role in these differences. Therefore, it is surprising that no work to date has investigated potential bias in AI models for cardiac image analysis. To the best of our knowledge, the closest related work was on assessing differences in radiomics features by gender [18].

In this paper, we perform the first analysis of the fairness of DL-based cardiac MR segmentation models, focusing in particular on the impact of gender or race imbalance in the training data. To the best of our knowledge, this is also the first analysis of the fairness of segmentation models in general. We discover significant bias in performance between racial groups when trained using one of the largest and widely used public databases of cardiac magnetic resonance (MR) data (the UK Biobank). We propose algorithms to mitigate this bias, resulting in a fairer segmentation model that ensures that no racial group will be disadvantaged when segmentations of their cardiac MR data are used to inform clinical management.

2   Background

In recent years, addressing fairness concerns in AI models has been an active research area. To date, most previous works have focused on classification tasks such as image recognition, although Hwang *et al.* [7] investigated bias in image-to-image translation. In an extensive search, we did not find any previous work on bias mitigation for image segmentation tasks.

At a high level, fairness techniques can be grouped into two categories: (i) 'fairness through awareness', also known as bias mitigation strategies or discrimination-aware classification, that aim to make the AI algorithms more aware of the protected attributes[1] by making predictions independently for each protected group [4]; and (ii) 'fairness through unawareness', which assumes that if the model is unaware of the protected attributes while making decisions, the decisions will be fair. This second approach has been the most common one used to train AI models. However, it has been shown to be unsuccessful in many cases [21,25] due to the correlation of protected attributes with other variables in the data.

Focusing now on bias mitigation algorithms (a.k.a. 'fairness through awareness'), DL pipelines contain three possible points of intervention to mitigate unwanted bias: the training data, the learning procedure, and the output predictions, and these are associated with three corresponding classes of bias mitigation strategy: pre-processing, in-processing, and post-processing:

1) **Pre-processing approaches** modify the training dataset to remove the discrimination before training an AI model. Common strategies used are under-sampling, over-sampling [21] or sample weighting [9] to neutralize discriminatory effects; data generation [14] or data augmentation [12] using generative adversarial networks to balance the training dataset; or training using a balanced dataset [20].

2) **In-processing approaches** try to modify state-of-the-art learning algorithms in order to remove discrimination during the model training process via model regularization. There are two main approaches: implicit regularization that adds implicit constraints which disentangle the association between model prediction and fairness sensitive attributes [3]; and explicit regularization that adds explicit constraints through updating the model's loss function to minimize the performance difference between different protected groups [14,23].

3) **Post-processing approaches** correct the output of an existing algorithm to satisfy the fairness requirements. Common strategies used are equalized odds post-processing [5] that solves a linear program to find probabilities with which to change output labels to optimize equalized odds; calibrated equalized odds post-processing [17] that optimizes over calibrated classifier score outputs to find probabilities with which to change output labels with an

---

[1] In fair AI, the *protected attribute(s)* are the ones for which fairness needs to be ensured, e.g. gender or race. A set of samples with the same value(s) for the protected attribute(s) are known as a *protected group*.

equalized odds objective; and reject option classification [10] that gives favorable outcomes to unprivileged protected groups and unfavorable outcomes to privileged protected groups in a confidence band around the decision boundary with the highest uncertainty.

3   Methods

To investigate whether training data imbalance can lead to bias between racial and/or gender groups in automated cardiac MR segmentation, we conducted a comparative study using five different approaches. The baseline approach was based on fairness through unawareness. To enable us to test the hypothesis that any observed bias was due to data imbalance, we also trained the segmentation network using a (smaller) race and gender balanced database. We subsequently investigated three approaches based on fairness through awareness (two preprocessing approaches: stratified batch sampling and protected group models and one in-processing method that we call a 'fair meta-learning for segmentation'). These approaches are illustrated in detail in Fig. 1.

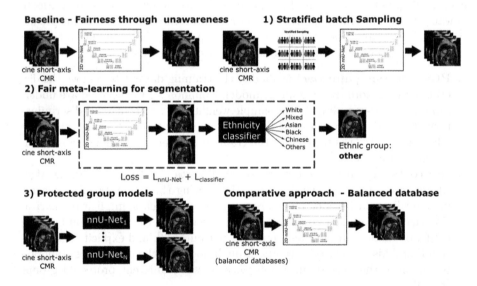

Fig. 1. Diagram showing the different strategies used for bias mitigation. Baseline is based on fairness through unawareness, and acts as the baseline approach. (1)–(3) are three approaches based on fairness through awareness: (1) stratified batch sampling, (2) 'fair meta-learning for segmentation' and (3) protected group models. The comparative approach of using a smaller balanced database allows us to test the hypothesis that observed bias is due to data imbalance.

Segmentation Network: For all of the approaches we used the nnU-Net network [8] for automatic segmentation of the left ventricle blood pool (LVBP), left ventricular myocardium (LVM) and right ventricle blood pool (RVBP) from cine short-axis cardiac MR slices at end diastole (ED) and end systole (ES). Details of the data used and experimental set up are provided in Sect. 4.

Baseline - Fairness Through Unawareness: We trained the nnU-net model blinded to the protected attributes (i.e. race and gender). This will be considered the baseline approach in the rest of the paper and its performance can be directly compared with other published cardiac MR segmentation techniques [1,19].

Comparative Approach - Balanced Database: This strategy aims to evaluate the scenario where the database is perfectly balanced. To this end, we randomly selected for each racial group the same number of subjects (N = 87), which corresponds to the number of subjects in the smallest racial group. In total, 522 subjects were used for training this model.

Approach 1 - Stratified Batch Sampling: This strategy aims to modify the training sampling strategy to remove the discrimination before training. For each training batch, the data are stratified by the protected attribute(s) and samples are selected to ensure that each protected group is equally represented. This approach has previously been used to train fair classifiers [9].

Approach 2 - Fair Meta-learning for Segmentation: This strategy aims to add a meta-fair classifier to the segmentation network to train a model that not only segments cardiac MR images but also performs classification of the protected attribute(s). As the classifier we used a DenseNet network [6] and the input of the classifier was the cardiac MR image as one channel and the output of the nnU-Net segmentation network as a second channel. We formulate the problem as a multi-task learning problem where both networks are jointly optimized. The idea is that the classification network prevents the learning of a dominant group from negatively impacting the learning of another one. Both networks were first trained for 500 epochs independently, and then jointly trained for another 500 epochs without deep supervision. This is a novel approach for segmentation fairness but is based on the work proposed by Xu *et al.,* [23] for classification fairness, which combined a ResNet-18 network for facial expression recognition with a fully connected network for protected attribute classification.

Approach 3 - Protected Group Models: In contrast to Approaches 1 and 2, this strategy assumes that the protected attributes are available at inference time (as well as training time) and aims to train a different segmentation model for each protected group. As the number of subjects for each protected group can vary, which could impact performance, we initially trained the nnU-Net model for 500 epochs using the full training database, and then fine-tuned independent nnU-Net models for each of the protected groups for another 500 epochs. This approach is similar in concept to [20] in which classifiers were first trained using an unbalanced database and then fine-tuned using balanced ones. However, we choose to fine-tune using only individual protected group data and we are not aware of previous work that has taken this approach.

Evaluation Metrics: In the literature, several fairness metrics have been proposed but because fairness research has so far focused on classification tasks all of them are tailored for such tasks. Because we deal with a segmentation task, we propose to evaluate our approaches using a segmentation overlap metric and two new segmentation fairness metrics. To measure segmentation overlap we use the average Dice similarity coefficient (DSC) over the three classes, i.e. LVBP, LVM and RVBP. As fairness metrics, we utilize the standard deviation (SD) and skewed error ratio (SER) of the average DSC values. The standard deviation reflects the amount of dispersion of the average DSC values between different protected groups. The SER is computed by the ratio of the highest error rate to the lowest error rate among different protected groups and it can be formulated as $SER = \frac{\max_g(1-DSC_g)}{\min_g(1-DSC_g)}$, where g are the protected groups. The SD and SER fairness metrics were adapted from [20], which used classification accuracy and classification error rate for SD and SER respectively.

4 Materials and Experiments

We demonstrated our approach for fairness in DL-based cardiac MR segmentation using data from the UK Biobank [16]. The dataset used consisted of ED and ES short-axis cine cardiac MR images of 5,903 subjects (61.5 ± 7.1 years). Data on race and gender were obtained from the UK Biobank database and their distribution is summarized in Fig. 2. For all subjects, the LV endocardial and epicardial borders and the RV endocardial border were manually traced at ED and ES frames using the cvi42 software (version 5.1.1, Circle Cardiovascular Imaging Inc., Calgary, Alberta, Canada). Each segmentation was generated by 1 of 10 experts who followed the same guidelines and were blinded to race and gender. Each expert contoured a random sample of images containing different races and genders. For all approaches (apart from the balanced database comparative approach), we used the same random split into training/validation/test sets of 4,723/590/590 subjects respectively. For the balanced database approach the training/validation/test set sizes were 417, 105 and 590. All models were trained on a NVIDIA GeForce GTX TITAN X. All networks were trained for 1,000 epochs and optimized using stochastic gradient descent with 'poly' learning rate policy (initial learning rate of 0.01 and Nesterov momentum of 0.99).

Bias Assessment: The first experiment aims to assess gender and racial bias for the baseline approach (i.e. 'fairness through unawareness'). Table 1 shows the DSC values for LVBP, LVM and RVBP at ED and ES as well as the overall average DSC. Unpaired Student's t−tests (significant difference reported for p−value < 0.01 with Scheffe post hoc test) were used for comparison between the full test

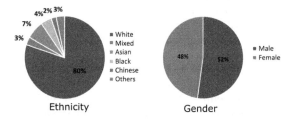

Fig. 2. The percentage of gender and different races in the UK Biobank dataset used in this paper.

set and each of the protected groups for the average DSC. From the results, we can make several important observations. First, there is a clear racial bias. For example, when the racial distribution is 80% white and 20% others, the nnU-net model reaches 93.51% average DSC for white subjects, but this dramatically decreases to less than 86% for black subjects and less than 85% for mixed race subjects. Second, the results show that there is no significant gender bias. This suggests that the accuracy of the baseline segmentation model for each group is correlated with its representation in the training set. The results of the comparative approach (Balanced database, see bottom row of Table 2) support this conclusion, achieving the lowest SD and SER. As expected, the smaller training database size led to a significant reduction in accuracy for all racial groups.

Bias Mitigation Strategies: Since Table 1 showed that the baseline segmentation model had a racial bias but no gender bias, in this second experiment we focus only on the use of bias mitigation techniques to reduce racial bias. Table 2 shows the comparison between the baseline approach and the three approaches for bias mitigation. The results show that all mitigation strategies reduce the racial bias but that Approach 3 (Protected group models) achieved the lowest SD and SER. Note, however, that Approach 3 requires knowledge of the protected attributes at inference time. The best-performing approach that does not require such knowledge is Approach 1 (Stratified batch sampling). For Approach 2 (Fair meta-learning for segmentation), the average accuracy of the protected attribute classifier was 0.77 and the precision and recall values per group varied between 0.69 and 0.89.

Table 1. Dice similarity coefficient (DSC) for the LV blood pool (LVBP), LV myocardium (LVM) and RV blood pool (RVBP) at end diastole (ED) and end systole (ES) for Baseline - fairness through unawareness. The 'Avg' column that represents the average DSC values across LVBP, LVM and RVBP at both ED and ES has been boldfaced for easier comparison with Table 2. The first row reports the DSC for the full database, the second and third rows report DSC by gender and the remaining rows report DSC by racial group. Asterisks indicate statistically significant differences between the full population and each protected group for the average DSC.

| | DSC (%) for Baseline—Fairness through unawareness | | | | | | |
| | ED | | | ES | | | Avg |
	LVBP	LVM	RVBP	LVBP	LVM	RVBP	
Total	93.48	83.12	89.37	89.37	86.31	80.61	**87.05**
Male	93.58	83.51	88.82	90.68	85.31	81.00	**87.02**
Female	93.39	82.71	89.90	89.59	86.60	80.21	**87.07**
White	97.33	93.08	94.09	95.06	90.58	90.88	**93.51***
Mixed	92.70	78.94	86.91	86.70	82.54	79.32	**84.52***
Asian	94.53	87.33	90.51	90.13	88.94	81.94	**88.90***
Black	92.77	85.93	89.49	89.42	85.74	71.91	**85.88***
Chinese	91.81	74.51	85.74	86.39	85.12	79.34	**83.82***
Others	91.74	78.94	89.50	88.53	84.96	80.27	**85.66***

Figure 3 shows some sample segmentation results for different racial groups with both high and low DSC.

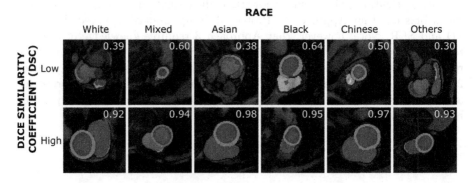

Fig. 3. Illustration of the segmentation results at ED for each racial group with high (top row) and low (bottom row) DSC.

Table 2. Comparison between baseline model (i.e. 'fairness through unawareness') and three approaches for bias mitigation. The values shown for each race represent the average DSC values across LVBP, LVM and RVBP at both ED and ES. The 'Avg' column is the average such value across all races (without weighting for the number of subjects in each group). We boldface the SD (lower is better) and SER (1 is the best) metrics since these are the important fairness criteria. The bottom row shows the results for using a smaller gender and race balanced training database.

Approach	Segmentation							Fairness	
	White	Mixed	Asian	Black	Chinese	Others	Avg	SD	SER
Baseline - Fairness through unawareness	93.51	84.52	88.90	85.88	87.63	85.66	87.68	**3.25**	**2.38**
1. Stratified batch sampling	90.88	93.84	93.65	93.07	94.35	93.50	93.22	**1.22**	**1.62**
2. Fair meta-learning for segmentation	92.75	88.03	90.64	89.60	88.18	88.27	89.58	**1.86**	**1.65**
3. Protected group models	91.03	93.17	93.34	92.15	93.04	93.08	92.64	**0.89**	**1.35**
Comparative approach - Balanced database	79.32	80.98	80.37	79.78	80.82	80.72	80.33	**0.65**	**1.09**

5   Discussion

To the best of our knowledge, this paper has presented the first study of fairness in AI-based image segmentation, focusing on the problem of data imbalance. We have shown, for the first time, that racial bias exists in DL-based cardiac MR segmentation models. Our hypothesis is that this bias is a result of the unbalanced nature of the training data, and this is supported by the results which show that there is racial bias but not gender bias when trained using the UK Biobank database, which is gender-balanced but not race-balanced. We also found very low bias when training using a balanced database. One reason why bias has yet to be investigated in DL-based cardiac MR segmentation is that in current clinical practice the outputs of such models are typically manually modified by a clinician. However, this modification process is time-consuming and prone to error, so does not eliminate the bias completely. Furthermore, in the future, as AI models become more clinically accepted, it may be that the role of clinician oversight decreases or is removed completely, in which case lack of bias in DL-based segmentation models will become a critical feature.

We have proposed three 'fairness through awareness' bias mitigation strategies. The first strategy (Stratified batch sampling) ensures that in each batch all protected groups are equally represented. The second strategy (Fair meta-learning for segmentation) aims to make the model more aware of the protected attribute(s) by simultaneously training a protected attribute classifier. The last strategy (Protected group models) aims to train independent segmentation models for each protected group. Our results show that all three strategies improve fairness, with lower SD and SER values compared to the baseline model. These results are in keeping with previous works where the strategy of 'fairness through unawareness' has shown to be insufficient in preventing bias in several problems [21,25].

It is important to note that the best-performing bias mitigation strategy (Approach 3) requires knowledge of the protected attribute at inference time. In general in medical image segmentation, such patient information is available via electronic health records. However, there are likely to be scenarios in which it is not available. In the case where the protected attribute(s) is not available at inference time it would be possible to use either Approach 1 or Approach 2. Approach 3 could be seen as a 'best achievable' model for such situations or where the protected attributes have been inadequately registered.

In conclusion, in this work we have highlighted the concerning issue of bias in DL-based image segmentation models. We have proposed three bias mitigation techniques, which were inspired by work from the literature on fairness in classification but which are all novel in the context of image segmentation. In the future, we aim to extend the current work to other kinds of existing fairness interventions, such as combinations of pre-processing, in-processing and post-processing strategies, as well as extending the analysis to other imaging modalities.

Acknowledgements. This work was supported by the EPSRC (EP/R005516/1 and EP/P001009/1), the Wellcome EPSRC Centre for Medical Engineering at the School of Biomedical Engineering and Imaging Sciences, King's College London (WT 203148/Z/16/Z) and has been conducted using the UK Biobank Resource under application numbers 17806 and 2964. SEP, SN and SKP acknowledge the BHF for funding the manual analysis to create a cardiac MR imaging reference standard for the UK Biobank imaging resource in 5000 CMR scans (PG/14/89/31194). We also acknowledge the following funding sources: the NIHR Biomedical Research Centre at Barts, EU's Horizon 2020 (grant no. 825903, euCanSHare project), the CAP-AI programme funded by the ERDF and Barts Charity, HDR UK, the Oxford NIHR Biomedical Research Centre, the Oxford BHF Centre of Research Excellence and the MRC eMed-Lab Medical Bioinformatics infrastructure (MR/L016311/1).

References

1. Bernard, O., et al.: Deep learning techniques for automatic MRI cardiac multi-structures segmentation and diagnosis: is the problem solved? IEEE Trans. Med. Imaging **37**(11), 2514–2525 (2018)
2. Buolamwini, J., Gebru, T.: Gender shades: intersectional accuracy disparities in commercial gender classification. In: Conference on Fairness, Accountability And Transparency, pp. 77–91. PMLR (2018)
3. Das, A., Dantcheva, A., Bremond, F.: Mitigating bias in gender, age and ethnicity classification: a multi-task convolution neural network approach. In: Leal-Taixé, L., Roth, S. (eds.) ECCV 2018. LNCS, vol. 11129, pp. 573–585. Springer, Cham (2019). https://doi.org/10.1007/978-3-030-11009-3_35
4. Dwork, C., et al.: Fairness through awareness. In: Proceedings of the 3rd Innovations in Theoretical Computer Science Conference, pp. 214–226 (2012)
5. Hardt, M., Price, E., Srebro, N.: Equality of opportunity in supervised learning. arXiv preprint arXiv:1610.02413 (2016)
6. Huang, G., et al.: Densely connected convolutional networks. In: Proceedings of the IEEE Conference on Computer Vision and Pattern Recognition, pp. 4700–4708 (2017)
7. Hwang, S., et al.: FairFaceGAN: fairness-aware facial image-to-image translation. arXiv preprint arXiv:2012.00282 (2020)

8. Isensee, F., et al.: nnU-Net: a self-configuring method for deep learning-based biomedical image segmentation. Nat. Methods **18**(2), 203–211 (2021)
9. Kamiran, F., Calders, T.: Data preprocessing techniques for classification without discrimination. Knowl. Inf. Syst. **33**(1), 1–33 (2012). https://doi.org/10.1007/s10115-011-0463-8
10. Kamiran, F., Karim, A., Zhang, X.: Decision theory for discrimination-aware classification. In: 2012 IEEE 12th International Conference on Data Mining, pp. 924–929. IEEE (2012)
11. Kishi, S., et al.: Race-ethnic and sex differences in left ventricular structure and function: the coronary artery risk development in young adults (CARDIA) study. J. Am. Heart Assoc. **4**(3), e001264 (2015)
12. Lu, K., Mardziel, P., Wu, F., Amancharla, P., Datta, A.: Gender bias in neural natural language processing. In: Nigam, V., et al. (eds.) Logic, Language, and Security. LNCS, vol. 12300, pp. 189–202. Springer, Cham (2020). https://doi.org/10.1007/978-3-030-62077-6_14
13. Mody, P., et al.: Most important articles on cardiovascular disease among racial and ethnic minorities. Circ.: Cardiovasc. Qual. Outcomes **5**(4), e33–e41 (2012)
14. Ngxande, M., Tapamo, J.R., Burke, M.: Bias remediation in driver drowsiness detection systems using generative adversarial networks. IEEE Access **8**, 55592–55601 (2020)
15. Obermeyer, Z., et al.: Dissecting racial bias in an algorithm used to manage the health of populations. Science **366**(6464), 447–453 (2019)
16. Petersen, S.E., et al.: UK Biobank's cardiovascular magnetic resonance protocol. J. Cardiovasc. Magn. Reson. **18**(1), 1–7 (2015)
17. Pleiss, G., et al.: On fairness and calibration. arXiv preprint arXiv:1709.02012 (2017)
18. Raisi-Estabragh, Z., et al.: Variation of cardiac magnetic resonance radiomics features by age and sex in healthy participants from the UK Biobank. Eur. Heart J. **41**(Supplement_2), ehaa946-0197 (2020)
19. Ruijsink, B., Puyol-Antón, E., et al.: Fully automated, quality-controlled cardiac analysis from CMR: validation and large-scale application to characterize cardiac function. Cardiovasc. Imaging **13**(3), 684–695 (2020)
20. Wang, M., Deng, W.: Mitigating bias in face recognition using skewness-aware reinforcement learning. In: Proceedings of the IEEE/CVF Conference on Computer Vision and Pattern Recognition, pp. 9322–9331 (2020)
21. Wang, Z., Qinami, K., et al.: Towards fairness in visual recognition: effective strategies for bias mitigation. In: Proceedings of the IEEE/CVF Conference on Computer Vision and Pattern Recognition, pp. 8919–8928 (2020)
22. Wilder, B., et al.: Clinical trial of an AI-augmented intervention for HIV prevention in youth experiencing homelessness. arXiv preprint arXiv:2009.09559 (2020)
23. Xu, T., White, J., Kalkan, S., Gunes, H.: Investigating bias and fairness in facial expression recognition. In: Bartoli, A., Fusiello, A. (eds.) ECCV 2020. LNCS, vol. 12540, pp. 506–523. Springer, Cham (2020). https://doi.org/10.1007/978-3-030-65414-6_35
24. Yoneyama, K., et al.: Cardiovascular magnetic resonance in an adult human population: serial observations from the multi-ethnic study of atherosclerosis. J. Cardiovasc. Magn. Reson. **19**(1), 1–11 (2017). https://doi.org/10.1186/s12968-017-0367-1
25. Zhang, B.H., Lemoine, B., Mitchell, M.: Mitigating unwanted biases with adversarial learning. In: Proceedings of the 2018 AAAI/ACM Conference on AI, Ethics, and Society, pp. 335–340 (2018)

An Interpretable Approach to Automated Severity Scoring in Pelvic Trauma

Anna Zapaishchykova[1,2], David Dreizin[3], Zhaoshuo Li[1], Jie Ying Wu[1], Shahrooz Faghihroohi[2], and Mathias Unberath[1(✉)]

[1] Johns Hopkins University, Baltimore, USA
mathias@jhu.edu
[2] Technical University of Munich, Munich, Germany
[3] School of Medicine, University of Maryland, College Park, USA

Abstract. Pelvic ring disruptions result from blunt injury mechanisms and are often found in patients with multi-system trauma. To grade pelvic fracture severity in trauma victims based on whole-body CT, the Tile AO/OTA classification is frequently used. Due to the high volume of whole-body trauma CTs generated in busy trauma centers, an automated approach to Tile classification would provide substantial value, e.g., to prioritize the reading queue of the attending trauma radiologist. In such scenario, an automated method should perform grading based on a transparent process and based on interpretable features to enable interaction with human readers and lower their workload by offering insights from a first automated read of the scan. This paper introduces an automated yet interpretable pelvic trauma decision support system to assist radiologists in fracture detection and Tile grade classification. The method operates similarly to human interpretation of CT scans and first detects distinct pelvic fractures on CT with high specificity using a Faster-RCNN model that are then interpreted using a structural causal model based on clinical best practices to infer an initial Tile grade. The Bayesian causal model and finally, the object detector are then queried for likely co-occurring fractures that may have been rejected initially due to the highly specific operating point of the detector, resulting in an updated list of detected fractures and corresponding final Tile grade. Our method is transparent in that it provides finding location and type using the object detector, as well as information on important counterfactuals that would invalidate the system's recommendation and achieves an AUC of 83.3%/85.1% for translational/rotational instability. Despite being designed for human-machine teaming, our approach does not compromise on performance compared to previous black-box approaches.

Keywords: Explainability · Explainable artificial intelligence · Deep learning · Machine learning · Human-computer interaction

Electronic supplementary material The online version of this chapter (https://doi.org/10.1007/978-3-030-87199-4_40) contains supplementary material, which is available to authorized users.

M. de Bruijne et al. (Eds.): MICCAI 2021, LNCS 12903, pp. 424–433, 2021.
https://doi.org/10.1007/978-3-030-87199-4_40

1   Introduction

Mortality in patients with pelvic fractures ranges from 4% to 15% [27] depending on fracture severity. Urgent surgical or endovascular treatment is necessary to ensure good patient outcome, however, pelvic ring disruption severity grading remains a subjective bottleneck in patient triage. In contemporary practice, whole-body CT is routinely performed upon admission of trauma victims with suspected pelvic fractures [9,13], which are then commonly graded with the unified AO Foundation/Orthopaedic Trauma Association (AO/OTA) system [24]. CT scans are interpreted by readers with widely disparate levels of experience in a time-sensitive environment resulting in limited accuracy and inter-observer agreement. An objective, automated approach to Tile AO/OTA grading of pelvic fractures could harmonize the accuracy of Tile grading across readers while accelerating the interpretation of whole-body CT to extract pelvic fracture-related risk scores, thus addressing an important unmet clinical need [27]. Previous work using automated deep learning methods to evaluate pelvic fracture severity relied on black-box models that predict severity from trauma imagery [12,21] but do not provide a transparent, interpretable association between Tile AO/OTA grade and detectable fractures. Plain radiographs have been used for pelvic fracture detection [8], but not classification tasks, and are severely hampered by bone and soft tissue overlap. CT remains the imaging gold standard.

Irrespective of the imaging modality used for fracture grading, black-box approaches are undesirable in a high-stakes decision making because they are designed to be run as a standalone application. They do not offer immediate opportunity for interaction and may be prone to automation bias. Interpretable models that offer insights into the decision-making process may promote human-machine teaming because they enable human interaction and verification, thus potentially bolstering clinicians' confidence in the recommendation. Prior methods [7,15,17] have been proposed to fill this gap by explaining causality through the identification of regions of interest that contribute to a predictor's outcome.

Contribution. We present an automated yet interpretable algorithm for first-order Tile AO/OTA grading from trauma CT. The system mimics a human expert read of the CT scan by first identifying fracture types and locations via a deep object detection algorithm, and then inferring Tile grade from presence or absence of fractures using a Bayesian network. An overview of the method is provided in Fig. 1. By establishing causal relationships between Tile grade and fracture presence, the framework provides a transparent inference pipeline that supplies fracture location and type, as well as information on counterfactuals, e.g. missed or misclassified fractures, that would invalidate the system's recommendation. As such, the system is designed to decrease the workload of the attending radiologist by facilitating validation and refinement.

2   Related Work

Specific fracture patterns and abnormal bony relationships in different parts of the pelvic ring on CT correspond with the degree of mechanical instability [24].

This section outlines pelvic fracture biomechanics and briefly reviews prior work on fracture detection and interpretability for medical decision making.

Fracture Detection on Plain Radiographs and CT. Previous studies have demonstrated the feasibility of deep learning for fracture detection in various anatomical regions from plain radiographs [8,16]. Developments in CT fracture detection have been presented for the rib cage [4], spine [5] and skull [16]. We use Faster-RCNN [20] for fracture localization and classification to extract pelvic ring disruptions from CT scans since it is a well-established architecture for both general purpose object and fracture detection [2,28].

Features of Pelvic Ring Disruption. The AO/OTA have adopted the Tile classification scheme for grading mechanical instability and severity of pelvic fractures [23]. First-order Tile grading (Grade **A** - translationally(T) and rotationally(R) stable, Grade **B** - T stable and R unstable, Grade **C** - both T and R *i.e. "globally"* unstable) has been shown to correlate with major arterial injury, need for surgical packing or catheter-based interventions, need for massive transfusion, and mortality [10,11]. Different degrees of instability manifest with different types of abnormal bony relationships on the CT scan, including pubic symphysis diastasis (PSD), divergent or parallel sacroischial (SI) joint diastasis, nondiastatic and diastatic sacral fractures, ischial spine avulsions (ISp), and innominate bone fractures involving the anterior and posterior pelvic ring [14,24]. Fragility in elderly patients is not expressly included in Tile's framework, but has come to be recognized as an important modifier of fracture severity [26].

Interpretable Artificial Intelligence for Medical Decision Support. There is increasing evidence that medical decision support systems that are interpretable or explainable are perceived as more trustworthy [3,22]. One way of making such models interpretable is to derive predictions with a task-specific causal model that is created from domain expertise and can model uncertainties between variables and the final prediction [25]. Using directed graphical models to encode the causal structure between variables [19], a complete causal graph can be constructed to induce the reasoning around diagnostic predictions [6]. Here, we combine the benefits of Bayesian structural causal models and deep learning-based feature extractors to create an automatic yet interpretable algorithm for pelvic Tile grading from CT scans.

3   Methods

To detect fractures and other pertinent findings (for brevity, we will summarize all of such findings under the term *fractures*) in a pelvic CT volume, we use a Faster-RCNN trained on annotated bounding boxes on axial maximum intensity projections (MIP) slices (Fig. 1a). After fracture locations and fracture types have been determined together with their corresponding confidence score, we obtain an initial Tile grade estimate from the Bayesian model (BM) applied to the detected fractures with high confidence (see Fig. 1b). We include the patient's age, which helps in clinical practice to determine overall severity

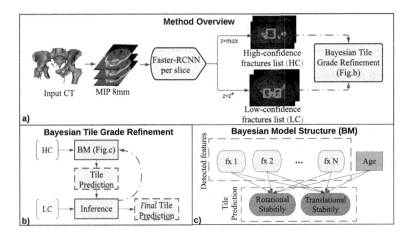

Fig. 1. a) Method overview. After training on maximum intensity projections (MIPs) and corresponding annotations generated from 3D CT segmentation maps of key fracture classes that contribute to Tile grading, Faster-RCNN is used to extract high- and lower-confidence findings, such as fractures. b) Tile grade refinement algorithm using causal Bayesian model. Detailed description can be found in Algorithm 1. c) Structure of the Bayesian model (BM): [fx 1 - fx N] are findings detected by the Faster-RCNN. Tile grade is represented by a combination of translational and rotational instability.

because fractures tend to propagate through the weaker bones in the elderly (see Fig. 1c for structure of the BM). By combining patient-specific information with fractures found by Faster-RCNN, we can compute inferences conditioned on the fractures that were detected with high confidence. Due to the highly specific operating point selected for Faster-RCNN, it is possible that some true fractures were erroneously rejected. To refine the Tile grade estimate based on these erroneously rejected fractures and simultaneously improve fracture detection, we intervene on the initial detected fractures to identify potential misses that would confirm or invalidate the current Tile estimate. We then check for presence of these candidate fractures identified by our causal model among the lower confidence fractures, and if present, include them in our high-confidence findings. Finally, we run our BM on the updated list of fractures to obtain the binding estimate of Tile grade.

The subset of pelvic fracture (fx)-related features that were determined a priori to be clinically meaningful for Tile grading by trauma radiologists includes: PSD; anteriorly divergent SI and parallel SI joint diastasis; non-diastatic and diastatic sacral fx; ISp, innominate bone fractures involving the anterior and posterior pelvic ring (henceforth "ring fx").

The proposed algorithm is formally described in Algorithm 1. An initial estimate of the Tile grade will be made by selecting the high-confidence features FX_{high} of pelvic ring disruption with confidence threshold set to $z = max$ to range [0.95-1.0], so that the feature detection network will have high specificity. By computing the marginal likelihood probabilities for each possible Tile state,

---

Algorithm 1: Proposed Tile refinement algorithm using a Bayesian network structural causal model (BM)

$BM(x)$ is a method for computing the marginal likelihood for nodes given x;
$p(x)$ is a probability of a variable x;
Input:
z is a confidence threshold setting;
FX_{high} is list of fractures detected with high confidence $> z*$;
FX_{low} is list of fractures detected with low confidence $< z*$;
Result: Tile grade prediction - A,B or C;
Compute BM on FX_{high} to get initial Tile grade;
for *Tile* **in** *[A,B,C]* **do**
    Compute $BM(FX_{high}, Tile)$ to get the p for features(fx);
    if *fx has $p > z*$ and is in FX_{low}* **then**
        | Update FX_{high} by adding fx;
    end
end
Compute BM on updated FX_{high};
argmax $(p(Tile))$;

---

we get a list of potentially missed fracture labels that are more likely to co-occur with the highly specific conditions found in the specific CT. For every fracture in this candidate list, we check its presence in a list of lower-confidence fracture detections FX_{low} that is generated using a lower threshold $z < z*$. If such matching between the FX_{low} and fractures suggested by BM are found, FX_{high} will be updated by considering the additional features found in FX_{low}, and the predicted Tile grade will be updated based on this new feature set.

While we consider our algorithm from a fully automated perspective in this manuscript, the refinement procedure based on a structural causal model is designed with immediate benefit for clinical workflows in mind. This is because, instead of setting a different threshold z for possibly missed fractures, our model could alert the attending radiologist for review of the fractures. Similarly, the BM could calculate counterfactuals to the current prediction to identify possible misses that would change treatment recommendations. Because the object detection network supplies bounding boxes for each fracture finding, our method allows for effective visualization of all findings. This may contribute to an effective hand-over from an autonomous to human-machine teaming operation.

4 Experiments and Results

4.1 Dataset and Metrics

We obtained approval for use of the dataset initially described in [12], that consists of 373 admission CT scans of adult patients from two level I trauma centers with bleeding pelvic fractures and varying degrees of instability with IRB-approved waiver of consent. Imaging was performed with intravenous contrast in the arterial phase on 40-, 64-, and dual source 128-row CT scanners and reconstructed with 0.7 – 5 mm slice thickness. Patient-level annotation using the Tile AO/OTA grading system was performed by three trauma-subspecialized radiologists and the consensus vote was used as gold standard for training and evaluation. Presence and location of fractures and fracture-related features such as pelvic ring disruption were performed on the 3D CTs by one radiologist with trauma expertise at the primary study site.

4.2   Implementation Details

The CT scans were interpolated, isotropically scaled and normalized on a per-volume basis. To preserve some volumetric information while reducing the complexity of the problem from 3D to 2D, MIP was used to generate 8 mm-thick slices along the axial plane. Reference standard segmentation of pelvic fractures were propagated to these MIPs as bounding boxes with corresponding fracture label. On all MIPs containing labels, we train a Faster-RCNN using TensorFlow for 4000 epochs with a ResNet-50 backbone pretrained on ImageNet to extract pelvic features. For evaluation, we used five-fold cross validation where, in each fold, 80% of the data was used for training and 20% of the data was used for testing. Random flip and random rotations were used for data augmentation. To reduce the number of false positives, we gather poorly predicted features during training and up-weigh them in the next training phase for 3000 epochs. The structure of BM is illustrated in Fig. 1c. The BM is developed using the Causal-Nex [1] package. We fit probabilities using Bayesian parameter estimation and learn conditional probability distributions (CPDs) w.r.t. Tile expert consensus from the same data split used during the fracture detection training.

4.3   Experimental Results and Discussion

To evaluate the Tile AO/OTA assessment, we split the Tile grade into component instabilities (*rotational* or *translational*), as in [11,14]. Human readers explicitly search for evidence of rotational and translational instability separately to synthesize the Tile grade and this represents the most intuitive real-world approach to the problem. We present all scores as an average over the 5 folds (Tab. 1) and use area under the receiver operating characteristic curve (AUC) for classification performance and Kappa scores [18] for reader agreement measurements. Inter-observer reliability between automated predictions and expert consensus improved using the proposed refinement technique to a level that is close to the inter-rater variability of human observers [29]. Since inter-observer reliability is only moderate to fair for many Tile grades [29], we hypothesize the lack of agreement is responsible for the imperfect performance of **BM**(*GT*) on expert-annotated fractures. Table 1 shows the full results.

   We evaluated Faster-RCNN for each fracture type separately (Fig. 2b). We then evaluate the performance of Faster-RCNN for fracture detection with and without BM-based refinement(Fig. 2a) and show that the proposed scheme not

Table 1. Average AUC and Kappa scores for rotational (R) and translational (T) instabilities. All columns are compared to the patient-level expert consensus Tile grade by three radiologists. **BM**(GT) represents Bayesian model (BM) fitness when using the reference standard fracture detections used to train the Faster-RCNN model; **BM**(FX_{low}) represents BM prediction on automatically detected fractures when lower confidence fractures (with $z = 0.5$) are immediately included in Tile grade inference; **BM**(FX_{high}) represents BM prediction on automatically detected fractures with high confidence ($z = max$); **BM refinement** represents predictions after the proposed refinement pipeline.

Metrics	BM(GT)		BM(FX_{low})		BM(FX_{high})		**BM refinement**	
	R	T	R	T	R	T	R	T
AUC	0.81	0.84	0.71	0.63	0.78	0.82	**0.85**	**0.83**
Kappa	0.32	0.48	0.1	0.18	0.15	0.38	**0.24**	**0.5**

only improves Tile grading but also positively affects fracture detection when the initial network operates at high specificity.

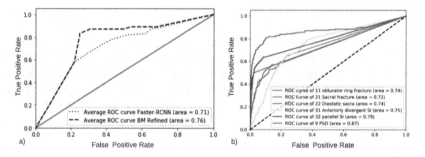

Fig. 2. a) Average RoC curve of Faster-RCNN-based fracture detection with and without BM-based refinement for $z = 0.5$ b) RoC curves for every fracture type using Faster-RCNN at $z = 0.5$

We found that the fracture with the lowest confidence by the object detector (Anteriorly Divergent SI) is often responsible for the Tile prediction update. Even though they were initially left out, anteriorly divergent SI fractures were included after the refinement. In a situation where a binder (a device used to stabilize suspected pelvic fractures in patients with traumatic injury) is present, the subset of visible fractures will appear less severe, even though presence of Anteriorly Divergent SI indicates a major high energy blunt trauma event. An example of a case is shown in the supplementary material.

To measure the impact of the confidence threshold z of the lower-confidence detections on Tile prediction, we compare the performance of BM on predicting Rotational (R)/Translational (T) instabilities based on fracture detections FX_{low} achieved with a Faster-RCNN operating at a confidence threshold in an

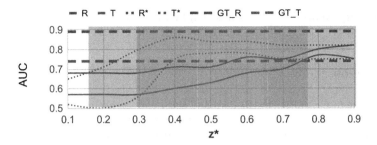

Fig. 3. Impact of the magnitude of z of FX_{low} on the AUC of Tile prediction for $FX_{high}(z = 0.95)$. **R/T** is BM model prediction without refinement on a list of fractures with confidence $> z*$, **R*/T*** is BM model prediction after refinement, **GT_R/GT_T** is BM model prediction on the ground truth semantic labels. The proposed method outperforms the BM without refinement when $z* \in [0.3 - 0.75]$ and therefore is less dependent on the initial parameter selection.

interval $z \in [0.1 - 0.9]$, while $z = 0.95$ is held constant for the high-confidence detections (Fig. 3). We find that the proposed refinement method improves upon direct inference on a less specific detection result without being overly dependent on hyperparameter selection.

5 Conclusion

In this work, by leveraging the respective strengths of deep learning for feature detection and a BM for causal inference, we achieve automatic and interpretable Tile grading of pelvic fractures from CT images. Using the proposed pipeline, Tile grade prediction had an AUC of 83.3% and 85.1% for translational and rotational instability comparable to the previous black-box method [12] but with added interpretability and potential for human-machine teaming.

The described methodology can benefit from interaction with experienced radiologists in place of the autonomous refinement, to further increase the inter-observer agreement. One limitation to be addressed in the future, either algorithmically or through interaction, are false-false detections, i.e., false positives that are detected with high confidence and then propagated unchanged through the refinement step. Because our BM-based refinement strategy cannot currently reject high-confidence fractures, these findings would irrevocably bias the Tile grade estimation. Another line of future work will include investigations on the effectiveness of the proposed system when working with attending trauma radiologists in a collaborative setting to see if the system can indeed decrease the time to triage. We anticipate that the core ideas of the presented framework may find application across various diagnostic image interpretation tasks. We understand our methods as a step towards machine-human teaming for quick and accurate pelvic fracture severity scoring using the Tile AO/OTA system to standardize and accelerate triage, and ultimately, reduce mortality rates.

Acknowledgements. S.F. was supported by the ICL-TUM Joint Academy for Doctoral (JADS) program. D.D. was funded by NIH K08 EB027141-01A1.

References

1. Welcome to CausalNex's API docs and tutorials! – causalnex 0.9.1 documentation (Feb 2021). https://causalnex.readthedocs.io/en/latest. Accessed 26 Feb 2021
2. Abbas, W., et al.: Lower leg bone fracture detection and classification using faster R-CNN for x-rays images. In: 2020 IEEE 23rd International Multitopic Conference (INMIC), pp. 1–6. IEEE (2020)
3. Ahmad, M.A., Eckert, C., Teredesai, A.: Interpretable machine learning in healthcare. In: Proceedings of the 2018 ACM International Conference on Bioinformatics, Computational Biology, and Health Informatics, pp. 559–560 (2018)
4. Blum, A., Gillet, R., Urbaneja, A., Teixeira, P.G.: Automatic detection of rib fractures: are we there yet? EBioMedicine **63** (2021). https://doi.org/10.1016/j.ebiom.2020.103158
5. Burns, J.E., Yao, J., Summers, R.M.: Artificial intelligence in musculoskeletal imaging: a paradigm shift. J. Bone Miner. Res. **35**(1), 28–35 (2020). https://doi.org/10.1002/jbmr.3849
6. Castro, D.C., Walker, I., Glocker, B.: Causality matters in medical imaging. Nat. Commun. **11**(3673), 1–10 (2020). https://doi.org/10.1038/s41467-020-17478-w
7. Chang, C.H., Creager, E., Goldenberg, A., Duvenaud, D.: Explaining image classifiers by counterfactual generation. arXiv preprint arXiv:1807.08024 (2018)
8. Cheng, C.T., et al.: A scalable physician-level deep learning algorithm detects universal trauma on pelvic radiographs. Nat. Commun. **12**(1066), 1–10 (2021). https://doi.org/10.1038/s41467-021-21311-3
9. Coccolini, F., et al.: Pelvic trauma: WSES classification and guidelines. World J. Emerg. Surg. **12**, 5 (2017). https://doi.org/10.1186/s13017-017-0117-6
10. Dreizin, D.: Commentary on multidetector CT in vascular injuries resulting from pelvic fractures. Radiographics **39**(7), 2130–2133 (2019)
11. Dreizin, D., et al.: CT prediction model for major arterial injury after blunt pelvic ring disruption. Radiology **287**(3), 1061–1069 (2018). https://doi.org/10.1148/radiol.2018170997
12. Dreizin, D., et al.: An automated deep learning method for tile AO/OTA pelvic fracture severity grading from trauma whole-body CT. J. Digit. Imaging **34**(1), 53–65 (2021)
13. Dreizin, D., Munera, F.: Blunt polytrauma: evaluation with 64-section whole-body CT angiography. Radiographics **32**(3), 609–631 (2012). https://pubs.rsna.org/doi/full/10.1148/rg.323115099
14. Dreizin, D., et al.: Can MDCT unmask instability in binder-stabilized pelvic ring disruptions? Am. J. Roentgenol. **207**(6), 1244–1251 (2016). https://doi.org/10.2214/AJR.16.16630
15. Fong, R., Patrick, M., Vedaldi, A.: Understanding deep networks via extremal perturbations and smooth masks. In: Proceedings of the IEEE/CVF International Conference on Computer Vision, pp. 2950–2958 (2019)
16. Kalmet, P.H., et al.: Deep learning in fracture detection: a narrative review. Acta Orthop. **91**(2), 215–220 (2020)

17. Lenis, D., Major, D., Wimmer, M., Berg, A., Sluiter, G., Bühler, K.: Domain aware medical image classifier interpretation by counterfactual impact analysis. In: Martel, A.L., et al. (eds.) MICCAI 2020. LNCS, vol. 12261, pp. 315–325. Springer, Cham (2020). https://doi.org/10.1007/978-3-030-59710-8_31

18. McHugh, M.L.: Interrater reliability: the kappa statistic. Biochem. Med. (Zagreb). **22**(3), 276 (2012). https://www.ncbi.nlm.nih.gov/pmc/articles/PMC3900052

19. Pearl, J.: Probabilistic Reasoning in Intelligent Systems: Networks of Plausible Inference. Elsevier, Amsterdam (2014)

20. Ren, S., He, K., Girshick, R., Sun, J.: Faster R-CNN: Towards Real-Time Object Detection with Region Proposal Networks. arXiv (June 2015). https://arxiv.org/abs/1506.01497v3

21. Sato, Y., et al.: A computer-aided diagnosis system using artificial intelligence for hip fractures-multi-institutional joint development research. arXiv e-prints pp. arXiv-2003 (2020)

22. Schölkopf, B., et al.: Towards Causal Representation Learning. arXiv (February 2021). https://arxiv.org/abs/2102.11107v1

23. Tile, M.: Pelvic ring fractures: should they be fixed? J. Bone Joint Surg. Br. **70**(1), 1–12 (1988). https://doi.org/10.1302/0301-620X.70B1.3276697

24. Tile, M.: Acute pelvic fractures: I. causation and classification. J. Am. Acad. Orthop. Surg. **4**(3), 143–151 (1996). https://doi.org/10.5435/00124635-199605000-00004, https://pubmed.ncbi.nlm.nih.gov/10795049/

25. Tonekaboni, S., Joshi, S., McCradden, M.D., Goldenberg, A.: What clinicians want: contextualizing explainable machine learning for clinical end use. In: Machine Learning for Healthcare Conference, pp. 359–380. PMLR (2019)

26. Vaidya, R., Scott, A.N., Tonnos, F., Hudson, I., Martin, A.J., Sethi, A.: Patients with pelvic fractures from blunt trauma. what is the cause of mortality and when? Am. J. Surg. **211**(3), 495–500 (2016). https://doi.org/10.1016/j.amjsurg.2015.08.038

27. Vaidya, R., Scott, A.N., Tonnos, F., Hudson, I., Sethi, A.: Patients with Pelvic Fractures from blunt trauma. what is the cause of mortality and when? Am. J. Surg. **211**(3), 495–500 (2015). https://doi.org/10.1016/j.amjsurg.2015.08.038

28. Yahalomi, E., Chernofsky, M., Werman, M.: Detection of distal radius fractures trained by a small set of X-ray images and Faster R-CNN. arXiv (December 2018). https://arxiv.org/abs/1812.09025v1

29. Zingg, T., et al.: Interobserver reliability of the tile classification system for pelvic fractures among radiologists and surgeons. Eur. Radiol. **31**(3), 1517–1525 (2021)

Scalable, Axiomatic Explanations of Deep Alzheimer's Diagnosis from Heterogeneous Data

Sebastian Pölsterl$^{(\boxtimes)}$ iD, Christina Aigner, and Christian Wachinger iD

Artificial Intelligence in Medical Imaging (AI-Med), Department of Child and Adolescent Psychiatry, Ludwig-Maximilians-Universität, Munich, Germany
`sebastian.poelsterl@med.uni-muenchen.de`

Abstract. Deep Neural Networks (DNNs) have an enormous potential to learn from complex biomedical data. In particular, DNNs have been used to seamlessly fuse heterogeneous information from neuroanatomy, genetics, biomarkers, and neuropsychological tests for highly accurate Alzheimer's disease diagnosis. On the other hand, their black-box nature is still a barrier for the adoption of such a system in the clinic, where interpretability is absolutely essential. We propose Shapley Value Explanation of Heterogeneous Neural Networks (SVEHNN) for explaining the Alzheimer's diagnosis made by a DNN from the 3D point cloud of the neuroanatomy and tabular biomarkers. Our explanations are based on the Shapley value, which is the unique method that satisfies all fundamental axioms for local explanations previously established in the literature. Thus, SVEHNN has many desirable characteristics that previous work on interpretability for medical decision making is lacking. To avoid the exponential time complexity of the Shapley value, we propose to transform a given DNN into a Lightweight Probabilistic Deep Network without re-training, thus achieving a complexity only quadratic in the number of features. In our experiments on synthetic and real data, we show that we can closely approximate the exact Shapley value with a dramatically reduced runtime and can reveal the hidden knowledge the network has learned from the data.

1 Introduction

In recent years, deep learning methods have become ubiquitous across a wide range of applications in biomedicine (see e.g. [6] for an overview). While these models may obtain a high predictive performance in a given task, a major obstacle for their routine use in the clinic is their black-box nature: the inner workings that lead to a particular prediction remain opaque to the user. Methods that try to open the black-box and make its decision process interpretable to the user, can be classified as explainable AI [2]. The goals of explainable AI are diverse and can be categorized into five groups [2]: to establish trust, discover causal relationships among inputs, guarantee fairness and privacy, and inform the user about the decision making process.

© Springer Nature Switzerland AG 2021
M. de Bruijne et al. (Eds.): MICCAI 2021, LNCS 12903, pp. 434–444, 2021.
https://doi.org/10.1007/978-3-030-87199-4_41

The focus of this paper is on *local explanation*: the ability to explain to the user how the prediction for one specific input came to be. We consider the input data to be heterogeneous, namely a combination of tabular biomarkers and brain morphology, both of which are common predictors for various neurodegenerative diseases. The shape of brain structures is rich in information and preferred over simple summary statistics such as volume, as demonstrated in recent deep learning approaches for neuroanatomical shape analysis [12,16,20]. However, due to the non-Euclidean geometry, deep learning on shapes requires dedicated network architectures that differ substantially from standard convolutional neural networks. When also incorporating tabular biomarkers, the network usually consists of two arms, each specific to one particular input type. Such networks can achieve impressive predictive performance, but to consider their deployment in the clinic, we need a principled methodology capable of explaining their predictions.

We require explanations to be easy to understand by a human and to accurately imitate the deep learning model. We attain these two properties by building upon the strong theoretical guarantees of the Shapley value [23], a quantity from cooperative game theory. The work in [1,18,25,26] established a set of fundamental axioms local explanations ought to satisfy, and proved that the Shapley value is the unique procedure that satisfies (1) completeness, (2) null player, (3) symmetry, (4) linearity, (5) continuity, (6) monotonicity, and (7) scale invariance. Unfortunately, computing the Shapley value scales exponential in the number of features, which necessitates developing an efficient approximation. To the best of our knowledge, the Shapley value has not been used to explain predictions made by a deep neural network from neuroanatomical shape and clinical information before.

In this work, we propose Shapley Value Explanation of Heterogeneous Neural Networks (SVEHNN): a scalable, axiomatic approach for explaining the decision of a deep neural network integrating neuroanatomical shape and tabular biomarkers. SVEHNN is based on the Shapley value [23] and thus inherits its strong theoretic guarantees. To overcome the cost, exponential in the number of features, we efficiently approximate it by converting standard layers into corresponding probabilistic layers without re-training. Thus, the computational complexity of explaining a model's prediction is quadratic instead of exponential in the number of features. We assess the approximation error of SVEHNN on synthetic data, for which the exact Shapley value can be computed, and demonstrate its ability to explain Alzheimer's disease diagnosis of patients on real clinical data.

Related Work. Early work on local explanations includes the occlusion method, which selectively occludes portions of the input and measures the change in the network's output [27]. It has been employed in [12] for Alzheimer's disease (AD) diagnosis based on shape representations of multiple brain structures, and in [17] for autism spectrum disorder (ASD) diagnosis based on fMRI. A second group of algorithms obtains local explanations by back-propagating a network's output through each layer of the network until it reaches individual features

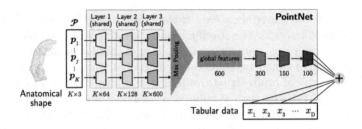

Fig. 1. Wide and Deep PointNet architecture (adapted from [20]).

of the input [3,22,24,26,28]. They differ in how the gradients are incorporated into the final feature relevance score. In [28], the authors incorporate forward activations and gradients in the relevance score and applied it to classification of cellular electron cryo-tomography. Different from the above, Zhuang et al. [29] proposed an invertible network for ASD diagnosis. By projecting the latent representation of an input on the decision boundary of the last linear layer and inverting this point, they obtain the corresponding representation in the input space. As shown in [1,18,25,26], the vast majority of the methods above do not satisfy all fundamental axioms of local explanations outlined above. Most relevant to our work is the efficient approximation of the Shapley value for multi-layer perceptrons and convolutional neural networks in [1]. However, it is not applicable to heterogeneous inputs we consider here. In fact, most methods were developed for image-based convolutional neural networks, and it remains unclear how they can be used for non-image data, e.g., when defining a suitable "non-informative reference input" for gradient-based methods [3,22,24,26,28].

2   Methods

We consider explaining predictions made by a Wide and Deep PointNet (WDPN, [20]) from the hippocampus shape and clinical variables of an individual for AD diagnosis by estimating the influence each input feature has on the model's output. Before describing our contribution, we will first briefly summarize the network's architecture and define the Shapley value.

2.1   Wide and Deep PointNet

Figure 1 depicts the architecture of our network to predict the probability of AD diagnosis. We represent the hippocampus shape as a point cloud $\mathcal{P}$, and known clinical variables associated with AD as tabular data $\boldsymbol{x} \in \mathbb{R}^D$ [20]. Point clouds are fed to a PointNet [21], which first passes each coordinate vector through a multilayer perceptron with shared weights among all points, before aggregating point descriptors using max pooling. We fuse $\boldsymbol{x}$ with the latent representation of $\mathcal{P}$ in a final linear layer. To account for non-linear effects, WDPN augments clinical markers by B-spline expansion or interactions.

2.2   Shapley Value

Given a trained WDPN, $f : \mathbb{R}^{K \times 3} \times \mathbb{R}^D \to [0; 1]$, we want to explain the predicted probability of AD for one particular input $\mathbf{z} = (\mathcal{P}, \mathbf{x})$ by estimating the contribution individual points in the point cloud and clinical features have on the prediction. In particular, we propose to estimate the Shapley value [23], which we will define next.

We denote by $\mathcal{F}$ the set of all features comprising the input, and $\mathbf{z}_{\mathcal{S}}$ the subset indexed by $\mathcal{S} \subseteq \mathcal{F}$. Let $g : \mathbf{P}(\mathcal{F}) \to \mathbb{R}$ be a set function defined on the power set of $\mathcal{F}$, with $g(\mathcal{S}) \neq 0$, $g(\emptyset) = 0$. The contribution of a set of features $\mathcal{S}$ on the prediction is given by the set function $g(\mathcal{S}) = f(\mathbf{z}_{\mathcal{S}}; \mathbf{z}^{\text{bl}}_{\mathcal{F} \setminus \mathcal{S}}) - f(\mathbf{z}^{\text{bl}})$, where the first term corresponds to the prediction of the model after replacing all features not in $\mathcal{S}$ with a baseline value from the vector $\mathbf{z}^{\text{bl}}$. The Shapley value for feature i is defined as

$$s_i(\mathbf{z} \mid f) = \frac{1}{|\mathcal{F}|!} \sum_{\mathcal{S} \subseteq \mathcal{F} \setminus \{i\}} |\mathcal{S}|! \cdot (|\mathcal{F}| - |\mathcal{S}| - 1)! \, (g(\mathcal{S} \cup \{i\}) - g(\mathcal{S})) \,. \tag{1}$$

Approximate Shapley Value. Computing (1) exactly would require exponential many evaluations of the difference $\Delta_i = g(\mathcal{S} \cup \{i\}) - g(\mathcal{S})$ for each feature i. Therefore, we employ the approximate Shapley value, first proposed for voting games in [8]. Let $\mathbb{E}_k(\Delta_i)$ denote the marginal contribution of feature i, where the expectation is over all $\mathcal{S}$ with $|\mathcal{S}| = k$. By writing (1) in an alternative form where we explicitly sum over all sets $\mathcal{S}$ of equal size and noting that there are $\binom{|\mathcal{F}|-1}{k}$ possible sets of size k, we arrive at the approximate Shapley value $\bar{s}_i$:

$$s_i(\mathbf{z} \mid f) = \frac{1}{|\mathcal{F}|!} \sum_{k=0}^{|\mathcal{F}|-1} \sum_{\substack{\mathcal{S} \subseteq \mathcal{F} \setminus \{i\} \\ |\mathcal{S}|=k}} k!(|\mathcal{F}| - k - 1)! \cdot \Delta_i \approx \frac{1}{|\mathcal{F}|} \sum_{k=0}^{|\mathcal{F}|-1} \mathbb{E}_k(\Delta_i) = \bar{s}_i(\mathbf{z} \mid f).$$

$$\tag{2}$$

We will now focus on our main contribution to efficiently estimate the expectation

$$\mathbb{E}_k(\Delta_i) = \mathbb{E}_k[f(\mathbf{z}_{\mathcal{S} \cup \{i\}}; \mathbf{z}^{\text{bl}}_{\mathcal{F} \setminus \mathcal{S} \cup \{i\}})] - \mathbb{E}_k[f(\mathbf{z}_{\mathcal{S}}; \mathbf{z}^{\text{bl}}_{\mathcal{F} \setminus \mathcal{S}})]. \tag{3}$$

2.3   Efficient Estimation of the Approximate Shapley Value

Our main contribution is based on the observation that we can treat $\mathbf{z}_{\mathcal{S}}$ over all sets $\mathcal{S}$ of size k as a source of aleatoric uncertainty, i.e., we are unsure about the input [1]. Then, the objective becomes propagating the aleatoric uncertainty through the network. This is a challenging task, because we need to create a version of the WDPN that is probabilistic and faithful to the original model. To this end, we propose a novel probabilistic WDPN, inspired by the Lightweight Probabilistic Deep Network (LPDN, [10]), that models aleatoric uncertainty by assuming inputs are comprised of independent univariate normal distributions, one for each feature. The resulting probabilistic WDPN directly outputs an estimate of $\mathbb{E}_k[f(\mathbf{z}_{\mathcal{S}}; \mathbf{z}^{\text{bl}}_{\mathcal{F} \setminus \mathcal{S}})]$ for a fixed k. Next, we will propose a new set of layers to transform inputs into distributions over sets $\mathcal{S}$ of size k.

Probabilistic PointNet. The first part of the network in Fig. 1 processes the point cloud. It consists of multiple fully-connected layers with ReLU activation and batch normalization, where weights are shared across points. For the initial linear layer with weights $\mathbf{W}$, the m-th output for the j-th point is $h_{jm} = \sum_{l=1}^{3} p_{jl} W_{lm}$. To make it probabilistic, we need to account for $\mathbf{p}_j$ being included in S randomly. Since each S is selected with probability $\binom{|\mathcal{F}|}{k}^{-1}$ and there are $\binom{|\mathcal{F}|-1}{k-1}$ sets S containing $\mathbf{p}_j$, we have

$$\mathbb{E}_k[h_{jm}] = \sum_{\substack{S \subseteq \mathcal{F} \\ |S|=k}} \binom{|\mathcal{F}|}{k}^{-1} \binom{|\mathcal{F}|-1}{k-1} h_{jm} = \frac{k}{|\mathcal{F}|} h_{jm}. \tag{4}$$

This is a well-known result from sampling theory [1,7], which also tells us that h_{jm} can be approximated with a normal distribution with mean (4) and variance

$$\mathbb{V}_k(h_{jm}) = k \frac{|\mathcal{F}| - k}{|\mathcal{F}| - 1} \left[\frac{1}{|\mathcal{F}|} \sum_{l=1}^{3} (p_{jl} W_{lm})^2 - \left(\frac{1}{|\mathcal{F}|} h_{jm} \right)^2 \right]. \tag{5}$$

After the first linear layer each output unit is approximated by a normal distribution, and we can employ a LPDN for all subsequent layers, by replacing ReLU, batch-norm, and max-pooling with their respective probabilistic versions [10]. The final output of the probabilistic PointNet is a global descriptor of the whole point cloud, where each feature has a mean and a variance attached to it.

Probabilistic Wide and Deep PointNet. The final step in estimating the expectation (3) is the integration of clinical markers with the latent point cloud representation (yellow node in Fig. 1). Both information is combined in a linear layer, for which we need to propagate uncertainty due to the distributions describing the point cloud, and due to S covering a subset of clinical markers. Since a linear layer is an associative operation, we can compute the sum over the latent point cloud features and the clinical features separately, before combining the respective results, thus we split the expectation (3) into $\mathbb{E}_k(\Delta_i) = \mathbb{E}_k(\Delta_i^x) + \mathbb{E}_k(\Delta_i^{\mathcal{P}})$. To compute the expectation with respect to tabular data $\mathbf{x}$, we utilize that for a linear layer with bias b and weights $\mathbf{w} \in \mathbb{R}^D$, $f(\mathbf{x}) = b + \mathbf{w}^\top \mathbf{x}$, thus $\mathbb{E}_k(\Delta_i^x) = w_i x_i$, if feature i is tabular and $\mathbb{E}_k(\Delta_i^x) = 0$ otherwise. The uncertainty due to the latent point cloud features can be propagated by a linear probabilistic layer [10] with mean μ_k^S, yielding $\mathbb{E}_k[f(\mathbf{z}_S; \mathbf{z}_{\mathcal{F} \setminus S}^{\mathrm{bl}})] = \mu_k^S$. Doing the same, but accounting for the inclusion of i in S, we have $\mathbb{E}_k[f(\mathbf{z}_{S \cup \{i\}}; \mathbf{z}_{\mathcal{F} \setminus S \cup \{i\}}^{\mathrm{bl}})] = \mu_k^{S \cup \{i\}}$. Finally, we subtract both means yielding $\mathbb{E}_k(\Delta_i^{\mathcal{P}}) = \mu_k^{S \cup \{i\}} - \mu_k^S$. As a result, we can estimate $\mathbb{E}_k(\Delta_i)$ with only two forward-passes, while keeping $\mathbf{z}$ and $k = |S|$ fixed.

Estimating the Approximate Shapley Value. Equipped with an efficient way to estimate the expectation over a fixed number of features in S, we can compute the difference in (3). For the first term, we only need a single forward pass through the probabilistic WDPN using the original input with all features. For the second term, we again need one forward pass, but without the contribution of the i-th feature, which is replaced by a baseline value. This value differs depending on whether the i-th feature is a point in the hippocampus or a clinical marker. Thus, the complexity to estimate the approximate Shapley value for all features in the input is $\mathcal{O}((D+|\mathcal{P}|)^2)$. For high-dimensional inputs, we can further reduce the complexity by estimating the average in (2) via M Monte Carlo samples, yielding $\mathcal{O}(M(D+|\mathcal{P}|))$.

Choosing a Baseline. Selection of an appropriate baseline $\mathbf{z}^{\mathrm{bl}}$ is crucial in obtaining semantically meaningful explanations, as the i-th Shapley value reflects the contribution of feature i to the difference $f(\mathbf{z}) - f(\mathbf{z}^{\mathrm{bl}}) = \sum_{i=1}^{|\mathcal{F}|} s_i(\mathbf{z}\,|\,f)$ [23]. Assuming no clinical marker has been zero-encoded, we can replace the original value with zero to eliminate its impact. If the i-th feature belongs to the point cloud, the situation is more nuanced. Using all-zeros would describe a point at the origin and Shapley values would explain the difference to the prediction made from the origin, which has no semantic meaning in the case of the hippocampus. We propose a better alternative by replacing it with a matching point from a hull containing all point clouds in the dataset. Our motivation for this approach is that it avoids erratic changes of the point cloud's surface when inserting a point at the origin, and that it provides a common, semantically meaningful reference for all patients, which eases interpretation of explanations across multiple patients.

3 Experiments

In this section, we are evaluating the computational efficiency, approximation error, and semantics of SVEHNN compared to the exact Shapley value [23], Occlusion [27], and Shapley sampling [5]. Note that except for Occlusion, these methods have not been used to explain predictions from shape data before. For each method, we compare three different approaches to choosing the baseline value in point clouds: (i) moving a point to the origin (zero), and (ii) replacing a point by its matching point from the common hull (hull). We quantitatively assess differences with respect to the exact Shapley value in terms of mean squared error (MSE), Spearman rank correlation (SRC), and normalized discounted cumulative gain (NDCG, [15]). MSE captures the overall approximation error, SRC is a measure of agreement between rankings of features, and NDCG is similar to SRC, but penalizes errors in important features – according to the ground truth – more. For NDCG, we use the absolute Shapley value as importance measure. We run Shapley sampling [5] with $M = 2,000$ Monte Carlo samples, which results in a runtime of $\mathcal{O}(|\mathcal{P}|M)$. Note that sampling usually requires $M \gg |\mathcal{P}|$ for a good approximation.

3.1 Synthetic Shape Data Set

To assess the approximation error compared to the exact Shapley value in a reasonable amount of time, we created a synthetic dataset of point clouds

Table 1. Mean difference to exact Shapley value across 100 synthetic point clouds with zero baseline. NE: Number of network evaluations for each example.

Method	MSE	SRC	NDCG	NE
Exact	0	1	1	65,538
Sampling	0.0008	0.9505	0.9986	32,000
Sampling	0.0340	0.5440	0.9481	512
Occlusion	16.1311	0.3180	0.8659	17
SVEHNN	0.0443	0.6918	0.9641	512

comprising 16 points ($|\mathcal{P}| = 16$). This allows calculation of the exact Shapley values across all 2^{16-1} subsets of $\mathcal{P}$ and use it as ground truth. We use a binary PointNet classifier to distinguish point clouds of the characters 'X' and 'I'.

Results in Table 1 show that occlusion fails in all categories, except runtime efficiency. SVEHNN closely approximates the exact Shapley value in terms of MSE with less than 0.8% of network evaluations. Running Shapley sampling until convergence (second row) ranks slightly higher, but at more than 60 times the cost of SVEHNN, it is a high cost for an improvement of less than 0.035 in NDCG. When considering Shapley sampling with the same runtime as SVEHNN (third row), rank metrics drop below that of SVEHNN; the MSE remains lower, which is due to the Gaussian approximation of activations in probabilistic layers of SVEHNN. Overall, we can conclude that SVEHNN is computationally efficient with only a small loss in accuracy.

3.2 Alzheimer's Disease Diagnosis

In this experiment, we use data from the Alzheimer's Disease Neuroimaging Initiative [13]. Brain scans were processed with FreeSurfer [9] from which we obtained the surface of the left hippocampus, represented as a point cloud with 1024 points. For tabular clinical data, we use age, gender, education (in orthogonal polynomial coding), APOE4, FDG-PET, AV45-PET, Aβ_{42}, total tau (t-tau), and phosphorylated tau (p-tau). We account for non-linear age effects using a natural B-spline expansion with four degrees of freedom. We exclude patients diagnosed with mild cognitive impairment and train a WDPN [20] to tell healthy controls from patients with AD. Data is split into 1308 visits for training, 169 for hyper-parameter tuning, and 176 for testing; all visits from one subject are included in the same split. The WDPN achieves a balanced accuracy of 0.942 on the test data. We run SVEHNN with 150 Monte Carlo samples and the hull baseline.

Figure 2 depicts the Shapley values of individual patients and features for all correctly classified test patients, sorted by the average feature relevance. For tabular biomarkers, colors indicate the range of feature values from low (blue) to high (red). For instance, APOE4 Shapley values of patients with no APOE4 allele are depicted in blue, with one allele in purple, and with two alleles in

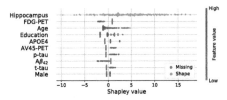

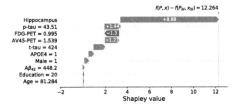

Fig. 2. Shapley values of biomarkers and hippocampus shape of 167 correctly classified patients (dots, 110 healthy, 57 AD), sorted by mean relevance.

Fig. 3. Shapley values of biomarkers of a single AD patient.

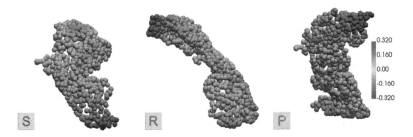

Fig. 4. Shapley values of hippocampus shape of same patient as in Fig. 3. Left: Superior view. Middle: Right view. Right: Posterior view.

red. It shows that for a few patients the hippocampus shape is very important, whereas for most patients its relevance is similar to that of tabular biomarkers. Overall, relevance scores echo clinically validated results: high concentrations of p-tau/t-tau and APOE4 are markers for AD, whereas low levels of $A\beta_{42}$ and many years of education are protective [4,11,19]. To provide an example of a local explanation by SVEHNN, we selected one AD patient (see Fig. 3). It shows that the AD diagnosis was driven by the hippocampus shape, followed by p-tau, FDG-PET, and AV45-PET. The explanation of the hippocampus reveals that points located in the CA1 subfield were most important for diagnosis (see Fig. 4). This is reassuring, because atrophy in the CA1 subfield is an established marker for AD [14]. By cross-referencing the explanation with clinical knowledge, we can conclude that the predicted AD diagnosis is likely trustworthy.

4   Conclusion

Obtaining comprehensible and faithful explanations of decisions made by deep neural networks are paramount for the deployment of such systems in the clinic. We proposed a principled methodology for explaining individual predictions, which can help to build trust among the users of such a system. In this work, we studied explaining networks integrating neuroanatomical shape and tabular biomarkers, which has not been studied before. We followed an axiomatic approach based on the Shapley value – the unique procedure that satisfies all fundamental axioms of local explanation methods. Our proposed method Shapley Value Explanation of Heterogeneous Neural Networks (SVEHNN) closely approximates the exact Shapley value while requiring only a quadratic instead of exponential number of network evaluations. Finally, we illustrated how SVEHNN can help to understand Alzheimer's diagnosis made by a network from heterogeneous data to reveal the hidden knowledge the network has learned from the data.

Acknowledgements. This research was supported by the Bavarian State Ministry of Science and the Arts and coordinated by the Bavarian Research Institute for Digital Transformation, and the Federal Ministry of Education and Research in the call for Computational Life Sciences (DeepMentia, 031L0200A).

References

1. Ancona, M., Oztireli, C., Gross, M.: Explaining deep neural networks with a polynomial time algorithm for shapley value approximation. In: Proceedings of the 36th International Conference on Machine Learning, vol. 97, pp. 272–281 (2019)
2. Arrieta, A.B., et al.: Explainable Artificial Intelligence (XAI): concepts, taxonomies, opportunities and challenges toward responsible AI. Inf. Fusion **58**, 82–115 (2020). https://doi.org/10.1016/j.inffus.2019.12.012
3. Bach, S., Binder, A., Montavon, G., Klauschen, F., Müller, K.R., Samek, W.: On pixel-wise explanations for non-linear classifier decisions by layer-wise relevance propagation. PLOS ONE **10**(7), e0130140 (2015). https://doi.org/10.1371/journal.pone.0130140
4. Blennow, K., Vanmechelen, E., Hampel, H.: CSF total tau, Aβ42 and phosphorylated tau protein as biomarkers for Alzheimer's disease. Mol. Neurobiol. **24**(1–3), 087–098 (2001). https://doi.org/10.1385/mn:24:1-3:087
5. Castro, J., Gómez, D., Tejada, J.: Polynomial calculation of the Shapley value based on sampling. Comput. Oper. Res. **36**(5), 1726–1730 (2009). https://doi.org/10.1016/j.cor.2008.04.004
6. Ching, T., et al.: Opportunities and obstacles for deep learning in biology and medicine. J. R. Soc. Interface **15**(141), 20170387 (2018). https://doi.org/10.1098/rsif.2017.0387
7. Cochran, W.G.: Sampling Techniques, 3rd edn. John Wiley & Sons, Hoboken (1977)
8. Fatima, S.S., Wooldridge, M., Jennings, N.R.: A linear approximation method for the Shapley value. Artif. Intell. **172**(14), 1673–1699 (2008). https://doi.org/10.1016/j.artint.2008.05.003
9. Fischl, B.: FreeSurfer. Neuroimage **62**(2), 774–781 (2012). https://doi.org/10.1016/j.neuroimage.2012.01.021
10. Gast, J., Roth, S.: Lightweight probabilistic deep networks. In: The IEEE Conference on Computer Vision and Pattern Recognition (CVPR), pp. 3369–3378 (2018)
11. Genin, E., et al.: APOE and Alzheimer disease: a major gene with semi-dominant inheritance. Mol. Psychiatry **16**(9), 903–907 (2011). https://doi.org/10.1038/mp.2011.52

12. Gutiérrez-Becker, B., Wachinger, C.: Deep multi-structural shape analysis: application to neuroanatomy. In: Medical Image Computing and Computer Assisted Intervention (MICCAI), pp. 523–531 (2018). https://doi.org/10.1007/978-3-030-00931-1_60

13. Jack, C.R., et al.: The Alzheimer's disease neuroimaging initiative (ADNI): MRI methods. J. Magn. Reson. Imaging **27**(4), 685–691 (2008). https://doi.org/10.1002/jmri.21049

14. Joie, R.L., et al.: Hippocampal subfield volumetry in mild cognitive impairment, Alzheimer's disease and semantic dementia. NeuroImage Clin. **3**, 155–162 (2013). https://doi.org/10.1016/j.nicl.2013.08.007

15. Järvelin, K., Kekäläinen, J.: Cumulated gain-based evaluation of IR techniques. ACM Trans. Inf. Syst. (TOIS) **20**(4), 422–446 (2002). https://doi.org/10.1145/582415.582418

16. Kopper, P., Pölsterl, S., Wachinger, C., Bischl, B., Bender, A., Rügamer, D.: Semi-structured deep piecewise exponential models. In: Proceedings of AAAI Spring Symposium on Survival Prediction - Algorithms, Challenges, and Applications 2021, vol. 146, pp. 40–53 (2021)

17. Li, X., Dvornek, N.C., Zhuang, J., Ventola, P., Duncan, J.S.: Brain biomarker interpretation in ASD using deep learning and fMRI. In: Medical Image Computing and Computer Assisted Intervention (MICCAI), pp. 206–214 (2018). https://doi.org/10.1007/978-3-030-00931-1_24

18. Lundberg, S.M., Lee, S.I.: A Unified Approach to Interpreting Model Predictions. Adv. Neural. Inf. Process. Syst. **30**, 4765–4774 (2017)

19. Meng, X., D'Arcy, C.: Education and dementia in the context of the cognitive reserve hypothesis: a systematic review with meta-analyses and qualitative analyses. PLoS ONE **7**(6), e38268 (2012). https://doi.org/10.1371/journal.pone.0038268

20. Pölsterl, S., Sarasua, I., Gutiérrez-Becker, B., Wachinger, C.: A wide and deep neural network for survival analysis from anatomical shape and tabular clinical data. In: Machine Learning and Knowledge Discovery in Databases, pp. 453–464 (2020). https://doi.org/10.1007/978-3-030-43823-4_37

21. Qi, C.R., Su, H., Mo, K., Guibas, L.J.: PointNet: deep learning on point sets for 3D classification and segmentation. In: The IEEE Conference on Computer Vision and Pattern Recognition (CVPR), pp. 652–660 (2017)

22. Selvaraju, R.R., Cogswell, M., Das, A., Vedantam, R., Parikh, D., Batra, D.: Grad-CAM: visual explanations from deep networks via gradient-based localization. In: The IEEE International Conference on Computer Vision (ICCV) (2017). https://doi.org/10.1109/iccv.2017.74

23. Shapley, L.S.: A value for n-person games. Contrib. Theory Games **2**(28), 307–317 (1953)

24. Shrikumar, A., Greenside, P., Kundaje, A.: Learning important features through propagating activation differences. In: Proceedings of the 34th International Conference on Machine Learning, vol. 70, pp. 3145–3153 (2017)

25. Sundararajan, M., Najmi, A.: The many Shapley values for model explanation. In: Proceedings of the 37th International Conference on Machine Learning, vol. 119, pp. 9269–9278 (2020)

26. Sundararajan, M., Taly, A., Yan, Q.: Axiomatic Attribution for Deep Networks. In: Proc. of the 34th International Conference on Machine Learning. vol. 70, pp. 3319–3328 (2017)

27. Zeiler, M.D., Fergus, R.: Visualizing and understanding convolutional networks. In: European Conference on Computer Vision (ECCV), pp. 818–833 (2014)

28. Zhao, G., Zhou, B., Wang, K., Jiang, R., Xu, M.: Respond-CAM: analyzing deep models for 3D imaging data by visualizations. In: Medical Image Computing and Computer Assisted Intervention (MICCAI), pp. 485–492 (2018). https://doi.org/10.1007/978-3-030-00928-1_55

29. Zhuang, J., Dvornek, N.C., Li, X., Ventola, P., Duncan, J.S.: Invertible network for classification and biomarker selection for ASD. In: Medical Image Computing and Computer Assisted Intervention (MICCAI), pp. 700–708 (2019). https://doi.org/10.1007/978-3-030-32248-9_78

SPARTA: An Integrated Stability, Discriminability, and Sparsity Based Radiomic Feature Selection Approach

Amir Reza Sadri[1], Sepideh Azarianpour Esfahani[1], Prathyush Chirra[1], Jacob Antunes[1], Pavithran Pattiam Giriprakash[1], Patrick Leo[1], Anant Madabhushi[1,2], and Satish E. Viswanath[1(✉)]

[1] Case Western Reserve University, Cleveland, OH 44106, USA
sev21@case.edu
[2] Louis Stokes Cleveland VA Medical Center, Cleveland, OH 44106, USA

Abstract. In order to ensure that a radiomics-based machine learning model will robustly generalize to new, unseen data (which may harbor significant variations compared to the discovery cohort), radiomic features are often screened for stability via test/retest or cross-site evaluation. However, as stability screening is often conducted independent of the feature selection process, the resulting feature set may not be simultaneously optimized for discriminability, stability, as well as sparsity. In this work, we present a novel radiomic feature selection approach termed SPARse sTable lAsso (SPARTA), uniquely developed to identify a highly discriminative and sparse set of features which are also stable to acquisition or institution variations. The primary contribution of this work is the integration of feature stability as a generalizable regularization term into a least absolute shrinkage and selection operator (LASSO)-based optimization function. Secondly, we utilize a unique non-convex sparse relaxation approach inspired by proximal algorithms to provide a computationally efficient convergence guarantee for our novel algorithm. SPARTA was evaluated on three different multi-institutional imaging cohorts to identify the most relevant radiomic features for distinguishing: (a) healthy from diseased lesions in 147 prostate cancer patients via T2-weighted MRI, (b) healthy subjects from Crohn's disease patients via 170 CT enterography scans, and (c) responders and non-responders to chemoradiation in 82 rectal cancer patients via T2w MRI. When compared to 3 state-of-the-art feature selection schemes, features selected via SPARTA yielded significantly higher classifier performance on unseen data in multi-institutional validation (hold-out AUCs of 0.91, 0.91, and 0.93 in the 3 cohorts).

Keywords: Feature selection · Feature stability · Convex optimization

---

Electronic supplementary material The online version of this chapter (https://doi.org/10.1007/978-3-030-87199-4_42) contains supplementary material, which is available to authorized users.

© Springer Nature Switzerland AG 2021
M. de Bruijne et al. (Eds.): MICCAI 2021, LNCS 12903, pp. 445–455, 2021.
https://doi.org/10.1007/978-3-030-87199-4_42

1   Introduction

In radiomics [1] applications, feature selection strategies such as maximum relevance minimum redundancy (mRMR) [2], least absolute shrinkage and selection operator (LASSO) [3], or Wilcoxon rank-sum testing (WLCX) [4], are widely used to find a relatively parsimonious (i.e. *sparse*) subset of features, which in combination are able to accurately distinguish between any classes of interest (i.e. are *discriminatory*). Critically, the ultimate validation of these radiomics models is in a multi-institutional setting where unseen datasets may suffer from significant variations compared to the original discovery cohort. Towards this, it has become critical to determine whether selected radiomic features are *stable* or *reproducible* within an institution (intra-site), between institutions (inter-site), as well as in repeated test/re-test evaluation [5]. The underlying hypothesis here is that stable radiomic features which are also discriminable will result in better model generalizability on new, unseen data.

2   Previous Work and Novel Contributions

While discriminatory radiomic features are often selected based on criteria such as mutual information (mRMR) or statistical testing (WLCX), the stability of these features is often ensured by initially removing features that are not reproducible in inter-site, intra-site, or repeat evaluation [6–10]. For instance, Aerts et al. [11] and Kim et al. [12] showed that selecting from among radiomic features that were stable between test/retest scans or annotations resulted in more generalizable classifier models for overall survival (head-and-neck and lung cancers) or tumor grade (brain tumors). Similarly, utilizing only those histomorphometric features that were stable between institutions resulted in significantly improved performance for prostate cancer detection and grading using digitized radical prostatectomy images [9]. However, this approach involves evaluating feature discriminability and stability independent of each other which can result in some marginally unstable but highly discriminable radiomic features being filtered out prior to feature selection; thus lowering overall model performance.

To address this, we present an approach that integrates feature stability screening and feature selection into a single optimization function; allowing for identification of a subset of radiomic features that are simultaneously sparse, discriminable, and stable in a fashion not attempted thus far. LASSO [3] is well-suited for this as it allows for easy incorporation of additional regularization terms in order to account for the curse of dimensionality (i.e. elastic net regularization [13]) or to allow predefined groups of features to be jointly selected (via sparse-group lasso [14]).

- The primary contribution of our SPARse sTable lAsso (SPARTA) approach is to integrate radiomic feature stability as an generalizable regularization term (which can capture test/retest, inter-site, or intra-site variations) into a LASSO-based optimization function.

– Our secondary contribution is to leverage a new class of proximal algorithms [15–17] to solve this novel optimization function in a computationally efficient fashion despite it being non-smooth (which can cause coordinate descent optimization algorithms to yield sub-optimal performance [18]).

We evaluate our new SPARTA feature selection scheme in the context of 3 clinical problems with multi-institutional data cohorts (n = 399 scans total) for distinguishing: (i) benign from malignant regions in the peripheral zone on prostate MRIs, (ii) healthy from Crohn's disease patients via CT enterography scans, and (iii) pathologic responders and non-responders to chemoradiation in rectal cancers via MRI.

3 Methodology

Problem Statement: A feature matrix is denoted as $\mathbf{X}_i = [\mathbf{x}_i^j] \in \mathbb{R}^{n_i \times m}$, where $\mathbf{x}_i^j$ is the feature vector, $i \in \{1, \ldots, s\}$ indicates the institution, and $j \in \{1, \ldots, m\}$ denotes the feature index. m and n_i represent the number of features and the number of samples in the ith institution, respectively. The complete set of feature values across all samples and institutions can be denoted as $\mathbf{X} \in \mathbb{R}^{n \times m}$ where $n = \sum_{i=1}^s n_i$. Let $\mathbf{y}_i \in \mathbb{R}^{n_i}$ be the corresponding label vector for n_i samples, such that the complete set of target class labels will be $\mathbf{y} \in \mathbb{R}^n$. The LASSO approach [3,19] for finding a sparse subset of discriminative features can be written as a constrained optimization problem with a convex relaxation:

$$\min_{\boldsymbol{\theta}} \left\{ J(\boldsymbol{\theta}) = \frac{1}{2} \sum_{i=1}^s \|\mathbf{y}_i - \mathbf{X}_i \boldsymbol{\theta}\|_2^2 + \lambda \|\boldsymbol{\theta}\|_1 \right\}, \tag{1}$$

where λ is the tuning parameter and $\boldsymbol{\theta} \in \mathbb{R}^m$ corresponds to a vector of coefficients reflecting the contribution of each feature.

Incorporating Feature Stability into LASSO: Feature stability can be quantified via measures including intra-class correlation coefficient (ICC) [6], concordance correlation coefficient (CCC) [10], coefficient of variation (CV) [20,21], or instability score (IS) [9]. Feature stability for every $\mathbf{x}^j$ is quantified via $\beta^j, j \in \{1, \ldots, m\}$, computed in a bootstrapped fashion for robustness (higher values of β^j correspond to more unstable features). Samples from across all s institutions are repeatedly and randomly split into L bootstrapped subsets, and pairwise comparisons of subsets are used to compute a statistical- or information-based stability function u as follows:

$$\beta^j = \frac{1}{Ls^2} \sum_{v=1}^s \sum_{w=1}^s \sum_{l=1}^L u\left(\mathbf{x}^j(D_v^l), \mathbf{x}^j(D_w^l)\right), \tag{2}$$

where D_v^l, D_w^l are the bootstrap subsets being compared in each iteration. The stability score vector for all m features can be defined as $\boldsymbol{\beta} \triangleq [\beta^1, \ldots, \beta^m]^T$, based on which a symmetric *Feature Stability Matrix* can be computed $\mathbf{R} \triangleq \boldsymbol{\beta}\boldsymbol{\beta}^T \in \mathbb{R}^{m \times m}$. $\mathbf{R}$ is incorporated as a penalty term in the LASSO cost function (Eq. (1)) as $\min_\theta J(\boldsymbol{\theta}) + \mu\boldsymbol{\theta}^T \mathbf{R}\boldsymbol{\theta}$ where μ is the regularization parameter. Note that $\mathbf{R}$ can be formulated to comprise multiple measures that individually capture intra-site, inter-site, or test/retest stability or take the form of a single integrated stability measure.

Optimization of SPARTA: The final equation can be written as:

$$\boldsymbol{\theta} = \underset{\theta}{\mathrm{argmin}}\left\{ \underbrace{\frac{1}{2}\sum_{i=1}^{s} \|\mathbf{y}_i - \mathbf{X}_i\boldsymbol{\theta}\|_2^2}_{\text{discriminability}} + \underbrace{\mu\boldsymbol{\theta}^T\boldsymbol{\beta}\boldsymbol{\beta}^T\boldsymbol{\theta}}_{\text{stability}} + \underbrace{\lambda\|\boldsymbol{\theta}\|_1}_{\text{sparsity}} \triangleq f(\boldsymbol{\theta}) + g(\boldsymbol{\theta}) \right\}. \quad (3)$$

In this, the discriminability and stability terms (denoted $f(\boldsymbol{\theta})$) are smooth functions while the sparsity term (denoted $g(\boldsymbol{\theta})$) is a non-smooth function. Hence, while $f + g$ in (3) is a convex objective function, it is still non-smooth and thus cannot be solved by regular optimization methods such as gradient descent. Rather than computationally expensive and complex alternatives [22,23], we utilize proximal algorithms [24] as they work under extremely general conditions (including cases where the functions are non-smooth), are much faster for challenging optimization problems, as well as being scalable and amenable to distributed optimization [17]. Proof that our problem meets the requirements of the general proximal algorithms [24] and that f in (3) is a Lipschitz continuous gradient function is presented as Lemma 1 in the Supplementary Materials.

Based on Lemma 2 (see Supplementary Materials), in order to minimize $f + g$, we can replace the non-smooth function f with its upper-bound (denoted $\bar{f}$) which results in the following iterative solution algorithm for (3),

$$\boldsymbol{\theta}_{k+1} = \underset{\theta}{\mathrm{argmin}}\left\{ \bar{f}(\boldsymbol{\theta}, \boldsymbol{\theta}_k) + g(\boldsymbol{\theta}) = f(\boldsymbol{\theta}_k) + \nabla^T f(\boldsymbol{\theta}_k)(\boldsymbol{\theta} - \boldsymbol{\theta}_k) + \frac{1}{2\gamma}\|\boldsymbol{\theta} - \boldsymbol{\theta}_k\|_2^2 + g(\boldsymbol{\theta}) \right\}. \quad (4)$$

This in turn is equivalent to

$$\boldsymbol{\theta}_{k+1} = \underset{\theta}{\mathrm{argmin}}\left\{ \frac{1}{2}\|\boldsymbol{\theta} - \bar{\boldsymbol{\theta}}_k\|_2^2 + \gamma g(\boldsymbol{\theta}) \right\} = \mathrm{prox}_{\gamma g}(\bar{\boldsymbol{\theta}}_k), \quad (5)$$

where $\bar{\boldsymbol{\theta}}_k = \boldsymbol{\theta}_k - \gamma\nabla f(\boldsymbol{\theta}_k)$ and $\mathrm{prox}_{\gamma g}$ is the proximal operator of the function γg [17] and γ is the learning rate. This base mapping of the proximal algorithm is a standard tool for solving non-smooth optimization problems [16]. Theorem 1 in the Supplementary Materials guarantees convergence of the iterations in Eq. (5) (even when the stability term comprises multiple measures). The final Algorithm 1 summarizes the overall approach to solve (5) within SPARTA.

Algorithm 1. Proximal Algorithm Solution to SPARTA

Input: y, X, β, μ, λ, K (number of inner-loop iterations), γ
    initialization : $\boldsymbol{\theta} = \mathbf{0}$, $\mathbf{R} = \beta\beta^T$
  1: **for** $k = 1, 2, \cdots, K$ **do**
  2:     $f(\boldsymbol{\theta}) = \frac{1}{2} \|\mathbf{y} - \mathbf{X}\boldsymbol{\theta}\|_2^2 + \mu\boldsymbol{\theta}^T \mathbf{R}\boldsymbol{\theta}$
  3:     $g(\boldsymbol{\theta}) = \lambda \|\boldsymbol{\theta}\|_1$
  4:     $\bar{\boldsymbol{\theta}} = \boldsymbol{\theta} - \gamma\nabla f(\boldsymbol{\theta})$
  5:     $\boldsymbol{\theta} = \mathrm{prox}_{\gamma g}(\bar{\boldsymbol{\theta}})$
  6: **end for**
Output: $\boldsymbol{\theta}$

---

4 Experimental Results and Discussion

4.1 Data and Implementation

All experiments were conducting using three different, multi-institutional, retrospectively accrued cohorts that had been segregated into independent discovery and validation sets (see Table 1).

1. C1 (prostate cancer) comprised 147 diagnostic T2-weighted (T2w) prostate MRIs from 4 institutions, with the goal of distinguishing benign from malignant lesions in the peripheral zone (discovery: 3 sites, validation: 1 site).
2. C2 (Crohn's disease) comprised 170 CT enterography scans from patients being screened for Crohn's disease with endoscopic confirmation of disease presence. The goal was to distinguish between healthy and diseased terminal ileum regions within this single institutional cohort harboring significant batch effects [25] due to changes in acquisition parameters.
3. C3 (rectal cancer) comprised 82 post-treatment T2w rectal MRIs from two institutions, where the goal is to distinguish pathologic complete response (pCR) or non-response to chemoradiation based on rectal wall regions (discovery: 1 site, validation: 1 site).

Implementation: As summarized in Fig. 1, total of 405 3D radiomic features were extracted on a voxel-wise basis from all imaging scans, after which the mean value of each feature was computed within expert annotated ROIs. Feature normalization was applied on a cohort-basis to ensure that all radiomic features lay within a comparable range of values (mean 0, std 1). SPARTA was implemented as Algorithm 1 (using IS [26] as the stability measure within $\mathbf{R}$), with $K = 100$ (empirically selected) and $\gamma = 0.5L$ (mid-point of convergence interval, see Eq. (4) of Supplementary Materials). Additionally, three popular FS schemes (mRMR, LASSO, and WLCX) in conjunction with stability screening [9] were also implemented. In the latter case, the IS score was computed for each of the 405 radiomic features across 100 bootstrap iterations (sampling 50% of the discovery cohorts) in each of C1, C2, and C3 separately. Features with an IS > 0.25 were excluded as being potentially unstable, resulting in 282, 294, and

Table 1. Data Description

Info	Split	C1 (Prostate cancer MRI)			C2 (Crohn's disease CTE)			C3 (Rectal cancer MRI)		
		Inst	Malignant ROI	Benign ROI	Inst	Healthy	Diseased	Inst	pCR	Non-pCR
Data	*Discovery*	S1	15	15	S5	56	56	S6	38	14
		S2	11	11						
		S3	65	65						
	Validation	S4	56	56		29	29	S7	4	26
Features	20 Grey, 152 Laws, 13 Gradient, 160 Gabor, and 60 Haralick									

297 "stable" features in each of C1, C2, and C3, respectively. This screened feature set was input to each of mRMR, LASSO, and WLCX for feature selection. All experiments were conducted in MATLAB 9.9 on a 64 bit Windows 10 PC with an Intel(R) Core(TM) i7 CPU 930 (3.60 GHz) and 32 GB RAM.

4.2 Experiment 1: Assessing Parameter Sensitivity and Radiomic Feature Contributions

Design: To evaluate differences between the top-ranked radiomic feature sets identified by each of SPARTA, mRMR, LASSO, and WLCX, the Shapley (SHAP) value was computed [27] for the top 5 (out of 10) ranked features identified by each FS scheme. The Shapley value is the average marginal contribution of a feature value over all possible coalitions [28], providing a natural way to compute how much each feature contributes to predictive performance. The trade-off between feature importance, stability, and discriminability was evaluated via a 3D scatter plot. The effect of the regularization parameters in SPARTA (μ and λ) was investigated by varying each parameter in the range $\{10^{-8}, 10^{-7}, \dots, 10^{2}\}$, and evaluating a Random Forests classifier (RF, with 50 trees, 50 maximum levels, 100 leaf samples) at each parameter combination.

Results: Figure 2 (top row) visualizes a 3D scatter plot for SHAP value vs IS vs AUC, for each of the 5 top-ranked radiomic features selected via each of SPARTA, mRMR, LASSO, and WLCX (in different colors), for each of C1, C2, and C3. SPARTA selected features (pink) appear to exhibit higher SHAP

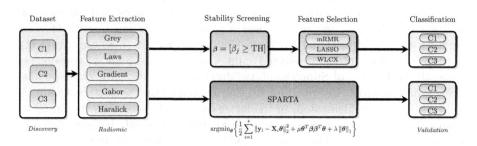

Fig. 1. Overall experimental design evaluating SPARTA for radiomic feature selection.

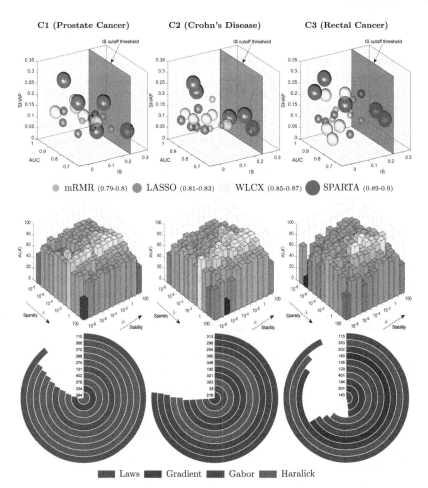

Fig. 2. Top row: 3D scatter plot of IS (X-axis), AUC (Y-axis), and SHAP (Z-axis) for 5 top-ranked features selected via each of mRMR, LASSO, WLCX, and SPARTA (in different colors). Spheres are sized based on the combined AUC value of each scheme. The IS screening threshold is visualized as a plane at IS = 0.25. Middle row: barplots of AUC values for SPARTA in the validation cohorts for each of C1, C2, and C3; when varying the regularization parameters μ (for stability) and λ (for sparsity). Bottom row: radial bar chart of 10 top-ranked features identified by SPARTA where outer to inner circles correspond to highest to lowest feature ranking. Shading of the perimeter denotes feature coefficient, while different colors correspond to different feature families.

values than the other FS schemes, indicating more informative features were selected via SPARTA. While a few SPARTA features do fall outside the stability screening threshold (visualized as a plane at IS = 0.25), their combination yields the highest AUC value compared to other approaches (denoted by the size of the spheres). The middle row in Fig. 2 shows a barplot of AUC values for SPARTA selected features at each combination of μ and λ when using the RF classifier on the validation set. The highest performance corresponds to $\mu = 0.1$ and $\lambda = 0.1$ with AUC $= 0.90, 0.91, 0.92$ in C1, C2, and C3, respectively. Note that increasing the weight of the stability term (via μ) results in an improved classifier performance in hold-out validation; albeit only for reasonably sparse feature sets (as governed by λ, similar to LASSO). The bottom row in Fig. 2 summarizes which feature families (different colors) comprise the top 10 radiomic features selected via SPARTA for each of C1, C2, and C3, via a radial bar chart where the perimeter of each circle is shaded based on their coefficient values. The top-ranked features in C1 primarily comprise Laws and Haralick features, while Gabor features largely comprise the top-ranked features in C2. An even mixture of Laws, Haralick, and Gabor features are seen to be top-ranked in C3.

4.3   Experiment 2: Comparative Strategies

Design: A RF classifier was used to evaluate the performance of different FS schemes (SPARTA, mRMR, LASSO, WLCX). In all cases, FS and classification was first done in an integrated setting within the discovery cohort over 100 iterations of 3-fold cross-validation. The best-performing model across all cross-validation runs was then evaluated in a hold-out fashion on the independent validation cohort for each of C1, C2, and C3. Classifier performance was evaluated via the area under the receiver-operator characteristic curve (AUC) in both the discovery (cross-validated) and validation (hold-out) cohorts. Pairwise Wilcoxon testing with multiple comparison correction was used to assess significant differences in feature set AUCs between SPARTA and each of the alternatives (mRMR, LASSO, WLCX); for each of C1, C2, and C3 separately. Computational efficiency was evaluated in terms of the average time (in seconds) taken to select features via each FS scheme across all cross-validation runs; reported for each cohort separately.

Results: Table 2 summarizes classifier performance for top-ranked radiomic features identified via each FS scheme being evaluated (SPARTA, stability-screened mRMR, LASSO, and WLCX). SPARTA yielded statistically significantly higher AUC values in each classification task in both discovery and hold-out validation, compared all 3 alternative strategies. This suggests the integration of feature stability directly into the selection scheme provides for optimal model performance even on new, unseen data in multi-institutional validation. SPARTA was also found to be more computationally efficient than both mRMR and WLCX (and marginally worse than LASSO) in all 3 cohorts.

Table 2. Classifier AUC and runtime (in seconds) evaluation of SPARTA feature set vs stability-screened mRMR, LASSO, WLCX-based feature sets; in distinguishing the 2 classes in each of C1, C2, and C3. Bolded values correspond to highest AUCs achieved in the discovery and validation sets for each cohort. * indicates $p < 0.005$ in pairwise Wilcoxon testing of SPARTA AUCs vs each of mRMR, LASSO, WLCX AUCs.

Set	Algorithm	C1 (Prostate cancer)	C2 (Crohn's Disease)	C3 (Rectal cancer)
Discovery	mRMR AUC	0.81 ± 0.11	0.80 ± 0.05	0.82 ± 0.01
	mRMR runtime	589 ± 8	351 ± 7	122 ± 4
	LASSO AUC	0.82 ± 0.06	0.81 ± 0.11	0.81 ± 0.12
	LASSO runtime	218 ± 4	165 ± 2	71 ± 1
	WLCX AUC	0.83 ± 0.06	0.85 ± 0.06	0.86 ± 0.07
	WLCX runtime	355 ± 5	193 ± 3	85 ± 2
	SPARTA AUC	**0.89 ± 0.02***	**0.90 ± 0.07***	**0.90 ± 0.14***
	SPARTA runtime	273 ± 3	182 ± 2	79 ± 3
Validation	mRMR	0.77	0.79	0.75
	LASSO	0.77	0.78	0.80
	WLCX	0.81	0.82	0.83
	SPARTA ($\mu = 0.1\ \lambda = 0.1$)	**0.91**	**0.91**	**0.93**

5   Concluding Remarks

In this study, we presented a novel radiomic feature selection scheme, SPARse sTable lAsso, (SPARTA), which represents a first effort at integrating feature stability as a generalizable regularization term directly into the optimization function used in feature selection. Radiomic features selected via SPARTA achieved significantly higher classification performance compared to popular feature selection strategies, across independent discovery and validation cohorts for (a) distinguishing benign from malignant regions in the peripheral zone of prostate T2w MRIs, (b) differentiating patients with and without Crohn's disease based on CT enterography scans, and (c) evaluating pathologic complete response from non-responsive rectal tumors on T2w MRIs. Future work will involve validating SPARTA as a feature selection strategy in other settings, incorporating additional variability measures into its formulation, as well as understanding its noise tolerance.

Acknowledgments. Research supported by NCI (1U24CA199374-01, 1R01CA2499 92-01A1, 1R01CA202752-01A1, 1R01CA208236-01A1, 1R01CA216579-01A1, 1R01CA 220581-01A1, 1R01CA257612-01A1, 1U01CA239055-01, 1U01CA248226-01, 1U54CA 254566-01, 1F31CA216935-01A1), NHLBI (R01HL15127701A1), NIBIB (1R43EB028 736-01), NCRR (1C06RR12463-01), DOD/CDMRP (W81XWH-19-1-0668, W81XWH-15-1-0558, W81XWH-20-1-0851, W81XWH-18-1-0440, W81XWH-20-1-0595, W81 XWH-18-1-0404, CA200789), VA (IBX004121A Merit Review Award), the KPMP Glue Grant, the Ohio Third Frontier Technology Validation Fund, the CTSC of Cleveland (UL1TR0002548), the Wallace H. Coulter Foundation Program in the Department of Biomedical Engineering at Case Western Reserve University, as well as sponsored research agreements from Bristol Myers-Squibb, Boehringer-Ingelheim, and Astrazeneca. Content solely responsibility of the authors and does not necessarily represent the official views of the NIH, USDVA, DOD, or the United States Government.

References

1. Parmar, C., Grossmann, P., Bussink, J., Lambin, P., Aerts, H.J.: Machine learning methods for quantitative radiomic biomarkers. Sci. Rep. **5**(1), 1–11 (2015)
2. Peng, H., Long, F., Ding, C.: Feature selection based on mutual information criteria of max-dependency, max-relevance, and min-redundancy. IEEE Trans. Pattern Anal. Mach. Intell. **27**(8), 1226–1238 (2005)
3. Tibshirani, R.: Regression shrinkage and selection via the lasso: a retrospective. J. R. Stat. Soc.: Ser. B (Stat. Methodol.) **73**(3), 273–282 (2011)
4. Wilcoxon, F., Katti, S., Wilcox, R.A.: Critical Values and Probability Levels for the Wilcoxon Rank Sum Test and the Wilcoxon Signed Rank Test. American Cyanamid Company, Pearl River (1963)
5. Traverso, A., Wee, L., Dekker, A., Gillies, R.: Repeatability and reproducibility of radiomic features: a systematic review. Int. J. Radiat. Oncol.* Biol.* Phys. **102**(4), 1143–1158 (2018)
6. Bologna, M., et al.: Assessment of stability and discrimination capacity of radiomic features on apparent diffusion coefficient images. J. Digit. Imaging **31**(6), 879–894 (2018)
7. Koo, T.K., Li, M.Y.: A guideline of selecting and reporting intraclass correlation coefficients for reliability research. J. Chiropractic Med. **15**(2), 155–163 (2016)
8. Li, L., Zeng, L., Lin, Z.J., Cazzell, M., Liu, H.: Tutorial on use of intraclass correlation coefficients for assessing intertest reliability and its application in functional near-infrared spectroscopy-based brain imaging. J. Biomed. Opt. **20**(5), 050801 (2015)
9. Leo, P., Elliott, R., Shih, N.N., Gupta, S., Feldman, M., Madabhushi, A.: Stable and discriminating features are predictive of cancer presence and Gleason grade in radical prostatectomy specimens: a multi-site study. Sci. Rep. **8**(1), 1–13 (2018)
10. van Timmeren, J.E., et al.: Test-retest data for radiomics feature stability analysis: generalizable or study-specific? Tomography **2**(4), 361 (2016)
11. Aerts, H.J., et al.: Decoding tumour phenotype by noninvasive imaging using a quantitative radiomics approach. Nat. Commun. **5**(1), 1–9 (2014)
12. Kim, M., et al.: Diffusion-and perfusion-weighted MRI radiomics model may predict isocitrate dehydrogenase (IDH) mutation and tumor aggressiveness in diffuse lower grade glioma. Eur. Radiol. **30**(4), 2142–2151 (2020)
13. Zou, H., Hastie, T.: Regularization and variable selection via the elastic net. J. R. Stat. Soc.: Ser. B (Stat. Methodol.) **67**(2), 301–320 (2005)
14. Simon, N., Friedman, J., Hastie, T., Tibshirani, R.: A sparse-group lasso. J. Comput. Graph. Stat. **22**(2), 231–245 (2013)
15. Sadri, A.R., Celebi, M.E., Rahnavard, N., Viswanath, S.E.: Sparse wavelet networks. IEEE Signal Process. Lett. **27**, 111–115 (2020)
16. Bejar, B., Dokmanic, I., Vidal, R.: The fastest l1, oo prox in the west. IEEE Trans. Pattern Anal. Mach. Intell. 1 (2021). https://doi.org/10.1109/tpami.2021.3059301
17. Parikh, N., Boyd, S.: Proximal algorithms. Found. Trends Optim. **1**(3), 127–239 (2014)
18. Friedman, J., Hastie, T., Tibshirani, R.: Regularization paths for generalized linear models via coordinate descent. J. Stat. Softw. **33**(1), 1 (2010)
19. Sohrab, H.H.: Basic Real Analysis. Birkhauser Basel (2014)
20. Saeedi, E., et al.: Radiomic feature robustness and reproducibility in quantitative bone radiography: a study on radiologic parameter changes. J. Clin. Densitomet. **22**(2), 203–213 (2019)

21. Chirra, P., et al.: Multisite evaluation of radiomic feature reproducibility and discriminability for identifying peripheral zone prostate tumors on mri. J. Med. Imaging **6**(2), 024502 (2019)

22. Hong, M., Luo, Z.Q.: On the linear convergence of the alternating direction method of multipliers. Math. Program. **162**(1–2), 165–199 (2017)

23. Li, H., Lin, Z.: Accelerated proximal gradient methods for nonconvex programming. Adv. Neural Inf. Process. Syst. **28**, 379–387 (2015)

24. Bolte, J., Sabach, S., Teboulle, M.: Proximal alternating linearized minimization for nonconvex and nonsmooth problems. Math. Program. **146**(1), 459–494 (2014)

25. Leek, J.T., et al.: Tackling the widespread and critical impact of batch effects in high-throughput data. Nat. Rev. Genet. **11**(10), 733–739 (2010)

26. Leo, P., Lee, G., Shih, N.N., Elliott, R., Feldman, M.D., Madabhushi, A.: Evaluating stability of histomorphometric features across scanner and staining variations: prostate cancer diagnosis from whole slide images. J. Med. Imaging **3**(4), 047502 (2016)

27. Pérez, E., Reyes, O., Ventura, S.: Convolutional neural networks for the automatic diagnosis of melanoma: an extensive experimental study. Med. Image Anal. **67**, 101858 (2021)

28. Shapley, L.S.: A value for n-person games. Contrib. Theory Games **2**(28), 307–317 (1953)

The Power of Proxy Data and Proxy Networks for Hyper-parameter Optimization in Medical Image Segmentation

Vishwesh Nath[✉], Dong Yang, Ali Hatamizadeh, Anas A. Abidin,
Andriy Myronenko, Holger R. Roth, and Daguang Xu

NVIDIA, Santa Clara, USA

Abstract. Deep learning models for medical image segmentation are primarily data-driven. Models trained with more data lead to improved performance and generalizability. However, training is a computationally expensive process because multiple hyper-parameters need to be tested to find the optimal setting for best performance. In this work, we focus on accelerating the estimation of hyper-parameters by proposing two novel methodologies: proxy data and proxy networks. Both can be useful for estimating hyper-parameters more efficiently. We test the proposed techniques on CT and MR imaging modalities using well-known public datasets. In both cases using one dataset for building proxy data and another data source for external evaluation. For CT, the approach is tested on spleen segmentation with two datasets. The first dataset is from the medical segmentation decathlon (MSD), where the proxy data is constructed, the secondary dataset is utilized as an external validation dataset. Similarly, for MR, the approach is evaluated on prostate segmentation where the first dataset is from MSD and the second dataset is PROSTATEx. First, we show higher correlation to using full data for training when testing on the external validation set using smaller proxy data than a random selection of the proxy data. Second, we show that a high correlation exists for proxy networks when compared with the full network on validation Dice score. Third, we show that the proposed approach of utilizing a proxy network can speed up an AutoML framework for hyper-parameter search by 3.3×, and by 4.4× if proxy data and proxy network are utilized together.

Keywords: Segmentation · AutoML · Hyper-parameter optimization · Proxy · Deep learning

1 Introduction

Data-driven methods have become the main approach for medical image segmentation based tasks for most imaging modalities such as computed tomography (CT) and magnetic resonance imaging (MRI) [8,14]. At the same time,

© Springer Nature Switzerland AG 2021
M. de Bruijne et al. (Eds.): MICCAI 2021, LNCS 12903, pp. 456–465, 2021.
https://doi.org/10.1007/978-3-030-87199-4_43

the drive for growing annotated data has been accelerating at an exponential rate [15]. However, the performance of deep learning [6] methods is critically dependent upon the hyper-parameters (learning rate, optimizers, augmentation probabilities, etc.) that are utilized while training the model. Hyper-parameter optimization (HPO) [1] is actively being researched by state-of-the-art (SOTA) approaches such as AutoML [3,10]. However, AutoML is usually an extremely computationally expensive process. Furthermore, the computational cost is exacerbated by medical imaging tasks where the data is often high-dimensional, i.e. 3D volumes.

Our work focuses on reducing the computational expense of HPO and AutoML approaches. In this paper, we propose the construction of a proxy dataset, which is representative of the full dataset yet relatively much smaller in size (see Fig. 1). Furthermore, we also propose the usage of proxy models, which are essentially smaller networks, yet representative of the larger/complete network structure. Utilizing the proposed proxy data and proxy networks, we show that computational burden of AutoML can be drastically reduced *from a few days to a few hours* and yet be able to estimate hyper-parameters that can lead to SOTA performance. For a robust validation of the approach of proxy data and proxy networks we utilize external validation data (completely withheld datasets) for testing of the trained models from the primary datasets. The approach is tested on CT and MR imaging modalities.

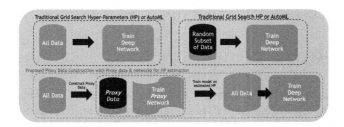

Fig. 1. The proposed pipeline of utilizing proxy data and proxy network to accelerate hyper-parameter estimation. Top row: Traditionally, hyper-parameters are either estimated on the full dataset or on a small subset of randomly selected data. Bottom row: Combining proxy networks with proxy data directly lead to speed ups in estimating hyper-parameters.

2   Related Work

Proxy Data and Networks: The term *proxy* implies computationally reduced settings [2]. Prior work has shown that proxy datasets can be useful in AutoML, specifically for sub-domains of Neural Architecture Search (NAS) and HPO [9,12]. However, they have only been tested for classification tasks by reduction of class labels or mining of easy and hard samples [12]. Similarly, proxy networks have been associated only with classification tasks [2]. To the best of our knowledge, there is no previous work which has proposed proxy data and proxy networks for medical image segmentation.

AutoML: Since the inception of AutoML [3], it has become the best way to estimate hyper-parameters and is effectively better than grid search. The prior literature indicates the popularity of AutoML [3,10] and it has also been effectively introduced for medical image segmentation [16,17]. While AutoML methods exist, their practical feasibility is of critical concern due to their heavy computational expense, which is a primary motivation for our work.

The proxy data and proxy networks serve as useful and much needed indicators of performance that can drastically speed up the estimation of hyper-parameters while reducing the computational expense.

2.1   Contributions

1. We propose the construction of proxy datasets via classical metrics of mutual information and normalized cross correlation. Our work also shows that the choice of metric needs to be paired with the spatial context of the task.
2. We show that proxy networks can be constructed systematically by reduction of residual blocks, channels and reducing number of levels in the U-net.
3. We show that our methods can be generalized across different imaging modalities, via evaluation on external validation datasets from which no data were utilized for training.

3   Method

Proxy Data Selection Strategy: Consider a dataset D containing a set of datapoints $\{x_1, x_2, ..x_n\}$, where x_i is a single data sample. To estimate the importance of a single datapoint x_i, we estimate its utility in relation to other datapoints x_j, resulting in a set of paired measures. An example of pairs for x_1 would be $\{(x_1, x_1), (x_1, x_2), (x_1, x_3)..(x_1, x_n)\}$. The mean of the measure is utilized as an indicator of the importance of the datapoint. There are multiple methods that can provide pair-wise measurements of the data. We explore mutual information (MI) (Eq. 1) on flattened vectors of the 3D images and normalized local cross-correlation (NCC) (Eq. 2) in local window size of $(9, 9, 9)$ [18] for each pair of data (x_i, x_j) as different variants.

$$\mathcal{MI}(x_i, x_j) = \sum_{x_i} \sum_{x_j} P(x_i, x_j) log \frac{P(x_i, x_j)}{P(x_i)P(x_j)}. \tag{1}$$

Here $P(x_i)$ & $P(x_j)$ are the marginal probability distributions while $P(x_i, x_j)$ is the joint probability distribution.

$$\mathcal{NCC}(x_i, x_j) = \frac{1}{\Omega} \sum_{p \in \Omega} \frac{(\sum_{p_i} (x_i(p_i) - \overline{x_i(p)})(x_j(p_i) - \overline{x_j(p)}))^2}{\sum_{p_i} (x_i(p_i) - \overline{x_i(p)})^2 \sum_{p_i} (x_j(p_i) - \overline{x_j(p)})^2} \tag{2}$$

Here, p_i is the 3D voxel position within a window around p and $\overline{x_i(p)}$ and $\overline{x_j(p)}$ are local means within the window surrounding the voxel position p_i in x_i, x_j correspondingly. Ω is the voxel coordinate space.

Task-Specific Region of Interest: The acquisition parameters (number of slices, resolution etc.) for different 3D volume scans vary. Therefore, when considering a pair (x_i, x_j), even if the x_i is re-sampled to x_j image size, there is misalignment for the region of interest (ROI) (the organ to be annotated by the model). Hence, we utilize only the task-specific ROI by utilizing the information from the existing label. The selected volume is cropped using the ROI and re-sampled to a cubic patch size.

The data points are ranked by their importance and the ones containing the lowest mutual information or lowest correlation are selected within a given budget B.

Proxy Network: U-net has become the go-to model for medical image segmentation tasks especially when deep learning methods are being utilized [4,11]. There are many variants of U-net that were proposed, we use a 5 level U-net with 2 residual blocks per level and skip connections between encoder and decoder blocks (this is our full model). The coarse hyper-parameters for a U-net are therefore: number of channels in first encoder block (successive encoder blocks are multiples of 2), number of residual blocks and the number of levels. To create a proxy network, we first reduce the number of channels to 4 and decrease the residual blocks from 2 to 1. The variants of proxy network are created by decreasing the number of levels to 5, 4 & 3 (can also be thought as reducing the number of encoding and decoding blocks).

AutoML: We use a recurrent neural network (RNN) that is part of the overarching reinforcement learning (RL) framework to estimate hyper-parameters [16].

4 Experiments and Data

4.1 Datasets

CT: The Spleen task dataset from MSD [13] with the segmentation annotation were used. Data was re-sampled to a resolution of $1.5 \times 1.5 \times 2.0$ mm^3 and the intensities were normalized in a HU window of $[-57, 164]$. Random patches of $96 \times 96 \times 96$ with or without the labels were used for training. For inference during validation and testing a patch size of $160 \times 160 \times 160$ was utilized. All 41 volumes were utilized for training and validation for ground truth.

External validation based testing was done on all the 30 volumes from Beyond the Cranial Vault (BTCV) challenge from MICCAI 2015 [5]. The pre-processing, patch-size for inference were kept consistent as with the first dataset. Please note that all other labels were ignored from BTCV as it is a multi-organ annotated dataset.

MR: The Prostate task from MSD [13] with the segmentation annotation were used. The original task includes the labels of transition zone and peripheral zone separately, for this work these two were combined to a single class, for consistency with the secondary dataset. Data was pre-processed using a resolution of $1.0 \times 1.0 \times 1.0 \ mm^3$, intensities were normalized, spatially padded to ensure consistency, random patches of $128 \times 128 \times 48$ were selected with or without the label using a ratio. For inference a patch size of $160 \times 160 \times 160$ were used. All 32 volumes were utilized for training and validation for ground truth.

External validation based testing was done on all 98 volumes from the PROSTATEx dataset [7]. The pre-processing, patch-size for inference were kept consistent as with the first dataset.

For all pre-processing and training of deep learning models, the MONAI library[1] was used.

4.2   Experimental Design

Hyper-parameter Space: We explore a hyper-parameter space of four different optimizers $\phi = \{$Adam, RMSProp, Adamax, Novograd$\}$ at varying learning rates in the set $\delta = \{0.001, 0.0006, 0.0004, 0.0001\}$ for validation of the proxy data selection strategy.

Ground Truth: The ground truth performance of different hyper-parameters was estimated on 7 random splits of training and validation for both MR and CT MSD datasets. For MSD Spleen dataset, training and validations splits are of 32 and 9 sample size. Similarly, for MSD Prostate, sample sizes were 26 and 6. Please note that no data samples from the secondary datasets were used for training. They were only used for testing as external validation datasets.

We aim to answer the following questions with our experimental design.

Which Proxy Dataset Construction Method is the Best and How Many Data Points are Sufficient for a Good Proxy? The metrics of MI and NCC with or without combinations of task-specific ROI selection are tested across 7 repeats (different initialization of the network, different data splits) where the selected B datapoints are randomly split into training set (50%) & validation set (50%). We also test the baseline of random selection of data. All data selection techniques are tested with $B = \{4, 6, 8$ and $10\}$ data points in total. To evaluate the best method, we study the correlation (Pearson correlation coefficient) of the Dice scores estimated on the secondary validation dataset by training on proxy data versus the Dice scores of the same validation set but trained with all training data (the "full" model). Meanwhile, through an ablative study, we show how much proxy data is needed for it to be a good proxy representing the entire dataset in terms of ground truth Dice scores.

How Shallow Can a Proxy Network Be? To evaluate, how small of a proxy network can be used, we systematically decrease the number of levels of the U-Net from 5 to 3 at steps of 1. The Pearson correlation is studied across seven

---

[1] https://monai.io.

different splits of data which are trained on seven different initializations of the network for proxy networks versus the full model.

Will Proxy Data and Proxy Network Lead to Relatively Closer Hyper-parameters When Estimated on All Data and a Full Model? We utilize the RL AutoML technique which is used to estimate hyper-parameters. Learning rate and the probability for the augmentation of a random shift in intensity are estimated via AutoML. The relative distance between the estimated parameters from proxy data and proxy networks is compared with hyper-parameters searched with a full model and all data. We also show comparisons when the hyper-parameters are estimated with a random selection of data. Once the hyper-parameters are estimated using a proxy technique, a full model with all data is trained for evaluation purposes.

5   Results

Proxy Data: It should be noted that the test Dice results should be compared relatively as they have been trained on reduced datasets and the performance is not expected to reflect the highest Dice scores. The best test Dice score was selected from the hyper-parameter search space for the ground truth and the corresponding settings were used for all the data selection methods (listed in Table 1). The method *MI+Labelcrop* (Labelcrop is referred to as the selected ROI based on the label) shows the highest test Dice score as compared to all other baselines and random selection of data for both spleen and prostate. With 24% of data with the proxy data selection method can achieve up to 90% of the performance of the full dataset.

Observing across the entire hyper-parameter space the proxy data selection method *MI+Labelcrop* shows a higher correlation of 0.37 versus 0.32 as compared to random selection of data on the external validation when being compared

Table 1. Summarized mean Dice score for internal validation of MSD spleen, external validation of BTCV, internal validation of MSD prostate, external validation of PROSTATEx across 7 random splits of selected data. NCC represents normalized cross-correlation and MI represents Mutual information. Spleen Dice scores are reported with 23% usage of full dataset for training. Prostate Dice scores are reported with 31% usage of full dataset

	CT—Spleen		MRI—Prostate	
Selection method	MSD Spln Dice	BTCV Dice	MSD Prst Dice	PRSTx Dice
NCC	0.8804 ± 0.0314	0.8012 ± 0.0247	0.6180 ± 0.1044	0.4059 ± 0.1326
NCC+Labelcrop	0.9223 ± 0.0194	0.6579 ± 0.0681	0.6647 ± 0.0955	0.4165 ± 0.1190
MI	0.8814 ± 0.0405	0.7446 ± 0.0469	0.5365 ± 0.0792	0.3401 ± 0.0706
MI+Labelcrop	0.8580 ± 0.0440	**0.8173 ± 0.0184**	0.6312 ± 0.0383	**0.5567 ± 0.0754**
Random baseline	0.8282 ± 0.0559	0.7678 ± 0.0391	0.5619 ± 0.1318	0.4298 ± 0.1406
Full Dataset	0.9450 ± 0.0106	0.8844 ± 0.0037	0.8534 ± 0.0121	0.7423 ± 0.0497

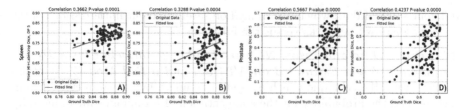

Fig. 2. Across 7 splits with different seeds. A) The "upper bound" Dice of BTCV external dataset is reported when trained on all MSD data. This is compared with Dice on BTCV when trained on a proxy dataset across the hyper-parameter space. B) Similarly, random data is used instead of proxy data. 24 % of full data were used respectively for A) & B). C) The "upper bound" Dice on PROSTATEx is reported when trained on all MSD data. This is compared with Dice on PROSTATEx when trained on a proxy dataset across the hyper-parameter space. D) Similarly, random data is used instead of proxy data. 31% of all data were used respectively for C) & D).

with the ground truth which is obtained by training with all the data (shown in Fig. 2).

Similarly, for prostate segmentation, the best test Dice score against the ground truth across the seven repeats of the hyper-parameter optimization was used to detect the best hyper-parameter setting. The test Dice scores reported in (listed in Table 1) belong to the same hyper-parameter setting. The best performing data selection method is *MI+Labelcrop* as also for Spleen. The correlation for *MI+Labelcrop* is in the high range as compared to random which is in the moderate range of correlation.

How Much Data is Proxy Enough? It can be observed that for spleen the cross-over for proxy enough begins to start showing at approximately 15% of data being used (Fig. 4A & 4B). A similar observation can be made for prostate as well.

Proxy Networks: Across the 7 different splits of data when the validation Dice score is compared for a proxy network versus the full network, a decreasing trend of correlation can be seen (shown in Fig. 3 top row). Given that the channels were fixed at 4 and only a single residual block was used, the correlation is highest for 5 levels in the U-net. Decreasing the number of levels of U-Net decreases the correlation. For all 3 variants a high degree (>0.5) of correlation is shown, suggesting that even smaller proxy networks could be utilized.

A similar observation as for spleen can be assessed for prostate across the 7 different splits of data. The validation Dice score is compared for a proxy network versus the full network, a decreasing trend of correlation can be seen (shown in Fig. 3 bottom row).

AutoML: The estimated hyper-parameters by RL using proxy data and proxy networks are closer to the full model with all data relatively as compared to

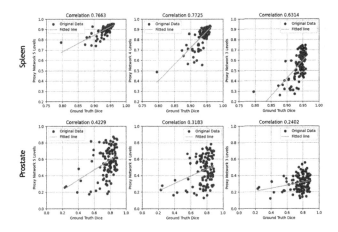

Fig. 3. Top row (Spleen): Proxy network correlation when the Dice score on the internal validation of the ground truth versus Dice score from the proxy network across the full data. From left to right reduce the number of levels of the U-net from 5 to 3 (Left to right corresponding levels are 5,4 and 3). All proxy U-nets have 4 channels and one residual block per level. Bottom row (Prostate): Number of levels of U-net are varied and the correlation is compared on the internal validation for prostate data similarly as for spleen.

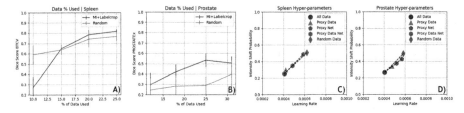

Fig. 4. A): For spleen BTCV dataset Dice score is plotted versus data usage. B): For PROSTATEx dataset Dice score is plotted versus data usage. C) For spleen the estimated hyper-parameter of learning rate and probability of intensity shift are shown with relative distance to the ground truth. D) Similarly for prostate

when using a random subset of data with the full model for spleen (Fig. 4C & 4D). A similar observation can be made for prostate.

For prostate, all proposed proxy methods perform similarly or higher (proxy data and network) vs. random selection of data or when all data is utilized for hyper-parameter estimation (Table 2). A similar observation can be made for spleen where proxy data provides the best results. The other methods perform similarly. Overall, a runtime improvement of 4.4× for spleen and 3.3× for prostate can be observed.

Table 2. Summarized mean Dice score for internal validation of MSD and external validation of PROSTATEx and BTCV across 5 repeats with the same split. The estimated hyperparameters (HP) are used to train full models with all data from MSD for comparison.

HP from	All data	Proxy data	Proxy net	Proxy data&Net	Random data
Prostate					
MSD Dice	0.8763 ± 0.02143	0.8798 ± 0.0182	0.8764 ± 0.01883	**0.8812 ± 0.0087**	0.8204 ± 0.0346
PRSTx Dice	0.7465 ± 0.0309	0.7462 ± 0.0511	0.7493 ± 0.0334	**0.7701 ± 0.0249**	0.7134 ± 0.0353
GPU Hours	772	480	432	**248**	480
Spleen					
MSD Dice	0.9532 ± 0.0010	**0.9546 ± 0.0011**	0.9537 ± 0.0010	0.9536 ± 0.0011	0.9534 ± 0.0011
BTCV Dice	0.8780 ± 0.0103	**0.8821 ± 0.0085**	0.8763 ± 0.0087	0.8798 ± 0.0048	0.8780 ± 0.0104
GPU Hours	1056	320	282	**240**	320

6   Discussion and Conclusions

We show that proxy data and proxy networks are a powerful tool to speed up the HPO estimation process. The results indicate that a maximum speedup of 4.4× can be obtained which can reduce days to a few hours. As a limitation the pairwise distance measures lead to a squared run-time, however the squared runtime is feasible for datasets in the size of thousands, which so far is uncommon in medical imaging segmentation tasks. While this work is a first step towards utilization of proxy techniques for basic hyper-parameter estimation in medical image segmentation, in future we plan to extend it for estimation of multiple hyper-parameters. The benefits can be extended towards multiple frameworks such as neural architecture search and federated learning which are both resource-critical settings.

References

1. Bergstra, J., Bengio, Y.: Random search for hyper-parameter optimization. J. Mach. Learn. Res. **13**(2) (2012)
2. Coleman, C., et al.: Selection via proxy: efficient data selection for deep learning. arXiv preprint arXiv:1906.11829 (2019)
3. Hutter, F., Kotthoff, L., Vanschoren, J.: Automated Machine Learning: Methods, Systems, Challenges. Springer, Heidelberg (2019). https://doi.org/10.1007/978-3-030-05318-5
4. Isensee, F., Jäger, P.F., Kohl, S.A., Petersen, J., Maier-Hein, K.H.: Automated design of deep learning methods for biomedical image segmentation. arXiv preprint arXiv:1904.08128 (2019)
5. Landman, B., Xu, Z., Igelsias, J., Styner, M., Langerak, T., Klein, A.: MICCAI multi-atlas labeling beyond the cranial vault-workshop and challenge. In: Proceedings of MICCAI Multi-Atlas Labeling Beyond Cranial Vault-Workshop Challenge (2015)
6. LeCun, Y., Bengio, Y., Hinton, G.: Deep learning. Nature **521**(7553), 436–444 (2015)

7. Litjens, G., Debats, O., Barentsz, J., Karssemeijer, N., Huisman, H.: Computer-aided detection of prostate cancer in MRI. IEEE Trans. Med. Imaging **33**(5), 1083–1092 (2014)

8. Litjens, G., et al.: A survey on deep learning in medical image analysis. Med. Image Anal. **42**, 60–88 (2017)

9. Park, M.: Data proxy generation for fast and efficient neural architecture search. arXiv preprint arXiv:1911.09322 (2019)

10. Real, E., Liang, C., So, D., Le, Q.: AutoML-zero: evolving machine learning algorithms from scratch. In: International Conference on Machine Learning, pp. 8007–8019. PMLR (2020)

11. Ronneberger, O., Fischer, P., Brox, T.: U-net: convolutional networks for biomedical image segmentation. In: Navab, N., Hornegger, J., Wells, W.M., Frangi, A.F. (eds.) MICCAI 2015. LNCS, vol. 9351, pp. 234–241. Springer, Cham (2015). https://doi.org/10.1007/978-3-319-24574-4_28

12. Shleifer, S., Prokop, E.: Using small proxy datasets to accelerate hyperparameter search. arXiv preprint arXiv:1906.04887 (2019)

13. Simpson, A.L., et al.: A large annotated medical image dataset for the development and evaluation of segmentation algorithms. arXiv preprint arXiv:1902.09063 (2019)

14. Tajbakhsh, N., Jeyaseelan, L., Li, Q., Chiang, J.N., Wu, Z., Ding, X.: Embracing imperfect datasets: a review of deep learning solutions for medical image segmentation. Med. Image Anal. **63**, 101693 (2020)

15. Wiesenfarth, M., et al.: Methods and open-source toolkit for analyzing and visualizing challenge results. Sci. Rep. **11**(1), 1–15 (2021)

16. Yang, D., Roth, H., Xu, Z., Milletari, F., Zhang, L., Xu, D.: Searching learning strategy with reinforcement learning for 3D medical image segmentation. In: Shen, D., et al. (eds.) MICCAI 2019. LNCS, vol. 11765, pp. 3–11. Springer, Cham (2019). https://doi.org/10.1007/978-3-030-32245-8_1

17. Yu, Q., et al.: C2FNAS: coarse-to-fine neural architecture search for 3d medical image segmentation. In: Proceedings of the IEEE/CVF Conference on Computer Vision and Pattern Recognition, pp. 4126–4135 (2020)

18. Zhu, W., et al.: NeurReg: neural registration and its application to image segmentation. In: Proceedings of the IEEE/CVF Winter Conference on Applications of Computer Vision, pp. 3617–3626 (2020)

Fighting Class Imbalance
with Contrastive Learning

Yassine Marrakchi[1,2(✉)], Osama Makansi[1], and Thomas Brox[1,2]

[1] Department of Computer Science, University of Freiburg,
Freiburg im Breisgau, Germany
{marrakch,makansio,brox}@cs.uni-freiburg.de
[2] CIBSS – Centre for Integrative Biological Signalling Studies,
University of Freiburg, Freiburg im Breisgau, Germany

Abstract. Medical image datasets are hard to collect, expensive to label, and often highly imbalanced. The last issue is underestimated, as typical average metrics hardly reveal that the often very important minority classes have a very low accuracy. In this paper, we address this problem by a feature embedding that balances the classes using contrastive learning as an alternative to the common cross-entropy loss. The approach is largely orthogonal to existing sampling methods and can be easily combined with those. We show on the challenging ISIC2018 and APTOS2019 datasets that the approach improves especially the accuracy of minority classes without negatively affecting the majority ones.

Keywords: Imbalance classification · Medical imaging · Contrastive learning

1 Introduction

Convolutional networks (CNN) have much to offer for computer-aided diagnostics (CAD) as they can potentially bring cheap pre-screening to people, who do not have regular access to medical experts, or they can decrease the screening intervals even for those who have this access. Several domain specific issues in medical image processing, like data scarcity, noisy labels, and low image quality, have been addressed. Another issue, yet with less attention for far, is the often imbalanced distribution of classes in medical datasets. Some classes are much more common than others, hence it is difficult to collect a dataset where all classes are represented equally. For example, among retinal diseases, diabetic retinopathy is more common than fundus pulverulentus and fundus albipunctatus [41]. In Fig. 1, we show the histogram of the classes in the ISIC dataset, a dermatoscopic dataset of common pigmented skin lesions, where few major classes have orders of magnitude larger frequency than others. Neural networks

This study was supported by the Excellence Strategy of the German Federal and State Governments, (CIBSS - EXC 2189).

M. de Bruijne et al. (Eds.): MICCAI 2021, LNCS 12903, pp. 466–476, 2021.
https://doi.org/10.1007/978-3-030-87199-4_44

trained on imbalanced datasets tend to perform worse on the minority classes – a problem that is well-known in machine learning and has been studied thoroughly in the last few years.

Few recent works in the medical domain apply the common technique of resampling the training data [44] or reweighting the loss function to give more attention to minority classes [2,44], hence improving their accuracy. Despite the marginal success of these methods, a thorough analysis of the problem and an effective approach towards a solution is still missing. In addition to the imbalance problem, medical image classification typically focuses on subtle, fine-grained differences. In Fig. 1, we show three samples from different classes which are hard to distinguish due to their subtle differences.

In this paper, we address the abovementioned problems by explicitly separating the feature space into different clusters by minimizing the distance between samples from the same class (intra-class) and maximizing the distance between samples from different classes (inter-class). We achieve this using a contrastive loss on the learned feature space. With this approach, the minority samples receive enough attention without negatively affecting performance on majority classes.

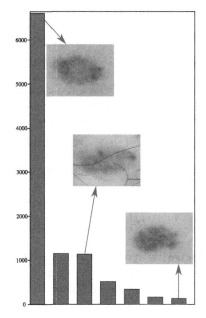

Fig. 1. Histogram of the ISIC dataset showing the majority (left) and minority classes (right). We also show a sample from three classes. These look very similar, yet should be classified differently.

In summary, we (1) emphasize the issue of imbalanced datasets in the medical domain, (2) propose a framework based on contrastive learning to better arrange the feature space for minority and majority classes, (3) show quantitatively on the challenging ISIC2018 [7] and APTOS2019 [1] datasets that our approach outperforms or performs on parwith existing techniques on all metrics, and (4) discuss the complementary of our method to existing techniques (resampling).

2   Related Work

Classification in Medical Imaging. Disease diagnosis and disease grading are major applications of computer-aided diagnosis that benefited from the flourishing era of medical image analysis [9]. The recent improvement is mainly due to the emergence of deep learning techniques at large scale as recent CAD methods are driven by learning-based classification methods employing neural networks [4]. Although adopting standard methods, such as finetuning pretrained networks, using ensembles and heavy data augmentation, led to decent improvement, more pronounced and domain specific issues have been tackled to gain further quantitative improvement. For instance, fighting against hard examples [14], inter-class correlation [42], and small inter-class differences [45], domain shift [13], catastrophic forgetting [29] and annotation scarcity [17,28] have been deeply studied in the last year and several new methods based on semi-supervised learning and meta-learning have been introduced. Unlike data for common diseases, which is available in large amounts with typically clean labels, it is hard to collect as much annotated data for rare diseases. As data for majority classes is not supposed to be thrown away (a.k.a. undersampling), this leads to a systematic data imbalance. Hence, training a neural network to classify all diseases at comparable accuracy is challenging, and so far, has seen very little attention in the medical imaging community. For instance, the winners of the ISIC-2018 challenge handled the data imbalance with a simple loss reweighting scheme [2,44]. The problem has been also implicitly tackled by splitting the data into two subsets consisting of majority classes and minority classes [28]. After training the network on the majority classes, it is finetuned to recognize the minority classes as well.

Learning from Imbalanced Datasets. There is a long line of works addressing the task of learning from datasets with class-imbalance. The most common technique is to manipulate the training data distribution by oversampling the data from minority classes [36,38] or undersampling the majority classes [12,20]. To avoid potential overfitting on the minority, recent methods proposed to generate new data from minority classes by simulation [5,18,26,33]. Instead of changing the data distribution, other works proposed to introduce a weight in the loss of samples based on their class frequency [8,22,31,39] or optimize a loss that is not sensitive to the class imbalance [15,35]. Recently, Kang et al. [23] proposed the two-stage approach of imbalanced feature learning and balanced classifier learning. In the latter, they use the common oversampling technique [38] for training the classifier. Others proposed to shape the feature space by explicitly designing a loss function to increase the inter-class distance while decreasing the intra-class one [22,43]. Recent works follow up on this idea by explicitly enlarging the margin for the minority to boost their performance [3,10,19,24]. We compare to the most relevant of these recent methods in Sect. 4.4.

Contrastive Learning. The concept of contrastive learning was first proposed to learn better feature representation in a self-supervised manner and was referred to as noise-contrastive learning [6,11,16,34]. In particular, for any input sample, a set of positives and negatives are constructed and the loss function is optimized to separate them. The positives are noisy versions of the input sample (e.x, augmentation) and the negatives are the other samples in the batch. Recently, [25] extended the positives to also include images from the same class yielding impressive results in image classification on ImageNet [37]. Despite the success of contrastive learning in both self-supervised and supervised settings, we are the first to adapt this concept in the task of learning from imbalance datasets.

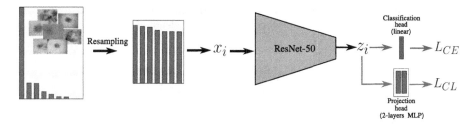

Fig. 2. Two-stages Framework. In the first stage, we learn the feature space via the contrastive loss (orange) by projecting the feature vector using 2-layers MLP. In the second stage, we through the projection head, freeze the backbone and only learn the classification head (blue). (Color figure online)

3   Supervised Representation Disentanglement via Contrastive Learning

Representation learning aims to disentangle the feature space, such that the resulting embedding can be mapped well to classes. However, when datasets are highly imbalanced, networks focus on the majority classes, as they dominate the loss. In order to overcome this limitation, we propose to separate the minority from the majority classes in feature space with contrastive learning. In particular, for every sample x_i in the batch, its positive samples, i.e., samples from the same class in the batch, are pushed closer while the negatives ones, i.e. all samples from different classes in the batch, are pushed away. Formally:

$$L_i^{CL} = -\frac{1}{N_{\mathbf{c}_i} - 1} \sum_{j \in \mathbf{c}_i} \mathbb{1}_{i \neq j} \cdot \log \frac{\exp(\mathbf{z}_i \cdot \mathbf{z}_j / \tau)}{\sum_{k=1}^{N} \mathbb{1}_{i \neq k} \cdot \exp(\mathbf{z}_i \cdot \mathbf{z}_k / \tau)}, \tag{1}$$

where $\mathbf{z}_i$ is the normalized feature vector of an image x_i, $\mathbf{c}_i$ is the positive set for sample i and corresponds to all samples in the batch with the same class label as i. $\mathbb{1}_{cond}$ is the indicator function that returns 1 if the condition *cond* returns true and 0 otherwise. N is the total number of samples in the batch, and $N_{\mathbf{c}_i}$ the size of the positive set of sample x_i. $\tau > 0$ is the temperature parameter, which controls the balance between positives and negatives. In the standard setting, cross-entropy is a function of the pseudo-probabilities generated by the Softmax function and the labels for each input image separately. Thus the resulting gradients for each single input depend solely on the considered image. Given the data unbalance, the network will prioritize learning major classes. Using contrastive loss mitigates this effect by computing the pseudo-probabilities based on the cosine similarity between feature embeddings of the full batch. Consequently, gradients depend on the feature embedding of all batch elements and minimizing the loss **explicitly** moves feature embeddings to form clusters.

After learning the feature space using the contrastive loss, a second stage is needed to learn to map from the feature space to the target class labels. To this end, we use the common cross-entropy loss:

Table 1. Quantitative evaluation on the ISIC2018 dataset. Our method using contrastive learning yields the best performance. Results denoted with [†] are taken from the respective papers.

	Accuracy	F-score
CE	$0.850 \pm 0.7e - 3$	$0.716 \pm 6e - 3$
Focal Loss [31]	$0.849 \pm 4.5e - 3$	$0.728 \pm 5e - 3$
LDAM [3]	$0.857 \pm 9.0e - 3$	$0.734 \pm 24e - 3$
CL (ours)	$\mathbf{0.865} \pm 4.0e - 3$	$\mathbf{0.739} \pm 13e - 3$
CE + resample	$0.861 \pm 0.9e - 3$	$0.735 \pm 6e - 3$
CL (ours) + resample	$\mathbf{0.868} \pm 7.0e - 3$	$\mathbf{0.751} \pm 16e - 3$
OHEM [40]	0.818 [†]	0.660 [†]
MTL [30]	0.811 [†]	0.667 [†]
DANIL [14]	0.825 [†]	0.674 [†]

$$L_i^{CE} = - \sum_{j=1}^{M} y_{i,j} \log p_{i,j} \, , \tag{2}$$

where M is the number of classes, $y_{i,j}$ is an indicator if j is the class label for the sample i, and $p_{i,j}$ is the predicted probability that the sample i is classified as class j. When training the second stage, we freeze the learned backbone and only learn the classification head as illustrated in Fig. 2. As the cross-entropy loss is not sensitive to the class distribution, it is important to sample the classes uniformly across batches as suggested by [23].

In Sect. 4.4, we compare and combine our approach with an oversampling strategy during the backbone training. We follow a simple oversampling strategy by extending the original data set with copies of samples from minority classes so that the artificially balanced dataset has the exact same number of items per class. For the second stage we always use artificially balanced data by following this simple scheme.

4   Experiments

4.1   Experimental Setup

We evaluate the proposed method on the ISIC2018 lesion diagnosis dataset [7] which consists of 10015 skin lesion images and 7 predefined categories and APTOS2019 [1] for diabetic retinopathy which has 5 classes and 3662 images. We split images randomly to a train and test set with a ratio of 7:3 as in [14]. Beside the average accuracy which is very sensitive to data imbalance, we report the F-score (also known as Dice similarity coefficient) which is the average of the classwise harmonic mean of precision and recall. Since the classwise F-score is normalized, this metric is particularly sensitive to the performance on minority classes. To evaluate the stability of our method, we report the means and standard deviations over 3 independent runs.

Table 2. Quantitative evaluation on the APTOS2019 dataset. Contrastive learning clearly outperforms the CE baseline and is on-par with the state of the art on this dataset. Results denoted with † are taken from the respective papers.

	Accuracy	F-score
CE	$0.812 \pm 7e-3$	$0.608 \pm 18e-3$
Focal Loss [31]	$0.815 \pm 1e-3$	$0.629 \pm 8e-3$
LDAM [3]	$0.813 \pm 3e-3$	$0.620 \pm 5e-3$
CL (ours)	$\mathbf{0.825} \pm 1e-3$	$\mathbf{0.652} \pm 3e-3$
CE + resample	$0.802 \pm 23e-3$	$0.583 \pm 55e-3$
CL (ours) + resample	$0.816 \pm 1e-3$	$0.608 \pm 4e-3$
CANet [27]	0.813 †	0.631 †
OHEM [40]	0.813 †	0.632 †
DANIL [14]	$\mathbf{0.825}$ †	$\mathbf{0.660}$ †

4.2 Baselines

We compare the proposed method to previous works that use the same experimental setup: **OHEM** [40]: a hard example mining method that samples training images according to a non-uniform distribution based on their current loss. **MTL** [30]: a deep multi-task learning framework that optimizes body location classification along with the skin lesion classification. **CANet** [27]: a CNN with an attention module to learn disease-specific features and a disease-dependent attention module to learn internal relationships between pairs of diseases. **DANIL** [14]: a method that synthesizes distractors in the feature space to learn stronger features. We also compare to popular model-based methods to learn from unbalanced data. Beside **focal loss** [31], a sophisticated way to balance the loss depending on the class frequency, we consider **LDAM** [3], a label-distribution-aware margin loss that applies stronger regularization to minority classes to improve the generalization of the model. We further investigate the effect of **Resampling**, which ensures having the same number of images per class per epoch by sampling images from the minority classes multiple times within the same epoch. We apply this method in combination with both the standard network training and with contrastive learning.

4.3 Implementation Details

Following the default data augmentation policy for training ResNet [21] on Imagenet [37], we use scaling, random flipping, color distortion, normalization and random cropping in training. We use Resnet50 as backbone and append a 2-layer perceptron of size 2048 and 128 respectively to apply the contrastive loss. The network is trained with SGD optimizer where the learning rate is initialized to 10^{-1} and decayed to 10^{-4} using the cosine schedule without restart [32]. The weight decay is set to 10^{-4}. For networks without data resampling and ISIC2018

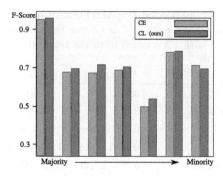

Fig. 3. F-score evaluation of our approach and the common cross-entropy baseline across different classes. This shows that our contrastive-based approach yields larger improvements on the minority (right) while the performance on the major class is not affected.

datasets, we train the backbone for 1000 epochs. For sake of fairness, we train networks with data resampling for 250 epochs only so that in both cases we run roughly 40k updates. For APTOS2019, we double the number of epochs. It takes about 8 h to train the backbone on 4 NVIDIA TITAN RTX GPUs with a batchsize of 192. Training the classification head follows the same settings except the batch size which is set to 512 and the learning rate which decays from 1 to 10^{-4}. This step takes less than an hour to train on a single GPU.

For testing, we resize the images to 256×256 and then process the normalized image. The code is implemented using PyTorch v1.5.1 and will be publicly available upon acceptance.

4.4   Results and Discussion

Tables 1 and 2 summarize the quantitative evaluation of our approach and several baselines. First we show a comparison of different model-based approaches for addressing the imbalance classification (top part). Here we clearly show that our approach based on contrastive learning is consistently superior to other techniques due to the explicit disentanglement of the feature space on both datasets. Next, we study the effect of the common data-based method (resampling) in conjunction with contrastive learning (middle part) and find that when resampling improves over standard cross-entropy, it is complementary to our approach and further improves its performance as in Table 1. In case resampling harms the performance, as in Table 2, contrastive learning helps stabilize the learning process and leads to clearly lower standard deviation. We think the drop in performance on APTOS19 is mainly due to the small size of the dataset. Thus, making copies from the very few samples results in overfitting to those. Finally, we compare to the best existing methods (lower part) and show that contrastive learning performance is on-par or even significantly better than those works without using their special modifications to standard classification. The gap in performance on ISIC shows that pretraining on ImageNet [37], used as

Table 3. Effect of contrastive loss temperature on the f-score of different classes without/with resampling

Temperature	Major	Middle	Minor
0.01	0.941/0.939	0.702/0.712	0.692/0.710
0.05	0.945/0.938	0.717/0.711	0.713/0.721
0.1	0.942/0.937	0.716/0.711	0.696/0.709
0.5	0.942/0.937	0.696/0.698	0.575/0.720
1.0	0.941/0.934	0.669/0.623	0.531/0.651

a common practice in existing methods, does not necessarily yield improvement due to the large domain shift. Figure 3 shows a break-down of the performance of our method and the cross-entropy baseline across all classes. This shows that the main source of improvement is due to better handling the minority while not affecting the performance of the major classes.

4.5   Ablation Study

We ran experiments on ISIC data to study the effect of the temperature used in contrastive loss without/with resampling. We let the temperature take values in $\{0.01, 0.05, 0.1, 0.5, 1.0\}$ for our experiments and report the averaged F-scores for 3 groups of classes: the major class (≥ 6000 images), the 2 medium size classes (~ 1100 images) and the 4 minor classes (≤ 550 images). Recall that the temperature controls the balance between positives and negatives and higher temperatures put more attention to pushing positives close to each other while lower temperatures focus more on pushing negatives away from each other which should theoretically lead to better performance on minority classes for lower temperatures and more stability using the resampling than without resampling. Experimentally, we show in Table 3 that the performance on the major class stays almost constant for different temperatures. We also notice a significant drop in performance for both middle and minor classes for increasing temperatures, which matches the theoretical explanation. We find out that the choice of the temperature for the contrastive loss with resampling is less critical as we recorded non-significant fluctuations in all classes for all temperatures in $\{0.01, 0.05, 0.1, 0.5\}$ while setting the temperature to 0.05 gives a performance boost for the plain contrastive learning. This fact aligns well with our expectation on the stability of both versions. Finally, we record a drop in performance for contrastive learning in conjunction with the lowest temperature. This drop can be explained by an excessive focus on pushing negatives away from each other, which slows down the learning process.

5   Conclusion

In this paper we highlighted the importance of learning from imbalanced datasets in medical image analysis. We proposed a new approach based on contrastive learning to better separate minority from majority classes in feature space. The approach consistently improved over the cross-entropy and oversampling baselines in our evaluation. Moreover, we showed that it is complementary to oversampling and sets a new state of the art for imbalanced datasets.

References

1. Aptos 2019 blindness detection (2019). https://www.kaggle.com/c/aptos2019-blindness-detection/data
2. Aleksey Nozdryn-Plotnicki, J.Y., Yolland, W.: Ensembling convolutional neural networks for skin cancer classification. ArXiv (2018)
3. Cao, K., Wei, C., Gaidon, A., Aréchiga, N., Ma, T.: Learning imbalanced datasets with label-distribution-aware margin loss. In: NeurIPS (2019)
4. Chan, H.P., Samala, R.K., Hadjiiski, L.M., Zhou, C.: Deep learning in medical image analysis (2020)
5. Chawla, N.V., Bowyer, K.W., Hall, L.O., Kegelmeyer, W.P.: SMOTE: synthetic minority over-sampling technique. JAIR **16**, 321–357 (2002)
6. Chen, T., Kornblith, S., Norouzi, M., Hinton, G.E.: A simple framework for contrastive learning of visual representations. In: ICML (2020)
7. Codella, N.C.F., et al.: Skin lesion analysis toward melanoma detection 2018: a challenge hosted by the international skin imaging collaboration (ISIC). CoRR abs/1902.03368 (2019)
8. Cui, Y., Jia, M., Lin, T., Song, Y., Belongie, S.J.: Class-balanced loss based on effective number of samples. In: CVPR (2019)
9. Doi, K.: Computer-aided diagnosis in medical imaging: historical review, current status and future potential. Comput. Med. Imaging Graph. **31**, 198–211 (2007)
10. Dong, Q., Gong, S., Zhu, X.: Imbalanced deep learning by minority class incremental rectification. IEEE TPAMI **41**, 1367–1381 (2019)
11. Dosovitskiy, A., Fischer, P., Springenberg, J.T., Riedmiller, M.A., Brox, T.: Discriminative unsupervised feature learning with exemplar convolutional neural networks. IEEE TPAMI, 1734–1747 (2016)
12. Drummond, C., Holte, R.: C4.5, class imbalance, and cost sensitivity: why undersampling beats oversampling. In: ICML Workshop (2003)
13. Fotedar, G., Tajbakhsh, N., Ananth, S., Ding, X.: Extreme consistency: overcoming annotation scarcity and domain shifts. In: Martel, A.L., et al. (eds.) MICCAI 2020. LNCS, vol. 12261, pp. 699–709. Springer, Cham (2020). https://doi.org/10.1007/978-3-030-59710-8_68
14. Gong, L., Ma, K., Zheng, Y.: Distractor-aware neuron intrinsic learning for generic 2D medical image classifications. In: Martel, A.L., et al. (eds.) MICCAI 2020. LNCS, vol. 12262, pp. 591–601. Springer, Cham (2020). https://doi.org/10.1007/978-3-030-59713-9_57
15. Guanjin Wang, K.W.W., Lu, J.: AUC-based extreme learning machines for supervised and semi-supervised imbalanced classification. IEEE Trans. Syst. Man Cybern.: Syst., 1–12 (2020)
16. Gutmann, M., Hyvärinen, A.: Noise-contrastive estimation: a new estimation principle for unnormalized statistical models. In: AISTATS (2010)
17. Gyawali, P.K., Ghimire, S., Bajracharya, P., Li, Z., Wang, L.: Semi-supervised medical image classification with global latent mixing. In: Martel, A.L., et al. (eds.) MICCAI 2020. LNCS, vol. 12261, pp. 604–613. Springer, Cham (2020). https://doi.org/10.1007/978-3-030-59710-8_59

18. Han, H., Wang, W.-Y., Mao, B.-H.: Borderline-SMOTE: a new over-sampling method in imbalanced data sets learning. In: Huang, D.-S., Zhang, X.-P., Huang, G.-B. (eds.) ICIC 2005. LNCS, vol. 3644, pp. 878–887. Springer, Heidelberg (2005). https://doi.org/10.1007/11538059_91

19. Hayat, M., Khan, S., Zamir, S.W., Shen, J., Shao, L.: Gaussian affinity for max-margin class imbalanced learning. In: ICCV (2019)

20. He, H., Garcia, E.A.: Learning from imbalanced data. IEEE TKDE **21**, 1263–1284 (2009)

21. He, K., Zhang, X., Ren, S., Sun, J.: Deep residual learning for image recognition. In: CVPR (2016)

22. Huang, C., Li, Y., Loy, C.C., Tang, X.: Learning deep representation for imbalanced classification. In: CVPR (2016)

23. Kang, B., et al.: Decoupling representation and classifier for long-tailed recognition. In: ICLR (2020)

24. Khan, S.H., Hayat, M., Zamir, S.W., Shen, J., Shao, L.: Striking the right balance with uncertainty. In: CVPR (2019)

25. Khosla, P., et al.: Supervised contrastive learning. In: NeurIPS (2020)

26. Kim, J., Jeong, J., Shin, J.: M2m: imbalanced classification via major-to-minor translation. In: CVPR (2020)

27. Li, X., Hu, X., Yu, L., Zhu, L., Fu, C.W., Heng, P.: CANet: cross-disease attention network for joint diabetic retinopathy and diabetic macular edema grading. IEEE Trans. Med. Imaging, 1483–1493 (2020)

28. Li, X., Yu, L., Jin, Y., Fu, C.-W., Xing, L., Heng, P.-A.: Difficulty-aware meta-learning for rare disease diagnosis. In: Martel, A.L., et al. (eds.) MICCAI 2020. LNCS, vol. 12261, pp. 357–366. Springer, Cham (2020). https://doi.org/10.1007/978-3-030-59710-8_35

29. Li, Z., Zhong, C., Wang, R., Zheng, W.-S.: Continual learning of new diseases with dual distillation and ensemble strategy. In: Martel, A.L., et al. (eds.) MICCAI 2020. LNCS, vol. 12261, pp. 169–178. Springer, Cham (2020). https://doi.org/10.1007/978-3-030-59710-8_17

30. Liao, H., Luo, J.: A deep multi-task learning approach to skin lesion classification. In: AAAI workshop (2017)

31. Lin, T., Goyal, P., Girshick, R., He, K., Dollár, P.: Focal loss for dense object detection. In: ICCV (2017)

32. Loshchilov, I., Hutter, F.: SGDR: stochastic gradient descent with warm restarts. In: ICLR (2017)

33. Mullick, S.S., Datta, S., Das, S.: Generative adversarial minority oversampling. In: ICCV (2019)

34. van den Oord, A., Li, Y., Vinyals, O.: Representation learning with contrastive predictive coding. CoRR abs/1807.03748 (2018)

35. Peilin Zhao, Steven C. H. Hoi, R.J., Yang, T.: Online AUC maximization. In: ICML (2011)

36. Peng, J., Bu, X., Sun, M., Zhang, Z., Tan, T., Yan, J.: Large-scale object detection in the wild from imbalanced multi-labels. In: CVPR (2020)

37. Russakovsky, O., et al.: ImageNet large scale visual recognition challenge. Int. J. Comput. Vis. **115**, 211–252 (2015)

38. Shen, L., Lin, Z., Huang, Q.: Relay backpropagation for effective learning of deep convolutional neural networks. In: Leibe, B., Matas, J., Sebe, N., Welling, M. (eds.) ECCV 2016. LNCS, vol. 9911, pp. 467–482. Springer, Cham (2016). https://doi.org/10.1007/978-3-319-46478-7_29

39. Shrivastava, A., Gupta, A., Girshick, R.B.: Training region-based object detectors with online hard example mining. In: CVPR (2016)
40. Shrivastava, A., Gupta, A., Girshick, R: Training region-based object detectors with online hard example mining. In: CVPR (2016)
41. Skorczyk-Werner, A., et al.: Fundus albipunctatus: review of the literature and report of a novel RDH5 gene mutation affecting the invariant tyrosine (p. Tyr175Phe). J. Appl. Genet. **56**, 317–327 (2015)
42. Wei, D., Cao, S., Ma, K., Zheng, Y.: Learning and exploiting interclass visual correlations for medical image classification. In: Martel, A.L., et al. (eds.) MICCAI 2020. LNCS, vol. 12261, pp. 106–115. Springer, Cham (2020). https://doi.org/10.1007/978-3-030-59710-8_11
43. Zhang, X., Fang, Z., Wen, Y., Li, Z., Qiao, Y.: Range loss for deep face recognition with long-tailed training data. In: ICCV (2017)
44. Zhuang, J.X., et al.: Skin lesion analysis towards melanoma detection using deep neural network ensemble (2018)
45. Zhuang, J., Cai, J., Wang, R., Zhang, J., Zheng, W.-S.: Deep kNN for medical image classification. In: Martel, A.L., et al. (eds.) MICCAI 2020. LNCS, vol. 12261, pp. 127–136. Springer, Cham (2020). https://doi.org/10.1007/978-3-030-59710-8_13

Interpretable Gender Classification from Retinal Fundus Images Using BagNets

Indu Ilanchezian[1,2], Dmitry Kobak[1], Hanna Faber[3], Focke Ziemssen[3], Philipp Berens[1,2], and Murat Seçkin Ayhan[1(✉)]

[1] Institute for Ophthalmic Research, Tübingen, Germany
{philipp.berens,murat-seckin.ayhan}@uni-tuebingen.de
[2] Tübingen AI Center, Tübingen, Germany
[3] University Eye Clinic, University of Tübingen, 72076 Tübingen, Germany

Abstract. Deep neural networks (DNNs) are able to predict a person's gender from retinal fundus images with high accuracy, even though this task is usually considered hardly possible by ophthalmologists. Therefore, it has been an open question which features allow reliable discrimination between male and female fundus images. To study this question, we used a particular DNN architecture called BagNet, which extracts local features from small image patches and then averages the class evidence across all patches. The BagNet performed on par with the more sophisticated Inception-v3 model, showing that the gender information can be read out from local features alone. BagNets also naturally provide saliency maps, which we used to highlight the most informative patches in fundus images. We found that most evidence was provided by patches from the optic disc and the macula, with patches from the optic disc providing mostly male and patches from the macula providing mostly female evidence. Although further research is needed to clarify the exact nature of this evidence, our results suggest that there are localized structural differences in fundus images between genders. Overall, we believe that BagNets may provide a compelling alternative to the standard DNN architectures also in other medical image analysis tasks, as they do not require post-hoc explainability methods.

Keywords: Retinal fundus image · Gender prediction · Interpretable deep neural networks · Bag-of-features models

1 Introduction

In recent years, deep neural networks (DNNs) have achieved physician-level accuracy in various image-based medical tasks, e.g. in radiology [21], dermatology [10], pathology [15] and ophthalmology [7,12]. Moreover, in some cases DNNs have been shown to have good performance in tasks that are not straightforward for physicians: for example, they can accurately predict the gender from retinal

© Springer Nature Switzerland AG 2021
M. de Bruijne et al. (Eds.): MICCAI 2021, LNCS 12903, pp. 477–487, 2021.
https://doi.org/10.1007/978-3-030-87199-4_45

images [25]. As this task is typically not clinically relevant, ophthalmologists are not explicitly trained for it. Nevertheless, the comparably poor performance of ophthalmologists at this task suggests that gender differences in fundus images are not obvious or salient. Even though saliency maps used by [25] and follow-up studies [5,9] have tentatively pointed at the optic disc, the macula, and retinal blood vessels as candidate regions for gender-related anatomical differences in fundus images, conclusive evidence is still lacking. Therefore the high gender prediction performance of DNNs has created lots of interest in the medical imaging community as one hope for DNNs is to unravel biomarkers that are not easily found by humans. Here, we performed a proof of principle study to make progress on the question of how DNNs are able to detect gender differences in retinal fundus. Our contribution is twofold: we (1) introduced BagNets [3]—a 'local' variant of the ResNet50 architecture [13]—as an interpretable-by-design architecture for image analysis in ophthalmology and (2) used them to narrow down the hypothesis space for question at hand.

We trained the BagNets on a large collection of retinal fundus images obtained from the UK Biobank [27] (Fig. 1a). BagNets use a linear classifier on features extracted from image patches to compute local evidence for each class, which is then averaged over space to form the final prediction, without considering any global relationships. Thus, BagNets resemble 'bag-of-features' models popular before deep learning [23]. Despite this simple bag-of-features approach, the BagNet performed on par with an Inception-v3 network in terms of gender prediction accuracy, indicating that gender can be determined from the local characteristics of the fundus image. Also, the BagNet architecture naturally allowed to construct saliency maps to highlight the most informative regions for gender prediction in the retina (Fig. 1b). We found that the macula contained most distinctive female patches, while the optic disk contained male ones. In addition, we showed that the decision of the BagNet was not simply caused by some exclusively female or male patches in the images, but rather by a change in both frequency and the degree of 'femaleness' or 'maleness' of individual patches. Overall, we argue that BagNets can be useful in medical imaging applications including both disease diagnosis and biomarker discovery, thanks to interpretability provided by their local architecture. Our code is available at https://github.com/berenslab/genderBagNets.

2   Related Work

Previous work on gender prediction from fundus images have used either standard DNN architectures or simple logistic regression on top of expert-defined features. For example, [25] trained Inception-v3 networks on the UK Biobank dataset to predict cardiovascular risk factors from fundus images and found that DNNs were also capable of predicting the patient's gender (AUC = 0.97). A similar network was used by [9]. In both studies, the authors computed post-hoc saliency maps to study the features driving the network's decisions. In a sample of 100 attention maps, [25] found that the optic disc, vessels, and other

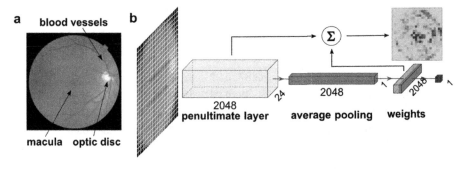

Fig. 1. A sketch of the gender prediction via BagNet33. (**a**) Example fundus image from the UK Biobank. The optic disc is the bright spot on the right, the macula is the slightly darker spot in the middle and the blood vessels are extending from the optic disc in darker red. (**b**) The BagNet33 extracts 2048-dimensional feature vectors from 33×33 patches and stores them in the penultimate layer. Via spatial average pooling and a linear classifier, it then forms the final predictions for the gender. The same linear classifier can be applied directly to the feature representation in the penultimate layer to compute the local evidence, which can be visualized as a saliency map. Plotted with PlotNeuralNet [14].

nonspecific parts of the images were frequently highlighted. However, this seems to be the case for almost all the dependent variables and it is very hard to derive testable hypotheses for gender specific differences. Likewise, [9] manually inspected a sample of occlusion maps and concluded that DNNs may use geometrical properties of the blood vessels at the optic disc for predicting gender. More recently, [5] demonstrated that DNNs can predict gender not only from retinal fundus images but also from OCT scans, where the foveal pit region seemed most informative based on gradient-based saliency maps. Taking a different approach, [29] used expert-defined image features in a simple logistic regression model. Although the performance of their model was worse ($AUC = 0.78$), they found various color-intensity-based metrics and the angle between certain retinal arteries to be significant predictors, but most effect sizes were small.

BagNets provide a compromise between linear classifiers operating on expert-defined features [29] and high-performing DNNs [5,9,25], which require complex post-hoc processing for interpretability [2,22]. In BagNets, a saliency map is also straightforward to compute by design, and it has been shown to provide more information about the location of class evidence than auxiliary interpretability methods [3]. Such native evidence-based maps returned by BagNets are interpretable as is, while standard saliency maps require fine-tuning and post-processing for compelling visualizations [2]. Thanks to these benefits, BagNets have also been used in the context of histopathological microscopy [24].

Table 1. Gender prediction performances of DNNs

	TRAINING		VALIDATION		TEST		CLINICAL	
	ACC.	AUC	ACC.	AUC	ACC.	AUC	ACC.	AUC
InceptionV3	92.99	0.98	83.87	0.92	82.97	0.91	62.07	0.78
BagNet33	93.44	0.98	85.48	0.93	85.26	0.93	72.41	0.70
BagNet17	86.66	0.94	82.30	0.90	82.11	0.90	37.93	0.51
BagNet9	82.41	0.92	79.95	0.89	80.57	0.90	41.38	0.45

3   Methods

3.1   Data and Preprocessing

The UK Biobank [27] offers a large-scale and multi-modal repository of health-related data from the UK. From this, we obtained records of over 84, 000 subjects with 174, 465 fundus images from both eyes and multiple visits per participant. Male and female subjects constituted 46% and 54% of the data, respectively. As a substantial fraction of the images were not gradable due to image quality issues (artefacts, high contrast, or oversaturation), we used the EyeQual networks [6] to filter out poor images. 47, 939 images (47% male, 53% female) passed the quality check by the EyeQual ensemble. We partitioned them into the training, validation and test sets with 75%, 10% and 15% of subjects, respectively, making sure that all images from each subject were allocated to the same set.

Additionally, we obtained 29 fundus images from patients (11 male, 18 female, all older than 47 years) at the University Eye Hospital with permission of the Institutional Ethics Board. We used these additional images as an independent test set. For all images, we applied a circular mask to capture the 95% central area and to remove camera artifacts at the borders.

3.2   Network Architecture and Training

We used BagNets [3] (Fig. 1b) and standard Inception-v3 [28] network as implemented in Keras [4]. In a BagNet, neurons in the final layer have a receptive field restricted to $q \times q$ pixels, where we used $q \in \{9, 17, 33\}$. The convolutional stack in the network extracts a 2048-dimensional feature vector for each $q \times q$ image patch. Patches were implicitly defined, with a stride for convolutions of 8 pixels for $q = 33$. Therefore local features were extracted for each patch on a 24 × 24 grid (Fig. 1b). A linear classifier combined these 2048 features to obtain the local class evidence which was then averaged across all image patches (average pooling layer).

All networks had been pretrained on ImageNet [26] by their respective developers. For our binary classification problem, we replaced the 1000-way softmax output layer with a single logistic output neuron (Fig. 1b). We initially trained only the output layer using the fundus images for 10 epochs. This was followed by fine-tuning all layers for 100 epochs. We used stochastic gradient descent (SGD) with the learning rate set to 0.01 and the batch size to 16. We used data augmentation via random rotations and flips, width and height shifts, random brightness, and random zooming operations. We picked the best epoch from the [95, 100] range based on the validation performance. We evaluated the final performance on both the test set and the data from the University Eye Hospital.

3.3 Generation of Saliency Maps

To compute saliency maps, we applied the weights $\mathbf{w}$ in the final classification layer of BagNet33 to the feature vectors, e.g. $\mathbf{x}$, in its penultimate layer (Fig. 1b), yielding the local evidence (logits) for each patch via $\mathbf{w} \cdot \mathbf{x} = \sum_i w_i x_i$. We clipped the resulting values to $[-75, 75]$ for visualization purposes. The resulting saliency maps were 24×24 (Fig. 2).

3.4 Embedding of Image Patches

To explore which image patches were informative for classification, we used t-Stochastic Neighborhood Embeddings (t-SNE) [20], a non-linear dimensionality reduction method. To embed the feature representations of $>1,000,000$ image patches extracted from the fundus images, we used FIt-SNE implementation [19] with uniform affinity kernel in the high-dimensional space across 15 nearest neighbours. We used PCA initialization to better preserve the global structure of the data and improve the reproducibility [16]. We used a heavy-tailed kernel $k(d) = 1/(1 + d^2/\alpha)^\alpha$ with $\alpha = 0.5$ to emphasize cluster structure [17].

4 Results

We trained BagNets with three different receptive field sizes to predict patient's gender from retinal fundus images based on the UK Biobank data. We evaluated their performances using prediction accuracy and the Area Under the Receiver Operating Characteristic curve (AUC) and compared to an Inception-v3 network (Table 1). BagNet33 and Inception-v3 performed on par with each other, while BagNet17 and BagNet9 performed worse. BagNet33 and Inception-v3 also generalized better to a new clinical dataset, albeit with a substantial drop in performance. Together, this suggests that the 33×33 patches captured the relevant information for gender prediction better than smaller patches. Thus, for the remainder of the paper, we will focus our analysis on the BagNet33 (referring to it simply as BagNet).

We inspected saliency maps for gender prediction computed by evaluating the classifier on each feature representation in the penultimate layer (Fig. 2, top). In a typical male example, we found that the optic disc provided high evidence for the male class, along with more scattered evidence around the major blood vessels. For a typical female example, high evidence was found for the female class in the macula. Averaging the saliency maps across all correctly classified

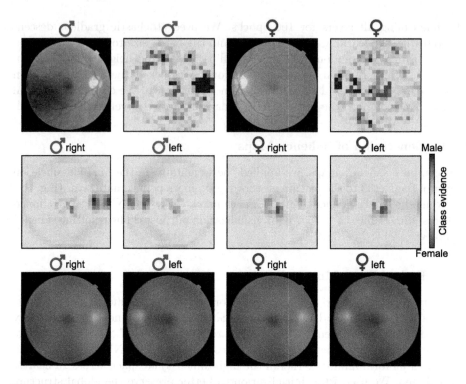

Fig. 2. Saliency maps obtained by BagNet showing class evidence for each of the image patches on a 24×24 grid. Top row shows exemplary test images along with their saliency maps. Middle row shows the average saliency maps for correctly classified male and female patients. Bottom row shows the average fundus images corresponding to the middle row.

male/female test images confirmed that the BagNet relied on the optic disc and the blood vessels to identify male images and on the macula to identify female ones (Fig. 2, middle).

Interestingly, the individual and the average saliency maps also showed that the optic disc patches tended to always provide male evidence, to some extent even in correctly classified female images. Similarly, the macula patches tended to provide female evidence, even in correctly classified male images. The BagNet could nevertheless achieve high classification performance after averaging the class evidence across all patches.

As a sanity check, we show the averaged fundus images across all correctly classified male/female images in the bottom row of Fig. 2. These average images are nearly identical across genders, demonstrating that it is not the location, the size, or the shape of the optic disc or macula that drive the BagNet predictions.

To further explore the structure of local image features informative about gender, we embedded the 2048-dimensional feature representation of each image patch into 2D using t-SNE and colored them by the provided class evidence

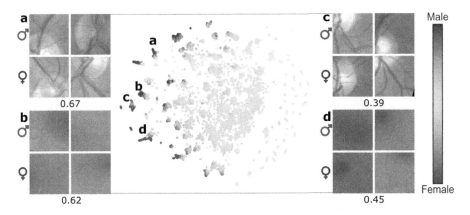

Fig. 3. Visualization of image patches and associated class evidence via t-SNE. 213,696 patches extracted from 371 correctly classified test images (using training set images yielded a similar embedding; not shown). Four patches with high evidence (two male, two female) are shown from each of the four highlighted clusters. The fraction of male patches in each of these clusters is given below the corresponding exemplary patches. The colors show the logit class evidence. Note that the color does not indicate the correct label of each patch.

(Fig. 3). We found that most image patches provided only weak evidence for either class, but some distinct clusters of patches had consistently high logits. We further explored these clusters and found that they consistently showed the optic disk with blood vessels (**a** and **c**) or the macula (**b** and **d**), in line with the saliency maps computed above (Fig. 2). However, even though the clusters **a** and **b** consistently provided evidence for the male class, patches in these clusters occurred in true female and male fundus images alike (67% and 62% patches from male images, respectively). Similarly, clusters **c** and **d** provided evidence for the female class but yet came from male and female fundus images (39% and 45% patches from male images, respectively).

This raised the question of whether the BagNet's decisions were mostly driven by (i) male/female images having individual patches with stronger male/female evidence; or (ii) male/female images having a larger number of patches with male/female evidence (Fig. 4). We found that both factors played a role in determining the final gender predictions, but the fraction of male/female patches seemed to be a stronger factor: Cohen's $d = 1.82$ and $d = 1.63$ for the difference in fraction of male (logit value >50) and female (logit value < -50) patches between genders, vs. $d = 0.77$ and $d = 0.76$ for the difference in the logit value of the most male and the most female patch. Thus, female images contained more patches providing strong female class evidence, and vice versa for male fundus images.

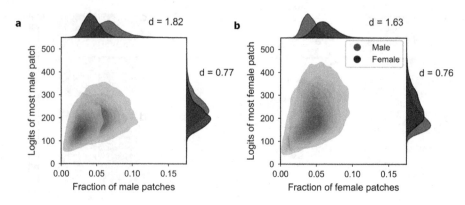

Fig. 4. Two factors determine the gender predictions of the BagNet: the maximal strength of evidence and the frequency of strong evidence. (**a**) Kernel density estimate of all male (red) and female (blue) test set images. Horizontal axis: fraction of male patches, defined as having logit values above 50. Vertical axis: the absolute logit value of the most male patch. (**b**) The same for patches providing female evidence (logit values below −50).

5   Discussion

In summary, we argued that the BagNet architecture is particularly suitable for medical image analysis, thanks to its built-in interpretability. Here we used Bag-Nets to investigate the high accuracy of DNNs in gender prediction from retinal fundus images. BagNet33 achieved a performance similar to Inception-v3 despite having a much simpler architecture and using only local image features for prediction. This suggested that local features are sufficient for gender prediction and the global arrangement of these features is not essential for this task.

In BagNets, saliency maps can be readily computed without auxiliary gradient-based methods or layer-wise relevance propagation [22]. We used the native saliency maps of BagNets and a two-dimensional t-SNE embedding of image patches to identify the most informative regions for the gender prediction task in fundus images. This allowed us to go beyond the previous reports [9,25] and for the first time to provide conclusive evidence that the optic disk region contains features used to inform a male prediction and the macula region for a female prediction. We found that both the frequency of informative male/female patches and—albeit to a lesser degree—the strength of the most informative male/female patches were important factors for gender prediction by BagNets.

It is, however, not the case that the optic disc in males is substantially larger than in females, as can be seen in the average fundus images shown in Fig. 2. The relative optic disc and macula sizes, shapes, brightness levels, etc. seem all to be roughly the same for both genders. Instead, our results suggest *structural but localized* differences in the male and female retinas, mainly within the optic disc and macula regions. This is supported by the previous findings showing that the retinal nerve fibre layer in the optic disk is slightly thicker in females [18] and that the macula is slightly thinner [5] and wider [8] in females. However, these previously reported gender differences have small to moderate effect sizes (Cohen's $d = 0.11$, $d = 0.52$, and $d = 0.17$ respectively for the comparisons referenced above; computed here based on reported means and standard deviations) and it is unclear if they alone can explain the BagNet performance.

Therefore, future work is needed to understand what exactly it is that allows the network to assign high male evidence to the optic disc patches from male patients and high female evidence to the optic disc patches from female patients. In this sense, the results presented here do not provide the final solution to the gender prediction mystery. Nevertheless, we believe that our results make a step in the right direction as they demonstrate structural but localized gender differences and reduce the problem complexity down to specific small patches of the fundus image that can be further analyzed separately.

We believe that BagNets may also be more widely applicable for clinically relevant diagnostic tasks involving medical images in ophthalmology and beyond, provided that they are coupled with reliable uncertainty estimation [1]. In many cases, pathologies often manifest in localized regions, which can be readily picked up by BagNets. For example, BagNets could be used to further explore clinically relevant changes underlying progressive diseases such as diabetic retinopathy. The interpretable architecture of BagNets may increase the trust of clinicians and patients, which is a critical issue for adoption of deep learning algorithms in medical practice [11].

Acknowledgements. We thank Wieland Brendel for his support with BagNets. This research was supported by the German Ministry of Science and Education (BMBF, 01GQ1601 and 01IS18039A) and the German Science Foundation (BE5601/4-2 and EXC 2064, project number 390727645). Hanna Faber received research funding from the Junior Clinician Scientist Program of the Faculty of Medicine, Eberhard Karls University of Tübingen, Germany (application number 463-0-0). Additional funding was provided by Novartis AG through a research grant. The funding bodies did not have any influence in the study planning and design. The authors thank the International Max Planck Research School for Intelligent Systems (IMPRS-IS) for supporting Indu Ilanchezian.

References

1. Ayhan, M.S., Kühlewein, L., Aliyeva, G., Inhoffen, W., Ziemssen, F., Berens, P.: Expert-validated estimation of diagnostic uncertainty for deep neural networks in diabetic retinopathy detection. Med. Image Anal. **64**, 101724 (2020)
2. Ayhan, M.S., et al.: Clinical validation of saliency maps for understanding deep neural networks in ophthalmology. medRxiv (2021)
3. Brendel, W., Bethge, M.: Approximating CNNs with bag-of-local-features models works surprisingly well on imagenet. In: International Conference on Learning Representations (2019)
4. Chollet, F., et al.: Keras (2015). https://github.com/fchollet/keras
5. Chueh, K.M., Hsieh, Y.T., Chen, H.H., Ma, I.H., Huang, S.L.: Prediction of sex and age from macular optical coherence tomography images and feature analysis using deep learning. medRxiv (2020)

6. Costa, P., et al.: EyeQual: accurate, explainable, retinal image quality assessment. In: 2017 16th IEEE International Conference on Machine Learning and Applications (ICMLA), pp. 323–330 (2017)
7. De Fauw, J., et al.: Clinically applicable deep learning for diagnosis and referral in retinal disease. Nat. Med. **24**(9), 1342 (2018)
8. Delori, F.C., Goger, D.G., Keilhauer, C., Salvetti, P., Staurenghi, G.: Bimodal spatial distribution of macular pigment: evidence of a gender relationship. JOSA A **23**(3), 521–538 (2006)
9. Dieck, S., et al.: Factors in color fundus photographs that can be used by humans to determine sex of individuals. Transl. Vis. Sci. Technol. **9**(7), 8–8 (2020)
10. Esteva, A., et al.: Dermatologist-level classification of skin cancer with deep neural networks. Nature **542**(7639), 115 (2017)
11. Grote, T., Berens, P.: On the ethics of algorithmic decision-making in healthcare. J. Med. Ethics **46**(3), 205–211 (2020)
12. Gulshan, V., et al.: Development and validation of a deep learning algorithm for detection of diabetic retinopathy in retinal fundus photographs. JAMA **316**(22), 2402–2410 (2016)
13. He, K., Zhang, X., Ren, S., Sun, J.: Deep residual learning for image recognition. In: Proceedings of the IEEE Conference on Computer Vision and Pattern Recognition, pp. 770–778 (2016)
14. Iqbal, H.: PlotNeuralNet (2018). https://github.com/HarisIqbal88/PlotNeuralNet. Accessed 26 Feb 2021
15. Kiani, A., et al.: Impact of a deep learning assistant on the histopathologic classification of liver cancer. npj Digit. Med. **3**(1), 1–8 (2020)
16. Kobak, D., Berens, P.: The art of using t-SNE for single-cell transcriptomics. Nat. Commun. **10**(1), 1–14 (2019)
17. Kobak, D., Linderman, G., Steinerberger, S., Kluger, Y., Berens, P.: Heavy-tailed kernels reveal a finer cluster structure in t-SNE visualisations. In: Brefeld, U., Fromont, E., Hotho, A., Knobbe, A., Maathuis, M., Robardet, C. (eds.) ECML PKDD 2019. LNCS (LNAI), vol. 11906, pp. 124–139. Springer, Cham (2020). https://doi.org/10.1007/978-3-030-46150-8_8
18. Li, D., et al.: Sex-specific differences in circumpapillary retinal nerve fiber layer thickness. Ophthalmology **127**(3), 357–368 (2020)
19. Linderman, G.C., Rachh, M., Hoskins, J.G., Steinerberger, S., Kluger, Y.: Fast interpolation-based t-SNE for improved visualization of single-cell RNA-seq data. Nat. Methods **16**, 243–245 (2019)
20. Maaten, L.V.D., Hinton, G.E.: Visualizing data using t-SNE. J. Mach. Learn. Res. **9**, 2579–2605 (2008)
21. McKinney, S.M., et al.: International evaluation of an AI system for breast cancer screening. Nature **577**(7788), 89–94 (2020)
22. Montavon, G., Samek, W., Müller, K.R.: Methods for interpreting and understanding deep neural networks. Digit. Signal Process. **73**, 1–15 (2018)
23. O'Hara, S., Draper, B.A.: Introduction to the bag of features paradigm for image classification and retrieval. arXiv preprint arXiv:1101.3354 (2011)
24. Paschali, M., Naeem, M.F., Simson, W., Steiger, K., Mollenhauer, M., Navab, N.: Deep learning under the microscope: improving the interpretability of medical imaging neural networks. arXiv preprint arXiv:1904.03127 (2019)
25. Poplin, R., et al.: Prediction of cardiovascular risk factors from retinal fundus photographs via deep learning. Nat. Biomed. Eng. **2**, 158–164 (2019)
26. Russakovsky, O., et al.: ImageNet large scale visual recognition challenge. Int. J. Comput. Vis. **115**(3), 211–252 (2015). https://doi.org/10.1007/s11263-015-0816-y

27. Sudlow, C., et al.: UK biobank: an open access resource for identifying the causes of a wide range of complex diseases of middle and old age. PLoS Med. **12**(3), e1001779 (2015)
28. Szegedy, C., Vanhoucke, V., Ioffe, S., Shlens, J., Wojna, Z.: Rethinking the inception architecture for computer vision. In: Proceedings of the IEEE Conference on Computer Vision and Pattern Recognition, pp. 2818–2826 (2016)
29. Yamashita, T., et al.: Factors in color fundus photographs that can be used by humans to determine sex of individuals. Transl. Vis. Sci. Technol. **9**(2), 4–4 (2020)

Explainable Classification of Weakly Annotated Wireless Capsule Endoscopy Images Based on a Fuzzy Bag-of-Colour Features Model and Brain Storm Optimization

Michael Vasilakakis⬤, Georgia Sovatzidi⬤, and Dimitris K. Iakovidis$^{(\boxtimes)}$ ⬤

University of Thessaly, Papasiopoulou str. 2-4, 35131 Lamia, Greece
{vasilaka,gsovatzidi,diakovidis}@uth.gr

Abstract. Wireless capsule endoscopy (WCE) constitutes a medical imaging technology developed for the endoscopic exploration of the gastrointestinal (GI) tract, whereas it provides a more comfortable examination method, in comparison to the conventional endoscopy technologies. In this paper, we propose a novel Explainable Fuzzy Bag-of-Words (XFBoW) feature extraction model, for the classification of weakly annotated WCE images. A comparative advantage of the proposed model over state-of-the-art feature extractors is that it can provide an explainable classification outcome, even with conventional classification schemes, such as Support Vector Machines. The explanations that can be derived are based on the similarity of the image content with the content of the training images, used for the construction of the model. The feature extraction process relies on data clustering and fuzzy sets. Clustering is used to encode the image content into visual words. These words are subsequently used for the formation of fuzzy sets to enable a linguistic characterization of similarities with the training images. A state-of-the-art Brain Storm Optimization algorithm is used as an optimizer to define the most appropriate number of visual words and fuzzy sets and also the fittest parameters of the classifier, in order to optimally classify the WCE images. The training of XFBoW is performed using only image-level, semantic labels instead of detailed, pixel-level annotations. The proposed method is investigated on real datasets that include a variety of GI abnormalities. The results show that XFBoW outperforms several state-of-the-art methods, while providing the advantage of explainability.

Keywords: Wireless capsule endoscopy · Feature extraction · Bag-of-words · Fuzzy sets · Swarm intelligence · Explainability

1 Introduction

Wireless Capsule Endoscopy (WCE) is a noninvasive diagnostic tool that enables the painless imaging of the whole gastrointestinal (GI) tract, without discomfort to patients, compared to other conventional diagnostic methods. A key weakness in clinical management of capsule endoscopy is that it is still demanding with respect to required review

© Springer Nature Switzerland AG 2021
M. de Bruijne et al. (Eds.): MICCAI 2021, LNCS 12903, pp. 488–498, 2021.
https://doi.org/10.1007/978-3-030-87199-4_46

time and effort per examination [1]. Thus, systems for the automatic detection and recognition of abnormalities are developed and contribute effectively in the reduction of the number of false negative diagnoses, while they reduce the time needed for an expert to be fully concentrated and review the endoscopy database of a patient.

In order to cooperate with this time-consuming problem, supervised learning with weakly annotated images is a promising solution, which does not require a detailed, pixel-wise annotation of the training images and involves the training of a learning machine [2]. One of the most widely used supervised learning methods with weakly annotated images is Bag-of-visual-Words (BoW); however, it has two main problems, which are the uncertainty and the plausibility of the words [3]. In [4] a color descriptor to characterize the image utilizing BoW model to build a vocabulary univocally describing each image for definition and recognition of unique characteristics for each colorectal polyp. In [5] a class-specific vocabulary utilizing supervised clustering and feature selection method was proposed for the generation of informative vocabularies. In addition, in [6] the Fuzzy Feature Encoding (FFE) method was proposed, where the construction of the image histogram is based on the contribution of the two closer and representative visual words to an image feature. Despite the efficiency of the recently proposed BoW methods, there is still a lack of understanding and explaining the result emerging from the image classification procedure.

Many efforts have been done, in order to develop robust and efficient methods to detect GI abnormalities [7]. Concerning the methods aiming at ulcer detection, in [2], a supervised learning classification approach with weakly annotated images was presented; the proposed scheme requires image-level, instead of pixel-level, annotations for training, in order to perform automated lesion detection in WCE videos, using color extracted features. Moreover, a Convolutional Neural Network (CNN) for the detection and localization of various GI abnormalities, using an iterative cluster unification algorithm, in weakly annotated images was proposed in [8]. In [9], a local color texture feature extraction approach for polyp detection was presented. This approach was applied on WCE images and compared to other conventional methods, it contains both color and texture information, whereas generation of a balanced codebook is achieved after using visual words based only on positive samples. In [10], a content-based image retrieval (CBIR) system was proposed aiming to the automatic recognition of colorectal polyps. In [11], a method for automated bleeding detection in WCE images was presented. In [12], the detection of abnormalities in gastrointestinal (GI) endoscopy images was performed, using Look-Behind Fully Convolutional Neural Network (LB-FCN).

In this paper, we propose an extension of the BoW method for image classification, named Explainable Fuzzy Bag-of-Words (XFBoW). The novel contribution of XFBoW is a fuzzy feature encoding method, enabling the linguistic interpretation of the image classification outcome. Feature encoding is performed using class-specific visual vocabularies from weakly annotated images, so that the visual content of WCE images is described by similarity histograms. These histograms express the similarity between the testing and training images based on visual words. Fuzzy sets are used to interpret the similarity information of the histograms using linguistic values, such as "Low", "Medium" or "High". The proposed XFBoW model is the first BoW extension able to provide explainable histogram features, where the similarity of the content of an image to every visual word is explained with a linguistic value. A comparative advantage of the Bag-of-Words approaches is that they resemble the way humans use specific vocabularies of words for the description of real-world concepts. Unlike current deep learning or BoW models that have been applied to date for WCE image classification, XFBoW is explainable, which means that it can explain why a WCE image is classified as abnormal or normal, depending on the existence of a lesion or not. The advantage of explaining the classification result is significant, especially in the content context of WCE, where the reasoning for the classification of an image could enhance the diagnostic yield. More importantly, XFBoW can be used for feature encoding with any feature extraction and classification scheme.

The rest of this paper is organized as follows: in Sect. 2, the proposed XFBoW feature extraction model is presented. The experimental results are presented in Sect. 3. Finally, in Sect. 4, conclusions derived from this study are summarized.

2   XFBoW Feature Extraction Model

The inspiration of BoW derives from text document analysis and is based on calculating the appearance frequencies of each word in the documents. When this technique is applied on images, a set of "visual words" captures the image content. XFBoW can be applied on different features, and it can be combined with any classifier, such as an SVM or a deep learning-based approach. In the proposed model, the feature extraction procedure of XFBoW includes parameters that are optimized along with the parameters of the classifier, by a state-of-the-art swarm optimization algorithm, called Determinative Brain Storm Optimization (DBSO) [13].

2.1   Explainable Fuzzy Bag-Of-Words Model

Let us consider a set of images $I_{d,m}$, $d \in [1, D_m]$, $m \in \{A, A'\}$, where A, A' represent the labels of two different classes, $e.g.$, in this paper, they represent the abnormal and normal class labels of the WCE images. From each image a set of F features $f_{d,j}$, $j \in [1, F]$, $d \in [1, D_m]$ are extracted, forming a respective feature vector, they are subsequently clustered and their centroids are considered as visual words. These words are used to construct a "vocabulary" $V_m = \{w_{n,m}\}$, $n \in [1, N_m]$, $m \in \{A, A'\}$, where N_m represents the number of clusters created for the vectors of class m. XFBoW is class-specific, since a vocabulary exists for each class, $i.e.$, abnormal and normal class vocabularies (Fig. 1). The union of all vocabularies V_m forms a global vocabulary $\mathcal{V}$, as it is presented in Eq. (1):

$$\mathcal{V} = \bigcup_m V_m, \ m \in \{A, A'\} \tag{1}$$

In this paper, a soft feature encoding method is used based on the mapping technique presented in [6]. The similarity $\varphi_{n,m}$ among the two closest visual words from $d \in$

$[1, D_m]$ is:

$$\varphi_{n,m}(f_{d,j}) = 1 - \frac{dist(f_{d,j}, w_{n,m})}{\sum_m \sum_{i=1}^{N_m} dist(f_{d,j}, w_{i,m})} \tag{2}$$

where $dist$ is the Euclidean distance and $w_{i,m}$ represents the ith visual word, $i \in [1, N_m]$. In addition, each $\varphi_{n,m}$ is normalized to the range $[0, 1]$. Thus, to describe the similarity between the input image and the existing visual words of every vocabulary, a histogram is constructed, by concatenating the histograms $H'_{d,A} = (\varphi_{1,A}, \ldots, \varphi_{N_A,A})$, $H'_{d,A'} = (\varphi_{1,A'}, \ldots, \varphi_{N'_A,A'})$ of the similarities of the visual words that belong to the abnormal and normal class:

$$H_{d,m} = \left[H'_{d,A}, H'_{d,A'} \right] = \left(\varphi_{1,A}, \ldots, \varphi_{N_A,A}, \varphi_{1,A'}, \ldots, \varphi_{N'_A,A'} \right) \tag{3}$$

where $d \in [1, D_m]$. Histograms extracted from different images, are schematically illustrated in Fig. 1. In that figure, the histogram bins corresponding to the abnormal visual words are colored red, whereas the bins corresponding to the normal visual words are colored green. Under each bin, an icon is provided as indicative example illustrating the content of the respective visual words.

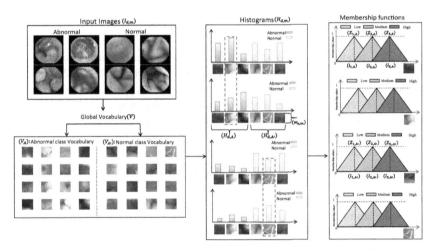

Fig. 1. Procedure of XFBoW model for the construction of the vocabularies.

In the sequel, linguistic values are derived from the calculated histograms to characterize, in a more intuitive way, the similarity between the visual words and the corresponding images that belong to the same class. In Fig. 1, the visual words that belong to the same class are represented by dashed lines. The k-means algorithm is then used to group histograms $H'_{d,A}$ and $H'_{d,A'}$ into $L_A, L_A < D_A$ and $L_{A'}, L_{A'} < D_{A'}$ clusters, respectively, with corresponding centroids $l_{b,n}, b \in [1, L_A], n = 1, \ldots, N_A$ and $l_{b,n} \in$

$[1, L_{A'}], n = 1, \ldots, N_A$. For each cluster centroid, a fuzzy set, $Z_{b,n}, b \in [1, L_A], n = 1, \ldots, N_A$ for abnormal class and $Z_{b,n}, b \in [1, L_{A'}], n = 1, \ldots, N_{A'}$ for normal class, is defined. Every fuzzy set is defined in terms of a membership function $\mu_{b,n} \in [0, 1]$, according to the category it belongs to, i.e., $\mu_{b,n}(\varphi_{n,A}), b \in [1, L_A], n = 1, \ldots, N_A$ and $\mu_{b,n}(\varphi_{n,A'}), b \in [1, L_{A'}], n = 1, \ldots, N_{A'}$ respectively. For simplicity, triangular membership functions are considered. These fuzzy sets express the respective degree of similarity between $f_{d,j}$ and $w_{n,m}$ and they are defined so that they are overlapping covering the whole range $[0, 1]$, with no gaps, as illustrated in Fig. 1. In that figure, the different color intensities of the triangles, indicate the similarity degrees they represent, e.g., "Low", "Medium" and "High" similarity. Considering the above, the visual content of an input image, $I_{d,m}$, can be described by the following feature vector, and used for classification by a classifier, such as an SVM:

$$\mathcal{H}_{d,m} = (\varphi_{1,A} \cdot \mu_{1,A}(\varphi_{1,A}), \ldots, \varphi_{N_A,A} \cdot \mu_{N_A,A}(\varphi_{N_A,A}), \varphi_{1,A} \cdot \\ \mu_{1,A}(\varphi_{1,A}), \ldots, \varphi_{N_A,A}\mu_{N_A,A}(\varphi_{N_A,A})) \tag{4}$$

In order to extract the desired visual descriptors "dense sampling" is used [14]. The color features are extracted and described with 9-dimensional vectors that are formed by the CIE-Lab values (L, a, b) and the minimum and maximum values of all three components within the entire patch $(min_L, min_a, min_b, max_L, max_a, max_b)$ [15].

2.2 Model Optimization

The proposed model is optimized using the state-of-the-art swarm-based algorithm, Determinative Brain Storm Optimization (DBSO) [13]. Specifically, DBSO is utilized for the purpose of finding the optimal number of visual words ($\{w_{n,m}\}$) for the vocabulary construction per class, as well as, the fittest number of fuzzy sets needed ($l_{1,n}$) and the fittest cost parameters c of the SVM classifier, in order to optimally classify the WCE images. DBSO is inspired from the effective brainstorming process of human beings, which is based on the consensus reached between the members of the brainstorming group that have similar ideas. To apply the DBSO algorithm, the following steps are performed.

Step 1 generates N potential solutions (individuals); in this case, the individuals represent the number of visual words ($\{w_{n,m}\}$) for the vocabulary construction per class, the number of fuzzy sets needed ($l_{1,n}$) and the cost parameter c. In **step 2**, the N individuals are clustered into $M < N$ groups. For each cluster, in **step 3**, the best individuals are recorded as cluster centers and they are sorted ascendingly, in order to facilitate the calculation of the Euclidean distance and the corresponding similarity in **step 4**. In **step 5**, the two most similar centers are detected and a merging strategy between their corresponding clusters is held. In **step 6**, new individuals are generated, based on one or two cluster and then they are compared with the existing ones. In **step 7**, the best variables $\{w_{n,m}\}, l_{1,n}, c$ are selected, to achieve the best possible image classification performance.

2.3 Explanation Extraction

During the XFBoW feature extraction process, explainable features are extracted enabling the interpretation of the classification outcome in an understandable way. Based on the methodology described in the previous subsection, the similarities $\varphi_{n,m}$ are characterized by a linguistic value represented by the fuzzy sets $Z_{b,n}, b \in [1, L_A], n = 1, \ldots, N_A$ for the abnormal class and $Z_{b,n}, b \in [1, L_{A'}], n = 1, \ldots, N_{A'}$ for the normal class. Thus, the linguistic values express the similarity degree between the content of an input image (expressed by the extracted feature vectors) and the visual words (Fig. 1). Consequently, for a fuzzy set corresponding to a centroid that is closer to 0, we chose the linguistic value "Low", and for the fuzzy set corresponding to a centroid that is closer to 1, we chose "High". Similarly, for the fuzzy sets corresponding to centroids in between, the linguistic values expressing intermediate similarity degrees, *e.g.,* "Medium", are chosen. Then, an explanation can be extracted for each f_j *e.g.,* if f_j is closer to an abnormal visual word with *"Medium"* similarity, then the content of the image in the location of f_j is abnormal. As a final step, the explanations of the entire number of f_j with respect to all the visual words, are averaged based on similarities $\varphi_{n,m}$ and they are weighted using their respective membership $\mu_{l,m}(\varphi_{n,m})$ to explain the classification outcome and the image content *e.g.,* for three features f_1, f_2, f_3 with respective similarities $\varphi_1(f_1) =$ *"High"*, $\varphi_1(f_2) =$ *"High"*, $\varphi_1(f_3) =$ *"Low"*, the total aggregated similarity is the average of $\varphi_1(f_1)$, $\varphi_1(f_2)$, $\varphi_1(f_3)$ and the linguistic values is *"High"*. A concise example of this process is provided to explain the classification results obtained on a real WCE dataset in the following section.

3 Experiment and Results

In this study, a subset of "Dataset 2" from the publicly available KID database is used [16]. This dataset is composed of WCE video frames obtained from the whole GI tract using a MiroCam capsule endoscope with a resolution of 360×360 pixels. The dataset includes 303 images of vascular, 44 images of polypoid abnormalities, 227 images of most common inflammatory lesions and 1778 normal images obtained from the esophagus, the stomach, the small bowel and the colon. All experiments were implemented with MATLAB and they were performed on a workstation with an Intel i5 2.5 GHz CPU, 4 GB RAM.

3.1 Performance Evaluation

XFBoW is compared with other state-of-the-art BoW and CNN methods in terms of Area Under receiver operating Characteristic (AUC) [17] Accuracy, Sensitivity, Specificity [18, 19] and time. The method proposed in [2] is considered as a baseline for the comparisons. Additional comparisons are performed with the Class-Specific BoW [5], the FFE [6], a combination of Class-Specific BoW with FFE, as well as with state-of-the-art CNN-based methods, namely WCNN [8] and LB-FCN [12]. In all cases of the

BoW-based methods, the color features of study [15] are used, k-means [20] is utilized for the construction of the vocabularies, and image classification is performed by an SVM [21]. The optimal number of visual words needed, the linguistic values per visual word and the SVM cost parameter c, are determined by DBSO. The number of linguistic values per visual word ranges between 3 to 7, whereas the examined vocabulary ranges between 50 to 150 words. To obtain statistically justified results, the experiments were executed for 15 independent runs, for a population of 20 individuals and 5 clusters. The rest of the parameters of DBSO were tuned according to the proposed values [13]. Moreover, a 10-fold cross validation (CV) evaluation scheme is adopted, *i.e.,* the dataset was randomly partitioned into 10 equally sized disjoint subsets, a single subset was retained as the validation data for testing the model and the remaining 9 subsets were used as training data were used for testing.

As it can be observed from Table 1, XFBoW has better results compared to all BoW-based methods. The focus should be mainly on the AUC value, due to the fact that unlike accuracy, it is more robust for datasets with imbalanced class distributions [17]. Furthermore, Table 2 presents the comparisons of the proposed model with the CNN-based approaches. XFBoW performs better than WCNN, in terms of AUC. LB-FCN has a higher performance than XFBoW; however, LB-FCN does not have the ability to provide explanations about the classification result, in contrast to XFBoW (as described in Sect. 3.3). Also, XFBoW is characterized by a significantly lower execution time, compared to all the CNN-based methods.

Table 1. Comparative results of XFBoW with state-of-the-art BoW-based methods.

Method	AUC	Accuracy	Sensitivity	Specificity	Time (sec)
Baseline	0.809	0.768	0.454	0.886	49
Class-Specific	0.798	0.778	0.447	0.885	50
FFE	0.790	0.781	0.461	0.884	48
Class-Specific FFE	0.791	0.772	0.449	0.876	53
Proposed	**0.841**	**0.811**	0.469	**0.901**	62

Table 2. Comparative results of XFBoW with state-of-the-art CNN-based methods.

Method	AUC	Accuracy	Sensitivity	Specificity	Time (sec)
WCNN	0.814	**0.899**	0.360	**0.913**	18,660
LB-FCN	**0.935**	0.882	**0.921**	0.764	7,200
Proposed	0.841	0.811	0.469	0.901	62

3.2 Statistical Analysis

In order to investigate the statistical significance of the performance of XFBoW with the compared methods in terms of AUC, the Student's t-test was used [22]. The significance level is set to be equal to 0.05. According to the results presented in Table 3, the difference between the compared mean AUC values is statistically significant, since the calculated p-values are smaller than 0.05, compared to Bow-based and CNN-based methods: Baseline [2], Class-Specific [5], FFE [6] and Class-Specific FFE, WCNN [8] and LB-FCN [12].

Table 3. T-test of mean AUC of XFBoW vs. state-of-the-art classification methods

Methods	Baseline	Class-Specific	FFE	Class-Specific FFE	WCNN	LB-FCN
---------	----------	----------------	-----	--------------------	----- -	--------
t-value	−6.281	−7.1764	−8.2075	−8.3329	−4.0234	13.1176
p-value	<.00001	<.00001	<.00001	<.00001	.000058	<.00001

3.3 Explainability Analysis

In order to demonstrate the explainability of the proposed XFBoW model, an indicative example is provided. The content of each visual word, per tissue type is presented in Fig. 2(b). In addition, Table 4 presents the calculated similarity of the extracted color features, between each image patch and the closer visual words. Specifically, the linguistic values that are assigned to each similarity are provided. For example, f_1 is closer to $w_{4,A'}$ with "*Medium*" similarity as it constitutes normal area of the image, while f_4 is closer to $w_{3,A}$, $w_{4,A}$ with "*High*" similarity to blood and angiectasias. Aggregating the average of all similarities, as described in Sect. 2.3 and their respective linguistic values, an explainable histogram is constructed, and it is presented as "**Total**" in the last row of Table 4. Thus, based on the final row of the given example, the explanation of the classification outcome of the examined image of Fig. 2(a) is the following:

"The image is classified as abnormal because there is "Low" similarity to a case of stenosis $w_{2,A}$, "High" similarity to a case of bleeding $w_{3,A}$ and angiectasias $w_{4,A}$ while a "Medium" similarity to normal cases can be observed."

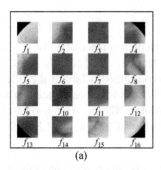
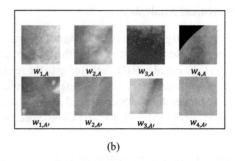

(a)   (b)

Fig. 2. (a) Subimages (f1–f16) sampled from an example image with bleeding (part of the vascular lesion images). (b) Representative illustrations of tissue type of visual words.

Table 4. Calculated similarity $(\varphi_{n,m})$ between color extracted features (f_j) and the closest visual words $(w_{n,m})$, where *"N"*, *"L"*, *"M"*, *"H"* stand for the linguistic values "None", "Low", "Medium", "High".

f_j	$w_{1,A}$	$w_{2,A}$	$w_{3,A}$	$w_{4,A}$	$w_{1,A'}$	$w_{2,A'}$	$w_{3,A'}$	$w_{4,A'}$
f_1	N	N	N	N	N	N	N	M
f_2	N	L	N	N	N	N	N	N
f_3	N	N	H	N	N	N	N	N
f_4	N	N	H	H	N	N	N	N
f_5	N	L	N	N	N	N	N	N
f_6	N	N	H	N	N	N	N	N
f_7	N	N	H	N	N	N	N	N
f_8	N	N	N	H	N	N	N	N
f_9	N	N	N	H	N	N	N	N
f_{10}	N	N	H	H	N	N	N	N
f_{11}	N	N	H	H	N	N	N	N
f_{12}	N	L	N	N	N	N	N	N
f_{13}	N	N	H	H	N	N	N	N
f_{14}	N	N	N	H	N	N	N	N
f_{15}	N	L	N	N	N	N	N	N
f_{16}	N	N	N	N	N	N	N	M
Total	N	L	H	H	N	N	N	M

4 Discussion and Conclusion

In this paper, XFBoW was presented as a novel approach to feature extraction for explainable supervised classification using weakly annotated WCE training images. The proposed model extends the concept of BoW by considering the similarity histograms that can explain the classification outcome of an image based on the similarity of its content with other images, from a weakly annotated training dataset. The performance of the proposed model was investigated on real datasets that include a variety of gastrointestinal abnormalities. XFBoW outperformed, in terms of AUC, relevant baseline and state-of-the-art methods, including Class-Specific, FFE, Class-Specific FFE and WCNN. The lower classification performance of XFBoW in comparison to LB-FCN, is offset by its significantly lower time-performance and its ability to provide explainable classification outcomes, even with conventional classification schemes, such as SVMs. Future research directions include further investigation of the effectiveness of XFBoW, with larger and diverse datasets, for potential applications in healthcare, such as the assessment of a disease severity. In addition, XFBoW is planned to be applied and combined with a variety of features and classifiers, including deep learning-based approaches.

Acknowledgment. This work was supported in part by the grant No. 5024 of the Special Account of Research Grants of the University of Thessaly, Greece.

References

1. Vasilakakis, M., Koulaouzidis, A., Yung, D.E., Plevris, J.N., Toth, E., Iakovidis, D.K.: Follow-up on: optimizing lesion detection in small bowel capsule endoscopy and beyond: from present problems to future solutions. Expert Rev. Gastroenterol. Hepatol. **13**, 129–141 (2019)
2. Vasilakakis, M., Iakovidis, D.K., Spyrou, E., Koulaouzidis, A.: Weakly-supervised lesion detection in video capsule endoscopy based on a bag-of-colour features model. In: Peters, T., et al. (eds.) CARE 2016. LNCS, vol. 10170, pp. 96–103. Springer, Cham (2017). https://doi.org/10.1007/978-3-319-54057-3_9
3. Van Gemert, J.C., Veenman, C.J., Smeulders, A.W., Geusebroek, J.-M.: Visual word ambiguity. IEEE Trans. Pattern Anal. Mach. Intell. **32**, 1271–1283 (2009)
4. Garca-Rodrguez, A., et al.: Polyp fingerprint: automatic recognition of unique features to univocally identify colorectal polyps. Endoscopy **51**, OP186V (2019)
5. Altintakan, U.L., Yazici, A.: Towards effective image classification using class-specific codebooks and distinctive local features. IEEE Trans. Multimedia **17**, 323–332 (2015)
6. Altintakan, U.L., Yazici, A.: A novel fuzzy feature encoding approach for image classification. In: 2016 IEEE International Conference on Fuzzy Systems (FUZZ-IEEE), pp. 1134–1139. IEEE (2016)
7. Vasilakakis, M.D., Koulaouzidis, A., Marlicz, W., Iakovidis, D.K.: The future of capsule endoscopy in clinical practice: from diagnostic to therapeutic experimental prototype capsules. Gastroenterology **15**, 179 (2020)
8. Iakovidis, D.K., Georgakopoulos, S.V., Vasilakakis, M., Koulaouzidis, A., Plagianakos, V.P.: Detecting and locating gastrointestinal anomalies using deep learning and iterative cluster unification. IEEE Trans. Med. Imaging **37**, 2196–2210 (2018)
9. Yang, J., Chang, L., Li, S., He, X., Zhu, T.: WCE polyp detection based on novel feature descriptor with normalized variance locality-constrained linear coding. Int. J. Comput. Assist. Radiol. Surg. **15**(8), 1291–1302 (2020). https://doi.org/10.1007/s11548-020-02190-3

10. García-Rodríguez, A., et al.: Polyp fingerprint: automatic recognition of colorectal polyps' unique features. Surg. Endosc. **34**(4), 1887–1889 (2020). https://doi.org/10.1007/s00464-019-07240-9

11. Patel, A., Rani, K., Kumar, S., Figueiredo, I.N., Figueiredo, P.N.: Automated bleeding detection in wireless capsule endoscopy images based on sparse coding. Multimedia Tools Appl. **80**, 30353–30366 (2021). https://doi.org/10.1007/s11042-020-09605-y

12. Diamantis, D.E., Iakovidis, D.K., Koulaouzidis, A.: Look-behind fully convolutional neural network for computer-aided endoscopy. Biomed. Signal Process. Control **49**, 192–201 (2019)

13. Sovatzidi, G., Iakovidis, D.K.: Determinative brain storm optimization. In: Tan, Y., Shi, Y., Tuba, M. (eds.) ICSI 2020. LNCS, vol. 12145, pp. 259–271. Springer, Cham (2020). https://doi.org/10.1007/978-3-030-53956-6_24

14. Tuytelaars, T.: Dense interest points. In: 2010 IEEE Computer Society Conference on Computer Vision and Pattern Recognition, pp. 2281–2288. IEEE (2010)

15. Vasilakakis, M.D., Iakovidis, D.K., Spyrou, E., Koulaouzidis, A.: DINOSARC: Color features based on selective aggregation of chromatic image components for wireless capsule endoscopy. Comput. Math. Methods Med. **2018** (2018). https://doi.org/10.1155/2018/2026962. Article ID 2026962

16. Koulaouzidis, A., et al.: KID Project: an internet-based digital video atlas of capsule endoscopy for research purposes. Endosc. Int. Open **5**, E477–E483 (2017)

17. Provost, F., Fawcett, T.: Analysis and visualization of classifier performance with nonuniform class and cost distributions. In: Proceedings of AAAI-97 Workshop on AI Approaches to Fraud Detection & Risk Management, pp. 57–63 (1997)

18. Jadon, S., Leary, O.P., Pan, I., Harder, T.J., Wright, D.W., Merck, L.H., Merck, D.L.: A comparative study of 2D image segmentation algorithms for traumatic brain lesions using CT data from the ProTECTIII multicenter clinical trial. In: Medical Imaging 2020: Imaging Informatics for Healthcare, Research, and Applications, p. 113180Q. International Society for Optics and Photonics (2020)

19. Jadon, S.: A survey of loss functions for semantic segmentation. In: 2020 IEEE Conference on Computational Intelligence in Bioinformatics and Computational Biology (CIBCB), pp. 1–7. IEEE (2020)

20. Drake, J., Hamerly, G.: Accelerated k-means with adaptive distance bounds. In: 5th NIPS Workshop on Optimization for Machine Learning (2012)

21. Theodoridis, S., Koutroumbas, K.: Pattern Recognition, 4th edn. Academic Press Inc., Orlando (2008)

22. Steel, R.G.D., Torrie, J.H., et al.: Principles and Procedures of Statistics (1960)

Towards Semantic Interpretation of Thoracic Disease and COVID-19 Diagnosis Models

Ashkan Khakzar[1], Sabrina Musatian[1], Jonas Buchberger[1],
Icxel Valeriano Quiroz[1], Nikolaus Pinger[1], Soroosh Baselizadeh[1],
Seong Tae Kim[2(✉)], and Nassir Navab[1,3]

[1] Technical University of Munich, Munich, Germany
[2] Kyung Hee University, Yongin-si, South Korea
st.kim@khu.ac.kr
[3] Johns Hopkins University, Baltimore, USA

Abstract. Convolutional neural networks are showing promise in the automatic diagnosis of thoracic pathologies on chest x-rays. Their black-box nature has sparked many recent works to explain the prediction via input feature attribution methods (aka saliency methods). However, input feature attribution methods merely identify the importance of input regions for the prediction and lack semantic interpretation of model behavior. In this work, we first identify the semantics associated with internal units (feature maps) of the network. We proceed to investigate the following questions; Does a regression model that is only trained with COVID-19 severity scores implicitly learn visual patterns associated with thoracic pathologies? Does a network that is trained on weakly labeled data (e.g. healthy, unhealthy) implicitly learn pathologies? Moreover, we investigate the effect of pretraining and data imbalance on the interpretability of learned features. In addition to the analysis, we propose semantic attribution to semantically explain each prediction. We present our findings using publicly available chest pathologies (CheXpert [5], NIH ChestX-ray8 [25]) and COVID-19 datasets (BrixIA [20], and COVID-19 chest X-ray segmentation dataset [4]). The Code (https://github.com/CAMP-eXplain-AI/CheXplain-Dissection) is publicly available.

Keywords: Interpretability · COVID-19 · Chest X-rays

1 Introduction

Convolutional neural networks (CNN) have demonstrated outstanding performance in automatic diagnosis on Chest X-rays [5,6,12,25]. There are reports of CNNs outperforming radiologists in chest x-ray pathology classification [17]. These diagnostic models can aid the clinicians and expedite the diagnosis resulting in more patients receiving the care they need. Such models can be especially beneficial in pandemic circumstances as the shortage of expert clinicians

S. T. Kim and N. Navab shared senior authorship.

M. de Bruijne et al. (Eds.): MICCAI 2021, LNCS 12903, pp. 499–508, 2021.
https://doi.org/10.1007/978-3-030-87199-4_47

becomes an issue [14,16,20]. Despite their performance, neural networks' lack of interpretability undermines their reliability. It is essential to understand the basis of the network predictions, and the networks' learned features to establish trust in the clinical domain. Therefore there have been efforts in explicitly making the models more interpretable during training [8,24], or interpreting neural network models after they are trained [7,10,17,25]. In this work, we investigate the latter case in order to see if these performant models trained without an infusion of interpretability, are learning human-interpretable concepts.

For post-hoc explanation of chest X-ray models, many works opt for feature attribution methodologies [1,9,18,21,23,25]. These works use feature attribution methods, such as Class Activation Maps (CAM) [18,25] to reveal which input regions are contributing to the output prediction. Albeit being insightful, the aforementioned methodology lacks semantic interpretation of the models and their predictions. Network Dissection [2] is a methodology for identifying the corresponding concept of internal units (feature maps) of the network.

Contributions - In this work, we first use Network Dissection [2] to quantify the interpretability of chest X-ray classification models. Then we proceed to investigate the following; Does a neural network regression model that is only trained on COVID-19 severity scores (on BrixIA) implicitly learn visual patterns associated with thoracic pathologies? We also study the effect of pretraining on CheXpert and ImageNet datasets, and the effect of considering data imbalance on the semantics of internal units. Does a network trained on a weakly labeled dataset (healthy/unhealthy labels) implicitly learn distinct pathologies? We combine NIH ChestX-ray8, CheXpert, and BrixIA datasets to generate a massive but weakly labeled ('healthy','unhealthy') dataset. In this case, we study the effect of considering data imbalance on the semantics of internal units. Moreover, for both cases, we observe the formation of semantic units during training. In all experiments, we use bounding boxes in the NIH ChestX-ray8 [25], and segmentation masks in COVID-19 chest X-ray segmentation dataset [4] for identifying the semantics of units. In addition to the analysis, we propose semantic attribution by combining feature attribution and network dissection to semantically explain the prediction for each chest x-ray.

Related Work - Interpreting Internal Units: There are two principal categories of methods for this purpose; Methods that generate images that maximally activate neurons/units [13,15,21], and methods that search over the dataset to find which images (also image regions) activates neurons/units [2,27]. Methods in the first category are prone to subjective interpretation as the generated images are ambiguous. **Network Dissection** [2] is a prominent method of the second category, that does quantitative analysis of the semantics of units. Though we use the same method (albeit on the different domain of chest x-rays rather than natural images), our experiments and insights differ. Effect of pretraining, imbalanced datasets, studying regression models and trained models on weakly-labeled datasets are exclusive to our work. We also propose semantic attribution. **DeepMiner** [26] is a methodology inspired by network dissection and applied to mammograms. The methodology differs from Network Dissection and our work. In DeepMiner, instead of automatic annotation, the most important units are annotated by experts. DeepMiner thus differs in methodology, discusses a different domain (mammograms), and does not address our research questions.

2 Methodology

2.1 Setup: Chest X-ray Models:

Classification Model: Each chest X-ray can contain multiple pathologies. Therefore we model the problem as a multi-label classification problem. For C pathologies, the network function is defined as $f_\Theta(x) : \mathbb{R}^{H \times W} \rightarrow \mathbb{R}^C$. The predicted probability for each category is $\hat{y} = sigmoid(f_\Theta(x))$. We use Binary Cross Entropy (BCE) loss on each output. Thus the loss is defined as:

$$\mathcal{L}_{BCE} = (\hat{y}, y) = -\sum_c \beta y_c \log(\hat{y}_c) + (1 - y_c) \log(1 - \hat{y}_c) \tag{1}$$

where β is a weighting factor to balance the positive labels, and defined as the ratio of the number of negative labels to the number of positive labels in a batch.

Regression Model: We consider two regression modelings on BrixIA dataset.

1) For each input j the label is one global severity score $y^{(j)} \in \{0, ..., 18\}$. For this case the neural network function is $f_\Theta(x) : \mathbb{R}^{H \times W} \rightarrow \mathbb{R}$. We use a weighted Mean Square Error (MSE) loss for a batch of size N:

$$\mathcal{L}_{MSE} = \frac{1}{N} \sum \beta(f_\Theta(x)^{(n)} - y^{(n)})^2 \tag{2}$$

2) In BrixIA dataset for each X-ray, the lung is divided into 6 regions, and a severity score $\in \{0, 1, 2, 3\}$ is assigned to each region. Thus the network is defined by $f_\Theta(x) : \mathbb{R}^{H \times W} \rightarrow \mathbb{R}^6$. Similar to [20] we use a mixed regression/classification loss by using Sparse Categorical Cross Entropy (SCCE) and diffentiable Mean Absolue Error (MAE^d), and for each sample and its corresponding network output $\hat{y} = [\hat{y}_c]^6$ it is defined as $\mathcal{L}_{SCCE} + \mathcal{L}_{MAE^d}$:

$$\mathcal{L}_{SCCE} = -\frac{1}{C} \sum_c y_c log(\hat{y}_c) \tag{3}$$

$$\mathcal{L}_{MAE^d} = \frac{1}{C} \left\| y - \sum_c \frac{e^{\hat{y}_c}}{\sum_c e^{\hat{y}_c}} c \right\| \tag{4}$$

2.2 Background: Network Dissection [2]

Network Dissection annotates individual units (feature maps) of neural networks with semantic concepts. We refer to a feature map that is associated with an individual concept (e.g. consolidation) as an individual concept detector/a

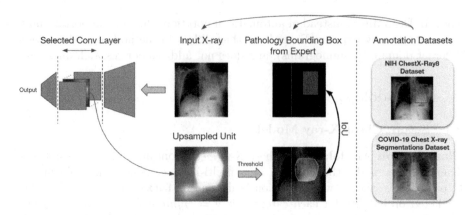

Fig. 1. Network Dissection on chest X-rays: To identify the corresponding concept of a unit in the network, the unit's activation for each input is compared with the ground truth mask/bounding-box of the concept in that input. For one unit the procedure is repeated for all inputs within the annotation datasets and average IoU with concepts in the dataset measures whether the unit is detecting a concept.

semantic unit. The method requires a dataset(s) where concepts are annotated with bounding box or segmentation masks. For each unit under investigation, the unit's feature map is computed for every image and is compared against the ground truth annotations (bounding boxes or segmentation masks) of the concepts in that image. The method compares the unit's feature map and the annotation in terms of their intersection over union. See Fig. 1 for a schematic overview. The activation map is first transformed in to a binary segmentation mask $S_k(\mathbf{x})$ via thresholding. For each unit k and its feature map $F_k(\mathbf{x})$, the threshold T_k is chosen such that $P(f_k > T_k) = 0.005$ [2] for every f_k given all images in the dataset. As the resolution of ground truth labels L_c for each concept c is different from S_k, bilinear interpolation is first applied to generate $S'_k(\mathbf{x})$. The binary segmentation mask is then derived by $M_k(\mathbf{x}) \doteq S'_k(\mathbf{x}) \geq T_k$. For each unit k and concept c pair, the intersection and the union between mask $M_k(\mathbf{x})$ derived from unit k and the label mask $L_c(\mathbf{x})$ is computed for all images containing concept c, and data-set-wide $IoU_{k,c}$ is defined as:

$$IoU_{k,c} = \frac{\sum |M_k(\mathbf{x}) \cap L_c(\mathbf{x})|}{\sum |M_k(\mathbf{x}) \cup L_c(\mathbf{x})|} \tag{5}$$

the sum is carried out over all the images with concept c, and $|.|$ denotes the cardinality of this set. The $IoU_{k,c}$ measures whether unit k detects concept c. We use a threshold of 0.04 similar to [2] for considering a unit as a detector. The threshold affects the number of concept detectors within a network, what is of interest is *comparing* the number of concept detectors between models (e.g. comparing trained and initial model). See Fig. 3 for units of different $IoU_{k,c}$.

Network Dissection for Chest X-ray Models: We use the NIH ChestX-ray8 [25] and COVID-19 chest X-ray segmentation [4] (Covid-CXR) dataset for annotating the chest X-ray models. NIH ChestX-ray8 contains bounding boxes for 8 pathologies, and we consider each pathology as a concept c in the Network Dissection framework. Covid-CXR contains segmentations for pathologies, lung components (e.g. right lung), and apparatus. we consider each one as a concept. The dataset contains 14 distinct concepts in total. For each model that we discuss in the paper, we use both NIH ChestX-ray8 and Covid-CXR datasets for annotating the models' internal units.

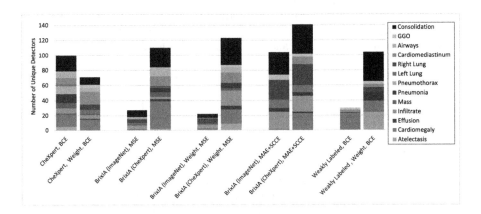

Fig. 2. The number/type of individual concept detectors: For all models under discussion, the number of concept detectors and the types of the detectors are presented. We can observe that pretraining on CheXpert is increasing the number and the variety of detectors. Moreover, considering data imbalance by using a weighted loss function also increases the number and variety of semantic detectors.

2.3 Semantic Attribution

For each prediction $\mathbf{f}_\Theta(\mathbf{x})$ on each input $\mathbf{x}$, we compute the importance of internal units A_k^L for the prediction. A_k^L denotes k_{th} feature map at layer L.

We follow a method, Integrated Gradients [23], that approximates the Aumann-Shapley value [22]. Shapley value [19] is the unique axiomatic definition of a feature's contribution to a final outcome. We first compute the individual neurons' importance a_k^i (activation i at channel k):

$$\mathbf{s}_k^i = \mathbf{a}_k^i \int_{\alpha=0}^1 \frac{\partial \mathbf{f}_\Theta(\alpha \mathbf{a}_k^i)}{\partial \mathbf{a}_k^i} d\alpha \tag{6}$$

We use $\sum_i |\mathbf{s}_k^i|$ as the contribution of the unit k. The absolute value is for considering both positive and negative contributions by taking the magnitude. We then select the top contributing unit(s) and use their annotations (if it exists) for semantic explanation. We visualize $\mathbf{s}_k^i$ of all neurons of A_k^L to highlight each semantic unit's most contributing areas to the output.

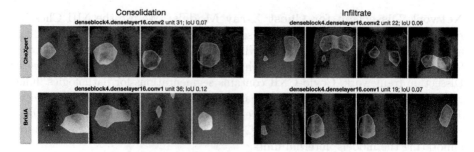

Fig. 3. Individual concept detectors: For CheXpert classification model (top) and BrixIA regression model (bottom) we visualize a Consolidation detector unit and an Infiltration detector unit for 4 different X-rays. The displayed IoU is not the IoU between the ground truth sample and the detector, it is the average from Eq. 5.

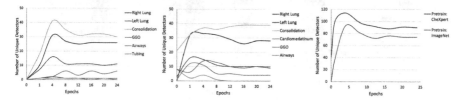

Fig. 4. Evolution of concepts: Number/Type of semantic detectors in two regression models at different epochs. (left): BrixIA (ImageNet), MAE+SCCE (middle): BrixIA (CheXpert), MAE+SCCE, (right) total number of semantic detectors for both models. We observe that the number of concept detectors is higher for trained models at different epochs compared to the initial model.

3   Results and Discussion

In all our experiments and for all models in Fig. 2, the network is a DenseNet121 [3]. All models are trained using Adam [11] optimizer without weight decay. If the loss function is a weighted one, it is indicated by "Weight". For BrixIA models, the model is trained on a pre-trained model on either ImageNet or CheXpert. The name in parentheses shows this. When the last part of the name shows the loss function. "MAE+SCCE" indicates using the 6 region prediction as in the original BrixIA paper. The weakly labeled models are binary classifiers, trained on the combination of NIH, CheXpert, and BrixIA datasets, deciding if the input is healthy or not. While NIH dataset has a healthy label for inputs, for CheXpert, the input was considered healthy when no pathology existed. For BriXIA, an input was considered healthy when all 6 regions were zero.

3.1   Semantics of Thoracic Classification Models

In this section, we analyze the semantics of detectors in classification models trained on CheXpert. Although these models are trained directly on the pathology labels, this does not imply that the individual detectors within the network are detecting features relevant to these pathologies. The models are denoted by CheXpert BCE, and CheXpert Weight. BCE on Fig. 2.

Balanced Training Effect. We see that the model trained with weighted BCE has a significantly higher number of concept detectors (100 vs 65). The number of detectors relevant to Pneumonia increases significantly in the weighted model. Pneumonia has a few positive instances in the dataset (0.027%), using a weighted loss causes the emergence of Pneumonia detectors. Consolidation is also imbalanced (0.066%), and we observe that the number of detectors is tripled. In addition, the weighted model shows more variety in terms of concepts than the unweighted version. This is particularly significant as we do not observe any improvement in terms of classification AUC, and also the F1 score. But the model interpretability increases with the weighted loss.

3.2   Semantics of COVID-19 Regression Models

In this section, we examine MSE and MAE SCCE segmentation losses for models trained on the BrixIA dataset. In addition, we investigate the effect of considering imbalance in the data in the MSE loss models. As the number of images in BrixIA is limited, we also investigate the effect of pretraining on the CheXpert dataset. The first observation in Fig. 2 is that training the model on 6 region regression loss (SCCE), results in a higher number of semantic detectors compared to global regression loss. The model has stronger supervision in the former case, and we observe that it also becomes more interpretable.

Pretraining Effect. For all training losses, we observe in Fig. 2 that pretraining the model on the CheXpert dataset is significantly increasing the number of individual concept detectors. This increase in the number of detectors is more pronounced in models trained with the global regression score labels. This implies that with weaker supervision with one global regression score, the model struggles in learning individual concepts, and pretraining becomes necessary.

Balanced Training Effect. We observe (Fig. 2) that using weighted MSE loss on regression models pre-trained on CheXpert datasets increases the number of unique concept detectors, specifically an increase in the number of detectors associated with Consolidation. For models that are not pretrained on CheXpert, the model struggles in both weighted and unweighted scenarios to learn concepts.

3.3   Semantics of Models Trained on Weakly Labeled Datasets

In this section we discuss models trained on a dataset of healthy/unhealthy labels. The dataset is comprised of CheXpert, NIH ChestX-ray8 and BrixIA.

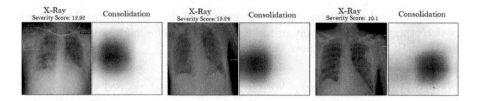

Fig. 5. Semantic attribution: X-rays with severe Covid condition from BrixIA. All have a unit with corresponding semantic annotation of Consolidation as their top contributing unit to the severity score. The contribution of corresponding unit is visualized.

Balanced Training Effect. We observe in Fig. 2 that the model that does not use a weighted loss, learns few semantic concepts. Moreover, the majority of individual detectors are related to Cardiomegaly and Cardiomediastinum, and the rest of the concepts are not learned. This implies that Cardiomegaly is an easy signal for the model to pick up during training. However, when we train the model with a weighted loss, we observe that the diversity of the semantic detectors and their numbers increases significantly. Specifically, there were not Consolidation, Pneumonia, Atelectasis detectors in the former case, but they emerge in high numbers in the weighted case. We also observe that the concept detectors indeed emerge in models trained on healthy/unhealthy labels, although it is relatively less than full supervision.

3.4   Evolution of Semantics

In this section, we observe the number/type of individual unit detectors for different epochs. We analyze the BrixIA regression models trained with MAE+SCCE. We observe (Fig. 4) an increase in the number of important concepts for COVID-19 during training (e.g. Consolidation and Ground Glass Opacity) and a decrease in less important concepts (e.g. left lung). The relative increase in the number of concepts of trained model vs. initial model points to the fact that the models are indeed learning patterns associated with pathologies.

3.5   Semantic Attribution

We observe that Consolidation is the most related concept to severe Covid scores as in Fig. 5 on BrixIA dataset. The results in Fig. 5 are for the last conv layer of the model described in Fig. 2 as BrixIA (CheXpert), Weight, MSE. To visualize, the final integrated gradients map of unit k is divided by the maximum s_k^i of its neurons to normalize and then upsampled to the input image size. Note the methodology here is different from using crude activation maps since we aim to discover the most *contributing* feature maps to the prediction. Highly activated maps do not necessarily correspond to contribution. Hence, we employ an axiomatic approach as explained in Sect. 2.3. Although this is a first step in this direction, more annotations can boost the recognition of other important concepts, and more semantic units for each prediction can be visualized.

4   Conclusion

In this work, we analyze the semantics of internal units of classification and regression models on chest X-rays. We observe that pretraining and considering data imbalance affect the number of semantic detectors. We observe how training on severity score regression labels results in the emergence of pathology-related detectors within the networks. We observe the same phenomenon when the model is trained on weakly labeled datasets. We also propose a semantic attribution method to semantically explain individual predictions.

Acknowledgement. This work is partially funded by the **Munich Center for Machine Learning (MCML)** and the **Bavarian Research Foundation** grant AZ-1429-20C. The computational resources for the study are provided by the **Amazon Web Services Diagnostic Development Initiative**. S.T. Kim is supported by the **Korean MSIT**, under the National Program for Excellence in SW (2017-0-00093), supervised by the IITP.

References

1. Bach, S., Binder, A., Montavon, G., Klauschen, F., Müller, K.R., Samek, W.: On pixel-wise explanations for non-linear classifier decisions by layer-wise relevance propagation. PLoS ONE **10**(7), e0130140 (2015)
2. Bau, D., Zhou, B., Khosla, A., Oliva, A., Torralba, A.: Network dissection: quantifying interpretability of deep visual representations. In: Proceedings of the IEEE Conference on Computer Vision and Pattern Recognition, pp. 6541–6549 (2017)
3. Huang, G., Liu, Z., Weinberger, K.Q.: Densely connected convolutional networks. CoRR abs/1608.06993 (2016). http://arxiv.org/abs/1608.06993
4. Inc, G.B.: Covid-19 chest x-ray segmentations dataset. https://github.com/GeneralBlockchain/covid-19-chest-xray-segmentations-dataset
5. Irvin, J., et al.: Chexpert: a large chest radiograph dataset with uncertainty labels and expert comparison. In: Proceedings of the AAAI Conference on Artificial Intelligence, vol. 33, pp. 590–597 (2019)
6. Johnson, A.E., et al.: MIMIC-CXR-JPG, a large publicly available database of labeled chest radiographs. arXiv preprint arXiv:1901.07042 (2019)
7. Karim, M., Döhmen, T., Rebholz-Schuhmann, D., Decker, S., Cochez, M., Beyan, O., et al.: Deepcovidexplainer: Explainable covid-19 predictions based on chest x-ray images. arXiv preprint arXiv:2004.04582 (2020)
8. Khakzar, A., Albarqouni, S., Navab, N.: Learning interpretable features via adversarially robust optimization. In: Shen, D., Liu, T., Peters, T.M., Staib, L.H., Essert, C., Zhou, S., Yap, P.-T., Khan, A. (eds.) MICCAI 2019. LNCS, vol. 11769, pp. 793–800. Springer, Cham (2019). https://doi.org/10.1007/978-3-030-32226-7_88
9. Khakzar, A., Baselizadeh, S., Khanduja, S., Rupprecht, C., Kim, S.T., Navab, N.: Neural response interpretation through the lens of critical pathways. In: Proceedings of the IEEE/CVF Conference on Computer Vision and Pattern Recognition (2021)
10. Khakzar, A., et al.: Explaining COVID-19 and thoracic pathology model predictions by identifying informative input features (2021)
11. Kingma, D.P., Ba, J.: Adam: a method for stochastic optimization. arXiv preprint arXiv:1412.6980 (2014)

12. Li, Z., et al.: Thoracic disease identification and localization with limited supervision. In: Proceedings of the IEEE Conference on Computer Vision and Pattern Recognition, pp. 8290–8299 (2018)
13. Nguyen, A., Dosovitskiy, A., Yosinski, J., Brox, T., Clune, J.: Synthesizing the preferred inputs for neurons in neural networks via deep generator networks. In: Advances in Neural Information Processing Systems, pp. 3387–3395 (2016)
14. Oh, Y., Park, S., Ye, J.C.: Deep learning COVID-19 features on CXR using limited training data sets. IEEE Trans. Med. Imaging 39(8), 2688–2700 (2020)
15. Olah, C., Mordvintsev, A., Schubert, L.: Feature visualization. Distill (2017). https://doi.org/10.23915/distill.00007, https://distill.pub/2017/feature-visualization
16. Punn, N.S., Agarwal, S.: Automated diagnosis of COVID-19 with limited posteroanterior chest X-ray images using fine-tuned deep neural networks. Appl. Intell. 51(5), 2689–2702 (2020). https://doi.org/10.1007/s10489-020-01900-3
17. Rajpurkar, P., et al.: Chexnet: radiologist-level pneumonia detection on chest x-rays with deep learning. arXiv preprint arXiv:1711.05225 (2017)
18. Selvaraju, R.R., Cogswell, M., Das, A., Vedantam, R., Parikh, D., Batra, D.: Grad-CAM: visual explanations from deep networks via gradient-based localization. In: Proceedings of the IEEE International Conference on Computer Vision, pp. 618–626 (2017)
19. Shapley, L.S.: A value for n-person games. Contrib. Theor. Games 2(28), 307–317 (1953)
20. Signoroni, A., et al.: End-to-end learning for semiquantitative rating of COVID-19 severity on chest X-rays. arXiv preprint arXiv:2006.04603 (2020)
21. Simonyan, K., Vedaldi, A., Zisserman, A.: Deep inside convolutional networks: Visualising image classification models and saliency maps. arXiv preprint arXiv:1312.6034 (2013)
22. Sundararajan, M., Najmi, A.: The many Shapley values for model explanation. In: 37th International Conference on Machine Learning, ICML 2020 (2020)
23. Sundararajan, M., Taly, A., Yan, Q.: Axiomatic attribution for deep networks. In: International Conference on Machine Learning, pp. 3319–3328. PMLR (2017)
24. Taghanaki, S.A., et al.: InfoMask: Masked Variational Latent Representation to Localize Chest Disease. In: Shen, D., et al. (eds.) MICCAI 2019. LNCS, vol. 11769, pp. 739–747. Springer, Cham (2019). https://doi.org/10.1007/978-3-030-32226-7_82
25. Wang, X., Peng, Y., Lu, L., Lu, Z., Bagheri, M., Summers, R.M.: Chestx-ray8: Hospital-scale chest X-ray database and benchmarks on weakly-supervised classification and localization of common thorax diseases. In: Proceedings of the IEEE Conference on Computer Vision and Pattern Recognition, pp. 2097–2106 (2017)
26. Wu, J., et al.: Deepminer: discovering interpretable representations for mammogram classification and explanation. arXiv preprint arXiv:1805.12323 (2018)
27. Zeiler, M.D., Fergus, R.: Visualizing and understanding convolutional networks. In: Fleet, D., Pajdla, T., Schiele, B., Tuytelaars, T. (eds.) ECCV 2014. LNCS, vol. 8689, pp. 818–833. Springer, Cham (2014). https://doi.org/10.1007/978-3-319-10590-1_53

A Principled Approach to Failure Analysis and Model Repairment: Demonstration in Medical Imaging

Thomas Henn[1]($\boxtimes$), Yasukazu Sakamoto[2], Clément Jacquet[1],
Shunsuke Yoshizawa[2], Masamichi Andou[2], Stephen Tchen[1], Ryosuke Saga[1],
Hiroyuki Ishihara[2], Katsuhiko Shimizu[2], Yingzhen Li[3], and Ryutaro Tanno[4]

[1] Rokken Inc., Osaka, Japan
thomas.henn@lab6.co.jp
[2] Corporate R&D Center, Terumo Corporation, Tokyo, Japan
yasukazu_sakamoto@terumo.co.jp
[3] Department of Computing, Imperial College London, London, UK
[4] Department of Computing, University College London, London, UK

Abstract. Machine learning models commonly exhibit unexpected failures post-deployment due to either data shifts or uncommon situations in the training environment. Domain experts typically go through the tedious process of inspecting the failure cases manually, identifying failure modes and then attempting to fix the model. In this work, we aim to standardise and bring principles to this process through answering two critical questions: (i) how do we know that we have identified meaningful and distinct failure types?; (ii) how can we validate that a model has, indeed, been repaired? We suggest that the quality of the identified failure types can be validated through measuring the intra- and inter-type generalisation after fine-tuning and introduce metrics to compare different subtyping methods. Furthermore, we argue that a model can be considered repaired if it achieves high accuracy on the failure types while retaining performance on the previously correct data. We combine these two ideas into a principled framework for evaluating the quality of both the identified failure subtypes and model repairment. We evaluate its utility on a classification and an object detection tasks. Our code is available at https://github.com/Rokken-lab6/Failure-Analysis-and-Model-Repairment.

Keywords: Failure analysis · Model repairment · Deep learning

1 Introduction

It is common for new failures to be discovered once a model has been deployed in the "wild". While recent lines of research in medical imaging have shown promising results in designing robust machine learning (ML) models [1–5], it may not be realistic to achieve perfect generalisation to every relevant environment.

© Springer Nature Switzerland AG 2021
M. de Bruijne et al. (Eds.): MICCAI 2021, LNCS 12903, pp. 509–518, 2021.
https://doi.org/10.1007/978-3-030-87199-4_48

Consequently, recently published guidelines for the reliable application of ML systems in healthcare [6,7] and recent work from Luke *et al.* [8] stress the importance of analyzing and reporting clinically relevant failure cases. However, there is a lack of standardised protocols to identify, validate and analyze those failure types. Typically, domain experts manually inspect the failure cases and make sense of them by identifying a set of failure modes. But this approach can be both expensive and biased by the human expertise. For example, a critical yet rare subgroup could be missed with such an approach and go unreported [8]. A notable recent work [9] recognises this issue and makes a first step towards data-driven approaches to failure subtyping through clustering of the learned features based on whether its presence or absence is predictive of poor performance. However to date, little attention has been gathered around the evaluation metrics of the identified failure types, hampering the development of new methods in this direction. Furthermore, even if a set of meaningful failure types could be identified, methods for *fixing* them and evaluating its success remain undeveloped.

In an attempt to bring principles to the process of failure analysis and model repairment, we introduce a framework for not only deriving subtypes of failure cases and measuring their quality, but also repairing the models and verifying their generalisation. We put forward a set of desirable properties that a meaningful set of failure subtypes should meet and design surrogate metrics. Moreover, we propose a data-driven method for identifying failure types based on clustering in feature or gradient spaces. This method was able not only to identify failure types with the highest quality according to our metrics but also to identify clinically important failures like undetected catheters close to the ultrasound probe in intracardiac echocardiography. Finally, we argue that model repairment should not only aim to fix each failure type in a generalizable manner but also to ensure the performance on previously successful cases is retained.

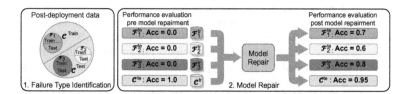

Fig. 1. An overview of the proposed failure analysis and model repairment framework which proceeds in two phases. Firstly, based on the evaluation of the model $\mathcal{M}$, the set of "failure cases" $\mathcal{F}$ (e.g., misclassified cases, examples with high errors, etc.) are clustered to identify failures types $\{\mathcal{F}_i\}_i$ such that $\mathcal{F} = \cup_i \mathcal{F}_i$. Secondly, the failure types $\{\mathcal{F}_i\}_i$ are "fixed" by a *repairment* algorithm (e.g., fine-tuning, continual learning methods, etc.) based on a "training split" of the failure sets $\{\mathcal{F}_i^{tr}\}$, and the correct set $\mathcal{C}^{tr}$. Finally, the success of the repairment is evaluated by measuring generalisation on the test sets of target failure types $\{\mathcal{F}_i^{te}\}$ and the correct cases $\mathcal{C}^{te}$.

2 Methods

We elaborate below the two phases (see Fig. 1) of our approach to model repair.

2.1 Phase I: Identification of Failure Types

It is not evident how to optimally separate failure cases into a set of distinct failure types. Domain experts could split the failure cases according to the visual appearance or consider the importance of different failures from a clinical perspective, for example according to stages or clinical signs of a disease. However, we suggest clinical relevance doesn't necessarily reflect the way the model distinctly fails on each failure type. Failure types should be specific not only to the model but the repairment methods that can be used to fix them. Moreover, objective metrics to quantify the quality of failure types for the purpose of model repairment are still lacking.

Two Desiderata of Failure Types: We postulate that a set of failure types should satisfy two desirable properties: *Independence* and *Learnability*. In addition, we propose two novel surrogate metrics to assess those properties in practice. They are measured by fine-tuning the model $\mathcal{M}$ on a subset $\mathcal{F}_i^{tr}$ of the given failure type $\mathcal{F}_i$ under a *Compatibility constraint*, and then calculating the performance on each $\mathcal{F}_j^{te}$. We assume a sufficiently large set of failures with no label noise or corrupted images.

(i) *Independence:* The subtypes should be as independent as possible from each other in the way the model fails. In other words, they can be fixed by comparatively distinct alterations to the decision boundary. If two types are independent, a model fine-tuned on one type should not be useful to the other. In practice, we suggest a continuous measure of *Independence* by calculating the average difference in the test performance between each failure type and the rest: $I(\mathcal{F}_i) := \text{mean}_{j \neq i} \left(m(\mathcal{M}_i, \mathcal{F}_i^{te}) - m(\mathcal{M}_i, \mathcal{F}_j^{te}) \right)$ where $m(\cdot)$ is a performance metric (e.g., accuracy).

(ii) *Learnability:* Each subtype should be homogeneous and consists of examples for which the model has failed in a similar way. In other words, a such failure type contains failure cases that can be fixed via a similar modification to the model's decision boundary. If a subtype is heterogeneous, fixing it would require more modifications to the model and would be more challenging. We thus posit that a more homogeneous failure type would be easier to learn, and measure *Learnability* as the generalisation of the fine-tuned model $\mathcal{M}_i$ on the chosen failure type $L(\mathcal{F}_i) := m(\mathcal{M}_i, \mathcal{F}_i^{te})$.

Compatibility Constraint: We argue in addition that the above surrogate metrics should be measured with the constraint of maintaining performance on correct data $\mathcal{C}$. This is necessary to avoid learning pathological discriminative rules just to solve a specific failure type. For example, given a failure type only containing images of a single class, the model could simply learn to ignore the input images and predict the same class everywhere, in which case the failure is fixed in a meaningless way. *Compatibility* ensures "locality" of the failure types by ensuring the required changes in the discriminative rules do not considerably influence the previous cases. *Compatibility* is achieved by fine-tuning on both a failure type $\mathcal{F}_i^{tr}$ and previously correct cases $\mathcal{C}^{tr}$ in equal proportions and early stopping when $\mathcal{C}^{tr}$ validation performance drops below 0.9.

Automatic Identification of Failure Types: Manual analysis can often be time-consuming but also suboptimal, potentially overlooking meaningful failure modes. Therefore, we wish to automatically uncover a set of failure types with good *Independence* and *Learnability* scores. We thus also explore two methods as specific instantiations of our framework; particularly, we experiment with clustering the failure cases $\mathcal{F}$ in the feature space (*Feature clustering*) and gradient space of a differentiable model $\mathcal{M}$ (*Gradient clustering*). The gradients of the loss with respect to the parameters are used (for object detection: the loss specific to the object). We expect that similar data will be close in the feature space and the failures whose correction requires similar changes to the model parameters will be close in the gradient space. Furthermore, features are averaged over spatial dimensions and both features and gradients are reduced by Gaussian random projection followed by UMAP [10] to 10 dimensions. Finally, the data is clustered through k-means with the highest Silhouette score between 3 and 10 clusters.

2.2   Phase II: Repairment of Failure Types

Once a set of target failure types has been identified, the model needs to be repaired. We argue that a successful repairment leads to a model that generalizes on unseen cases of the target failure types while maintaining performance on the cases $\mathcal{C}$ where the model previously performed well. First, a target set of failure types $\mathcal{F}^T \subset \{\mathcal{F}_i\}_i$ is selected by the end-users, and then each type is split into two sets, where one is used to first repair the model (e.g. $\mathcal{F}_i^{tr}$) and then the other is used to evaluate generalization (e.g. $\mathcal{F}_i^{te}$). For repairment procedures, we experiment with fine-tuning on the failure types (with or without the correct cases) and *elastic weight consolidation* (EWC), a popular continual learning approach [11]. We note, however, that more recent model adaptation approaches are also applicable in our framework such as more recent variants of continual learning [12,13], meta-learning [14] and domain adaptation [15].

3   Datasets and Implementation Details

We evaluate the efficacy of the proposed model repairment framework on two medical imaging datasets. The details of the respective datasets along with the specification of models/optimisation are provided below.

Binary PathMNIST (BP-MNIST) is a publicly available classification dataset consisting of colorectal cancer histology slides patches [16,17] derived originally from the NCT-CRC-HE-100K [17] dataset and resized to $3 \times 28 \times 28$ as a part of the MedMNIST benchmark [16]. We simplify the original 9-way classification task into a binary task of discriminating benign and malignant classes (*cancer-associated stroma* and *colorectal adenocarcinoma epithelium*) and use the original classes of granular tissue types as metadata to interpret the discovered failure types. Moreover, 40% of the dataset was put into the test set to increase the sample size for the evaluation of both subtyping and model repairment. Finally, the model was trained with Adam with a learning rate of 10^{-4} in combination with early stopping on the validation accuracy. The architecture is a version of VGG [18] with 6 convolutional layers starting at 16 channels and the fully connected layer replaced by a 1×1 convolution to two output channels and a spatial average.

ICE Catheter Detection (ICE-CD) is a private real-world object detection dataset comprised of ultrasound images of intra-cardiac catheters made by a Intracardiac Echocardiography (ICE) device on pigs. Furthermore, for the purpose of evaluating the performance of catheter detection models, each catheter image was classified into different types representing known difficult situations based on catheter appearance or position. In addition, information about the rough anatomical locations of the probe is available as metadata. The architecture is composed of 5 residual blocks of two convolutional layers (starting at 8 channels and doubling up to 128 channels) followed by two 1×1 convolutions branches: a classification and a center position regression branch. This dataset has been acquired in accordance with animal experiment regulations.

4   Experiments and Results

4.1   Comparison of Failure Subtyping Methods

Baselines and Experiments: we aim to quantify the quality of the proposed automatic subtyping methods (see Table 1 and Fig. 2) which we compare against several baselines: *Random* (random clusters), *False positives and negatives (FP/FN)* (two clusters) and *Metadata* (BP-MNIST: the original 9 *tissue types*; ICE-CD: *image types* as identified by the clinicians and in addition *anatomical locations* of the images). For each failure type $\mathcal{F}_i$, the model was fine-tuned on both $\mathcal{F}_i^{tr}$ and the correct cases $\mathcal{C}^{tr}$ (to satisfy *Compatibility*). Early stopping is performed with the best validation score (accuracy on BP-MNIST and F_1 on ICE-CD) on $\mathcal{F}_i^{tr}$ while maintaining 0.9 validation accuracy on $\mathcal{C}^{tr}$. Table 1 displays the average metrics for the respective methods while Fig. 2 shows granular results i.e., matrix $m(\mathcal{M}_i, \mathcal{F}_j^{te})$ which denotes the test accuracy on $\mathcal{F}_j^{te}$ of the model fine-tuned on $\mathcal{F}_i^{tr}$.

Analysis: first of all, *Gradient clustering* reached better scores than any other method with the exception of *Independence* for *FP/FN* clustering on BP-MNIST. However, it had a 18% higher Learnability score and is more informative with more identified failure types. Remarkably, *Gradient clustering* was better than using *Metadata*, including the ICE-CD metadata made through very time consuming visual inspection. In the case of BP-MNIST, the lack of independence of *Metadata* subtyping was clearly visible in Fig. 2(a). This implies that *Gradient clustering* might be able to identify independent failure types which are not obvious through human eyes but are relevant to the model. On the other hand, *Feature clustering* seemed to achieve lower scores than metadata. Furthermore, *Gradient clustering* might achieve higher *Learnability* and *Independence* due to being more aligned with the repairment method. Finally, *Random* resulted in by far the lowest independence scores as is apparent in Fig. 2(c) where all types had the same score. This shows that the *Independence* metric is effective in detecting when failure types are mixed together. Moreover, *Random* and *FP/FN* clustering showed lower *Learnability* which may be explained by the diversity of tasks to be learned within each cluster. However, *FP/FN* had the highest *Independence* due to matching the two classes.

Table 1. Comparison of methods for failure types identification on both BP-MNIST and ICE-CD. The best and the second best results are shown in red and blue.

Method	BP-MNIST (ACC)		Catheter Detection (ACC)	
	Learnability	Independence	Learnability	Independence
Random	0.42±0.01	0.00±0.02	0.65±0.02	0.01±0.01
False positives and negatives	0.74±0.22	**0.74±0.22**	0.69±0.14	0.14±0.37
BP-MNIST: *tissue type*	0.92±0.05	0.58±0.16	-	-
ICE-CD: *image type*	-	-	0.77±0.17	0.46±0.16
ICE-CD: *anatomical location*	-	-	0.75±0.22	0.37±0.20
Feature clustering	0.79±0.25	0.52±0.23	**0.83±0.06**	0.36±0.07
Gradient clustering	**0.92±0.04**	0.69±0.04	**0.83±0.12**	**0.48±0.18**

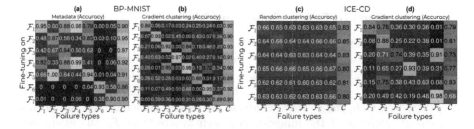

Fig. 2. Accuracy of fine-tuning on each failure type on BP-MNIST according to tissue types (a) and gradient clustering (b) and on ICE-CD according to random clustering (c) and gradient clustering (d). The lack of independence is apparent for tissue types (a) and random clustering (c) while a diagonal pattern is noticeable in (b) and (d).

4.2 Analysis of Automatically Discovered Failure Subtypes

We aim to inspect the automatically discovered subtypes on BP-MNIST and ICE-CD by *Gradient clustering* which achieved the best scores.

Binary PathMNIST (BP-MNIST): first, we observe that False positives and false negatives were mostly separated into two sets of clusters (i.e., each cluster contains mostly either circles or crosses as shown in Fig. 3(a)). Moreover, the two malignant tissue types were recovered separately: Cluster 2 and 4 for *cancer-associated stroma* and cluster 1 for *colorectal adenocarcinoma epithelium.* Secondly, even within one tissue type, *Gradient clustering* was able to discover independent failure types. Cluster 2 and 4 both focused on *cancer-associated stroma* but were relatively independent and differed when evaluating on $\mathcal{F}_1$, $\mathcal{F}_2$ and $\mathcal{F}_4$ (See Fig. 2.). In addition, clusters 2 and 4 were visually different as seen in Fig. 3 with cluster 4 corresponding to darker less textured images. Finally, only cluster 8 seemed to contain *normal colon mucosa* (in addition to *Debris*) and does seem to contain darker textured images than other false positive clusters.

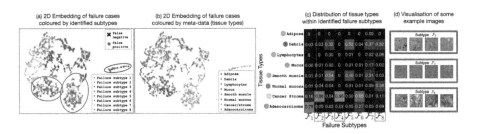

Fig. 3. Inspection of failure types obtained through gradient space clustering for BP-MNIST. (a) UMAP embedding coloured by *failure types.* (b) UMAP embedding coloured by *tissue types.* (c) Distribution of tissue types within each failure type. (d) Example images for remarkable failure types. Failures $\mathcal{F}_2$ and $\mathcal{F}_4$ which contain the same tissue but different visual appearances are identified as two different types.

Catheter detection (ICE-CD): *Gradient clustering* was able to recover a known and important but under-represented failure type: Cluster 4 (red cluster in Fig. 4(a)) focused on *Near-probe* catheters which are close to the ultrasound probe. Indeed, these catheters are hard to detect due to noise in this region of the images. Secondly, *Gradient clustering* was able to automatically discover some of the anatomical locations. Indeed, Cluster 6 (brown cluster in Fig. 4(a)) focused on the *LAA* and Cluster 3 (green cluster in Fig. 4(a)) focused mostly on the *SVC/IVC*. Finally, *Gradient clustering* was able to separate false positives (see the orange cluster in Fig. 4(a)) from false negatives (the others).

4.3   Model Repairment

Experiments: we aim to evaluate how much failure types performance can be improved while retaining performance on correct cases (See Table 2). We compare several repairment approaches on both datasets based on fine-tuning and EWC [11]. The fine-tuning is done on either a single failure type: $\mathcal{F}_i^{tr}$ or all: $\mathcal{F} = \cup_i \mathcal{F}_i^{tr}$. Also, we compare to using $\mathcal{C}^{tr}$ with a 50% sample ratio. For

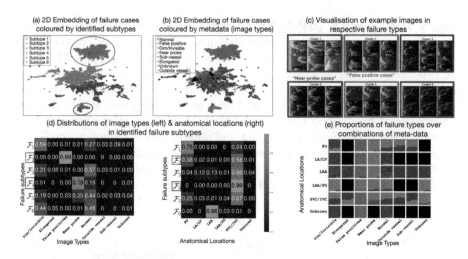

Fig. 4. Inspection of the failure types obtained through gradient space clustering for ICE-CD. UMAP embedding coloured by failure types (a) and by image types (b). (c) Example images for each subtype. (d) Distribution of image types and anatomical locations within each subtype. (e) Proportions of failure types over each combination of anatomical location and image types coloured by failure type. Two outstanding failure types are $\mathcal{F}_2$ focusing on "false positives" and $\mathcal{F}_4$ focusing on "Near-probe" cases.

all methods, early stopping is performed by selecting the best accuracy on $\mathcal{F}_i^{tr}$. Models were fine-tuned with a learning rate of 10^{-6} on BP-MNIST and 10^{-4} on ICE-CD and weight decay of 10^{-3}. Table 2 reports the accuracy on the test set of each failure type $\mathcal{F}_i^{te}$, the previously correct cases $\mathcal{C}^{te}$ and the overall test set.

Analysis: first, fine-tuning on a single failure type $\mathcal{F}_i^{tr}$ generalized more on that specific failure type than fine-tuning on all incorrect cases $\mathcal{F}^{tr}$ at once (see Table 2). This may indicate that the failure types are conflicting during fine-tuning and it is more difficult to simultaneously learn a diverse set of cases than simple ones. Therefore, if learning unimportant failures is conflicting with critical ones, it makes sense to first start by repairing a carefully selected subset of the failures. Secondly, fine-tuning on the failures only couldn't maintain performance on $\mathcal{C}^{te}$ while including the correct cases $\mathcal{C}^{tr}$ helped to preserve performance. Fine-tuning on $\mathcal{F}_i^{tr} \cup \mathcal{C}^{tr}$ for ICE-CD dropped to 0.73 but this was still higher than 0.32 if using only $\mathcal{F}_i^{tr}$ for fine-tuning. Finally, while for EWC [11] the performance on correct cases didn't drop as much as simple fine-tuning, EWC wasn't able to maintain correct cases accuracy to more than 0.27 and 0.51.

Table 2. Comparison of model repairment methods on BP-MNIST (top) and ICE-CD (bottom) evaluated on failure types obtained in gradient space (ACC). Methods retaining at least 0.85 accuracy on $\mathcal{C}$ are shown in green and those that do not in red.

Method (BP-MNIST)	$\mathcal{F}_1^{te}$	$\mathcal{F}_2^{te}$	$\mathcal{F}_3^{te}$	$\mathcal{F}_4^{te}$	$\mathcal{F}_5^{te}$	$\mathcal{F}_6^{te}$	$\mathcal{F}_7^{te}$	$\mathcal{F}_8^{te}$	$\mathcal{C}^{te}$	All
Pre-repairment	0	0	0	0	0	0	0	0	1.0	0.91
Fine-tuning on a single $\mathcal{F}_i^{tr}$	0.92	0.94	0.97	1.00	0.99	0.94	0.96	0.92	0.81	0.77
Fine-tuning on $\mathcal{F}_i^{tr} \cup \mathcal{C}^{tr}$	0.91	0.93	0.92	0.97	0.99	0.89	0.88	0.89	0.91	0.86
Fine-tuning on $\mathcal{F} = \cup_i \mathcal{F}_i^{tr}$	0.4	0.79	0.72	0.83	0.87	0.71	0.82	0.78	0.19	0.24
EWC [11] on $\mathcal{F}^{tr}$	0.34	0.65	0.79	0.79	0.85	0.64	0.79	0.76	0.27	0.30
Fine-tuning on $\mathcal{F}^{tr} \cup \mathcal{C}^{tr}$	0.25	0.65	0.71	0.66	0.80	0.48	0.69	0.68	0.9	0.88
Method (ICE-CD)	$\mathcal{F}_1^{te}$	$\mathcal{F}_2^{te}$	$\mathcal{F}_3^{te}$	$\mathcal{F}_4^{te}$	$\mathcal{F}_5^{te}$	$\mathcal{F}_6^{te}$	–	–	$\mathcal{C}^{te}$	All
Pre-repairment	0	0	0	0	0	0	–	–	1	0.59
Fine-tuning on a single $\mathcal{F}_i^{tr}$	0.85	0.99	0.77	0.69	0.64	0.96	–	–	0.32	0.36
Fine-tuning on $\mathcal{F}_i^{tr} \cup \mathcal{C}^{tr}$	0.81	0.97	0.79	0.96	0.67	0.97	–	–	0.73	0.64
Fine-tuning on $\mathcal{F} = \cup_i \mathcal{F}_i^{tr}$	0.61	0.9	0.71	0.28	0.57	0.72	–	–	0.64	0.66
EWC [11] on $\mathcal{F}^{tr}$	0.41	0.89	0.59	0.26	0.47	0.94	–	–	0.51	0.56
Fine-tuning $\mathcal{F}^{tr} \cup \mathcal{C}^{tr}$	0.41	0.8	0.65	0.39	0.61	0.82	–	–	0.86	0.78

5 Conclusion

We have introduced a principled framework to address the problems of failure identification, analysis and model repairment. Firstly, we put forward a set of desirable properties for meaningful failure types and novel surrogate metrics to assess those properties in practice. Secondly, we argued that model repairment should not only aim to fix the failures but also to retain performance on the previously correct data. Finally, we showed specific instantiations of our framework and demonstrated that clustering in feature and gradient space can automatically identify clinically important failures and outperform manual inspection.

References

1. Panfilov, E., Tiulpin, A., Klein, S., Nieminen, M.T., Saarakkala, S.: Improving robustness of deep learning based knee MRI segmentation: Mixup and adversarial domain adaptation. In: Proceedings of the IEEE/CVF International Conference on Computer Vision Workshops (2019)
2. Bdair, T., Navab, N., Albarqouni, S.: Roam: Random layer mixup for semi-supervised learning in medical imaging. arXiv preprint arXiv:2003.09439 (2020)
3. Billot, B., Greve, D., Van Leemput, K., Fischl, B., Iglesias, J.E., Dalca, A.V.: A learning strategy for contrast-agnostic mri segmentation. arXiv preprint arXiv:2003.01995 (2020)
4. Liu, Q., Dou, Q., Yu, L., Heng, P.A.: Ms-net: multi-site network for improving prostate segmentation with heterogeneous MRI data. IEEE Trans. Med. Imaging **39**(9), 2713–2724 (2020)
5. Dou, Q., de Castro, D.C., Kamnitsas, K., Glocker, B.: Domain generalization via model-agnostic learning of semantic features. Adv. Neural Inf. Process. Syst. **32**, 6450–6461 (2019)
6. Collins, G.S., Moons, K.G.M.: Reporting of artificial intelligence prediction models. Lancet **393**(10181), 1577–1579 (2019)

7. Liu, X., Rivera, S.C., Moher, D., Calvert, M.J., Denniston, A.K.: Reporting guidelines for clinical trial reports for interventions involving artificial intelligence: the consort-AI extension. BMJ, 370 (2020)
8. Oakden-Rayner, L., Dunnmon, J., Carneiro, G., Ré, C.: Hidden stratification causes clinically meaningful failures in machine learning for medical imaging. In: Proceedings of the ACM conference on health, inference, and learning, pp. 151–159 (2020)
9. Singla, S., Nushi, B., Shah, S., Kamar, E., Horvitz, E.: Understanding failures of deep networks via robust feature extraction. In: The IEEE Conference on Computer Vision and Pattern Recognition (CVPR) (2021)
10. McInnes, L., Healy, J., Saul, N., Grossberger, L.: UMAP: Uniform Manifold Approximation and Projection. J. Open Source Softw. 3(29), 861 (2018)
11. Kirkpatrick, J., et al.: Overcoming catastrophic forgetting in neural networks. Proc. Nat. Acad. Sci. 114(13), 3521–3526 (2017)
12. Karani, N., Chaitanya, K., Baumgartner, C., Konukoglu, E.: A lifelong learning approach to brain MR segmentation across scanners and protocols. In: Frangi, A.F., Schnabel, J.A., Davatzikos, C., Alberola-López, C., Fichtinger, G. (eds.) MICCAI 2018. LNCS, vol. 11070, pp. 476–484. Springer, Cham (2018). https://doi.org/10.1007/978-3-030-00928-1_54
13. Hofmanninger, J., Perkonigg, M., Brink, J.A., Pianykh, O., Herold, C., Langs, G.: Dynamic memory to alleviate catastrophic forgetting in continuous learning settings. In: Martel, A.L., Abolmaesumi, P., Stoyanov, D., Mateus, D., Zuluaga, M.A., Zhou, S.K., Racoceanu, D., Joskowicz, L. (eds.) MICCAI 2020. LNCS, vol. 12262, pp. 359–368. Springer, Cham (2020). https://doi.org/10.1007/978-3-030-59713-9_35
14. Karani, N., Erdil, E., Chaitanya, K., Konukoglu, E.: Test-time adaptable neural networks for robust medical image segmentation. Med. Image Anal. 68, 101907 (2021)
15. Kamnitsas, K., et al.: Unsupervised domain adaptation in brain lesion segmentation with adversarial networks. In: Niethammer, M., et al. (eds.) IPMI 2017. LNCS, vol. 10265, pp. 597–609. Springer, Cham (2017). https://doi.org/10.1007/978-3-319-59050-9_47
16. Yang, J., Shi, R., Ni, B.: Medmnist classification decathlon: a lightweight automl benchmark for medical image analysis. arXiv preprint arXiv:2010.14925 (2020)
17. Kather, J.N., et al.: Predicting survival from colorectal cancer histology slides using deep learning: a retrospective multicenter study. PLoS Med. 16(1), e1002730 (2019)
18. Simonyan, K., Zisserman, A.: Very deep convolutional networks for large-scale image recognition. arXiv preprint arXiv:1409.1556 (2014)

Using Causal Analysis for Conceptual Deep Learning Explanation

Sumedha Singla[1][(✉)], Stephen Wallace[2], Sofia Triantafillou[3],
and Kayhan Batmanghelich[3]

[1] Computer Science Department, University of Pittsburgh, Pittsburgh, USA
[2] University of Pittsburgh School of Medicine, University of Pittsburgh,
Pittsburgh, USA
[3] Department of Biomedical Informatics, University of Pittsburgh, Pittsburgh, USA

Abstract. Model explainability is essential for the creation of trustworthy Machine Learning models in healthcare. An ideal explanation resembles the decision-making process of a domain expert and is expressed using concepts or terminology that is meaningful to the clinicians. To provide such explanation, we first associate the hidden units of the classifier to clinically relevant concepts. We take advantage of radiology reports accompanying the chest X-ray images to define concepts. We discover sparse associations between concepts and hidden units using a linear sparse logistic regression. To ensure that the identified units truly influence the classifier's outcome, we adopt tools from Causal Inference literature and, more specifically, mediation analysis through counterfactual interventions. Finally, we construct a low-depth decision tree to translate all the discovered concepts into a straightforward decision rule, expressed to the radiologist. We evaluated our approach on a large chest x-ray dataset, where our model produces a global explanation consistent with clinical knowledge.

1 Introduction

Machine Learning, specifically, Deep Learning (DL) methods are increasingly adopted in healthcare applications. Model explainability is essential to build trust in the AI system [5] and to receive clinicians' feedback. Standard explanation methods for image classification delineates regions in the input image that significantly contribute to the model's outcome [13,17,19]. However, it is challenging to explain *how* and *why* variations in identified regions are relevant to the model's decision. Ideally, an explanation should resemble the decision-making process of a domain expert. This paper aims to map a DL model's neuron activation patterns to the radiographic features and constructs a simple rule-based model that partially explains the Black-box.

Methods based on feature attribution have been commonly used for explaining DL models for medical imaging [1]. However, an alignment between feature attribution and radiology concepts is difficult to achieve, especially when a single region may correspond to several radiographic concepts. Recently, researchers

© Springer Nature Switzerland AG 2021
M. de Bruijne et al. (Eds.): MICCAI 2021, LNCS 12903, pp. 519–528, 2021.
https://doi.org/10.1007/978-3-030-87199-4_49

have focused on providing explanations in the form of human-defined concepts [2,12,23]. In medical imaging, such methods have been adopted to derive an explanation for breast mammograms [22], breast histopathology [6] and cardiac MRIs [4]. A major drawback of the current approach is their dependence on explicit concept-annotations, either in the form of a representative set of images [12] or semantic segmentation [2], to learn explanations. Such annotations are expensive to acquire, especially in the medical domain. We use weak annotations from radiology reports to derive concept annotations. Furthermore, these methods measure correlations between concept perturbations and classification predictions to quantify the concept's relevance. However, the neural network may not use the discovered concepts to arrive at its decision. We borrow tools from causal analysis literature to address that drawback [21].

In this work, we used radiographic features mentioned in radiology reports to define concepts. Using a National Language Processing (NLP) pipeline, we extract weak annotations from text and classify them based on their positive or negative mention [9]. Next, we use sparse logistic regression to identify sets of hidden-units correlated with the presence of a concept. To quantify the causal influence of the discovered concept-units on the model's outcome, we view concept-units as a *mediator* in the treatment-mediator-outcome framework [8]. Using measures from mediation analysis, we provide an effective ranking of the concepts based on their causal relevance to the model's outcome. Finally, we construct a low-depth decision tree to express discovered concepts in simple decision rules, providing the global explanation for the model. The rule-based nature of the decision tree resembles many decision-making procedures by clinicians.

2   Method

We consider a pre-trained *black-box* classifier $f : \mathbf{x} \to \mathbf{y}$ that takes an image $\mathbf{x}$ as input and process it using a sequence of hidden layers to produce a final output $\mathbf{y} \in \mathbb{R}^D$. Without loss of generality, we decompose function f as $\Phi_2 \circ \Phi_1(\mathbf{x})$, where $\Phi_1(\mathbf{x}) \in \mathbb{R}^L$ is the output of the initial few layers of the network and Φ_2 denotes the rest of the network. We assume access to a dataset $\mathcal{X} = \{(\mathbf{x}_n, \mathbf{y}_n, \mathbf{c}_n)\}^N$, where $\mathbf{x}_n$ is input image, $\mathbf{y}_n$ is a d-dimensional one-hot encoding of the class labels and $\mathbf{c}_n \in \mathbb{R}^K$ is a k-dimensional concept-label vector. We define concepts as the radiographic observations mentioned in radiology reports to describe and provide reasoning for a diagnosis. We used a NLP pipeline [9] to extract concept annotations. The NLP pipeline follows a rule-based approach to extract and classify observations from the free-text radiology report. The extracted k^{th} concept-label $\mathbf{c}_n[k]$ is either 0 (negative-mention), 1(positive-mention) or -1 (uncertain or missing-mention). An overview of our method is shown in Fig. 1. Our method consists of three sequential steps:

(1) *Concept associations*: We seek to discover sparse associations between concepts and the hidden-units of $f(\cdot)$. We express k^{th} concept as a sparse vector $\beta_k \in \mathbb{R}^L$ that represents a linear direction in the intermediate space $\Phi_1(\cdot)$.

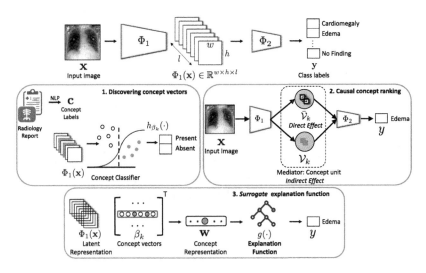

Fig. 1. Method overview: We provide explanation for the black-box function $f(\mathbf{x})$ in terms of concepts, that are radiographic observations mentioned in radiology reports. 1) The intermediate representation $\Phi_1(\mathbf{x})$ is used to learn a sparse logistic regression $h_{\beta_k}(\cdot)$ to classify k^{th} concept. 2) The non-zero coefficients of β_k represents a set of concept units $\mathcal{V}_k$ that serves as a mediator in the causal path connecting input $\mathbf{x}$ and outcome y. 3) A decision tree function is learned to map concepts to class labels.

(2) *Causal concept ranking*: Using tools from causal inference, we find an effective ranking of the concepts based on their relevance to the classification decision. Specifically, we consider each concept as a mediator in the causal path between the input and the outcome. We measure concept relevance as the effect of a counterfactual intervention on the outcome that passes indirectly through the concept-mediator.

(3) *Surrogate explanation function*: We learn an easy-to-interpret function $g(\cdot)$ that mimics function $f(\cdot)$ in its decision. Using $g(\cdot)$, we seek to learn a global explanation for $f(\cdot)$ in terms of the concepts.

2.1 Concept Associations

We discover concept associations with intermediate representation $\Phi_1(\cdot)$ by learning a binary classifier that maps $\Phi_1(\mathbf{x})$ to the concept-labels [12]. We treat each concept as a separate binary classification problem and extract a representative set of images $\mathcal{X}^k$, in which concept $c_n[k]$ is present and a random negative set. We define concept vector (β_k) as the solution to the logistic regression model $c_n[k] = \sigma(\beta_k^T \text{vec}(\Phi_1(\mathbf{x}_n))) + \epsilon$, where $\sigma(\cdot)$ is the sigmoid function. For a convolutional neural network, $\Phi_1(\mathbf{x}) \in \mathbb{R}^{w \times h \times l}$ is the output activation of a convolutional layer with width w, height h and number of channels l. We experimented with two vectorization for Φ_1. In first, we flatten $\Phi_1(\mathbf{x})$ to be a whl-dimensional vector. In second, we applied a spatial aggregation by max-pooling along the width and height to obtain l-dimensional vector. Unlike TCAV [12] that uses linear regression, we used lasso regression to enable sparse feature selection and minimize the following loss function,

$$\min_{\beta_k} \sum_{\mathbf{x}_n \in \mathcal{X}_k} \ell(h_{\beta_k}(\mathbf{x}), c_n[k]) + \lambda ||\beta_k||_1 \tag{1}$$

where $\ell(\cdot, \cdot)$ is the cross entropy loss, $h_{\beta_k}(\mathbf{x}) = \sigma(\beta_k^T \text{vec}(\Phi_1(\mathbf{x}_n)))$ and λ is the regularization parameter. We performed 10-fold nested-cross validation to find λ with least error. The non-zero elements in the concept vector β_k forms the set of hidden units $(\mathcal{V}_k)$ that are most relevant to the kth concept.

2.2 Causal Concept Ranking

Concept associations identified hidden units that are strongly correlated with a concept. However, the neural network may or may not use the discovered concepts to arrive at its decision. We use tools from causal inference, to quantify what fraction of the outcome is mediated through the discovered concepts.

To enable causal inference, we first define *counterfactual* $\mathbf{x}'$ as a perturbation of the input image $\mathbf{x}$ such that the decision of the classifier is flipped. Following the approach proposed in [20], we used a conditional generative adversarial network (cGAN) to learn the counterfactual perturbation. We conditioned on the output of the classifier, to ensure that cGAN learns a classifier-specific perturbation for the given image $\mathbf{x}$. Next, we used theory from causal mediation analysis to causally relate a concept with the classification outcome. Specifically, we consider concept as a mediator in the causal pathway from the input $\mathbf{x}$ to the outcome $\mathbf{y}$. We specify following effects to quantify the causal effect of the counterfactual perturbation and the role of a mediator in transferring such effect,

1. Average treatment effect (ATE): ATE is the total change in the classification outcome $\mathbf{y}$ as a result of the counterfactual perturbation.
2. Direct effect (DE): DE is the effect of the counterfactual perturbation that comprises of any causal mechanism that *do not* pass through a given mediator. It captures how the perturbation of input image changes classification decision directly, without considering a given concept.
3. Indirect effect (IE): IE is the effect of the counterfactual perturbation which is mediated by a set of mediators. It captures how the perturbation of input image changes classification decision indirectly through a given concept.

Following the potential outcome framework from [18,21], we define the ATE as the proportional difference between the factual and the counterfactual classification outcome,

$$\mathbf{ATE} = \mathbb{E}\left[\frac{f(\mathbf{x}')}{f(\mathbf{x})} - 1\right]. \tag{2}$$

To enable causal inference through a mediator, we borrow Pearl's definitions of natural direct and indirect effects [16] (*ref* Fig. 2). We consider set of concept-units $\mathcal{V}_k$ as a mediator, representing the kth concept. We decompose the latent representation $\Phi_1(\mathbf{x})$ as concatenation of response of concept-units $\mathcal{V}_k(\mathbf{x})$ and rest of the hidden units $\bar{\mathcal{V}}_k(\mathbf{x})$ *i.e.*, $\Phi_1(\mathbf{x}) = [\mathcal{V}_k(\mathbf{x}), \bar{\mathcal{V}}_k(\mathbf{x})]$. We can re-write classification outcome as $f(\mathbf{x}) = \Phi_2(\Phi_1(\mathbf{x})) = \Phi_2([\mathcal{V}_k(\mathbf{x}), \bar{\mathcal{V}}_k(\mathbf{x})])$. To disentangle the direct effect from the indirect effect, we use the concept of *do*-operation on the unit level of the learnt network. Specifically, we use $do(\mathcal{V}_k(\mathbf{x}))$ to denote that we set the value of the concept-units to the value obtained by using the original image as input. By intervening on the network and setting the value of the concept units, we can compute the direct effect as the proportional difference between the factual and the counterfactual classification outcome, while holding mediator *i.e.*, $\mathcal{V}_k$ fixed to its value before the perturbation,

$$\mathbf{DE} = \mathbb{E}\Big[\frac{\Phi_2([do(\mathcal{V}_k(\mathbf{x})), \bar{\mathcal{V}}_k(\mathbf{x}')])}{\Phi_2([\mathcal{V}_k(\mathbf{x}), \bar{\mathcal{V}}_k(\mathbf{x})])} - 1\Big]. \tag{3}$$

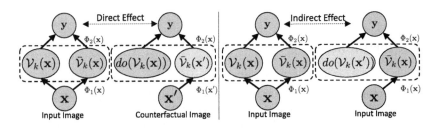

Fig. 2. Illustration of direct and indirect effects in causal mediation analysis.

We compute indirect effect as the expected change in the outcome, if we change the mediator from its original value to its value using counterfactual, while holding everything else fixed to its original value,

$$\mathbf{IE} = \mathbb{E}\Big[\frac{\Phi_2([do(\mathcal{V}_k(\mathbf{x}')), \bar{\mathcal{V}}_k(\mathbf{x})])}{\Phi_2([\mathcal{V}_k(\mathbf{x}), \bar{\mathcal{V}}_k(\mathbf{x})])} - 1\Big]. \tag{4}$$

If the perturbation has no effect on the mediator, then the causal indirect effect will be zero. Finally, we use the indirect effect associated with a concept, as a measure of its relevance to the classification decision.

2.3 Surrogate Explanation Function

We aim to learn a surrogate function $g(\cdot)$, such that it reproduces the outcome of the function $f(\cdot)$ using an interpretable and straightforward function. We formulated $g(\cdot)$ as a decision tree as many clinical decision-making procedures follow a rule-based pattern. We summarize the internal state of the function $f(\cdot)$ using output of k concept regression functions $h_{\beta_k}(\cdot)$ as follows,

$$\mathbf{w}_n = [\text{logit}(h_{\beta_1}(\mathbf{x}_n)), \text{logit}(h_{\beta_2}(\mathbf{x}_n)), \cdots]. \tag{5}$$

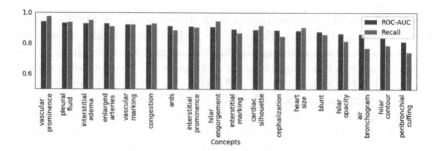

Fig. 3. AUC-ROC and recall metric for different concept classifiers.

Next, we fit a decision tree function, $g(\cdot)$, to mimic the outcome of the function $f(\cdot)$ as,

$$g^* = \arg\min_g \sum_n \mathcal{L}(g(\mathbf{w}_n), f(\mathbf{x}_n)), \qquad (6)$$

where $\mathcal{L}$ is the splitting criterion based on minimizing entropy for highest information gain from every split.

3   Experiments

We first evaluated the concept classification performance and visualized concept-units to demonstrate their effectiveness in localizing a concept. Next, we summarized the indirect effects associated with different concepts across different layers of the classifier. We evaluated a proposing ranking of the concepts based on their causal contribution to the classification decision. Finally, we used the top-ranked concepts to learn a surrogate explanation function in the form of a decision tree.

Data preprocessing: We perform experiments on the MIMIC-CXR [10] dataset, which is a multi-modal dataset consisting of 473K chest X-ray images and 206K reports. The dataset is labeled for 14 radiographic observations, including 12 pathologies. We used state-of-the-art DenseNet-121 [7] architecture for our classification function [9]. DenseNet-121 architecture is composed of four dense blocks. We experimented with three versions of $\Phi_1(\cdot)$ to represent the network until the second, third, and fourth dense block. For concept annotations, we considered radiographic features that are frequently mentioned in radiology reports in the context of labeled pathologies. Next, we used Stanford CheXpert [9] to extract and classify these observations from free-text radiology reports.

3.1   Evaluation of Concept Classifiers

The intermediate representations from third dense-block consistently outperformed other layers in concept classification. In Fig. 3, we show the testing-ROC-AUC and recall metric for different concept classifiers. All the concept classifiers achieved high recall, demonstrating a low false-negative (type-2) error.

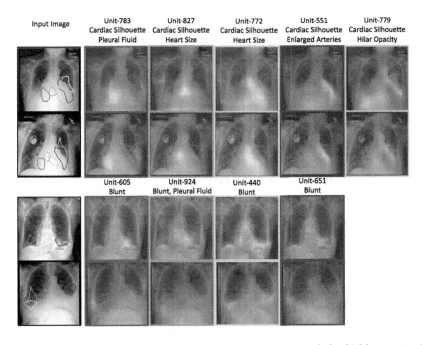

Fig. 4. A qualitative demonstration of the activation maps of the hidden units that act as visual concept detectors. Each column represents one hidden unit identified as part of concept vector $\mathcal{V}_k$. Top two rows show $k = cardiac\text{-}silhouette$ and bottom rows have $k = blunt\ costophrenic\ angle$.

In Fig. 4, we visualize the activation map of hidden units associated with the concept vector $\mathcal{V}_k$. For each concept, we visualize hidden units that have large logistic regression-coefficient (β_k). To highlight the most activated region for a unit, we threshold activation map by the top 1% quantile of the distribution of the selected units' activations [2]. Consistent with prior work [3], we observed that several hidden units have emerged as concept detectors, even though concept labels were not used while training f. For $cardiac\text{-}silhouette$, different hidden units highlight different regions of the heart and its boundary with the lung. For localized concept such as $blunt\ costophrenic\ angle$, multiple relevant units were identified that all focused on the lower-lobe regions. Same hidden unit can be relevant for multiple concepts. The top label in Fig. 4. shows the top two important concepts for each hidden unit.

3.2 Evaluating Causal Concepts Using Explanation Function

We evaluate the success of the counterfactual intervention by measuring ATE. High values for ATE confirms that counterfactual image generated by [20] successfully flips the classification decision. We achieved an ATE of 0.97 for cardiomegaly, 0.89 for pleural effusion and 0.96 for edema. In Fig. 5 (heat-map), we show the distribution of the indirect effect associated with concepts, across

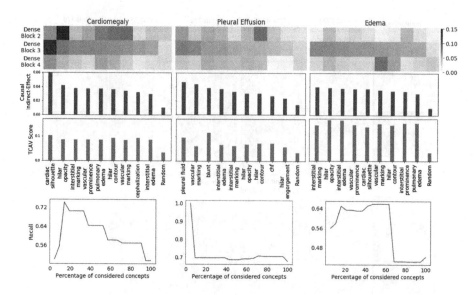

Fig. 5. Indirect effects of the concepts, calculated over different layers of the DenseNet-121 architecture (heat-map). The derived ranking of the concepts based on their causal relevance to the diagnosis (bar-graph). A comparative ranking based on concept sensitivity score from TCAV [12]. The trend of recall metric for the decision tree function $g(\cdot)$, while training using top x% of top-ranked concepts (trend-plot).

different layers. The middle layer demonstrates a large indirect effect across all concepts. This shows that the hidden units in dense-block 3 played a significant role in mediating the effect of counterfactual intervention.

In Fig. 5 (bar-graph), we rank the concepts based on their indirect effect. The top-ranked concepts recovered by our ranking are consistent with the radiographic features that clinicians associates with the examined three diagnoses [11,14,15]. Further, we used the concept sensitivity score from TCAV [12] to rank concepts for each diagnosis. The top-10 concepts identified by our indirect effect and TCAV are the same, while their order is different. The top-3 concepts are also the same, with minor differences in ranking. Both the methods have low importance score for random concept. This confirms that the trend in importance score is unlikely to be caused by chance. For our approach, random concept represents an ablation of the concept-association step. Here, rather than performing lasso regression to identify relevant units, we randomly select units.

To quantitatively demonstrate the effectiveness of our ranking, we iteratively consider x% of top-ranked concepts and retrain the explanation function $g(\mathbf{w})$. In Fig. 5 (bottom-plot), we observe the change in recall metric for the classifier $g(\cdot)$ as we consider more concepts. In the beginning, as we add relevant concepts, the true positive rate increases resulting in a high recall. However, as less relevant concepts are considered, the noise in input features increased, resulting in a lower recall. Figure 6 visualize the decision tree learned for the best performing model.

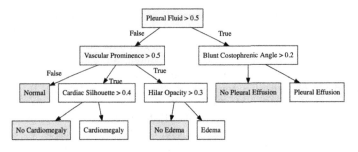

Fig. 6. The decision tree for the three diagnosis with best performance on recall metric.

4   Conclusion

We proposed a novel framework to derive global explanation for a black-box model. Our explanation is grounded in terms of clinically relevant concepts that are causally influencing the model's decision. As a future direction, we plan to extend our definition of concepts to include a broader set of clinical metrics.

Acknowledgement. This work was partially supported by NIH Award Number 1R01HL141813-01, NSF 1839332 Tripod+X, SAP SE, and Pennsylvania's Department of Health. We are grateful for the computational resources provided by Pittsburgh SuperComputing grant number TG-ASC170024.

References

1. Basu, S., Mitra, S., Saha, N.: Deep learning for screening COVID-19 using chest X-ray images. In: IEEE Symposium Series on Computational Intelligence (SSCI) (2020)
2. Bau, D., Zhou, B., Khosla, A., Oliva, A., Torralba, A.: Network dissection: quantifying interpretability of deep visual representations. In: IEEE Computer Vision and Pattern Recognition (CVPR), pp. 6541–6549 (2017)
3. Bau, D., Zhu, J.Y., Strobelt, H., Lapedriza, A., Zhou, B., Torralba, A.: Understanding the role of individual units in a deep neural network. Nat. Acad. Sci. **117**(48), 30071–30078 (2020)
4. Clough, J.R., Oksuz, I., Puyol-Antón, E., Ruijsink, B., King, A.P., Schnabel, J.A.: Global and local interpretability for cardiac MRI classification. In: Medical Image Computing and Computer-Assisted Intervention (MICCAI), pp. 656–664 (2019)
5. Glass, A., McGuinness, D.L., Wolverton, M.: Toward establishing trust in adaptive agents. In: International Conference on Intelligent User Interfaces (2008)
6. Graziani, M., Andrearczyk, V., Marchand-Maillet, S., Müller, H.: Concept attribution: explaining CNN decisions to physicians. Comput. Biol. Med. **123**, 103865 (2020)
7. Huang, G., Liu, Z., Van Der Maaten, L., Weinberger, K.Q.: Densely connected convolutional networks. In: IEEE Computer Vision and Pattern Recognition (CVPR), pp. 4700–4708 (2017)

8. Imai, K., Jo, B., Stuart, E.A.: Commentary: using potential outcomes to understand causal mediation analysis. Multivar. Behav. Res. **46**(5), 861–873 (2011)
9. Irvin, J., et al.: Chexpert: a large chest radiograph dataset with uncertainty labels and expert comparison. AAAI Conf. Artif. Intell. **33**, 590–597 (2019)
10. Johnson, A.E., et al.: Mimic-CXR, a de-identified publicly available database of chest radiographs with free-text reports. Sci. Data **6**(1), 1–8 (2019)
11. Karkhanis, V.S., Joshi, J.M.: Pleural effusion: diagnosis, treatment, and management. Open Access Emerg. Med. (OAEM) **4**, 31 (2012)
12. Kim, B., Wattenberg, M., Gilmer, J., Cai, C., Wexler, J., Viegas, F., et al.: Interpretability beyond feature attribution: quantitative testing with concept activation vectors (TCAV). In: International Conference on Machine Learning (ICML), pp. 2668–2677 (2018)
13. Lundberg, S.M., Lee, S.I.: A unified approach to interpreting model predictions. Adv. Neural. Inf. Process. Syst. **30**, 4765–4774 (2017)
14. Milne, E., Pistolesi, M., Miniati, M., Giuntini, C.: The radiologic distinction of cardiogenic and noncardiogenic edema. Am. J. Roentgenol. **144**(5), 879–894 (1985)
15. Nakamori, N., MacMahon, H., Sasaki, Y., Montner, S., et al.: Effect of heart-size parameters computed from digital chest radiographs on detection of cardiomegaly. potential usefulness for computer-aided diagnosis. Invest. Radiol. **26**(6), 546–550 (1991)
16. Pearl, J.: Direct and indirect effects. In: Conference on Uncertainty and Artificial Intelligence (UAI), pp. 411–420 (2001)
17. Ribeiro, M.T., Singh, S., Guestrin, C.: Why should i trust you? explaining the predictions of any classifier. In: ACM SIGKDD International Conference on Knowledge Discovery and Data Mining, pp. 1135–1144 (2016)
18. Rubin, D.B.: Estimating causal effects of treatments in randomized and nonrandomized studies. J. Educ. Psychol. **66**(5), 688 (1974)
19. Selvaraju, R.R., Cogswell, M., Das, A., Vedantam, R., Parikh, D., Batra, D.: Gradcam: Visual explanations from deep networks via gradient-based localization. In: International Conference on Computer Vision (ICCV), pp. 618–626 (2017)
20. Singla, S., Pollack, B., Chen, J., Batmanghelich, K.: Explanation by progressive exaggeration. In: International Conference on Learning Representations (ICLR) (2019)
21. Vig, J., et al.: Investigating gender bias in language models using causal mediation analysis. In: Larochelle, H., Ranzato, M., Hadsell, R., Balcan, M.F., Lin, H. (eds.) Advances in Neural Information Processing Systems, vol. 33, pp. 12388–12401 (2020)
22. Yeche, H., Harrison, J., Berthier, T.: UBS: a dimension-agnostic metric for concept vector interpretability applied to radiomics. In: Suzuki, K., et al. (eds.) MLCDS/IMIMIC -2019. LNCS, vol. 11797, pp. 12–20. Springer, Cham (2019). https://doi.org/10.1007/978-3-030-33850-3_2
23. Zhou, B., Sun, Y., Bau, D., Torralba, A.: Interpretable basis decomposition for visual explanation. In: European Conference on Computer Vision (ECCV), pp. 119–134 (2018)

A Spherical Convolutional Neural Network for White Matter Structure Imaging via dMRI

Sara Sedlar[(✉)], Abib Alimi, Théodore Papadopoulo, Rachid Deriche, and Samuel Deslauriers-Gauthier

Inria, Université Côte d'Azur, Valbonne, France
{sara.sedlar,abib.alimi,theodore.papadopoulo,
rachid.deriche,samuel.deslauriers-gauthier}@inria.fr

Abstract. Diffusion Magnetic Resonance Imaging (dMRI) is a powerful non-invasive and in-vivo imaging modality for probing brain white matter structure. Convolutional neural networks (CNNs) have been shown to be a powerful tool for many computer vision problems where the signals are acquired on a regular grid and where translational invariance is important. However, as we are considering dMRI signals that are acquired on a sphere, rotational invariance, rather than translational, is desired. In this work, we propose a spherical CNN model with fully spectral domain convolutional and non-linear layers. It provides rotational invariance and is adapted to the real nature of dMRI signals and uniform random distribution of sampling points. The proposed model is positively evaluated on the problem of estimation of neurite orientation dispersion and density imaging (NODDI) parameters on the data from Human Connectome Project (HCP).

Keywords: Spherical CNN · Diffusion MRI · White matter micro-structures

1 Introduction

Diffusion MRI (dMRI) is an imaging modality that exploits the interactions of diffusing water molecules with the surrounding tissue micro-structures to create contrast. As such, it is well suited for in-vivo non-invasive white matter structure imaging. Progress in dMRI acquisition, from diffusion tensor imaging (DTI) [1] to high angular resolution diffusion imaging (HARDI), allowed application of more insightful mathematical tools in dMRI analysis. As a consequence, a number of biophysically inspired models which make finer link between underlying tissue microstructures and observed signals have been proposed [2–6]. Several studies have shown that the parameters of these models can be used in the evaluation of several neurological diseases [6–9]. However, non-linear optimizers employed for the models' parameter estimation, such as Markov

© Springer Nature Switzerland AG 2021
M. de Bruijne et al. (Eds.): MICCAI 2021, LNCS 12903, pp. 529–539, 2021.
https://doi.org/10.1007/978-3-030-87199-4_50

chain Monte Carlo (MCMC) [2], Levenberg–Marquardt [3,4], Gauss-Newton [5], exhibit high computational complexity and often require high number of acquisition points [3–5], which is not clinically applicable.

Given that a single voxel of dMRI volume contains a high dimensional q-space signal, entire brain scan contains hundreds of thousands dMRI signals. This large quantity of data opens the door to deep learning (DL) approaches for various problems where voxel-wise estimation is required. In [10], a fully connected neural network (FCN) was proposed for the estimation of diffusion kurtosis and micro-structure parameters from dMRI data obtained with a significantly reduced acquisition protocol. For a similar problem, a sparse reconstruction framework has been unfolded into a DL approach in [11] and was further improved by incorporating long-short term memory (LSTM) units in [12]. In [13], a model inspired by [11] was developed for single-shell microstructure estimation. Furthermore, DL was used for learning features related to certain neurological disease. In [14] a CNN model adjusted to the dMRI data domain which combines inter- and intra-voxel features, was proposed for classification into Parkinson disease and control group. Rotation and translation equivariant network was developed and successfully applied on the problem of multiple sclerosis lesion segmentation in [15]. Furthermore, as dMRI acquisition protocols significantly vary over different research centers and studies, a comparison of harmonization techniques was conducted and showed potential of deep CNN based approaches over parametric models [16]. A number of approaches was proposed for the estimation of fiber orientation distribution functions (fODF). In [17,18], estimation of fODFs was achieved with planar CNNs applied on dMRI data in spectral and signal domain. For the same task, spherical U-nets which exhibit rotational equivariance were proposed in [19,20].

Prior to the models tailored to the spherical nature of dMRI [14,15,19,20], a number of rotationally equivariant DL models were proposed for arbitrary S^2 signals. One of the prominent spherical CNNs was proposed in [21], where convolutions are performed in spectral instead of signal domain in order to avoid computationally expensive interpolations. Since convolution of two S^2 signals gives a signal in $SO(3)$ manifold, apart from the first layer, in all the following, convolution is performed between $SO(3)$ signals and kernels [21]. In the same work and in accordance with planar CNNs, non-linearity is applied in signal domain, which requires computationally demanding transformation from Fourier to $SO(3)$ domain. To address this problem, in [22], a spherical CNN model with zonal kernels was proposed. Another issue that arises from the non-linearity in signal domain is the introduction of high frequency components, which might introduce aliasing. In the work presented by [23], a fully Fourier space CNN was proposed. This model applies non-linearities of quadratic nature in spectral domain, so it completely eliminates conversions from spectral to signal space and the distortions introduced by aliasing.

In this work, we propose a spherical CNN with rotation equivariant Fourier domain convolutional and non-linear layers, where trainable kernels and biases are represented in Fourier domain of S^2 and $SO(3)$ manifold. Whereas the spherical CNN models proposed in [21–23] assume that input is bandlimited S^2 signal acquired at equiangular grid as one defined in [27], our model is tailored to the spherical dMRI signals acquired at randomly uniformly distributed points [28] and which are affected by significant Rician noise. The proposed model is evaluated on the problem of estimation of neurite orientation dispersion and density imaging (NODDI) [5] parameters from the data from the Human Connectome Project (HCP) [29].

2 Theory

In this section, we provide theoretical grounds related to S^2 and $SO(3)$ signals which are used to define our spherical CNN model described in the section Methods. The presented theory is, to a certain extent, common to the multiple rotation equivariant DL models proposed in [15,21–23].

An arbitrary square integrable ($\mathbb{L}^2$) bandlimited signal $s : S^2 \to \mathbb{C}$ can be represented as a linear combination of spherical harmonic (SH) basis elements as

$$s(\mathbf{r}) = \sum_{l=0}^{B} \sum_{m=-l}^{m=l} \hat{s}_l^m Y_l^m(\mathbf{r}) \tag{1}$$

where $\mathbf{r} \in \mathbb{R}^3, ||\mathbf{r}|| = 1$. $Y_l^m : S^2 \to \mathbb{C}$ is the SH basis element of degree l and order m. $\hat{s}_l^m$ is the corresponding SH coefficient and B is signal's bandwidth. As the SH basis are orthogonal, $\hat{s}_l^m$ can be obtained as

$$\hat{s}_l^m = \int_{S^2} s(\mathbf{r}) Y_l^{m*}(\mathbf{r}) d\mathbf{r}. \tag{2}$$

Given two $\mathbb{L}^2$ signals $f, g : S^2 \to \mathbb{C}$ of bandwidth B their convolution is defined as

$$[f * g](R) = \int_{S^2} f(\mathbf{r}) g^*(R^{-1}\mathbf{r}) \, d\mathbf{r} = \sum_{l=0}^{B} \sum_{m=-l}^{l} \sum_{n=-l}^{l} D_l^{mn}(R) \hat{f}_l^m \hat{g}_l^{n*} , \tag{3}$$

where $R = R(\alpha, \beta, \gamma) \in SO(3)$, $\hat{f}_l^m$ and $\hat{g}_l^n$ are SH coefficients of degree l and orders m and n of the signals $f(\mathbf{r})$ and $g(\mathbf{r})$ [24]. $D_l^{mn} : SO(3) \to \mathbb{C}$ is an element of Wigner-D matrix of degree l and orders m and n. Consequently, the convolution of two spherical signals results in a signal whose domain is the $SO(3)$ manifold. An arbitrary $\mathbb{L}^2$ bandlimited signal $h : SO(3) \to \mathbb{C}$ can be represented as a linear combination of the elements of Wigner-D matrices (referred here to as rotational harmonics (RH)) as

$$h(R) = \sum_{l=0}^{B} \sum_{m=-l}^{m=l} \sum_{n=-l}^{n=l} \hat{H}_l^{mn} D_l^{mn}(R). \tag{4}$$

where $\hat{H}_l^{mn}$ is the RH coefficient associated with the element of Wigner-D matrix $D_l^{mn}(R)$ and B is signal's bandwidth [24]. If $h(R) = [f * g](R)$, then from Eqs. 3

and 4, $\hat{H}_l = \hat{\mathbf{f}}_l\,\hat{\mathbf{g}}_l^H$ where $\hat{\mathbf{f}}_l$ and $\hat{\mathbf{g}}_l \in \mathbb{C}^{2l+1}$ contain the SH coefficients of degree l of signals $f(\mathbf{r})$ and $g(\mathbf{r})$. $\hat{H}_l \in \mathbb{C}^{(2l+1)\times(2l+1)}$ contains the RH coefficients of degree l of signal $h(R)$. Convolution of two $\mathbb{L}^2$ signals $f, g : SO(3) \rightarrow \mathbb{C}$ of bandwidth B, $h(R) = [f * g](R)$, is defined as

$$h(R) = \int_{SO(3)} f(Q)g^*(R^{-1}Q)\,dQ = \sum_{l=0}^{B} \frac{8\pi^2}{2l+1} \sum_{m=-l}^{l} \sum_{n=-l}^{l} D_l^{mn}(R) \sum_{k=-l}^{l} \hat{F}_l^{mk}\hat{G}_l^{nk*},$$
(5)

where $\hat{F}_l^{mk}$ and $\hat{G}_l^{kn}$ are the RH coefficients of degree l and orders m, k and k, n of signals $f(R)$ and $g(R)$ [24]. From Eqs. 4 and 5, we have $\hat{H}_l = \frac{8\pi^2}{2l+1}\hat{F}_l\hat{G}_l^H$ where $\hat{H}_l, \hat{F}_l, \hat{G}_l \in \mathbb{C}^{(2l+1)\times(2l+1)}$ contain the RH coefficients of degree l of functions $h(R), f(R)$ and $g(R)$. Product of two $\mathbb{L}^2$ signals $f, g : SO(3) \rightarrow \mathbb{C}$ of bandwidths B_f and B_g, $h(R) = f(R) \times g(R)$, is defined as

$$h(R) = \sum_{l'=0}^{B_f} \sum_{m'=-l'}^{l'} \sum_{n'=-l'}^{l'} \sum_{l''=0}^{B_g} \sum_{m''=-l''}^{l''} \sum_{n''=-l''}^{l''} \hat{F}_{l'}^{m'n'}\hat{G}_{l''}^{m''n''}D_{l'}^{m'n'}(R)D_{l''}^{m''n''}(R)$$
(6)

where product of elements of Wigner-D matrices can be represented as a linear combination of elements of Wigner-D matrices using Clebsch-Gordan coefficients $< l'k'l''k''|lk >$ which are non-zero only if $k = k' + k''$ as

$$D_{l'}^{m'n'}(R)D_{l''}^{m''n''}(R) = \sum_{l=|l'-l''|}^{l'+l''} < l'm'l''m''|lm >< l'n'l''n''|ln > D_l^{mn}(R) \quad (7)$$

where $m = m' + m''$ and $n = n' + n''$ [26]. This means that $h(R)$ is an $SO(3)$ signal of bandwidth $B_g + B_f$ whose RH coefficients, $\hat{H}_l$, in vector-matrix notation can be represented as

$$\hat{H}_l = \sum_{l',l''} C_{l',l''}^{l}{}^T \left[\hat{F}_{l'} \otimes \hat{G}_{l''}\right] C_{l',l''}^{l} \quad \text{s.t.} \quad |l' - l''| \leq l \leq l' + l'' \quad (8)$$

where matrix $C_{l',l''}^{l} \in \mathbb{R}^{(2l'+1)(2l''+1)\times(2l+1)}$ is a sparse matrix containing Clebsch-Gordan coefficients. As we are dealing with real S^2 signals, we use real SH and Wigner-D basis and accordingly defined Clebsch-Gordan matrices using the unitary matrices as in [25].

3   Methods

Architecture of the model proposed in this work is illustrated in Fig. 1. It is composed of denoising layers, S^2 and $SO(3)$ convolutional layers, layers with quadratic non-linearity and fully connected layers at the end. From denoised input signals and each feature map, after non-linearity is applied, rotation invariant power spectrum features are extracted. They are concatenated and fed into fully connected layers which perform final inference.

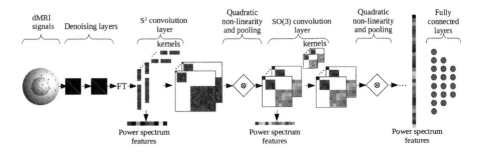

Fig. 1. Illustration of proposed spherical CNN architecture

3.1 Denoising Layers

Since dMRI signals used in our experiments are acquired at random uniformly distributed points [28] we cannot use quadrature rule for computation of SH coefficients as in [21–23]. In addition, the signals are affected by significant Rician noise. As dMRI signals are positive, we define denoising layer as $ReLU((I + \lambda W)s)$ where I is identity matrix, W are trainable weights, s is input signal and λ is parameter which should ensure that matrix $(I + \lambda W)$ is not too far from identity matrix.

3.2 Convolutional Layers

Convolutions are performed in spectral domain as firstly proposed in [21]. Given input denoised dMRI signal of bandwidth L acquired at K shells (channels) $\{s_k : \{(\theta_i, \phi_i)\}_{i=1}^{N_k} \to R\}_{k=1}^{K}$, SH coefficients $\{\{\hat{\mathbf{s}}_l^k \in \mathbb{R}^{2l+1}\}_{l=0}^{L}\}_{k=1}^{K}$ are estimated using real SH basis [25], orthonormalized using Gram-Schmidt process. First convolutional layer contains weights and biases expressed in terms of SH coefficients $\{\{\{\hat{\mathbf{w}}_l^{ki} \in \mathbb{R}^{2l+1}\}_{l=0}^{L}\}_{k=1}^{K}\}_{i=1}^{M}, \{\hat{\mathbf{b}}^i \in \mathbb{R}^1\}_{i=1}^{M}$, where M is the number of output channels. Convolution is given by

$$\hat{F}_l^i = \sum_{k=1}^{K} \hat{\mathbf{s}}_l^k \hat{\mathbf{w}}_l^{kiT} \quad \text{if} \quad l \neq 0, \quad \hat{F}_l^i = \sum_{k=1}^{K} \hat{\mathbf{s}}_l^k \hat{\mathbf{w}}_l^{kiT} + \hat{\mathbf{b}}^i \quad \text{if} \quad l = 0, \quad (9)$$

where $\{\hat{F}_l^i \in \mathbb{R}^{(2l+1)\times(2l+1)}\}_{i=1}^{M}$ are the real RH coefficients of the resulting $SO(3)$ signals. As in [21], all the following convolutional layers contain convolutions between $SO(3)$ feature maps and kernels. For a layer $n > 1$, weights and biases are expressed in terms of real RH coefficients $\{\{\{\hat{W}_l^{pq} \in \mathbb{R}^{(2l+1)\times(2l+1)}\}_{l=0}^{L_n}\}_{p=1}^{P}\}_{q=1}^{Q}, \{\hat{B}^q \in \mathbb{R}^1\}_{q=1}^{Q}$, where P and Q are the number of input and output channels and L_n is bandwidth of input feature maps. Given the RH coefficients of the input feature maps $\{\{\hat{S}_l^p \in \mathbb{R}^{(2l+1)\times(2l+1)}\}_{l=0}^{L_n}\}_{p=1}^{P}$ convolution is given by

$$\hat{F}_l^q = \frac{8\pi^2}{2l+1} \sum_{p=1}^{P} \hat{S}_l^p \hat{W}_l^{pqT} \quad \text{if} \quad l \neq 0, \quad \hat{F}_l^q = \frac{8\pi^2}{2l+1} \sum_{p=1}^{P} \hat{S}_l^p \hat{W}_l^{pqT} + \hat{B}^q \quad \text{if} \quad l = 0$$

$$(10)$$

where $\{\hat{F}_l^q \in \mathbb{R}^{(2l+1)\times(2l+1)}\}_{q=1}^Q$ are RH coefficients of the resulting $SO(3)$ signals.

3.3 Non-linearity and Pooling

In the domain of rotation equivariant neural networks, non-linear activations of quadratic nature were firstly introduced in [23], where Clebsch-Gordan decomposition is used to decompose tensor product of $SO(3)$ covariant vectors into irreducible fragments. In our work, we propose a non-linearity $f^2(R)$ which is rotationally equivariant and allows to preserve $SO(3)$ feature maps. Given real RH coefficients $\{\hat{F}_l^q\}_{l=0}^{L_n}$ of feature map $f(R)^q$, using Eq. 8, RH coefficients of $h(R)^q = f(R)^q \times f(R)^q$ can be expressed as

$$\hat{H}_l^q = \sum_{l',l''} \mathcal{C}_{l',l''}^l{}^T [\hat{F}_{l'}^q \otimes \hat{F}_{l''}^q] \mathcal{C}_{l',l''}^l \quad \text{s.t.} \quad |l'-l''| \le l \le l'+l'' \tag{11}$$

where $\mathcal{C}_{l',l''}^l$ are corresponding real Clebsch-Gordan matrices [25]. Although the bandwidth of the function $h(R)^q$ is $2L_n$ to achieve effect of pooling (reducing feature map resolution) RH coefficients of degree higher then some L_m ($L_m \le L_n$) are discarded as in [21,23].

3.4 Rotation Invariant Feature Vector

As goal of the cascade of convolutional and non-linear layers is to extract rotation invariant features, in [23] fragments of degree $l = 0$ at the output of each non-linear layer are concatenated (including $l = 0$ fragment of input signals) and represent rotation invariant vector that is further fed to a FCN. In our model, RH coefficients of degree $l = 0$ are rotation invariant as well. In addition, from the generalization of Parseval's theorem to S^2 and $SO(3)$ signals, we can notice that angular and rotation power spectra defined as

$$P_{ff}(l) = \sum_{m=-l}^{l} |\hat{f}_l^m|^2 \quad \text{and} \quad P_{hh}(l) = \frac{8\pi^2}{2l+1} \sum_{m=-l}^{l} \sum_{n=-l}^{l} |\hat{H}_l^{mn}|^2 \tag{12}$$

where $\hat{f}_l^m$ and $\hat{H}_l^{mn}$ are SH and RH coefficients of arbitrary $\mathbb{L}^2$ functions f and h on S^2 and $SO(3)$, are rotation invariant as well. Given this, we create rotation invariant feature vectors by concatenating the power spectra values of all degrees l of outputs of each non-linear layer and input signals. As in [23], they are fed to a FCN which performs the final inference.

4 Experiments

We used in our experiments real data of 50 subjects from Human Connectome Project (HCP) [29], 30 for training, 10 for validation and 10 testing. Scans are composed of three shells with b-values of 1000, 2000 and 3000 s/mm^2, each with

90 gradient directions and 18 $b = 0$ images. Scans were previously registered to T1w images. To select relevant white matter voxels, brain tissue segmentation computed from T1w images using FAST algorithm [31] implemented in *mrtrix* library [32] was used. NODDI gold standard parameters, namely intra-axonal volume fraction ν_{IC}, isotropic volume fraction ν_{iso} and orientation dispersion OD, were estimated using *brute2fine* optimizer from *dmipy* toolbox [30] on full acquisition scheme, while the methods are compared on significantly downsampled scheme with 30 sampling points per two shells - 1000 and 2000 s/mm^2.

We compared our method with *brute2fine* optimizer and two deep learning approaches, a model developed specifically for NODDI parameter estimation MEDN [11] and a FCN which was initially used for microstructure parameter estimation in [10]. Our model is composed of: (1) two denoising layers of size 60×60, (2) three convolutional layers of bandwidths 6, 4 and 2 with convolutional kernels of sizes $2 \times 8 \times 28$, $8 \times 16 \times 165$, $16 \times 32 \times 35$, followed by quadratic non-linearity, and (3) four fully connected layers of sizes 128×128, 128×64, 64×32, 32×3 , followed by ReLU, except the last one which is followed by sigmoid activation. FCN model has seven fully connected layers of size 60×256, 256×192, 192×128, 128×64, 64×32, 32×16 and 16×3. All models are trained over 200 epochs. In each epoch 12 800 dMRI signals were randomly selected from white matter region from each of 10 randomly selected training subjects. Models are trained in a way that they see same amount of data during training. For all three DL approaches, loss function is defined as mean squared error (MSE). For MEDN and FCN initial learning rate is 0.0001 and for our model it is 0.001, after 50 epochs they are reduced by factor 5 and after 100 by factor 10 of initial values. Batch size is 128. Total number of trainable parameters is 100 579, 109 865 and 73 691 for FCN, MEDN and our model respectively. Codes and trained models are provided at https://gitlab.inria.fr/ssedlar/fourier_s2cnn.

Table 1. Mean absolute error and std over 10 testing subjects

Model	ν_{IC}	ν_{ISO}	OD
brute2fine	$(5.06 \pm 0.36) \times 10^{-2}$	$(3.87 \pm 0.24) \times 10^{-2}$	$(2.97 \pm 0.27) \times 10^{-2}$
FCN	$(3.86 \pm 0.28) \times 10^{-2}$	$(3.14 \pm 0.21) \times 10^{-2}$	$(2.26 \pm 0.15) \times 10^{-2}$
MEDN	$(3.81 \pm 0.28) \times 10^{-2}$	$(3.16 \pm 0.20) \times 10^{-2}$	$(2.17 \pm 0.16) \times 10^{-2}$
Ours	$(3.77 \pm 0.28) \times 10^{-2}$	$(3.00 \pm 0.21) \times 10^{-2}$	$(2.08 \pm 0.16) \times 10^{-2}$

Table 2. Mean absolute error and std over 10 testing subjects with direction restricted training

Model	ν_{IC}	ν_{ISO}	OD
FCN	$(4.73 \pm 0.16) \times 10^{-2}$	$(3.43 \pm 0.18) \times 10^{-2}$	$(18.31 \pm 0.77) \times 10^{-2}$
MEDN	$(5.09 \pm 0.21) \times 10^{-2}$	$(3.76 \pm 0.19) \times 10^{-2}$	$(13.46 \pm 0.50) \times 10^{-2}$
Ours	$(3.99 \pm 0.31) \times 10^{-2}$	$(3.19 \pm 0.20) \times 10^{-2}$	$(4.10 \pm 0.15) \times 10^{-2}$

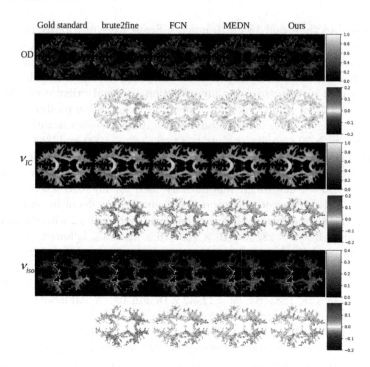

Fig. 2. Qualitative comparison of NODDI parameter estimation and differences with respect to gold standard. Red color indicates underestimation and blue color overestimation. (Color figure online)

5   Results and Conclusions

We compared results in terms of mean absolute value and its standard deviation over 10 testing subjects. Quantitative comparison of the results is given in Table 1, where we can see that our proposed approach yields lower average errors. In Fig. 2, qualitative comparison between methods is provided for one axial slice of a testing subject. Whereas, the differences between *brute2fine* and DL approaches are noticeable, some differences between DL approaches can be seen in OD estimation, where MEDN and FCN tend to overestimate OD values. Furthermore, to investigate rotation invariance of DL approaches, we have trained models on data whose diffusion tensor fit direction is in range $[0, \frac{\pi}{6})$ (or $(\frac{5\pi}{6}, \pi)$) and the quantitative results of the experiments are provided in Table 2 clearly indicating rotation invariance of our model.

In this work we have proposed a rotation invariant model for signals acquired at randomly distributed points on sphere and affected by considerable noise such as dMRI signals. We have demonstrated that the proposed model can be used successfully in the domain of dMRI analysis such as microstructure parameter estimation. In the future work we will investigate how computational complexity of quadratic activation function can be decreased and how the model performance can be improved by incorporating neighbourhood information.

Acknowledgment. This work was supported by the ERC under the European Union's Horizon 2020 research and innovation program (ERC Advanced Grant agreement No 694665:CoBCoM : Computational Brain Connectivity Mapping).

This work has been partly supported by the French government, through the 3IA Côte d'Azur Investments in the Future project managed by the National Research Agency (ANR) with the reference number ANR-19-P3IA-0002.

Data were provided [in part] by the Human Connectome Project, WU-Minn Consortium (Principal Investigators: David Van Essen and Kamil Ugurbil; 1U54MH091657) funded by the 16 NIH Institutes and Centers that support the NIH Blueprint for Neuroscience Research; and by the McDonnell Center for Systems Neuroscience at Washington University.

The authors are grateful to Inria Sophia Antipolis - Méditerranée https://wiki.inria.fr/ClustersSophia/Usage_policy "Nef" computation cluster for providing resources and support.

The authors are grateful to the OPAL infrastructure from Université Côte d'Azur for providing resources and support.

References

1. Le Bihan, D., et al.: Diffusion tensor imaging: concepts and applications. J. Magn. Reson. Imaging: an Off. J. Int. Soc. Magn. Reson. Med. **13**(4), 534-546 (2001)
2. Behrens, T.E.J., et al.: Characterization and propagation of uncertainty in diffusion-weighted MR imaging. Magn. Reson. Med.: An Off. J. Int. Soc. Magn. Reson. Med. **50**(5), 1077–1088 (2003)
3. Assaf, Y., Basser, P.J.: Composite hindered and restricted model of diffusion (CHARMED) MR imaging of the human brain. Neuroimage **27**(1), 48–58 (2005)
4. Assaf, Y., et al.: AxCaliber: a method for measuring axon diameter distribution from diffusion MRI. Magn. Reson. Med.: An Off. J. Int. Soc. Magn. Reson. Med. **59**(6), 1347–1354 (2008)
5. Zhang, H., et al.: NODDI: practical in vivo neurite orientation dispersion and density imaging of the human brain. Neuroimage **61**(4), 1000–1016 (2012)
6. Panagiotaki, E., et al.: Noninvasive quantification of solid tumor microstructure using VERDICT MRI. Cancer Res. **74**(7), 1902–1912 (2014)
7. De Santis, S., et al.: Early axonal damage in normal appearing white matter in multiple sclerosis: novel insights from multi-shell diffusion MRI. In: 2017 39th Annual International Conference of the IEEE Engineering in Medicine and Biology Society (EMBC). IEEE (2017)
8. Schneider, T., et al.: Sensitivity of multi-shell NODDI to multiple sclerosis white matter changes: a pilot study. Funct. Neurol. **32**(2), 97 (2017)
9. Broad, R.J., et al.: Neurite orientation and dispersion density imaging (NODDI) detects cortical and corticospinal tract degeneration in ALS. J. Neurol. Neurosurg. Psychiatry **90**(4), 404–411 (2019)
10. Golkov, V., et al.: Q-space deep learning: twelve-fold shorter and model-free diffusion MRI scans. IEEE Trans. Med. Imaging **35**(5), 1344–1351 (2016)

11. Ye, Chuyang: Estimation of tissue microstructure using a deep network inspired by a sparse reconstruction framework. In: Niethammer, M., et al. (eds.) IPMI 2017. LNCS, vol. 10265, pp. 466–477. Springer, Cham (2017). https://doi.org/10.1007/978-3-319-59050-9_37

12. Ye, C., Li, X., Chen, J.: A deep network for tissue microstructure estimation using modified LSTM units. Med. Image Anal. **55**, 49–64 (2019)

13. Faiyaz, A., et al.: DLpN: Single-Shell NODDI Using Deep Learner Estimated Isotropic Volume Fraction. arXiv preprint arXiv:2102.02772 (2021)

14. Banerjee, M., et al.: DMR-CNN: a CNN tailored for DMR scans with applications to PD classification. In: 2019 IEEE 16th International Symposium on Biomedical Imaging (ISBI 2019). IEEE (2019)

15. Müller, P., et al.: Rotation-Equivariant Deep Learning for Diffusion MRI. arXiv preprint arXiv:2102.06942 (2021)

16. Ning, L., et al.: Muti-shell Diffusion MRI harmonisation and enhancement challenge (MUSHAC): progress and results. In: Bonet-Carne, E., et al. (eds.) MICCAI 2019. MV, pp. 217–224. Springer, Cham (2019). https://doi.org/10.1007/978-3-030-05831-9_18

17. Lin, Z., et al.: Fast learning of fiber orientation distribution function for MR tractography using convolutional neural network. Med. Phys. **46**(7), 3101–3116 (2019)

18. Koppers, S., Merhof, D.: Direct Estimation of Fiber Orientations using Deep Learning in Diffusion Imaging. International Workshop on Machine Learning in Medical Imaging, Springer, Cham (2016). https://doi.org/10.1007/978-3-319-47157-0_7

19. Sedlar, S., et al.: Diffusion MRI fiber orientation distribution function estimation using voxel-wise spherical U-net. In: Computational Diffusion MRI, MICCAI Workshop (2020)

20. Elaldi, A., et al.: Equivariant Spherical Deconvolution: Learning Sparse Orientation Distribution Functions from Spherical Data. arXiv preprint arXiv:2102.09462 (2021)

21. Cohen, T.S., Geiger, M., Köhler, J., Welling, M.: Spherical CNNs. In: International Conference on Learning Representations (ICLR) (2018)

22. Esteves, C., et al.: Learning so (3) equivariant representations with spherical CNNS. In: Proceedings of the European Conference on Computer Vision (ECCV) (2018)

23. Kondor, R., Lin, Z., Trivedi, S.: Clebsch-gordan nets: a fully fourier space spherical convolutional neural network. Adv. Neural. Inf. Process. Syst. **31**, 10117–10126 (2018)

24. Sugiura, M.: Unitary Representations and Harmonic Analysis: an Introduction. Elsevier, Amsterdam (1990)

25. Homeier, H.H.H., Steinborn, E.O.: Some properties of the coupling coefficients of real spherical harmonics and their relation to Gaunt coefficients. J. Mol. Struct.: THEOCHEM **368**, 31–37 (1996)

26. Rose, M.E.: Elementary Theory of Angular Momentum. Courier Corporation (1995)

27. Driscoll, J.R., Healy, D.M.: Computing Fourier transforms and convolutions on the 2-sphere. Adv. Appl. Math. **15**(2), 202–250 (1994)

28. Caruyer, E., et al.: Design of multishell sampling schemes with uniform coverage in diffusion MRI. Magn. Reson. Med. **69**(6), 1534–1540 (2013)

29. Van Essen, D.C., et al.: The WU-Minn human connectome project: an overview. Neuroimage **80**, 62–79 (2013)

30. Fick, R.H.J., Wassermann, D., Deriche, R.: The dmipy toolbox: Diffusion MRI multi-compartment modeling and microstructure recovery made easy. Front. Neuroinformatics **13**, 64 (2019)
31. Zhang, Y., Brady, M., Smith, S.: Segmentation of brain MR images through a hidden Markov random field model and the expectation-maximization algorithm. IEEE Trans. Med. Imaging **20**(1), 45–57 (2001)
32. Tournier, J.-D., et al.: MRtrix3: A fast, flexible and open software framework for medical image processing and visualisation. NeuroImage **202**, 116137 (2019)

Sharpening Local Interpretable Model-Agnostic Explanations for Histopathology: Improved Understandability and Reliability

Mara Graziani[1,2(✉)], Iam Palatnik de Sousa[3], Marley M.B.R. Vellasco[3], Eduardo Costa da Silva[3], Henning Müller[1,2], and Vincent Andrearczyk[1]

[1] University of Applied Sciences of Western Switzerland (Hes-so Valais), 3960 Sierre, Switzerland
mara.graziani@hevs.ch
[2] University of Geneva, 1227 Carouge, Switzerland
[3] Pontifical Catholic University of Rio de Janeiro, Gávea, Rio de Janeiro 22541-041, Brazil

Abstract. Being accountable for the signed reports, pathologists may be wary of high-quality deep learning outcomes if the decision-making is not understandable. Applying off-the-shelf methods with default configurations such as Local Interpretable Model-Agnostic Explanations (LIME) is not sufficient to generate stable and understandable explanations. This work improves the application of LIME to histopathology images by leveraging nuclei annotations, creating a reliable way for pathologists to audit black-box tumor classifiers. The obtained visualizations reveal the sharp, neat and high attention of the deep classifier to the neoplastic nuclei in the dataset, an observation in line with clinical decision making. Compared to standard LIME, our explanations show improved understandability for domain-experts, report higher stability and pass the sanity checks of consistency to data or initialization changes and sensitivity to network parameters. This represents a promising step in giving pathologists tools to obtain additional information on image classification models. The code and trained models are available on GitHub.

Keywords: Histopathology · Interpretable AI · Reliable AI

1 Introduction

Convolutional Neural Networks (CNNs) can propose with very high accuracy regions of interest and their relative tumor grading in Whole Slide Images (WSIs), gigapixel scans of pathology glass slides [24]. This can support pathologists in clinical routine by reducing the size of the areas to analyze in detail and

---

M. Graziani and I.P. de Sousa—Equal contribution (a complex randomization process was employed to determine the order of the first and second authors).

M. de Bruijne et al. (Eds.): MICCAI 2021, LNCS 12903, pp. 540–549, 2021.
https://doi.org/10.1007/978-3-030-87199-4_51

eventually highlighting missed or underestimated anomalies [3]. Without justifications for the decision-making, there is an opaque barrier between the model criteria and the clinical staff. Reducing such opaqueness is important to ensure the uptake of CNNs for sustained clinical use [22]. An already wide variety of off-the-shelve toolboxes has been proposed to facilitate the explanation of CNN decisions while keeping the performance untouched [2,5,14,17,19]. Among these, Local Interpretable Model-agnostic Explanations (LIME) are widely applied in radiology [16] and histopathology [15,20].

As argued by Sokol and Flach [19], enhancements of existing explainability tools are needed to provide machine learning consumers with more accessible and interactive technologies. Existing visualization methods present pitfalls that urge for improvement, as pointed out by the unreliability shown in [1,11]. LIME outputs for histopathology, for example, do not indicate any alignment of the explanations to clinical evidence and show high instability and scarce reproducibility [6]. Optimizing and reformulating this existing approach is thus a necessary step to promote its realistic deployment in clinical routines.

In this work, we propose to employ a better segmentation strategy that leads to sharper visualizations, directly highlighting relevant nuclei instances in the input images. The proposed approach brings improved understandability and reliability. Sharp-LIME heat maps appear more understandable to domain experts than the commonly used LIME and GradCAM techniques [18]. Improved reliability is shown in terms of result consistency over multiple seed initializations, robustness to input shifts, and sensitivity to weight randomizations. Finally, Sharp-LIME allows for direct interaction with pathologists, so that areas of interest can be chosen for explanations directly. This is desirable to establish trust [19]. In this sense, we propose a relevant step towards reliable, understandable and more interactive explanations in histopathology.

2 Methods

2.1 Datasets

Three publicly available datasets are used for the experiments, namely Camelyon 16, Camelyon 17 [13] and the breast subset of the PanNuke dataset [4][1]. Camelyon comprises 899 WSIs of the challenge collection run in 2017 and 270 WSIs of the one in 2016. Slide-level annotations of metastasis type (i.e. negative, macro-metastases, micro-metastases, isolated tumor cells) are available for all training slides, while a few manual segmentations of tumor regions are available for 320 WSIs. Breast tissue scans from the PanNuke dataset are included in the analysis. For these images, the semi-automatic instance segmentation of multiple nuclei types is available, allowing to identify neoplastic, inflammatory, connective, epithelial, and dead nuclei in the images. No dead nuclei are present, however, in the breast tissue scans [4]. Image patches of 224 × 224 pixels are extracted at the highest magnification level from the WSIs to build training, validation and test splits as in Table 1. To balance the under-representation,

---

[1] camelyon17.grand-challenge.org and jgamper.github.io/PanNukeDataset.

Table 1. Summary of the train, validation, internal and external test splits.

	Label	Cam16	C. 0	C. 1	C. 2	C. 3	C. 4	F. 1	F. 2	F. 3
			Cam17 (5 Centers)					PanNuke (3 Folds)		
Train	Neg.	12954	31108	25137	38962	25698	0	1425	1490	0
	Pos.	6036	8036	5998	2982	1496	0	2710	2255	0
Val.	Neg.	0	325	0	495	0	0	0	0	0
	Pos.	0	500	0	500	0	0	0	0	0
Int. Test	Neg.	0	0	274	483	458	0	0	0	1475
	Pos.	0	500	999	0	0	0	0	0	2400
Ext. Test	Neg.	0	0	0	0	0	500	0	0	0
	Pos.	0	0	0	0	0	500	0	0	0

PanNuke input images were oversampled by five croppings, namely in the center, upper left, upper right, bottom left and bottom right corners. The pre-existing PanNuke folds were used to separate the patches in the splits. Reinhard normalization is applied to all the patches to reduce the stain variability.

2.2 Network Architectures and Training

Inception V3 [21] with ImageNet pre-trained weights is used for the analysis. The network is fine-tuned on the training images to classify positive patches containing tumor cells. The fully connected classification block has four layers, with 2048, 512, 256 and 1 neurons. A dropout probability of 0.8 and L2 regularization were used to avoid overfitting. This architecture was trained with mini-batch Stochastic Gradient Descent (SGD) optimization with standard parameters (learning rate of $1e^{-4}$, Nesterov momentum of 0.9). For the loss function, class-weighted binary cross-entropy was used. Network convergence is evaluated by early stopping on the validation loss with patience of 5 epochs. The model performance is measured by the average Area Under the ROC Curve (AUC) over ten runs with multiple initialization seeds, reaching 0.82 ± 0.0011 and 0.87 ± 0.005 for the internal and external test sets respectively.

Nuclei contours of the Camelyon input are extracted by a Mask R-CNN model [7] fine-tuned from ImageNet weights on the Kumar dataset for the nuclei segmentation task [12]. The R-CNN model identifies nuclei entities and then generates pixel-level masks by optimizing the Dice score. ResNet50 [7] is used for the convolutional backbone as in [10]. The network is optimized by SGD with standard parameters (learning rate of 0.001 and momentum of 0.9).

2.3 LIME and Sharp-LIME

LIME for Image Classifiers Defined by Ribeiro et al. [17] for multiple data classifiers, a general formulation of LIME is given by:

$$\xi(x) = \underset{g \in G}{\operatorname{argmin}} \quad \mathcal{L}(f, g, \pi_x) + \Omega g \tag{1}$$

Eq. (1) represents the minimization of the explanatory infidelity $\mathcal{L}(f, g, \pi_x)$ of a potential explanation g, given by a surrogate model G, in a neighborhood defined by $\pi_x(z)$ around a given sample of the dataset (x). The neighborhood is obtained by perturbations of x around the decision boundary.

For image classifiers, that are the main focus of this work, an image x is divided into representative image sub-regions called super-pixels using a standard segmentation algorithm, e.g. Quickshift [23]. Perturbations of the input image are obtained by filling random super-pixels with black pixels. The surrogate linear classifier G is a ridge regression model trained on the perturbed instances weighed by the cosine similarity $(\pi_x(z))$ to approximate the prediction probabilities. The coefficients of this linear model (referred to as *explanation weights*) explain the importance of each super-pixel to the model decision-making. Explanation weights are displayed in a symmetrical heatmap where super-pixels in favor of the classification (positive explanation weights) are in blue, and those against (negative weights) in red.

Previous improvements of LIME for histopathology proposed a systematic manual search for parameter heuristics to obtain super-pixels that visually correspond to expert annotations [20]. Consistency and super-pixel quality were further improved by genetic algorithms in [15]. Both solutions are impractical for clinical use, being either too subjective or too expensive to compute.

Sharp-LIME The proposed implementation of Sharp-LIME, as illustrated in Fig. 1, uses nuclei contours as input super-pixels for LIME rather than other segmentation techniques. Pre-existing nuclei contour annotations may be used. If no annotations are available, the framework suggests automatic segmentation of nuclei contours by the Mask R-CNN. Manual annotations of regions of interest may also be drawn directly by end-users to probe the network behavior for specific input areas. For the super-pixel generation, the input image is split into nuclei contours and background. The background is further split into 9 squares of fixed size. This splitting reduces the difference between nuclei and background areas, since overly large super-pixels may achieve large explanation weights by sheer virtue of their size. The code to replicate the experiments (developed with Tensorflow > 2.0 and Keras 2.4.0) is available

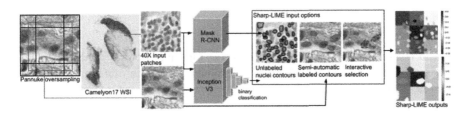

Fig. 1. Overview of the approach. An InceptionV3 classifies tumor from non-tumor patches at high magnification sampled from the input WSIs. Manual or automatically suggested nuclei contours (by Mask R-CNN) are used as input to generate the Sharp-LIME explanations on the right.

at github.com/maragraziani/sharp-LIME, alongside the trained CNN weights. Experiments were run using a GPU NVIDIA V100. A single Sharp-LIME explanation takes roughly 10 s to generate in this setting. 200 perturbations were used, as it already showed low variability in high explanation weight super-pixels, as further discussed in Sect. 3.

2.4 Evaluation

Sharp-LIME is evaluated against the state-of-the-art LIME by performing multiple quantitative evaluations. Not having nuclei type labels for Camelyon, we focused on the PanNuke data. We believe, however, that the results would also apply to other inputs. Sanity checks are performed, testing for robustness to constant input shifts and sensitivity to network parameter changes as in [1,11]. Spearman's Rank Correlation Coefficient (SRCC) is used to evaluate the similarity of the ranking of the most important super-pixels. The cascading randomization test in [1] is performed by assigning random values to the model weights starting from the top layer and progressively descending to the bottom layer. We already expect this test to show near-zero SRCC for both techniques, since by randomizing the network weights, the network output is randomized as well as LIME and Sharp-LIME explanations. The repeatability and consistency for multiple seed initializations are evaluated by the SRCC, the Intraclass Correlation Coefficient (ICC) (two-way model), and the coefficient of variation (CV) of the explanation weights.

Additionally, we quantify domain appropriateness as the alignment of the explanations with relevant clinical factors [22]. The importance of a neoplastic nucleus, an indicator of a tumor [4], is measured by the sign and magnitude of the explanation weight. Descriptive statistics of the explanation weights are compared across the multiple types of nuclei in PanNuke. Pairwise non-parametric Kruskal tests for independent samples are used for the comparisons. A paired t-test is used to compare LIME weights obtained from a randomly initialized and a trained network, as suggested in [6].

3 Results

3.1 Improved Understandability

Qualitative Evaluation By Domain Experts. Figure 2 shows a qualitative comparison of LIME and Sharp-LIME for PanNuke and Camelyon inputs. For conciseness, only two examples are provided. An extended set of results can be inspected in the GitHub repository[2].

Five experts in the digital pathology domain with experience in CNN-based applications for clinical research purposes compared LIME, Sharp-LIME and Gradient Weighted Class Activation Mapping (Grad-CAM) [18] for a few images in this work. The experts generally use these visualizations to improve their

---

[2] (github.com/maragraziani/sharp-LIME).

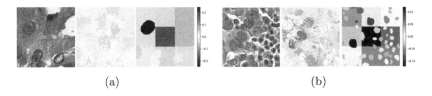

Fig. 2. From left to right, input image with overlayed nuclei contours, standard LIME and sharp LIME for a) a PanNuke and b) a Camelyon input image.

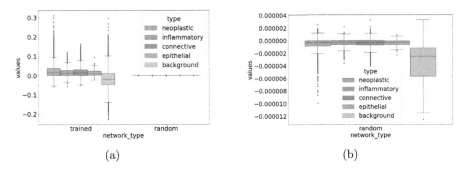

Fig. 3. a) Comparison between Sharp-LIME explanation weights for a trained and a randomly initialized CNN; b) Zoom on the random CNN in a). These results can be compared to those obtained for standard LIME in [6].

model understanding, particularly if the suggested diagnosis is different from theirs. Sharp-LIME was assessed as easier to understand than Grad-CAM and LIME by 60% of them. Two of the five experts further confirmed that these explanations help increasing their confidence in the model's decision-making. While it is difficult to obtain quantitative comparisons, we believe this expert feedback, although subjective, is an essential evaluation.

3.2   Improved Reliability

Quantification of Network Attention. We quantify the Sharp-LIME explanation weights for each of the functionally diverse nuclei types of the PanNuke dataset in Fig. 3. As Fig. 3a shows, the explanation weights of the neoplastic nuclei, with average value 0.022 ± 0.03, are significantly larger than those of the background squared super-pixels, with average value -0.018 ± 0.05. Explanation weights of the neoplastic nuclei are also significantly larger than those of inflammatory, neoplastic and connective nuclei (Kruskal test, p-value < 0.001 for all pairings). Sharp-LIME weights are compared to those obtained by explaining a random CNN, that is the model with randomly initialized parameters. The Sharp-LIME explanation weights for the trained and random CNN present significant differences (paired t-test, p-value< 0.001), with the explanations for the latter being almost-zero values as shown by the boxplot in Fig. 3b.

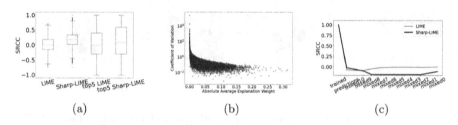

Fig. 4. a) SRCC of the entire and top-5 super-pixel rankings obtained over three re-runs with changed initialization. The means of the distributions are significantly different (paired t-test, p-value< 0.001); b) CV against average explanation weight for three re-runs with multiple seeds; c) SRCC of the super-pixel rankings obtained in the cascading randomization test

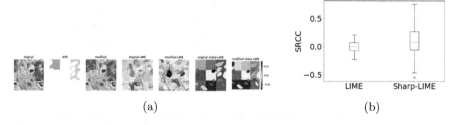

Fig. 5. Robustness to constant input shift. a) Qualitative evaluation for one PanNuke input image; b) SRCC of the super-pixel rankings for all PanNuke inputs.

Consistency. The consistency of Sharp-LIME explanations for multiple seed initialization is shown in Figs. 4a and 4b. The mean of LIME SRCC is significantly lower than that of Sharp-LIME, 0.015 against 0.18 (p-value< 0.0001). As Fig. 4b shows, super-pixels with large average absolute value of the explanation weight are more consistent across re-runs of Sharp-LIME, with lower CV. We compare the SRCC of the five super-pixels with the highest ranking, obtaining average LIME explanation weights 0.029 and 0.11 for Sharp-LIME. The ICC of the most salient super-pixel in the image, i.e. first in the rankings, for different initialization seeds, further confirms the largest agreement of Sharp-LIME, with ICC 0.62 against the 0.38 of LIME. As expected, the cascading randomization of network weights shows nearly-zero SRCC in Fig. 4c. A visual example of LIME robustness to constant input shifts is given in Fig. 5a. The SRCC of LIME and Sharp-LIME is compared for original and shifted inputs with unchanged model prediction in Fig. 5b. Sharp-LIME is significantly more robust than LIME (t-test, p-value< 0.001).

4   Discussion

The experiments evaluate the benefits of the Sharp-LIME approach against the standard LIME, showing improvements in the understandability and reliability of the explanations. This improvement is given by the choice of a segmentation algorithm that identifies regions with a semantic meaning in the images. Differently from standard LIME, Sharp-LIME justifies the model predictions by the relevance of image portions that are easy to understand as shown in Fig. 2. Our visualizations have higher explanation weights and show lower variability than standard LIME. The feedback from the domain-experts is encouraging (Sect. 3.2). Despite being only qualitative, it reinforces the importance of a feature often overseen in explainability development, namely considering the target of the explanations during development to provide them with intuitive and reliable tools. The quantitative results in Sect. 3.2 show the improved reliability of Sharp-LIME. Neoplastic nuclei appear more relevant than other nuclei types, aligning with clinical relevance. Since these nuclei are more frequent than other types in the data, the results are compared to a randomly initialized CNN to confirm that their importance is not due to hidden biases in the data (Fig. 3). The information contained in the background, often highlighted as relevant by LIME or Grad-CAM [6], seems to rather explain the negative class, with large and negative explanation weights on average. Large Sharp-LIME explanation weights point to relevant super-pixels with little uncertainty, shown by low variation and high consistency in Figs. 4b and 4a. The instability of LIME reported in [6] can therefore be explained by the choice of the segmentation algorithm, an observation in line with the work in [20].

The simplicity of this approach is also its strength. Our super-pixel choice of nuclei segmentation adds little complexity to the default LIME, being a standard data analysis step in various histopathology applications [8,9]. Extensive annotations of nuclei contours are not needed since automated contouring can be learned from small amounts of labeled data [8] (Fig. 2b). Additionally, the users may directly choose the input super-pixels to compare, for example, the relevance of one image area against the background or other areas. Requiring only a few seconds to be computed, Sharp-LIME is faster than other perturbation methods that require a large number of forward passes to find representative super-pixels. For this reason, the technique represents a strong building-block to develop interactive explainability interfaces where users can visually query the network behavior and quickly receive a response.

The small number of available experts is a limitation of this study, which does not propose quantitative estimates of user confidence and satisfaction in the explanations. We will address this point in future user-evaluation studies.

5   Conclusions

This work shows important points in the development of explainability for healthcare. Optimizing existing methods to the application requirements and user satisfaction promotes the uptake and use of explainability techniques.

Our proposed visualizations are sharp, fast to compute and easy to apply to black-box histopathology classifiers by focusing the explanations on nuclei contours and background portions. Other image modalities may benefit from this approach. The relevance of the context surrounding tumor regions, for example, can be evaluated in radiomics. Further research should focus on the specific demands of the different modalities.

Acknowledgements. We thank MD. Dr. Filippo Fraggetta for his relevant feedback on our methods. We also thank Lena Kajland Wilén, Dr. Francesco Ciompi and the Swiss Digital Pathology consortium for helping us getting in contact with the pathologists that participated in the user studies. This work is supported by the European Union's projects ExaMode (grant 825292) and AI4Media (grant 951911).

References

1. Adebayo, J., Gilmer, J., Muelly, M., Goodfellow, I., Hardt, M., Kim, B.: Sanity checks for saliency maps. In: Proceedings of the 32nd International Conference on Neural Information Processing Systems, p. 9525–9536. NIPS 2018, Curran Associates Inc., Red Hook, NY, USA (2018)
2. Alber, M., Let al.: Innvestigate neural networks! J. Mach. Learn. Res. **20**(93), 1–8 (2019). http://jmlr.org/papers/v20/18-540.html
3. Fraggetta, F.: Clinical-grade computational pathology: alea iacta est. J. Pathol. Inf. **10** (2019)
4. Gamper, J., Alemi Koohbanani, N., Benet, K., Khuram, A., Rajpoot, N.: Pan-Nuke: an open pan-cancer histology dataset for nuclei instance segmentation and classification. In: Reyes-Aldasoro, C.C., Janowczyk, A., Veta, M., Bankhead, P., Sirinukunwattana, K. (eds.) ECDP 2019. LNCS, vol. 11435, pp. 11–19. Springer, Cham (2019). https://doi.org/10.1007/978-3-030-23937-4_2
5. Graziani, M., Andrearczyk, V., Müller, H.: Regression concept vectors for bidirectional explanations in histopathology. In: Stoyanov, D., et al. (eds.) MLCN/DLF/IMIMIC -2018. LNCS, vol. 11038, pp. 124–132. Springer, Cham (2018). https://doi.org/10.1007/978-3-030-02628-8_14
6. Graziani, M., Lompech, T., Müller, H., Andrearczyk, V.: Evaluation and comparison of CNN visual explanations for histopathology. In: Explainable Agency in Artificial Intelligence at AAAI21, pp. 195–201 (2020)
7. He, K., Gkioxari, G., Dollar, P., Girshick, R.: Mask r-CNN. In: Proceedings of the IEEE International Conference on Computer Vision (ICCV) (2017)
8. Janowczyk, A., Madabhushi, A.: Deep learning for digital pathology image analysis: a comprehensive tutorial with selected use cases. J. Pathol. Inf. **7** (2016)
9. Janowczyk, A., Zuo, R., Gilmore, H., Feldman, M., Madabhushi, A.: Histoqc: an open-source quality control tool for digital pathology slides. JCO Clin. Cancer Inf. **3**, 1–7 (2019)
10. Jung, H., Lodhi, B., Kang, J.: An automatic nuclei segmentation method based on deep convolutional neural networks for histopathology images. BMC Biomed. Eng. **1**, 1–2 (2019). https://doi.org/10.1186/s42490-019-0026-8
11. Kindermans, P.J., et al.: The (un) Reliability of Saliency Methods. Interpreting, Explaining and Visualizing Deep Learning, Springer International Publishing, Explainable AI (2019)
12. Kumar, N., et al.: A dataset and a technique for generalized nuclear segmentation for computational pathology. IEEE Trans. Med. Imaging **36**(7), 1550–1560 (2017). https://doi.org/10.1109/TMI.2017.2677499

13. Litjens, G., et al.: 1399 H&E-stained sentinel lymph node sections of breast cancer patients: the CAMELYON dataset. GigaScience **7**(6), giy065 (2018)

14. Lundberg, S.M., Lee, S.I.: A unified approach to interpreting model predictions. In: Guyon, I., et al. (eds.) Advances in Neural Information Processing Systems 30, pp. 4765–4774. Curran Associates, Inc. (2017). http://papers.nips.cc/paper/7062-a-unified-approach-to-interpreting-model-predictions.pdf

15. Palatnik de Sousa, I., Bernardes Rebuzzi Vellasco, M.M., Costa da Silva, E.: Evolved Explainable Classifications for Lymph Node Metastases. arXiv e-prints arXiv:2005.07229, May 2020

16. Reyes, M., et al.: On the interpretability of artificial intelligence in radiology: challenges and opportunities. Radiol.: Artif. Intell. **2**, e190043 (2020). https://doi.org/10.1148/ryai.2020190043

17. Ribeiro, M.T., Singh, S., Guestrin, C.: Why Should I Trust You?: explaining the predictions of any classifier. In: Proceedings of the 22nd ACM SIGKDD International Conference on Knowledge Discovery and Data Mining, San Francisco, CA, USA, 13–17 August 2016, pp. 1135–1144 (2016)

18. Selvaraju, R.R., Cogswell, M., Das, A., Vedantam, R., Parikh, D., Batra, D.: Gradcam: visual explanations from deep networks via gradient-based localization. In: Proceedings of the IEEE International Conference on Computer Vision, pp. 618–626 (2017)

19. Sokol, K., Flach, P.: One explanation does not fit all. KI-Künstliche Intelligenz, pp. 1–16 (2020)

20. Palatnik de Sousa, I., Bernandes Rebuzzi Vellasco, M.M., Costa da Silva, E.: Local interpretable model-agnostic explanations for classification of lymph node metastases. Sensors (Basel, Switzerland) **19** (2019)

21. Szegedy, C., Vanhoucke, V., Ioffe, S., Shlens, J., Wojna, Z.: Rethinking the inception architecture for computer vision. In: Proceedings of the IEEE Conference on Computer Vision and Pattern Recognition, pp. 2818–2826 (2016)

22. Tonekaboni, S., Joshi, S., McCradden, M.D., Goldenberg, A.: What clinicians want: contextualizing explainable machine learning for clinical end use. In: Machine Learning for Healthcare Conference, pp. 359–380. PMLR (2019)

23. Vedaldi, A., Soatto, S.: Quick Shift and Kernel Methods for Mode Seeking. In: Forsyth, D., Torr, P., Zisserman, A. (eds.) ECCV 2008. LNCS, vol. 5305, pp. 705–718. Springer, Heidelberg (2008). https://doi.org/10.1007/978-3-540-88693-8_52

24. Wang, D., Khosla, A., Gargeya, R., Irshad, H., Beck, A.H.: Deep learning for identifying metastatic breast cancer. arXiv preprint arXiv:1606.05718 (2016)

Improving the Explainability of Skin Cancer Diagnosis Using CBIR

Catarina Barata[✉] [ID] and Carlos Santiago[ID]

Institute for Systems and Robotics, Instituto Superior Técnico, Lisbon, Portugal
ana.c.fidalgo.barata@tecnico.ulisboa.pt

Abstract. Explainability is a key feature for computer-aided diagnosis systems. This property not only helps doctors understand their decisions, but also allows less experienced practitioners to improve their knowledge. Skin cancer diagnosis is a field where explainability is of critical importance, as lesions of different classes often exhibit confounding characteristics. This work proposes a deep neural network (DNN) for skin cancer diagnosis that provides explainability through content-based image retrieval. We explore several state-of-the-art approaches to improve the feature space learned by the DNN, namely contrastive, distillation, and triplet losses. We demonstrate that the combination of these regularization losses with the categorical cross-entropy leads to the best performances on melanoma classification, and results in a hybrid DNN that simultaneously: i) classifies the images; and ii) retrieves similar images justifying the diagnosis. The code is available at https://github.com/catarina-barata/CBIR_Explainability_Skin_Cancer.

Keywords: Skin cancer · CBIR · Explainable AI

1 Introduction

During the last five years we have witnessed an impressive transformation in the field of dermoscopy image analysis. The development of increasingly more complex DNN approaches to segment [12,20] and diagnose [2,5,11] skin lesions led to outstanding performances, on par with those of dermatologists with different levels of expertise [4]. Recently, an extensive study with more than 300 dermatologists has demonstrated that a DNN assistant is a useful second opinion tool, particularly for less experienced doctors [18]. However, this only holds if the

This work was supported by the FCT project and multi-year funding [CEECIND/00326/2017] and LARSyS - FCT Plurianual funding 2020–2023; and by a Google Research Award'21. The Titan Xp used in this project were donated by the NVIDIA Corporation.

Electronic supplementary material The online version of this chapter (https://doi.org/10.1007/978-3-030-87199-4_52) contains supplementary material, which is available to authorized users.

M. de Bruijne et al. (Eds.): MICCAI 2021, LNCS 12903, pp. 550–559, 2021.
https://doi.org/10.1007/978-3-030-87199-4_52

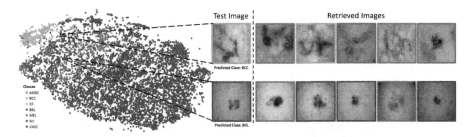

Fig. 1. T-SNE of the global average pooling (GAP) embedding space from a DenseNet121. Applying a CBIR module to these embeddings leads to poor retrieval results, where most retrieved images are visually dissimilar and belong to a different class (represented by the border color).

DNN assistant is i) reliable, *i.e.*, it consistently predicts the correct class for a given data set; and ii) equipped with a good visualization interface that justifies the predicted outcome. The second property becomes particularly relevant to dermatologists if we consider that these assistants may influence their diagnosis and also teach less experienced ones.

Incorporating a content-based image retrieval (CBIR) module into a DNN assistant for skin cancer diagnosis has the potential to improve its explainability, since CBIR consists of finding images on an previously collected data set that share similar features to the one being diagnosed. As a matter of fact, this mimics the medical procedure, where sometimes doctors rely on their past experience and similar cases to reach the final diagnosis. However, the adoption of CBIR in skin cancer diagnosis is not only poorly studied, but also presents conflicting results [13,14,18]. One source of controversy is the number of images presented to the user. In [18], the model retrieved k similar images for each type of skin lesion, offering a lot of variability to support the user's final decision. Other works opted to present only the set of k most similar dermoscopy images, irrespective of their class, significantly reducing the number of images to be inspected and, consequently, the time to answer [13,14].

Besides the number of images, the actual set of retrieved images also affects the CBIR module utility. Figure 1 (right) shows the diagnosis predicted by a DNN and the $k = 5$ most similar images from the training set, as well as their classes. Surprisingly, the predominant class retrieved by the CBIR often does not match the one predicted by the DNN. Moreover, CBIR returns several classes. These issues are bound to decrease the explainability of the DNN and would probably influence negatively a dermatologist, particularly a less experienced one. When inspecting the t-SNE visualization [10] of the features used to compute image similarities (see Fig. 1, left), it is clear that the feature space learned by the DNN lacks structure. Features from different lesion classes are close to each other and mixed. This suggests that, although the classification layer is able to handle this feature space, it is not appropriate for CBIR.

This work aims to improve the structure of the feature space learned by a DNN for skin cancer diagnosis. We demonstrate that with appropriate regularization strategies imposed by contrastive, distillation, and triplet losses, the DNN

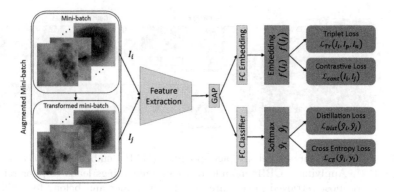

Fig. 2. Overview of the proposed two-headed model comprising a feature extraction module with GAP, which feeds a FC layer for the embedding space and another FC layer for the classification.

model becomes suitable for both the diagnosis and CBIR tasks. A recent work [1] has proposed the use of triplet-loss [15] to develop a CBIR model for skin cancer diagnosis. However, the CBIR model was not implemented together with the classification one, as is performed in our work. Moreover, we observed that the triplet loss seems to become unstable when a small batch size is used. Finally, by using contrastive and distillation losses, we also improve the robustness of our CBIR model to geometric transformations.

The remaining of the paper is organized as follows. Section 2 provides a description of the proposed approach. The experimental setup and results are presented in Sect. 3, while Sect. 4 concludes the paper with some insights on future work.

2 Proposed Approach

2.1 Classification-CBIR Model Overview

The proposed approach follows a standard DNN architecture for skin cancer diagnosis: a backbone of convolutional and pooling layers for feature extraction, followed by a global average pooling (GAP) operation and a fully connected (FC) layer with softmax activation to predict the lesion class probabilities. We found that to make this model more suitable for CBIR tasks, it should comprise a second FC head with linear activation. This head is responsible for creating a feature embedding. The general overview of the model is shown in Fig. 2. Regarding the feature extraction backbone, any architecture can be used. In this work, we validate our approach using three convolutional architectures that have been extensively used in dermoscopy image analysis: VGG16 [16], ResNet101 [7], and DenseNet121 [9] (without the fully connected (FC) layers).

Typically, DNN for image classification is trained using a dataset of D dermatoscopy images that are randomly divided into mini-batches of size N. During training, the DNN will learn to classify the dermatoscopy images in each mini-batch through the minimization of the categorical cross-entropy loss function

$$\mathcal{L}_{CE} = -\sum_{c=1}^{C} y_i^c \log(\hat{y}_i^c), \tag{1}$$

where C is the number of classes, y_i is the ground truth class of the i-th element in the mini-batch, and $\hat{y}_i$ is the corresponding softmax probability.

As shown in Fig. 1, the cross-entropy loss is not suitable to train a system that not only classifies images, but is also capable of performing CBIR. Thus, in this work we propose to improve the CBIR capabilities of a DNN through the use of an augmented loss function. The global loss function used to train the classification-CBIR model is the following

$$\mathcal{L} = \mathcal{L}_{CE} + \alpha\mathcal{L}_{tr} + \beta\mathcal{L}_{Cont} + \gamma\mathcal{L}_{Dist}, \tag{2}$$

where α, β, and γ are hyperparameters that define the contribution of each term to the overall loss, $\mathcal{L}_{Tr}$ is the triplet loss, $\mathcal{L}_{Cont}$ is the contrastive loss, and $\mathcal{L}_{Dist}$ is the distillation loss. These additional loss functions are used to regularize the feature-space learned by the DNN, as we will discuss in the following section. Training details associated with each loss are also introduced.

The inference phase of the classification-CBIR model works as follows. The classification head predicts a diagnosis (FC Classifier in Fig. 2), while retrieval can be performed using the secondary head (FC Embedding in Fig. 2) or the output of the GAP layer. The similarity metric used for retrieval is the cosine similarity $x^T y / \|x\| \|y\|$. CBIR is also used to predict a diagnosis, by assessing the most voted class amongst the retrieved ones.

2.2 Feature-Space Regularization

Triplet-Loss: This loss function is added as a regularization term to yield more organized feature spaces, since a structured space is expected to improve the performance of CBIR [15]. Let us define an embedding $f(I_i)$ for image I_i, which is the output of the feature extraction backbone of the DNN, normalized to have unitary norm. Then, the triplet-loss can be defined as

$$\mathcal{L}_{Tr}(I_i, I_p, I_n) = \max(\text{sim}(f(I_i), f(I_p)) - \text{sim}(f(I_i), f(I_n)) + m, 0), \tag{3}$$

where (I_p, I_n) are, respectively, positive and negative pairs of I_i, m is the desired margin between negative and positive pairs, and $\text{sim}(x, y)$ is the cosine similarity between feature vectors. It is common to set $f(I_i)$ as the output of the global average pooling (GAP) operation [1]. However, this may disrupt the classification performance. A recent work showed that using a two-headed network, where one of the heads corresponds to the classification and the other is used for CBIR, leads to better results [17]. Here, we followed the same strategy, using $f(I_i)$ as the embedding produced by a second head, as exemplified in Fig. 2.

Negative pairs were selected from the mini-batch following a semi-hard sampling, which as been shown to lead to better scores [17].

Contrastive Loss: An ideal CBIR should be robust to multiple instances of the same image, *i.e.*, its performance should not change dramatically when presented with transformed versions of the query image. Inspired by recent works on self-supervised learning, we propose to explicitly model the previous constraint using a contrastive loss that compares features from different instances of the same image [3,6].

Following the strategy proposed in [3], we augment each mini-batch presented to the DNN during the training phase, by applying an additional random geometric transformation, such that each image I_i will have a transformed version, denoted by I_j. This means that the actual size of the mini-batch is $2N$. To compare the features of I_i and I_j, we use the normalized-temperature cross entropy. For each positive image pair (I_i, I_j), this loss function is given by

$$\mathcal{L}_{Cont}(I_i, I_j) = -\log \frac{\exp(\text{sim}(f(I_i), f(I_j))/\tau_c)}{\sum_{k=1}^{2N} \mathbb{1}_{[k \neq i]} \exp(\text{sim}(f(I_k), f(I_i))/\tau_c)}, \tag{4}$$

where τ_c is a temperature hyperparameter and $\text{sim}(x, y)$ is the cosine similarity defined above. Similarly to the triplet-loss, the contrastive loss is applied to the embeddings produced by the second head in our two-headed model (see Fig. 2).

Distillation Loss: This loss is commonly used in knowledge distillation approaches, where the knowledge is transferred from a teacher to a student DNN [8]. Nonetheless, we believe that it can also be used to improve the robustness of a DNN to different instances of the same images.

Similarly to the case of contrastive loss, we augment a mini-batch with an additional random geometric transformation per image. Then, we compute the distillation loss between each image I_i and its transformed version I_j

$$\mathcal{L}_{Dist} = -\sum_{c=1}^{C} \frac{\hat{y}_i^c}{\tau_d} \log \frac{\hat{y}_j^c}{\tau_d}, \tag{5}$$

where τ_d is a temperature hyperparameter that smooths softmax probabilities.

3   Experimental Results

3.1   Experimental Setup

The proposed classification-CBIR model is trained and evaluated using the ISIC 2018 dermoscopy dataset [19], which contains 10,015 images for training and 1,512 for testing, comprising seven skin lesion classes: nevu (NV), melanoma (MEL), actinic (AKIEC), basal cell carciona (BCC), dermatofibroma (DF), benign keratosis (BKL), and vascular (VASC). The training set was further split into training (80%) and validation (20%), to select model hyperparameters.

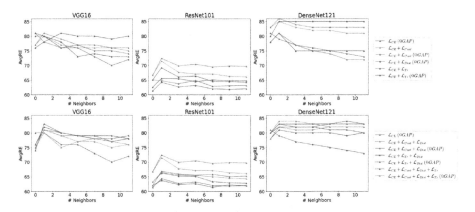

Fig. 3. Average recall of the CBIR model for various number of neighbors k, where $k = 0$ corresponds to the performance of the classification layer. The top row shows models trained with a single regularization function, while the bottom row shows the results of combining different loss functions. @GAP refers to performing retrieval using the output of the GAP layer, instead of the second head (if available).

To evaluate the performance of our classification-CBIR model we compute the following statistics for the validation set: a) *recall* for the diagnosis predicted by the model and by the one predicted by CBIR; and b) *precision@K*, to specifically evaluate the quality of the CBIR module. The ground truth labels of the test set are unavailable. The evaluation was performed using the ISIC platform[1], which scores the models based on the balanced accuracy score (average recall).

We validate our approach on three CNN architectures: VGG16, ResNet101, and DenseNet121. The size of the FC layers is 7 for classification and 256 for embedding (recall Fig. 2). We train several models for each CNN architecture, using different configurations of hyperparameters: i) $\alpha \in \{0,1\}$, $\beta \in \{0,1\}$, and $\gamma \in \{0, \tau_d^2\}$ in (2); ii) $\tau_d = \{2, 5, 10\}$ in (5); iii) $\tau_c = 0.07$ in (4); and iv) $m = 0.2$ in (3). VGG16 models are trained for 150 epochs with a mini-batch of $N = 20$, while ResNet101 and DenseNet121 models are both trained for 80 epochs with a mini-batch of $N = 10$. The validation set is used to select the best set of hyperparameters, considering the average recall. To improve the generalization of the model, we use online data augmentation (prior to the additional mini-batch augmentation described in Sect. 2) and drouput with $p = 0.5$ after the GAP layer. The weights of the CNN architectures are also initialized using models pre-trained on ImageNet.

3.2   Results

Figure 3 shows the average recall curves obtained by the various models. Each of the represented curves corresponds to a different training configuration.

---

[1] https://challenge.isic-archive.com/.

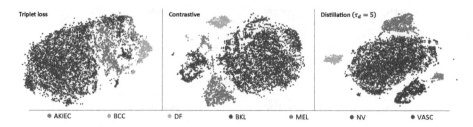

Fig. 4. T-SNE of the embedding space obtained after GAP with $\mathcal{L}_{CE}$ plus each loss. The colors represent the different classes. These visualizations correspond to DenseNet121.

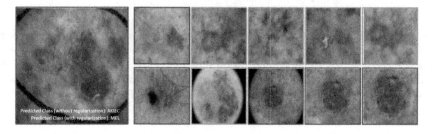

Fig. 5. Example of CBIR@GAP without regularization (top row), $\mathcal{L}_{CE}$, and with regularization (bottom row), $\mathcal{L}_{CE} + \mathcal{L}_{Cont} + \mathcal{L}_{Dist}$. The GT class is shown with the same color scheme as Fig. 4. These results were obtained using DenseNet161.

As expected, the average recall of the models trained using only $\mathcal{L}_{CE}$ (blue line in all plots) decreases rapidly, as we use more neighbors to perform the CBIR. When we incorporate at least one of the regularization losses, it generally has a strong and positive impact on the model, improving the average recall for a larger number of neighbors. Additionally, the performance of the CBIR module is often better than that of the classification layer and, in the cases of ResNet101 and DenseNet121, the use of the additional loss functions even improves the performance of the classification layer over the baseline. This improvement may be justified by a reduction of the model overfitting to the training data, since both DenseNet and ResNet are significantly bigger models than VGG16 and the latter does not exhibit the same performance improvement. Finally, it is noteworthy that the best scores are achieved when the retrieval is performed using the embeddings generated by the GAP layer. Therefore, we will adopt this strategy in the remaining of the results.

Surprisingly, the best overall performances seem to be achieved with contrastive and distillation losses, as well as their combination. Since both losses were not specifically proposed for retrieval purposes, contrary to triplet loss, this is a very interesting result. An inspection of the t-SNE visualization obtained for each of these losses (see Fig. 4) proves that both contrastive and distillation losses lead to more structured feature spaces than the triplet loss. The worse performance of triplet loss may be related with the batch size used in our exper-

Table 1. Evaluation scores for lesion classification using CBIR with $k = 1$ and $k = 11$. These scores were obtained using DenseNet121 and with $\tau_d = 5$ in $\mathcal{L}_{Dist}$.

| | | $\mathcal{L}_{CE}$(@GAP) | | | | | | | |
| | | − | | +$\mathcal{L}_{Dist}$ | | +$\mathcal{L}_{Cont}$ | | +$\mathcal{L}_{Cont}$ + $\mathcal{L}_{Dist}$ | |
Lesion class	k	Re	Pr@k	Re	Pr@k	Re	Pr@k	Re	Pr@k
AKIEC	1	61.5%	61.5%	**75.4%**	**75.4%**	69.2%	69.2%	70.8%	70.8%
	11	63.1%	56.4%	**73.8%**	72.2%	69.2%	65.9%	**73.8%**	69.7%
BCC	1	85.4%	85.4%	89.3%	89.3%	90.3%	90.3%	**91.3%**	**91.3%**
	11	88.3%	78.0%	88.3%	88.2%	90.3%	89.3%	**92.2%**	**91.8%**
DF	1	82.6%	82.6%	**91.3%**	**91.3%**	91.3%	91.3%	82.6%	82.6%
	11	60.9%	49.8%	**91.3%**	**89.3%**	87.0%	86.2%	73.9%	76.3%
BKL	1	71.8%	71.8%	77.7%	77.7%	**78.6%**	**78.6%**	73.6%	73.6%
	11	64.5%	55.6%	**76.8%**	**76.7%**	73.6%	71.9%	74.1%	72.9%
MEL	1	71.2%	71.2%	66.7%	66.7%	**77.9%**	**77.9%**	74.3%	74.3%
	11	55.4%	51.2%	67.1%	67.4%	**72.5%**	70.1%	**72.5%**	**72.6%**
NV	1	96.3%	96.3%	96.3%	96.3%	96.2%	96.2%	95.7%	95.7%
	11	96.9%	94.2%	95.8%	95.7%	96.0%	95.7%	95.5%	95.4%
VASC	1	82.8%	82.8%	**100%**	**100%**	96.6%	96.6%	93.1%	93.1%
	11	79.3%	71.1%	**100%**	**99.1%**	89.7%	89.0%	93.1%	92.8%
AVG	1	78.8%	78.8%	85.2%	85.2%	**85.7%**	**85.7%**	83.1%	83.1%
	11	72.6%	65.2%	**84.7%**	**84.1%**	82.6%	81.2%	82.2%	81.6%
		AvgRE							
Test set score	0	66.8%		65.8%		66.2%		65.5%	
	1	59.4%		**70.1%**		62.6%		65.6%	
	11	58.6%		**69.1%**		65.5%		68.1%	

iments. This loss has been shown to have some limitations when the batch size is small, since it requires the comparison of positive and negative pairs of images. The results in Fig. 4 must also be compared with those in Fig. 1(left), since all of them correspond to t-SNE visualizations of the GAP layer. It is noteworthy that, although contrastive and triplet losses are applied to a second FC head, they are still able to impose some structure on the GAP layer.

Table 1 shows a more detailed analysis of the scores obtained for the best models, as well as an evaluation of the models on the held-out test set. For additional results please refer to the supplementary material. When the models are only trained using $\mathcal{L}_{CE}$, the CBIR performance drops abruptly for the following lesion classes: MEL, DF, and BKL. When the regularization losses are incorporated, the performance for these classes becomes much more stable. This is particularly important in the case of MEL, since this is the most aggressive form of skin cancer. Although the evaluation on the test set does not allow us to individually analyze each lesion class, it is still possible to observe the same trend of the validation set: the performance of the classification layer is similar, but the performance of CBIR improves significantly with the regularization functions, in particular with the incorporation of $\mathcal{L}_{Dist}$. Moreover, the performance of CBIR is even better than that of classification, which can be very promising for future research on this topic.

Figure 5 shows a comparison between a model without and with regularization. It is clear that using regularization improves the performance of CBIR. Moreover, this example shows that our model is robust to geometric transformations, as three of the retrieved images are different instances of the same lesion.

4 Conclusions

This paper introduced a new formulation to train a classification-CBIR DNN for dermoscopy image analysis. Our approach is based on the augmentation of the loss function with regularization terms that improve the structure of the feature space. The extensive experimental validation demonstrated the potential of the proposed approach. Future work will be devoted to perform a user-study, in order to better assess the impact of our approach on the clinical practice.

References

1. Allegretti, S., Bolelli, F., Pollastri, F., Longhitano, S., Pellacani, G., Grana, C.: Supporting skin lesion diagnosis with content-based image retrieval. In: 2020 25th International Conference on Pattern Recognition (ICPR) (2020)
2. Barata, C., Celebi, M.E., Marques, J.S.: Explainable skin lesion diagnosis using taxonomies. Pattern Recogn. **110**, 107413 (2021)
3. Chen, T., Kornblith, S., Norouzi, M., Hinton, G.: A simple framework for contrastive learning of visual representations. In: International Conference on Machine Learning, pp. 1597–1607 (2020)
4. Esteva, A., et al.: Dermatologist-level classification of skin cancer with deep neural networks. Nature **542**(7639), 115–118 (2017)
5. Gessert, N., et al.: Skin lesion classification using CNNs with patch-based attention and diagnosis-guided loss weighting. IEEE Trans. Biomed. Eng. **67**(2), 495–503 (2020)
6. He, K., Fan, H., Wu, Y., Xie, S., Girshick, R.: Momentum contrast for unsupervised visual representation learning. In: Proceedings of the IEEE/CVF Conference on Computer Vision and Pattern Recognition, pp. 9729–9738 (2020)
7. He, K., Zhang, X., Ren, S., Sun, J.: Identity mappings in deep residual networks. In: Leibe, B., Matas, J., Sebe, N., Welling, M. (eds.) ECCV 2016. LNCS, vol. 9908, pp. 630–645. Springer, Cham (2016). https://doi.org/10.1007/978-3-319-46493-0_38
8. Hinton, G., Vinyals, O., Dean, J.: Distilling the knowledge in a neural network. arXiv preprint arXiv:1503.02531 (2015)
9. Huang, G., Liu, Z., Van Der Maaten, L., Weinberger, K.Q.: Densely connected convolutional networks. In: Proceedings of the IEEE Conference on Computer Vision and Pattern Recognition, pp. 4700–4708 (2017)
10. Van der Maaten, L., Hinton, G.: Visualizing data using t-SNE. J. Mach. Learn. Res. **9**(11) (2008)
11. Mahbod, A., Schaefer, G., Wang, C., Dorffner, G., Ecker, R., Ellinger, I.: Transfer learning using a multi-scale and multi-network ensemble for skin lesion classification. Comput. Methods Program. Biomed. **193**, 105475 (2020)

12. Mirikharaji, Z., Hamarneh, G.: Star shape prior in fully convolutional networks for skin lesion segmentation. In: Frangi, A.F., Schnabel, J.A., Davatzikos, C., Alberola-López, C., Fichtinger, G. (eds.) MICCAI 2018. LNCS, vol. 11073, pp. 737–745. Springer, Cham (2018). https://doi.org/10.1007/978-3-030-00937-3_84

13. Sadeghi, M., Chilana, P., Yap, J., Tschandl, P., Atkins, M.S.: Using content-based image retrieval of dermoscopic images for interpretation and education: a pilot study. Skin Res. Technol. **26**(4), 503–512 (2020)

14. Sadeghi, M., Chilana, P.K., Atkins, M.S.: How users perceive content-based image retrieval for identifying skin images. In: Stoyanov, D., et al. (eds.) MLCN/DLF/IMIMIC -2018. LNCS, vol. 11038, pp. 141–148. Springer, Cham (2018). https://doi.org/10.1007/978-3-030-02628-8_16

15. Schroff, F., Kalenichenko, D., Philbin, J.: FaceNet: a unified embedding for face recognition and clustering. In: Proceedings of the IEEE Conference on Computer Vision and Pattern Recognition, pp. 815–823 (2015)

16. Simonyan, K., Zisserman, A.: Very deep convolutional networks for large-scale image recognition. arXiv preprint arXiv:1409.1556 (2014)

17. Taha, A., Chen, Y.T., Misu, T., Shrivastava, A., Davis, L.: Boosting standard classification architectures through a ranking regularizer. In: Proceedings of the IEEE/CVF Winter Conference on Applications of Computer Vision, pp. 758–766 (2020)

18. Tschandl, P., et al.: Human-computer collaboration for skin cancer recognition. Nat. Med. **26**(8), 1229–1234 (2020)

19. Tschandl, P., Rosendahl, C., Kittler, H.: The ham10000 dataset, a large collection of multi-source dermatoscopic images of common pigmented skin lesions. Sci. Data **5**(1), 1–9 (2018)

20. Wu, H., Pan, J., Li, Z., Wen, Z., Qin, J.: Automated skin lesion segmentation via an adaptive dual attention module. IEEE Trans. Med. Imaging **40**(1), 357–370 (2020)

PAC Bayesian Performance Guarantees for Deep (Stochastic) Networks in Medical Imaging

Anthony Sicilia[1], Xingchen Zhao[2], Anastasia Sosnovskikh[2], and Seong Jae Hwang[1,2(✉)]

[1] Intelligent Systems Program, University of Pittsburgh, Pittsburgh, USA
{anthonysicilia,sjh95}@pitt.edu
[2] Department of Computer Science, University of Pittsburgh, Pittsburgh, USA
{xiz168,anastasia}@pitt.edu

Abstract. Application of deep neural networks to medical imaging tasks has in some sense become commonplace. Still, a "thorn in the side" of the deep learning movement is the argument that deep networks are prone to overfitting and are thus unable to generalize well when datasets are small (as is common in medical imaging tasks). One way to bolster confidence is to provide mathematical guarantees, or bounds, on network performance after training which explicitly quantify the possibility of overfitting. In this work, we explore recent advances using the PAC-Bayesian framework to provide bounds on generalization error for large (stochastic) networks. While previous efforts focus on classification in larger natural image datasets (e.g., MNIST and CIFAR-10), we apply these techniques to both classification and segmentation in a smaller medical imagining dataset: the ISIC 2018 challenge set. We observe the resultant bounds are competitive compared to a simpler baseline, while also being more explainable and alleviating the need for holdout sets.

1 Introduction

Understanding the generalization of learning algorithms is a classical problem. Practically speaking, verifying whether a fixed method of inference will generalize may not seem to be a challenging task. Holdout sets are the tool of choice for most practitioners – when sample sizes are large, we can be confident the measured performance is representative. In medical imaging, however, sample sizes are often small and stakes are often high. Thus, mathematical guarantees[1] on the performance of our algorithms are of paramount importance. Yet, it is

---

[1] Guarantees in this paper are probabilistic. Similar to confidence intervals, one should interpret with care: the guarantees hold with high probability prior to observing data.

Electronic supplementary material The online version of this chapter (https://doi.org/10.1007/978-3-030-87199-4_53) contains supplementary material, which is available to authorized users.

M. de Bruijne et al. (Eds.): MICCAI 2021, LNCS 12903, pp. 560–570, 2021.
https://doi.org/10.1007/978-3-030-87199-4_53

not abundantly common to provide such guarantees in medical imaging research on deep neural networks; we are interested in supplementing this shortage.

A simple (but effective) guarantee on performance is achieved by applying a Hoeffding Bound to the error of an inference algorithm reported on a holdout set. In classification tasks, Langford provides a useful tutorial on these types of high probability bounds among others [32]. The bounds are easily extended to any bounded performance metrics in general, and we use this methodology as a baseline in our own experimentation (Sect. 3). While effective, Hoeffding's Inequality falls short in two regards: (1) use of a holdout set requires that the model does not see all available data and (2) the practitioner gains no insight into *why* the model generalized well. Clearly, both short-comings can be undesirable in a medical imaging context: (1) access to the entire dataset can be especially useful for datasets with rare presence of a disease and (2) understanding *why* can both improve algorithm design and give confidence when deploying models in a clinical setting. Thus, we desire practically applicable bounds – i.e., competitive with Hoeffding's Inequality – which avoid the aforementioned caveats.

Unfortunately, for deep neural networks, *practically applicable* guarantees of this nature can be challenging to produce. Traditional PAC bounds based on the Vapnik-Chervonenkis (VC) dimension [8,46,48,49] accomplish our goals to some extent, but require (much) more samples than parameters in our network to produce a good guarantee.[2] When our networks are large – e.g., a ResNet-18 [23] with more than 10M parameters – our datasets are thus required to be unreasonably sized to ensure generalization, especially, in medical imaging contexts. In effect, these bounds are *vacuous*; they are logically meaningless for the sample sizes we observe in practice. Specifically, vacuous describes any bound on error which is larger than 1, and therefore, gives us no new insight on a model's generalization ability because error rates lie in the range $[0, 1]$. The term was coined by Dziugaite & Roy [16] who observed that most PAC bounds on error for deep networks were in fact vacuous (when computed). The authors demonstrate a technique to compute non-vacuous bounds for some deep stochastic networks trained on standard datasets (e.g., MNIST) using the *PAC-Bayesian* framework.[3] As discussed later, the motivation for using PAC-Bayesian bounds is the hypothesis that stochastic gradient descent (SGD) trained networks generalize well when their solutions lie in large, flat minima.[4] Thus, these PAC-Bayes bounds give us exactly what we desire. They provide practically applicable guarantees on the performance of deep networks, while allowing the model access to all data and also giving insight on what network properties lead to good generalization in practice. To borrow an analogy from Arora [2], this approach is

---

[2] We (very roughly) estimate this by Thm. 6.11 of Shalev-Shwartz & Ben-David [46]. Bartlett et al. [7] provide tight bounds on VC dimension of ReLU networks. Based on these, the sample size must be magnitudes larger than the parameter count for a small generalization gap. See Appendix for additional details and a plot.

[3] PAC-Bayes is attributed to McAllester [37]; also, Shawe-Taylor & Williamson [47].

[4] Early formulations of this hypothesis are due to Hochreiter & Schmidhuber [25].

"prescriptive". It is similar to a doctor's orders to resolve high blood pressure: cut the salt, or in our case, look for flat minima.

In this context, the end goal of this paper is to validate whether the "flat minima prescription" – with observed evidence in traditional computer vision – may find similar success in explaining generalization for the small-data, non-traditional tasks that are common to medical imaging. Our contributions, in this sense, are primarily experimental. We demonstrate non-vacuous PAC-Bayesian performance guarantees for deep stochastic networks applied to the classification and segmentation tasks within the ISIC 2018 Challenge [13]. Importantly, our results show that PAC-Bayesian bounds are competitive against Hoeffding's Inequality, offering a practical alternative which avoids the aforementioned caveats. We employ much the same strategies used by Dziugaite et al. [15,16] as well those used by Pérez-Ortiz et al. [43]. With that said, our different setting yields novel experimental results and poses some novel challenges. Specifically, in segmentation, we compute non-vacuous bounds for a fully-sized U-Net using a medical imaging dataset with about 2.3K training samples. To our knowledge, for deep stochastic networks, we are the first to compute non-vacuous bounds in segmentation on such small datasets. Along the way, we offer some practical insights for the medical imaging practitioner including a (mathematically sound) trick to handle batch normalization layers in PAC-Bayes bounds and an experimental technique to "probe" parameter space to learn about the generalization properties of a particular model and dataset. We hope these results promote continued research on the important topic of guarantees in medical imaging.

2   PAC-Bayesian Theory and Generalization

2.1   Formal Setup

In the PAC-Bayes setting, we consider a hypothesis space $\mathcal{H}$ and a distribution Q over this space. Specific to our context, $\mathcal{H} = \mathbb{R}^d$ will represent the space of deep networks with some fixed architecture, and Q will be a distribution over $\mathcal{H}$. Typically, we will set $Q = \mathcal{N}(\mu, \Sigma)$, a multivariate normal. For some fixed space $\mathcal{Z} = \mathcal{X} \times \mathcal{Y}$, the hypothesis $h \in \mathcal{H}$ defines a mapping $x \mapsto h(x)$ with $x \in \mathcal{X}$, and $h(x) \in \mathcal{Y}$. Given a $[0,1]$-bounded loss function $\ell : \mathcal{H} \times \mathcal{Z} \to [0,1]$ and a distribution D over $\mathcal{Z}$, the risk of h is defined $\mathcal{R}_\ell(h, D) = \mathbf{E}_{(x,y) \sim D}\, \ell(h, (x,y))$. Given instead a sample $S \sim D^m$ over $\mathcal{Z}$, the empirical risk is denoted $\hat{\mathcal{R}}_\ell(h, S)$ and is computed as usual by averaging. In all of our discussions, the data distribution D or sample S is usually easily inferred from context. Therefore, we typically write $\mathcal{R}_\ell(h) = \mathcal{R}_\ell(h, D)$ and $\hat{\mathcal{R}}_\ell(h) = \hat{\mathcal{R}}_\ell(h, S)$. With these definitions, we are interested in quantifying the risk of a stochastic model.[5] In the context of neural networks, one can imagine sampling the distribution Q over $\mathbb{R}^d$ and setting the weights before performing inference on some data-point x. Often, we will refer

---

[5] Sometimes, in classification, this may be called the Gibbs classifier. Not to be confused with the "deterministic", majority vote classifier. An insightful discussion on the relationship between risk in these distinct cases is provided by Germain et al. [19].

to Q itself as the stochastic predictor. The associated risk for such a model is defined as $\mathcal{R}_\ell(Q) = \mathbf{E}_{h \sim Q}\, \mathcal{R}_\ell(h)$ with $\hat{\mathcal{R}}_\ell(Q)$ similarly defined. Typically, we cannot exactly compute expectations over Q. For this reason, we also define $\hat{R}_\ell(\hat{Q})$ for a sample $\hat{Q} \sim Q^n$ as an empirical variant, computed by averaging. The last components for any PAC-Bayesian bound come from two notions of the Kullback-Leibler (KL) divergence between two distributions Q and P written $KL(Q||P)$ and defined as usual. For numbers $q, p \in [0,1]$, we write $kl(q||p)$ as shorthand for the KL divergence between Bernoulli distributions parameterized by q, p. This is typically used to quantify the difference between the risk on a sample and the true risk. In the next section, we put the discussed pieces in play.

2.2   The PAC-Bayesian Theorem

The PAC-Bayesian theory begins primarily with the work of McAllester [37] with similar conceptualizations given by Shawe-Taylor & Williamson [47]. Besides what is discussed in this section, for completeness, readers are also directed to the work of Catoni [11], McAllester [36], Germain et al. [19,20], and the primer by Guedj [21]. We start by stating the main PAC-Bayesian Theorem as given by Maurer [35]. See also Langford & Seeger [33] for the case where ℓ is a 01 loss.

Theorem 1 *(Maurer). Let ℓ be a $[0,1]$-bounded loss function, D be a distribution over $\mathcal{Z}$, and P be a probability distribution over $\mathcal{H}$. Then, for $\delta \in (0,1)$*

$$\Pr_{S \sim D^m} \left(\forall\, Q\; :\; kl\left(\hat{\mathcal{R}}_\ell(Q) || \mathcal{R}_\ell(Q) \right) \leq \frac{KL(Q||P) + \ln\frac{1}{\delta} + \ln\sqrt{4m}}{m} \right) \geq 1 - \delta. \quad (1)$$

By way of Pinsker's inequality, the above may be loosened for the purpose of interpretation [19]

$$\mathcal{R}_\ell(Q) \leq \hat{\mathcal{R}}_\ell(Q) + \sqrt{\frac{KL(Q||P) + \ln\frac{1}{\delta} + \ln\sqrt{4m}}{2\,m}}. \quad (2)$$

In Sect. 3, we compute a much tighter formulation of the bound given in Eq. (2) which handles the term $kl\left(\hat{\mathcal{R}}_\ell(Q) || \mathcal{R}_\ell(Q) \right)$ directly. We provide the derivation of this bound in the Appendix. Various insights and results used to build the final bound are of course due to Langford & Caruana [31]; Dziugaite et al. [15,16]; and Pérez-Ortiz et al. [43] who have all computed similar (or identical) bounds on stochastic neural networks before us. In Sect. 3, we compute this bound for classification and segmentation tasks. For classification, we take ℓ to be the 01-loss defined $\ell_{01}(h, (x, y)) = \mathbf{1}[h(x) \neq y]$ where $\mathbf{1}$ is the indicator function. Precisely, $\mathcal{R}_{\ell_{01}}(h)$ is equal to 1 minus the accuracy. For segmentation, we pick ℓ to be $\ell_{DSC}(h, (x, y)) = 1 - DSC(h, (x, y))$ where DSC is the $[0,1]$-valued Dice similarity coefficient. These upperbounds trivially yield corresponding lowerbounds for both the accuracy and the Dice similarity coefficient, respectively.

Selecting the Prior. Often P is referred to as the prior and Q as the posterior. Still, it is important to note that no restriction on the form of P and Q is required (e.g., as it is in Bayesian parameter estimation). What is required is that P be fixed before observing the sample S that is used to compute the bound. Albeit, P can depend on D and samples independent of S. In fact, it is not uncommon for the prior P to be **data-dependent**.[6] That is, P may be trained on a sample which is disjoint from that which is used to compute the bound; i.e., disjoint from S. On the other hand, the bound holds for all posteriors Q regardless of how Q is selected. So, the datasets used to train Q and P may actually intersect. All in all, we can train Q with all available data without violating assumptions. We must only ensure the datasets used to train P and compute the bound do not intersect. In effect, we avoid the first caveat of Hoeffding's Inequality.

Interpretation. The bound also offers insight into *why* the model generalizes. Intuitively, we quantify the complexity of the stochastic predictor Q in so much as it deviates from some prior knowledge we have on the solution space (e.g., from the data-dependent prior). This is captured in the term $KL(Q||P)$. Dziugaite & Roy [16] also relate PAC-Bayesian bounds to the flat-minima hypothesis. To understand their observation, we consider the case where Q is a normal distribution $\mathcal{N}(\mu, \sigma^2 I)$ with σ a constant and I the identity matrix. The model Q on which we bound the error is stochastic: each time we do inference, we sample from $\mathcal{N}(\mu, \sigma^2 I)$. Because the distribution has some variance (dictated by σ), we sample network weights in a region *around* the mean. Thus, when performance of the stochastic model Q is good, there must be a non-negligible area around the mean where most networks perform well, i.e., a flat minimum around the mean. We know the variance is non-negligible in the posterior network Q because a small upperbound implies small KL divergence with the prior P which itself has non-negligible variance (we pick this value). So, to reiterate, small KL divergence and small empirical risk imply a flat-minimum of appropriate size around the mean of Q. In this sense the bound is explainable: Q generalizes well because it does not deviate much from prior knowledge and it lies in a flat minimum.

Additional Context. Dziugaite & Roy [16] provide a nice synopsis of the history behind the flat-minima hypothesis including the work of Hochreiter & Schmidhuber [25], Hinton & Van Camp [24], Baldassi et al. [3,4], Chaudhari et al. [12], and Keskar et al. [27]. Since then, large scale empirical studies – e.g., Jiang et al. [26]; Dziugaite, Drouin, et al. [14] – have continued to indicate that measures of sharpness of the minimum may be good indicators of neural network generalization in practice. For completeness, we also point out some other theoretically plausible indicators of deep network generalization. These include small weight norms – e.g., Bartlett [5,6]; Neyshabur et al. [40,41] – and the notion of algorithmic stability proposed by Bousquet & Elisseeff [10] which focuses instead on the SGD algorithm – e.g., Hardt et al. [22]; Mou et al. [39]; Kuzborskij & Lampert [29].

---

[6] For example, see Ambroladze et al. [1], Parrado-Hernández et al. [42], Pérez-Ortiz et al. [43], and Dziugaite et al. [15,17].

3   Experiments

In this section, we first evaluate PAC-Bayesian bounds within a self-bounded learning setting. Specifically, a self-bounded learner must both learn and provide a guarantee using the same dataset.[7] As noted, providing guarantees with our trained networks can bolster confidence in small data regimes. We compare the PAC-Bayesian bounds discussed in Sect. 2 to a simple baseline for producing performance guarantees: application of Hoeffding's Inequality to a holdout set.[8] We show PAC-Bayesian bounds are competitive with Hoeffding's Inequality, while also alleviating some caveats discussed in the previous sections. This result (in medical imaging) compliments those previously shown on natural image datasets by Pérez-Ortiz et al. [43]. Specifically, we compute bounds on the **Lesion Segmentation** and the **Lesion Classification Tasks** in the ISIC 2018 Skin Lesion Challenge Dataset [13]. The data in these sets used for training (and bound computation) totals 2.3K and 9K labeled examples, respectively, which is much smaller than previous computation [15,16,43] using MNIST [34] or CIFAR-10 [28].[9] Our second contribution in this section comes from tricks and tools which we hope prove useful for the medical imaging practitioner. We demonstrate an experiment to probe the loss landscape using PAC-Bayesian bounds and also devise a strategy to handle batch normalization layers when computing PAC-Bayesian bounds. Our code is available at: https://github.com/anthonysicilia/PAC-Bayes-In-Medical-Imaging.

3.1   Setup

Models. For segmentation, we use U-Net (UN) [45] and a light-weight version of U-Net (LW) with 3% of the parameters and no skip-connections. For classification, we use ResNet-18 (RN) [23]. Probabilistic models use the same architecture but define a multivariate normal distribution over network weights $\mathcal{N}(\mu, \Sigma)$ with a diagonal covariance matrix Σ. The distribution is sampled to do inference.

Losses. For segmentation, we train using the common Dice Loss [38] which is a smooth surrogate for $1 - DSC$. For classification, we use the negative log-likelihood. Probabilistic models used modified losses we describe next.

Training Probabilistic Models. Recall, in the PAC-Bayes setting we define both the prior P and the posterior Q as both are needed to compute bounds and Q is needed for inference. The prior P is a probabilistic network defined by $\mathcal{N}(\mu_{\mathrm{p}}, \sigma_{\mathrm{p}}^2 I)$ where I is the identity matrix and σ_{p} is a constant. In this text, we use a data-dependent prior unless otherwise noted (see Sect. 2). To pick the prior, μ_{p} is learned by performing traditional optimization on a dataset disjoint from that which is used to compute the bound (see **Data Splits**). The parameter $\sigma_{\mathrm{p}} = 0.01$ unless otherwise noted. The posterior Q is initialized identically to P before

---

[7] See Freund [18] or Langford & Blum [30].

[8] We provide additional details on this procedure in the Appendix.

[9] These datasets have 60K and 50K labeled examples, respectively.

it is trained using PAC-Bayes with Backprop (PBB) as proposed by Rivasplata et al. [43,44]. This training technique may be viewed as (mechanically) similar to Bayes-by-Backprop (BBB) [9]. In particular, it uses a re-parameterization trick to optimize a probabilistic network through SGD. Where PBB and BBB differ is the motivation, and subsequently, the use of PAC-Bayes upperbounds as the objective to optimize.[10] Note, PAC-Bayes bounds are valid for all $[0,1]$-bounded losses, and thus, are valid for the Dice Loss or normalized negative log-likelihood. The upperbound used for our PBB objective is the Variational bound of Dziugaite et al. [15].

Probabilistic Batch Norm Layers. While generally each weight in the probabilistic networks we consider is sampled independently according to a normal distribution, batch norm layers[11] must be handled with special care. We treat the parameters of batch norm layers as point mass distributions. Specifically, the parameter value has probability 1 and 0 is assigned to all other values. The posterior distribution for these parameters is made identical to the prior by "freezing" the posterior batch norm layers during training and inference. In effect, we avoid sampling the means and variances in our batch norm layers, and importantly, the batch norm layers do not contribute to the KL-divergence computation. In the Appendix, we provide a derivation to show this strategy is (mathematically) correct; it relies primarily on the independence of the weight distributions.

Optimization Parameters. Optimization is done using SGD with momentum set to 0.95. For classification, the batch size is 64, and the initial learning rate is 0.5. For segmentation, the batch size is 8, and the initial learning rate is 0.1 for LW and 0.01 for U-Net. All models are initialized at the same random location and are trained for 120 epochs with the learning rate decayed by a factor of 10 every 30 epochs. In the PAC-Bayesian setting, the data-dependent prior mean μ_{p} is randomly initialized (as other models) and trained for 30 epochs. The posterior Q is initialized at the prior P and trained for the remaining 90 epochs. The learning rate decay schedule is not reset for posterior training.

Bound Details. Note, in all cases, PAC-Bayes bounds are computed using a data-dependent prior. Bounds are computed with 95% confidence ($\delta = 0.05$) with data sample size given in the next section. For PAC-Bayes bounds, the number of models sampled is either 1000 (in Fig. 1a) or 100 (in Fig. 1b, c, d).

Data Splits for Self-bounded Learning. Each method is given access to a base training set (90% of the data) and is expected to both learn a model and provide a performance guarantee for this model when applied to unseen data. To evaluate both the model and performance guarantee in a more traditional fashion, each method is also tested on a final holdout set (10% of the data) which no model sees. Splits are random but identically fixed for all models. For probabilistic networks trained using PBB, we split the base training data into a 50%-prefix set[12] used to train the prior and a disjoint 50%-bound set. Both

---

[10] See Pérez-Ortiz et al. [43] for more detailed discussion.

[11] We refer here to both the running statistics and any learned weights.

[12] See Dziugaite et al. [15] who coin the term "prefix".

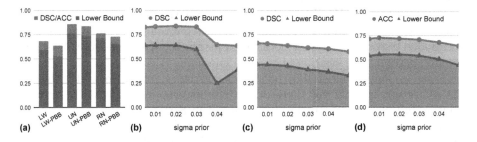

Fig. 1. (a) DSC/ACC (red) and lowerbounds (blue). (b, c, d) Modulation of prior variance for U-Net (b) and LW (c) and ResNet-18 (d). (Color figure online)

the prefix-set and bound-set are used to train the posterior, but recall, the PAC-Bayes bound can only be computed on the bound-set (Sect. 2). For the baseline non-probabilistic networks, we instead train the model using a traditional training set (composed of 90% of the base training set) and then compute a guarantee on the model performance (i.e., a Hoeffding bound) using an independent holdout set (the remaining 10% of the base training set). In this sense, all models are on an equal footing with respect to the task of a self-bounded learner. All models have access *only* to the base training set for both training and computation of valid performance guarantees. In relation to Fig. 1, performance metrics such as DSC are computed using the final holdout set. Lowerbound computation and training is done using the base training set.

3.2 Results

Comparison to Hoeffding's Inequality. As promised, we observe in Fig. 1a performance guarantees by both bounds are comparable and performance on the final holdout set is also comparable. Hoeffding's Inequality does have a slight advantage with regards to these metrics, but as mentioned, PAC-Bayes bounds possess some desirable qualitative properties which make it an appealing alternative. For PAC-Bayes bounds the posterior Q sees all training data, while for the Hoeffding Bound, one must maintain an unseen holdout set to compute a guarantee. Further, we may explain the generalization of the PBB trained model through our interpretation of the PAC-Bayes bound (see Sect. 2). These qualities make PAC-Bayes appealing in medical imaging contexts where explainability is a priority and we often need to maximize the utility of the training data.

Flat Minima and Their Size. As discussed, the application of PAC-Bayesian bounds may be motivated by the flat minima hypothesis (see Sect. 2). We explore this idea in Fig. 1b, c, d by modulating the prior variance σ_p across runs. Informally, these plots can give us insight into our loss landscape. The reasonably tight lowerbounds – which are slightly looser than in Fig. 1a only due to fewer model samples – imply small KL-Divergence and indicate the prior and posterior variances are of a similar magnitude. Likewise, the difference between the prior and posterior means should not be too large, relative to the variance. A fixed random-seed ensures priors are identical, so each data-point within a plot should correspond to roughly the same location in parameter space; i.e., we will assume we are analyzing the location of a single minimum.[13] For U-Net, we see stable performance and a sudden drop as the prior variance grows. Before the drop at $\sigma_p = 0.04$, consistently high DSC and a high lowerbound indicate the network solution lies in a flat minimum (as discussed in Sect. 2). So, we may conclude a flat minimum proportional in size to $\sigma_p = 0.03$. For LW and ResNet-18, we instead see consistent performance degradation as the prior variance grows. For these networks, the minima may not be as flat. Informally, such sensitivity analysis can tell us "how flat" the minima are for a particular network and dataset as well as "how large". Practically, information like this can be useful to practitioners interested in understanding the generalization ability of their models. Namely, it is hypothesized "larger" flat minima lead to better generalization because less precision (fewer bits) is required to specify the weights [25].

4   Conclusion

As a whole, our results show how PAC-Bayes bounds can be practically applied in medical imaging contexts – where theoretical guarantees (for deep networks) would appear useful but not commonly discussed. With this said, we hope for this paper to act primarily as a conversation starter. At 2.3K examples, the segmentation dataset we consider is still larger than commonly available in some application domains (e.g., neuroimaging applications). It remains to be considered how effective these bounds can be in ultra-low resource settings.

Acknowledgment. This work is supported by the University of Pittsburgh Alzheimer Disease Research Center Grant (P30 AG066468).

References

1. Ambroladze, A., Parrado-Hernández, E., Shawe-Taylor, J.: Tighter PAC-Bayes Bounds (2007)
2. Arora, S.: Generalization Theory and Deep Nets, An introduction (2017). https://www.offconvex.org/2017/12/08/generalization1/
3. Baldassi, C., et al.: Unreasonable effectiveness of learning neural networks: from accessible states and robust ensembles to basic algorithmic schemes. PNAS **113**, E7655–E7662 (2016)
4. Baldassi, C., Ingrosso, A., Lucibello, C., Saglietti, L., Zecchina, R.: Subdominant dense clusters allow for simple learning and high computational performance in neural networks with discrete synapses. Phys. Rev. Lett. **115**, 128101 (2015)

---

[13] Notice, another approach might be to the fix the posterior mean at the result of, say, the run with $\sigma_p = 0.01$ and then modulate the variance from this fixed location. We are not guaranteed this run will be near the center of a minimum, and so, may underestimate the minimum's size by this procedure. Our approach, instead, allows the center of the posterior to change (slightly) when the variance grows.

5. Bartlett, P.L.: For valid generalization, the size of the weights is more important than the size of the network (1997)
6. Bartlett, P.L.: The sample complexity of pattern classification with neural networks: the size of the weights is more important than the size of the network. IEEE Trans. Inform. Theory **44**, 525–536 (1998)
7. Bartlett, P.L., Harvey, N., Liaw, C., Mehrabian, A.: Nearly-tight VC-dimension and pseudodimension bounds for piecewise linear neural networks. JMLR **20**, 2285–2301 (2019)
8. Blumer, A., Ehrenfeucht, A., Haussler, D., Warmuth, M.K.: Learnability and the Vapnik-Chervonenkis dimension. J. ACM **36**, 929–965 (1989)
9. Blundell, C., Cornebise, J., Kavukcuoglu, K., Wierstra, D.: Weight uncertainty in neural network. In: ICML (2015)
10. Bousquet, O., Elisseeff, A.: Stability and generalization. JMLR **2**, 499–526 (2002)
11. Catoni, O.: PAC-Bayesian supervised classification: the thermodynamics of statistical learning. arXiv:0712.0248v1 (2007)
12. Chaudhari, P., et al.: Entropy-SGD: biasing gradient descent into wide valleys. arXiv:1611.01838v5 (2016)
13. Codella, N., et al.: Skin lesion analysis toward melanoma detection 2018: a challenge hosted by the international skin imaging collaboration (ISIC). arXiv:1902.03368v2 (2019)
14. Dziugaite, G.K., et al.: In search of robust measures of generalization. arXiv:2010.11924v2 (2020)
15. Dziugaite, G.K., Hsu, K., Gharbieh, W., Roy, D.M.: On the role of data in PAC-Bayes bounds. arXiv:2006.10929v2 (2020)
16. Dziugaite, G.K., Roy, D.M.: Computing nonvacuous generalization bounds for deep (stochastic) neural networks with many more parameters than training data. arXiv:1703.11008v2 (2017)
17. Dziugaite, G.K., Roy, D.M.: Data-dependent PAC-Bayes priors via differential privacy. In: NeurIPS (2018)
18. Freund, Y.: Self bounding learning algorithms. In: COLT (1998)
19. Germain, P., Lacasse, A., Laviolette, F., March, M., Roy, J.F.: Risk bounds for the majority vote: from a PAC-Bayesian analysis to a learning algorithm. JMLR **16**, 787–860 (2015)
20. Germain, P., Lacasse, A., Laviolette, F., Marchand, M.: PAC-Bayesian learning of linear classifiers. In: ICML (2009)
21. Guedj, B.: A primer on PAC-Bayesian learning. arXiv:1901.05353v3 (2019)
22. Hardt, M., Recht, B., Singer, Y.: Train faster, generalize better: stability of stochastic gradient descent. In: ICML (2016)
23. He, K., Zhang, X., Ren, S., Sun, J.: Deep residual learning for image recognition. In: CVPR (2016)
24. Hinton, G.E., Van Camp, D.: Keeping the neural networks simple by minimizing the description length of the weights. In: COLT (1993)
25. Hochreiter, S., Schmidhuber, J.: Flat minima. Neural Comput. **9**, 1–42 (1997)
26. Jiang, Y., Neyshabur, B., Mobahi, H., Krishnan, D., Bengio, S.: Fantastic generalization measures and where to find them. arXiv:1912.02178v1 (2019)
27. Keskar, N.S., Mudigere, D., Nocedal, J., Smelyanskiy, M., Tang, P.T.P.: On large-batch training for deep learning: generalization gap and sharp minima. arXiv:1609.04836v2 (2016)
28. Krizhevsky, A., Hinton, G., et al.: Learning multiple layers of features from tiny images (2009)

29. Kuzborskij, I., Lampert, C.: Data-dependent stability of stochastic gradient descent. In: ICML (2018)
30. Langford, J., Blum, A.: Microchoice bounds and self bounding learning algorithms. Mach. Learn. **51**, 165–179 (2003)
31. Langford, J., Caruana, R.: (Not) bounding the true error. In: NeurIPS (2002)
32. Langford, J., Schapire, R.: Tutorial on practical prediction theory for classification. JMLR **6**, 273–306 (2005)
33. Langford, J., Seeger, M.: Bounds for averaging classifiers (2001)
34. LeCun, Y., Cortes, C.: MNIST handwritten digit database (2010). http://yann.lecun.com/exdb/mnist/
35. Maurer, A.: A note on the PAC Bayesian theorem. arXiv:cs/0411099v1 (2004)
36. McAllester, D.: A PAC-Bayesian tutorial with a dropout bound. arXiv:1307.2118v1 (2013)
37. McAllester, D.A.: Some PAC-Bayesian theorems. Mach. Learn. **37**, 355–363 (1999)
38. Milletari, F., Navab, N., Ahmadi, S.A.: V-net: fully convolutional neural networks for volumetric medical image segmentation. In: 3DV (2016)
39. Mou, W., Wang, L., Zhai, X., Zheng, K.: Generalization bounds of SGLD for non-convex learning: two theoretical viewpoints. In: COLT (2018)
40. Neyshabur, B., Bhojanapalli, S., McAllester, D., Srebro, N.: Exploring generalization in deep learning. arXiv:1706.08947v2 (2017)
41. Neyshabur, B., Tomioka, R., Srebro, N.: In search of the real inductive bias: on the role of implicit regularization in deep learning. arXiv:1412.6614v4 (2014)
42. Parrado-Hernández, E., Ambroladze, A., Shawe-Taylor, J., Sun, S.: PAC-Bayes bounds with data dependent priors. JMLR **13**, 3507–3531 (2012)
43. Pérez-Ortiz, M., Rivasplata, O., Shawe-Taylor, J., Szepesvári, C.: Tighter risk certificates for neural networks. arXiv:2007.12911v2 (2020)
44. Rivasplata, O., Tankasali, V.M., Szepesvari, C.: PAC-Bayes with backprop. arXiv:1908.07380v5 (2019)
45. Ronneberger, O., Fischer, P., Brox, T.: U-net: convolutional networks for biomedical image segmentation. In: Navab, N., Hornegger, J., Wells, W.M., Frangi, A.F. (eds.) MICCAI 2015. LNCS, vol. 9351, pp. 234–241. Springer, Cham (2015). https://doi.org/10.1007/978-3-319-24574-4_28
46. Shalev-Shwartz, S., Ben-David, S.: Understanding Machine Learning: From Theory to Algorithms. Cambridge University Press, Cambridge (2014)
47. Shawe-Taylor, J., Williamson, R.C.: A PAC analysis of a Bayesian estimator. In: COLT (1997)
48. Valiant, L.G.: A theory of the learnable. Commun. ACM **27**, 1134–1142 (1984)
49. Vapnik, V.N., Chervonenkis, A.Y.: On uniform convergence of the frequencies of events to their probabilities. Teoriya Veroyatnostei Primeneniya **16** (1971)

Machine Learning – Uncertainty

Medical Matting: A New Perspective on Medical Segmentation with Uncertainty

Lin Wang[1,2,3], Lie Ju[2,3], Donghao Zhang[2], Xin Wang[3], Wanji He[3],
Yelin Huang[3], Zhiwen Yang[3], Xuan Yao[3], Xin Zhao[3], Xiufen Ye[1],
and Zongyuan Ge[2,3]

[1] Harbin Engineering University, Harbin 150001, Heilongjiang, China
`yexiufen@hrbeu.edu.cn`
[2] Monash Medical AI Group, Monash University, Clayton, VIC 3800, Australia
`zongyuan.ge@monash.edu`
[3] Airdoc Co., Ltd., Beijing 100089, China

Abstract. In medical image segmentation, it is difficult to mark ambiguous areas accurately with binary masks, especially when dealing with small lesions. Therefore, it is a challenge for radiologists to reach a consensus by using binary masks under the condition of multiple annotations. However, these uncertain areas may contain anatomical structures that are conducive to diagnosis. Uncertainty is introduced to study these situations. Nevertheless, the uncertainty is usually measured by the variances between predictions in a multiple trial way. It is not intuitive, and there is no exact correspondence in the image. Inspired by image matting, we introduce matting as a soft segmentation method and a new perspective to deal with and represent uncertain regions into medical scenes, namely medical matting. More specifically, because there is no available medical matting dataset, we first labeled two medical datasets with alpha matte. Secondly, the matting methods applied to the natural image are not suitable for the medical scene, so we propose a new architecture to generate binary masks and alpha matte in a row. Thirdly, the uncertainty map is introduced to highlight the ambiguous regions from the binary results and improve the matting performance. Evaluated on these datasets, the proposed model outperformed state-of-the-art matting algorithms by a large margin, and alpha matte is proved to be a more efficient labeling form than a binary mask.

Keywords: Uncertainty quantification · Soft segmentation · Image matting

1 Introduction

Due to the limitation of imaging methods, ambiguities are common in medical images. The pathological or anatomical structures, especially those around

M. de Bruijne et al. (Eds.): MICCAI 2021, LNCS 12903, pp. 573–583, 2021.
https://doi.org/10.1007/978-3-030-87199-4_54

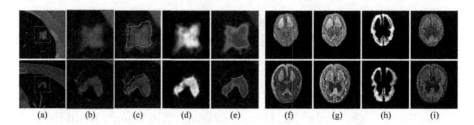

Fig. 1. Examples of the ambiguity in the medical image. (a) (b) and (f) are the original or enlarged images of lung nodule [2] and newborn brain myelination process [26]. (c) and (g) are the contours of the human binary labels. Inaccuracy and inconsistency are more likely to be located in the ambiguity regions. (d) and (h) shows the targets labeled by the alpha matte. It is more explainable than the binary labels in ambiguity regions. (e) and (i) reveals the original images mixed with the alpha matte in pseudo-color for better display. Better view in zoom and color.

Fig. 2. The most descriptive label (the alpha matte or one of the binary masks) for each medical image is evaluated by clinicians. (a) and (b) shows 100 randomly selected cases of the LIDC-IDRI and cases of the Brain-growth, respectively. Rows denote votes of clinicians, and columns denote different image cases in the two datasets. The alpha mattes (red blocks) are more preferred in expressing the anatomical structures than the binary masks (blocks in other colors). (Color figure online)

their boundaries, can be blurred and difficult to be segmented. Hence the study of predictive uncertainty of segmentation results is as essential as improving segmentation accuracy because the trustworthiness of model output is vital for clinicians.

Generally, the uncertainty can be categorized into two types, aleatoric uncertainty and epistemic uncertainty [16,27,31]. Aleatoric uncertainty is often related to intrinsic noise in the data that hard to reduce. In contrast, epistemic uncertainty is associated with suboptimal parameters of a model due to insufficient data [9,18], which can be alleviated by providing more data. However, the difficulty in data collection and the high cost of annotation make lack of data common in medical image study, which urge us to use the insufficient data to the fullest. After observing the annotations from multiple radiologists, we found a large proportion of the discrepancies in segmentation tasks, which was thought hard to reduce, is due to the inadequate expressiveness of binary masks and the predicament of existing labeling techniques for tiny structures (see Fig. 1) and could be mitigated by a more effective labeling method.

Many research works focus on quantifying the uncertainty and alleviate its impact on segmentation for disease diagnosis and prognosis [3,12,15,17,18,20, 28]. However, most of them measure the uncertainty by learning the predictions' discrepancies in various forms, such as cross-entropy or variance. It is unintuitive and hard to be evaluated visually. Besides, the binary masks may lose information for diagnosing.

Matting [1,6,8,11,21–23,25] is a specific segmentation approach widely used in picture editing, green screen, virtual conference, etc. It can obtain fine-grained details by introducing a blending coefficient named alpha matte α, between the foreground $\mathcal{F}$ and background $\mathcal{B}$ of an image $\mathcal{I}$, which makes $\mathcal{I} = \alpha\mathcal{F} + (1-\alpha)\mathcal{B}$. Analogically, the uncertainty can be considered the degree of mixing the pathological tissue and its surrounding environment. However, matting methods are not widely used in medical images and are limited as an auxiliary of segmentation [7,10,34,35]. Since uncertainty is highly coupled with the challenging regions of segmentation results and there is no clear definition of uncertainty, we introduce alpha matte into medical scenarios as calibration of uncertainty and a more accurate segmentation method.

Our contribution is summarized as follows: 1) Alpha matte is first introduced in medical scenarios, which is more expressive than binary masks by experiments. Valuable information in such regions can be reserved. 2) The uncertainty is defined more intuitively by the alpha matte. 3) A compact multi-task network was proposed, producing the alpha matte and binary masks simultaneously. 4) Uncertainty map is proposed as an analog to the trimap in image matting based on the predictions of binary masks and improves the matting network's performance. 5) The datasets labeled for uncertainty learning and medical matting are publicly available as benchmark datasets for the research community.

2   Datasets

Two datasets with alpha mattes reviewed by clinicians, a subset of LIDC-IDRI [2] and the Brain-growth of QUBIQ [26], are used in this work. LIDC-IDRI dataset consists of thoracic CT scans for lung nodules diagnosis. The scans are cropped and centered into size 128×128 patches as practice [15,20]. Each patch is labeled out the region of the pulmonary nodules by four binary masks. To better focus on the uncertainty study, we select 1609 patches in which an identical nodule is labeled in the corresponding masks. The Brain-growth dataset consists of 39 low-intensity contrast T2-W MR images for the newborn brain's white matter tissue myelination process. Each image was labeled by seven binary masks.

Alpha mattes use continuous values such that they are more capable of depicting the anatomical structures in ambiguous areas. Specifically, the uncertain region in the LIDC dataset can describe the indistinct border and ground-glass shadow around a lesion better, which is vital for nodules staging. In the Brain-growth dataset, the newborn white tissue undergoes a rapid myelination process. Thus it is hard to tag the white matter in the shifting regions myelinated or non-myelinated with a binary label.

The alpha mattes are labeled in a semi-automatic way. i.e., rough mattes are generated by matting methods and refined manually by image editors to fit the anatomical structure, which is efficient and widely used in natural image matting [29]. Here Information-Flow [1], a laplacian-based matting method,

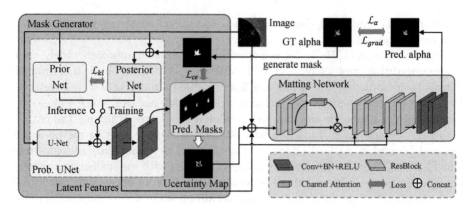

Fig. 3. The schematic diagram of Medical Matting. The Mask Generator outputs various segmentation predictions under the target distribution. Such intermediate results are merged to create an Uncertainty Map, which assists the following Matting Network like the trimap.

was selected to create the rough alpha mattes as its comparatively better performance in our scenario, while other methods are also possible. A trimap is required in these matting methods, which denotes the foreground, background, and unknown region as prior information. We generate it by the manual labeled masks. In particular, pixels are categorized as foreground or background only if they are tagged the same label in all the binary masks. The left inconsistent pixels are tagged as the unknown region.

Three qualified clinicians, including pediatricians and oncologists, were invited to review the various ground truth, including the proposed alpha matte and the conventional binary masks. Each of them picked the labeling to describe best the anatomical structures, shown in Fig. 2. It is demonstrated that the alpha mattes are significantly more favorable than the binary masks. The datasets are available at https://github.com/wangsssky/MedicalMatting.

3   Methodology

Labeling with continuous values can accurately depict the anatomical structure and provide an intuitive way to quantify uncertainty, as shown earlier. Still, the binary mask is the mainstream segmentation method in the medical scene. Therefore, we design a multi-task network predicting the alpha matte and binary masks simultaneously to expand the application scope. Overall, the proposed medical matting framework consists of Mask Generator and Matting Network. The mask generator predicts multiple binary masks of each lesion, which can be considered a simulation of clinicians' labeling process. Then, we use the predicted score maps to build a map indicating the unsure regions, namely the uncertainty map. In the matting network, the uncertainty map, input image, and latent feature from the mask generator are merged to predict the alpha matte. Figure 3 provides a schematic view of the framework.

3.1   Mask Generator

Mask generator can produce a bunch of binary masks, and the intermediate score maps are used to build an uncertainty map for the following matting network as assistance.

Probabilistic UNet[1] is selected as the network to generate a set of binary masks under the target distribution. The body structures are continuous, which will also reflect in the corresponding alpha mattes. Therefore unlike the original method, which randomly samples a mask from multiple labeled masks in each training iteration, we generate a binary mask by random thresholding the ground truth alpha matte. Thus we can generate more abundant masks with structural continuity. Moreover, masks generated by different thresholds correspond to different tolerance of uncertainty. The generated mask can be formulated as Eq. 1.

$$\mathcal{M}ask = Threshold(\alpha_{gt}, \tau), \tau \in [a, b] \tag{1}$$

where α_{gt} denotes the ground truth alpha matte, τ stands for the threshold level, a and b are practically set to 0.2 and 0.7 of the maximum of α_{gt} to get reasonable masks.

Uncertainty Map. Matting methods generally introduce a priori trimap as a restriction of the foreground, background, and unknown regions, which vastly reduces the task complexity. Unlike the trimap, in medical images, it can be even difficult to tell the definite foreground, the lesions, from its surrounding structures. Inspired by the Monte Carlo dropout [18] approximation, we create a score map named uncertainty map, which indicates challenging areas to identify, playing a similar role to trimap. The uncertainty map is defined as the entropy:

$$\mathcal{U}(x) = -\sum_{c=1}^{m} \mathcal{P}_c(x) log \mathcal{P}_c(x) \tag{2}$$

where $\mathcal{P}_c(x)$ is the probability of the pixel x in class c of the average score map of the probabilistic UNet predictions, and m is the number of class. Figure 4(d) shows examples of the generated uncertain regions.

3.2   Matting Network

The matting network outputs the alpha matte with the help of the aforementioned uncertainty map. It consists of three propagation units, and each consists of two Residual Blocks [14]. Between the first two units, a channel attention module [32] was inserted to help the network focus on effective features. The output block contains two convolution layers at the end of the pipeline. The input image, latent features from the probabilistic UNet, and the uncertainty map are concatenated as the matting network's input. The uncertainty map is also injected at the last two propagation units as a guidance of the information flow [5].

---

[1] We reference to the Pytorch implementation from https://github.com/stefanknegt/Probabilistic-Unet-Pytorch.

3.3 Multi-task Loss

Multi-task learning is employed in our network for binary masks and alpha matte prediction, as it simplifies the training procedure and achieves better performance by sharing the information of the interrelated tasks. Each task is guided by its corresponding losses and balanced by an uncertainty-weighted way.

For the segmentation, following the probabilistic UNet [20] practice, Kullback-Leibler loss $\mathcal{L}_{kl}$ and cross-entropy loss $\mathcal{L}_{ce}$ are applied. The former is for minimizing the divergence of the prior distribution and the posterior distribution, and the latter is for matching the generated mask and the ground truth mask.

For the matting, the absolute difference and the gradient difference between the predicted alpha matte and the ground truth alpha matte are both considered by $\mathcal{L}_\alpha$ and $\mathcal{L}_{grad}$, respectively. The gradient expresses the correlation of one pixel to its surrounding pixels, significant to the medical structure's continuity. Moreover, a mask based on the uncertainty map is applied to make the gradient loss concentrate on the uncertain regions. The losses are defined as:

$$\mathcal{L}_\alpha(\tilde{\alpha}, \alpha_{gt}) = \frac{1}{|\tilde{\alpha}|} \sum_{i \in \tilde{\alpha}} \|\tilde{\alpha}(i) - \alpha_{gt}(i)\|_1 \tag{3}$$

$$\mathcal{L}_{grad}(\{\tilde{\alpha}, \mathcal{U}\}, \alpha_{gt}) = \frac{1}{|\mathcal{R}|} \sum_{i \in \mathcal{R}} \|\nabla_{\tilde{\alpha}}(i) - \nabla_{\alpha_{gt}}(i)\|_1, \mathcal{R} = \mathcal{U} > thresh \tag{4}$$

where $\tilde{\alpha}$, α_{gt} denotes the predicted and ground truth alpha matte, respectively. $\mathcal{U}$ stands for the uncertainty map. A sub-region $\mathcal{R}$ of $\mathcal{U}$ selected by thresholding is used as a mask for gradient loss, which makes the loss more focused on the uncertain regions.

The outputs of each task can be assumed as each following a Gaussian distribution with observation noise. Therefore, we introduced the uncertainty weighted loss [19] to balance them. Finally, our loss $\mathcal{L}$ is defined as:

$$\mathcal{L} = \frac{\mu\mathcal{L}_{kl} + \upsilon\mathcal{L}_{ce}}{\sigma_1^2} + \frac{\zeta\mathcal{L}_\alpha + \xi\mathcal{L}_{grad}}{2\sigma_2^2} + \log \sigma_1\sigma_2 \tag{5}$$

where σ_1 and σ_2 are trainable parameters, μ, υ and ζ, ξ are parameters balancing the $\mathcal{L}_{kl}$, $\mathcal{L}_{ce}$ and $\mathcal{L}_\alpha$, $\mathcal{L}_{grad}$ for binary mask and alpha matte prediction, respectively.

4   Experiments

Implementation Details. Our method is evaluated on the datasets above with 4-fold cross-validation. The base learning rate l_r is set to 0.0001. The cosine annealing schedule [4,24] was used after a 1-epoch long steady increasing warm-up from 0 to l_r. The ADAM optimizer was used with momentum 0.9 and weight decay 5×10^{-5}. The σ_1 and σ_2 of multi-task losses are initialized to 4, and the loss weight μ is set to 10, υ, ζ, ξ are set to 1. The threshold for the mask in $\mathcal{L}_{grad}$

Table 1. Quantitative comparisons with state-of-the-art matting algorithms.

Model	LIDC-IDRI				Brain-growth			
	SAD↓	MSE↓	Grad.↓	Conn.↓	SAD↓	MSE↓	Grad.↓	Conn.↓
Bayesian [8]	0.0778	0.0819	0.1535	0.0724	0.8435	0.1662	1.5921	0.8683
Closed-form [21]	0.3040	0.4736	0.7584	0.3188	1.5419	0.4410	2.6960	1.6258
KNN [6]	0.0737	0.0451	0.1381	0.0731	0.6534	0.1073	1.1548	0.6945
Information-flow [1]	0.0663	0.0351	0.1001	0.0652	0.6819	0.1056	1.5007	0.7209
Learning based [36]	0.0554	0.0286	0.0826	0.0508	0.6061	0.0898	1.0559	0.6441
FBA [11]	0.0598	0.0395	0.1143	0.0557	0.7711	0.1390	1.2350	0.8220
Ours (w/o uncertainty map)	0.0448	0.0180	0.0414	0.0401	0.4926	0.0611	0.7692	0.5189
Ours	**0.0422**	**0.0166**	**0.0401**	**0.0363**	**0.4675**	**0.0557**	**0.7123**	**0.4948**

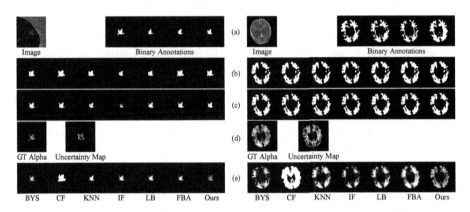

Fig. 4. Qualitative comparisons on samples from LIDC-IDRI and Brain-growth datasets. (a) original images and labeled binary masks. (b) predicted masks by Probabilistic U-net with manual binary masks. (c) predicted masks by our model with uncertainty map, which reveals more details in the ambiguity regions. (d) the target alpha mattes and the uncertainty maps in which the hard areas to recognize are highlighted. (e) the results of different matting methods. BYS: Bayesian [8], CF: Closed-Form [21], KNN [6], IF: Information-Flow [1], LB: Learning Based [36], FBA [11]. Better view in zoom.

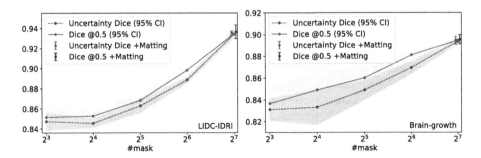

Fig. 5. The chart reveals the relationship between the number provided when training the segmentation network and the model performance. The performance grows as more masks are provided.

is set to 0.1. Sixteen masks are generated for uncertainty map generation. The LIDC-IDRI and the Brain-growth are trained 50, 100 epochs with batch-size 1, respectively. The input image is resized to 128 × 128 and elastic transformation [30] is deployed. All the models are trained from scratch with convolution parameters initialized by He initialization [13]. The number of generalized masks is kept the same as that of the ground truth masks during evaluation.

Results and Analysis. We compared the predicted alpha mattes, i.e., the calibrated uncertainty, with six state-of-the-art matting methods. The trained models of the deep-learning-based methods for the natural image are used as the data augmentation mechanism [11,33], which generate training images by randomly compose foreground and background, may not be anatomically correct in medical images. Moreover, we provide the methods which need a trimap with the one we used in generating the ground truth alpha mattes, mentioned in Sect. 2.

Table 1 shows that our model outperforms all the other methods in both datasets, which illustrates that our method is more applicable to the medical scenarios. It also demonstrated that the performance has a further improvement in all four metrics with the uncertainty map, indicating that the uncertainty map benefits the matting task.

The visual results are shown in Fig. 4. Our method can better express the edge of the fuzzy transition zone and subtle structural features in the matting results. Moreover, binary masks generated by our final model have a better ability to depict the details. It may be due to using the alpha matte to generate masks so that the masks also inherit the better ability of detail expression.

Ablation Study. Thanks to the continuous value, alpha matte can describe the lesion structure more precisely than binary masks and quantify uncertainty intuitively. Moreover, a bunch of binary masks in the same distribution can be generated by thresholding a single alpha matte, which is advantageous over the limited manually labeled masks, reducing the labeling cost and improving efficiency. To prove it quantitatively, we investigate how the number of labeled masks used in training increase can improve the model performance. To ensure the target masks share the same distribution, we generate 128 binary masks by using Eq. 1 on alpha matte with equidistant thresholds, from which a specific number of masks is obtained by evenly sampling.

Two metrics are used to evaluate the similarity between the predicted and the ground truth masks. Uncertainty Dice $Dice_{\mathcal{U}}$ is proposed to evaluate the uncertainty with multiple annotations [26] quantitatively. It can reveal the similarity of the predicted mask distribution to the target distribution by calculating the average Dice scores at different score map levels. We also adapt conventional dice with a threshold at 0.5 to the multi-label scenario, denoted as $Dice_{@0.5}$, to reveal the similarity to a specific target mask instead of overall performance. The metrics are formulated as:

$$Dice_{\mathcal{U}}(\tilde{\mathcal{P}}, \mathcal{D}_{gt}) = \frac{1}{|\mathcal{T}|} \sum_{\tau \in \mathcal{T}} \frac{1}{|\tilde{\mathcal{P}}|} \sum_{\tilde{p} \in \tilde{\mathcal{P}}} Dice\left(Threshold\left(\tilde{p}, \tau\right), \mathcal{D}_{gt}\right) \quad (6)$$

$$Dice_{@0.5}(\tilde{\mathcal{M}}, \mathcal{M}_{gt}) = \frac{1}{|\tilde{\mathcal{M}}|} \sum_{\tilde{m} \in \tilde{\mathcal{M}}} max\left\{\mathcal{D}ice(\tilde{m}, m_{gt}), m_{gt} \in \mathcal{M}_{gt}\right\} \quad (7)$$

where $\tilde{\mathcal{P}}$ is the set of predicted score maps. $\mathcal{D}_{gt}$ denotes the ground truth distribution, calculated by averaging the target masks. $\mathcal{T}$ is a set of thresholds, that $\mathcal{T} = \{\tau | 0.05, 0.15, ..., 0.95\}$. $\tilde{\mathcal{M}}$ and $\mathcal{M}_{gt}$ denote the predicted and target masks, respectively.

We deployed experiments on probabilistic UNet with different numbers of masks, ranging from 8 to 128. Figure 5 shows the results. It is demonstrated that both metrics' results grow as more masks are provided. It also reveals that alpha matte is a promising annotation form in medical segmentation that can achieve comparable performance with much more binary labels. Moreover, an additional experiment was deployed to compare the performance of the Probabilistic UNet and our model. We find the performance can be improved with our model, which indicates training with the matting network may improve the performance, and the matting task benefits the binary segmentation task (Fig. 5 +Matting).

5   Conclusions

In this work, we creatively calibrate the uncertainty by alpha matte introduced from image matting, which has a better ability to reveal tiny and ambiguous structures and has a potential for diagnosis. A well-designed multi-task network was proposed to predict binary masks and alpha matte simultaneously. The uncertainty map, an analogy to trimap, is generated by the intermediate outputs and improves the matting network performance. The binary masks, uncertainty map, and alpha matte express the target with uncertainty in different aspects, so the sub-tasks can benefit each other by sharing the latent information during training. Experiments reveal our model outperforms the other state-of-the-art matting methods on all four metrics with a considerable margin and demonstrate that alpha matte is a more powerful annotation method than the binary mask. We labeled two datasets with alpha matte, including CT and MRI images, and they are released to the public to promote the study on uncertainty learning and matting in medical scenarios. More modalities will be examined with medical matting, and also new experiments will be carried out to test its value in diagnosis in the future.

Acknowledgements. The authors would like to thank the clinicians for their hard work in evaluating the alpha matte datasets. They are Yi Luo of Chongqing hospital of traditional Chinese medicine, Huan Luo of Chongqing Renji Hospital of Chinese Academy of Sciences, and Feng Jiang of the First Affiliated Hospital of Wenzhou Medical University.

References

1. Aksoy, Y., Ozan Aydin, T., Pollefeys, M.: Designing effective inter-pixel information flow for natural image matting. In: Proceedings of the IEEE Conference on Computer Vision and Pattern Recognition (CVPR), pp. 29–37 (2017)
2. Armato III, S.G., et al.: Lung image database consortium: developing a resource for the medical imaging research community. Radiology **232**(3), 739–748 (2004)
3. Baumgartner, C.F., et al.: PHiSeg: capturing uncertainty in medical image segmentation. In: Shen, D., et al. (eds.) MICCAI 2019. LNCS, vol. 11765, pp. 119–127. Springer, Cham (2019). https://doi.org/10.1007/978-3-030-32245-8_14
4. Bochkovskiy, A., Wang, C.Y., Liao, H.Y.M.: YOLOv4: optimal speed and accuracy of object detection. arXiv preprint arXiv:2004.10934 (2020)
5. Cai, S., et al.: Disentangled image matting. In: Proceedings of the IEEE International Conference on Computer Vision (ICCV), pp. 8819–8828 (2019)
6. Chen, Q., Li, D., Tang, C.K.: KNN Matting. IEEE Trans. Pattern Anal. Mach. Intell. (TPAMI) **35**(9), 2175–2188 (2013)
7. Cheng, J., Zhao, M., Lin, M., Chiu, B.: AWM: adaptive weight matting for medical image segmentation. In: Medical Imaging 2017: Image Processing, vol. 10133, p. 101332P. International Society for Optics and Photonics (2017)
8. Chuang, Y.Y., Curless, B., Salesin, D.H., Szeliski, R.: A Bayesian approach to digital matting. In: Proceedings of the IEEE Conference on Computer Vision and Pattern Recognition (CVPR), vol. 2, p. II. IEEE (2001
9. Der Kiureghian, A., Ditlevsen, O.: Aleatory or epistemic? Does it matter? Struct. Saf. **31**(2), 105–112 (2009)
10. Fan, Z., Lu, J., Wei, C., Huang, H., Cai, X., Chen, X.: A hierarchical image matting model for blood vessel segmentation in fundus images. IEEE Trans. Image Process. (TIP) **28**(5), 2367–2377 (2018)
11. Forte, M., Pitié, F.: F, B, Alpha matting. arXiv preprint arXiv:2003.07711 (2020)
12. Gantenbein, M., Erdil, E., Konukoglu, E.: RevPHiSeg: a memory-efficient neural network for uncertainty quantification in medical image segmentation. In: Sudre, C.H., et al. (eds.) UNSURE/GRAIL -2020. LNCS, vol. 12443, pp. 13–22. Springer, Cham (2020). https://doi.org/10.1007/978-3-030-60365-6_2
13. He, K., Zhang, X., Ren, S., Sun, J.: Delving deep into rectifiers: surpassing human-level performance on ImageNet classification. In: Proceedings of the IEEE International Conference on Computer Vision (CVPR), pp. 1026–1034 (2015)
14. He, K., Zhang, X., Ren, S., Sun, J.: Deep residual learning for image recognition. In: Proceedings of the IEEE Conference on Computer Vision and Pattern Recognition (CVPR), pp. 770–778 (2016)
15. Hu, S., Worrall, D., Knegt, S., Veeling, B., Huisman, H., Welling, M.: Supervised uncertainty quantification for segmentation with multiple annotations. In: Shen, D., et al. (eds.) MICCAI 2019. LNCS, vol. 11765, pp. 137–145. Springer, Cham (2019). https://doi.org/10.1007/978-3-030-32245-8_16
16. Hüllermeier, E., Waegeman, W.: Aleatoric and epistemic uncertainty in machine learning: a tutorial introduction. arXiv preprint arXiv:1910.09457 (2019)
17. Kendall, A., Badrinarayanan, V., Cipolla, R.: Bayesian segNet: model uncertainty in deep convolutional encoder-decoder architectures for scene understanding. In: British Machine Vision Conference (BMVC) (2017)
18. Kendall, A., Gal, Y.: What uncertainties do we need in Bayesian deep learning for computer vision? In: Advances in Neural Information Processing Systems (NIPS), pp. 5574–5584 (2017)

19. Kendall, A., Gal, Y., Cipolla, R.: Multi-task learning using uncertainty to weigh losses for scene geometry and semantics. In: Proceedings of the IEEE Conference on Computer Vision and Pattern Recognition (CVPR), pp. 7482–7491 (2018)

20. Kohl, S., et al.: A probabilistic U-Net for segmentation of ambiguous images. In: Advances in Neural Information Processing Systems (NIPS), pp. 6965–6975 (2018)

21. Levin, A., Lischinski, D., Weiss, Y.: A closed-form solution to natural image matting. IEEE Trans. Pattern Anal. Mach. Intell. (TPAMI) **30**(2), 228–242 (2007)

22. Li, Y., Lu, H.: Natural image matting via guided contextual attention. In: Proceedings of the AAAI Conference on Artificial Intelligence (AAAI), vol. 34, pp. 11450–11457 (2020)

23. Li, Y., Xu, Q., Lu, H.: Hierarchical opacity propagation for image matting. arXiv preprint arXiv:2004.03249 (2020)

24. Loshchilov, I., Hutter, F.: SGDR: Stochastic gradient descent with warm restarts. Learning **10**, 3

25. Lutz, S., Amplianitis, K., Smolic, A.: AlphaGAN: generative adversarial networks for natural image matting. arXiv preprint arXiv:1807.10088 (2018)

26. Menze, B., Joskowicz, L., Bakas, S., Jakab, A., Konukoglu, E., Becker, A.: Quantification of uncertainties in biomedical image quantification challenge. [EB/OL]. https://qubiq.grand-challenge.org/Home/. Accessed 22 Oct 2020

27. Monteiro, M., et al.: Stochastic segmentation networks: modelling spatially correlated aleatoric uncertainty. arXiv preprint arXiv:2006.06015 (2020)

28. Rupprecht, C., et al.: Learning in an uncertain world: Representing ambiguity through multiple hypotheses. In: Proceedings of the IEEE International Conference on Computer Vision (ICCV), pp. 3591–3600 (2017)

29. Shen, X., Tao, X., Gao, H., Zhou, C., Jia, J.: Deep automatic portrait matting. In: Leibe, B., Matas, J., Sebe, N., Welling, M. (eds.) ECCV 2016. LNCS, vol. 9905, pp. 92–107. Springer, Cham (2016). https://doi.org/10.1007/978-3-319-46448-0_6

30. Simard, P.Y., Steinkraus, D., Platt, J.C., et al.: Best practices for convolutional neural networks applied to visual document analysis. In: Icdar, vol. 3 (2003)

31. Wang, G., Li, W., Aertsen, M., Deprest, J., Ourselin, S., Vercauteren, T.: Aleatoric uncertainty estimation with test-time augmentation for medical image segmentation with convolutional neural networks. Neurocomputing **338**, 34–45 (2019)

32. Woo, S., Park, J., Lee, J.Y., So Kweon, I.: CBAM: convolutional block attention module. In: Proceedings of the European Conference on Computer Vision (ECCV), pp. 3–19 (2018)

33. Xu, N., Price, B., Cohen, S., Huang, T.: Deep image matting. In: Proceedings of the IEEE Conference on Computer Vision and Pattern Recognition (CVPR), pp. 2970–2979 (2017)

34. Zeng, Z., Wang, J., Shepherd, T., Zwiggelaar, R.: Region-based active surface modelling and alpha matting for unsupervised tumour segmentation in pet. In: IEEE International Conference on Image Processing (ICIP), pp. 1997–2000. IEEE (2012)

35. Zhao, H., Li, H., Cheng, L.: Improving retinal vessel segmentation with joint local loss by matting. Pattern Recogn. (PR) **98**, 107068 (2020)

36. Zheng, Y., Kambhamettu, C.: Learning based digital matting. In: 2009 IEEE 12th International Conference on Computer Vision (ICCV), pp. 889–896. IEEE (2009)

Confidence-Aware Cascaded Network for Fetal Brain Segmentation on MR Images

Xukun Zhang[1,2], Zhiming Cui[3], Changan Chen[4], Jie Wei[5], Jingjiao Lou[6],
Wenxin Hu[2], He Zhang[4(✉)], Tao Zhou[7], Feng Shi[8(✉)],
and Dinggang Shen[1,8(✉)]

[1] School of Biomedical Engineering, ShanghaiTech University, Shanghai, China
dgshen@shanghaitech.edu.cn
[2] School of Data Science and Engineering, East China Normal University,
Shanghai, China
[3] Department of Computer Science, The University of Hong Kong, Hong Kong, China
[4] Department of Radiology, Obstetrics and Gynecology Hospital of Fudan
University, Shanghai, China
[5] School of Computer Science and Engineering, Northwestern Polytechnical
University, Xi'an, Shaanxi, China
[6] School of Control Science and Engineering, Shandong University, Jinan, China
[7] School of Computer Science and Engineering, Nanjing University of Science
and Technology, Nanjing, China
[8] Shanghai United Imaging Intelligence Co., Ltd.,
Shanghai, China
feng.shi@united-imaging.com

Abstract. Fetal brain segmentation from Magnetic Resonance (MR) images is a fundamental step in brain development study and early diagnosis. Although progress has been made, performance still needs to be improved especially for the images with motion artifacts (due to unpredictable fetal movement) and/or changes of magnetic field. In this paper, we propose a novel confidence-aware cascaded framework to accurately extract fetal brain from MR image. Different from the existing coarse-to-fine techniques, our two-stage strategy aims to segment brain region and simultaneously produce segmentation confidence for each slice in 3D MR image. Then, the image slices with high-confidence scores are leveraged to guide brain segmentation of low-confidence image slices, especially on the brain regions with blurred boundaries. Furthermore, a slice consistency loss is also proposed to enhance the relationship among the segmentations of adjacent slices. Experimental results on fetal brain MRI dataset show that our proposed model achieves superior performance, and outperforms several state-of-the-art methods.

Keywords: Fetal brain segmentation · Confidence-aware cascaded network · Slice consistency loss

---

X. Zhang, Z. Cui and C. Chen—Contributed equally.

© Springer Nature Switzerland AG 2021
M. de Bruijne et al. (Eds.): MICCAI 2021, LNCS 12903, pp. 584–593, 2021.
https://doi.org/10.1007/978-3-030-87199-4_55

1   Introduction

With the development of brain science and neuroimaging technology, Magnetic Resonance (MR) imaging plays an increasingly important role in brain analysis. Fetal brain segmentation from MR images is a fundamental step for various subsequent tasks, including early diagnosis and fetal brain development study [1, 2]. In clinical practice, doctors usually localize and extract fetal brain by manual annotation, which is rather tedious and time-consuming. Thus, designing an automatic method to accurately segment fetal brain from MR images is greatly demanded and has attracted tremendous research attention.

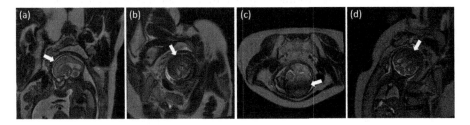

Fig. 1. Four typical examples of complex appearances include: (a) low contrast between tissues, (b) motion artifacts, (c) and (d) bias fields.

However, automatic and accurate fetal brain segmentation from MR images is a challenging task. The low image contrast, such as boundary between amniotic fluid and cerebrospinal fluid (CSF) (Fig. 1(a)), often greatly affects the segmentation performance. Previous works [3–5] mainly relying on handcrafted features or atlas models often failed to robustly handle those low-contrast regions, and under- or over-segmented some important brain structures. Recently, with the advancement of deep learning [6–8], many learning-based algorithms have been developed, including one-stage neural networks [9] and coarse-to-fine network [10–14]. Unfortunately, all these methods are applied to 2D slice independently without considering the relationship among adjacent slices. Actually, the relationship is very important in fetal brain segmentation, especially on the slices with severe artifacts due to unpredictable fetal motion (Fig. 1(b)) and magnetic field changes [1,15] (Fig. 1(c)(d)).

In this paper, to tackle the aforementioned challenges, we present a novel confidence-aware cascaded framework to accurately segment fetal brain from MR images. In the first stage, we utilize a segmentation network to coarsely localize the brain region on MR image, and simultaneously generate the segmentation confidence for each slice based on the predicted segmentation probability map. Next, in the second stage, with the foreground brain region cropped from the original MR image, a confidence-guided segmentation network is designed to obtain the final results, using two novel strategies. 1) Each slice with low segmentation confidence is guided by two nearest adjacent slices with higher segmentation confidence to generate better fetal brain segmentation. 2) A slice consistency loss is proposed to encourage the consistent segmentations of neighboring slices. To evaluate our method, we have conducted extensive experiments on the fetal brain MRI dataset collected from real-world clinics. Experimental results show our proposed framework achieves superior performance, and outperforms various state-of-the-art methods, qualitatively and quantitatively.

2   Methods

Since the fetal brain only occupies a small proportion of pixels in the fetal MRI image, we follow a coarse-to-fine strategy by first localizing brain region based on a coarse segmentation model and then conducting fine brain segmentation. The framework of our proposed method for fetal brain segmentation is shown in Fig. 2, which consists of the first stage of coarse segmentation (Fig. 2(a)) and the second stage of fine brain segmentation (Fig. 2(b)) (using both slice confidence guidance and slice segmentation consistency). The details of our proposed framework are elaborated below.

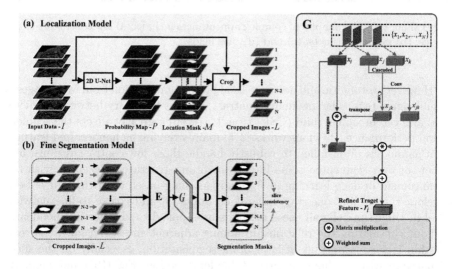

Fig. 2. The overall framework of our proposed method for fetal brain segmentation, including (a) brain localization with coarse segmentation model and (b) fine brain segmentation with another segmentation model (using both slice confidence guidance and slice segmentation consistency).

2.1   Coarse Segmentation with Slice Confidence

Considering the fact that the 3D fetal brain MR image has a large inter-slice interval and thus anatomical shapes vary significantly across consecutive slices, we employ a 2D U-Net [16] to coarsely segment and localize fetal brain slice-by-slice. In the meanwhile, the segmentation confidence of each slice is also produced. Then, high-confidence slices can be utilized to guide the segmentation of low-confidence slices with initial poor segmentation (as described in Sect. 2.2).

Given an input image I, the output of coarse segmentation model is a probability map P of the fetal brain. We directly adopt a threshold λ ($\lambda = 0.5$) to obtain the binary segmentation mask M. Hence, the foreground fetal brain area can be localized and cropped based on the mask M. To train the segmentation network, we use a multi-scale Dice loss to encourage the similarity between the probability map P and the ground-truth Y at multiple non-local scales, which can be defined as:

$$\mathcal{L}_{mDice}(P,Y) = \frac{1}{S} \sum_{s=1}^{S} (1 - \frac{2 \sum P_s Y_s}{\sum P_s^2 + \sum Y_s^2}), \tag{1}$$

where P_s and Y_s refer to the downscale versions of P and Y, respectively. S indicates the number of downscales, which is set to 3 in all our experiments.

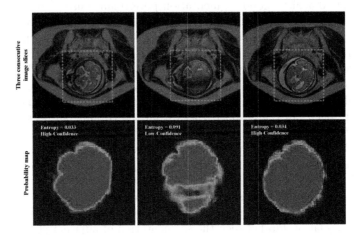

Fig. 3. Due to fetal movement or magnetic field changes, one middle slice obtains poor segmentation with low confidence.

Slice Confidence Level Based on Localization Probability Map. The segmentation quality for the images cropped from the segmentation mask M varies from slice to slice, due to acquisition issues and motion artifacts as shown in Fig. 1. Clear image slices with obvious contours are easy to produce good segmentation masks, while low-contrast and fuzzy image slices will confuse the segmentation model and produce uncertain predictions. One typical example is shown in Fig. 3. It can be seen that the intensity of the middle slice may be distorted due to bias field, which usually produces poor segmentation results with low confidence. Inspired by Pan et al.'s work on autonomous driving [17], we introduce the concept of slice confidence to measure the segmentation quality of each slice, by calculating the entropy of the corresponding cropped probability map P in each slice, as defined below:

$$R(t) = avg(\sum_{h,w} -P_t^{(h,w)} \cdot \ln P_t^{(h,w)}), \tag{2}$$

where $P_t^{(h,w)}$ denotes the prediction of the pixel coordinate (h, w) corresponding to the probability map of the t-th slice in an MR image.

Based on Eq. (2), we calculate the entropy for each slice in an MR image. A lower $R(t)$ indicates a higher confidence, while a higher $R(t)$ indicates a lower confidence. In this regard, given a ranking of scores from $R(t)$, we mark the top 50% as the cropped image slices with high confidence, and others as those with low confidence.

2.2   Confidence-Aware Fine Fetal Brain Segmentation

Intuitively, due to the existence of image slices with quality problems (Fig. 1), it is difficult to achieve good segmentation performance using common cascading strategy. Different from the existing coarse-to-fine techniques, we propose a confidence-aware fine fetal brain segmentation model (Fig. 2(b)), which uses the slices with high-confidence to guide more reliable segmentation for the slices with low-confidence. Specifically, we first enhance the foreground fetal brain regions for the cropped slices with high confidence, i.e., increasing their local intensity twice, while keeping intensities of the cropped slices with low confidence being unchanged. Then, all the cropped slices from the training subject are used as a batch of input for the fine segmentation model. The features of each slice extracted by the encoder E pass through a novel guiding module G, in which the features of the low-confidence slice will be guided by two nearest adjacent high-confidence slices features to produce refined features. Finally, the refined features are passed through the decoder D to produce a fine segmentation mask. In the training process of the model, we also propose a slice consistency loss to enhance the relationship between the segmentation of adjacent slices. Below we elaborate on the key components of the fine model.

Guidance Module - G. As shown in Fig. 3, it is difficult for the segmentation model to give accurate segmentation predictions for image slices with quality problems. Thus, to tackle this issue, we design a novel guidance module G, which applies guidance to the target image feature by using the features of the adjacent teacher images with high confidence, for improving the segmentation results of low-quality image slices.

Specifically, as shown in Fig. 2, the encoder E of the second stage network extracts the corresponding features $x_i \in \{x_1, x_2, .., x_N\}$ from the input slice sequence $L_i \in \{L_1, L_2, ..., L_N\}$ in a subject. For the feature x_i extracted from the slice L_i, G finds the high-confidence feature maps x_j and x_k with the closest slice distance, and uses them as teacher features. Note that the high confidence is used to ensure the teacher features have high quality, and the closest slice distance aims to preserve the shape similarity between target and teacher features. Then, we cascade the two teacher features and generate two new teacher features x_{jk} and x'_{jk} by using different convolution operations. The relationship W between the teacher features x_{jk} and the target feature x_i is obtained through matrix transpose, matrix multiplication and softmax operation, as shown in the following equation:

$$W = softmax(x_{jk}^T * x_i),$$
(3)

where T and $*$ represent transposition and multiplication operation, respectively. Based on this, we apply the obtained relationship W to another teacher features x'_{jk} and perform a weighted sum with the target feature x_i to obtain the refined target feature r_i:

$$r_i = W * x'_{jk} + x_i.$$
(4)

Finally, the decoder D outputs the fine segmentation mask by using the feature $r_i \in \{r_1, r_2, .., r_N\}$ refined by G.

Segmentation Consistency Loss. Considering that the relationship between adjacent slices is very important in fetal brain segmentation, especially on slices with severe artifacts due to unpredictable fetal movement (Fig. 1(b)) and magnetic field changes (Fig. 1(c)(d)). We propose a segmentation consistency loss function (*i.e.*, $\mathcal{L}_C$) to exploit the consistent relationship among adjacent image slices, which is defined as follows:

$$\mathcal{L}_C = \frac{1}{hw} \sum_{t=2}^{N-1} \sum_{h,w} (P_t^{(h,w)} - P_{t-1}^{(h,w)})^2 + (P_t^{(h,w)} - P_{t+1}^{(h,w)})^2,$$
(5)

where $P_t^{(h,w)}$ denotes the prediction of the pixel coordinate (h, w) corresponding to the t-th image slice of a subject. The total loss function is a combination of the segmentation consistency loss $\mathcal{L}_C$ and the multi-scale Dice loss $\mathcal{L}_{mDice}$ as shown below:

$$\mathcal{L} = \mathcal{L}_{mDice} + \beta \mathcal{L}_C,$$
(6)

where β is an introduced hyper-parameter, which is set to 0.01 in the experiment.

3   Experiments

3.1   Dataset and Evaluation Metrics

We evaluated the proposed framework on 239 fetal MRI scans, and we randomly split the dataset with 120 scans for training, 30 scans for validation and 89 scans for testing. The ground truth of all subjects were annotated by doctors. Each subject was scanned and acquired by Single Shot Fast Spin Echo (SSFSE), with image size of 256×256 and the voxel spacing of $3.5 \times 1 \times 1$ (mm). The gestational week (GW) ranges from 23 GWs to 34 GWs. We used the bias field

Table 1. Comparison results of different segmentation methods.

Method	Dice (%)	Jaccard (%)	ASD (mm)	Hausdorff (mm)
2D U-Net [16]	91.05	84.35	1.23	9.87
U-Net3Plus [19]	91.39	85.61	2.10	19.0
Salehi *et al.* [9]	92.42	86.05	1.40	10.67
Lou *et al.* [10]	93.34	89.51	0.93	5.96
Ours	**95.77**	**91.99**	**0.46**	**3.99**

correction method to preprocess all images [18], and normalized them via a z-score operation.

In this study, we use four popular metrics, *i.e.*, Dice (%), Jaccard coefficient (%), average surface distance (ASD) (mm), and 95% Hausdoff (mm) to evaluate the performance of the fetal brain segmentation.

3.2 Implementation Details

The proposed framework was implemented on a RTX2080Ti GPU using PyTorch. For coarse segmentation stage, the random rotations of 90°, 180°, and 270° were performed on each input slice as data augmentation. For fine segmentation, we first rotated the original MR images at any angle within the range of 0 to 360, and then cropped a 128×128 patch as an augmentation approach for producing the training data. Besides, we used SGD to optimize the network parameters, and the learning rate was set to 0.0001.

3.3 Comparison with State-of-the-art Methods

We compared our framework with several state-of-the-art approaches, including 2D U-Net [16], U-Net3Plus [19], and the methods of Salehi *et al.* [9] and Lou *et al.* [10]. The first two are well-known deep learning methods for medical image segmentation, and have achieved the remarkable performance in many public segmentation datasets. The last two are specialized methods for fetal brain segmentation. Specifically, Salehi *et al.* developed a 2D U-Net with automatic context. Lou *et al.* proposed a multi-stage model based on a coarse-to-fine strategy to achieve leading performance in this specific task. The quantitative and qualitative comparison are shown in Table 1 and Fig. 4, respectively.

Quantitative Comparison. The overall comparison results under the four metrics are summarized in Table 1. Comparing the first two rows, we can see that 2D U-Net and U-Net3Plus have relatively low Dice scores, which demonstrates that common one-stage methods is difficult to achieve acceptable performance in the fetal brain segmentation task. On this basis, the 2D U-Net with automatic context proposed by Salehi *et al.* has a Dice improvement of about 1.0%. Although Lou *et al.* achieves the state-of-the-art performance in this specific task by multi-stage models, our framework still outperforms it over the four

Table 2. Ablation study on two key components.

Methods	Dice (%)	Jaccard (%)	ASD (mm)	Hausdorff (mm)
bNet	92.93	86.97	0.97	7.53
bNet-G	94.95	90.53	0.61	5.57
bNet-S	93.40	88.56	0.63	4.70
FullNet(Ours)	**95.77**	**91.99**	**0.46**	**3.99**

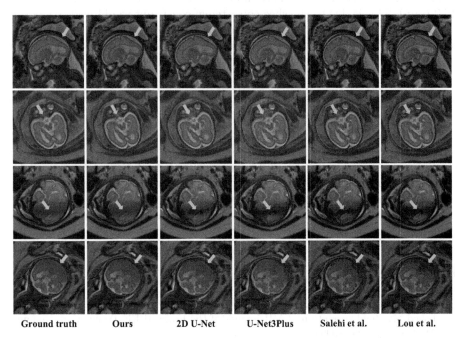

<table>
<tr><td>Ground truth</td><td>Ours</td><td>2D U-Net</td><td>U-Net3Plus</td><td>Salehi et al.</td><td>Lou et al.</td></tr>
</table>

Fig. 4. Comparison segmentation results of our model and other methods. Extracted brain masks of different methods are overlaid on the cropped fetal brain MR images.

segmentation metrics (2.43% improvements of Dice score), demonstrating the effectiveness of confidence-aware cascaded module.

Qualitative Comparison. To further evaluate the effectiveness of our method, we also provide qualitative results, as shown in Fig. 4. It can be found that the segmentation results produced by our method have more reliable fetal brain boundaries. One typical example is shown in the third row, where the MR image has a relatively low quality due to the bias field. It can be seen that our method obtains accurate segmentation results, while all compared methods suffer from severe under-segmentation on the area with low image intensity contrast.

3.4   Ablation Study

To validate the effectiveness of the key components proposed in our method, we build a baseline network (bNet) by directly utilizing cascaded network to segment the fetal brain. Then, we respectively add the guidance module and slice consistence loss on bNet, which is denoted as bNet-G and bNet-S. Finally, in the FullNet, these two components are all included. The statistic results are presented in Table 2. It can seen that bNet-G consistently improves the segmentation accuracy (2.02% improvements in terms of Dice accuracy), indicating that the guidance module pays more attention to the slice with low confidence in the coarse segmentation. Additionally, with the slice consistency loss, the distance error between the predicted and ground truth brain shape decreases from 7.53 to 4.70 (mm). This is due to the slice consistency constraint would lead to smooth segmentation results among slices, which can efficiently remove the segmentation artifacts. At last, the FullNet achieves the best performance in the fetal brain segmentation task.

4   Conclusion

We propose a novel confidence-aware cascaded framework to accurately segment fetal brain from MR image. Our method utilizes high-segmentation confidence slices to guide low-segmentation confidence slices through the guidance module. Besides, we propose a consistence constraint to exploit the relationship among the adjacent slices. Experimental results demonstrate the effectiveness of the proposed method against other state-of-the-art methods. In the future, we will carry out neuroimaging analysis based on our fetal brain segmentation results.

References

1. Van Leemput, K., Maes, F., Vandermeulen, D., Suetens, P.: Automated model-based bias field correction of MR images of the brain. IEEE Trans. Med. Imaging **18**(10), 885–896 (1999)
2. Makropoulos, A., Counsell, S.J., Rueckert, D.: A review on automatic fetal and neonatal brain MRI segmentation. NeuroImage **170**, 231–248 (2018)
3. Anquez, J., Angelini, E.D., Bloch, I.: Automatic segmentation of head structures on fetal MRI. In: IEEE International Conference on Symposium on Biomedical Imaging: From Nano to Macro, pp. 109–112 (2009)
4. Tourbier, S., et al.: Automatic brain extraction in fetal MRI using multi-atlas-based segmentation, SPIE, vol. 9413, pp. 94130Y–94130Y-7 (2015)
5. Wright, R., et al.: Automatic quantification of normal cortical folding patterns from fetal brain MRI. Neuroimage **91**(2), 21–32 (2014)
6. Ching, T., et al.: Opportunities and obstacles for deep learning in biology and medicine. J. R. Soc. Interface **15**, 20170387 (2018)
7. Litjens, G., et al.: A survey on deep learning in medical image analysis. Med. Image Anal. **42**, 60–88 (2017)
8. Shen, D., Wu, G., Suk, H.I.: Deep learning in medical image analysis. Ann. Rev. Biomed. Eng. **19**(1), 221–248 (2017)
9. Salehi, S.S.M., et al.: Real-time automatic fetal brain extraction in fetal MRI by deep learning. In: 15th IEEE International Symposium on Biomedical Imaging, ISBI 2018, Washington, DC, USA, 4–7 April 2018, pp. 720–724 (2018)

10. Lou, J., et al.: Automatic fetal brain extraction using multi-stage U-Net with deep supervision. In: Suk, H.-I., Liu, M., Yan, P., Lian, C. (eds.) MLMI 2019. LNCS, vol. 11861, pp. 592–600. Springer, Cham (2019). https://doi.org/10.1007/978-3-030-32692-0_68

11. Ebner, M., et al.: An automated framework for localization, segmentation and super-resolution reconstruction of fetal brain MRI. NeuroImage **206**, 116324 (2020)

12. Xing, J., et al.: Lesion segmentation in ultrasound using semi-pixel-wise cycle generative adversarial nets. IEEE/ACM Trans. Comput. Biol. Bioinform. (2020)

13. Li, J., et al.: Automatic fetal brain extraction from 2D in utero fetal MRI slices using deep neural network. Neurocomputing **378**, 335–349 (2020)

14. Chen, J., et al.: Automatic brain extraction from 3D fetal MR image with deep learning-based multi-step framework. Comput. Med. Imaging Graph. **88**, 101848 (2021)

15. Sled, J.G., Pike, G.B.: Understanding intensity non-uniformity in MRI. In: Wells, W.M., Colchester, A., Delp, S. (eds.) MICCAI 1998. LNCS, vol. 1496, pp. 614–622. Springer, Heidelberg (1998). https://doi.org/10.1007/BFb0056247

16. Ronneberger, O., Fischer, P., Brox, T.: U-Net: convolutional networks for biomedical image segmentation. In: Navab, N., Hornegger, J., Wells, W.M., Frangi, A.F. (eds.) MICCAI 2015. LNCS, vol. 9351, pp. 234–241. Springer, Cham (2015). https://doi.org/10.1007/978-3-319-24574-4_28

17. Pan, F., Shin, I., Rameau, F., Lee, S., Kweon, I.S.: Unsupervised intra-domain adaptation for semantic segmentation through self-supervision. In: 2020 IEEE/CVF Conference on Computer Vision and Pattern Recognition, CVPR 2020, Seattle, WA, USA, 13–19 June 2020, pp. 3763–3772. IEEE (2020)

18. Yushkevich, P.A., et al.: User-guided 3D active contour segmentation of anatomical structures: significantly improved efficiency and reliability. Neuroimage **31**(3), 1116–1128 (2006)

19. Huang, H., et al.: UNet 3+: a full-scale connected UNet for medical image segmentation. In: 2020 IEEE International Conference on Acoustics, Speech and Signal Processing, ICASSP 2020, Barcelona, Spain, 4–8 May 2020, pp. 1055–1059 (2020)

Orthogonal Ensemble Networks for Biomedical Image Segmentation

Agostina J. Larrazabal[1(✉)], César Martínez[1], Jose Dolz[2], and Enzo Ferrante[1]

[1] Research Institute for Signals, Systems and Computational Intelligence, Sinc(i),
FICH-UNL/CONICET, Santa Fe, Argentina
alarrazabal@sinc.unl.edu.ar
[2] Laboratory for Imagery, Vision and Artificial Intelligence, École de Technologie
Supérieure, Montreal, Canada

Abstract. Despite the astonishing performance of deep-learning based approaches for visual tasks such as semantic segmentation, they are known to produce miscalibrated predictions, which could be harmful for critical decision-making processes. Ensemble learning has shown to not only boost the performance of individual models but also reduce their miscalibration by averaging independent predictions. In this scenario, model diversity has become a key factor, which facilitates individual models converging to different functional solutions. In this work, we introduce Orthogonal Ensemble Networks (OEN), a novel framework to explicitly enforce model diversity by means of orthogonal constraints. The proposed method is based on the hypothesis that inducing orthogonality among the constituents of the ensemble will increase the overall model diversity. We resort to a new pairwise orthogonality constraint which can be used to regularize a sequential ensemble training process, resulting on improved predictive performance and better calibrated model outputs. We benchmark the proposed framework in two challenging brain lesion segmentation tasks –brain tumor and white matter hyper-intensity segmentation in MR images. The experimental results show that our approach produces more robust and well-calibrated ensemble models and can deal with challenging tasks in the context of biomedical image segmentation.

Keywords: Image segmentation · Ensemble networks · Orthogonal constraints

1 Introduction

In the past few years, deep learning-based methods have become the *de facto* solution for many computer vision and medical imaging tasks. Nevertheless, despite their success and great ability to learn highly discriminative features, they are shown to be poorly calibrated [1], often resulting in over-confident predictions. This results in a major problem, which can have catastrophic consequences in critical decision-making systems, such as medical diagnosis, where the downstream decision depends on predicted probabilities.

M. de Bruijne et al. (Eds.): MICCAI 2021, LNCS 12903, pp. 594–603, 2021.
https://doi.org/10.1007/978-3-030-87199-4_56

Ensemble learning is a simple strategy to improve both the robustness and calibration performance of predictive models [2,3]. In this scenario, a common approach is to train the same model under different conditions, which can foster the model convergence to different functional solutions. Techniques to produce ensembles include dataset shift [4], Monte-Carlo Dropout [5], batch-ensemble [6] or different model hyperparameters [7], among others. Then, by averaging the predictions, individual mistakes can be dismissed leading to a reduced miscalibration. In this context, ensuring *diversity* across models is a key factor to build a robust ensemble. To promote model diversity in ensembles many mechanisms have been proposed. These include using latent variables [8], integrating attention in the embeddings to enforce different learners to attend to different parts of the object [9] or isolating the adversarial vulnerability in sub-models by distilling non-robust features to induce diverse outputs against a transfer attack [10].

Nevertheless, despite the relevance of obtaining well-calibrated models in clinical applications, relatively few works have studied this problem. Particularly, in the context of medical image segmentation, it was suggested that models trained with the well-known soft Dice loss [11] produce miscalibrated models [12], which tend to be highly overconfident. Furthermore, the recent work in [13] proposed the use of ensembles to improve confidence calibration. However, the importance of model diversity was not assessed in this work. Thus, given the negative impact of miscalibrated models in health-related tasks, and the current practices in medical image segmentation of systematically employing the Dice loss as an objective function, we believe it is of paramount importance to investigate the effect of ensemble learning in image segmentation, and how to enforce model diversity to generate high-performing and well-calibrated models.

Contributions. In this work, we propose a novel learning strategy to boost model diversity in deep convolutional neural networks (DCNN) ensembles, which improves both segmentation accuracy and model calibration in two challenging brain lesion segmentation scenarios. The main hypothesis is that inducing orthogonality among the constituents of the ensemble will increase the overall model diversity. We resort to a novel pairwise orthogonality constraint which can be used to regularize a sequential ensemble training process, resulting on improved predictive performance and better calibrated model outputs. In this context, our contributions are 3-fold: (1) we propose a novel filter orthogonality constraint for ensemble diversification, (2) we show that diversified ensembles improve not only segmentation accuracy but also confidence calibration and (3) we showcase the proposed framework in two challenging brain lesion segmentation tasks, including tumor and white-matter hyperintensity (WMH) segmentation on magnetic resonance images.

2   Related Works

Diversifying ensembles has been used to improve classification and segmentation performance of DCNNs in several contexts. In [14] authors propose an explicit way to construct diverse ensembles bringing together multiple CNN models and architectures. Although they obtain successful results, this approach requires to manually design and train various architectures. An ensemble of 3D U-Nets with different hyper-parameters for brain tumor segmentation is proposed in [15], where authors point out that using different hyper-parameters reduces the correlations of random errors with respect to homogeneous configurations. However, no study on the diversity of the models and its influence on performance is presented. In [16] authors present a different view, highlighting that many automatic segmentation algorithms tend to exhibit asymmetric errors, typically producing more false positives than false negatives. By modifying the loss function, they train a diverse ensemble of models with very high recall, while sacrificing their precision, with a sufficiently high threshold to remove all false positives. While the authors achieve a significant increase in performance no study on the final calibration of the ensemble is carried out.

Following the success of ensemble methods at improving discriminative performance, its capability to improve confidence calibration has begun to be explored. [2] uses a combination of independent models to reduce confidence uncertainty by averaging predictions over multiple models. In [13] authors achieve an improvement in both segmentation quality and uncertainty estimation by training ensembles of CNNs with random initialization of parameters and random shuffling of training data. While these results are promising, we believe that confidence calibration can be further improved by directly enforcing diversity into the models instead of randomly initializing the weights.

As pointed out in [17] over-sized DNNs often result in a high level of overfitting and many redundant features. However, when filters are learned to be as orthogonal as possible, they become decorrelated and their filter responses are no longer redundant, thereby fully utilizing the model capacity. [18] follows a very similar approach but they regularize both negatively and positively correlated features according to their differentiation and based on their relative cosine distances. Differently from these works where ortogonality constraints are used to decorrelated the filters within a single model, here we propose to enforce filter orthogonality among the constituents of the ensemble to boost model diversity.

3 Orthogonal Ensemble Networks for Image Segmentation

Given a dataset $\mathcal{D} = \{(\mathbf{x}, \mathbf{y})_i\}_{0 \leq i \leq |\mathcal{D}|}$ composed of images $\mathbf{x}$ and corresponding segmentation masks $\mathbf{y}$, we aim at training a model which approximates the underlying conditional distribution $p(\mathbf{y}|\mathbf{x})$, mapping input images $\mathbf{x}$ into segmentation maps $\mathbf{y}$. Thus, $p(y_j = k|\mathbf{x})$ will indicate the probability that a given pixel (or voxel) j is assigned class $k \in \mathcal{C}$ from a set of possible classes $\mathcal{C}$. The distribution is commonly approximated by a neural network $f_{\mathbf{w}}$ parameterized by weights $\mathbf{w}$. In other words, $f_{\mathbf{w}}(\mathbf{x}) = p(\mathbf{y}|\mathbf{x}; \mathbf{w})$. Parameters $\mathbf{w}$ are learnt so that they minimize a particular loss function over the training dataset. Given a set of segmentation networks $\{f_{\mathbf{w}^1}, f_{\mathbf{w}^2}...f_{\mathbf{w}^N}\}$, a simple strategy to build an ensemble network $f_{\mathbf{E}}$ is to average their predictions as:

$$f_{\mathbf{E}}(\mathbf{x}) = \frac{1}{N} \sum_{i=1}^{N} f_{\mathbf{w}^i}(\mathbf{x}). \tag{1}$$

Under the hypothesis that diversifying the set of models $f_{\mathbf{w}}^i$ will lead to more accurate and calibrated ensemble predictions, we propose to boost its overall performance by incorporating pairwise orthogonality constraints during training.

Inducing Model Diversity via Orthogonal Constraints. Modern deep neural networks are parameterized by millons of learnable weights, resulting in redundant features that can be either a shifted version of each other or be very similar with almost no variation [18]. Inducing orthogonality between convolutional filters from the same layer of a given network has shown to be a good way to reduce filter redundancy [17]. Here we exploit this principle not only to avoid redundancy within a single neural model, but among the constituents of a neural ensemble.

Given two vectors $\mathbf{x}$ and $\mathbf{y}$, cosine similarity quantifies orthogonality (or decorrelation), ranging from -1 (i.e., exactly opposite) to 1 (i.e., exactly the same), with 0 indicating orthogonality. It can be defined as:

$$\mathrm{SIM}_C(\mathbf{x}, \mathbf{y}) = \frac{<\mathbf{x}, \mathbf{y}>}{||\mathbf{x}|| \, ||\mathbf{y}||}. \tag{2}$$

Following [18], we consider the squared cosine similarity to induce orthogonality between filters through a new regularization term in the loss function. An advantage of this measure is that it takes into account both negative and positive correlations.

In order to enforce diversity within and between the ensemble models, we propose to include two regularization terms into the overall learning objective. The first one, referred to as self-orthogonality loss ($\mathcal{L}_{\mathrm{SelfOrth}}$), aims at penalizing the correlation between filters in the same layer, for a given model. Thus, for a given convolutional layer l, this term is calculated as follows:

$$\mathcal{L}_{\mathrm{SelfOrth}}(\mathbf{w}_l) = \frac{1}{2} \sum_{i=1}^{n} \sum_{j=1, j \neq i}^{n} \mathrm{SIM}_C(\mathbf{w}_{l,i}, \mathbf{w}_{l,j})^2, \tag{3}$$

where $\mathbf{w}_{l,i}$ and $\mathbf{w}_{l,j}$ are vectorized versions of each of the n convolutional kernels from layer l. We also define an inter-orthogonality loss term ($\mathcal{L}_{\mathrm{InterOrth}}$) which penalizes correlation between filters from different models in the ensemble. To this end, following a sequential training scheme, the inter-orthogonality loss for layer l of model N_e is estimated as follows:

$$\mathcal{L}_{\mathrm{InterOrth}}(\mathbf{w}_l; \{\mathbf{w}_l^e\}_{0 \leq e < N_e}) = \frac{1}{N_e} \sum_{e=0}^{N_e-1} \sum_{i=1}^{n} \sum_{j=1}^{n} \mathrm{SIM}_C(\mathbf{w}_{l,i}, \mathbf{w}_{l,j}^e)^2, \tag{4}$$

where $\{\mathbf{w}_l^e\}_{0 \leq e < N_e}$ are the parameters of the previous $N_e - 1$ models trained during the sequential ensemble construction.

Thus, the learning objective to train the proposed OEN amounts to:

$$\mathcal{L} = \mathcal{L}_{Seg} + \lambda \sum_l \left(\mathcal{L}_{\text{SelfOrth}}(\mathbf{w}_l) + \mathcal{L}_{\text{InterOrth}}(\mathbf{w}_l; \{\mathbf{w}_l^e\}) \right), \quad (5)$$

where $\mathcal{L}_{Seg}$ is the segmentation loss (e.g. soft Dice loss or cross entropy) and λ is a hyperparamether controlling the influence of the orthogonality terms.[1]

4   Experimental Framework

Database Description. We benchmark the proposed method in the context of brain tumor and WMH segmentation in MR images. For brain tumor we use the BraTS 2020 dataset [19–21] which contains 369 images with expert segmentation masks (including GD-enhancing tumor, peritumoral edema, and the necrotic and non-enhancing tumor core). Each patient was scanned with FLAIR, T1ce, T1, and T2. The images were re-sampled to an isotropic 1.0 mm voxel spacing, skull-striped and co-registered by the challenge organizers. The provided training set, we divide the database in training (315), validation (17) and test (37). The second dataset [22] consists of 60 MR images with binary masks indicating the presence of WMH lesions. For each subject, co-registered 3D T1-weighted and a 2D multi-slice FLAIR images were provided. We split the dataset in training (42), validation (3) and test (15). All images have 3 mm spacing in the z dimension, and approximately 1 mm × 1 mm in the axial plane.

Segmentation Network. For all the experiments, the backbone segmentation network was a state-of-the-art ResUNet architecture [23] implemented in Keras 2.3 with TensorFlow as backend, with soft Dice [11] as segmentation loss $\mathcal{L}_{Seg}$. For the BraTS dataset, the input was a four-channel tensor (FLAIR, T1ce, T1, and T2) and a softmax activation was used as output, whereas a two-channel input (T1, FLAIR) was employed in the WMH, with a sigmoid activation function in the output. During training, patches of size 64 × 64 × 64 were extracted from each volume, and networks were trained until convergence by sampling the patches randomly, with equal probability for each class in the case of tumour segmentation, and 0.9 probability in the case of WMH. We used Adam optimizer with a batch size of 64. The initial learning rate was set to 0.001 for BraTS and 0.0001 for WMH, and it was reduced by a factor of 0.85 every 10 epochs. Hyper-parameters were chosen using the validation split, and results reported on the hold-out test set.

Baselines and Ensemble Training. We trained two different baselines to benchmark the proposed method. In the first one (*random* ensemble) each model was randomly initialized and trained to reduce only the segmentation error $\mathcal{L}_{Seg}$. Therefore, its main source of diversity comes from the initialization of the weights. The second approach (*self-orthogonal* ensemble) includes the

---

[1] Our code associated to the orthogonal ensemble networks training is publicly available at: https://github.com/agosl/Orthogonal_Ensemble_Networks.

$\mathcal{L}_{\text{SelfOrth}}$ term in the learning objective, creating and ensemble of models individually trained with the self-orthogonality constraint. Thus, while each model learns orthogonal filters, orthogonality between different models in the ensemble was not imposed. We compared these two models with the proposed orthogonal ensemble network which also encourages inter-model diversity by minimizing the full objective defined in Eq. 5 (referred as *inter-orthogonal*). Note that in our approach models are trained sequentially. For each of the proposed settings we trained 10 models. During evaluation, we assembled groups of 1, 3 and 5 models from each setting by averaging the individual probability outputs. To provide better statistics, we repeated this process 10 times, each with different model selection. We empirically observed that beyond 5 models, the performance of the ensemble did not improve. Furthermore, λ was set to 0.1 and 1 for the WMH and brain tumour segmentation task, respectively.

Measuring Calibration for Image Segmentation. Given a segmentation network $f_{\mathbf{w}}$, if the model is well-calibrated its output for a single pixel j can be interpreted as the probability $p(y_j = k|\mathbf{x}; \mathbf{w})$ for a given class $k \in \mathcal{C}$. In this case, the class probability can be seen as the model confidence or probability of correctness, and can be used as a measure for predictive uncertainty at the pixel level [13]. A common metric used to measure calibration performance is the Brier score [24], a proper scoring rule whose optimal value corresponds to a perfect prediction. In other words, a system that is both perfectly calibrated and perfectly discriminative will have a Brier score of zero. In the context of image segmentation, for an image with N pixels (voxels), the Brier score can be defined as:

$$Br = \frac{1}{N} \sum_{i=1}^{N} \frac{1}{|\mathcal{C}|} \sum_{k=1}^{|\mathcal{C}|} \left(p(y_i = k|\mathbf{x}; \mathbf{w}) - \mathbb{1}[\bar{y}_i = k] \right)^2, \tag{6}$$

where $\mathbb{1}[\bar{y}_i = k]$ is the indicator function whose value is 1 when $\bar{y}_i$ (the ground truth class for pixel i) is equal to k, and 0 otherwise.

Stratified Brier Score. In problems with highly imbalanced classes (such as brain lesion segmentation where most of the pixels are background), calibration may be good overall but poor for the minority class. In this case, the majority class will dominate and miscalibration in the class of interest will not be reflected in the standard Brier score. In [25], the authors proposed the stratified Brier score to measure calibration in binary classification problems with high imbalance. Here, we extend this concept to the segmentation task and propose to measure the stratified Brier score individually per-class, treating every structure of interest as a binary segmentation problem, to account for mis-calibration in the minority classes. For a given image with ground truth segmentation $\bar{\mathbf{y}}$, we construct the *stratified* Brier score for the class k, Br^k, by computing it only in the subset of pixels $\mathcal{P}_k = \{p : \bar{y}_p = k\}$, i.e. pixels whose ground truth label is k. The problem is therefore binarized considering all the other classes within a single background class. The formulation of the stratified Brier score Br^k is given by:

$$Br^k = \frac{1}{|\mathcal{P}_k|} \sum_{i \in \mathcal{P}_k} \left(p(y_i = k | \mathbf{x}; \mathbf{w}) - \mathbb{1}[\bar{y}_i = k] \right)^2. \tag{7}$$

Segmentation Evaluation. In addition to the metrics presented to measure the model miscalibration, we resort to the common Dice Similarity coefficient (DSC) to assess the quality of the segmentations.

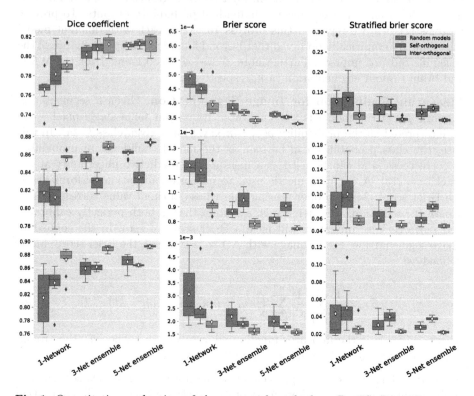

Fig. 1. Quantitative evaluation of the proposed method on BraTS: Rows from top to bottom show results for: (i) enhanced tumor; (ii) tumor core; (iii) whole tumor. Boxplots show mean and standard deviation for predictions obtained with individual models, 3-networks ensembles and 5-networks ensembles.

5   Results and Discussion

We present quantitative results for brain tumor and WMH segmentation in Fig. 1 and Fig. 2, respectively. We can observe that the model just integrating *self-orthogonality* outperforms the baseline model across groups and metrics. This improvement is further stressed when explicitly enforcing model diversity by incorporating the *inter − orthogonality* term computed between pairs of models during sequential training. In particular, our proposed learning strategy consistently leads to improvement on both model calibration and segmentation performance and across the two different segmentation tasks. This demonstrates the benefits of the proposed learning strategy to generate well-calibrated and highly performing segmentation models.

Another important observation is related to differences between Brier and stratified Brier scores. Given the small Brier value reported for all the models (less than 1^{-3}), one could think that these models are well calibrated. However, when having a closer look at the stratified Brier score, the higher value (more than 0.1 in most of the cases) reflects calibration issues. This results from the majority class dominating the traditional Brier score. Thus, studying the stratified Brier score allows us to better appreciate the improvements obtained by the inter-orthogonal ensemble with respect to the other models.

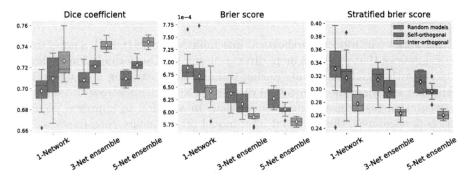

Fig. 2. Quantitative evaluation of the proposed method for WMH segmentation. Boxplots show mean and standard deviation for predictions obtained with individual models, 3-networks ensembles and 5-networks ensembles.

In addition, we depict in Fig. 3 the variance in the predictions across the components of the ensemble trained with and without the orthogonal losses, demonstrating that the orthogonal constraints bring diversity to the ensemble. As expected, we found that integrating the inter-orthogonal objective term leads to an increase in the variance of the predictions compared to the baseline models.

Last but not least, it is surprising to see that the inter-orthogonal regularization term boosts the performance even when considering the individual models. We believe that this is due to a regularization effect of the inter-orthogonal term, which implicitly reduces the complexity of the model by adding orthogonality constraints with respect to specific points in the parameter space, i.e. the weights of the previously trained models.

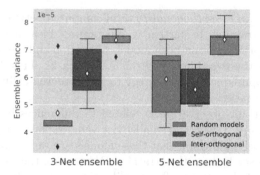

Fig. 3. Quantitative evaluation of the ensembles diversity. Boxplots depict the mean and standard deviation of the variance in the predictions when training the ensemble with and without the proposed orthogonal losses.

6   Conclusions

In this work we introduced Orthogonal Ensemble Networks (OEN), a novel training framework that produces more diverse ensembles. Our formulation explicitly imposes orthogonal constraints during training by integrating a regularization term that enhances the inter-model diversity. Experiments across two different segmentation tasks have demonstrated that, in addition to improved segmentation performance, the proposed inter-model orthogonality constraints reduce miscalibration, leading to more reliable predictions.

Acknowledgments. The authors gratefully acknowledge NVIDIA Corporation with the donation of the GPUs used for this research, and the support of UNL (CAID-0620190100145LI, CAID-50220140100084LI) and ANPCyT (PICT 2018-03907). This research was enabled in part by support provided by Calcul Québec and Compute Canada.

References

1. Guo, C., Pleiss, G., Sun, Y., Weinberger, K.Q.: On calibration of modern neural networks. In: ICML, pp. 1321–1330 (2017)
2. Lakshminarayanan, B., Pritzel, A., Blundell, C.: Simple and scalable predictive uncertainty estimation using deep ensembles. In: NeurIPS (2017)
3. Stickland, A.C., Murray, I.: Diverse ensembles improve calibration. In: ICML 2020 Workshop on Uncertainty and Robustness in Deep Learning (2020)
4. Ovadia, Y., et al.: Can you trust your model's uncertainty? Evaluating predictive uncertainty under dataset shift. In: NeurIPS (2019)
5. Gal, Y., Ghahramani, Z.: Dropout as a Bayesian approximation: Representing model uncertainty in deep learning. In: International Conference on Machine Learning, pp. 1050–1059 (2016)
6. Wen, Y., Tran, D., Ba, J.: Batchensemble: an alternative approach to efficient ensemble and lifelong learning. In: ICLR (2020)

7. Wenzel, F., Snoek, J., Tran, D., Jenatton, R.: Hyperparameter ensembles for robustness and uncertainty quantification. In: NeurIPS (2020)
8. Sinha, S., Bharadhwaj, H., Goyal, A., Larochelle, H., Garg, A., Shkurti, F.: Dibs: diversity inducing information bottleneck in model ensembles. In: AAAI (2020)
9. Kim, W., Goyal, B., Chawla, K., Lee, J., Kwon, K.: Attention-based ensemble for deep metric learning. In: ECCV, pp. 736–751 (2018)
10. Yang, H., et al.: DVERGE: diversifying vulnerabilities for enhanced robust generation of ensembles. arXiv preprint arXiv:2009.14720 (2020)
11. Milletari, F., Navab, N., Ahmadi, S.A.: V-Net: fully convolutional neural networks for volumetric medical image segmentation. In: 2016 Fourth International Conference on 3D Vision (3DV), pp. 565–571 IEEE (2016)
12. Sander, J., de Vos, B.D., Wolterink, J.M., Išgum, I.: Towards increased trustworthiness of deep learning segmentation methods on cardiac MRI. In: Medical Imaging 2019: Image Processing, vol. 10949, p. 1094919. International Society for Optics and Photonics (2019)
13. Mehrtash, A., Wells, W.M., Tempany, C.M., Abolmaesumi, P., Kapur, T.: Confidence calibration and predictive uncertainty estimation for deep medical image segmentation. IEEE Trans. Med. Imaging 39(12), 3868–3878 (2020)
14. Kamnitsas, K., et al.: Ensembles of multiple models and architectures for robust brain tumour segmentation. In: Crimi, A., Bakas, S., Kuijf, H., Menze, B., Reyes, M. (eds.) BrainLes 2017. LNCS, vol. 10670, pp. 450–462. Springer, Cham (2018). https://doi.org/10.1007/978-3-319-75238-9_38
15. Feng, X., Tustison, N.J., Patel, S.H., Meyer, C.H.: Brain tumor segmentation using an ensemble of 3D U-Nets and overall survival prediction using radiomic features. Front. Comput. Neurosci. 14, 25 (2020)
16. Ma, T., et al.: Ensembling low precision models for binary biomedical image segmentation. In: Proceedings of the IEEE/CVF Winter Conference on Applications of Computer Vision, pp. 325–334 (2021)
17. Wang, J., Chen, Y., Chakraborty, R., Yu, S.X.: Orthogonal convolutional neural networks. In: Proceedings of the IEEE/CVF Conference on Computer Vision and Pattern Recognition, pp. 11505–11515 (2020)
18. Ayinde, B.O., Inanc, T., Zurada, J.M.: Regularizing deep neural networks by enhancing diversity in feature extraction. IEEE Trans. Neural Networks Learning Syst. 30(9), 2650–2661 (2019)
19. Bakas, S., et al.: Advancing the cancer genome atlas glioma MRI collections with expert segmentation labels and radiomic features. Sci. Data 4, 170117 (2017)
20. Bakas, S., et al.: Identifying the best machine learning algorithms for brain tumor segmentation, progression assessment, and overall survival prediction in the brats challenge. arXiv preprint arXiv:1811.02629 (2018)
21. Menze, B.H., et al.: The multimodal brain tumor image segmentation benchmark (brats). IEEE Trans. Med. Imaging 34(10), 1993–2024 (2014)
22. Kuijf, H.J., et al.: Standardized assessment of automatic segmentation of white matter hyperintensities and results of the WMH segmentation challenge. IEEE Trans. Med. Imaging 38(11), 2556–2568 (2019)
23. Zhang, Z., Liu, Q., Wang, Y.: Road extraction by deep residual U-Net. IEEE Geosci. Remote Sens. Lett. 15(5), 749–753 (2018)
24. Brier, G.W.: Verification of forecasts expressed in terms of probability. Mon. Weather Rev. 78(1), 1–3 (1950)
25. Wallace, B.C., Dahabreh, I.J.: Improving class probability estimates for imbalanced data. Knowl. Inf. Syst. 41(1), 33–52 (2013). https://doi.org/10.1007/s10115-013-0670-6

Learning to Predict Error for MRI Reconstruction

Shi Hu[1(✉)], Nicola Pezzotti[2,3], and Max Welling[1]

[1] University of Amsterdam, Amsterdam, The Netherlands
s.hu@uva.nl
[2] Philips Research, Eindhoven, The Netherlands
[3] Eindhoven University of Technology, Eindhoven, The Netherlands

Abstract. In healthcare applications, predictive uncertainty has been used to assess predictive accuracy. In this paper, we demonstrate that predictive uncertainty estimated by the current methods does not highly correlate with prediction error by decomposing the latter into random and systematic errors, and showing that the former is equivalent to the variance of the random error. In addition, we observe that current methods unnecessarily compromise performance by modifying the model and training loss to estimate the target and uncertainty jointly. We show that estimating them separately without modifications improves performance. Following this, we propose a novel method that estimates the target labels and magnitude of the prediction error in two steps. We demonstrate this method on a large-scale MRI reconstruction task, and achieve significantly better results than the state-of-the-art uncertainty estimation methods.

Keywords: Deep learning · Uncertainty · MRI reconstruction

1 Introduction

Healthcare has been increasingly facilitated by artificial intelligence technologies [3]. Uncertainty is ubiquitous in these technologies, and it can arise due to randomness or imperfect knowledge [15], such as the disagreement among human annotators, missing entries in electronic health records, or occlusions in MRIs. Kennedy and O'Hagan [13] lists six sources of uncertainty that affect predictive outcomes, which include data noise, input variability, model structure and parameters, optimization, and interpolation. Unfortunately, the current methods in deep learning [12,19] quantify only two of them, which are data noise and model parameters; in addition, they use the sum of the two uncertainties to estimate predictive uncertainty. Lastly, they need to modify the model structure and training objective to estimate the target and uncertainty jointly.

Electronic supplementary material The online version of this chapter (https://doi.org/10.1007/978-3-030-87199-4_57) contains supplementary material, which is available to authorized users.

M. de Bruijne et al. (Eds.): MICCAI 2021, LNCS 12903, pp. 604–613, 2021.
https://doi.org/10.1007/978-3-030-87199-4_57

In this paper, we show that their estimated predictive uncertainty cannot highly correlate with prediction error; in addition, if we estimate the target and uncertainty separately, the performance improves since the model and training loss are unchanged, and regularization schemes such as early stopping can have separate effects on the two estimates. Following this, we propose a novel two-step method where we train one deep model to estimate the target, and another the magnitude of the prediction error. We demonstrate this method on a large-scale MRI reconstruction task, and achieve significantly better results than the state-of-the-art uncertainty estimation methods.

2   Notation

We denote an input by x, a noisy target by $y(x)$, and the true target by $h(x)$. The noise $\epsilon(x)$ is assumed to be additive and Gaussian, i.e. $y(x) = h(x) + \epsilon(x)$ where $\epsilon(x) \sim \mathcal{N}(0, \sigma^2(x))$, and we will refer to the noise level $\sigma(x)$ as "sigma". Further, the estimates are marked with the caret symbol, e.g., $\hat{h}(x)$ is an estimate of $h(x)$. Unless otherwise stated, the model structure $\mathcal{M}$, training data $\mathcal{D}$ and optimizer $\mathcal{O}$ are all fixed. The expectation of any estimate is taken over the random seed s. For clarity of notation, we omit these symbols when possible, e.g., the expected estimate of the true target $\mathbb{E}_s\left[\hat{h}(x; s, \mathcal{M}, \mathcal{D}, \mathcal{O})\right]$ is abbreviated to $\mathbb{E}[\hat{h}(x)]$.

3   An Anatomy of Prediction Error

To estimate the magnitude of the prediction error (or squared error) on unseen data, we first decompose the prediction error into systematic and random errors, and analyze the two separately:

$$\underbrace{y(x) - \hat{h}(x)}_{\text{prediction error}} = \underbrace{h(x) - \mathbb{E}[\hat{h}(x)]}_{\text{systematic error}} + \underbrace{\left[y(x) - h(x)\right] + \left[\mathbb{E}[\hat{h}(x)] - \hat{h}(x)\right]}_{\text{random error}}. \quad (1)$$

Systematic error is the difference between the true target and the expected estimate of the true target [11]. In deep learning, it can be reduced if we have better knowledge regarding the model structure, or can diminish the gap between the global optimum and the local optimum obtained by the optimizer. The lack of knowledge of the model structure and optimization process reflects two types of epistemic uncertainties, which are called structural and algorithmic uncertainties [13]. Unfortunately, to the best of our knowledge, neither has been quantified by the current deep learning methods.

The total random error equals the prediction error minus the systematic error [11]. The first random error $y(x) - h(x) = \epsilon(x)$ represents the label noise, which is unpredictable. However, its conditional variance is equivalent to that of the noisy targets, i.e., $\sigma^2(x) = \text{Var}[\epsilon(x)] = \text{Var}[y(x) - h(x)] = \text{Var}[y(x)]$. If we can access multiple noisy targets per x, we can predict their variance through supervised learning [10]. Otherwise, to estimate $\sigma(x)$, we need to assume that

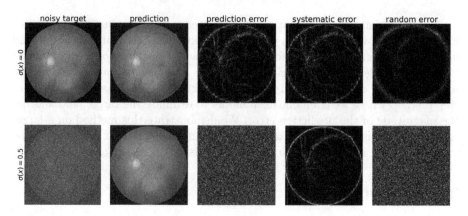

Fig. 1. The error decomposition for a super-resolution prediction (the retina image is shown in the luminance channel in YCbCr colour space). The target noise is part of the random error, which does not affect the systematic error.

it is smooth over x, then predict it along with the true target [20]. Likewise, the second random error $\mathbb{E}[\hat{h}(x)] - \hat{h}(x)$ is also unpredictable, but its variance $\mathrm{Var}\left[\mathbb{E}[\hat{h}(x)] - \hat{h}(x)\right] = \mathrm{Var}[\hat{h}(x)]$ can be estimated using Monte Carlo dropout [5,12] or ensemble methods [19]. Previous works [5,12] refer to the two variances as aleatoric and model (or epistemic) uncertainty.

In summary, systematic error is deterministic, and predictable if the true target is a smooth function [2]. On the other hand, random error is unpredictable, so it can be best estimated using its expectation. An illustration of the error decomposition on a retina image is shown in Fig. 1.

4   What Does the Predictive Uncertainty Quantify?

Due to the random error, the magnitude of the prediction error cannot be precisely known, so it can be best estimated with the expected squared error that integrates out the randomness in the label noise and random seed. This error can be decomposed using the bias-variance decomposition [9] as the following:

$$\underbrace{\mathbb{E}\left[(y(x) - \hat{h}(x))^2\right]}_{\text{expected squared error}} = \underbrace{\sigma^2(x) + \mathrm{Var}[\hat{h}(x)]}_{\text{variance of random error}} + \underbrace{(h(x) - \mathbb{E}[\hat{h}(x)])^2}_{\text{squared systematic error}}, \qquad (2)$$

which means it is influenced by both random and systematic errors. However, several recent works [4,10,12,19] estimate predictive uncertainty by combining the uncertainties in data noise and model parameters under the law of total variance as follows:

$$\underbrace{\mathrm{Var}[\hat{y}(x)]}_{\text{predictive uncertainty}} = \underbrace{\mathbb{E}[\hat{\sigma}^2(x)]}_{\text{aleatoric uncertainty}} + \underbrace{\mathrm{Var}[\hat{h}(x)]}_{\text{model uncertainty}}. \qquad (3)$$

This predictive uncertainty is equivalent to the variance of the random error, and therefore does not contain the systematic error. As a result, it underestimates the squared error on average.

5   A Two-Step Estimation Method

We have shown that the current predictive uncertainty cannot highly correlate with the squared error, especially when the systematic error is high. In this section, we introduce a simple method that estimates the true target and expected squared error in two steps. We assume that for a given task, the predictive model and its training objective are known, and our method is as follows. First, we train the model to estimate only the target. After training, for each training input x, we compute the squared error $e^2(x) = (y(x) - \hat{h}(x))^2$ as an unbiased estimator of its expectation $\mathbb{E}[e^2(x)]$. Then, we train the same model from scratch to estimate $e^2(x)$, where the loss function can simply be the L_1 or L_2 loss.

This two-step method has two benefits. First, to estimate aleatoric uncertainty, the current methods [12,19] need to add a second branch to the model to output $\hat{\sigma}^2(x)$, and incorporate it into the training objective (we refer to this as a "two-head" model). However, it is challenging to incorporate $\hat{\sigma}^2(x)$ into complex objective functions, such as the structural similarity index measure (SSIM) loss [29] commonly used in the MRI reconstruction models [22,23]. Furthermore, if the original training objective is a combination of multiple losses, it is non-trivial to include $\hat{\sigma}^2(x)$ without affecting the target prediction accuracy. In comparison, the two-step method does not need to modify the model or training loss. Second, the true target and aleatoric uncertainty are independent quantities, but the two-head model estimates them jointly using the same set of hyperparameters, including the number of epochs. Since the two estimands can have different magnitudes, there is no guarantee that they will reach their best estimates simultaneously. The two-step method can avoid this problem by applying early stopping, which prevents overfitting in each step.

6   Experiments

6.1   Datasets

Retina. We use the diabetic retinopathy dataset[1] for a synthetic single image super-resolution task, where we predict the high resolution (HR) image from its low resolution (LR) counterpart. We randomly sample 500 good-quality square images and resize each to 255×255 as the HR image, and use the downsampled 85×85 image as the LR counterpart (i.e., the upscaling factor is 3). We split this dataset into 200/100/200 training/validation/test images.

---

[1] https://www.kaggle.com/c/diabetic-retinopathy-detection.

FastMRI. The fastMRI dataset [30] contains fully anonymized clinical MR images and raw MR measurements. We use the multi-coil knee dataset for a reconstruction task, where we predict the fully sampled MR image from its undersampled image with 4- or 8-time acceleration. The dataset contains a training, validation, test and challenge set, but only the first two provide fully sampled data, which are used for evaluation. Hence, we randomly split the validation set into a validation and test set. After the split, there are 973/59/140 training/validation/test MRI volumes.

6.2   Single Image Super-Resolution

We run a set of synthetic experiments on a single image super-resolution task using the retina dataset, where the HR images used as the training targets are corrupted by the pixel-wise Gaussian noise $\epsilon(x) \sim \mathcal{N}(0, \sigma^2(x))$. For ease of comparison, we use the efficient sub-pixel convolutional neural network (ESPCN) [25], which is a feedforward network with the mean squared error (MSE) loss: $(y(x) - \hat{h}(x))^2$, and it processes the images in the luminance channel in YCbCr colour space.

We compare the prediction accuracies of the true target and sigma estimates among the original, two-head, and two-head ($2\times$) models. The first two use the same number of parameters (excluding the extra branch in the two-head), and the last uses twice as many proportional to the model structure. To implement the two-head models, we duplicate the last layer of the ESPCN and append both to the penultimate layer. The uncertainty estimate is incorporated into the MSE loss, which becomes the Gaussian negative log-likelihood (NLL) loss: $\frac{(y(x)-\hat{h}(x))^2}{\hat{\sigma}^2(x)} + \log \hat{\sigma}^2(x)$. To avoid the division by zero error, we use the numerical stable implementation as in [12]. We train 1000 epochs using the Adam optimizer [14] with learning rate 10^{-4}, though we observe that learning rates do not affect the relative performance.

The first row of Fig. 2 compares the test errors in the true target estimate by the three models. When the target noise is negligible, all three have relatively stable error curves, and the original model consistently outperforms the others. But when the target noise is high, the original model starts to overfit within 10 epochs, and the accuracy declines faster than the two-heads. This confirms that the uncertainty estimate $\hat{\sigma}^2(x)$ regularizes the training loss, as suggested by [12]. However, before overfitting, the original model achieved the lowest error among all. Therefore, rather than training it for a predetermined number of epochs, we apply early stopping with the validation MSE criterion. As shown in these plots, early stopping effectively prevents overfitting and leads to better estimates. Further, the second row of Fig. 2 shows the test errors in the target and sigma estimates by the two-head ($2\times$) model. As indicated by the two vertical lines, the best epochs are substantially separated, and the estimand with the smaller magnitude gets to the optimal faster. Therefore, even with early stopping, we cannot simultaneously obtain both optimal estimates. As a comparison, we train a second ESPCN with the same NLL loss to estimate only sigma (i.e., the model

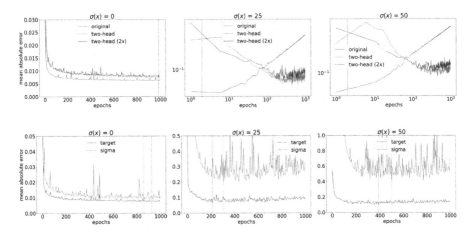

Fig. 2. First row: test errors in the true target estimate. The stopping epoch suggested by early stopping for the original model is shown by the vertical dotted line. **Second row**: test errors in the target and sigma estimates by the two-head (2×) model, and best epochs are indicated by the two vertical dotted lines.

structure is unchanged). In this loss, the target estimates are obtained using the first (original) ESPCN in the previous experiment. For fairness, we use the same learning rate in the second model, and apply early stopping with the validation NLL criterion to all models. Table 1 compares the test mean absolute error (MAE) in the sigma estimate, and the original model achieves the best results at all noise levels (each result is computed with 4 random seeds).

We have shown that estimating the true target and aleatoric uncertainty in two steps outperforms the joint model with parameter sharing; in addition, doubling the number of parameters for the two-head model does not have a significant impact on the results.

Table 1. Test MAE in the sigma estimate (⋆ means statistically significant).

$\sigma(x)$	0	25	50
Original	**0.0085 ± 0.0001***	**0.1896 ± 0.0114**	**0.3880 ± 0.0109**
Two-head	0.0108 ± 0.0004	0.1909 ± 0.0053	0.3983 ± 0.0375
Two-head (2×)	0.0101 ± 0.0002	0.1966 ± 0.0072	0.4088 ± 0.0167

6.3   MRI Reconstruction

We run the MRI reconstruction experiment on the fastMRI dataset, where no synthetic noise is added to the targets. We use the Adaptive-CS-Net (ACSNet)

[22], which won the multi-coil track of the 2019 fastMRI challenge[2]. It uses a recurrent neural network to iteratively improve the target estimate in k-space, and its training loss is a combination of the L_1 and SSIM losses in the image space. There are two output channels in each iteration, which represent the real and imaginary components of the k-space estimate. A diagram of the model is shown in Fig. 1 of their paper.

We compare with three state-of-the-art predictive uncertainty quantification methods, which are Monte Carlo dropout (MC-D) [5], Kendall and Gal (K&G) [12] and deep ensembles (DE) [19]. Since dropout [26] is not used in the ACSNet, we follow [30] and insert a dropout layer after each convolution block. To implement the two-head model, it is not feasible to duplicate the last layer due to the recurrent model structure; therefore, we use the model's last two outputs as the image and uncertainty estimates, and train with the NLL loss. For MC-D and K&G, we use dropout rates 0.05, 0.1, 0.2 and 0.5, and draw 50 samples for evaluation. For DE, we follow [19] and train an ensemble of 5 models. We use the 16GB Nvidia Tesla V100 GPUs, and training one model takes 62 h and 12 GB memory. Due to hardware limits, we cannot implement the fine-tuning step in the original paper; in addition, we cannot double the number of parameters for K&G and DE, or run the adversarial training [7] for DE as it needs to store the gradients twice per input. However, we have run these experiments on a U-net [24] model[3] and found the memory-intensive experiments bring little extra benefits. For our method, we follow [19] and train an ensemble of 5 models to estimate the target, and 1 model to estimate the absolute error with the L_1 loss. For an ablation study, we also train 1 model to estimate the target and 1 model the absolute error (results are marked with "Ours (one)" in Table 2). For ease of comparison, we use Adam with learning rate 10^{-4} for 40 epochs, 10^{-5} for 10 more epochs, and apply early stopping with the validation L_1 criterion to all models. Some of the fully sampled test MR images are noisy, but we cannot access the true targets (i.e. *clean* and fully sampled images), so we use these noisy targets as ground truths for target evaluation. We note this makes the error prediction very challenging, since the noise levels greatly vary among test images. We follow [30] and report the NMSE, PSNR and SSIM results. For error evaluation, we report the absolute error prediction accuracy in L_1 and MSE. Lastly, since baseline methods do not estimate the systematic error, for fairness, we calibrate their uncertainty estimates on the validation set by optimizing: $\alpha_\star = \min_\alpha \sum_x (|y(x) - \hat{h}(x)| - \alpha \hat{u}(x))^2$, where $\hat{u}(x)$ is the square root of the estimated predictive uncertainty, then the calibrated estimate is $\alpha_\star \hat{u}(x)$.

Table 2 shows our method achieves the best results in all metrics[4]. Additionally, we have three remarks. First, MC-D estimates only model uncertainty, and the accuracy in error estimates improves as the dropout rate decreases (same for K&G). This suggests the second random error in Eq. 1 is very small, which

---

[2] https://fastmri.org.

[3] The original U-net code is from: https://github.com/facebookresearch/fastMRI.

[4] The high standard deviations (SD) are due to high image noise (we note that most prior works on this dataset do not report SD, including [30]).

Table 2. Target and absolute error estimation results for 4× (top) and 8× (bottom) accelerated MRIs. Dropout rates are indicated after "MC-D" and "K&G". Best results are in bold.

Method	Target (4×)			Absolute Error (4×)	
	NMSE (10^{-3}) ↓	PSNR ↑	SSIM ↑	L_1 (10^{-6}) ↓	MSE (10^{-12}) ↓
MC-D 0.05	7.12 ± 5.97	38.46 ± 2.97	0.912 ± 0.062	1.98 ± 0.92	9.62 ± 14.91
MC-D 0.1	7.73 ± 6.14	38.07 ± 2.97	0.909 ± 0.063	2.06 ± 0.89	9.90 ± 13.83
MC-D 0.2	8.28 ± 6.29	37.58 ± 2.76	0.905 ± 0.061	2.25 ± 0.99	11.92 ± 15.67
MC-D 0.5	16.60 ± 17.88	35.91 ± 3.83	0.888 ± 0.072	2.81 ± 1.05	17.69 ± 19.84
K&G 0.05	7.38 ± 5.78	38.11 ± 2.74	0.907 ± 0.063	2.51 ± 0.96	14.55 ± 2.25
K&G 0.1	8.69 ± 6.72	37.24 ± 2.53	0.898 ± 0.063	2.90 ± 1.10	19.70 ± 30.72
K&G 0.2	10.47 ± 7.21	36.22 ± 2.28	0.889 ± 0.062	3.39 ± 1.30	27.51 ± 40.92
K&G 0.5	43.13 ± 11.33	29.67 ± 2.34	0.778 ± 0.070	9.85 ± 4.14	174.60 ± 254.10
DE	7.58 ± 5.41	37.99 ± 2.74	0.905 ± 0.063	1.98 ± 1.01	10.36 ± 17.22
Ours (one)	6.56 ± 6.00	38.87 ± 3.03	0.915 ± 0.063	1.77 ± 0.82	7.77 ± 13.10
Ours	**6.34** ± 5.92	**39.08** ± 3.11	**0.917** ± 0.062	**1.74** ± 0.79	**7.32** ± 12.59

Method	Target (8×)			Absolute Error (8×)	
	NMSE (10^{-2}) ↓	PSNR ↑	SSIM ↑	L_1 (10^{-6}) ↓	MSE (10^{-11}) ↓
MC-D 0.05	1.23 ± 0.67	35.51 ± 2.43	0.878 ± 0.064	2.74 ± 1.53	2.31 ± 3.49
MC-D 0.1	1.32 ± 0.70	35.22 ± 2.40	0.875 ± 0.063	2.84 ± 1.56	2.34 ± 3.35
MC-D 0.2	1.52 ± 0.76	34.55 ± 2.38	0.867 ± 0.063	3.10 ± 1.65	2.93 ± 4.02
MC-D 0.5	2.92 ± 1.96	32.09 ± 2.83	0.834 ± 0.074	4.20 ± 2.26	4.91 ± 6.45
K&G 0.05	1.54 ± 0.69	34.41 ± 2.30	0.863 ± 0.064	4.50 ± 1.79	5.11 ± 7.94
K&G 0.1	1.91 ± 0.86	33.40 ± 2.31	0.849 ± 0.062	5.24 ± 2.02	6.95 ± 10.39
K&G 0.2	2.23 ± 0.91	32.70 ± 2.31	0.838 ± 0.061	5.81 ± 2.33	8.79 ± 13.26
K&G 0.5	7.37 ± 2.84	27.48 ± 2.66	0.720 ± 0.074	12.82 ± 5.76	35.86 ± 56.94
DE	1.55 ± 0.69	34.35 ± 2.30	0.861 ± 0.063	3.01 ± 1.83	3.04 ± 4.52
Ours (one)	1.21 ± 0.70	35.62 ± 2.48	0.880 ± 0.065	2.54 ± 1.46	2.28 ± 3.95
Ours	**1.13** ± 0.68	**35.98** ± 2.52	**0.884** ± 0.065	**2.44** ± 1.38	**2.00** ± 3.44

makes sense since we use 50 samples for evaluation. Nevertheless, there is a gap between their results and ours, as we also estimate the systematic error and target noise. Second, K&G and DE do not perform well in this task, since they modified the training objective and used the last two outputs for image and uncertainty estimates, which changes the dynamics of the model; in addition, they do not estimate the systematic error. Finally, the calibration improves all baseline results, except for MC-D 0.05 (4×). In the supplementary material, we show the uncalibrated baseline results in Table S1, and compare the *calibrated* baseline error plots and ours on a random 8× test MRI in Fig. S1.

7   Related Work

To assess the predictive quality without ground truth, the reverse classification accuracy framework [28], and a regression algorithm using shape and appearance features [16] have been proposed, though these methods are limited to the segmentation task. Predictive quality can also be evaluated by the calibrated confidence, which estimates the frequency of the target falling in a given interval [8,17,21]. In addition, the generalization error, which measures the prediction accuracy on unseen data, can be estimated by cross-validation, but this is done on a separate test set [9]. Further, when the estimands are related to each other, the joint model with parameter sharing can be effective [6]. Lastly, other interesting uncertainty estimation methods in medical imaging include [1,18,27], etc.

8   Conclusion

Current methods in deep learning estimate predictive uncertainty by the sum of data and model uncertainties. In this work, we show this estimate cannot highly correlate with prediction error; in addition, estimating the target and uncertainty separately outperforms the joint model by the current methods. Following this, we propose a novel two-step method that can accurately estimate the target and magnitude of the prediction error on unseen in-distribution data. For future work, we would like to extend this method to tackle out-of-distribution detection.

Acknowledgements. We thank Tony O'Hagan, Yoshua Bengio and the anonymous reviewers for helpful discussions. This research was supported by the NWO Perspective Grant DLMedIA and in-cash and in-kind contributions by Philips.

References

1. Adler, J., Öktem, O.: Deep Bayesian inversion. arXiv preprint arXiv:1811.05910 (2018)
2. Cybenko, G.: Approximation by superpositions of a sigmoidal function. Math. Control Signals Syst. **2**, 303–314 (1989)
3. Davenport, T., Kalakota, R.: The potential for artificial intelligence in healthcare. Future Healthc. J. **6**, 94 (2019)
4. Depeweg, S., Hernández-Lobato, J.M., Doshi-Velez, F., Udluft, S.: Decomposition of uncertainty in Bayesian deep learning for efficient and risk-sensitive learning. In: ICML (2018)
5. Gal, Y., Ghahramani, Z.: Dropout as a Bayesian approximation: representing model uncertainty in deep learning. In: ICML (2016)
6. Goodfellow, I., Bengio, Y., Courville, A.: Deep Learning. MIT Press, Cambridge (2016)
7. Goodfellow, I.J., Shlens, J., Szegedy, C.: Explaining and harnessing adversarial examples. In: ICLR (2015)
8. Guo, C., Pleiss, G., Sun, Y., Weinberger, K.Q.: On calibration of modern neural networks. In: ICML (2017)
9. Hastie, T., Tibshirani, R., Friedman, J.: The Elements of Statistical Learning. SSS, Springer, New York (2009). https://doi.org/10.1007/978-0-387-84858-7
10. Hu, S., Worrall, D., Knegt, S., Veeling, B., Huisman, H., Welling, M.: Supervised uncertainty quantification for segmentation with multiple annotations. In: Shen, D., et al. (eds.) MICCAI 2019. LNCS, vol. 11765, pp. 137–145. Springer, Cham (2019). https://doi.org/10.1007/978-3-030-32245-8_16

11. Joint Committee for Guides in Metrology: JCGM 100:2008. evaluation of measurement data - guide to the expression of uncertainty in measurement. Technical report (2008)
12. Kendall, A., Gal, Y.: What uncertainties do we need in Bayesian deep learning for computer vision? In: NIPS (2017)
13. Kennedy, M.C., O'Hagan, A.: Bayesian calibration of computer models. J. Roy. Stat. Soc. Ser. B (Stat. Methodol.) **63**(3), 425–464 (2001)
14. Kingma, D.P., Ba, J.: Adam: a method for stochastic optimization. In: ICLR (2015)
15. Kiureghian, A.D., Ditlevsen, O.: Aleatory or epistemic? does it matter? Struct. Saf. **31**, 105–112 (2009)
16. Kohlberger, T., Singh, V., Alvino, C., Bahlmann, C., Grady, L.: Evaluating segmentation error without ground truth. In: Ayache, N., Delingette, H., Golland, P., Mori, K. (eds.) MICCAI 2012. LNCS, vol. 7510, pp. 528–536. Springer, Heidelberg (2012). https://doi.org/10.1007/978-3-642-33415-3_65
17. Kuleshov, V., Fenner, N., Ermon, S.: Accurate uncertainties for deep learning using calibrated regression. In: ICML (2018)
18. Kwon, Y., Won, J.H., Kim, B.J., Paik, M.C.: Uncertainty quantification using Bayesian neural networks in classification: application to biomedical image segmentation. Comput. Stat. Data Anal. **142** (2019). Article no. 106816
19. Lakshminarayanan, B., Pritzel, A., Blundell, C.: Simple and scalable predictive uncertainty estimation using deep ensembles. In: NIPS (2017)
20. Nix, D.A., Weigend, A.S.: Estimating the mean and variance of the target probability distribution. In: IEEE International Conference on Neural Networks (1994)
21. Ovadia, Y., et al.: Can you trust your model's uncertainty? Evaluating predictive uncertainty under dataset shift. In: NeurIPS (2019)
22. Pezzotti, N., et al.: An adaptive intelligence algorithm for undersampled knee MRI reconstruction. IEEE Access **8**, 204825–204838 (2020)
23. Putzky, P., et al.: i-RIM applied to the fastMRI challenge. arXiv preprint arXiv:1910.08952 (2019)
24. Ronneberger, O., Fischer, P., Brox, T.: U-Net: convolutional networks for biomedical image segmentation. In: Navab, N., Hornegger, J., Wells, W.M., Frangi, A.F. (eds.) MICCAI 2015. LNCS, vol. 9351, pp. 234–241. Springer, Cham (2015). https://doi.org/10.1007/978-3-319-24574-4_28
25. Shi, W., et al.: Real-time single image and video super-resolution using an efficient sub-pixel convolutional neural network. In: CVPR (2016)
26. Srivastava, N., Hinton, G., Krizhevsky, A., Sutskever, I., Salakhutdinov, R.: Dropout: a simple way to prevent neural networks from overfitting. JMLR **15**, 1929–1958 (2014)
27. Tanno, R., et al.: Bayesian image quality transfer with CNNs: exploring uncertainty in dMRI super-resolution. In: Descoteaux, M., Maier-Hein, L., Franz, A., Jannin, P., Collins, D.L., Duchesne, S. (eds.) MICCAI 2017. LNCS, vol. 10433, pp. 611–619. Springer, Cham (2017). https://doi.org/10.1007/978-3-319-66182-7_70
28. Valindria, V.V., et al.: Reverse classification accuracy: predicting segmentation performance in the absence of ground truth. IEEE Trans. Med. Imaging **36**, 1597–1606 (2017)
29. Wang, Z., Bovik, A.C., Sheikh, H.R., Simoncelli, E.P.: Image quality assessment: from error visibility to structural similarity. IEEE Trans. Image Process. **13**, 600–612 (2004)
30. Zbontar, J., et al.: fastMRI: an open dataset and benchmarks for accelerated MRI. arXiv preprint arXiv: 1811.08839 (2018)

Uncertainty-Guided Progressive GANs for Medical Image Translation

Uddeshya Upadhyay[1]([⊠]), Yanbei Chen[1], Tobias Hepp[1,2], Sergios Gatidis[1,2], and Zeynep Akata[1,2]

[1] University of Tübingen, Tübingen, Germany
[2] Max Planck Institute for Intelligent Systems, Tübingen, Germany

Abstract. Image-to-image translation plays a vital role in tackling various medical imaging tasks such as attenuation correction, motion correction, undersampled reconstruction, and denoising. Generative adversarial networks have been shown to achieve the state-of-the-art in generating high fidelity images for these tasks. However, the state-of-the-art GAN-based frameworks do not estimate the uncertainty in the predictions made by the network that is essential for making informed medical decisions and subsequent revision by medical experts and has recently been shown to improve the performance and interpretability of the model. In this work, we propose an uncertainty-guided progressive learning scheme for image-to-image translation. By incorporating aleatoric uncertainty as attention maps for GANs trained in a progressive manner, we generate images of increasing fidelity progressively. We demonstrate the efficacy of our model on three challenging medical image translation tasks, including PET to CT translation, undersampled MRI reconstruction, and MRI motion artefact correction. Our model generalizes well in three different tasks and improves performance over state of the art under full-supervision and weak-supervision with limited data. Code is released here: https://github.com/ExplainableML/UncerGuidedI2I.

Keywords: Image-to-image translation · Uncertainty estimation · Progressive GANs · PET · CT · MRI · Artefact correction

1 Introduction

In the medical domain, each imaging modality reflects particular physical properties of the tissue under examination. This results in images with different dimensionality, spatial resolution, and contrast. Various imaging modalities provide a complimentary stream of information for clinical diagnostics or technical pre and post-processing steps. Moreover, acquiring medical images is susceptible to various kinds of noise and modality-specific artefacts. To remedy these issues, translating images between different domains is of great importance.

Inter-modal image-to-image translation can potentially replace additional acquisition procedures, reducing examination costs and time. Besides, intra-modality image-to-image translation enables complex artefact and noise correction. For example, attenuation correction of positron emission tomography

© Springer Nature Switzerland AG 2021
M. de Bruijne et al. (Eds.): MICCAI 2021, LNCS 12903, pp. 614–624, 2021.
https://doi.org/10.1007/978-3-030-87199-4_58

(PET) data is challenging in situations where no density distribution is available from computed tomography (CT) data, as in the case for stand-alone PET scanners or combined PET/magnetic resonance imaging (MRI). In these situations, the generation of pseudo-CTs from PET data can be helpful. Further examples are related to image reconstruction and/or correction in MRI: Reconstruction of undisturbed artifact-free images is hard to achieve with traditional methods; deep-learning-based image-to-image translation can solve this challenge. In particular, generative adversarial networks (GAN) based on convolutional neural networks (CNN) have proven to provide a high visual quality of the generated synthetic images. However, predictions of GANs can be unreliable, and particularly in medical applications, the quantification of uncertainty is of high importance for the interpretation of the results. In this work, we propose a generic end-to-end model that introduces high-capacity conditional progressive GANs to synthesize high-quality images, using aleatoric uncertainty estimates as the guide to focus on improving image quality in regions where the network is highly uncertain about the prediction. We perform experiments on three challenging and vital medical imaging tasks: PET to CT translation, undersampled MRI reconstruction, and motion correction in MRI. Moreover, we empirically demonstrate the efficacy of our model under weak supervision with limited data.

2   Related Works

Traditional machine learning techniques for medical image translation rely on explicit feature representations [6,12,17,35]. More recently, convolutional neural networks have been proposed for various image translation tasks [3–5,8,13,19] and state-of-the-art performance is achieved by generative adversarial networks [1,2,7,9,15,21,22,29–32]. The existing methods propose conditional GAN architectures with deterministic outputs that typically uses $\mathcal{L}_1/\mathcal{L}_2$-based fidelity loss for the generator assumes a pixel-wise *homoscedasticity* and also assumes the pixel-wise error (i.e., residual) to be *independent and identically distributed* (i.i.d) following a Laplace or Gaussian distribution. This is a limiting assumption as explained in [10,23,25]. While these methods can provide synthetic images of high visual quality, the image content may still deviate significantly from the corresponding ground-truth. This results in overconfidence or misinterpretation with negative consequences, particularly in the medical domain. There have been recent works on quantifying aleatoric and epistemic uncertainty in task-specific medical imaging algorithms like classification, segmentation, super-resolution etc. [14,20,25–27] quantifying it for general image-to-image translation problem largely remains unexplored. Thus, the central motivation of our work is to provide measures of uncertainty for image-to-image translation tasks that can contribute to safe applications of results.

Moreover, recent work has shown that high-capacity generators that are progressive in nature lead to high-quality results as described in [1,2,9]. However, the progressive generation of high-quality images remains unguided without specifically attending to poorly translated regions. Prior works indicate a correlation between estimated uncertainty and prediction error [20,23,33]. We exploit

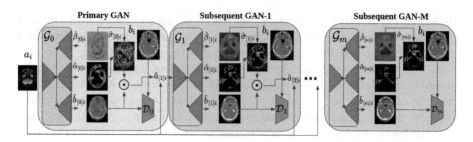

Fig. 1. Uncertainty-guided Progressive GANs (UP-GAN): The primary GAN takes the input image from domain A, while subsequent GANs absorb outputs from the preceding GAN (see Eq. 3 and 4). Explicitly guided by the attention maps, the uncertainty maps are estimated from the preceding GAN.

this relationship for the progressive enhancement of synthetic images, which has not been investigated by prior work before.

3   Uncertainty-Guided Progressive GAN (UP-GAN)

Let A and B be two image domains with a set of images $S_A := \{a_1, a_2...a_n\}$ and $S_B := \{b_1, b_2...b_m\}$ where a_i and b_i represent the ith image from domain A and B respectively. Let each image drawn from an underlying *unknown* probability distribution $\mathcal{P}_{AB}$, i.e., $(a_i, b_i) \sim \mathcal{P}_{AB} \forall i$ have K pixels, and u_{ik} represent the kth pixel of a particular image u_i. Our goal is to learn a mapping from domain A to B $(A \rightarrow B)$ in a paired manner, i.e., learning the underlying conditional distribution $\mathcal{P}_{B|A}$ from the set of given samples $\{(a_i, b_i)\}$, following the distribution $\mathcal{P}_{AB}$. For a given image a_i in domain A, the estimated image in domain B is called $\hat{b}_i$. The pixel wise error is defined as $\epsilon_{ij} = \hat{b}_{ij} - b_{ij}$. While the existing framework models the residual as the i.i.d as described above, we relax that assumption by modelling the residual as non i.i.d variables and learning the optimal distribution from the dataset, as described in the following.

Figure 1 shows our model that consists of cascaded GANs, where each generator is capable of estimating the aleatoric uncertainty, along with generating images. Our solution alleviates the aforementioned limitations of recent methods by modelling the underlying per-pixel residual distribution as *independent* but *non-identically* distributed *zero-mean generalized Gaussian distribution* (GGD) as in [23], where the network learns to predict the optimal *scale* (α) and *shape* (β) of the GGD for every pixel, Therefore, $\hat{b}_{ij} = b_{ij} + \epsilon_{ij}$ with, $\epsilon_{ij} \sim GGD(\epsilon; 0, \alpha_{ij}, \beta_{ij}) \equiv \beta_{ij}(2\alpha_{ij}\Gamma(\beta_{ij}^{-1}))^{-1} \exp\left(-\alpha_{ij}^{-1}|\epsilon|^{\beta_{ij}}\right)$. We generate images in multiple phases, with each phase generating output images along with the aleatoric uncertainty estimates. The outputs from one phase serve as the input to the subsequent GAN in the next phase, explicitly guided by the attention map derived from uncertainty estimates. Importantly, this uncertainty-based guidance enforces the model to focus on refining the uncertain regions that are likely to be poorly synthesized, resulting in progressively improving quality.

Our framework is composed of a sequence of M GANs, where the mth GAN is represented by a pair of networks, generator and discriminator, given by, $(\mathcal{G}_m(\cdot;\theta_m), \mathcal{D}_m(\cdot;\phi_m))$. Both the generator and discriminator can have arbitrary network architecture as long as generator can estimate *aleatoric uncertainty* as described in [23]. We choose all the discriminators to be the patch discriminators from [7] and generators to be modified U-Net [16], where the head is split into three to estimate the parameters of the GGD as shown in Fig. 1 and in [23].

Primary GAN. We train the first GAN $(\mathcal{G}_0)$ using the dataset S_A and S_B. The predictions of the generator are given by $(\hat{\alpha}_{[0]i}, \hat{\beta}_{[0]i}, \hat{b}_{[0]i})$. The network is trained with an adaptive fidelity loss function $\mathcal{L}_{\alpha\beta}^G$ [23] and an adversarial loss $\mathcal{L}_{adv}^G$ [36], combined as $\mathcal{L}_{tot}^G$ for the generator $(\mathcal{G}_0(\cdot;\theta_0) : A \to B)$:

$$\mathcal{L}_{\alpha\beta}^G(\hat{b}_{[0]i}, \hat{\alpha}_{[0]i}, \hat{\beta}_{[0]i}, b_i) = \frac{1}{K}\sum_j \left(\frac{|\hat{b}_{[0]ij} - b_{ij}|}{\hat{\alpha}_{[0]ij}}\right)^{\hat{\beta}_{[0]ij}} - \log\frac{\hat{\beta}_{[0]ij}}{\hat{\alpha}_{[0]ij}} + \log\Gamma(\hat{\beta}_{[0]ij}^{-1}) \quad (1)$$

$$\mathcal{L}_{adv}^G = \mathcal{L}_2(\mathcal{D}_1(\hat{b}_{[0]i}), 1) \text{ and } \mathcal{L}_{tot}^G = \lambda_1\mathcal{L}_{\alpha\beta}^G + \lambda_2\mathcal{L}_{adv}^G. \quad (2)$$

The patch discriminator $(\mathcal{D}_1)$ is trained using the adversarial loss from [36] given by $\mathcal{L}_{adv}^D = \mathcal{L}_2(\mathcal{D}^A(b_i), 1) + \mathcal{L}_2(\mathcal{D}^A(\hat{b}_{[0]i}), 0)$.

Subsequent GANs. The mth GAN (where $m > 0$) takes the output produced by the $(m-1)^{th}$ GAN, i.e. $(\hat{\alpha}_{[m-1]i}, \hat{\beta}_{[m-1]i}, \hat{b}_{[m-1]i})$, along with the original sample a_i from domain A as its input and generates a refined output. The image estimated by the $(m-1)^{th}$ GAN along with its uncertainty map learns to create the input feature $f_{[m]i}$ for the mth GAN, where the uncertainty map serves as an attention mechanism to highlight the uncertain regions in the image. The input $a_{[m]i}$ for the mth generator is given by concatenating a_i and $f_{[m]i}$, i.e.,

$$\hat{\sigma}_{[m-1]i} = \hat{\alpha}_{[m-1]i}\sqrt{\frac{\Gamma(3/\hat{\beta}_{[m-1]i})}{\Gamma(1/\hat{\beta}_{[m-1]i})}}, \text{ and } f_{[m]i} = \hat{b}_{[m-1]i} \odot \frac{\hat{\sigma}_{[m-1]i}}{\sum_j \hat{\sigma}_{[m-1]ij}} \quad (3)$$

$$a_{[m]i} = \texttt{concat}(f_{[m]i}, a_i) \quad (4)$$

The input $a_{[m]i}$ for the mth GAN encourages the generator to further refine the highly uncertain regions in the image given the original input context. The generator and the discriminator are trained using $\mathcal{L}_{tot}^G$ and $\mathcal{L}_{adv}^D$, respectively.

Progressive Training Scheme. We initialize the parameters $\theta \cup \phi$ sequentially. First, we initialize $\theta_1 \cup \phi_1$ using the training set (S_A, S_B) to minimize the loss function given by $\mathcal{L}_{tot}^G$ and $\mathcal{L}_{adv}^D$. Then, for the subsequent GANs, we initialize the $\theta_m \cup \phi_m$ $(m > 1)$ by fixing the weights of all the previous generators and training the mth GAN alone (see Eq. 3 and 4 with losses $\mathcal{L}_{tot}^G$ and $\mathcal{L}_{adv}^D$). Once all the parameters have been initialized (i.e., $\theta_m \cup \phi_m \forall m$), we do further fine tuning by training all the networks end-to-end by combining the loss functions of all the intermediate phases and a significantly smaller learning-rate.

4 Experiments

In this section, we first detail the experimental setup and comparative methods in Sect. 4.1, and present the corresponding results in Sect. 4.2.

4.1 Experimental Setup

Tasks and Datasets. We evaluate our method on the following three tasks.

(i) PET to CT translation: We synthesize CT images from PET scans to be used for the attenuation correction, e.g. for PET-only scanners or PET/MRI. We use paired data sets of non-attenuation-corrected PET and the corresponding CT of the head region of 49 patients acquired on a state-of-the-art PET/CT scanner (Siemens Biograph mCT), approved by ethics committee of the Medical Faculty of the University of Tübingen. Data is split into 29/5/15 for training/val/test sets. Figure 2 shows exemplary slices for co-registered PET and CT.

(ii) Undersampled MRI reconstruction: We translate undersampled MRI images to fully-sampled MRI images. We use MRI scans from the open-sourced IXI[1] dataset that consists of T1-weighted (T1w) MRI scans. We use a cohort of 500 patients split into 200/100/200 for training/val/test, and retrospectively create the undersampled MRI with an acceleration factor of 12.5×, i.e., we preserve only 8% of the fully-sampled k-space measurement (from the central region) to obtain the undersampled image.

(iii) MRI Motion correction: We generate sharp images from motion corrupted images. We retrospectively create the motion artefacts in the T1w MRI from IXI following the transformations in the *k-space* as described in [18]. Figure 3-(ii) shows the input MRI scan with artefacts and ground-truth.

Training Details and Evaluation Metrics. All GANs are first initialized using the aforementioned progressive learning scheme with (λ_1, λ_2) in Eq. 2 set to $(1, 0.001)$. We use Adam [11], with the hyper-parameters $\beta_1 := 0.9$, $\beta_2 := 0.999$, an initial learning rate of 0.002 for initialization and 0.0005 post-initialization that decays based on cosine annealing over 1000 epochs, using a batch size of 8. We use three widely adopted metrics to evaluate image generation quality: PSNR measures $20 \log \mathrm{MAX}_I / \sqrt{\mathrm{MSE}}$, where MAX_I is the highest possible intensity value in the image and MSE is the mean-squared-error between two images. SSIM computes the structural similarity between two images [28]. MAE computes the mean absolute error between two images. Higher PSNR, SSIM, and lower MAE indicate a higher quality of the generated images (wrt ground-truth).

Compared Methods. We compare our model to representative state-of-the-art methods for medical image translation, including Pix2pix [7], a baseline conditional adversarial networks for image-to-image translation tasks using GANs,

---

[1] From https://brain-development.org/ixi-dataset/.

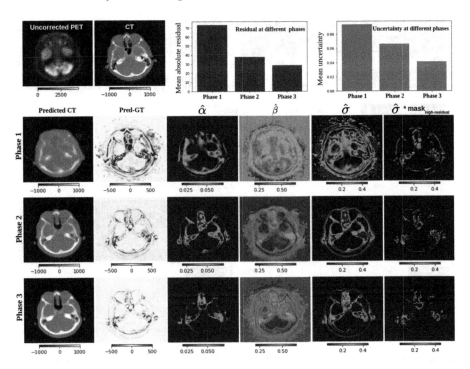

Fig. 2. Outputs from different phases of UP-GAN (with M = 3). (Top) The input (uncorrected PET), the corresponding ground-truth CT, mean residual values over different phases, mean uncertainty values over different phases. (Bottom) Each row shows the predicted output, the residual between the prediction and the ground-truth, the predicted scale (α) map, the predicted shape (β) map, the uncertainty map, and the uncertainty in high residual regions.

PAN [24], and MedGAN [2], a GAN-based method that relies on *external-pre-trained feature extractors*, with a generator that refines the generated images progressively. MedGAN is shown to perform superior to methods like, Fila-sGAN [34], ID-cGAN [32], and achieve state-of-the-art performance for several medical image-to-image translation problems.

4.2 Results and Analysis

Qualitative Results. Figure 2 visualizes the (intermediate) outputs of the generators at different phases of the framework. The visual quality of the generated image content increasingly improves along the network phases (as shown in the first column, second row onward). At the same time, prediction error and uncertainty decrease continuously (second column and fifth column, second row onward, respectively). High uncertainty values are found in anatomical regions with fine osseous structures, such as the nasal cavity and the inner ear in the

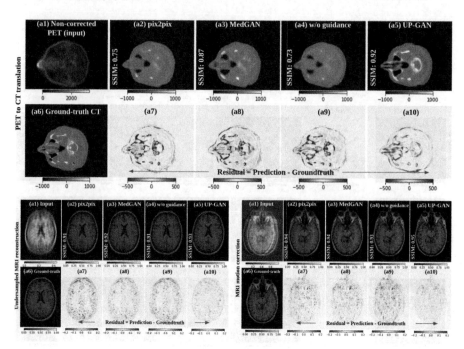

Fig. 3. Qualitative results. (Top) PET to CT translation. (Bottom) Undersampled MRI reconstruction (left), and MRI motion correction (right).

petrous portion of the temporal bone. Particularly in such regions of high uncertainty, we achieve a progressive improvement in the level of detail.

Figure 3-(Top) visualizes the generated CT images from the PET for all the compared methods along with our methods. We observe that more high-frequency features are present in our prediction compared to the previous state-of-the-art model (MedGAN). We also observe that the overall residual is significantly lower for our method compared to the other baselines. MedGAN performs better than pix2pix in synthesizing high-frequency features and sharper images. Figure 3-(Bottom) shows similar results for the undersampled MRI reconstruction task and MRI motion correction task. In both cases, our model yields superior images, as can be seen via relatively neutral residual maps.

Quantitative Results. Table 1 shows the quantitative performance of all the methods on the three tasks; for all the tasks, our method outperforms the recent models. In particular, for the most challenging task, PET to CT translation, our method with uncertainty-based guide outperforms the previous state-of-the-art method, MedGAN (that relies on task-specific external feature extractor), *without using any external feature extractor.* Therefore, the uncertainty guidance reduces the burden of having an externally trained task-specific feature extractor to achieve high fidelity images. The same trend holds for undersampled MRI reconstruction and motion correction in MRI. The statistical tests on

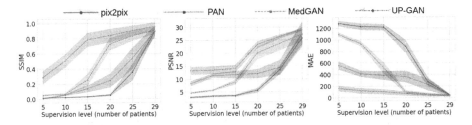

Fig. 4. Quantitative results in the presence of limited labeled training data.

Table 1. Evaluation of various methods on three medical image translation tasks.

Methods	PET to CT			Undersampled MRI Recon			MRI Motion Correction		
	SSIM	PSNR	MAE	SSIM	PSNR	MAE	SSIM	PSNR	MAE
pix2pix [7]	0.89±0.04	26.0±2.0	38.5±10.7	0.92±0.03	28.5±0.9	27.6±9.3	0.94±0.06	29.6±1.4	26.3±8.2
PAN [24]	0.90±0.08	26.5±4.5	37.2±15.6	0.93±0.05	28.8±0.7	26.2±10.4	0.95±0.10	30.1±2.8	24.9±9.7
MedGAN [2]	0.90±0.04	27.1±2.5	35.4±11.8	0.94±0.02	**29.7±1.9**	24.2±8.7	0.95±0.04	30.8±1.8	23.6±9.1
UP-GAN	**0.95±0.05**	**28.9±0.4**	**24.7±12.9**	**0.97±0.07**	29.4±2.1	**24.1±7.5**	**0.96±0.03**	**32.1±0.3**	**22.8±11.1**

SSIM values of MedGAN and our UP-GAN gives us a *p-value* of 0.016 for PET-to-CT translation, 0.021 for undersampled MRI reconstruction, and 0.036 for MRI motion correction. As all the *p-values* are < 0.05, results are statistically significant.

Ablation Study. We study the model that does not utilize the estimated uncertainty maps as attention maps and observe that the model without the uncertainty as the guide performs inferior to the UP-GAN with a performance (SSIM/PSNR/MAE) of (0.87/25.4/40.7), (0.93/27.3/38.7), and (0.92/26.2/35.1) for PET to CT translation, undersampled MRI reconstruction, and MRI motion correction, respectively. UP-GAN model leverages the uncertainty map to refine the predictions where the model is uncertain, which is also correlated to the regions where the translation is poor. The model without uncertainty-based guidance does not focus on the regions mentioned above in the prediction and is unable to perform as well as UP-GAN.

Evaluating Models with Weak Supervision. We evaluate all the models for PET to CT synthesis by limiting the number of paired image samples used for training. We define *five* supervision levels corresponding to different amounts of cross-domain pairwise training sample slices. For this experiment, we train the recent state-of-the-art models with a varying number of patients in the training stage, i.e., we use 5, 10, 15, 20, and 29 patients, respectively. Figure 4 shows the performance of all the models at varying supervision levels. We observe that our model with uncertainty guidance outperforms all the baselines at full supervision (with 29 patients). Moreover, our model sharply outperforms the baselines with limited training data (with < 29 patients). UP-GAN produces intermediate uncertainty maps that have higher values under weak supervision (compared to the full supervision case), but this still allows UP-GAN to focus on highly uncertain regions, that the current state-of-the-art models do not have access to, hence are not able to leverage that to refine the predicted images.

5   Conclusion

In this work, we propose a new generic model for medical image translation using uncertainty-guided progressive GANs. We demonstrate how uncertainty can serve as an attention map in progressive learning schemes. We demonstrate the efficacy of our method on three challenging medical image translation tasks, including PET to CT translation, undersampled MRI reconstruction, and motion correction in MRI. Our method achieves state-of-the-art in various tasks. Moreover, it allows the quantification of uncertainty and shows better generalizability with smaller sample sizes than recent approaches.

Acknowledgements. This work has been partially funded by the ERC (853489 - DEXIM) and by the DFG (2064/1 – Project number 390727645). The authors thank the International Max Planck Research School for Intelligent Systems (IMPRS-IS) for support.

References

1. Armanious, K., Jiang, C., Abdulatif, S., Küstner, T., Gatidis, S., Yang, B.: Unsupervised medical image translation using cycle-MedGAN. In: European Signal Processing Conference (EUSIPCO) (2019)
2. Armanious, K., et al.: MedGAN: medical image translation using GANs. Comput. Med. Imaging Graph. **79**, 101684 (2020)
3. Chen, H., et al.: Low-dose CT with a residual encoder-decoder convolutional neural network. IEEE TMI **36**, 2524–2535 (2017)
4. Dou, Q., Chen, H., Yu, L., Qin, J., Heng, P.A.: Multilevel contextual 3-D CNNs for false positive reduction in pulmonary nodule detection. IEEE Trans. Biomed. Eng. **64**, 1558–1567 (2016)
5. Havaei, M., et al.: Brain tumor segmentation with deep neural networks. Med. Image Anal. **35**, 18–31 (2017)
6. Huynh, T., et al.: Estimating CT image from MRI data using structured random forest and auto-context model. IEEE TMI **35**, 174–183 (2015)
7. Isola, P., Zhu, J.Y., Zhou, T., Efros, A.A.: Image-to-image translation with conditional adversarial networks. In: IEEE CVPR (2017)
8. Kamnitsas, K., et al.: Efficient multi-scale 3D CNN with fully connected CRF for accurate brain lesion segmentation. Med. Image Anal. **36**, 61–78 (2017)
9. Karras, T., Aila, T., Laine, S., Lehtinen, J.: Progressive growing of GANs for improved quality, stability, and variation. In: ICLR (2018)
10. Kendall, A., Gal, Y.: What uncertainties do we need in Bayesian deep learning for computer vision? In: NIPS (2017)
11. Kingma, D.P., Ba, J.: Adam: a method for stochastic optimization. preprint arXiv:1412.6980 (2014)
12. Küstner, T., et al.: MR-based respiratory and cardiac motion correction for pet imaging. Med. Image Anal. **42**, 129–144 (2017)

13. Litjens, G., et al.: A survey on deep learning in medical image analysis. Med. Image Anal. **42**, 60–88 (2017)
14. Nair, T., Precup, D., Arnold, D.L., Arbel, T.: Exploring uncertainty measures in deep networks for multiple sclerosis lesion detection and segmentation. Med. Image Anal. **59**, 101557 (2020)
15. Nie, D., et al.: Medical image synthesis with deep convolutional adversarial networks. IEEE Trans. Biomed. Eng. **65**, 2720–2730 (2018)
16. Ronneberger, O., Fischer, P., Brox, T.: U-Net: convolutional networks for biomedical image segmentation. In: Navab, N., Hornegger, J., Wells, W.M., Frangi, A.F. (eds.) MICCAI 2015. LNCS, vol. 9351, pp. 234–241. Springer, Cham (2015). https://doi.org/10.1007/978-3-319-24574-4_28
17. Rueda, A., Malpica, N., Romero, E.: Single-image super-resolution of brain MR images using overcomplete dictionaries. Med. Image Anal. **17**, 113–132 (2013)
18. Shaw, R., Sudre, C., Ourselin, S., Cardoso, M.J.: MRI k-space motion artefact augmentation: model robustness and task-specific uncertainty. In: MIDL (2019)
19. Shin, H.C., et al.: Deep convolutional neural networks for computer-aided detection: CNN architectures, dataset characteristics and transfer learning. IEEE TMI **35**, 1285–1298 (2016)
20. Tanno, R., et al.: Bayesian image quality transfer with CNNs: exploring uncertainty in dMRI super-resolution. In: Descoteaux, M., Maier-Hein, L., Franz, A., Jannin, P., Collins, D.L., Duchesne, S. (eds.) MICCAI 2017. LNCS, vol. 10433, pp. 611–619. Springer, Cham (2017). https://doi.org/10.1007/978-3-319-66182-7_70
21. Upadhyay, U., Awate, S.P.: A mixed-supervision multilevel GAN framework for image quality enhancement. In: Shen, D., et al. (eds.) MICCAI 2019. LNCS, vol. 11768, pp. 556–564. Springer, Cham (2019). https://doi.org/10.1007/978-3-030-32254-0_62
22. Upadhyay, U., Awate, S.P.: Robust super-resolution GAN, with manifold-based and perception loss. In: 2019 IEEE 16th International Symposium on Biomedical Imaging (ISBI 2019), pp. 1372–1376. IEEE (2019)
23. Upadhyay, U., Chen, Y., Akata, Z.: Uncertainty-aware generalized adaptive Cycle-GAN. preprint arXiv:2102.11747 (2021)
24. Wang, C., Xu, C., Wang, C., Tao, D.: Perceptual adversarial networks for image-to-image transformation. IEEE TIP **27**, 4066–4079 (2018)
25. Wang, G., Li, W., Aertsen, M., Deprest, J., Ourselin, S., Vercauteren, T.: Aleatoric uncertainty estimation with test-time augmentation for medical image segmentation with convolutional neural networks. Neurocomputing **338**, 34–45 (2019)
26. Wang, G., Li, W., Ourselin, S., Vercauteren, T.: Automatic brain tumor segmentation using convolutional neural networks with test-time augmentation. In: Crimi, A., Bakas, S., Kuijf, H., Keyvan, F., Reyes, M., van Walsum, T. (eds.) BrainLes 2018. LNCS, vol. 11384, pp. 61–72. Springer, Cham (2019). https://doi.org/10.1007/978-3-030-11726-9_6
27. Wang, G., Li, W., Ourselin, S., Vercauteren, T.: Automatic brain tumor segmentation based on cascaded convolutional neural networks with uncertainty estimation. Front. Comput. Neurosci. **13**, 56 (2019)
28. Wang, Z., Bovik, A.C., Sheikh, H.R., Simoncelli, E.P.: Image quality assessment: from error visibility to structural similarity. IEEE TIP **13**, 600–612 (2004)
29. Wolterink, J.M., Dinkla, A.M., Savenije, M.H.F., Seevinck, P.R., van den Berg, C.A.T., Išgum, I.: Deep MR to CT synthesis using unpaired data. In: Tsaftaris, S.A., Gooya, A., Frangi, A.F., Prince, J.L. (eds.) SASHIMI 2017. LNCS, vol. 10557, pp. 14–23. Springer, Cham (2017). https://doi.org/10.1007/978-3-319-68127-6_2

30. Yang, Q., et al.: Low-dose CT image denoising using a generative adversarial network with Wasserstein distance and perceptual loss. IEEE TMI **37**, 1348–1357 (2018)
31. Yi, X., Walia, E., Babyn, P.: Generative adversarial network in medical imaging: a review. Med. Image Anal. **58**, 101552 (2019)
32. Zhang, H., Sindagi, V., Patel, V.M.: Image de-raining using a conditional generative adversarial network. IEEE Trans. Circuits Syst. Video Technol. **30**, 3943–3956 (2019)
33. Zhang, Z., Romero, A., Muckley, M.J., Vincent, P., Yang, L., Drozdzal, M.: Reducing uncertainty in undersampled MRI reconstruction with active acquisition. In: IEEE CVPR (2019)
34. Zhao, H., Li, H., Cheng, L.: Synthesizing filamentary structured images with GANs. preprint arXiv:1706.02185 (2017)
35. Zhong, L., et al.: Predict CT image from MRI data using KNN-regression with learned local descriptors. In: IEEE ISBI (2016)
36. Zhu, J.Y., Park, T., Isola, P., Efros, A.A.: Unpaired image-to-image translation using cycle-consistent adversarial networks. In: IEEE ICCV (2017)

Variational Topic Inference for Chest X-Ray Report Generation

Ivona Najdenkoska[1($\boxtimes$)], Xiantong Zhen[1,2], Marcel Worring[1], and Ling Shao[2]

[1] University of Amsterdam, Amsterdam, The Netherlands
{i.najdenkoska,x.zhen,m.worring}@uva.nl
[2] Inception Institute of Artificial Intelligence, Abu Dhabi, UAE
ling.shao@ieee.org

Abstract. Automating report generation for medical imaging promises to reduce workload and assist diagnosis in clinical practice. Recent work has shown that deep learning models can successfully caption natural images. However, learning from medical data is challenging due to the diversity and uncertainty inherent in the reports written by different radiologists with discrepant expertise and experience. To tackle these challenges, we propose *variational topic inference* for automatic report generation. Specifically, we introduce a set of topics as latent variables to guide sentence generation by aligning image and language modalities in a latent space. The topics are inferred in a conditional variational inference framework, with each topic governing the generation of a sentence in the report. Further, we adopt a visual attention module that enables the model to attend to different locations in the image and generate more informative descriptions. We conduct extensive experiments on two benchmarks, namely Indiana U. Chest X-rays and MIMIC-CXR. The results demonstrate that our proposed variational topic inference method can generate novel reports rather than mere copies of reports used in training, while still achieving comparable performance to state-of-the-art methods in terms of standard language generation criteria.

Keywords: Chest X-ray · Radiology report generation · Latent variables · Variational topic inference

1 Introduction

Chest X-rays are one of the most frequently used imaging modalities in clinical practice. However, interpreting X-ray images and writing reports is laborious and creates an extensive workload for radiologists. Automated radiology report generation using machine learning techniques has thus arisen to potentially alleviate the burden and expedite clinical workflows. Fundamentally, this can be

Electronic supplementary material The online version of this chapter (https://doi.org/10.1007/978-3-030-87199-4_59) contains supplementary material, which is available to authorized users.

M. de Bruijne et al. (Eds.): MICCAI 2021, LNCS 12903, pp. 625–635, 2021.
https://doi.org/10.1007/978-3-030-87199-4_59

regarded as translating visual input into textual output, which is broadly known as image captioning [1,21,28,30]. It is non-trivial and challenging to transfer this to X-ray reports though, as we need to learn their complex structure and diversity, as well as, to model the uncertainty induced by the varying expertise and experience of radiologists.

Successful chest X-ray report generation methods mainly follow the neural encoder-decoder architecture [3,10,11,17,19,20,31–33], where a convolutional neural network (CNN) encodes the image into a fixed-size representation and then, sentence by sentence, a recurrent neural network decodes the representation into a report. To enhance this architecture, additional techniques have been introduced. For instance, [11] incorporate a co-attention mechanism to exploit the relationships between visual features and medical tags and uses hierarchical LSTMs [7] to generate multiple sentences. Furthermore, to generate reports with high clinical correctness, [19] proposes to optimize a clinical coherence reward by reinforcement learning. To use the information encoded in both the frontal and lateral views, [33] explores the fusion of multi-view chest X-rays. Another relevant approach exploits the structure of reports by modeling the relationship between findings and impression sections [10]. More recent works [3,20] leverage the Transformer [27] as a more powerful language model to better capture long-term dependencies for sentence generation.

Despite being the state of the art in terms of benchmark measures, these deterministic encoder-decoder models tend to overfit to the data, producing generic results and making them unable to represent the inherent uncertainty in the reports. This uncertainty arises from the fact that the reports are written by radiologists with different levels of expertise, experience and expressive styles. Naturally, this can yield diversity when several radiologists interpret an X-ray image into a report. In order to improve their generalizability, it is thus highly important to capture the uncertainty when designing algorithms for report generation. Probabilistic modeling is able to handle the uncertainty, diversity and complex structure of reports [15,22] in a well founded way. Instead of simply compressing inputs into fixed-sized deterministic representations, which could cause information loss, adopting stochastic latent variables [15] allows the holistic characteristics of sentences, such as topic, style and high-level patterns, to be explicitly modeled [2], which enables more diverse but controllable text generation [23,29].

In this paper, we propose variational topic inference (VTI), which addresses report generation for chest X-ray images with a probabilistic latent variable model. In particular, we introduce a set of latent variables, each defined as a topic governing the sentence generation. The model is optimized by maximizing an evidence lower bound objective (ELBO) [26]. During training, the topics are inferred from visual and language representations, which are aligned by minimizing the Kullback-Leibler (KL) divergence between them. By doing so, at test time the model is able to infer topics from the visual representations to generate the sentences and maintain coherence between them. Finally, we adopt visual attention which enables the model to attend to different local image regions when generating specific words.

Our main contributions can be summarized as follows: (1) We propose a variational topic inference framework to address the radiology report generation problem, which enables diversity and uncertainty to be better handled when generating reports. (2) We adopt Transformers to aggregate local visual features with each attention head producing a specific representation for each sentence, which encourages diverse sentences to provide informative and comprehensive descriptions. (3) We demonstrate that our method achieves comparable performance to the state of the art on two benchmark datasets under a broad range of evaluation criteria.

2   Methodology

2.1   Problem Formulation

Given the input image $\mathbf{x}$, we aim to generate a report that consists of multiple sentences $\{\mathbf{y}_i\}_{i=1}^N$, which are assumed to be conditionally independent. From a probabilistic perspective, we aim to maximize the conditional log-likelihood:

$$\theta^* = \arg\max_{\theta} \sum_{i=1}^N \log p_\theta(\mathbf{y}_i|\mathbf{x}), \tag{1}$$

where θ contains the model parameters and N is the number of sentences in each report. To solve the model, we formulate the report generation as a conditional variational inference problem.

2.2   Variational Topic Inference

In order to encourage diversity and coherence between the generated sentences in a report, we introduce a set of latent variables to represent topics $\mathbf{z}$, each of which governs the generation of one sentence $\mathbf{y}$ in the final report (note that the subscript i is omitted for brevity). By incorporating $\mathbf{z}$ into the conditional probability $p_\theta(\mathbf{y}|\mathbf{x})$, we have:

$$\log p_\theta(\mathbf{y}|\mathbf{x}) = \int_z \log p_\theta(\mathbf{y}|\mathbf{x}, \mathbf{z}) p_\theta(\mathbf{z}|\mathbf{x}) d\mathbf{z}, \tag{2}$$

where $p_\theta(\mathbf{z}|\mathbf{x})$ is the conditional prior distribution. We define a variational posterior $q_\phi(\mathbf{z})$ to approximate the intractable true posterior $p_\theta(\mathbf{z}|\mathbf{y}, \mathbf{x})$ by minimizing the KL divergence between them: $D_{\mathrm{KL}}[q_\phi(\mathbf{z})||p_\theta(\mathbf{z}|\mathbf{x}, \mathbf{y})]$. we arrive at:

$$D_{\mathrm{KL}}[q_\phi(\mathbf{z})||p_\theta(\mathbf{z}|\mathbf{x}, \mathbf{y})] = \mathbb{E}[\log q_\phi(\mathbf{z}) - \log \frac{p_\theta(\mathbf{y}|\mathbf{z}, \mathbf{x}) p_\theta(\mathbf{z}|\mathbf{x})}{p_\theta(\mathbf{y}|\mathbf{x})}] \geq 0, \tag{3}$$

which gives rise to the ELBO of the log-likelihood:

$$\log p_\theta(\mathbf{y}|\mathbf{x}) \geq \mathbb{E}[\log p_\theta(\mathbf{y}|\mathbf{z}, \mathbf{x})] - D_{\mathrm{KL}}[q_\phi(\mathbf{z})||p_\theta(\mathbf{z}|\mathbf{x})] = \mathcal{L}_{\mathrm{ELBO}}(\theta, \phi), \tag{4}$$

where the variational posterior $q(\mathbf{z})$ can be designed in various forms to approximate the true posterior.

To leverage the language modality during training, we design the variational posterior as $q_\phi(\mathbf{z}|\mathbf{y})$ conditioned on the ground-truth sentence. Based on the ELBO, we derive the objective function w.r.t. a report of N sentences as follows:

$$\mathcal{L}_{\mathrm{ELBO}}(\theta, \phi) = \sum_{i=1}^{N} \Big[\sum_{\ell=1}^{L} \log p_\theta(\mathbf{y}_i | \mathbf{z}^{(\ell)}, \mathbf{x}) - \beta D_{\mathrm{KL}}[q_\phi(\mathbf{z}|\mathbf{y}_i) || p_\theta(\mathbf{z}|\mathbf{x})] \Big], \qquad (5)$$

where $\mathbf{z}^{(\ell)}$ is the ℓ-th of L Monte Carlo samples, and β is a weighting parameter that controls the behavior of the KL divergence. In order to efficiently draw samples and conduct backpropagation, we use the reparametrization trick to draw samples from the variational posterior [14]. During training, the samples are drawn from the variational posterior distribution $\mathbf{z}^{(l)} \sim q_\phi(\mathbf{z}|\mathbf{y})$, whereas during inference the samples are drawn from the prior distribution $\mathbf{z}^{(l)} \sim p_\theta(\mathbf{z}|\mathbf{x})$.

2.3   Implementation Using Neural Networks

For efficient optimization, we implement the model with deep neural networks using amortization techniques [14]. $p_\theta(\mathbf{z}|\mathbf{x})$ and $q_\phi(\mathbf{z}|\mathbf{y})$ are parameterized as fully factorized Gaussian distributions and inferred by multi-layer perceptrons (MLPs), which we refer to as the visual prior net and the language posterior net, respectively. The log-likelihood is implemented as a cross entropy loss based on the output of the sentence generator net and the ground-truth sentence. Figure 1 illustrates our proposed VTI model.

Visual Prior Net. To establish more holistic visual representations, we leverage a Transformer to aggregate local visual features from a pre-trained CNN. Specifically, the convolutional feature maps are flattened along the spatial dimensions to obtain a set of k local visual features $\mathbf{V} = \{\mathbf{v}_1, \mathbf{v}_2, \cdots, \mathbf{v}_k\}$, where $\mathbf{v}_i \in \mathbb{R}^{d_v}$ and d_v is the dimension of a visual feature vector. To explore the similarity among local features, we adopt the Transformer to encode them into a special visual token $\mathbf{v}_{[\mathrm{IMG}]}$ as the holistic representation of the image. To encourage diversity among topics in a report, we employ a multi-head attention in the Transformer and use each attention head to generate a specific representation for each topic governing the generation of a sentence.

Language Posterior Net. Each sentence is represented as a sequence of word tokens including a special language token [SENT]. Each word token is embedded by an embedding matrix W_e, which yields a sequence of n word embeddings $\{\mathbf{e}_1, \mathbf{e}_2, ..., \mathbf{e}_n\}$, where $\mathbf{e}_i \in \mathbb{R}^{d_e}$ and d_e is the dimension of the embedding matrix W_e. A Transformer with positional embedding encodes the relationships between the word embeddings, which are aggregated into the special token $\mathbf{e}_{[\mathrm{SENT}]}$ as the holistic representation of the sentence. This net takes ground-truth sentences as input to aid the generation of latent topics, which is used during training only.

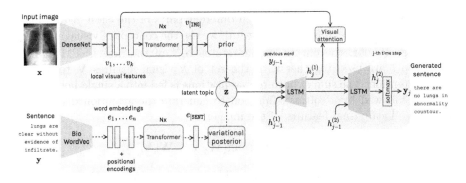

Fig. 1. Architecture of the proposed variational topic inference model. Note that the language stream is only used at training time, when we infer the distributions of latent topics from both the visual and language modalities and minimize their KL divergence. At test time, we infer topics from the visual modality only to generate sentences.

Sentence Generator Net. The sentences in a report are generated jointly, where the generation of each sentence $\mathbf{y}$ is formulated as $p_\theta(\mathbf{y}|\mathbf{x}, \mathbf{z})$. $\mathbf{y}$ is a sequence of word tokens $\mathbf{y}_0, \mathbf{y}_1, \cdots, \mathbf{y}_t$ and it is common to use the joint probability over the tokens to formulate the generation process:

$$p_\theta(\mathbf{y}|\mathbf{x}, \mathbf{z}) = \prod_{t=1}^{T} p_\theta(\mathbf{y}_t|\mathbf{x}, \mathbf{z}, \mathbf{y}_t). \tag{6}$$

The sentence generator net is designed in an auto-regressive manner and it is implemented by two consecutive LSTMs [1] with injected latent topic variables and enhanced by visual attention:

$$\mathbf{h}_t^{(1)} = \text{LSTM}^{(1)}\left(\mathbf{y}_t, \mathbf{h}_{t-1}^{(1)}, \mathbf{c}_{t-1}^{(1)}\right), \tag{7}$$

where $\mathbf{c}^{(1)}$ is the memory cell state initialized by the latent topic $\mathbf{z}$ for the first time step. The next word in the sequence is predicted by the second LSTM, which takes as input the concatenation of the attentive visual representation and the hidden state $\mathbf{h}_t^{(1)}$ of the first LSTM:

$$\mathbf{h}_t^{(2)} = \text{LSTM}^{(2)}([\mathbf{v}_a; \mathbf{h}_t^{(1)}], \mathbf{h}_{t-1}^{(2)}, \mathbf{c}_{t-1}^{(2)}), \tag{8}$$

where $\mathbf{v}_a$ is obtained by a visual attention module, as described in the next subsection. The output $\mathbf{h}_t^{(2)}$ of the second LSTM$^{(2)}$ is used to predict the probability distribution p_t of the next word, as in [1]:

$$p_t = \text{softmax}(\mathbf{W}_p \mathbf{h}_t^{(2)}), \tag{9}$$

where $\mathbf{W}_p \in \mathbb{R}^{d_h \times d_{vocab}}$ is a learnable linear layer that projects $\mathbf{h}_t^{(2)} \in R^{d_h}$ to a probability distribution p_t over the vocabulary of size d_{vocab}.

Visual Attention. To place focus on different parts of the chest X-ray image while decoding the sentence word by word, we use the concept of visual attention [30]. In particular, the output hidden states $\mathbf{h}_t^{(1)}$ of the first LSTM at each time step t are used together with the set of k visual features $\mathbf{V}$ to achieve visual attention. The sum of both representations is fed into a single-layer neural network followed by a softmax function to generate the attention distribution over the k local visual features of the image:

$$\alpha_t = \text{softmax}\big(\mathbf{w}_a^\top \tanh(\mathbf{W}_v \mathbf{V} + \mathbf{W}_h \mathbf{h}_t^{(1)})\big), \tag{10}$$

where $\mathbf{w}_a^T \in \mathbb{R}^k$, $\mathbf{W}_v, \mathbf{W}_h \in \mathbb{R}^{k \times d_h}$ are all learnable parameters. Once the attention distribution α_t is obtained, we can compute the weighted visual representation as follows:

$$\mathbf{v}_a = \sum_{t=0}^{k} \alpha_t \cdot \mathbf{v}_t, \tag{11}$$

which is essentially the aggregated visual representation specific to each word at a given time step t.

3   Experiments

3.1   Datasets and Implementation Details

We evaluate our VTI model on the Indiana University Chest X-Ray collection [4] and MIMIC-CXR [12] dataset. Following standard procedure, images are normalized and resized to 224×224, making them appropriate for extracting visual features from a pre-trained DenseNet-121 [8]. Data entries with missing or incomplete reports are discarded. The impressions and findings sections of the reports are concatenated, lower-cased and tokenized. Non-alphabetical words and words that occur less than a pre-defined threshold are filtered out and replaced with a [UNK] token. Shorter sentences and reports are padded to obtain squared batches. After pre-processing, Indiana U. Chest X-Ray consists of 3,195 samples, which are split into training, validation and test sets with a ratio of 7:1:2. MIMIC-CXR consists of 218,101 samples and is split according to the official splits.

The word embeddings are initialized with the pre-trained biomedical embeddings BioWordVec [34], which represent 200-dimensional contextualized vectors. All hyperparameters are set through cross-validation. The linear layers are initialized from a uniform distribution [6] and each one has a hidden dimension of 512, followed by ReLU non-linearity and a dropout with a rate of 0.5. The Transformers in both streams use a hidden dimension of 512. The model is trained end-to-end on four NVIDIA GTX 1080Ti GPUs using the Adam optimizer [13] with a learning rate of 3e-05 and early stopping with a patience of five epochs. We use cyclical annealing [5] to deal with the notoriously difficult training with KL divergence in the objective function. To further improve the readability and coherence, we use a temperature hyperparameter to skew the output distribution towards higher probability events and then apply top-k sampling.

Table 1. Results on Indiana U. Chest X-ray and MIMIC-CXR using NLG metrics.

Indiana U. X-Ray

Method	BLEU-1	BLEU-2	BLEU-3	BLEU-4	METEOR	ROUGE
HRGR-Agent [17]	0.438	0.298	0.208	0.151	–	0.322
Clinical-NLG [19]	0.369	0.246	0.171	0.115	–	0.359
MM-Att [32]	0.464	0.358	0.270	**0.195**	**0.274**	0.366
MvH [33]	0.478	0.334	0.277	0.191	0.265	0.318
CMAS-RL [10]	0.464	0.301	0.210	0.154	-	0.362
Memory-Transformer [3]	0.470	0.304	0.219	0.165	0.187	0.371
VTI (Ours)	**0.493**	**0.360**	**0.291**	0.154	0.218	**0.375**
MIMIC-CXR						
Clinical-NLG [19]	0.352	0.223	0.153	0.104	-	0.307
Memory-Transformer [3]	0.353	0.218	0.145	0.103	0.142	0.277
CC-Transformer [20]	0.415	0.272	**0.193**	**0.146**	0.159	**0.318**
VTI (Ours)	**0.418**	**0.293**	0.152	0.109	**0.177**	0.302

Table 2. Results on MIMIC-CXR using the clinical efficacy metrics.

Method	Micro			Macro		
	F1	Precision	Recall	F1	Precision	Recall
Clinical-NLG [19]	–	0.419	0.360	–	0.225	0.209
CC-Transformer [20]	**0.411**	0.475	**0.361**	**0.228**	0.333	**0.217**
VTI (Ours)	0.403	**0.497**	0.342	0.210	**0.350**	0.151

3.2 Results and Discussion

We adopt commonly used evaluation metrics for natural language generation (NLG), including BLEU [24], METEOR [16] and ROUGE [18]. We compare to several other neural network based state-of-the-art methods: [3,10,17,19,32,33] for Indiana U. X-Rays, and [3,19,20] for MIMIC-CXR. As shown in Table 1, our VTI achieves comparable performance or yields higher scores in terms of BLEU-1-2-3, ROUGE (for Indiana U. Chest X-ray) and METEOR (for MIMIC-CXR). The probabilistic nature of our approach, which imposes diversity, prevents the model from generating longer n-grams similar to the ground-truth, which is important when computing the NLG metrics. Our approach is able to maintain a better trade-off between accuracy and diversity, which is desirable when generating descriptions for images, as pointed out in [22]. The benefit of using Transformer encoders to learn holistic representations and BioWordVec for pretrained word embeddings is empirically observed in the experimental results, details of which are provided in the supplementary material.

As an additional evaluation in terms of the clinical coherence and correctness, we employ clinical efficacy metrics, i.e., precision, recall and F1 score [19] to compare the extracted labels by the rule-based CheXpert labeler [9] for the

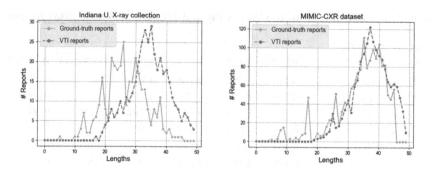

Fig. 2. Length distributions of the ground-truth reports and the VTI generated reports.

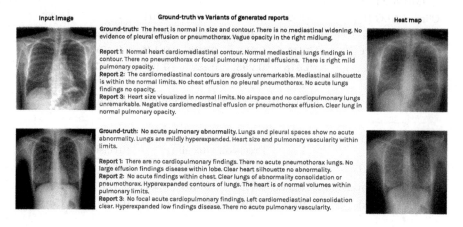

Fig. 3. Examples of the reports generated for Indiana U. Chest X-rays by our VTI model, and heat maps highlighting relevant image regions using Grad-CAM [25].

ground-truth and generated reports. As shown in Table 2, our model scores higher in precision due to the diversity of the generated reports, which can capture additional information in the image, demonstrating the advantage of probabilistic modeling. Moreover, we plot the length distributions of the generated and ground-truth reports, following [3], in Fig. 2. The generated reports tend to be longer for both datasets, suggesting that more detailed information is captured during decoding. They also follow similar distributions, indicating that our VTI is general and not biased towards a particular dataset or simply replicating the exact ground-truth.

We further examine the results from a qualitative perspective. Specifically, for each image we show three report variants in Fig. 3, in which we draw one topic sample per sentence, demonstrating that different Monte Carlo samples yield variation in the sentence generation process. We provide heat maps, which show that VTI can focus on relevant image regions while generating the reports. Additionally, visualizations of the attention maps for the generation of each word are presented in the supplementary material.

We notice that the variants describe similar topics with different sentence structures, indicating that the VTI model is aware of more than one correct combination of sentences. Some sentences have variability in their topics owing to the probabilistic modeling. This tackles the uncertainty in the chest X-ray interpretation process. For instance, report 1 in the first image describes the *cardiomediastinal contour* as *normal*, whereas report 2 describes it as *grossly unremarkable*, both with similar semantics. One limitation is that some sentences may have missing words, due to the difficulty of LSTMs in handling long-term dependencies in sequences. This could be alleviated by using a more powerful language decoder, such as Transformer. Nevertheless, VTI can generate reports not necessarily limited to the ground-truth, indicating its generalization potential, which is a major challenge for report generation [31, 33]. In clinical scenarios, it is often relevant to have a single best report among a variety. The VTI model produces such a report by combining the most probable sentences in terms of Bayesian model averaging in a principled way under the probabilistic framework [14, 26].

4   Conclusion

In this paper we present a probabilistic latent variable model for automated report generation for chest X-ray images. We formulate the report generation as a variational inference problem. We introduce topics as latent variables to guide the sentence generation by aligning the image and language modalities in a latent space. Our approach allows the diversity and uncertainty that exist in the chest X-ray interpretation process to be modeled. Moreover, it provides a new, theoretically well-grounded framework of probabilistic modeling to deal with common issues in report generation, such as generic, incoherent and repetitive sentences. We perform extensive experiments on two benchmark datasets, namely Indiana U. Chest X-rays and MIMIC-CXR, and provide a qualitative analysis demonstrating the effectiveness of the proposed model for chest X-ray radiology report generation.

References

1. Anderson, P., et al.: Bottom-up and top-down attention for image captioning and visual question answering. In: IEEE Conference on Computer Vision and Pattern Recognition (2018)
2. Bowman, S.R., Vilnis, L., Vinyals, O., Dai, A., Jozefowicz, R., Bengio, S.: Generating sentences from a continuous space. In: Proceedings of The 20th SIGNLL Conference on Computational Natural Language Learning, pp. 10–21 (2016)
3. Chen, Z., Song, Y., Chang, T.H., Wan, X.: Generating radiology reports via memory-driven transformer. In: Proceedings of the 2020 Conference on Empirical Methods in Natural Language Processing (2020)
4. Demner-Fushman, D., et al.: Preparing a collection of radiology examinations for distribution and retrieval. J. Am. Med. Inform. Assoc.: JAMIA **23**, 304–10 (2016)

5. Fu, H., Li, C., Liu, X., Gao, J., Celikyilmaz, A., Carin, L.: Cyclical annealing schedule: a simple approach to mitigating KL vanishing. In: North American Chapter of the Association for Computational Linguistics, pp. 240–250 (2019)
6. He, K., Zhang, X., Ren, S., Sun, J.: Delving deep into rectifiers: surpassing human-level performance on ImageNet classification. In: IEEE International Conference on Computer Vision, pp. 1026–1034 (2015)
7. Hochreiter, S., Schmidhuber, J.: Long short-term memory. Neural Comput. **9**(8), 1735–1780 (1997)
8. Huang, G., Liu, Z., Van Der Maaten, L., Weinberger, K.Q.: Densely connected convolutional networks. In: IEEE Conference on Computer Vision and Pattern Recognition, pp. 4700–4708 (2017)
9. Irvin, J., et al.: CheXpert: a large chest radiograph dataset with uncertainty labels and expert comparison. In: 33rd AAAI Conference on Artificial Intelligence (2019)
10. Jing, B., Wang, Z., Xing, E.: Show, describe and conclude: on exploiting the structure information of chest X-ray reports. In: Proceedings of the 57th Annual Meeting of the Association for Computational Linguistics, pp. 6570–6580. Association for Computational Linguistics, July 2019
11. Jing, B., Xie, P., Xing, E.: On the automatic generation of medical imaging reports. In: Proceedings of the 56th Annual Meeting of the Association for Computational Linguistics (2018)
12. Johnson, A.E., et al.: MIMIC-CXR-JPG, a large publicly available database of labeled chest radiographs. arXiv preprint arXiv:1901.07042 (2019)
13. Kingma, D.P., Ba, J.: Adam: a method for stochastic optimization. In: International Conference on Learning Representations (2015)
14. Kingma, D.P., Welling, M.: Auto-encoding variational Bayes. arXiv preprint arXiv:1312.6114 (2013)
15. Kohl, S.A., et al.: A probabilistic U-Net for segmentation of ambiguous images. arXiv preprint arXiv:1806.05034 (2018)
16. Lavie, A., Denkowski, M.J.: The Meteor metric for automatic evaluation of machine translation. Mach. Transl. **23**, 105–115 (2009). https://doi.org/10.1007/s10590-009-9059-4
17. Li, Y., Liang, X., Hu, Z., Xing, E.P.: Hybrid retrieval-generation reinforced agent for medical image report generation. In: Advances in Neural Information Processing Systems, vol. 31 (2018)
18. Lin, C.Y.: ROUGE: a package for automatic evaluation of summaries. Association for Computational Linguistics (ACL) (2004)
19. Liu, G., et al.: Clinically accurate chest x-ray report generation. In: Machine Learning for Healthcare Conference, pp. 249–269 (2019)
20. Lovelace, J., Mortazavi, B.: Learning to generate clinically coherent chest X-ray reports. In: Findings of the Association for Computational Linguistics: EMNLP, pp. 1235–1243 (2020)
21. Lu, J., Xiong, C., Parikh, D., Socher, R.: Knowing when to look: adaptive attention via a visual sentinel for image captioning. In: IEEE Conference on Computer Vision and Pattern Recognition, pp. 375–383 (2017)
22. Luo, R., Shakhnarovich, G.: Analysis of diversity-accuracy tradeoff in image captioning (2020)
23. Mahajan, S., Roth, S.: Diverse image captioning with context-object split latent spaces. In: Advances in Neural Information Processing Systems (NeurIPS) (2020)
24. Papineni, K., Roukos, S., Ward, T., Zhu, W.J.: BLEU: a method for automatic evaluation of machine translation. In: Association for Computational Linguistics, pp. 311–318 (2002)

25. Selvaraju, R.R., Cogswell, M., Das, A., Vedantam, R., Parikh, D., Batra, D.: Grad-CAM: visual explanations from deep networks via gradient-based localization. In: IEEE International Conference on Computer Vision, pp. 618–626 (2017)
26. Sohn, K., Lee, H., Yan, X.: Learning structured output representation using deep conditional generative models. In: Cortes, C., Lawrence, N., Lee, D., Sugiyama, M., Garnett, R. (eds.) Advances in Neural Information Processing Systems (2015)
27. Vaswani, A., et al.: Attention is all you need. In: Advances in Neural Information Processing Systems, vol. 30 (2017)
28. Vinyals, O., Toshev, A., Bengio, S., Erhan, D.: Show and tell: a neural image caption generator. In: IEEE Conference on Computer Vision and Pattern Recognition, pp. 3156–3164 (2015)
29. Wang, W., et al.: Topic-guided variational auto-encoder for text generation. North American Chapter of the Association for Computational Linguistics (2019)
30. Xu, K., et al.: Show, attend and tell: neural image caption generation with visual attention. In: International Conference on Machine Learning, pp. 2048–2057 (2015)
31. Xue, Y., Huang, X.: Improved disease classification in chest X-rays with transferred features from report generation. In: Chung, A.C.S., Gee, J.C., Yushkevich, P.A., Bao, S. (eds.) IPMI 2019. LNCS, vol. 11492, pp. 125–138. Springer, Cham (2019). https://doi.org/10.1007/978-3-030-20351-1_10
32. Xue, Y., et al.: Multimodal recurrent model with attention for automated radiology report generation. In: Frangi, A.F., Schnabel, J.A., Davatzikos, C., Alberola-López, C., Fichtinger, G. (eds.) MICCAI 2018. LNCS, vol. 11070, pp. 457–466. Springer, Cham (2018). https://doi.org/10.1007/978-3-030-00928-1_52
33. Yuan, J., Liao, H., Luo, R., Luo, J.: Automatic radiology report generation based on multi-view image fusion and medical concept enrichment. In: Shen, D., et al. (eds.) MICCAI 2019. LNCS, vol. 11769, pp. 721–729. Springer, Cham (2019). https://doi.org/10.1007/978-3-030-32226-7_80
34. Zhang, Y., Chen, Q., Yang, Z., Lin, H., Lu, Z.: BioWordVec: improving biomedical word embeddings with subword information and MeSH ontology (2018)

Uncertainty Aware Deep Reinforcement Learning for Anatomical Landmark Detection in Medical Images

James Browning[1]([⊠]), Micha Kornreich[1], Aubrey Chow[1], Jayashri Pawar[1], Li Zhang[1], Richard Herzog[1,2], and Benjamin L. Odry[1]

[1] Covera Health, New York, NY, USA
james.browning@coverahealth.com
[2] Hospital for Special Surgery, New York, NY, USA

Abstract. Deep reinforcement learning (DRL) is a promising technique for anatomical landmark detection in 3D medical images and a useful first step in automated medical imaging pathology detection. However, deployment of landmark detection in a pathology detection pipeline requires a self-assessment process to identify out-of-distribution images for manual review. We therefore propose a novel method derived from the full-width-half-maxima of q-value probability distributions for estimating the uncertainty of a distributional deep q-learning (dist-DQN) landmark detection agent. We trained two dist-DQN models targeting the locations of knee fibular styloid and intercondylar eminence of the tibia, using 1552 MR sequences (Sagittal PD, PDFS and T2FS) with an approximate 75%, 5%, 20% training, validation, and test split. Error for the two landmarks was 3.25 $\pm$ 0.12 mm and 3.06 $\pm$ 0.10 mm respectively (mean $\pm$ standard error). Mean error for the two landmarks was 28% lower than a non-distributional DQN baseline (3.16 $\pm$ 0.11 mm vs 4.36 $\pm$ 0.27 mm). Additionally, we demonstrate that the dist-DQN derived uncertainty metric has an AUC of 0.91 for predicting out-of-distribution images with a specificity of 0.77 at sensitivity 0.90, illustrating the double benefit of improved error rate and the ability to defer reviews to experts.

Keywords: Reinforcement learning · Uncertainty · Anatomical landmark detection

1 Introduction and Related Work

Representing uncertainty is an important problem and ongoing research question in the field of deep learning. Practical use of the resulting models in risk-sensitive domains, including medical diagnosis, requires reliable quantification of uncertainty in model predictions for the purposes of identifying critical mistakes or out-of-distribution data for manual review.

Deep Q networks (DQN) have been shown to perform at above-human level on sequential decision making tasks such as simulated Atari environments in discrete action spaces (Mnih et al. 2015). DQN has been used to train single agents to locate anatomical

© Springer Nature Switzerland AG 2021
M. de Bruijne et al. (Eds.): MICCAI 2021, LNCS 12903, pp. 636–644, 2021.
https://doi.org/10.1007/978-3-030-87199-4_60

structures within 3D medical images (Ghesu et al. 2019; Alansary et al. 2019) as well as multiple-agents for multiple landmark detection within single images (Vlontzos et al. 2019). In order to improve performance in stochastic environments, Distributional DQN (dist-DQN) also learns probability distributions over q-values (Bellemare et al. 2017), and similar techniques have been used to improve agent performance via risk-aware and uncertainty-aware exploration and decision making in uncertain and stochastic environments (Chua et al. 2018; Nikolov et al. 2018; Clements et al. 2020). Kahn et al. demonstrated that uncertainty derived from a combination of bootstrapping and dropout could be used to flag high-risk transitions for collision avoidance of autonomous robots using a model-based reinforcement learning algorithm.

In this work we present a novel anatomical landmark detection method, with improved uncertainty estimation. We trained two dist-DQN models targeting the locations of knee fibular styloid and the intercondylar eminence of the tibia, and demonstrate that the resulting model has higher accuracy than a non-distributional baseline. Our proposed uncertainty measure for landmark location, U^{FWHM}, is derived from the full-width-half-maxima (FWHM) of q-value probability distributions and can be used to identify out-of-distribution images and landmark prediction errors.

2   Methods

2.1   Landmark Annotation

1,600 knee MRI images from unique studies were annotated for 3D point location of fibular styloid and intercondylar eminence by annotators trained in the task by a radiologist. Annotators were directed to select 1) the tip of the fibular styloid in a sagittal image at the slice level where the lateral meniscus becomes visible and 2) the intercondylar eminence (tibial spike) at the level of the anterior cruciate ligament.

2.2   Agent Architecture

Distributional DQN modifies the DQN Bellman equation to form the distributional Bellman equation (Bellman 1957; Bellemare et al. 2017):

$$Z^{\pi}(x, a) = R(x, a) + \gamma Z^{\pi}\left(x', a^*\right) \tag{1}$$

where $Z^{\pi}(x, a)$ is the predicted value distribution of taking action a in state x given model parameters π, $R(x, a)$ is the distribution of immediate rewards given state-action pair (x, a), and $\gamma Z^{\pi}\left(x', a'\right)$ is the value distribution of the successive state-action pair $\left(x', a'\right)$ discounted by factor γ. Our dist-DQN training implementation follows that of Bellemare et al. in which the q-value distribution for each allowable action in a given state is learned as a discrete random variable.

A deep reinforcement agent was trained to find a landmark within a 3D MR image volume based on the landmark detection methods of Alansary et al. and Ghesu et al. (Alansary et al. 2019; Ghesu et al. 2019). Landmark detection was framed as a markov decision process (MDP) wherein the agent moves through the image and is tasked with centering itself on the landmark of interest. The agent's state consists of a 42 × 42 × 42 voxel isentropic grayscale image volume field of view (FOV) centered on the agent and the agent's allowable actions at each step in the MDP are the six signed orthogonal image directions (ie. up, down, left, right, forward, backward).

The agent's learned model is a neural network consisting of four convolutional + maxpooling blocks (the encoder network) followed by four dense layers and a final softmax layer (the action-value network). The final softmax layer has dimensions of 6 × 64 which encodes 64-bin discrete q-value probability distributions evenly distributed over the range $[-10, 20]$ for each of the six allowable actions. The predicted q-value of each action for a given state is the expected value of the q-value probability distribution. A schematic of the agent is shown in Fig. 1.

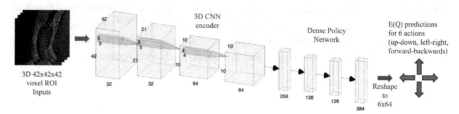

Fig. 1. Schematic of the distributional DQN medical image anatomical landmark detection agent. Note that schematic shows a simplified 2D convolutional encoder for illustrative purposes, whereas the proposed method uses a 3D convolutional encoder analog.

During training, the agent receives a reward of +2 each time it centers itself at the image voxel corresponding to the landmark. In order to reduce reward sparsity, the agent also receives a reward at each time step equal to the signed decrease in Euclidean distance (in units of voxels) resulting from the action taken.

Multi-scale landmark detection is used as described by Ghesu et al., but with two agents training simultaneously in a single image, one for each landmark. The two agents move through the same image, but each has its own unique location and FOV and there is no communication between agents. Resolution changes occur when both agents oscillate within a small region for a predetermined number of steps, and the episode terminates when oscillation occurs within the highest resolution image. The final predicted landmark location is the state reached within the final oscillation period having the lowest maximum expected q-value. This is justified by the intuition that any state reached from the correct state (i.e. the landmark location) should have lower immediate reward, and thus lower expected q-value, than an adjacent state.

2.3 Inference and Uncertainty Estimation

During inference, the agent's optimal action in a given state was calculated as the maximum expected value of the discrete q-value distribution over all actions, and the proposed uncertainty metric was derived from the normalized (i.e. normalized for 0th moment to equal 1) q-value probability distribution. Operating under the intuition that a wider probability distribution reflects uncertainty in q-value estimation, two metrics were investigated to quantify uncertainty during an agent's episode; $H(Z^\pi(x, a))$ (Shannon entropy), and full-width-half-maximum $FWHM(Z^\pi(x, a))$, (Virtanen et al. 2020) from the discrete q-value distributions. These metrics are averaged over all steps in the inference episode for each agent and the maximum over the two agents is taken as the uncertainty metric for a given input image. The uncertainty-aware dist-DQN inference algorithm is summarized in Algorithm 1 and Table 1.

Algorithm 1: Uncertainty-aware dist-DQN inference algorithm

Require: number of agents $K \in \mathbb{R}^+$, distributional NN policy models Z^{π_k}
Initialize step counter $n = 0$
do until episode end
    for $k = 1, \dots, K$ **do**
        Take optimal action $a_k^* = \text{argmax}_a(E(Z^{\pi_k}(x_k, a)))$
        Calculate and save $FWHM_{k,n}(Z^{\pi_k}(x_k, a_k^*))$ and $H_{k,n}(Z^{\pi_k}(x_k, a_k^*))$
    end for
    $n = n + 1$
    for $k = 1, \dots, K$ **do**
        calculate mean $FWHM_k = \frac{1}{n'}\sum_{n'} FWHM_{k,n'}$ and mean $H_k = \frac{1}{n'}\sum_{n'} H_{k,n'}$
    return: final agent locations,
        $U^{FWHM} = \text{max}_k(FWHM_k)$ over all agents,
        $U^H = \text{max}_k(H_k)$ over all agents

Table 1. Training and agent architecture parameters

Parameter	Value	Parameter	Value	Parameter	Value
multi-scale voxel spacing	(6.75 mm, 2.25 mm, 0.75 mm)	number of workers	16	optimizer	Adam
Agent FOV	42x42x42	number of worker gpus	3	learning rate	0.0001
inference epsilon	0.05	number of trainer gpus	1	batch size	128
maximum steps	400	reward for find	2	gamma (discount factor)	0.95
oscillation voxel bounds	3x3x3	number of distribution bins	64	network update frequency	50,000
oscillation step length	20	distribution bin bounds	(-10, 20)	training exploration policy	per worker epsilon greedy
				replay buffer size	200,000

3 Experiment and Results

529 Sagittal Proton Density (SagPD), 665 Sagittal Proton Density with Fat Saturation (SagPDFS), and 259 Sagittal T2 with Fat Saturation (SagT2FS) images were annotated for anatomical landmarks.

Two agents, one for each knee landmark, were trained using the Ape-X DQN algorithm in the ray/rllib 0.8.5 reinforcement learning framework (Horgan et al. 2018; Moritz et al. 2018) and a custom OpenAI Gym environment (Brockman et al. 2016). A unique agent was trained for each of the landmarks, rather than a single agent tasked with finding both landmarks, in order to decrease the complexity of the task and reward structure and increase the robustness of the resulting landmark location predictions.

The training set consisted of 1152 total images; 410 SagPD, 504 SagPDFS, and 238 SagT2FS. To improve generalization, random augmentation on the 3D image volumes was performed during training including superposition of random noise, rotation and translation in the sagittal image plane, and image stack reverse-ordering (i.e. creating an anatomically correct left knee from a right knee image and vice-versa). Clip limited adaptive histogram equalization followed by normalization for zero mean and standard deviation = 1 was performed on all images.

Agents were trained for a maximum of 2×10^7 steps and mean landmark error was calculated every 100,000 steps on a 100 image validation set. The network with the lowest mean landmark location error on the validation set was checkpointed and used for testing. As a baseline for landmark detection accuracy comparison, an additional agent was trained for each of the two landmarks without distributional DQN but with other training and inference parameters held the same. Double DQN and Noisy Nets were employed for training both distributional and baseline models. (van Hasselt et al. 2016; Fortunato et al. 2019). Training took approximately 15 h on an Amazon AWS g4dn.12xlarge server with 16 rollout workers sharing 3 GPUs and 1 GPU dedicated to asynchronous model training.

Three experiments were performed to test the accuracy and uncertainty estimation of the proposed dist-DQN landmark detection agent with U^{FWHM} and U^H; 1) error of landmark detection was calculated on a 300 image annotated test set without out-of-distribution images, 2) performance of the U^{FWHM} and U^H uncertainty metrics were calculated on an expanded 377 image test set that included 77 out-of-distribution images, and 3) performance of the U^{FWHM} as a method for flagging landmark detection errors was calculated on an additional 602 image unlabeled test set.

The secondary test set of 77 out-of-distribution images was curated to assess model uncertainty and automatic identification of out-of-distribution images. Accordingly, this set contained low quality or post-operative images for which annotators could not find landmarks, a synthetically generated noise image, and lumbar spine and shoulder MR images. Examples of these secondary test images are shown in Fig. 2.

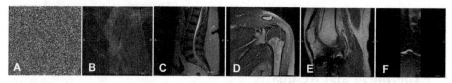

Fig. 2. Images from the secondary out-of-distribution test set illustrating a) synthetic noise image, b) imaging artifacts, c) lumbar spine, d) shoulder, e) post-operative, and f) incomplete field of view (does not contain fibular styloid due to overly restricted left-right FOV as shown in the coronal view).

3.1 Experiment 1: Landmark Detection Accuracy

Landmark detection accuracy was measured on the primary 300 image annotated test set that did not contain out-of-distribution images. dist-DQN accuracy for detecting the location of the fibular styloid and intercondylar eminence in 3D knee MRI volumes (N = 168 SagPD, 236 SagPDFS, and 196 SagT2FS) was 3.25 ± 0.12 mm and 3.06 ± 0.10 mm respectively (mean $\pm$ standard error). Posterior-anterior, superior-inferior, and left-right (out-of-plane) accuracy was 1.22 ± 0.05, 1.37 ± 0.05, and 2.09 ± 0.07 respectively. Mean accuracy over both landmarks was significantly lower than the non-distributional baseline (3.16 ± 0.08 vs. 4.36 ± 0.27 mm, $p < 0.001$).

3.2 Experiment 2: Uncertainty Estimation

During inference, q-value probability distributions for each allowable action at each timestep were saved. Randomly selected example q-value distributions marked with FWHM and final tibial intercondylar eminence landmark location for an image that passed manual QC (left) and failed manual QC (right) are shown in Fig. 3. Note that the failed image is an axial shoulder image and was intentionally included to test the U^{FWHM} method on detecting out-of-distribution images. In general, inspection of q-value distributions from the pass and fail cases showed a trend of narrower distribution for pass cases corresponding to lower uncertainty.

Candidate metrics, U^{FWHM} and U^{H}, were evaluated for use as binary classifiers for detecting out-of-distribution images or inaccurate landmark predictions in the combined primary 300 image and secondary 77 image test sets. Landmark prediction results were manually reviewed, with the reviewer blind to U^{FWHM} and U^{H} values, and categorized into two classes; 1) successes, defined as clearly visible landmarks with landmark prediction error <10 mm and 2) failures, defined as either landmark predictions with error >10 mm or out-of-distribution image characteristics as illustrated in Fig. 2. In general, failures and successes were intended to illustrate the usefulness of the resulting image or landmark prediction as a component in an automated pathology detection pipeline. The resulting AUC for the U^{FWHM} error flagging metric was greater than that of the U^{H} entropy metric (0.91 vs 0.76) and had higher specificity in the high-sensitivity operating regime. At 90% sensitivity to out-of-distribution images or landmark detection errors, the proposed U^{FWHM} metric had 77% specificity.

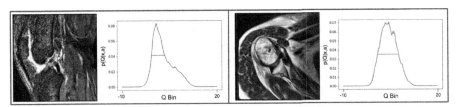

Fig. 3. Value distributions marked with peaks (yellow x's) and FWHM (green lines) and final landmark locations (red dots) for tibial intercondylar eminence landmark detection agent for an image that passed manual QC (left) and failed manual QC (right). (Color figure online)

3.3 Experiment 3: Landmark Prediction Error Detection

As a final experiment, the dist-DQN algorithm with U^{FWHM} metric was deployed "in the wild" as a knee landmark detection agent on studies that did not have manual landmark annotations. To detect failure cases, we used binary automatic quality control flagging based on a U^{FWHM} threshold, 9.6, corresponding to the 90% sensitivity operating point derived from the 377 image expanded test set experiment. The algorithm was run on 602 unique studies without patient overlap in the training or validation sets and without landmark annotations. Inference was performed on each Sagittal PD, T2FS, or PDFS image available in the study (in that order) until the U^{FWHM} uncertainty metric was below the 9.6 threshold, indicating low uncertainty in the landmark results. If no image in the study produced a U^{FWHM} metric below 9.6, the results were flagged as low confidence for that study and the result with lowest U^{FWHM} was saved. The 602 landmark prediction results were manually reviewed, and landmark prediction failures were defined as images in which at least one of the two landmark predictions had error greater than 10 mm. Of the 602 images, there were four such landmark prediction failures for a failure rate of 0.7%. Three of these four failures were correctly flagged by the proposed U^{FWHM} auto-QC method and four of the 598 successful landmark detection results were erroneously flagged by the proposed method. For this experiment, with a limited number of true positives, this corresponds to a U^{FWHM} error detection sensitivity and specificity of 75% and 99.3% respectively.

3.4 Failure Modes

To investigate potential failure modes of U^{FWHM} as an auto-QC method, we inspected erroneous U^{FWHM} filtering decisions in the expanded 377 image test set. Auto-QC binary filtering was based on U^{FWHM} values, with larger values corresponding to greater uncertainty. Cases where U^{FWHM} auto-QC exceeded the uncertainty threshold, but passed manual QC, are considered false positives. Correspondingly, cases that passed U^{FWHM} auto-QC, but failed manual QC, are false negatives. Figure 4 shows results for the top three false positives and top three false negatives from these sets along with their U^{FWHM} uncertainty metrics.

The false positives include two knee images with apparently normal pathology, normal imaging characteristics, and accurate landmark detection results (A and C). The third false positive, panel B, shows abnormal T2 hyperintensities in subcortical bone marrow of tibia. However, it is unclear if this warrants a designation as an out-of-distribution as it may be common enough to occur in the training set. Among the three false negatives for out-of-distribution detection are two lumbar spine images and a corrupted knee image. Note that in panels D and E, the agent has found local visual features in lumbar spine with horizontal spatial gradient similar to the intercondylar eminence in T2FS and PDFS images (see the red markers at the top of the pelvis and the top of the 5th lumbar disc).

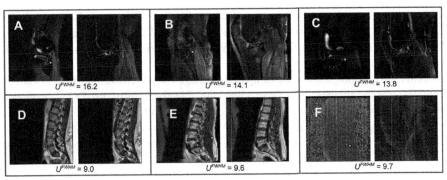

Fig. 4. dist-DQN U^{FWHM} confidence estimation failure cases. Top three results with lowest confidence that passed manual review (top panels A-C) and top three results with highest confidence that failed manual QC are shown. Yellow and red markers show fibular styloid and intercondylar eminence location predictions respectively in the sagittal view. (Color figure online)

4 Discussion

We presented a dist-DQN model for MRI anatomical landmark detection, and proposed a corresponding novel uncertainty metric, U^{FWHM}. Our proposed method and uncertainty metric improves landmark detection accuracy, and successfully identifies out-of-distribution images and errors. At a 90% sensitivity and 77% specificity operating point, application of the U^{FWHM} method of error flagging is expected to correctly identify 90% of errors while incorrectly identifying 23% of correct landmark predictions and in-distribution images in this two-landmark detection task. The method has the potential to greatly reduce the workload of landmark detection compared to a fully manual process while maintaining robustness by detecting critical errors and out-of-distribution images.

Investigation of U^{FWHM} measured uncertainty failures, when used to flag out-of-distribution images or inaccurate results, indicates that the false positives consist of images without apparent anatomical irregularities or abnormal imaging characteristics. However, review of false negatives reveals confusion with local visual features in lumbar spine images and corrupt images that are not caught with the proposed uncertainty method. The uncertainty metric may better reflect true uncertainty if the agent is simplified to a single-scale agent to avoid confusion between low-resolution knee images and corrupt images, or if the U^{FWHM} measurements are confined to the tail end (high resolution) portion of the episode.

Deploying the dist-DQN U^{FWHM} metric in the wild as an automatic error flagging method in an unannotated 602 study experiment resulted in very promising results, with sensitivity and specificity of 75% and 99.3% respectively and a landmark detection error rate (defined as at least one landmark in an image with error >10 mm) of 0.7%. Although the number of true positives (i.e. true errors) was low due to the high general performance of the algorithm in this experiment, it is further evidence that dist-DQN landmark detection with U^{FWHM} uncertainty algorithm can be a useful tool in semi or fully-automated radiology pipelines.

In addition to enabling identification of out-of-distribution images or landmark detection failures, a limited comparison to a non-distributional DQN baseline indicates that the dist-DQN landmark detection method may decrease error on some anatomical landmark detection tasks. Medical images may be thought of as a stochastic environment wherein each test image is unique compared to images on which the algorithm is trained. It appears that the agent is able to take advantage of the benefits of dist-DQN, developed for improved performance in stochastic environments, leading to lower error.

The double benefit of improved error rate and the ability to defer reviews to experts makes dist-DQN with U^{FWHM} uncertainty estimation a promising algorithm for anatomical landmark detection in semi or fully-automated clinical pipelines.

References

Alansary, A., et al.: Evaluating reinforcement learning agents for anatomical landmark detection. Med. Image Anal. **53**, 156–164 (2019)

Bellemare, M.G., Dabney, W., Munos, R.: A distributional perspective on reinforcement learning. In: International Conference on Machine Learning, pp. 449–458. PMLR (2017)

Bellman, R.E.: Dynamic Programming. Princeton University Press, Princeton (1957)

Brockman, G., et al.: OpenAI Gym. arXiv:1606.01540 [cs] (2016)

Chua, K., Calandra, R., McAllister, R., Levine, S.: Deep reinforcement learning in a handful of trials using probabilistic dynamics models. In: Advances in Neural Information Processing Systems 31 (2018)

Clements, W.R., Van Delft, B., Robaglia, B.-M., Slaoui, R.B., Toth, S.: Estimating risk and uncertainty in deep reinforcement learning. arXiv:1905.09638 [cs, stat] (2020)

Fortunato, M., et al.: Noisy networks for exploration. arXiv:1706.10295 [cs, stat] (2019)

Ghesu, F., et al.: Multi-scale deep reinforcement learning for real-time 3D-landmark detection in CT scans. IEEE Trans. Pattern Anal. Mach. Intell. **41**, 176–189 (2019)

van Hasselt, H., Guez, A., Silver, D.: Deep reinforcement learning with double Q-learning. In: Proceedings of the Thirtieth AAAI Conference on Artificial Intelligence 2094–2100. AAAI Press (2016)

Horgan, D., et al.: Distributed prioritized experience replay. In: International Conference on Learning Representations. ICLR (2018)

Kahn, G., Villaflor, A., Pong, V., Abbeel, P., Levine, S.: Uncertainty-aware reinforcement learning for collision avoidance. arXiv:1702.01182 [cs] (2017)

Mnih, V., et al.: Human-level control through deep reinforcement learning. Nature **518**, 529–533 (2015)

Moritz, P., et al.: Ray: a distributed framework for emerging AI applications. arXiv:1712.05889 [cs, stat] (2018)

Nikolov, N., Kirschner, J., Berkenkamp, F., Krause, A.: Information-directed exploration for deep reinforcement learning. In: International Conference on Learning Representations. ICLR (2018)

Virtanen, P., et al.: SciPy 1.0: fundamental algorithms for scientific computing in Python. Nat. Methods **17**, 261–272 (2020)

Vlontzos, A., Alansary, A., Kamnitsas, K., Rueckert, D., Kainz, B.: Multiple landmark detection using multi-agent reinforcement learning. arXiv:1907.00318 [cs] (2019)

Author Index

Printed in the United States
by Baker & Taylor Publisher Services